Heat Shock Proteins

Volume 19

Series editors
Alexzander A. A. Asea
Professor, Department of Medicine and Director
Precision Therapeutics Proteogenomics Diagnostic Center
Eleanor N. Dana Cancer Center
University of Toledo, College of Medicine and Life Sciences
Toledo, USA

Stuart K. Calderwood
Professor and Director, Division of Molecular and Cellular Radiation Oncology
Department of Radiation Oncology
Beth Israel Deaconess Medical Center and Harvard Medical School
Boston, USA

Heat Shock Proteins: key mediators of Health and Disease. Heat shock proteins (HSP) are essential molecules conserved through cellular evolution required for cells to survive the stresses encountered in the environment and in the tissues of the developing and aging organism. These proteins play the essential roles in stress of preventing the initiation of programmed cell death and repairing damage to the proteome permitting resumption of normal metabolism. Loss of the HSP is lethal either in the short-term in cases of acute stress or in the long-term when exposure to stress is chronic. Cells appear to walk a fine line in terms of HSP expression. If expression falls below a certain level, cells become sensitive to oxidative damage that influences aging and protein aggregation disease. If HSP levels rise above the normal range, inflammatory and oncogenic changes occur. It is becoming clear that HSP are emerging as remarkably versatile mediators of health and disease. The aim of this series of volumes is to examine how HSP regulation and expression become altered in pathological states and how this may be remedied by pharmacological and other interventions.

More information about this series at http://www.springer.com/series/7515

Alexzander A. A. Asea • Punit Kaur
Editors

Heat Shock Protein 90 in Human Diseases and Disorders

Springer

Editors
Alexzander A. A. Asea
Department of Medicine and Director,
Precision Therapeutics Proteogenomics
Diagnostic Center, Eleanor N. Dana
Cancer Center
University of Toledo, College of Medicine
and Life Sciences
Toledo, OH, USA

Punit Kaur
Department of Experimental
Radiation Oncology
MD Anderson Cancer Center
Houston, TX, USA

ISSN 1877-1246 ISSN 1877-1254 (electronic)
Heat Shock Proteins
ISBN 978-3-030-23157-6 ISBN 978-3-030-23158-3 (eBook)
https://doi.org/10.1007/978-3-030-23158-3

This Springer imprint is published by the registered company Springer Nature Switzerland AG.
The registered company address is: Gewerbestrasse 11, 6330 Cham, Switzerland

Preface

The ninety-kilo Dalton heat-shock protein (HSP90) regulates the stability, activation, and degradation of a diverse array of proteins associated with growth, proliferation, and survival. Thus, it is core to regulation of protein stability and protein-degradation pathways and modulating transcription factors, signaling transduction networks, and kinases. It facilitates the survival of cells during stress response and exhibits a pronounced anti-apoptotic and stabilization effect. Thus, HSP90 has been associated with development and progression of a wide range of pathological conditions including cancers, diabetes, Gaucher disease, neurodegenerative diseases, and autoimmune dysfunction.

The book ***Heat Shock Protein 90 in Human Diseases and Disorders*** provides the most comprehensive review on contemporary knowledge on the role of HSP90. Using an integrative approach, the contributors provide a synopsis of novel mechanisms, previously unknown signal transduction pathways. To enhance the ease of reading and comprehension, this book has been subdivided into various section including; Part I, reviews current progress on our understanding Oncogenic Aspects of HSP90; Part II, focuses on Bimolecular Aspects of HSP90; Part III, emphasizes and HSP90 in Natural Products Development and Part IV; give the most up to date reviews on Clinical Aspects of HSP90.

Key basic and clinical research laboratories from major universities, academic medical hospitals, biotechnology and pharmaceutical laboratories around the world have contributed chapters that review present research activity and importantly project the field into the future. The book is a must read for graduate students. medical students, basic science researchers and postdoctoral scholars in the fields of Translational Medicine, Clinical Research, Human Physiology, Biotechnology, Natural Products, Cell & Molecular Medicine, Pharmaceutical Scientists and Researchers involved in Drug Discovery.

Toledo, OH, USA — Alexzander A. A. Asea
Houston, TX, USA — Punit Kaur

Contents

About the Editors

Prof. Dr. Alexzander A. A. Asea is a highly innovative and accomplished world renowned clinical and basic research scientist and visionary executive leader who has exceptional experience spearheading clinical and basic science research, training, education, and commercialization initiatives within top ranked academic biomedical institutes. Prof. Dr. Asea's initial findings studying the effects of Hsp72 on human monocytes lead to the proposal of a novel paradigm that Hsp72, previously known to be an intracellular molecular chaperones, can be found in the extracellular milieu where it has regulatory effects on immuno-competent cells – a term now called chaperokine. Prof. Asea has authored over 255 scientific publications including peer-reviewed articles, reviews, books, book chapters, editorials, and news headliners in a wide range of biomedical-related disciplines. Prof. Asea is the series editor of the widely successful book series *Heat Shock Proteins* (Springer Nature Publishing) and is an editorial board member of numerous scientific peer-reviewed journals. Currently, Prof. Dr. Asea is at the University of Toledo College of Medicine and Life Sciences in Toledo, USA.

Dr. Punit Kaur is an expert in onco-proteogenomics, with extensive training and experience in quantitative mass spectrometry imaging, protein chemistry and biomarker discovery. Dr. Kaur's main research focus is on the use of heat-induced nanotechnology in combination with radiotherapy and chemotherapy in the cancer stem cell therapy. Dr. Kaur has published more than 40 scientific articles, book chapters, and reviews, and currently serves as editorial board member for the European *Journal of* Cancer Prevention and the *Journal of Proteomics and Bioinformatics*. Dr. Kaur is an editor of 10 books in the highly successful *Heat Shock Proteins* book series by Springer Nature Publishers. Currently, Dr. Kaur is a Visiting Scientist Professor at the University of Texas MD Anderson Cancer Center in Houston, USA.

Part I
Oncogenic Aspects of HSP90

Chapter 1
Regulatory Roles of HSP90-Rich Extracellular Vesicles

Takanori Eguchi, Kisho Ono, Kazumi Kawata, Kuniaki Okamoto, and Stuart K. Calderwood

Abstract HSP90 is an essential protein in protein folding, cancer progression and wound healing. Originally, most studies were focused on the intracellular molecular chaperone role of HSP90. However, more recent studies, including ours, have reported the secretion of HSP90 and novel functions for this protein in the extracellular space (ex-HSP90). Additionally, HSP90 has been found to be a major cargo contained in extracellular vesicles (EV) such as exosomes. HSP90 can directly bind to and promote functions of CD91/LRP1 and receptor tyrosine kinases such as EGF receptor. HSP90 also regulates the recycling of Rab proteins that control the secretion of exosomes. This chapter reviews current knowledge and the future potential of ex-HSP90 and EV-HSP90.

T. Eguchi (✉)
Department of Dental Pharmacology, Graduate School of Medicine, Dentistry and Pharmaceutical Sciences, Okayama University, Okayama, Japan

Advanced Research Center for Oral and Craniofacial Sciences, Graduate School of Medicine, Dentistry and Pharmaceutical Sciences, Okayama University, Okayama, Japan
e-mail: eguchi@okayama-u.ac.jp; eguchi.takanori@gmail.com

K. Ono
Department of Dental Pharmacology, Graduate School of Medicine, Dentistry and Pharmaceutical Sciences, Okayama University, Okayama, Japan

Department of Oral and Maxillofacial Surgery, Okayama University Hospital, Okayama, Japan

K. Kawata
Department of Biochemistry and Molecular Dentistry, Graduate School of Medicine, Dentistry and Pharmaceutical Sciences, Okayama University, Okayama, Japan

K. Okamoto
Department of Dental Pharmacology, Graduate School of Medicine, Dentistry and Pharmaceutical Sciences, Okayama University, Okayama, Japan

S. K. Calderwood
Department of Radiation Oncology, Beth Israel Deaconess Medical Center, Harvard Medical School, Boston, MA, USA

A. A. A. Asea, P. Kaur (eds.), *Heat Shock Protein 90 in Human Diseases and Disorders*, Heat Shock Proteins 19,
https://doi.org/10.1007/978-3-030-23158-3_1

Keywords Ectosome · Epithelial-mesenchymal transition · Exosome · Extracellular matrix · Extracellular vesicle · HSP90 · LRP1/CD91 · Rab proteins

Abbreviations

ECM	Extracellular matrix
EGFR	Epidermal growth factor receptor
EMT	Epithelial-mesenchymal transition
EV	Extracellular vesicle
ex-HSP90	Extracellular HSP90
FN	Fibronectin
HIF-1	Hypoxia-inducible factor-1
HNC	Head and neck cancer
HSP	Heat shock protein
ic-HSP90	Intracellular HSP90
LRP1	Lipoprotein receptor-related protein 1
MV	Microvesicle
MVB	Multivesicular body
OSCC	Oral squamous cell carcinoma
RTK	Receptor tyrosine kinase
TM	Transmembrane
TSP1	Thrombospondin 1

1.1 Introduction

Heat Shock Protein (HSP) is a protein universally present in every cell, while the HSP family consists of two types; a cell stress-response type of HSP and a constitutively expressed housekeeping type of HSP. The stress-responsive HSP is expressed when cells are exposed to stress such as heat, cold, and hypoxia. Among members of the HSP family, HSP90 is one of the most abundant molecular chaperones playing key roles in proteostasis in the cells. The HSP90 family consists of four members; the proteotoxic stress-inducible HSP90α encoded by *HSP90AA1*, the constitutively expressed HSP90β encoded by *HSP90AB1*, the mitochondria-localized TRAP1, and an ER resident paralog GRP96/HSP90B1. Of note, HSP90α is often highly expressed in cancer cells and is secreted to extracellular space as a soluble protein so-called chaperokine (Eguchi et al. 2018) or as a cargo protein of extracellular vesicles (EVs) (Ono et al. 2018). Additionally, HSP90β, TRAP1, and some members of HSP70 were recently found to be secreted with EVs from cancer cells (Ono et al. 2018). However, the mechanism by which HSP is incorporated with EVs and their significance are unknown.

1.2 Extracellular HSP90 in Skin Wound Healing and Cancer

Intracellular HSP90 (ic-HSP90) impacts many components of the cellular proteostasis network, cytoplasmic protein quality control, and the stress response (Neckers et al. 2018). More recently the roles and clinical applications of extracellular HSP90 (ex-HSP90) have been established by Drs Wei Li and David Woodley since 2007. Dr. Li's studies have shown a novel homeostatic mechanism involving the pathway: hypoxia>HIF-1>ex-Hsp90α secretion>skin cell migration>wound healing, and identified ex-Hsp90α as a potential therapeutic agent for the healing of skin wounds (Li et al. 2007; Jayaprakash et al. 2015). This group subsequently showed that TGFα, a member of the EGF ligand family, also stimulates secretion of HSP90α (Cheng et al. 2008). In addition, the low-density lipoprotein receptor-related protein 1 (LRP1/CD91) was identified as a key receptor for ex-HSP90 to promote human skin cell migration (Cheng et al. 2008). ex-HSP90 was demonstrated to bind to the subdomain II of LRP1 and it was shown that the intracellular NPVY motif of LRP1 was essential for activation of Akt1/2 signaling (Tsen et al. 2013). These studies regarding wound healing were also relevant to cancer studies. Secreted ex-HSP90α and ex-HSP90β were found in the conditioned media (CM) of breast cancer cell lines such as MDA-MB-231, MDA-MB-468, MDA-MB-361, BT474, T47D, and Skbr3, in which HIF-1α is also constitutively active, but not in HBL100 or HS-578-T cells (Dong et al. 2016). In breast cancer MDA-MB-231 cells, the secreted ex-HSP90 increased cancer cell survival in a hostile hypoxic environment via LRP1-mediated activation of Akt, a kinase that is known to mediate cell survival (Tsen et al. 2013). The interiors of large tumors are hypoxic due outgrowing the local capillary system (LePage 1948; Najafi et al. 2019). It is noteworthy that hypoxia induces LRP1 (Koong et al. 2000; Kawata et al. 2012) and HSP90 expression (Eguchi et al. 2018). LRP1-HSP90 interaction on the surface of cells and of exosomes might therefore promote tumor growth.

1.3 Extracellular Vesicles (EVs)

Recent studies have demonstrated the significance of extracellular vesicles (EVs) in many biological and medical phenomena, including: cancer (Peinado et al. 2011; Fujita et al. 2016; Kalluri 2016; Fujiwara et al. 2018a, b), the immune system (Carstens et al. 2017), tissue development and repair (Barile and Vassalli 2017; Mathiyalagan et al. 2017), bone metabolism (Taverna et al. 2017), microbiology (Beveridge 1999), and amyloidogenesis (van Niel 2016) as well as EVs as drug delivery system (DDS) (Fais et al. 2016; Ha et al. 2016) and carriers of biomarkers (Kalluri and LeBleu 2016; Minciacchi et al. 2017b; Mendt et al. 2018; Ono et al. 2018). In a brief classification, EVs include exosomes (50–200 nm), ectosomes (100–1000 nm, a.k.a. shed microvesicles, MVs), large EVs such as large oncosomes (1–10 um) (Minciacchi et al. 2017a), exophers (~4 um) generated upon neurotoxic

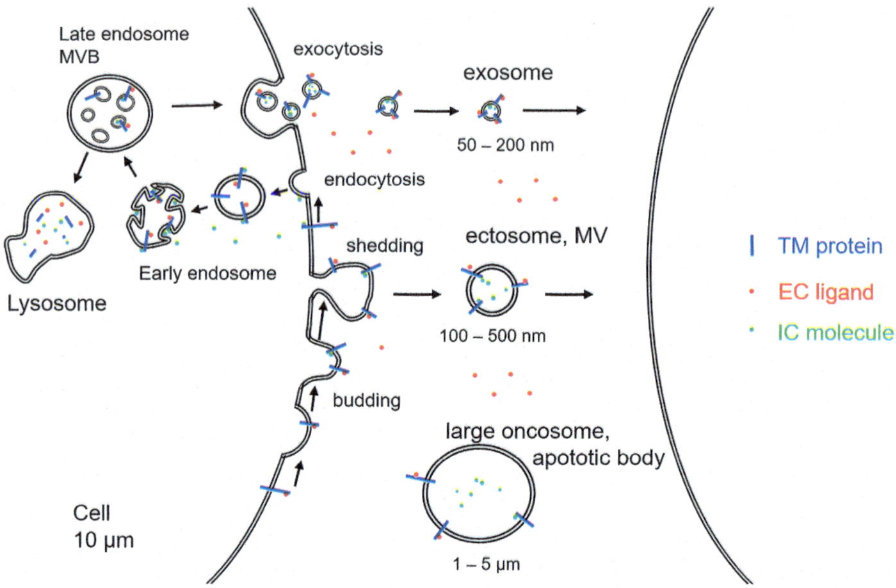

Fig. 1.1 Roles of HSP90 and LRP1 within EVs. EVs are a heterogeneous mixture of structures classified respectively as exosomes, ectosomes (a.k.a. shed MVs), large oncosomes and apoptotic bodies as shown at the center in this figure. Exosomes are secreted via exocytosis of late endosomes a.k.a. multi-vesicular bodies (MVBs) (top). Distinctively, budding and shedding of plasma membrane generate ectosomes (center). Transmembrane (TM) proteins such as LRP1 (blue) can localize on the surface of EVs. Extracellular (EC) ligands such as ex-HSP90 (red) bind to the extracellular domain of LRP1 on the surface of EVs. LRP1-positive EVs can gather ex-HSP90. Intracellular HSP90 (green) can be kept bound to the intracellular domains of the TM proteins such as receptor tyrosine kinases (RTKs) including EGFR, which is incorporated into EVs with HSP90. In this case, HSP90 can enter into EVs

stress (Melentijevic et al. 2017) and apoptotic bodies (Fig. 1.1, at the center), migrasomes associated with cilia (Ma et al. 2015), and exomeres (~35 nm) (Zhang et al. 2018; Zijlstra and Di Vizio 2018). Differences in the generation mechanisms involved in their production define these EVs rather than their size. Exosomes are secreted via exocytosis of late endosomes a.k.a. multivesicular bodies (MVBs) (Fig. 1.1, upper left). By contrast, the budding and shedding of plasma membrane lead to the generation of ectosomes (Fig. 1.1, center).

1.4 EV-Associated HSP90

We recently reported that the secretion of ex-HSP90α was boosted along with the formation of tumor organoids (tumoroids/cell aggregates/spheroids) of prostate cancer PC-3 cells, in which intra-tumoral hypoxic milieu was reconstituted (Eguchi et al. 2018). ex-HSP90α receptor LRP1 was robustly expressed in the PC-3 cells

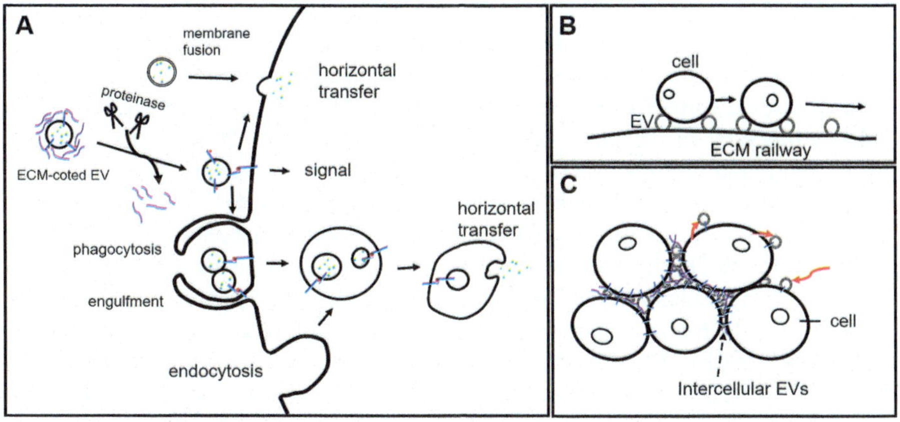

Fig. 1.2 The multiple actions of EVs on/to the cells. The action of EVs on cells can be classified as: (i) horizontal transfer of EV cargos (**a**, top), which can largely change the recipient cells, (ii) signal transduction to the recipient cells using the EV-surface molecules such as ex-HSP90 and cell surface receptor such as LRP1 (**a**, center), which can trigger subsequent membrane fusion, phagocytosis, macropinocytosis (Nakase et al. 2015) or endocytosis (**a**, bottom) (Horibe et al. 2018). After the phagocytosis and endocytosis, EV cargos can be processed in lysosomes, horizontally transferred into the cytoplasm (**a**, right) or recycled in recycling endosomes. EVs can stay on the surface of cells via interaction between cell surface molecules and extracellular matrix (ECM). EVs can also be coated by ECM (purple), which may physically interfere with membrane fusion and horizontal transfer from the EVs to the cells (**a**, left). EVs can be used as wheel-like adaptors that enable cells to drive (migrate) on the cell migration highway made by ECM in tissues, most notably in cancer metastasis (**b**). EVs can also be kept on the surface of cells and at intercellular space, where EVs could mediate cell aggregation (**c**). ECM molecules such as fibronectin (FN) and TM proteins such as LRP1, EpCAM (Eguchi et al. 2018) and integrins that associate with both cells and EVs could mediate such new roles of EVs in cell aggregation

(unpublished data) and thus a mechanism of tumoral hypoxia>HIF-1α>secretion of ex-HSP90>LRP-1>hypoxia-resistant survival signal (e.g. Akt) can be generalized in many types of tumors. However, we showed that HSP90 and LRP1 were carried by EVs secreted by the prostate cancer PC-3 cells (Eguchi et al. 2018), indicating that LRP1-HSP90 on the surface of EVs can act on recipient cells (Figs. 1.1 and 1.2). ex-HSP90 species including EV-HSP90 and EV-free ex-HSP90, are also secreted by oral squamous cell carcinoma (OSCC) cells (Ono et al. 2018). These cells are classified among head and neck cancers (HNC). HSP family members such as EV-HSP90α and EV-HSP90β were more significantly detected in the lymph node (LN)-metastatic OSCC-derived EV compared with low-metastatic OSCC-EVs (Ono et al. 2018). We thereafter demonstrated that high-expression of HSP90s, including HSP90α, HSP90β, TRAP1 (mitochondrial HSP90) and HSPH1/HSP105 were correlated with increased metastatic tendencies and poor prognosis of patients suffering from HNCs (Ono et al. 2018), indicating that high HSP90 in tumors and EVs are potential prognostic biomarkers of HNCs. Double targeting of HSP90α and HSP90β using siRNA reduced the survival of OSCC cells (Ono et al. 2018), suggesting potential usefulness of small RNA medications that target HSP90 mRNA in cells and EVs.

Although many papers have reported that HSP90 is included in exosome or EV fractions, it is still unclear whether HSP90 is located inside the vesicles or outside (on the surface of) vesicles. Such sub-vesicular localization of HSP90 is important for their functional properties. Cytoplasmic HSP90 and its cochaperone CDC37 play key roles in functions of receptor tyrosine kinases (RTK). Many RTKs including EGFR have been found in exosomes (Fujiwara et al. 2018a, b; Ono et al. 2018). Therefore, RTK-HSP90 complex is thought to be incorporated into vesicles. By this mechanism, HSP90 can be incorporated inside EVs, whose membrane fusion with recipient cells enable HSP90/RTKs transfer into recipient cytoplasm. Intra-exosomal HSP90 could be horizontally transferred into recipient cells where it could exert chaperone functions (Fig. 1.2). Horizontal transfer of HSP90 from the EVs to the recipient cells could thus increase the cellular proteostasis networks, including autophagy, and the UPR in the ER, as well as cytoplasmic protein quality control and stress response in tumoral immune cells such as tumor-associated macrophages (TAMs), T cells, cancer-associated fibroblasts (CAFs), tumor endothelial cells (TECs), normal subtumoral epithelial cells, and other cells in the local and distant milieu.

1.5 Roles of LRP1 in Vesicle Traffic

LRP1/CD91 is a macromolecule that is composed of the extracellular α-subunit (approx. 500kD) and the transmembrane β-subunit (approx. 85kD). Although LRP1 is one of the receptors for ex-HSP90, this macromolecule receptor binds with many ligands that control signal transduction (Misra et al. 1995, 1999; Zilberberg et al. 2004; Yang and Williams 2017), endocytosis (Marynen et al. 1982; Actis Dato and Chiabrando 2018), transcytosis (Burgess and Stanley 1997; Fillebeen et al. 1999; Kawata et al. 2012; Jarosz-Griffiths et al. 2016), and exocytosis (Meng et al. 2011; Roy et al. 2015; Leca et al. 2016). Interestingly, a recent study has shown that LRP1 expression is significantly elevated on exosomes, especially in lung adenocarcinoma patients' sera (Ueda et al. 2014). Strong staining patterns of LRP1 were observed in stromal cells surrounding cancer cells in tissue sections from lung adenocarcinoma patients with poor clinical outcomes, while minimal expression of LRP1 in lung cancer cells was observed in the normal tissues (Meng et al. 2011). These reports suggested that the high level of serum LRP1-expressing exosomes might be secreted from stromal cells surrounding lung adenocarcinoma cells (Ueda et al. 2014). LRP1-rich EVs were also found in studies that model tumor milieu as follows. CAFs under physiopathologic conditions such as coculture with macrophages, under hypoxia and lipid deprivation, were enriched in EVs that contain annexin A6 (ANXA6)/LRP1/thrombospondin 1 (TSP1) complex (Leca et al. 2016). In this study, the ANXA6/LRP1/TSP1 complex was necessary for the uptake of these EVs by cancer cells and for the increased tumor cell aggressiveness (Leca et al. 2016). It is noteworthy that hypoxia induces LRP1 expression (Kawata et al. 2012) as well as HSP90α (Eguchi et al. 2018). Elevated expression of LRP1-HSP90α in tumor milieu might promote further tumor growth via LRP1-mediated control of vesicle molecular traffic.

In addition to roles for LRP1 in vesicle traffic, this receptor plays key roles in intracellular signaling including the Akt1/2 pathways (Tsen et al. 2013), STAT3 (Signal transducers and activator of transcription), and β-catenin signaling. It was shown that glioblastoma-derived plasminogen activator inhibitor 1 (PAI1) binding to LRP1 increased STAT3 phosphorylation and subsequent exocytosis in mast cells (Roy et al. 2015). It was also shown that LRP1 is a receptor for lipoproteins that alter canonical Wnt/β-catenin signals and sterol signals (Willnow et al. 2007). We recently showed lipoprotein transport to be a key for tumorigenesis (Namba et al. 2018).

1.6 ex-HSP90 Promotes Epithelial-Mesenchymal Transition (EMT)

TGFα/EGFR signaling is another inducer of secretion of ex-HSP90 (Cheng et al. 2011). We recently reported that EGF-EGFR signaling is essential for secretion of EGFR/HSP90-contained exosomes from OSCC cells (Fujiwara et al. 2018a, b). Interestingly, OSCC-derived EGFR/HSP90-containing exosomes had an ability to induce carcinogenic epithelial-mesenchymal transition (EMT) in the oral epithelial cell line RT7 (Fujiwara et al. 2018a, b). HSP90 is often found in cancer EVs and EV-free CMs (Eguchi et al. 2018; Ono et al. 2018). We hypothesize that ex-HSP90 may be enclosed in EVs and displayed on the outer surface of the EV membrane via its receptor EV-LRP1. The HSP90 inside EVs could also be horizontally transferred into recipient cells, which subsequently acquire multiple functions of cytoplasmic HSP90 as mentioned above. The HSP90 outside EVs could bind to cell surface receptors such as LRP1 on the surface of recipient cells leading to signal transduction and ligand-dependent endocytosis. As ex-HSP90 has been shown to promote EMT, EV-associated HSP90 can also promote EMT potentially via recipient cell surface receptor LRP1.

ex-HSP90-driven promotion of EMT was first reported by a group of Dr. Jennifer Isaacs in 2012 (Hance et al. 2012). This report showed that ex-HSP90 binding of LRP1 promoted EMT in prostate cancer cells via activation of ERK signaling. This group subsequently added a mechanism underlying ex-HSP90>LRP1>p-ERK>repression of E-cadherin gene by showing that polycomb group repressor EZH2 is induced under ex-HSP90>LRP1>pERK and the induced EZH2 repressed E-cadherin gene by histone H3 K27 methylation (Nolan et al. 2015). Alternatively, intracellular HSP90 was also shown to promote EMT, motility, and invasion of colorectal cancer cells via activation of HIF-1α and NF-κB (Nagaraju et al. 2015). Therefore, a positive feedback loop of hypoxia>HIF-1α>HSP90 (ic-HSP90 and ex-HSP90)>multiple signaling>EMT may play a key role in the promotion of cancer. Secreted ex-HSP90 promotes not only EMT but also heterogeneity of cancer stem cells or cancer-initiating cells (CSC/CIC) (Nolan et al. 2017). We recently reported that tumor organoids with CSC/CIC/EMT traits of prostate cancer cells profoundly secreted ex-HSP90α that may play autocrine and paracrine roles in tumor progression (Eguchi et al. 2018).

A new role of HSP90β was recently shown that promotes aggressive vasculogenic mimicry (VM) via EMT in hepatocellular carcinoma (HCC), a typical hypervascular solid tumor (Meng et al. 2019). In this study, HSP90β interacted with Twist1 and promoted its deubiquitination, stabilization, and nuclear translocation and enhanced the vascular endothelial (VE)-cadherin promoter activity. An HSP90 inhibitor NVP-BEP800 suppressed VM formation by releasing the HSP90β and Twist1 interaction. Such a new role of HSP90 in VM could also be mediated by HCC cell-derived HSP90-contained EVs.

1.7 LRP1- HSP90 Complexes Can Gather Extracellular Matrix (ECM) on the Surface of Cells and EVs

We have also identified many ECM proteins in the fraction of EVs (Ono et al. 2018). It was shown that fibronectin (FN) on the surface of myeloma cell-derived exosomes mediates exosome-cell interaction (Purushothaman et al. 2016). Dr. Adrienne Edkins group has reported that FN is a stress-responsive gene regulated by HSF1 (Dhanani et al. 2017) and is anchored to the plasma membrane by LRP1-HSP90 complex (Hunter et al. 2014; Boel et al. 2018). We reported that cancer EVs often carry abundant levels of FN, LRP1, and HSP90 (Ono et al. 2018). FN often coats EVs (Purushothaman et al. 2016). Therefore, EV-coating FNs can be anchored to the EV membrane via HSP90-LRP1. Not only FNs but also many species of ECM are found in the EV fractions (Ono et al. 2018). EV-coating ECM can physically interfere with membrane fusion and horizontally transfer from the EVs to recipient cells (Fig. 1.2a, left). EVs can be used as wheel-like adaptors that enable cells to drive (migrate) on the "metastasis highways" made by ECM in tissues, most notably cancer cells (Sung et al. 2015) (Fig. 1.2b). EVs can also be kept on the surface of cells and at intercellular space, where EVs mediate cell aggregation (Fig. 1.2c) (Eguchi et al. 2018). ECM proteins and transmembrane proteins such as LRP1, EpCAM (Eguchi et al. 2018), and integrins were found to be associated with both cells and EVs and might thus mediate such new roles of EVs in cell aggregation.

However, we showed that HSP-rich metastatic OSCC-EVs lost ECM whereas low-metastatic HSP-poor OSCC-EVs were ECM-rich. Therefore, HSP90 might not connect EVs with ECM. We also identified members of matricellular CCN protein family in cancer exosome fractions (unpublished data). CCNs have been shown to bind with ECM, transmembrane (TM) proteins such as LRP1 and integrins, and growth factors. Therefore, we prospect not HSP90 but CCN proteins are essential adaptors between EVs, TM proteins, and ECM and thus contribute for generation of ECM-coating of EVs. Importantly, CCN2/CTGF plays a key role in bone metastasis (Shimo et al. 2006), indicating that CCN-positive, ECM-positive EVs may play a key role in pre-metastatic niche formation in bone metastatic breast cancer and prostate cancer.

1.8 Roles of Rab Proteins in Exosome Secretion

MVBs are formed by endocytic budding from an endosomal membrane into the lumen side of the compartment such as late endosome (Fig. 1.1). After vesicular accumulation, the MVBs are either sorted for cargo degradation in the lysosome or released from the cells into the extracellular space as exosomes by fusion with the plasma membrane. Recently, a large number of Rab proteins have been associated with the exocytic pathway, including Rab3, Rab11, Rab26, Rab27, Rab37 and Rab38 (Masuda et al. 2000; Nashida et al. 2006; Wasmeier et al. 2006; Rupnik et al. 2007; Tolmachova et al. 2007; Takahashi et al. 2012). Rab proteins belonging to Ras GTPase superfamily are small GTPases (20–25 kDa) comprising more than 60 proteins in *homo sapiens*. They play pivotal roles in regulating intercellular membrane trafficking including endocytosis and exocytosis such as exosome secretion and vesicles delivery between organelles (Chavrier and Goud 1999; Pereira-Leal and Seabra 2000; Stenmark 2009). The Hsp90 chaperone complex regulates GDI-dependent Rab (Rab1 and Rab3A) recycling (Chen and Balch 2006). Among Rab members, Rab11, Rab27, and Rab35 have been shown to play crucial roles in exosome secretion. Rab11 exists as two isoforms, Rab11A and Rab11B, and involved in recycling from an endosome to the plasma membrane, so-called slow recycling. Overexpression of the wild-type of Rab11 slightly stimulated exosome secretion (Savina et al. 2002). On the other hand, the inhibition of Rab11 function by overexpression of a dominant-negative mutant decreased exosome release (Savina et al. 2002). Similar to Rab11, Rab27 is widely conserved and are existed two isoforms, Rab27A and Rab27B. Their regulation of exosome secretion was confirmed by using breast cancer cell lines. Their inhibition of both Rab27 A and B was observed the fewer exosomes into the culture medium (Zheng et al. 2013). Rab35 regulates a fast endocytic recycling pathway for a lot of proteins to the plasma membrane (Kouranti et al. 2006). The inhibition of Rab35 activity in oligodendrocytes leads to intracellular accumulation of endosomal vesicles and reduces exosome secretion (Hsu et al. 2010). Such roles of Rab proteins in vesicle molecular trafficking might play roles in controlling cell differentiation. Indeed, Rab27A regulates the transport of cell surface receptors, modulating multinucleation and lysosome-related organelles in osteoclasts (Shimada-Sugawara et al. 2015). Additionally, Rab44, a novel large Rab GTPase, negatively regulates osteoclast differentiation by modulating intracellular calcium levels followed by NFATc1 activation (Yamaguchi et al. 2018). Although it is likely that more Rab family proteins are involved in secretion of exosomes and cell differentiation, the detailed mechanism remains to be clarified.

1.9 Conclusions

The significance and potential of EV-HSP90 and ex-HSP90 in the progression of cancer and wound healing is becoming apparent (Table 1.1). HSP90-LRP1 binding to the surface of EVs and cells appears to impact both cellular and tissue homeostasis and tumor progression at local regions and distant milieu. On-going hypotheses and theories of HSP90/LRP1-EVs were also explored in the review and suggested new areas of investigation particularly in the cancer field.

Table 1.1 Classification and functions of intracellular and extracellular HSP90

Alias	Localization of HSP90	Mechanism of generation	Function
HSP90, ic-HSP90	Cytoplasm, mitochondria, ER	1. Translation of cellular mRNA 2. Translation of EV-derived mRNA upon horizontal transfer	Impact on cellular proteostasis network, including; 1. Autophagy 2. The UPR in the ER 3. Cytoplasmic protein quality control 4. The cytoplasmic stress response
ex-HSP90 (EV-free)	Extracellular space	1. Hypoxia and HIF-1alpha 2. TGFα-EGFR signal 3. Stemness 4. Membrane damages of cells and EVs upon injuries, inflammation and cell stress	1. Chaperokine role via binding to its receptor LRP1, whose signaling promote wound healing, cancer cell survival, and EMT 2. Activity control of extracellular proteins, e.g. MMP (Eustace et al. 2004)
EV-HSP90	Inside EVs	1. Intracellular HSP90 is engulfed and incorporated into exosomes when MVB is generated 2. Intracellular HSP90 is engulfed and incorporated into ectosomes, apoptotic bodies or large oncosomes 3. Heat shock stress increased EV-HSP90 (unpublished data)	1. EV-HSP90 can be released into cytoplasm of recipient cells after membrane fusion between EVs and cells 2. New MVBs can be generated in recipient cells after endocytosis or phagocytosis of EVs
	On the outer surface of EVs	1. ex-HSP90 are bound to LRP1 on the surface of EVs after/upon secretion 2. HSP90 kept bound to LRP1-EVs before secretion	1. Chaperokine role via binding to its receptor LRP1, whose signaling promote wound healing 2. EV-cell interaction mediated by HSP90 and/or LRP1 triggers membrane fusion, endocytosis or phagocytosis 3. Activity control of extracellular proteins, e.g. MMP

Acknowledgements This paper is dedicated to the memory of one of our mentors, Professor Ken-ichi Kozaki, who passed away on May 29, 2016. The authors thank Yuka Okusha, Chiharu Sogawa, Masaharu Takigawa, Sati Kubota, Akira Sasaki, Ayesha Murshid, Thomas Prince, and Benjamin Lang for useful discussion and helpful suggestion. This work was supported by JSPS KAKENHI, grant numbers JP17K11642 (TE), JP17K11643 (TE), JP17K11669 (TE), JP16K11863 (KOk), JP18K09789 (TE) and by SUZUKEN memorial foundation (TE).

References

Actis Dato V, Chiabrando GA (2018) The role of low-density lipoprotein receptor-related protein 1 in lipid metabolism, glucose homeostasis and inflammation. Int J Mol Sci 19(6)

Barile L, Vassalli G (2017) Exosomes: therapy delivery tools and biomarkers of diseases. Pharmacol Ther 174:63–78

Beveridge TJ (1999) Structures of gram-negative cell walls and their derived membrane vesicles. J Bacteriol 181:4725–4733

Boel NM, Hunter MC, Edkins AL (2018) LRP1 is required for novobiocin-mediated fibronectin turnover. Sci Rep 8:11438

Burgess JW, Stanley KK (1997) Estrogen-stimulated transcytosis of desialylated ligands and alpha2 macroglobulin in rat liver. Biochim Biophys Acta 1359:48–58

Carstens JL, Correa de Sampaio P, Yang D, Barua S, Wang H, Rao A, Allison JP, LeBleu VS, Kalluri R (2017) Spatial computation of intratumoral T cells correlates with survival of patients with pancreatic cancer. Nat Commun 8:15095

Chavrier P, Goud B (1999) The role of ARF and Rab GTPases in membrane transport. Curr Opin Cell Biol 11:466–475

Chen CY, Balch WE (2006) The Hsp90 chaperone complex regulates GDI-dependent Rab recycling. Mol Biol Cell 17:3494–3507

Cheng CF, Fan J, Fedesco M, Guan S, Li Y, Bandyopadhyay B, Bright AM, Yerushalmi D, Liang M, Chen M, Han YP, Woodley DT, Li W (2008) Transforming growth factor alpha (TGFalpha)-stimulated secretion of HSP90alpha: using the receptor LRP-1/CD91 to promote human skin cell migration against a TGFbeta-rich environment during wound healing. Mol Cell Biol 28:3344–3358

Cheng CF, Sahu D, Tsen F, Zhao Z, Fan J, Kim R, Wang X, O'Brien K, Li Y, Kuang Y, Chen M, Woodley DT, Li W (2011) A fragment of secreted Hsp90alpha carries properties that enable it to accelerate effectively both acute and diabetic wound healing in mice. J Clin Invest 121:4348–4361

Dhanani KCH, Samson WJ, Edkins AL (2017) Fibronectin is a stress responsive gene regulated by HSF1 in response to geldanamycin. Sci Rep 7:17617

Dong H, Zou M, Bhatia A, Jayaprakash P, Hofman F, Ying Q, Chen M, Woodley DT, Li W (2016) Breast Cancer MDA-MB-231 cells use secreted heat shock protein-90alpha (Hsp90alpha) to survive a hostile hypoxic environment. Sci Rep 6:20605

Eguchi T, Sogawa C, Okusha Y, Uchibe K, Iinuma R, Ono K, Nakano K, Murakami J, Itoh M, Arai K, Fujiwara T, Namba Y, Murata Y, Shimomura M, Okamura H, Takigawa M, Nakatsura T, Kozaki K, Okamoto K, Calderwood S (2018) Organoids with cancer stem cell-like properties secrete exosomes and HSP90 in a 3D nanoenvironment. PLoS One 13:e0191109

Eustace BK, Sakurai T, Stewart JK, Yimlamai D, Unger C, Zehetmeier C, Lain B, Torella C, Henning SW, Beste G, Scroggins BT, Neckers L, Ilag LL, Jay DG (2004) Functional proteomic screens reveal an essential extracellular role for hsp90 alpha in cancer cell invasiveness. Nat Cell Biol 6:507–514

Fais S, O'Driscoll L, Borras FE, Buzas E, Camussi G, Cappello F, Carvalho J, Cordeiro da Silva A, Del Portillo H, El Andaloussi S, Ficko Trcek T, Furlan R, Hendrix A, Gursel I, Kralj-Iglic V, Kaeffer B, Kosanovic M, Lekka ME, Lipps G, Logozzi M, Marcilla A, Sammar M, Llorente A,

Nazarenko I, Oliveira C, Pocsfalvi G, Rajendran L, Raposo G, Rohde E, Siljander P, van Niel G, Vasconcelos MH, Yanez-Mo M, Yliperttula ML, Zarovni N, Zavec AB, Giebel B (2016) Evidence-based clinical use of nanoscale extracellular vesicles in nanomedicine. ACS Nano 10:3886–3899
Fillebeen C, Descamps L, Dehouck MP, Fenart L, Benaissa M, Spik G, Cecchelli R, Pierce A (1999) Receptor-mediated transcytosis of lactoferrin through the blood-brain barrier. J Biol Chem 274:7011–7017
Fujita Y, Yoshioka Y, Ochiya T (2016) Extracellular vesicle transfer of cancer pathogenic components. Cancer Sci 107:385–390
Fujiwara T, Eguchi T, Sogawa C, Ono K, Murakami J, Ibaragi S, Asaumi J, Okamoto K, Calderwood S, Kozaki K (2018a) Anti-EGFR antibody cetuximab is secreted by oral squamous cell carcinoma and alters EGF-driven mesenchymal transition. Biochem Biophys Res Commun 503:1267–1272
Fujiwara T, Eguchi T, Sogawa C, Ono K, Murakami J, Ibaragi S, Asaumi J-i, Calderwood SK, Okamoto K, Kozaki K-i (2018b) Carcinogenic epithelial-mesenchymal transition initiated by oral cancer exosomes is inhibited by anti-EGFR antibody cetuximab. Oral Oncol 86:251–257
Ha D, Yang N, Nadithe V (2016) Exosomes as therapeutic drug carriers and delivery vehicles across biological membranes: current perspectives and future challenges. Acta Pharm Sin B 6:287–296
Hance MW, Dole K, Gopal U, Bohonowych JE, Jezierska-Drutel A, Neumann CA, Liu H, Garraway IP, Isaacs JS (2012) Secreted Hsp90 is a novel regulator of the epithelial to mesenchymal transition (EMT) in prostate cancer. J Biol Chem 287:37732–37744
Horibe S, Tanahashi T, Kawauchi S, Murakami Y, Rikitake Y (2018) Mechanism of recipient cell-dependent differences in exosome uptake. BMC Cancer 18:47
Hsu C, Morohashi Y, Yoshimura S, Manrique-Hoyos N, Jung S, Lauterbach MA, Bakhti M, Gronborg M, Mobius W, Rhee J, Barr FA, Simons M (2010) Regulation of exosome secretion by Rab35 and its GTPase-activating proteins TBC1D10A-C. J Cell Biol 189:223–232
Hunter MC, O'Hagan KL, Kenyon A, Dhanani KC, Prinsloo E, Edkins AL (2014) Hsp90 binds directly to fibronectin (FN) and inhibition reduces the extracellular fibronectin matrix in breast cancer cells. PLoS One 9:e86842
Jarosz-Griffiths HH, Noble E, Rushworth JV, Hooper NM (2016) Amyloid-beta receptors: the good, the bad, and the prion protein. J Biol Chem 291:3174–3183
Jayaprakash P, Dong H, Zou M, Bhatia A, O'Brien K, Chen M, Woodley DT, Li W (2015) Hsp90alpha and Hsp90beta together operate a hypoxia and nutrient paucity stress-response mechanism during wound healing. J Cell Sci 128:1475–1480
Kalluri R (2016) The biology and function of exosomes in cancer. J Clin Invest 126:1208–1215
Kalluri R, LeBleu VS (2016) Discovery of double-stranded genomic DNA in circulating exosomes. Cold Spring Harb Symp Quant Biol 81:275–280
Kawata K, Kubota S, Eguchi T, Aoyama E, Moritani NH, Kondo S, Nishida T, Takigawa M (2012) Role of LRP1 in transport of CCN2 protein in chondrocytes. J Cell Sci 125:2965–2972
Koong AC, Denko NC, Hudson KM, Schindler C, Swiersz L, Koch C, Evans S, Ibrahim H, Le QT, Terris DJ, Giaccia AJ (2000) Candidate genes for the hypoxic tumor phenotype. Cancer Res 60:883–887
Kouranti I, Sachse M, Arouche N, Goud B, Echard A (2006) Rab35 regulates an endocytic recycling pathway essential for the terminal steps of cytokinesis. Curr Biol 16:1719–1725
LePage GA (1948) Phosphorylated intermediates in tumor glycolysis; effects of anoxia and hyperglycemia. Cancer Res 8:201
Leca J, Martinez S, Lac S, Nigri J, Secq V, Rubis M, Bressy C, Serge A, Lavaut MN, Dusetti N, Loncle C, Roques J, Pietrasz D, Bousquet C, Garcia S, Granjeaud S, Ouaissi M, Bachet JB, Brun C, Iovanna JL, Zimmermann P, Vasseur S, Tomasini R (2016) Cancer-associated fibroblast-derived annexin A6+ extracellular vesicles support pancreatic cancer aggressiveness. J Clin Invest 126:4140–4156

Li W, Li Y, Guan S, Fan J, Cheng CF, Bright AM, Chinn C, Chen M, Woodley DT (2007) Extracellular heat shock protein-90alpha: linking hypoxia to skin cell motility and wound healing. EMBO J 26:1221–1233

Ma L, Li Y, Peng J, Wu D, Zhao X, Cui Y, Chen L, Yan X, Du Y, Yu L (2015) Discovery of the migrasome, an organelle mediating release of cytoplasmic contents during cell migration. Cell Res 25:24–38

Marynen P, Van der Schueren B, Van Leuven F, Cassiman JJ, Van den Berghe H (1982) Absence of specific binding of receptor-mediated endocytosis, and of secretion of alpha-2-macroglobulin by cultured endothelial cells. Haemostasis 11:210–214

Masuda ES, Luo Y, Young C, Shen M, Rossi AB, Huang BC, Yu S, Bennett MK, Payan DG, Scheller RH (2000) Rab37 is a novel mast cell specific GTPase localized to secretory granules. FEBS Lett 470:61–64

Mathiyalagan P, Liang Y, Kim D, Misener S, Thorne T, Kamide CE, Klyachko E, Losordo DW, Hajjar RJ, Sahoo S (2017) Angiogenic mechanisms of human CD34+ stem cell exosomes in the repair of ischemic hindlimb. Circ Res 120:1466–1476

Melentijevic I, Toth ML, Arnold ML, Guasp RJ, Harinath G, Nguyen KC, Taub D, Parker JA, Neri C, Gabel CV, Hall DH, Driscoll M (2017) *C. elegans* neurons jettison protein aggregates and mitochondria under neurotoxic stress. Nature 542:367–371

Mendt M, Kamerkar S, Sugimoto H, McAndrews KM, Wu CC, Gagea M, Yang S, Blanko EVR, Peng Q, Ma X, Marszalek JR, Maitra A, Yee C, Rezvani K, Shpall E, LeBleu VS, Kalluri R (2018) Generation and testing of clinical-grade exosomes for pancreatic cancer. JCI Insight 3:8

Meng H, Chen G, Zhang X, Wang Z, Thomas DG, Giordano TJ, Beer DG, Wang MM (2011) Stromal LRP1 in lung adenocarcinoma predicts clinical outcome. Clin Cancer Res 17:2426–2433

Meng J, Chen S, Lei YY, Han JX, Zhong WL, Wang XR, Liu YR, Gao WF, Zhang Q, Tan Q, Liu HJ, Zhou HG, Sun T, Yang C (2019) Hsp90beta promotes aggressive vasculogenic mimicry via epithelial-mesenchymal transition in hepatocellular carcinoma. Oncogene 38:228–243

Minciacchi VR, Spinelli C, Reis-Sobreiro M, Cavallini L, You S, Zandian M, Li X, Mishra R, Chiarugi P, Adam RM, Posadas EM, Viglietto G, Freeman MR, Cocucci E, Bhowmick NA, Di Vizio D (2017a) MYC mediates large oncosome-induced fibroblast reprogramming in prostate cancer. Cancer Res 77:2306–2317

Minciacchi VR, Zijlstra A, Rubin MA, Di Vizio D (2017b) Extracellular vesicles for liquid biopsy in prostate cancer: where are we and where are we headed? Prostate Cancer Prostatic Dis 20:251–258

Misra UK, Gawdi G, Pizzo SV (1995) Ligation of the alpha 2-macroglobulin signalling receptor on macrophages induces protein phosphorylation and an increase in cytosolic pH. Biochem J 309:151–158

Misra UK, Gawdi G, Pizzo SV (1999) Ligation of low-density lipoprotein receptor-related protein with antibodies elevates intracellular calcium and inositol 1,4, 5-trisphosphate in macrophages. Arch Biochem Biophys 372:238–247

Nagaraju GP, Long TE, Park W, Landry JC, Taliaferro-Smith L, Farris AB, Diaz R, El-Rayes BF (2015) Heat shock protein 90 promotes epithelial to mesenchymal transition, invasion, and migration in colorectal cancer. Mol Carcinog 54:1147–1158

Najafi M, Goradel NH, Farhood B, Salehi E, Solhjoo S, Toolee H, Kharazinejad E, Mortezaee K (2019) Tumor microenvironment: interactions and therapy. J Cell Physiol 234:5700–5721

Nakase I, Kobayashi NB, Takatani-Nakase T, Yoshida T (2015) Active macropinocytosis induction by stimulation of epidermal growth factor receptor and oncogenic Ras expression potentiates cellular uptake efficacy of exosomes. Sci Rep 5:10300

Namba Y, Sogawa C, Okusha Y, al e (2018) Depletion of lipid efflux pump ABCG1 triggers the intracellular accumulation of extracellular vesicles and reduces aggregation and tumorigenesis of metastatic cancer cells. Front Oncol 8:376

Nashida T, Imai A, Shimomura H (2006) Relation of Rab26 to the amylase release from rat parotid acinar cells. Arch Oral Biol 51:89–95

Neckers L, Blagg B, Haystead T, Trepel JB, Whitesell L, Picard D (2018) Methods to validate Hsp90 inhibitor specificity, to identify off-target effects, and to rethink approaches for further clinical development. Cell Stress Chaperones 23:467–482

Nolan KD, Franco OE, Hance MW, Hayward SW, Isaacs JS (2015) Tumor-secreted Hsp90 subverts polycomb function to drive prostate tumor growth and invasion. J Biol Chem 290:8271–8282

Nolan KD, Kaur J, Isaacs JS (2017) Secreted heat shock protein 90 promotes prostate cancer stem cell heterogeneity. Oncotarget 8:19323–19341

Ono K, Eguchi T, Sogawa C, Calderwood SK, Futagawa J, Kasai T, Seno M, Okamoto K, Sasaki A, Kozaki KI (2018) HSP-enriched properties of extracellular vesicles involve survival of metastatic oral cancer cells. J Cell Biochem 119:7363–7376

Peinado H, Lavotshkin S, Lyden D (2011) The secreted factors responsible for pre-metastatic niche formation: old sayings and new thoughts. Semin Cancer Biol 21:139–146

Pereira-Leal JB, Seabra MC (2000) The mammalian Rab family of small GTPases: definition of family and subfamily sequence motifs suggests a mechanism for functional specificity in the Ras superfamily. J Mol Biol 301:1077–1087

Purushothaman A, Bandari SK, Liu J, Mobley JA, Brown EE, Sanderson RD (2016) Fibronectin on the surface of myeloma cell-derived exosomes mediates exosome-cell interactions. J Biol Chem 291:1652–1663

Roy A, Coum A, Marinescu VD, Polajeva J, Smits A, Nelander S, Uhrbom L, Westermark B, Forsberg-Nilsson K, Ponten F, Tchougounova E (2015) Glioma-derived plasminogen activator inhibitor-1 (PAI-1) regulates the recruitment of LRP1 positive mast cells. Oncotarget 6:23647–23661

Rupnik M, Kreft M, Nothias F, Grilc S, Bobanovic LK, Johannes L, Kiauta T, Vernier P, Darchen F, Zorec R (2007) Distinct role of Rab3A and Rab3B in secretory activity of rat melanotrophs. Am J Phys Cell Phys 292:C98–C105

Savina A, Vidal M, Colombo MI (2002) The exosome pathway in K562 cells is regulated by Rab11. J Cell Sci 115:2505–2515

Shimada-Sugawara M, Sakai E, Okamoto K, Fukuda M, Izumi T, Yoshida N, Tsukuba T (2015) Rab27A regulates transport of cell surface receptors modulating multinucleation and lysosome-related organelles in osteoclasts. Sci Rep 5:9620

Shimo T, Kubota S, Yoshioka N, Ibaragi S, Isowa S, Eguchi T, Sasaki A, Takigawa M (2006) Pathogenic role of connective tissue growth factor (CTGF/CCN2) in osteolytic metastasis of breast cancer. J Bone Miner Res 21:1045–1059

Stenmark H (2009) Rab GTPases as coordinators of vesicle traffic. Nat Rev Mol Cell Biol 10:513–525

Sung BH, Ketova T, Hoshino D, Zijlstra A, Weaver AM (2015) Directional cell movement through tissues is controlled by exosome secretion. Nat Commun 6:7164

Takahashi S, Kubo K, Waguri S, Yabashi A, Shin HW, Katoh Y, Nakayama K (2012) Rab11 regulates exocytosis of recycling vesicles at the plasma membrane. J Cell Sci 125:4049–4057

Taverna S, Pucci M, Giallombardo M, Di Bella MA, Santarpia M, Reclusa P, Gil-Bazo I, Rolfo C, Alessandro R (2017) Amphiregulin contained in NSCLC-exosomes induces osteoclast differentiation through the activation of EGFR pathway. Sci Rep 7:3170

Tolmachova T, Abrink M, Futter CE, Authi KS, Seabra MC (2007) Rab27b regulates number and secretion of platelet dense granules. Proc Natl Acad Sci U S A 104:5872–5877

Tsen F, Bhatia A, O'Brien K, Cheng CF, Chen M, Hay N, Stiles B, Woodley DT, Li W (2013) Extracellular heat shock protein 90 signals through subdomain II and the NPVY motif of LRP-1 receptor to Akt1 and Akt2: a circuit essential for promoting skin cell migration in vitro and wound healing in vivo. Mol Cell Biol 33:4947–4959

Ueda K, Ishikawa N, Tatsuguchi A, Saichi N, Fujii R, Nakagawa H (2014) Antibody-coupled monolithic silica microtips for highthroughput molecular profiling of circulating exosomes. Sci Rep 4:6232

van Niel G (2016) Study of exosomes shed new light on physiology of amyloidogenesis. Cell Mol Neurobiol 36:327–342

Wasmeier C, Romao M, Plowright L, Bennett DC, Raposo G, Seabra MC (2006) Rab38 and Rab32 control post-Golgi trafficking of melanogenic enzymes. J Cell Biol 175:271–281
Willnow TE, Hammes A, Eaton S (2007) Lipoproteins and their receptors in embryonic development: more than cholesterol clearance. Development 134:3239–3249
Yamaguchi Y, Sakai E, Okamoto K, Kajiya H, Okabe K, Naito M, Kadowaki T, Tsukuba T (2018) Rab44, a novel large Rab GTPase, negatively regulates osteoclast differentiation by modulating intracellular calcium levels followed by NFATc1 activation. Cell Mol Life Sci 75:33–48
Yang T, Williams BO (2017) Low-density lipoprotein receptor-related proteins in skeletal development and disease. Physiol Rev 97:1211–1228
Zhang H, Freitas D, Kim HS, Fabijanic K, Li Z, Chen H, Mark MT, Molina H, Martin AB, Bojmar L, Fang J, Rampersaud S, Hoshino A, Matei I, Kenific CM, Nakajima M, Mutvei AP, Sansone P, Buehring W, Wang H, Jimenez JP, Cohen-Gould L, Paknejad N, Brendel M, Manova-Todorova K, Magalhaes A, Ferreira JA, Osorio H, Silva AM, Massey A, Cubillos-Ruiz JR, Galletti G, Giannakakou P, Cuervo AM, Blenis J, Schwartz R, Brady MS, Peinado H, Bromberg J, Matsui H, Reis CA, Lyden D (2018) Identification of distinct nanoparticles and subsets of extracellular vesicles by asymmetric flow field-flow fractionation. Nat Cell Biol 20:332–343
Zheng Y, Campbell EC, Lucocq J, Riches A, Powis SJ (2013) Monitoring the Rab27 associated exosome pathway using nanoparticle tracking analysis. Exp Cell Res 319:1706–1713
Zijlstra A, Di Vizio D (2018) Size matters in nanoscale communication. Nat Cell Biol 20:228–230
Zilberberg A, Yaniv A, Gazit A (2004) The low density lipoprotein receptor-1, LRP1, interacts with the human frizzled-1 (HFz1) and down-regulates the canonical Wnt signaling pathway. J Biol Chem 279:17535–17542

Chapter 2
HSP90-Based Heterocomplex as Essential Regulator for Cancer Disease

Mario D. Galigniana

Abstract Heat-shock protein 90 (HSP90) is a molecular chaperone that represents the most abundant soluble protein of the cell. This chaperone shows intrinsic ATPase activity and associates to a great number of client factors such as steroid receptors, tyrosine-kinases, transcription factors, enzymes, cytoskeletal proteins, channels, histones, etc. Because of these interactions, HSP90 is implicated in diverse biological processes that require critical and coordinated regulatory mechanisms to modulate its activity. HSP90 forms heterocomplexes with other chaperones and co-chaperones such as TPR-domain immunophilins that modulate HSP90 properties and are themselves subject of regulation. In malignancy, HSP90 is essential to preserve the metastable forms of oncoproteins that are usually mutated, overamplified, and sometimes translocated from their normal subcellular compartments. Thus, HSP90 helps to buffer the proteostasis of the cell that is assaulted by the onset of the stress generated by the malignant condition. Consequently, HSP90 is an attractive pharmacologic target, and the pharmaceutical industry has generated over the last years several advances in both the basic biology and the translational drug development around HSP90. This chapter is focused on the multiple aspects by which the HSP90 activity and that related to HSP90-binding proteins, particularly immunophilins, are regulated in malignancy.

Keywords Cancer · HSP90 inhibitors · Immunophilins · Steroid receptors · Stress response · TPR domain proteins

---

M. D. Galigniana (✉)
Instituto de Biología y Medicina Experimental (IBYME)-CONICET & Departamento de Química Biológica, Facultad de Ciencias Exactas y Naturales, Universidad de Buenos Aires, Buenos Aires, Argentina
e-mail: mgaligniana@conicet.gov.ar

A. A. A. Asea, P. Kaur (eds.), *Heat Shock Protein 90 in Human Diseases and Disorders*, Heat Shock Proteins 19,
https://doi.org/10.1007/978-3-030-23158-3_2

Abbreviations

CyP	Cyclophilin
FKBP	FK506-Binding Protein
GA	Geldanamycin
HSP	Heat shock protein
hTERT	Reverse transcriptase subunit of human telomerase
TPR	Tetratricopeptide repeats

2.1 Introduction

From a social perspective, the term chaperone referred to a person, usually a matron, who used to accompany young people at a social gathering to supervise and ensure their proper behavior in public. By analogy, proteins that assist others in their correct folding and proper biological actions are also referred to as 'chaperones' (e.g. HSP90 or HSP70) (Quinta et al. 2011; Ellis 1996; Pearl and Prodromou 2006). The term 'molecular chaperone' was first coined for nucleoplasmin, a protein that shows the ability to prevent the aggregation of histones with DNA during the nucleosome assembly. Then, the term was also extended to those proteins that mediate post-translational assembly of other proteins (see (Ellis 1993) for an appealing review). On the other hand, the term 'heat-shock protein' (HSP) stems from the fact that upon the onset of heat-stress, the production of a specific class of proteins subsequently recognized as molecular chaperones is greatly enhanced. This implies that all HSP are molecular chaperones, but the opposite is not true since there are proteins showing the ability to function as molecular chaperones that are not induced by heat stress, for example TRiC/CCT. Finally, the term co-chaperone refers to proteins that assist chaperones in their biological function or are associated to them influencing the properties of the client-protein (e.g. p23 or FKBP52). The main biological function for all these proteins is to assist the proper folding of their client-proteins preventing both newly synthesized polypeptide chains and assembled subunits from aggregating into non-functional structures. Based on their molecular masses, the HSP family of molecular chaperones has been classified into six major subfamilies—HSP100, HSP90, HSP70, HSP60, HSP40, and the small HSP, the latter being a group characterized by the presence of subunit of 12–43 kDa and the presence of a highly conserved stretch of 80–100 amino acids in their C-terminal domains called the "α-crystallin domain" (Bakthisaran et al. 2015).

In view of the fact that even at physiological temperatures the stability of cellular proteins is relatively low, and aggregation competes with their efficient folding, the need for chaperone assistance during de novo protein folding and refolding of non-native polypeptide chains is almost permanent (Mayer 2010; Richter et al. 2010). The exposure of hydrophobic residues is a cardinal condition to discriminate a native protein from its partially or globally unfolded, non-native counterpart.

Normally, binding of chaperones to hydrophobic surfaces and structural regions of the non-native protein occurs with the purpose to protect these proteins from denaturation, aggregation and/or degradation. It is accepted that molecular chaperones do not contribute to the structural information for protein folding, but they prevent the formation of unwanted intermolecular interactions. The mechanism by which chaperones do this work is through the controlled binding and release of non-native proteins, an event that is usually accomplished by affinity changes for the client substrate of the chaperone. Normally, such exchange between two or more affinity states is regulated by the binding of the chaperone followed by ATP hydrolysis (except in the case of sHSP that lack ATPase intrinsic activity).

2.2 Structural Properties of HSP90

HSP90 is the major soluble protein present under both normal (1–3% of the total) and stressful conditions (6–10% of the total) (Hoter et al. 2018; Echeverria and Picard 2010). Mammals express two isoforms, HSP90α and HSP90β, which are encoded by two different genes originated ~500 Myrs ago. While HSP90α is inducible under stress conditions, HSP90β is a constitutively expressed isoform (Langer et al. 2003). HSP90 exists as homodimer, each protomer containing three flexibly linked and quite conserved regions from bacteria to mammals—the N-terminal ATP-binding domain (N-domain), the middle domain (M-domain), and the C-terminal dimerization domain (C-domain). An interesting structural property of the N-terminal end is the presence of various conserved amino acids that form a sort of lid closing over the ATP-binding pocket in the ATP-bound isoform, but it is open in the ADP-bound isoform state of the chaperone (Ali et al. 2006). The middle M-domain shows the crucial catalytic residues of the ATPase site responsible for ATP hydrolysis. This domain also contributes to the binding sites of client proteins and some co-chaperones (Meyer et al. 2004). The C-terminal domain is responsible for HSP90 dimerization, it is anticorrelated to the closing of the N-terminal domain (Ratzke et al. 2010) and shows a highly conserved MEEVD motif at the very C-terminal end that functions as the docking site for co-chaperones showing a tetratricopeptide repeat (TPR) clamp (Scheufler et al. 2000). Mutations of this sequence may abolish the recognition of TPR domain co-chaperones such as immunophilins, PP5 or Hop/p60 (Ramsey et al. 2009).

The homodimers are primarily cytosolic, although a small fraction of the chaperone is located in the nucleus, and that amount increases upon the onset of stress. In the absence of nucleotide, HSP90 adopts a V-shaped open conformation (Fig. 2.1), whereas ATP binding triggers a series of sequential conformational changes including repositioning of the N-terminal lid region and a drastic modification in the orientation of the N-M region. At the final point, HSP90 reaches a compacted conformation usually named "closed isoform" where the N-terminal ends associate generating the closed dimer (for a detailed review, see Li and Buchner 2013).

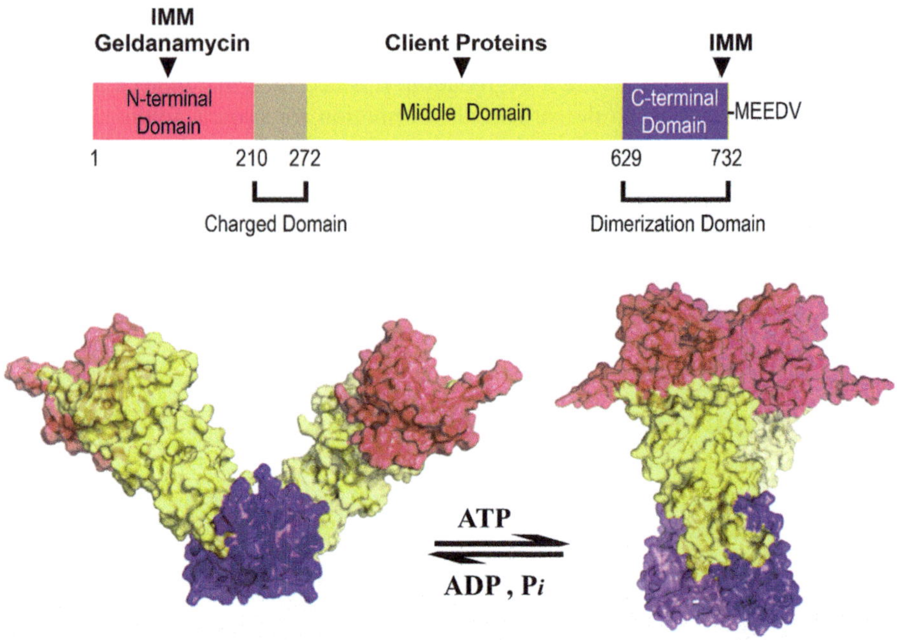

FIG. 1

Fig. 2.1 Structure of HSP90. The N-terminal domain is depicted in red. It binds ATP and inhibitors such as geldanamycin (GA). The middle M-domain is shown in yellow. It is the main region (but not the only one) where client proteins bind. The C-terminal domain is depicted in purple. It shows the MEEVD motif where HSP90 and HSP70 bind, and also the TPR acceptor site where immunophilins (IMM) and other TPR proteins are recruited. The three-dimensional crystal structures show the V-shaped open conformation that closes when ATP binds to the N-terminal triggering sequential conformational changes

HSP90 is a phosphoprotein containing a number of phosphorylated serine, threonine, and tyrosine residues (see Mollapour and Neckers 2012 for an updated revision). This post-translational modification is thought to enhance HSP90 function and its interaction with both client proteins and co-chaperones. Interestingly, both the dephosphorylated and hyperphosphorylated states seem to impair HSP functions (Scroggins and Neckers 2007). A number of protein-kinases are able to phosphorylate HSP90 such as CK2 protein kinase, double-stranded DNA protein kinase, protein kinase A (PKA), c-Src kinase, Swe1Wee1 kinase, etc. (Mollapour and Neckers 2012; Li and Buchner 2013). Interestingly, many of the kinases are HSP90 client proteins. It should be noted that, in addition to phosphorylation, HSP90 is also affected by other post-translational modifications such as acetylation by p300 (Yang et al. 2008) and deacetylation by HDAC6 (Kovacs et al. 2005), as well as S-nitrosylation by endothelial nitric oxide synthase (Martinez-Ruiz et al. 2005).

Both modifications (acetylation and S-nitrosylation) weakens the chaperone activity of HSP90 over the client-protein. Under oxidative stress conditions, HSP90 also undergoes thiol oxidation, which inhibits its chaperone function and targets the client-protein to proteasomal degradation (Chen et al. 2008).

Hsp90 shows intrinsic ATPase activity, and most drugs designed to inhibit HSP90 function target the ATP-binding domain preventing this activity. However, no direct correlation between the inhibitory action of the drug and ATPase activity has been demonstrated. Regarding the chaperoning action of HSP90, one particular property of this chaperone is the fact that it is highly selective with respect to other conventional chaperones and shows a relatively low affinity for unfolded proteins. Actually, the principal role of HSP90 in the cell is to provide biological activity to properly folded client proteins with a preserved tertiary structure, such that HSP90 functions as a delicate and refined sensor of protein function rather than a gross folding enhancer. For example, the presence of HSP90 in steroid receptors determine their ability to bind hormone, and a similar observation is also valid for protein-kinases for their enzymatic activity, but this does not mean that both inactive client-proteins are unfolded without HSP90 (Mazaira et al. 2014, 2018).

2.3 The Steroid Receptor•HSP90•Immunophilin Heterocomplex

Steroid receptors bind their cognate ligands only if they are assembled with HSP90 in oligomeric structures (Pratt 1997). This is particularly notorious and sensitive for some members of the steroid receptor family such as GR, MR, PR and AR (Pratt and Toft 1997). These receptors are assembled with molecular chaperones and co-chaperones, including HSP90, HSP70, HSP40, the HSP90 co-chaperone p23 and a TPR-domain protein, usually a high molecular weight immunophilin such as FKBP51, FKBP52, CyP40 or PP5. TPR domains consist of at least three motifs in tandem that are highly degenerate repeats of 34 amino acids (Scheufler et al. 2000). The interaction of TPR domains with HSP90 and HSP70 are conserved in nature and broadly distributed in the animal and plant kingdoms (Harrell et al. 2002; Pratt et al. 2001; Erlejman et al. 2014b; Bohne et al. 2016).

Some of the TPR-domain co-chaperones that form complexes with steroid receptors belong to the immunophilin family of proteins (e.g. FKBP51, FKBP52, Cyp40, etc.). Immunophilins are characterized by the presence of an additional, signature domain, the PPIase domain, that shows peptidyl-prolyl-(cis/trans)-isomerase activity, i.e., the reversible cis↔trans interconversion of Xaa-Pro bonds. Also, this domain is the binding site for exogenous ligands with immunosuppressant properties (Erlejman et al. 2013; Barik 2006). These compounds define two main subfamilies of immunophilins—the cyclophilins (or CyPs) when they bind the undecapeptide cyclosporine A, and the FK506-binding proteins (or FKBPs) when they bind the macrolide FK506 (tacrolimus). Normally, FKBP proteins also recognize another macrolide compound, rapamycin (sirolimus). The ability to function as immunosuppressive drugs led to a great success in the transplantation field to avoid pharmacologically the rejection of foreign tissue. However not all immunophilins show the property to be intracellular receptors of these drugs and mediate the immunosuppression effect, but the smallest members of each subfamily only, i.e. CyPA (17-kDa) and FKBP12 (12-kDa) (Sigal and Dumont 1992; Wiederrecht et al. 1995). Immunosuppression occurs when the corresponding drug binds the proper immunophilin (cyclosporine A for CyPA or FK506 for FKBP12), such that the drug•protein complex inhibits the activity of PP2B/calcineurin (Dumont 2000). This prevents the dephosphorylation of the transcription factor NFAT (Nuclear Factor of Activated T cells), its subsequent nuclear translocation, and the production of interleukines and interferon-γ. On the other hand, the immunosuppressive FKBP12•rapamycin complex shows a different mechanism of action (Wiederrecht et al. 1995) since it targets mTOR (Mammalian Target of Rapamycin), a Ser/Thr protein kinase able to regulate cell proliferation, cell growth, cell motility, and protein synthesis (Zhou and Huang 2010).

In the absence of steroid, some members of the steroid receptor family such as GR reside primarily in the cytoplasm, whereas others are constitutively nuclear, for example ER. Regardless of their primary location, receptors are not confined to any particular cell compartment and they continuously shuttle between cytoplasm and nucleus (Galigniana et al. 2010a). For years, it was assumed that simple diffusion was the driving force for receptor movement, and that the molecular chaperone HSP90 must be dissociated from the receptor upon steroid binding to release the transcription factor from cytoplasmic anchorage sites. However, the observation that motor proteins of the dynein/dynactin motor complex co-immunoprecipitate with GR (Galigniana et al. 2001) and MR (Galigniana et al. 2010b) via the immunophilin FKBP52 suggested a role for FKBP52 to facilitate the retrograde movement of the receptor (Galigniana 2012; Storer et al. 2011). The highly homologous partner FKBP51, an immunophilin that shares more than 70% similitude with FKBP52, antagonizes the effect of the later since FKBP51 does not bind dynein (Wochnik et al. 2005). In line with these properties, it has been demonstrated that FKBP51 is present in unliganded receptor, whereas it is exchanged by FKBP52 when the steroid binds (Gallo et al. 2007; Galigniana et al. 2010b)(Fig. 2.2). Moreover, it was proved that the dissociation of the HSP90-based complex and the subsequent dimerization of the receptor takes place in the nucleus rather than in the cytoplasm, as it has always been believed (Galigniana et al. 2010b; Grossmann et al. 2012).

The trafficking model depicted in Fig. 2.2 was also proved for other models (Galigniana et al. 2010b; Echeverria and Picard 2010; Wochnik et al. 2005; Ebong et al. 2016; Vandevyver et al. 2012; Tatro et al. 2009) changing substantially the classic molecular model for the mechanism of action of steroid receptors posed in the literature since the '80s, and was also expanded to other soluble proteins that share the same molecular chaperone machinery (Vafopoulou and Steel 2012; Schuster et al. 2017; Erlejman et al. 2014a; Galigniana et al. 2004; Colo et al. 2008;

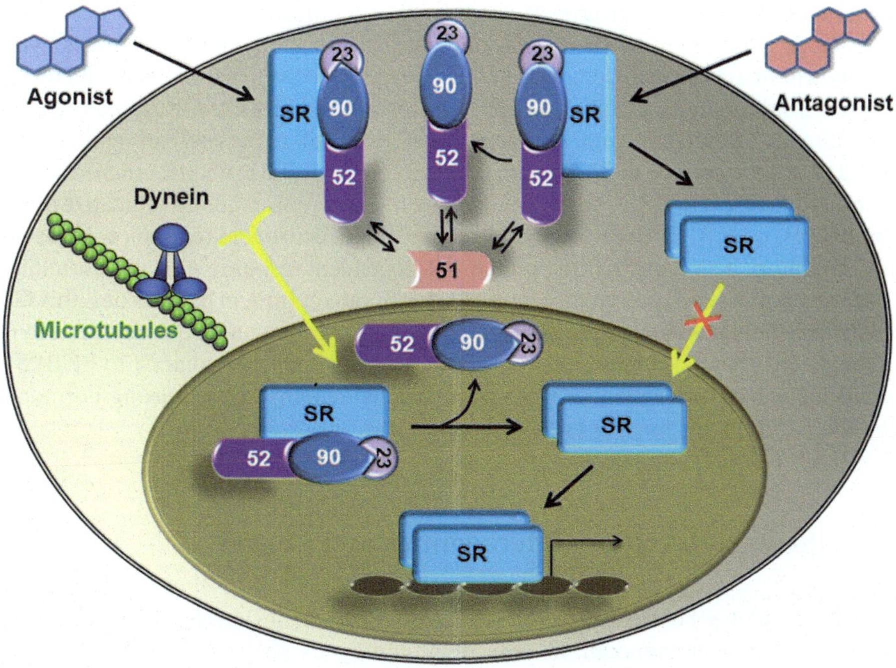

Fig. 2.2 Model of steroid receptor activation. In the absence of ligand, steroid receptors (SR) such as GR, MR or AR are primarily located in the cytoplasm associated to the HSP90•FKBP51 heterocomplex. Upon steroid binding, the immunophilin FKBP51 is exchanged by FKBP52, an immunophilin able to bind dynein/dynactin. Thus, the SR heterocomplex is transported towards the nucleus via the motor protein complex. Then, SR translocates to the nucleoplasm still associated to the HSP90 heterocomplex, such that SR transformation (i.e. the dissociation of the HSP90 complex) occurs in the nucleoplasmic milieu followed by SR homodimerization. The receptor is finally targeted to the promoter binding-sites to trigger the proper biological response. When an antagonist binds (for example RU486 for GR or spironolactone for MR), SR transformation and homodimerization can also take place, but in the cytoplasm rather than in the nucleus. This implies that the steroid-receptor complex translocation to the nucleus is impaired. FKBP51 is a very weak dynein-interacting protein. Yellow arrows represent the active transport of the SR complex

Lagadari et al. 2016; Jeong et al. 2016). Because the association of dynein with FKBPs has also been demonstrated in plant systems (Harrell et al. 2002; Pratt et al. 2001), the functional role of this complex seems to be preserved along the evolution. Importantly, the disruption of HSP90 function is critical to abrogate receptor transport, which is not surprising if we consider that this chaperone is the gravity centre of the trafficking machinery. The most relevant extrapolation of this property is the fact that drugs able to interfere with this system impair the biological action of the client-protein. This is the basis of ongoing clinical trials that are testing HSP90 inhibitors in cancer treatment (Jhaveri et al. 2014; Chatterjee and Burns 2017).

Another important finding was related to the hormone-dependent biological activity of steroid receptors, which is itself affected by immunophilins. It was known for many years the property of New World primates to show glucocorticoid resistance with high levels of plasma cortisol, but the molecular bases were unknown. In 2001 it was demonstrated in squirrel monkeys that FKBP51 decreases both steroid binding capacity and transcriptional activity of the GR, properties that are directly correlated with the very high levels of expression of endogenous FKBP51 in the cells of these primates (Scammell et al. 2001) On the other hand, the expression of FKBP52 is significantly lower. These observations indicated that a high level of FKBP51 expression contributes to glucocorticoid resistance. Later on, the inhibitory action of FKBP51 on the GR-dependent response was also correlated with the expression of certain isoforms of this immunophilin in the neurons, this GR resistance being associated to the development of stress-related post-traumatic syndrome and other psychiatric disorders (Binder et al. 2004). In contrast to FKBP51, FKBP52 selectively enhances GR transcriptional activity by increasing hormone-binding affinity and nuclear retrotransport.

2.4 Steroid Receptor Heterocomlex and Cancer

According to these observations, one of the obvious conclusions is that, even though HSP90-binding immunophilins are ubiquitously expressed, alterations in the expression balance in a given tissue could impact on receptor function. In this regard, this is particularly important for the cases of hormone-dependent breast and prostate cancers, where ERα and AR are the main targets for anti-hormonal therapy. Interestingly, the expression of both immunophilins CyP40 and FKBP52 is up-regulated in breast cancer compared to normal tissue (Ward et al. 1999; Ratajczak 2015; Gougelet et al. 2005). Accordingly, both immunophilins are the main co-chaperones associated to HSP90 in these complexes. A contrasting pattern of protein expression was observed in a panel of breast cancer cell lines, where HSP90•FKBP52 complexes are more frequent and abundant than HSP90•CyP40 complexes (Ward et al. 1999). The expression of ERα is strongly correlated with higher levels of FKBP52, and it is accepted that relative loss of CyP40 compromises ERα function by altering immunophilin composition in receptor•HSP90 heterocomplexes. This impact anti-oestrogen resistance resulting in phenotypic changes in breast cancer.

In prostate cancer, FKBP52 is also highly expressed (Periyasamy et al. 2007; Lin et al. 2007) and favours AR signalling. This may be of particular importance in cases of androgen ablation therapy, where hormone levels are reduced, but they can still stimulate AR•HSP90 complexes (Mostaghel et al. 2007). Studies where needle prostate biopsy specimens were analysed concluded that FKBP52 is a reliable biomarker of prostate cancer (Lin et al. 2007). Its close-related partner FKBP51 is also overexpressed in this type of cancers and shows the ability to stimulate AR activity (Periyasamy et al. 2010), the immunophilin being itself a product of the AR activity.

This generates a harmful feedback circuit that is worsened by the fact that FKBP51 impairs the biological activity of the GR. Thus, it is accepted that ligand-activated GR attenuates androgen-activated AR gene expression, suggesting an AR- dependent mechanism through which GR could act as a tumour suppressor in prostate cancer (Sahu et al. 2013; Kach et al. 2015). Importantly, the macrolide FK506 inhibits androgen-induced cell proliferation, an effect abolished by the FKBP51 knock-down, but not by FKBP52 knock-down, suggesting that no other TPR-domain factor could be an efficient replacement for the function of these immunophilins in these HSP90 heterocomplexes. Accordingly, the overexpression of FKBP51 favours prostate cancer cell growth and severely affects the efficacy of antiandrogens such as bicalutamide, a frequent therapy used in patients undergoing androgen ablation therapy (Ni et al. 2010).

Several studies have reported noticeable changes in the expression levels of both FKBP51 and FKBP52 in various types of hepatocellular cancer, oral cell squamous carcinomas, and oesophageal adenocarcinoma, suggesting that these genes are useful prognosis biomarkers of the disease (Solassol et al. 2011; Russo et al. 2017; Bonner and Boulianne 2017). In a recent study, it was reported that FKBP51 is up-regulated 100-fold in taxol-resistant treatments (Huang and Chao 2015). Moreover, FKBP51 silencing sensitized taxol-resistant cells to taxol, while overexpression of FKBP51 increased resistance to the drug, a mechanism that involves the AR in prostate and ovarian cells. This may be related to the fact that several of the newly identified taxol resistance genes are transcriptionally regulated by FKBP51 and silencing of these genes sensitized cells to taxol (Huang and Chao 2015). Consequently, it was suggested that taxol should not be used against ovarian cancer where the AKT/FKBP51/AR axis is activated (Huang and Chao 2015). In colorectal adenocarcinoma tissues, the expression of FKBP51 protein was associated with an immature phenotype of stromal fibroblasts and with the epithelial-to-mesenchymal transition phenotype, suggesting a role for this protein in the process (Rotoli et al. 2016). Due to the fact that only certain cells of the stroma express FKBP51 protein, a potential role for this immunophilin as a stroma cell subtype marker has been postulated (Rotoli et al. 2017).

Even though the expression of FKBP51 appears to be increased in most types of cancer, curiously, its expression in pancreatic cancer has been reported alternatively as absent or down-regulated (Pei et al. 2009). FKBP51 has also been related to melanoma development (Romano et al. 2010), and its down-regulation favoured the induction of massive apoptosis after irradiation both in vitro and in vivo, by a mechanism controlled by the activation of NF-κB. These observations suggest that FKBP51 is a promising candidate target for radiosensitizing strategies against malignant melanoma. Interestingly, FKBP51 is also described to work as tumour suppressor in pancreatic cancer (Wang 2011), where FKBP51 shows low or lost levels of expression correlating with increased AKT phosphorylation and downstream genes of the AKT pathway.

2.5 The hTERT•HSP90•Immunophilin Complex

Recent studies showed that some immunophilins are also mitochondrial factors. Among them, it is particularly interesting the case of FKBP51 (Gallo et al. 2011; Lagadari et al. 2016) because it is highly abundant in this organelle (~50% of the cell pool), but it lacks a known mitochondrial localization signal. Nonetheless, TPR domain mutants of FKBP51 are constitutively nuclear rather cytoplasmic or mitochondrial, suggesting that HSP90 is responsible for the extra-nuclear localization of the immunophilin. Moreover, it is quite reasonable to postulate that the HSP90-dependent mechanism for mitochondrial import of proteins (Fan and Young 2011) is the one used by FKBP51 to be translocated into the organelle. It was demonstrated (Gallo et al. 2011; Lagadari et al. 2016) that FKBP51 shows antiapoptotic action and undergoes nuclear-mitochondrial shuttling upon the onset of stress conditions such as high production of reactive oxygen species or deficit of nutrients. These are frequent situations in cancer tissues, where FKBP51 is also overexpressed (Lagadari et al. 2016). Cancer cells are also characterized by possessing high telomerase activity, an enzyme that is essential for their rapid clonal expansion of the cells (Eisenstein 2011). Telomerase is a ribonucleoprotein that compensates for the loss of telomeric DNA by adding repeated sequences to the chromosome ends using its intrinsic RNA component as a template for DNA synthesis. Cell generations are counted by telomeres, which constitute a particular molecular device composed of thousand repeats of a short sequence element located at the termini of chromosomes (Gomes et al. 2010). The chromosome end undergoes the loss of DNA fragments during each cell cycle, leading to the progressive shortening of the telomere. This phenomenon is assigned to failure of the DNA polymerases to completely replicate those sequences at the ends of the chromosomes during the S phase (Shore and Bianchi 2009). This mechanism prevents cancer development because the number of cell divisions become limited. When cells lose the control of the cell cycle and become cancerous, they divide more frequently than normal cells, such that their telomeres become shorter. This explains why a mechanism for telomere elongation becomes essential for cancer cells to divide indefinitely, to the point that they may reach an immortalization state (Calado and Young 2009).

While the protein expression and enzymatic activity of telomerase are very low and sometimes absent in most multicellular eukaryotic organisms, it is quite active in germ cells, stem cells, and certain blood cells, telomerase being up-regulated in the vast majority of human cancer cells (Harley 2008). The reverse transcriptase subunit of telomerase, hTERT, is an HSP90 client that contains the catalytic activity of the enzyme, whereas the associated RNA component, hTR, serves as the template for synthesis of telomeric sequences. The HSP90-based chaperone heterocomplex is required for assembly of hTERT (Holt et al. 1999) following similar assembly mechanism to that already demonstrated for steroid receptors. Therefore, it is not surprising that FKBP51 and FKBP52 belong to the hTERT•HSP90 heterocomplex (Lagadari et al. 2016). Both immunophilins are strong activators of the telomerase activity, which represents an important observation for cancer cells where both immunophilins are overexpressed and concentrated in the nuclei, as it is expected for cells that undergo several types of stress. Such trafficking influences the amount of immunophilins associated to the enzyme, favouring the enhancing action of the co-chaperones.

Importantly, the telomerase enzymatic activity is abrogated by the macrolide FK506 or by transfection of inactive immunophilins carrying point mutations in the PPIase domain, suggesting that the prolyl-cis/trans-isomerase activity is involved in this biological response (Holt et al. 1999). In other words, immunophilins are novel pharmacologic targets to prevent the clonal expansion of cancer cells simply by impairing the stimulant action over the catalytic subunit of telomerase. In this regard, it is important to note the recent synthesis of new highly selective ligands that prevent immunophilin enzymatic activity without showing immunosuppressive action (Gaali et al. 2015).

2.6 Chaperoning p53

Under normal conditions, cells use an intrinsic cell cycle program that suppresses their limitless proliferation. Cancer cells show exacerbated proliferation signals, but they also have mechanisms able to circumvent uncontrolled proliferation. This permit a reasonable expansion of tumours up to an inconvenient large life-threatening size. Apoptosis is the control system for programmed cell death that mostly affect the multiplication rate of cells and promotes anti-proliferative mechanisms that become natural barriers for uncontrolled tumour development. Consequently, cancer cells should also evolve a number of strategies to circumvent apoptosis and acquired resistance against this programme. The development and progression of tumours is stimulated by the inactivation of the pro-apoptotic machinery. This is one of the hallmarks of cancer cells (Hanahan and Weinberg 2000). The pivotal factor of the apoptotic system is the tumour suppressor transcription factor p53 (Wawrzynow et al. 2018). Therefore, a prerequisite for tumour development is often the disruption of normal p53, in most of the cases (>70%) due to a single amino acid substitution clustered in the DNA-binding domain of p53 (Hollstein et al. 1991).

As a consequence of the onset of cellular stress due to failure of DNA damage-repair system, exposure to ionizing radiation, hypoxia conditions, presence of toxic compounds, etc., p53 elicits either cell cycle arrest, a condition that may persist within a certain period of time until the conditions are back to normal, or trigger the apoptosis mechanism if the stress stimulus is overwhelming, prolonged and/or the damage is irreversible. Dysregulation of p53 by mutations suppress its normal anti-proliferative function, this functional inactivation being highly frequent in human cancers (Brady and Attardi 2010). Usually, these mutated forms of p53 are chaperoned by HSP90 (Calderwood et al. 2006), whose transcriptional rate is increased in tumour cells. This is related to the loss of p53 function as well as to higher expression of some proto-oncogenes, in particular HER2 and c-Myc. In this sense, it is important to emphasize that HSP90 is not only required for the stability and activity of HER2 itself, but also for its downstream proteins that show cardinal roles in cell survival such as AKT, c-Src, Raf-1 (Neckers and Ivy 2003), protein kinases that play key roles in cell growth and survival, as well as due to the effect of the proto-oncogene c-Myc, which up-regulates the expression of both HSP70 (Taira et al. 1999) and HSP90 (Teng et al. 2004) through activation of heat-shock regulatory elements (or HSE) present in the promoter regions of HSP genes. Normally, massive increases in HSP gene expression take place in cells under stress condition, an event that favours the recruitment of the transcription factor HSF1 (Heat-Shock Factor-1) to those HSEs. Interestingly, the tumourigenic factor heregulin-β1 binds to receptors in the cell surface of breast cancer cells leading to increased HSP expression. This enhances both cell survival and cell transformation through induced stabilization of HSF1 in a mechanism that involves the activity of PI3K (phosphatidyl-inositol-3-kinase) (Hoang et al. 2000), a key enzyme related to malignant progression. In addition, the c-Myc proto-oncogene is also able to activate the HSP90 promoter. The inhibition of this induction with the specific HSP90 inhibitor geldanamycin (GA) (Whitesell et al. 1994) decreases the transforming effects of c-Myc expression (Teng et al. 2004).

The HSP90-binding immunophilins are also related to p53 action following the same trafficking mechanism described for steroid receptors (Galigniana et al. 2010a). Thus, FKBP52 associates to p53•HSP90 cytoplasmic complexes in cells carrying p53 mutants, and the heterocomplex is responsible for the retrotransport mechanism of the pro-oncogenic factor via dynein motors (Galigniana et al. 2004). HSP90 is required to maintain the folded and transcriptionally active state of p53. Moreover, many of tumour-inducing p53 mutations increase the p53 binding affinity for HSP90. It is accepted that wild-type p53 also interacts transiently with HSP90 in normal cells, whereas in cancer cells p53 mutants associate to the chaperone in a more stably fashion. This makes tumour p53 'addicted' to HSP90.

2.7 The Evolutional Capacitor Function of HSP90

In a unique and original research project, Susan Lindquist and colleagues used Drosophila, S.cerevisiae, and Arabidopsis as organism model to demonstrate that HSP90 plays a reservoir role in the evolutionary steps of species working as a 'moderator factor' able to suppress the phenotypical emergence caused by genomic mutations (Rutherford and Lindquist 1998; Queitsch et al. 2002; Cowen and Lindquist 2005; Sangster et al. 2008). After a mutation, proteins may become structurally and/or functionally unstable unless HSP90 binds to the mutant proteins suppressing the phenotypic expression of that mutation. Consequently, that originally modified and aberrant factor may become normal. Analogously, there is a large number of mutations in proteins expressed in cancer cells as an expected result of their rapid proliferation and high genome instability. It is reasonable to think that not all mutations must be harmful, but some of them may be advantageous for the tumour itself allowing its growth and subsistence. Therefore, cell autonomous properties of isolated monotypic cells do not represent the biology of the whole tumour. Actually, solid tumours are composed of an assemblage of different cell types (Hanahan and Weinberg 2011) producing subpopulations of specialized cells with divergent capabilities that are beneficial for the tumour. In recent years, it was discovered the existence of a previously unrecognized type of neoplastic cells, the cancer stem cells (Clevers 2011), characterized by their self-renewal capability and differentiation capacity into multiple undifferentiated or tumour cell types. Consequently, new tumour populations are seeded. The discovery of cancer stem cells in tumours strengthens the notion that tumours are constituted by phenotypically heterogeneous cells in distinct states of differentiation, such that during the progression of the disease, the high genetic instability could be responsible for driving diversification by a selective Darwinian process responsible for the subsequent genetic variability. In summary, beneficial genetic modifications may lead to the progressive conversion of normal cells into malignant cancer cells.

Within this fascinating and complex picture, the repertoire for HSP90 in cancer cells is to work as a molecular reservoir capable to buffer lethal phenotypes due to non-viable gene mutations. Therefore, HSP90 could play a crucial role in the progression of tumours composed by highly proliferating cells possessing an instable genome. In short, the biological function of HSP90 in this type of system could be exactly the same as that already demonstrated by the Lindquist lab in the evolution process of species, where the accumulation of gene mutations favours beneficial phenotypic differentiation. A good example is the reported case of small-cell lung cancer where oncogenic mutations of the EGFR (Epidermal Growth Factor Receptor) responsible for drug resistance are stabilized by HSP90 binding. Consistent with this role, the use of HSP90 inhibitors targets the receptor to proteasomal degradation without affecting the stability of the wild-type EGFR (Shimamura et al. 2005), and clinical trials have recently demonstrated the efficiency of the non-geldanamycin inhibitor luminespib to favour the degradation of EGFR exon-20 isoform mutations inducing the apoptosis of lung adenocarcinoma cells even in a patient that was insensitive to other conventional treatments (Jorge et al. 2018).

In line with this model, the relatively novel discovery of regulatory microRNAs has led to careful revision of our understanding of the mechanisms of genetic control. Today, it is well known that microRNAs play a relevant role to profile cancer cells, leading them to tumour formation (van Kouwenhove et al. 2011; Bracken et al. 2016). In recent years, it was demonstrated that HSP90 is associated with protein complexes related to RNA processing (Miyoshi et al. 2010; Iwasaki et al. 2010; Aziz 2016; Pan et al. 2018; Lampis et al. 2018; Naruse et al. 2018). Nonetheless, many of the potential roles postulated for HSP90 in the miRNA world associated to tumour progression remain to be demonstrated.

2.8 Extracellular HSP90

Recent studies have demonstrated that HSP90 is not confined to the intracellular milieu. Many cell types express HSP90 on the cell surface and secrete HSP90 into the extracellular space (Li et al. 2013; Calderwood 2018; Eustace and Jay 2004; Eustace et al. 2004). In addition, the amounts of HSP90 on the cell surface is higher in cancer cells than in normal cells, its presence being correlated with metastasic activity (Becker et al. 2004) and grater invasiveness capacity. Normal cells do not secrete large amounts of HSP90 or directly do not secrete the chaperone (unless they suffer environmental stress), whereas tumour cells constitutively secrete large amounts of HSP90 to the medium. There are upstream regulators of HSP90 secretion. Some interesting cases are: (a) the Hectd1-ubiquitin ligase, that negatively regulates the intracellular localization and secretion of HSP90 (Sarkar and Zohn 2012); (b) p53, which is involved in exosome formation (Yu et al. 2006); and (c) HIF-1α that mediates hypoxia-triggered secretion (Li et al. 2007), as shown by the fact that a dominant negative mutant of HIF-1α blocks HSP90 secretion and a constitutively active mutant of HIF-1α preserves active secretion of the chaperone even under normoxia. It is thought that the same types of mechanism as a whole take place in tumour cells (Li et al. 2013).

Cell motility can be prevented in vitro by blocking or neutralizing secreted HSP90 assays, and in vivo this neutralization inhibits the generation of tumour metastasis (Tsutsumi et al. 2008). Nevertheless, the precise molecular mechanism underlying HSP90 secretion is not entirely understood at the present time. It is accepted that secretion is stimulated by environmental stresses and growth factors and is also affected by HSP90 post-translational modifications (Trepel et al. 2010). Importantly, specific inhibition of extracellular HSP90 does not appear to affect cancer cell proliferation in vitro or tumour xenograft growth in vivo (Tsutsumi et al. 2008). As it was previously detailed, HSP90 forms heterocomplexes with other chaperones and co-chaperones. Interestingly, some of these proteins have also been detected extracellularly, including Hop/p60 and p23. The co-chaperones can be found in conditioned media. These findings support the notion that HSP90 complexes function in invasiveness.

2.9 Targeting HSP90 in Cancer Therapy

The number of HSP90 client-proteins reported to date is more than 400 (Taipale et al. 2012, 2014; da Silva and Ramos 2012), including receptors, transcription factors, kinases, proteins related to the control of cell survival, chromatin remodelation, trafficking, cell proliferation, apoptosis, enzymes, etc. On the other hand, the amounts of HSP90 in tumour cells ranged between twice and ten-fold higher than in normal cells (Ferrarini et al. 1992; Li and Buchner 2013; Pennisi et al. 2015;

Clarke et al. 2000; Hao et al. 2010). Moreover, even those cancer cells where the expression of HSP90 is not increased show higher activity of HSP90. All these properties make HSP90 an attractive therapeutic target for cancer treatment since its inhibition inactivates client-proteins and also induces their degradation. At the end, these events favour the attack of NK cells and the killing of the tumour.

The first HSP90 inhibitor was a natural benzoquinone ansamycin antibiotic produced by Streptomyces hygroscopicus, the macrocyclic polyketide geldanamycin (GA). It was first reported that GA reverts the transformed phenotype of v-src-transformed cells via binding to HSP90 and disrupting its chaperone function (Whitesell et al. 1994). The subsequent identification of the nucleotide binding site in the N-terminal of HSP90 demonstrated GA competitively binds to this domain and prevents ATP binding, thus disrupting the ATP-dependent conformational cycling reactions (Fig. 2.1) for a broad range of signal transduction factors. The inactivation and subsequent degradation of these client-proteins constitutes the mechanistic reason for the anti-tumour effect of GA. Even though GA shows potent anti-tumour action in vitro, its low solubility in an aqueous medium and its severe hepatotoxicity made this drug not useful for human treatment (Supko et al. 1995).

Like GA, the resorcinol derivative radicicol is another natural product with similar activity first isolated by the fungus Monosporium bonorden (this is why it is also named monorden). Radicicol also binds to the N-terminal ATP-binding pocket of HSP90 (Schulte et al. 1998), although its affinity is greater than that shown by GA. The first in vitro studies already demonstrated that radicicol is also a potent anti-proliferative agent (Kwon et al. 1992), although reducing agents inactivate its biological action. Curiously, radicicol showed no inhibitory effect on PKC and PKA activities (Kwon et al. 1992), and in vivo studies failed to demonstrate efficient anti-tumour activity, an observation that was assigned to its low biological stability (Sreedhar et al. 2004). Interestingly, it has been demonstrated (Prodromou et al. 2009) that a single point mutation in the N-terminal domain of HSP90 in the fungus *Humicola fuscoatra* reduces its affinity for radicicol protecting the organism from the production of the compound. Neither GA binding nor ATP binding to the N-terminal end are affected, demonstrating that the appearance of a drug resistance-conferring mutation with no effect on HSP90 function is possible in nature. Also, it shows the need of having a vast drug armamentarium to prevent resistance to the treatment. Although clinical trials demonstrated that both GA and radicicol are unstable and/or toxic for clinical use, both pioneer inhibitors have provided the structural basis for drug derivatives that subsequently entered the clinic. One of the first to progress to clinical trials was the better tolerated and more hydrophilic GA analogous 17-AAG (17-allylamino-17-demethoxy-geldanamycin), which showed some success in preclinical and clinical studies (Schulte and Neckers 1998). 17-AAG showed promising anticancer action both in vitro and in vivo, and it has currently completed phase I to III in clinical trials for several malignancies such as metastatic melanoma, breast and ovarian cancer (Solit and Chiosis 2008; Mellatyar et al. 2018). However, its poor water solubility, low stability, potential liver toxicity, short biological half-life, and the requirement to be dissolved in dimethyl sulfoxide or Cremophor-EL formulation has limited the progress for its clinical application.

These setbacks led to the development of a semi-synthetic derivative of GA, 17-dimethyl-aminoethyl-amino-17-demethoxygeldanamycin (17-DMAG), which shows several advantages over 17-AAG. Among them, it shows higher water solubility, good bioavailability, reduced metabolism (i.e., reduced potential to generate toxic metabolites), and greater anti-tumour capability. In addition, 17-DMAG can be administered orally (Haque et al. 2016; Mellatyar et al. 2018). 17-DMAG has been studied in preclinical and clinical trials, in both conditions alone and in conjunction with other anti-cancer compounds (see an updated revision in (Mellatyar et al. 2018). Results are encouraging and suggests that 17-DMAG, particularly in combination with anticancer drugs, may be suitable to achieve a clinically significant anti-cancer response. Several other compounds were and are being tested at the present time. Among the most interesting: (1) Radamide (a radanamycin amide) that was designed based on the analysis of the co-crystallization structures of GA (Stebbins et al. 1997) and radicicol (Clevenger and Blagg 2004) bound to the N-domain of HSP90. The design created a sort of chimeric compound containing the resorcinol ring that characterizes radicicol and the quinine ring typical from GA. Radamide shows strong inhibitory action in breast cancer cells in the low μM range (Clevenger and Blagg 2004). (2) Novobiocin, a coumarin antibiotic first isolated from Streptomyces spheroids that targets the HSP90 C-terminal ATP binding domain of Hsp90 inducing degradation of client-proteins (Soti et al. 2002). Its low affinity for the chaperone precluded its use in clinical trials and forced to obtain more potent novobiocin analogues, for example F4, which shows good efficacy to induce apoptosis in prostate cancer cell lines (Matthews et al. 2010), or Ku135, a good inhibitor of Jurkat T cells and T lymphocyte proliferation and a more potent pro-apoptotic drug than 17-AAG (Shelton et al. 2009). (3) Diaryl-pyrazole resorcinols such as CCT018159, VER-49009 and NVP-AUY922 inhibit cell proliferation in a wide range of human cancer cell lines with an average efficacy of 10 nM (reviewed by Hao et al. 2010), and xenograft tumour models showed pharmacologic efficiency to supress the tumour and a very good half-life (~9.5 h) (Brough et al. 2008). (4) PU3, which has become a classic purine-based inhibitor that shows potent inhibitory action on the growth of breast cancer cells (Chiosis et al. 2001). (5) A hybrid inhibitor was designed using the trimethoxy-phenyl group of PU3 that interacts with amino acids of the ATP-binding pocket of HSP90, and the resorcinol ring of the inhibitor NVP-AUY922 (Brough et al. 2008), both embedded in a chalcone (aromatic ketone) scaffold. The resultant compound named '1 g' shows strong inhibitory action of cell proliferation and downregulates the expression of HSP90 client proteins such as EGFR, MET, Her2 and AKT, whereas HSP70 expression is up-regulated (Jeong et al. 2017). (6)- EGCG [(−)Epigallocatechin-3-gallate] is flavonoid found in natural products, among them green tea. It is a potent chemoprevention and anticancer component (104). EGCG inhibits the transcriptional activity of the aryl hydrocarbon receptor (AhR) via direct binding to the C-terminal region of HSP90 (Palermo et al. 2005). It also affects the association of HSP90 with its co-chaperones including AKT, Cdk4, Raf-1, Her-2 and pERK, etc., which in turn, favours the degradation of these proteins and result in anti-proliferating effects (Li et al. 2009). Several compounds with good inhibitory activity and acceptable toxicity have been designed and synthesized. Table I shows a classification of these compounds according to their natural or synthetic origin. The practical results in the clinic are promising, but still several key questions need to be answered. Nevertheless, prospects are quite optimistic, and the expectation is that major advances will be achieved in the near future.

2.10 Targeting HSP90-Binding Immunophilins in Cancer Therapy

The developing of ligands able to specifically bind to HSP90-binding immunophilins shows promising pharmacological perspectives in the near future. The ability to regulate the functions of a specific protein using cell-permeable small molecules is an unquestionable powerful method not only to study biological systems from the mechanistic perspective, but also a desired alternative to be used in therapeutic treatments. In this sense, HSP90-binding immunophilins are novel targets that could offer new therapeutic opportunities in many fields, most likely in cancer therapy, as it is inferred from the previously discussed features of these proteins, but also in neurodegenerative diseases and other neurological disorders such as depression (Galigniana et al. 2012). Following the isolation of rapamycin and cyclosporine A, it was FK506 the most used drug for the prevention of liver transplant rejection and since then, its use expanded rapidly into the transplantation of other organs (for a recent review, see Erlejman et al. 2013). Strong attempts to synthesize new selective immunophilin ligands are in course of action (Liu et al. 2013a; Ashida and Kikuchi 2013; Wang et al. 2013; Gaali et al. 2015; Feng et al. 2015). Nonetheless, drug discovery has always been hampered because the failure to pharmacologically differentiate against the highly homologous members of the family, in particular for the case of FKBP51 and FKBP52. Most of the novel ligands are still unselective in this regard.

Targeting immunophilins by treatments with FK506 (tacrolimus) inhibited the proliferation of prostate cancer cells, a fact that was assigned to blockade of the enhancing effect of FKBP51 on the AR in these cells (Periyasamy et al. 2010; Ni et al. 2010). It has also been reported a physical association of FKBP51 with the HSP90 bound kinase upstream of IκB, IKK, via its IKKα subunit (Jiang et al. 2008). Consistent with this, the FKBP ligand rapamycin blocked IκB/NF-κB/mTOR signalling cascade (Giordano et al. 2006). Because the drug concentration used in the study is also able to inhibit FKBP52, it results unclear whether this effect on NF-κB signalling was due the inhibition of the stimulant FKBP52 or due to prevention of the overall inhibitory action recently reported for FKBP51 (Erlejman et al. 2014a). The development of selective drugs for each immunophilin such as SAFit, a recently developed ligand for FKBP51 (Gaali et al. 2015), will help to think about novel and specific pharmacologic approaches for cancer treatment, in particular by combination of therapies.

2.11 Conclusions

HSP90 plays significant roles in the regulation of several biological processes including cell differentiation, proliferation, differentiation and transformation. In tumours, HSP90 is usually up-regulated ten-fold the basal expression level, as well as in normal cells experiencing various types of stress, such as deficit of nutrients, fluctuations of oxygen supply (particularly in solid tumours), acidic pH, exposure to oxidants or toxics, etc. It is consequently reasonable to detect high levels of HSP90 expression in several systems such as acute leukaemias (Yufu et al. 1992; Reikvam et al. 2009), ovarian cancers (Liu et al. 2013b), breast cancer (Dimas et al. 2018), Ras-transformed cells (Lebeau et al. 1991), malignant cell lines (Wu et al. 2017), melanomas (Protti et al. 1994), pancreatic carcinomas (Kim et al. 2008), endometrial carcinomas (Witkin 2001), gastrointestinal cancers (Ehrenfried et al. 1995), colon adenocarcinoma (Ansa-Addo et al. 2016). Moreover, overexpression of HSP90α is usually associated with poor prognosis of the disease (Yano et al. 1996; Shi et al. 2014), and has been closely correlated with the overall proliferative potential of malignant cells. Elevated levels of heat-shock proteins may participate in drug resistance to treatments and also in the reorganization of chromatin structure and confer a significant advantage on tumour cells to survive in a hostile environment favouring genome rearrangements and gene amplifications that are often involved in the tumorigenesis process. In view of the properties of HSP90, a priori it sounds reasonable that HSP90 inhibitors should block the oncogenic function of client-proteins. The reality is somehow unsatisfactory since most tested drugs have produced disappointing results. The failures are related to the fact that in vivo, most of these drugs bind to HSP90 protein in the micromolar range and/or show a short half-life. Success of HSP90 inhibition also depends on other factors sometimes not considered such as the real 'selectivity' of the drug for tumour cells over non-malignant cells. Moreover, most clinical inhibitors have been designed to target the N-terminus of the chaperone, but the fact that classical inhibitors are significantly more effective at blocking cell growth than they are at binding to the target domain implies that it is entirely possible that HSP90 is not the primary target of these novel drugs, as it is thought. This is one of the reasons why it is also convenient to think about targeting HSP90-binding factors, for example, immunophilins. Although our basic understanding of the role of HSP90 in cancer biology has come a long way in a short time, the next years hold the hope of similar progress in effectively translating these findings to the improved treatment of cancer (Table 2.1).

Table 2.1 Most common inhibitors of HSP90 activity

Chemical or conventional name	Source
Geldanamycin	Natural
17-AAG	Semisynthetic
17-DMAG	Semisynthetic
IPI-504	Synthetic
C-11	Synthetic
Radicicol	Natural
Radicicol derivates	Semisynthetic
PochoninA-F	Natural
Radamide	Synthetic
Novobiocin	Natural
Novobiocin derivates	Semisynthetic
EGCG	Natural
Derrubone	Natural
Gedunin	Natural
Celastrol	Natural
Pu3	Synthetic
BIIB021	Synthetic
CCTO18159	Synthetic
NVP-AUY922	Synthetic
SNX-2112	Synthetic
STA9090	Synthetic

Acknowledgements The author acknowledges the financial support of the National Research Council of Argentina (CONICET), The University of Buenos Aires (UBACyT programme), and the National Agency for Scientific & Technological Programming (ANPCyT).

References

Ali MM, Roe SM, Vaughan CK, Meyer P, Panaretou B, Piper PW, Prodromou C, Pearl LH (2006) Crystal structure of an Hsp90-nucleotide-p23/Sba1 closed chaperone complex. Nature 440:1013–1017

Ansa-Addo EA, Thaxton J, Hong F et al (2016) Clients and oncogenic roles of molecular chaperone gp96/grp94. Curr Top Med Chem 16:2765–2778

Ashida T, Kikuchi T (2013) Estimation of relative binding free energy based on a free energy variational principle for the FKBP-ligand system. J Comput Aided Mol Des 27:479–490

Aziz F (2016) The emerging role of miR-223 as novel potential diagnostic and therapeutic target for inflammatory disorders. Cell Immunol 303:1–6

Bakthisaran R, Tangirala R, Rao Ch M (2015) Small heat shock proteins: role in cellular functions and pathology. Biochim Biophys Acta 1854:291–319

Barik S (2006) Immunophilins: for the love of proteins. Cell Mol Life Sci 63:2889–2900

Becker B, Multhoff G, Farkas B, Wild PJ, Landthaler M, Stolz W, Vogt T (2004) Induction of Hsp90 protein expression in malignant melanomas and melanoma metastases. Exp Dermatol 13:27–32

Binder EB, Salyakina D, Lichtner P, Wochnik GM, Ising M, Pütz B, Papiol S, Seaman S, Lucae S, Kohli MA, Nickel T, Künzel HE, Fuchs B, Majer M, Pfennig A, Kern N, Brunner J, Modell S, Baghai T, Deiml T, Zill P, Bondy B, Rupprecht R, Messer T, Köhnlein O, Dabitz H, Brückl T, Müller N, Pfister H, Lieb R, Mueller JC, Lõhmussaar E, Strom TM, Bettecken T, Meitinger T, Uhr M, Rein T, Holsboer F, Muller-Myhsok B (2004) Polymorphisms in FKBP5 are associated with increased recurrence of depressive episodes and rapid response to antidepressant treatment. Nat Genet 36:1319–1325

Bohne AV, Schwenkert S, Grimm B, Nickelsen J (2016) Roles of tetratricopeptide repeat proteins in biogenesis of the photosynthetic apparatus. Int Rev Cell Mol Biol 324:187–227

Bonner JM, Boulianne GL (2017) Diverse structures, functions and uses of FK506 binding proteins. Cell Signal 38:97–105

Bracken CP, Scott HS, Goodall GJ (2016) A network-biology perspective of microRNA function and dysfunction in cancer. Nat Rev Genet 17:719–732

Brady CA, Attardi LD (2010) p53 at a glance. J Cell Sci 123:2527–2532

Brough PA, Aherne W, Barril X et al (2008) 4,5-diarylisoxazole Hsp90 chaperone inhibitors: potential therapeutic agents for the treatment of cancer. J Med Chem 51:196–218

Calado RT, Young NS (2009) Telomere diseases. N Engl J Med 361:2353–2365

Calderwood SK (2018) Heat shock proteins and cancer: intracellular chaperones or extracellular signalling ligands? Philos Trans R Soc Lond B Biol Sci 373(1738). pii: 20160524. https://doi.org/10.1098/rstb.2016.0524

Calderwood SK, Khaleque MA, Sawyer DB, Ciocca DR (2006) Heat shock proteins in cancer: chaperones of tumourigenesis. Trends Biochem Sci 31:164–172

Chatterjee S, Burns TF (2017) Targeting heat shock proteins in cancer: a promising therapeutic approach. Int J Mol Sci 18, 18(9). pii: E1978. https://doi.org/10.3390/ijms18091978

Chen WY, Chang FR, Huang ZY, Chen JH, Wu YC, Wu CC (2008) Tubocapsenolide-A, a novel withanolide, inhibits proliferation and induces apoptosis in MDA-MB-231 cells by thiol oxidation of heat shock proteins. J Biol Chem 283:17184–17193

Chiosis G, Timaul MN, Lucas B, Munster PN, Zheng FF, Sepp-Lorenzino L, Rosen N (2001) A small molecule designed to bind to the adenine nucleotide pocket of Hsp90 causes Her2 degradation and the growth arrest and differentiation of breast cancer cells. Chem Biol 8:289–299

Clarke PA, Hostein I, Banerji U, Stefano FD, Maloney A, Walton M, Judson I, Workman P (2000) Gene expression profiling of human colon cancer cells following inhibition of signal transduction by 17-allylamino-17-demethoxygeldanamycin, an inhibitor of the hsp90 molecular chaperone. Oncogene 19:4125–4133

Clevenger RC, Blagg BS (2004) Design, synthesis, and evaluation of a radicicol and geldanamycin chimera, radamide. Org Lett 6:4459–4462

Clevers H (2011) The cancer stem cell: premises, promises and challenges. Nat Med 17:313–319

Colo GP, Rubio MF, Nojek IM, Werbajh SE, Echeverria PC, Alvarado CV, Nahmod VE, Galigniana MD, Costas MA (2008) The p160 nuclear receptor co-activator RAC3 exerts an anti-apoptotic role through a cytoplasmatic action. Oncogene 27:2430–2444

Cowen LE, Lindquist S (2005) Hsp90 potentiates the rapid evolution of new traits: drug resistance in diverse fungi. Science 309:2185–2189

da Silva VC, Ramos CH (2012) The network interaction of the human cytosolic 90 kDa heat shock protein Hsp90: a target for cancer therapeutics. J Proteome 75:2790–2802

Dimas DT, Perlepe CD, Sergentanis TN, Misitzis I, Kontzoglou K, Patsouris E, Kouraklis G, Psaltopoulou T, Nonni A (2018) The prognostic significance of Hsp70/Hsp90 expression in breast cancer: a systematic review and meta-analysis. Anticancer Res 38:1551–1562

Dumont FJ (2000) FK506, an immunosuppressant targeting calcineurin function. Curr Med Chem 7:731–748

Ebong IO, Beilsten-Edmands V, Patel NA, Morgner N, Robinson CV (2016) The interchange of immunophilins leads to parallel pathways and different intermediates in the assembly of Hsp90 glucocorticoid receptor complexes. Cell Discov 2:16002

Echeverria PC, Picard D (2010) Molecular chaperones, essential partners of steroid hormone receptors for activity and mobility. Biochim Biophys Acta 1803:641–649

Ehrenfried JA, Herron BE, Townsend CM Jr, Evers BM (1995) Heat shock proteins are differentially expressed in human gastrointestinal cancers. Surg Oncol 4:197–203

Eisenstein M (2011) Telomeres: all's well that ends well. Nature 478:S13–S15

Ellis RJ (1993) The general concept of molecular chaperones. Philos Trans R Soc Lond Ser B Biol Sci 339:257–261

Ellis RJ (1996) Discovery of molecular chaperones. Cell Stress Chaperones 111:155–160

Erlejman AG, Lagadari M, Galigniana MD (2013) Hsp90-binding immunophilins as a potential new platform for drug treatment. Future Med Chem 5:591–607

Erlejman AG, De Leo SA, Mazaira GI, Molinari AM, Camisay MF, Fontana V, Cox MB, Piwien-Pilipuk G, Galigniana MD (2014a) NF-kappaB transcriptional activity is modulated by FK506-binding proteins FKBP51 and FKBP52: a role for peptidyl-prolyl isomerase activity. J Biol Chem 289:26263–26276

Erlejman AG, Lagadari M, Toneatto J, Piwien-Pilipuk G, Galigniana MD (2014b) Regulatory role of the 90-kDa-heat-shock protein (Hsp90) and associated factors on gene expression. Biochim Biophys Acta 1839:71–87

Eustace BK, Jay DG (2004) Extracellular roles for the molecular chaperone, hsp90. Cell Cycle 3:1098–1100

Eustace BK, Sakurai T, Stewart JK et al (2004) Functional proteomic screens reveal an essential extracellular role for hsp90 alpha in cancer cell invasiveness. Nat Cell Biol 6:507–514

Fan AC, Young JC (2011) Function of cytosolic chaperones in Tom70-mediated mitochondrial import. Protein Pept Lett 18:122–131

Feng X, Pomplun S, Hausch F (2015) Recent progress in FKBP ligand development. Curr Mol Pharmacol 9:27–36

Ferrarini M, Heltai S, Zocchi MR, Rugarli C (1992) Unusual expression and localization of heat-shock proteins in human tumour cells. Int J Cancer 51:613–619

Gaali S, Kirschner A, Cuboni S et al (2015) Selective inhibitors of the FK506-binding protein 51 by induced fit. Nat Chem Biol 11:33–37

Galigniana MD (2012) Steroid receptor coupling becomes nuclear. Chem Biol 19:662–663

Galigniana MD, Radanyi C, Renoir JM, Housley PR, Pratt WB (2001) Evidence that the peptidylprolyl isomerase domain of the hsp90-binding immunophilin FKBP52 is involved in both dynein interaction and glucocorticoid receptor movement to the nucleus. J Biol Chem 276:14884–14889

Galigniana MD, Harrell JM, O'Hagen HM, Ljungman M, Pratt WB (2004) Hsp90-binding immunophilins link p53 to dynein during p53 transport to the nucleus. J Biol Chem 279:22483–22489

Galigniana MD, Echeverria PC, Erlejman AG, Piwien-Pilipuk G (2010a) Role of molecular chaperones and TPR-domain proteins in the cytoplasmic transport of steroid receptors and their passage through the nuclear pore. Nucleus 1:299–308

Galigniana MD, Erlejman AG, Monte M, Gomez-Sanchez C, Piwien-Pilipuk G (2010b) The hsp90-FKBP52 complex links the mineralocorticoid receptor to motor proteins and persists bound to the receptor in early nuclear events. Mol Cell Biol 30:1285–1298

Galigniana NM, Ballmer LT, Toneatto J, Erlejman AG, Lagadari M, Galigniana MD (2012) Regulation of the glucocorticoid response to stress-related disorders by the Hsp90-binding immunophilin FKBP51. J Neurochem 122:4–18

Gallo LI, Ghini AA, Piwien Pilipuk G, Galigniana MD (2007) Differential recruitment of tetratricorpeptide repeat domain immunophilins to the mineralocorticoid receptor influences both heat-shock protein 90-dependent retrotransport and hormone-dependent transcriptional activity. Biochemistry 46:14044–14057

Gallo LI, Lagadari M, Piwien-Pilipuk G, Galigniana MD (2011) The 90-kDa heat-shock protein (Hsp90)-binding immunophilin FKBP51 is a mitochondrial protein that translocates to the nucleus to protect cells against oxidative stress. J Biol Chem 286:30152–30160

Giordano A, Avellino R, Ferraro P, Romano S, Corcione N, Romano MF (2006) Rapamycin antagonizes NF-kappaB nuclear translocation activated by TNF-alpha in primary vascular smooth muscle cells and enhances apoptosis. Am J Physiol Heart Circ Physiol 290:H2459–H2465

Gomes NM, Shay JW, Wright WE (2010) Telomere biology in Metazoa. FEBS Lett 584:3741–3751

Gougelet A, Bouclier C, Marsaud V, Maillard S, Mueller SO, Korach KS, Renoir JM (2005) Estrogen receptor alpha and beta subtype expression and transactivation capacity are differentially affected by receptor-, hsp90- and immunophilin-ligands in human breast cancer cells. J Steroid Biochem Mol Biol 94:71–81

Grossmann C, Ruhs S, Langenbruch L, Mildenberger S, Stratz N, Schumann K, Gekle M (2012) Nuclear shuttling precedes dimerization in mineralocorticoid receptor signaling. Chem Biol 19:742–751

Hanahan D, Weinberg RA (2000) The hallmarks of cancer. Cell 100:57–70

Hanahan D, Weinberg RA (2011) Hallmarks of cancer: the next generation. Cell 144:646–674

Hao H, Naomoto Y, Bao X et al (2010) HSP90 and its inhibitors. Oncol Rep 23:1483–1492

Haque A, Alam Q, Alam MZ, Azhar EI, Sait KH, Anfinan N, Mushtaq G, Kamal MA, Rasool M (2016) Current understanding of HSP90 as a novel therapeutic target: an emerging approach for the treatment of cancer. Curr Pharm Des 22:2947–2959

Harley CB (2008) Telomerase and cancer therapeutics. Nat Rev Cancer 8:167–179

Harrell JM, Kurek I, Breiman A, Radanyi C, Renoir JM, Pratt WB, Galigniana MD (2002) All of the protein interactions that link steroid receptor-hsp90.immunophilin heterocomplexes to cytoplasmic dynein are common to plant and animal cells. Biochemistry 41:5581–5587

Hoang AT, Huang J, Rudra-Ganguly N, Zheng J, Powell WC, Rabindran SK, Wu C, Roy-Burman P (2000) A novel association between the human heat shock transcription factor 1 (HSF1) and prostate adenocarcinoma. Am J Pathol 156:857–864

Hollstein M, Sidransky D, Vogelstein B, Harris CC (1991) p53 mutations in human cancers. Science 253:49–53

Holt SE, Aisner DL, Baur J et al (1999) Functional requirement of p23 and Hsp90 in telomerase complexes. Genes Dev 13:817–826

Hoter A, El-Sabban ME, Naim HY (2018) The HSP90 family: structure, regulation, function, and implications in health and disease. Int J Mol Sci 19(9). pii: E2560. doi: https://doi.org/10.3390/ijms19092560

Huang SL, Chao CC (2015) Silencing of taxol-sensitizer genes in cancer cells: lack of sensitization effects. Cancer 7:1052–1071

Iwasaki S, Kobayashi M, Yoda M, Sakaguchi Y, Katsuma S, Suzuki T, Tomari Y (2010) Hsc70/Hsp90 chaperone machinery mediates ATP-dependent RISC loading of small RNA duplexes. Mol Cell 39:292–299

Jeong YY, Her J, Oh SY, Chung IK (2016) Hsp90-binding immunophilin FKBP52 modulates telomerase 1069 activity by promoting the cytoplasmic retrotransport of hTERT. Biochem J 473:3517–3532

Jeong JH, Oh YJ, Kwon TK, Seo YH (2017) Chalcone-templated Hsp90 inhibitors and their effects on gefitinib resistance in non-small cell lung cancer (NSCLC). Arch Pharm Res 40:96–105

Jhaveri K, Ochiana SO, Dunphy MP et al (2014) Heat shock protein 90 inhibitors in the treatment of cancer: current status and future directions. Expert Opin Investig Drugs 23:611–628

Jiang W, Cazacu S, Xiang C, Zenklusen JC, Fine HA, Berens M, Armstrong B, Brodie C, Mikkelsen T (2008) FK506 binding protein mediates glioma cell growth and sensitivity to rapamycin treatment by regulating NF-kappaB signaling pathway. Neoplasia 10:235–243

Jorge S, Lucena-Araujo AR, Yasuda H et al (2018) EGFR exon 20 insertion mutations display sensitivity to Hsp90 inhibition in preclinical models and lung adenocarcinomas. Clin Cancer Res 24:6548–6555

Kach J, Conzen SD, Szmulewitz RZ (2015) Targeting the glucocorticoid receptor in breast and prostate cancers. Sci Transl Med 7:305ps319. https://doi.org/10.1126/scitranslmed.aac7531

Kim HL, Cassone M, Otvos L Jr, Vogiatzi P (2008) HIF-1alpha and STAT3 client proteins interacting with the cancer chaperone Hsp90: therapeutic considerations. Cancer Biol Ther 7:10–14

Kovacs JJ, Murphy PJ, Gaillard S et al (2005) HDAC6 regulates Hsp90 acetylation and chaperone-dependent activation of glucocorticoid receptor. Mol Cell 18:601–607

Kwon HJ, Yoshida M, Fukui Y, Horinouchi S, Beppu T (1992) Potent and specific inhibition of p60v-src protein kinase both in vivo and in vitro by radicicol. Cancer Res 52:6926–6930

Lagadari M, Zgajnar NR, Gallo LI, Galigniana MD (2016) Hsp90-binding immunophilin FKBP51 forms complexes with hTERT enhancing telomerase activity. Mol Oncol 10:1086–1098

Lampis A, Carotenuto P, Vlachogiannis G et al (2018) MIR21 drives resistance to heat shock protein 90 inhibition in cholangiocarcinoma. Gastroenterology 154:1066–1079 e1065

Langer T, Rosmus S, Fasold H (2003) Intracellular localization of the 90 kDA heat shock protein (HSP90alpha) determined by expression of a EGFP-HSP90alpha-fusion protein in unstressed and heat stressed 3T3 cells. Cell Biol Int 27:47–52

Lebeau J, Le Chalony C, Prosperi MT, Goubin G (1991) Constitutive overexpression of a 89 kDa heat shock protein gene in the HBL100 human mammary cell line converted to a tumourigenic phenotype by the EJ/T24 Harvey-ras oncogene. Oncogene 6:1125–1132

Li J, Buchner J (2013) Structure, function and regulation of the hsp90 machinery. Biom J 36:106–117

Li W, Li Y, Guan S, Fan J, Cheng CF, Bright AM, Chinn C, Chen M, Woodley DT (2007) Extracellular heat shock protein-90alpha: linking hypoxia to skin cell motility and wound healing. EMBO J 26:1221–1233

Li Y, Zhang T, Jiang Y, Lee HF, Schwartz SJ, Sun D (2009) Epigallocatechin-3-gallate inhibits Hsp90 function by impairing Hsp90 association with cochaperones in pancreatic cancer cell line Mia Paca-2. Mol Pharm 6:1152–1159

Li W, Tsen F, Sahu D, Bhatia A, Chen M, Multhoff G, Woodley DT (2013) Extracellular Hsp90 (eHsp90) as the actual target in clinical trials: intentionally or unintentionally. Int Rev Cell Mol Biol 303:203–235

Lin JF, Xu J, Tian HY, Gao X, Chen QX, Gu Q, Xu GJ, Song JD, Zhao FK (2007) Identification of candidate prostate cancer biomarkers in prostate needle biopsy specimens using proteomic analysis. Int J Cancer 121:2596–2605

Liu F, Wang YQ, Meng L, Gu M, Tan RY (2013a) FK506-binding protein 12 ligands: a patent review. Expert Opin Ther Pat 23:1435–1449

Liu H, Xiao F, Serebriiskii IG et al (2013b) Network analysis identifies an HSP90-central hub susceptible in ovarian cancer. Clin Cancer Res 19:5053–5067

Martinez-Ruiz A, Villanueva L, Gonzalez de Orduna C, Lopez-Ferrer D, Higueras MA, Tarin C, Rodriguez-Crespo I, Vazquez J, Lamas S (2005) S-nitrosylation of Hsp90 promotes the inhibition of its ATPase and endothelial nitric oxide synthase regulatory activities. Proc Natl Acad Sci U S A 102:8525–8530

Matthews SB, Vielhauer GA, Manthe CA, Chaguturu VK, Szabla K, Matts RL, Donnelly AC, Blagg BS, Holzbeierlein JM (2010) Characterization of a novel novobiocin analogue as a putative C-terminal inhibitor of heat shock protein 90 in prostate cancer cells. Prostate 70:27–36

Mayer MP (2010) Gymnastics of molecular chaperones. Mol Cell 39:321–331

Mazaira GI, Lagadari M, Erlejman AG, Galigniana MD (2014) The emerging role of TPR-domain immunophilins in the mechanism of action of steroid receptors. Nucl Recept Res 1:1–17. https://doi.org/10.11131/2014/101094. ID:101094

Mazaira GI, Zgajnar NR, Lotufo CM, Daneri-Becerra C, Sivils JC, Soto OB, Cox MB, Galigniana MD (2018) The nuclear receptor field: a historical overview and future challenges. Nucl Recept Res 5:101320. https://doi.org/10.11131/2018/101320

Mellatyar H, Talaei S, Pilehvar-Soltanahmadi Y, Barzegar A, Akbarzadeh A, Shahabi A, Barekati-Mowahed M, Zarghami N (2018) Targeted cancer therapy through 17-DMAG as an Hsp90 inhibitor: overview and current state of the art. Biomed Pharmacother = Biomedecine Pharmacotherapie 102:608–617

Meyer P, Prodromou C, Liao C et al (2004) Structural basis for recruitment of the ATPase activator Aha1 to the Hsp90 chaperone machinery. EMBO J 23:1402–1410

Miyoshi T, Takeuchi A, Siomi H, Siomi MC (2010) A direct role for Hsp90 in pre-RISC formation in Drosophila. Nat Struct Mol Biol 17:1024–1026
Mollapour M, Neckers L (2012) Post-translational modifications of Hsp90 and their contributions to chaperone regulation. Biochim Biophys Acta 1823:648–655
Mostaghel EA, Page ST, Lin DW et al (2007) Intraprostatic androgens and androgen-regulated gene expression persist after testosterone suppression: therapeutic implications for castration-resistant prostate cancer. Cancer Res 67:5033–5041
Naruse K, Matsuura-Suzuki E, Watanabe M, Iwasaki S, Tomari Y (2018) In vitro reconstitution of chaperone-mediated human RISC assembly. RNA 24:6–11
Neckers L, Ivy SP (2003) Heat shock protein 90. Curr Opin Oncol 15:419–424
Ni L, Yang CS, Gioeli D, Frierson H, Toft DO, Paschal BM (2010) FKBP51 promotes assembly of the Hsp90 chaperone complex and regulates androgen receptor signaling in prostate cancer cells. Mol Cell Biol 30:1243–1253
Palermo CM, Westlake CA, Gasiewicz TA (2005) Epigallocatechin gallate inhibits aryl hydrocarbon receptor gene transcription through an indirect mechanism involving binding to a 90 kDa heat shock protein. Biochemistry 44:5041–5052
Pan J, Jiang F, Zhou J, Wu D, Sheng Z, Li M (2018) HSP90: a novel target gene of miRNA-628-3p in A549 cells. Biomed Res Int 2018:1–10. ID:4149707. https://doi.org/10.1155/2018/4149707
Pearl LH, Prodromou C (2006) Structure and mechanism of the Hsp90 molecular chaperone machinery. Annu Rev Biochem 75:271–294
Pei H, Li L, Fridley BL, Jenkins GD, Kalari KR, Lingle W, Petersen G, Lou Z, Wang L (2009) FKBP51 affects cancer cell response to chemotherapy by negatively regulating Akt. Cancer Cell 16:259–266
Pennisi R, Ascenzi P, di Masi A (2015) Hsp90: a new player in DNA repair? Biomol Ther 5:2589–2618
Periyasamy S, Warrier M, Tillekeratne MP, Shou W, Sanchez ER (2007) The immunophilin ligands cyclosporin A and FK506 suppress prostate cancer cell growth by androgen receptor-dependent and -independent mechanisms. Endocrinology 148:4716–4726
Periyasamy S, Hinds T Jr, Shemshedini L, Shou W, Sanchez ER (2010) FKBP51 and Cyp40 are positive regulators of androgen-dependent prostate cancer cell growth and the targets of FK506 and cyclosporin A. Oncogene 29:1691–1701
Pratt WB (1997) The role of the hsp90-based chaperone system in signal transduction by nuclear receptors and receptors signaling via MAP kinase. Annu Rev Pharmacol Toxicol 37:297–326
Pratt WB, Toft DO (1997) Steroid receptor interactions with heat shock protein and immunophilin chaperones. Endocr Rev 18:306–360
Pratt WB, Krishna P, Olsen LJ (2001) Hsp90-binding immunophilins in plants: the protein movers. Trends Plant Sci 6:54–58
Prodromou C, Nuttall JM, Millson SH, Roe SM, Sim TS, Tan D, Workman P, Pearl LH, Piper PW (2009) Structural basis of the radicicol resistance displayed by a fungal hsp90. ACS Chem Biol 4:289–297
Protti MP, Heltai S, Bellone M, Ferrarini M, Manfredi AA, Rugarli C (1994) Constitutive expression of the heat shock protein 72 kDa in human melanoma cells. Cancer Lett 85:211–216
Queitsch C, Sangster TA, Lindquist S (2002) Hsp90 as a capacitor of phenotypic variation. Nature 417:618–624
Quinta HR, Galigniana NM, Erlejman AG, Lagadari M, Piwien-Pilipuk G, Galigniana MD (2011) Management of cytoskeleton architecture by molecular chaperones and immunophilins. Cell Signal 23:1907–1920
Ramsey AJ, Russell LC, Chinkers M (2009) C-terminal sequences of hsp70 and hsp90 as non-specific anchors for tetratricopeptide repeat (TPR) proteins. Biochem J 423:411–419
Ratajczak T (2015) Steroid receptor-associated Immunophilins: candidates for diverse drug-targeting approaches in disease. Curr Mol Pharmacol 9:66–95

Ratzke C, Mickler M, Hellenkamp B, Buchner J, Hugel T (2010) Dynamics of heat shock protein 90 C-terminal dimerization is an important part of its conformational cycle. Proc Natl Acad Sci U S A 107:16101–16106

Reikvam H, Ersvaer E, Bruserud O (2009) Heat shock protein 90 – a potential target in the treatment of human acute myelogenous leukemia. Curr Cancer Drug Targets 9:761–776

Richter K, Haslbeck M, Buchner J (2010) The heat shock response: life on the verge of death. Mol Cell 40:253–266

Romano S, D'Angelillo A, Pacelli R et al (2010) Role of FK506-binding protein 51 in the control of apoptosis of irradiated melanoma cells. Cell Death Differ 17:145–157

Rotoli D, Morales M, Del Carmen Maeso M, Del Pino Garcia M, Morales A, Avila J, Martin-Vasallo P (2016) Expression and localization of the immunophilin FKBP51 in colorectal carcinomas and primary metastases, and alterations following oxaliplatin-based chemotherapy. Oncol Lett 12:1315–1322

Rotoli D, Morales M, Avila J, Maeso M DC, Garcia MDP, Mobasheri A, Martin-Vasallo P (2017) Commitment of scaffold proteins in the onco-biology of human colorectal cancer and liver metastases after oxaliplatin-based chemotherapy. Int J Mol Sci 18, pii: E891.https://doi.org/10.3390/ijms18040891

Russo D, Merolla F, Mascolo M et al. (2017) FKBP51 immunohistochemical expression: a new prognostic biomarker for OSCC? Int J Mol Sci 18, pii: E443. https://doi.org/10.3390/ijms18020443

Rutherford SL, Lindquist S (1998) Hsp90 as a capacitor for morphological evolution. Nature 396:336–342

Sahu B, Laakso M, Pihlajamaa P, Ovaska K, Sinielnikov I, Hautaniemi S, Janne OA (2013) FoxA1 specifies unique androgen and glucocorticoid receptor binding events in prostate cancer cells. Cancer Res 73:1570–1580

Sangster TA, Salathia N, Lee HN et al (2008) HSP90-buffered genetic variation is common in *Arabidopsis thaliana*. Proc Natl Acad Sci U S A 105:2969–2974

Sarkar AA, Zohn IE (2012) Hectd1 regulates intracellular localization and secretion of Hsp90 to control cellular behavior of the cranial mesenchyme. J Cell Biol 196:789–800

Scammell JG, Denny WB, Valentine DL, Smith DF (2001) Overexpression of the FK506-binding immunophilin FKBP51 is the common cause of glucocorticoid resistance in three New World primates. Gen Comp Endocrinol 124:152–165

Scheufler C, Brinker A, Bourenkov G, Pegoraro S, Moroder L, Bartunik H, Hartl FU, Moarefi I (2000) Structure of TPR domain-peptide complexes: critical elements in the assembly of the Hsp70-Hsp90 multichaperone machine. Cell 101:199–210

Schulte TW, Neckers LM (1998) The benzoquinone ansamycin 17-allylamino-17-demethoxygeldanamycin binds to HSP90 and shares important biologic activities with geldanamycin. Cancer Chemother Pharmacol 42:273–279

Schulte TW, Akinaga S, Soga S, Sullivan W, Stensgard B, Toft D, Neckers LM (1998) Antibiotic radicicol binds to the N-terminal domain of Hsp90 and shares important biologic activities with geldanamycin. Cell Stress Chaperones 3:100–108

Schuster M, Schnell L, Feigl P et al (2017) The Hsp90 machinery facilitates the transport of diphtheria toxin into human cells. Sci Rep 7:613

Scroggins BT, Neckers L (2007) Post-translational modification of heat-shock protein 90: impact on chaperone. Expert Opin Drug Discovery 2:1403–1414

Shelton SN, Shawgo ME, Matthews SB et al (2009) KU135, a novel novobiocin-derived C-terminal inhibitor of the 90-kDa heat shock protein, exerts potent antiproliferative effects in human leukemic cells. Mol Pharmacol 76:1314–1322

Shi Y, Liu X, Lou J, Han X, Zhang L, Wang Q, Li B, Dong M, Zhang Y (2014) Plasma levels of heat shock protein 90 alpha associated with lung cancer development and treatment responses. Clin Cancer Res 20:6016–6022

Shimamura T, Lowell AM, Engelman JA, Shapiro GI (2005) Epidermal growth factor receptors harboring kinase domain mutations associate with the heat shock protein 90 chaperone and are destabilized following exposure to geldanamycins. Cancer Res 65:6401–6408

Shore D, Bianchi A (2009) Telomere length regulation: coupling DNA end processing to feedback regulation of telomerase. EMBO J 28:2309–2322

Sigal NH, Dumont FJ (1992) Cyclosporin A, FK-506, and rapamycin: pharmacologic probes of lymphocyte signal transduction. Annu Rev Immunol 10:519–560

Solassol J, Mange A, Maudelonde T (2011) FKBP family proteins as promising new biomarkers for cancer. Curr Opin Pharmacol 11:320–325

Solit DB, Chiosis G (2008) Development and application of Hsp90 inhibitors. Drug Discov Today 13:38–43

Soti C, Racz A, Csermely P (2002) A nucleotide-dependent molecular switch controls ATP binding at the C-terminal domain of Hsp90. N-terminal nucleotide binding unmasks a C-terminal binding pocket. J Biol Chem 277:7066–7075

Sreedhar AS, Soti C, Csermely P (2004) Inhibition of Hsp90: a new strategy for inhibiting protein kinases. Biochim Biophys Acta 1697:233–242

Stebbins CE, Russo AA, Schneider C, Rosen N, Hartl FU, Pavletich NP (1997) Crystal structure of an Hsp90-geldanamycin complex: targeting of a protein chaperone by an antitumour agent. Cell 89:239–250

Storer CL, Dickey CA, Galigniana MD, Rein T, Cox MB (2011) FKBP51 and FKBP52 in signaling and disease. Trends Endocrinol Metab 22:481–490

Supko JG, Hickman RL, Grever MR Malspeis L (1995) Preclinical pharmacologic evaluation of geldanamycin as an antitumor agent. Cancer Chemother Pharmacol 36:305–315

Taipale M, Krykbaeva I, Koeva M, Kayatekin C, Westover KD, Karras GI, Lindquist S (2012) Quantitative analysis of HSP90-client interactions reveals principles of substrate recognition. Cell 150:987–1001

Taipale M, Tucker G, Peng J et al (2014) A quantitative chaperone interaction network reveals the architecture of cellular protein homeostasis pathways. Cell 158:434–448

Taira T, Sawai M, Ikeda M, Tamai K, Iguchi-Ariga SM, Ariga H (1999) Cell cycle-dependent switch of up-and down-regulation of human hsp70 gene expression by interaction between c-Myc and CBF/NF-Y. J Biol Chem 274:24270–24279

Tatro ET, Everall IP, Kaul M, Achim CL (2009) Modulation of glucocorticoid receptor nuclear translocation in neurons by immunophilins FKBP51 and FKBP52: implications for major depressive disorder. Brain Res 1286:1–12

Teng SC, Chen YY, Su YN, Chou PC, Chiang YC, Tseng SF, Wu KJ (2004) Direct activation of HSP90A transcription by c-Myc contributes to c-Myc-induced transformation. J Biol Chem 279:14649–14655

Trepel J, Mollapour M, Giaccone G, Neckers L (2010) Targeting the dynamic HSP90 complex in cancer. Nat Rev Cancer 10:537–549

Tsutsumi S, Scroggins B, Koga F, Lee MJ, Trepel J, Felts S, Carreras C, Neckers L (2008) A small molecule cell-impermeant Hsp90 antagonist inhibits tumour cell motility and invasion. Oncogene 27:2478–2487

Vafopoulou X, Steel CG (2012) Cytoplasmic travels of the ecdysteroid receptor in target cells: pathways for both genomic and non-genomic actions. Front Endocrinol 3:43. https://doi.org/10.3389/fendo.2012.00043

van Kouwenhove M, Kedde M, Agami R (2011) MicroRNA regulation by RNA-binding proteins and its implications for cancer. Nat Rev Cancer 11:644–656

Vandevyver S, Dejager L, Libert C (2012) On the trail of the glucocorticoid receptor: into the nucleus and back. Traffic 13:364–374

Wang L (2011) FKBP51 regulation of AKT/protein kinase B phosphorylation. Curr Opin Pharmacol 11:360–364

Wang Y, Kirschner A, Fabian AK et al (2013) Increasing the efficiency of ligands for FK506-binding protein 51 by conformational control. J Med Chem 56:3922–3935

Ward BK, Mark PJ, Ingram DM, Minchin RF, Ratajczak T (1999) Expression of the estrogen receptor-associated immunophilins, cyclophilin 40 and FKBP52, in breast cancer. Breast Cancer Res Treat 58:267–280

Wawrzynow B, Zylicz A, Zylicz M (2018) Chaperoning the guardian of the genome. The two-faced role of molecular chaperones in p53 tumour suppressor action. Biochim Biophys Acta Rev Cancer 1869:161–174

Whitesell L, Mimnaugh EG, De Costa B, Myers CE, Neckers LM (1994) Inhibition of heat shock protein HSP90-pp60v-src heteroprotein complex formation by benzoquinone ansamycins: essential role for stress proteins in oncogenic transformation. Proc Natl Acad Sci U S A 91:8324–8328

Wiederrecht GJ, Sabers CJ, Brunn GJ, Martin MM, Dumont FJ, Abraham RT (1995) Mechanism of action of rapamycin: new insights into the regulation of G1-phase progression in eukaryotic cells. Prog Cell Cycle Res 1:53–71

Witkin SS (2001) Heat shock protein expression and immunity: relevance to gynecologic oncology. Eur J Gynaecol Oncol 22:249–256

Wochnik GM, Ruegg J, Abel GA, Schmidt U, Holsboer F, Rein T (2005) FK506-binding proteins 51 and 52 differentially regulate dynein interaction and nuclear translocation of the glucocorticoid receptor in mammalian cells. J Biol Chem 280:4609–4616

Wu J, Liu T, Rios Z, Mei Q, Lin X, Cao S (2017) Heat shock proteins and cancer. Trends Pharmacol Sci 38:226–256

Yang Y, Rao R, Shen J, Tang Y, Fiskus W, Nechtman J, Atadja P, Bhalla K (2008) Role of acetylation and extracellular location of heat shock protein 90alpha in tumour cell invasion. Cancer Res 68:4833–4842

Yano M, Naito Z, Tanaka S, Asano G (1996) Expression and roles of heat shock proteins in human breast cancer. Japan J Cancer Res: Gann 87:908–915

Yu X, Harris SL, Levine AJ (2006) The regulation of exosome secretion: a novel function of the p53 protein. Cancer Res 66:4795–4801

Yufu Y, Nishimura J, Nawata H (1992) High constitutive expression of heat shock protein 90 alpha in human acute leukemia cells. Leuk Res 16:597–605

Zhou H, Huang S (2010) mTOR signaling in cancer cell motility and tumour metastasis. Crit Rev Eukaryot Gene Expr 20:1–16

Chapter 3
Therapeutic Potential of Heat Shock Protein 90 Inhibitors in Colorectal Cancer

Reyhaneh Moradi-Marjaneh, Seyed Mahdi Hassanian, Gordon A. Ferns, Amir Avan, and Majid Khazaei

Abstract Colorectal cancer (CRC) is one of the most common causes of cancer-related death globally. Despite recent advances in diagnosis and treatment, CRC remains one of the leading cause mortality, with 5 years survival rate of approximately 60%. Hence, there is an urgent need for new therapeutic approaches. A large number of Hsp90 client proteins have importance in the stability and function of crucial oncogenic proteins and play critical roles in the development of CRC. Hence, Hsp90 inhibitors may have therapeutic benefits in the treatment of CRC, and they are currently in clinical trials for various malignancies. In this chapter, we have

R. Moradi-Marjaneh
Torbat Heydariyeh University of Medical Sciences, Torbat Heydariyeh, Iran

S. M. Hassanian
Metabolic Syndrome Research Center, Mashhad University of Medical Sciences, Mashhad, Iran

Department of Medical Biochemistry, Faculty of Medicine, Mashhad University of Medical Sciences, Mashhad, Iran

G. A. Ferns
Division of Medical Education, Brighton & Sussex Medical School, Brighton, Sussex, UK

A. Avan (✉)
Metabolic Syndrome Research Center, Mashhad University of Medical Sciences, Mashhad, Iran

Cancer Research Center, Mashhad University of Medical Sciences, Mashhad, Iran

Department of Modern Sciences and Technologies, Faculty of Medicine, Mashhad University of Medical Sciences, Mashhad, Iran
e-mail: avana@mums.ac.ir

M. Khazaei (✉)
Metabolic Syndrome Research Center, Mashhad University of Medical Sciences, Mashhad, Iran

Department of Physiology, Faculty of Medicine, Mashhad University of Medical Sciences, Mashhad, Iran
e-mail: Khazaeim@mums.ac.ir

A. A. A. Asea, P. Kaur (eds.), *Heat Shock Protein 90 in Human Diseases and Disorders*, Heat Shock Proteins 19,
https://doi.org/10.1007/978-3-030-23158-3_3

reviewed the potential role of Hsp90 in the pathogenesis of CRC, and have summarized recent preclinical and clinical studies of Hsp90 inhibitors as a potential therapeutic in CRC.

Keywords Cancer · Clinical · Colorectal · Hsp90 inhibitors · Treatment

Abbreviations

15-PGDH	15-hydroxyprostaglandin dehydrogenase
Akt	Protein kinase B
BIM	Bcl-2 interacting mediator of cell death
Cdk2	Cyclin-dependent kinase 2
COX-2	Cyclooxygenase-2
ECM	Cell-extracellular matrix
ERK	Extracellular signal–regulated kinases
FDG	Fluorodeoxyglucose
FGFR	Fibroblast growth factor receptor
HER2/neu	Human epidermal growth factor receptor 2
IKK	IκB kinase
LRP-1	Lipoprotein receptor-related protein 1
MMP9	Matrix metalloproteinase 9
NF-κB	Nuclear factor kappa-light-chain-enhancer of activated B cells
NNM	Non-neoplastic mucosa
PET	Positron emission tomography
PGE2	Prostaglandin E2
PI3K	Phosphoinositide 3 – kinase
PKC	Protein kinase C
RIG-I	Retinoic acid-inducible gene I
STK33	Serine/threonine kinase 33
TCF12	Transcription factor 12
TRAP1	Tumor necrosis factor associated protein
Tyk2	Tyrosine kinase 2

3.1 Introduction

Colorectal cancer (CRC) is the third most common cancer world wild. The incidence rates of CRC in 2030 have been estimated to more than 2.2 million new cases and approximately 1.1 million deaths (Arnold et al. 2016; Moradi-Marjaneh et al. 2018a). The treatments usually used for CRC include a combination of surgery, radiotherapy, chemotherapy, and targeted therapy. The most common chemotherapeutic regimens used include: 5-fluorouracil (5-FU) in combination with leucovorin and oxaliplatin or irinotecan, known as FOLFOX and FOLFIRI respectively. However a large number of patients do not benefit from the current regimens that are being used, and local recurrence and metastatic disease still often occur, so more effective therapeutic approaches are warranted (Marjaneh et al. 2018; Moradi-Marjaneh et al. 2018b).

Heat shock proteins 90 (Hsp90) are molecular chaperones that mediate intracellular process involved in cell proliferation and survival. Hsp90 expression is elevated in CRC, and this is linked with poor prognosis (Ge et al. 2018). There is accumulating evidence for the beneficial effects of Hsp90 inhibitors in cancer treatment, and that they may augment the effect of other anticancer drugs through various mechanisms. The efficacy of Hsp90 inhibitors is under investigating in clinical trials. Furthermore, a large number of compounds have been synthetized to increase their efficiency and bioavailability, such as oral compound (Garcia-Carbonero et al. 2013). In most trials to date, CRC patients have not comprised the entire studied, but they have evaluated a part of the studies population in the treatment of the advanced solid tumor or gastrointestinal stromal tumors. In this review, we provide an overview of preclinical and clinical studies evaluating the therapeutic potential of Hsp90 inhibitors in patients with CRC.

3.2 Hsp90 and Colorectal Cancer

The Hsp90 superfamily consists of cytoplasmic chaperones Hsp90α (inducible) (84 kDa), and Hsp90ß (constitutively expressed) (86 kDa) (Pearl and Prodromou 2001), membrane associated homologue Hsp 90 N (75 kDa) (Grammatikakis et al. 2002), endoplasmic reticulum homologue GRP94, also called endoplasmin (94 kDa), and mitochondrial homologue TRAP1 (Tumor necrosis factor associated protein (75 kDa) (Pearl and Prodromou 2001). Hsp90 plays an important role in the pathogenesis of CRC. Hsp90α levels in CRC tumor tissue is significantly greater than healthy adjacent tissue (Zhang et al. 2009). Its overexpression has been related to metastasis and poor prognosis in CRC patients (Chen et al. 2011). Higher expression levels of Hsp90α on the surface of CRC cells may enable Hsp90α to act as a facilitator of invasion, suggesting its role in cell migration (Milicevic et al. 2008). In addition, the expression of the Hsp90ß isoform is higher in poorly-differentiated carcinomas compared to well differentiated CRC, indicating its involvement in the inhibition of cancer cells differentiation (Milicevic et al. 2008). There is a strong relationship between high levels of Hsp90N expression and neoplastic transformation. Thus it might play a role in tumor formation (Grammatikakis et al. 2002).

Hsp90α induce tumor-genesis through various mechanisms in CRC cells. It induces TCF_{12} expression in a CD_{91}/IKK/NF-κB-dependent signaling cascade to result in E-cadherin down-regulation and increased migration of CRC cells (Chen et al. 2013). In addition, Hsp90α bind to CD_{91}/LRP-1 and HER_2/neu and induce integrinαV expression in an NF-κB dependent-signaling pathway, leading to CRC cell invasion (Chen et al. 2010). It has also been shown that the histone methyltransferase activity of $SMYD_3$

is increased in the presence of Hsp90α (Hamamoto et al. 2004). Hsp90α associates with RIG-I, an intracellular pattern recognition receptor that plays key roles during innate immune responses and protects RIG-I from proteasomal degradation (Matsumiya et al. 2009). It has been reported that Hsp90α is an IMH-2 epitope-associated protein suggesting IMH-2 applications as a tumor marker or therapeutic target (Chen et al. 2011).

Hsp90β also induces tumorogenesis through various mechanisms in CRC cells. It has been shown that Hsp90β regulates TGF-β signaling. It binds to LAP of TGF-β leading to TGF-β negative regulation (Suzuki and Kulkarni 2010). In addition, Hsp90β binds to both TGF-β receptors, TGF-βRI and TGF-βRII, protecting them from degradation in the non-cancerous human cell (Wrighton et al. 2008). There is a synergic activity between TGF-β1 and Hsp90β that stimulates the migratory behavior of CRC cells when signaling occurs via αvβ6 integrin, a protein that facilitates ECM adhesion, compared to the canonical TGF-β1 pathway (de la Mare et al. 2017). Furthermore, it seems that GRP94 play a role in tumorogenesis and progression of CRC. GRP94 up-regulated from NNM, low-grade adenoma, high-grade adenoma, to carcinoma (Takahashi et al. 2011).

3.3 Hsp90 Inhibitors

Hsp90 have attracted recent attention as a potential target in the treatment of cancer. Indeed Hsp90 inhibitors exert their anticancer function through degradation of oncoproteins that are Hsp90 dependent clients. Hsp90 inhibitors bind to ATP binding pocket of Hsp90, inhibit the dissociation of Hsp90 client proteins from the chaperone complex, and so trap proteins and inhibit their stable conformation leading to complex degradation within the proteasome (Powers and Workman 2007; Koga et al. 2009) (Fig. 3.1). There is evidence that Hsp90 inhibitors selectively target tumor cells, which may be due to heteroprotein complexes present in cancer cells, versus the homodimer complexes found in normal cells. Also in cancer cells Hsp90 exists in an activated form with increased affinity for ATP or inhibitors (Kamal et al. 2003). Various classes of Hsp90 inhibitors have been discovered. They consist of

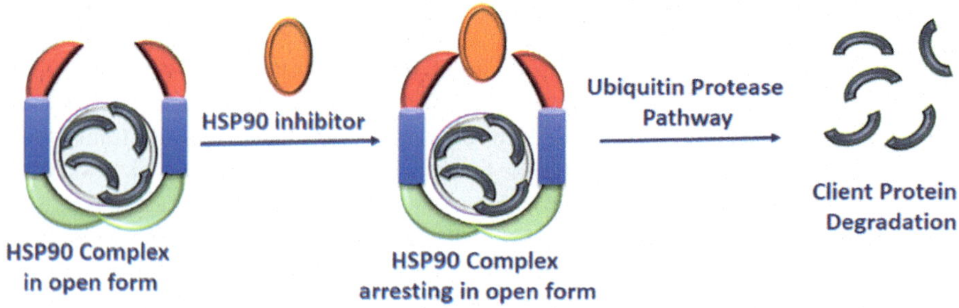

Fig. 3.1 Biding of Hsp90 inhibitors with Hsp90 multiprotein complex inhibit the formation of the closed conformation of Hsp90 complex which is essential for maturation of client proteins

natural products including geldanamycin and radicicol and geldanamycin derivatives such as 17-AAG (17-allylamino-17-demethoxygeldanamycin), and 17-DMAG (17- dimethylamino-noethylamino-17-demethoxygeldanamycin), purine derivatives including PU3, PU24FCl and PU-H58, pyrazoles, naphthol and quinazolinone (Doddareddy et al. 2011). Despite the striking advances in production of novel Hsp90 inhibitors, however the therapeutic potential of them remains yet to be fully realized in the clinical setting to treat CRC patients.

3.3.1 Geldanamycin and Its Derivatives

The geldanamycin (GA) is a natural product, and the first Hsp90 inhibitor identified (Fig. 3.2). It obtained from the bacteria Streptomyces hygroscopicus (Taldone et al. 2009) and competes for the N-terminal ATP binding site of Hsp90 (Mimnaugh et al. 1996). Geldanamycin exerts a negative effect on the cyclin E (an essential component to initiate DNA replication) gene promoter activity (Bedin et al. 2004). In addition, geldanamycin enhances the ubiquitinylation of cyclin E which leading to its active degradation and inhibits the cyclin E/Cdk2 kinase activity via the proteasome pathway without direct binding in CRC cell lines (Bedin et al. 2009; Hashemzehi et al. 2018). Despite promising anticancer activities of GA in vitro (Messaoudi et al. 2008), it was shown to be hepatotoxic in preclinical toxicology studies. In addition to hepatotoxicity, gastrointestinal toxicity and nephrotoxicity were also observed, making it unsuitable for clinical trials (Supko et al. 1995). GA toxic activity has prompted the development of less toxic analogues including Tanespimycin (17-AAG), Alvespimycin (17-DMAG), Retaspimycin (IPI-504) and IPI-493.

3.3.1.1 Tanespimycin (17-AAG)

Tanespimycin (17-allyl-amino-geldanamycin (17-AAG); NSC 330507/KOS-953) is an analog of GA which has a substitution at the 17-position of the ansamycin quinone (Taldone et al. 2009) (Fig. 3.3). Tanespimycin is less toxic and more stable

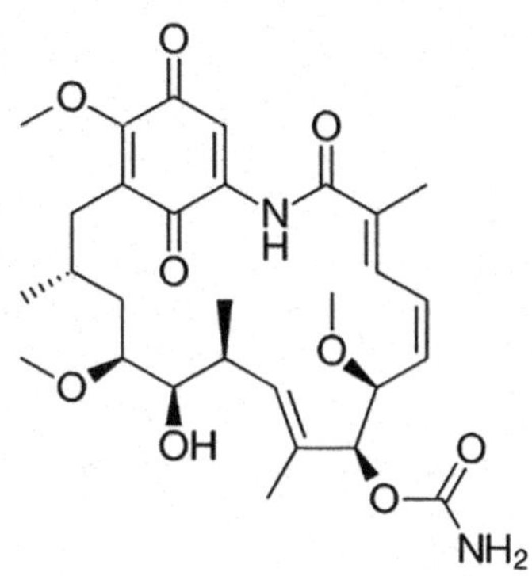

Fig. 3.2 The molecular structure of geldanamycin (GA)

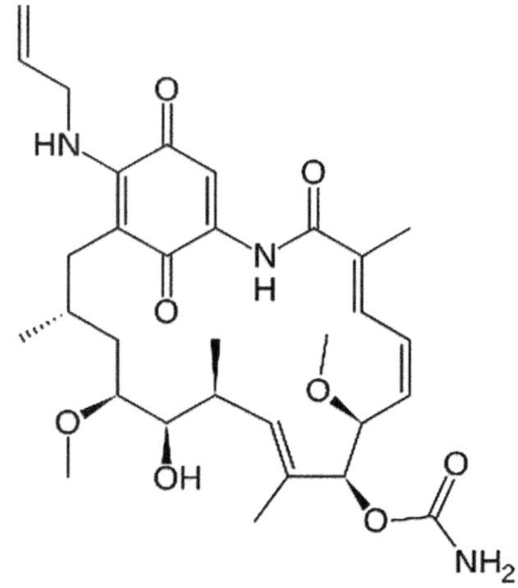

Fig. 3.3 The molecular structure of tanespimycin (17-AAG)

compared with GA. Even though 17-AAG/Hsp90 binding is weaker compared with GA/Hsp90 binding, 17-AAG functions as GA and exerts stronger toxicity profile in cancer cells (Agnew et al. 2001). 17-AAG metabolized to 17-AG (17-aminogeldanamycin), the major active metabolite of 17-AAG, by cytochrome P450 (CYP3A5) and widely distributed in body tissues but not the central nervous system (Egorin et al. 1998, 2001). However CYP3A5 polymorphisms do not have an effect on the disposition or toxicity of 17-AAG (Goetz et al. 2005).

17-AAG in combination with celecoxib increases the anti-proliferative effect of celecoxib in CRC cells (Mohammadi et al. 2016a). Also, it was shown to have a potent synergistic effect in combination with capecitabine and oxaliplatin in CRC cell lines (Mohammadian et al. 2017) suggesting Hsp90 inhibitors in combination with common chemotherapeutic regimes may be a useful therapeutic strategy for CRC treatment. 17-AAG exerts anti-proliferative activity in CRC cells through various mechanisms. For example, a study showed that 17-AAG degrades STK33 (an important enzyme for the viability and proliferation of KRAS mutant cell) in KRAS mutant CRC and induce apoptosis (Azoitei et al. 2012). In addition, mitochondrial proteins accumulate in CRC cells following treated by 17-AAG. So it may be able to induce the mitochondrial apoptosis pathways by 17-AAG (Margineantu et al. 2007). It has been also showed that 17-AAG decreases the levels of PGE2 (major mediators of inflammation in CRC) levels in CRC cells by modulating COX-2 and 15-PGDH expression (Mohammadi et al. 2016b).

Although there is several evidence showing the safety of 17-AAG, a striking significant anticancer activity of 17-AAG has not been observed. A phase I study of 17-AAG was performed by Mayo clinical investigators and evaluated its pharmacokinetic, response, maximum tolerated dose, and toxicity in 21 patients with advanced cancer. Dose-limiting toxicities (DLTs) at 431 mg/m^2 was grade 3 bilirubin, aspartate transaminase, anemia, nausea, vomiting, and myalgia. They determined 308 mg/m^2 for the maximum tolerated dose (MTD) of weekly 17-AAG. No objective responses were observed (Goetz et al. 2005). Another phase I pharmacokinetic-pharmacodynamic study of 17-AAG was conducted in 45 patients with refractory advanced cancers. Escalating doses between 10 and 395 mg/m^2 were given intravenously. 17-AAG was well tolerated until the dose of 295 mg/m^2. Common toxic effects were fatigue, anorexia, diarrhea, nausea, vomiting and the limited number of hematologic toxicity. Among 44 patients, 16 patients had reversible elevations of liver enzymes. Consistent with the results of Mayo Clinical investigators study no objective response was detected (NCT00019708) (Ramanathan et al. 2005).

The effect of 17-AAG in combination with paclitaxel was evaluated in phase 1 clinical study for the treatment of patients with metastatic or unresectable solid tumors progressing on standard therapy. The results of anticancer activity were disappointing. However, the recommended phase II dose (RP2D) of this combination were 175 mg/m^2 17-AAG twice-weekly and 80 mg/m^2 paclitaxel weekly for phase II trials. Also, any drug-drug interactions between 17-AAG and paclitaxel was not observed (NCT00087217) (Ramalingam et al. 2008). The combination of 17-AAG and sorafenib was also evaluated in a phase I trial with the primary objective of defining a phase II dose in 27 patients with renal cancer, melanoma or CRC. DLT was observed at 450 mg/m^2 of 17-AAG. RP2D was declared at the dose of 400 mg of sorafenib twice daily and 400 mg/m^2 17-AAG on days 1, 8, and 15, every 28 days. In addition, clinical efficacy was promising so that 9% showed a partial response (PR) and 61% had stable disease (SD). Time to progression (TTP) was also 2.9 months for all patients (NCT00121264) (Vaishampayan et al. 2010). The combination of 17-AAG with irinotecan was also used in 27 patients with solid tumors in a dose escalation phase. DLT was observed in four patients at the dose of 100 mg/m^2 irinotecan and 375 mg/m^2 17-AAG and was including nausea, vomiting, diarrhea, and pulmonary embolism. The two compounds did not affect another pharmacokinetic and the authors suggested that although no PR was observed the 17-AAG/irinotecan combination could be given with acceptable toxicity (NCT00119236) (Archie et al. 2008). Anti-cancer activity was observed in the combination of 17-AAG with gemcitabine and cisplatin, but it was not clinically tolerable due to significant hematologic toxicities (NCT00047047) (Hubbard et al. 2011). In another study, the DLTs of 17-AAG were included dehydration, diarrhea, hyperglycemia, and liver toxicity and no tumor response was observed. Interestingly the level of Hsp70 was increased following treatment with 17-AAG which may be a reason for resistance (NCT00003969) (Nowakowski et al. 2006) (Table 3.1).

17-AAG has entered into phase II clinical trials to evaluate its efficacy combined with gemcitabine in patients with advanced ovarian and peritoneal cancer. 17-AAG was administered by the dose of 154 mg/m^2 on days 1 and 9 of cycle 1 and days 2 and 9 of subsequent cycles. Gemcitabine was also administered intravenously by the dose of 750 mg/m^2 on day 8 of the first treatment cycle and days 1 and 8 of subsequent cycles. Although 17-AAG/gemcitabine combination was well tolerated, however a limited clinical efficacy was observed (NCT00093496) (Hendrickson et al. 2012). However, against this finding 17-AAG in combination with trastuzumab showed a significant anti-cancer activity in patients with HER2 positive, metastatic breast cancer, so that the overall response rate was 22%, the clinical benefit was 59%, the median progression-free survival (PFS) was 6 months and the

Table 3.1 Clinical studies of phase I Tanespimycin (17-AAG)

Clinical trials identifier	Intervention	Condition	Recommended dose for phase II	Clinical efficacy	References
NCT00119236	Tanespimycin	27 patients with advanced solid tumors, (6 patients had CRC)	17-AAG: 300 mg/m^2 QW	No complete or PR was observed	Archie et al. (2008)
	Irinotecan hydrochloride		Irinotecan: 100 mg/m^2 QW	11 patients achieved SD with median duration of 3 months	
NCT00121264	Tanespimycin	27 patients with advanced solid tumors, (4 patients had CRC)	17-AAG: 400 mg/m^2 on days 1, 8, and 15 of a 4 W cycle	2 patients achieved PR	Vaishampayan et al. (2010)
				14 patients achieved SD	
	Sorafenib		Sorafenib: 400 mg twice QD	Clinical benefit was observed in 70% of patients	
				Median TTP for all patients was 2.9 months	
				Median OS was 12.1 months	
NCT00087217	Tanespimycin	25 patients with advanced solid tumors, (2 patients had CRC)	17-AAG: 175 mg/m^2 BIW	5 patients achieved SD	Ramalingam et al. (2008)
	Paclitaxel		Paclitaxel: 80 mg/m^2 QW	Disease progression led to treatment discontinuation in 16 patients	
NCT00047047	Tanespimycin	39 patients with advanced solid tumors, (4 patients had CRC)	17-AAG: 154 mg/m^2 QW	6 patients achieved PR	Hubbard et al. (2011)
			Gemcitabine:750 mg/m^2 QW	13 patients achieved SD	
	Gemcitabine hydrochloride		No RP2D for the cisplatin-containing combinations		
	Cisplatin				
NCT00003969	Tanespimycin	13 patients with advanced solid tumors, (7 patients had CRC)	17-AAG: 220 mg/m^2 on days 1, 4, 8, and 11 of a 3 W cycle	No tumor responses were seen	Nowakowski et al. (2006)

median overall survival (OS) was 17 months. These patients received trastuzumab by the dose of 2 mg/kg and tanespimycin by the dose of 450 mg/m^2 weekly (NCT00773344) (Modi et al. 2011).

3.3.1.2 Alvespimycin (17-DMAG)

Alvespimycin (17-DMAG) is another analog of GA with approximately two times potency against human Hsp90 than 17-AAG and exhibit better solubility and less toxicity to normal cells (Fig. 3.4). In addition, a new form of this drug is orally bioavailable (Hertlein et al. 2010). It has been shown that 17- DMAG and oxaliplatin exert a synergistic effect on inhibition of proliferation and invasion of CRC cells through increasing the level of Bax (an apoptosis promoting gene) and decreasing the levels of MMP9 and integrin β3 (Zhou et al. 2015). Furthermore, the 17-DMAG ability to function as an NF-κB inhibitor is of great interest in treatment of CRC (Hertlein et al. 2010).

In a Phase I dose escalation study, Pacey evaluated pharmacokinetic profile and tumor response of 17-DMAG in patients with advanced solid tumors. DLT was observed at the dose of 106 mg/m^2 in 2/4 patients and one treatment-related death occurred. In this trial, frequent adverse events (AEs) included nausea, vomiting, fatigue and liver enzyme disturbances. Intriguingly anti-cancer efficacy was observed across a range of dose levels. Pharmacokinetic studies showed a narrow therapeutic window of 17-DMAG. Plasma Hsp72 was increased during the therapeutic period and was more significant in patients who experienced DLT. The RP2D of 17-DMAG was 80 mg/m^2 weekly, IV (Pacey et al. 2011).

The effect of 17-DMAG in combination with trastuzumab was evaluated in phase I clinical study for the treatment of patients with advanced solid tumors who were refractory to proven therapy. 17-DMAG was administered in three dose (60 mg/m^2, 80 mg/m^2, 100 mg/m^2), IV, once weekly in a 4-week cycle. Trastuzumab was also injected at a dose of 4 mg/kg on 1st week. In contrast to the study of Pacey et al., DLT was observed at the dose of 80 mg/m^2 and was keratitis in two patients. Congestive heart failure was also observed in one patient at the dose of 100 mg/m^2.

Fig. 3.4 The molecular structure of Alvespimycin (17-DMAG)

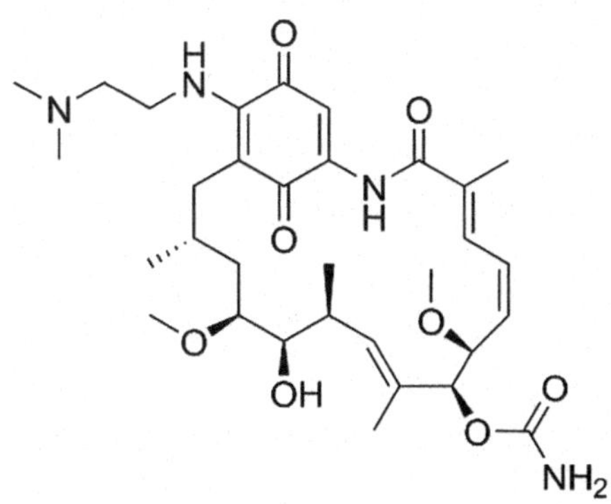

Table 3.2 Clinical studies of Alvespimycin (17-DMAG)

Clinical trials identifier	Intervention	Condition	Phase	Main results	References
NCT00248521	17-DMAG	25 patients with advanced solid tumors, (2 patients had CRC)	I	3 patients achieved SD for a duration of >6 months	Pacey et al. (2011)
	QIW (2.5–106 mg/m^2)			1 patient with melanoma achieved PR	
				1 patient with castration refractory prostate cancer achieved CR	
				RP2D: 80 mg/m^2 QW, IV	
NCT00803556	17-DMAG	28 patients with advanced solid tumors	I	1 patient achieved PR	Jhaveri et al. (2012)
	QIW (60, 80, 100 mg/m^2)			7 patients achieved SD (range; 4–10 months)	
	Trastuzumab			RP2D in combination with trastuzumab: 80 mg/m^2, QW, IV	

A promising considerable anti-cancer efficacy was observed in HER2 metastatic breast cancer patients so that among 23 patients one achieved PR and 7 patients achieved SD (range; 4–10 months). In addition, pharmacokinetics study showed that dose or dosing day did no effect on the volume of distribution (range; 142–155 L) and half-lives (range; 16–20 h) (Jhaveri et al. 2012) (Table 3.2). Currently, there are no phase II or III trials of 17-DMAG that we are aware of. Future studies of 17-DMAG should consider finding its therapeutic capacity in various type of cancer.

3.3.1.3 Retaspimycin Hydrochloride (IPI-504)

Retaspimycin hydrochloride (IPI-504) is an analogue derivative of 17-AAG which is designed to improve the water-solubility activity of 17-AAG and enable to exert high concentrations without the need for organic solvents (Hanson and Vesole 2009) (Fig. 3.5). IPI-504 passed phase Ib trial in combination with docetaxel in patients with a solid tumor. Toxicities were reversible and similar to those seen with docetaxel or IPI-504 alone. In this study the regimen of IPI-504 450 mg/m^2 with docetaxel 75 mg/m^2 once every 3 weeks was RP2D (NCT00606814) (Riely et al. 2009). A low toxicity profile of IPI-504 was confirmed in several clinical trial studies showed. In addition, it showed promising anti-cancer activity in gastrointestinal stromal tumors (GIST) and other soft tissue sarcomas. In a phase I dose-escalation trial, 54 patients

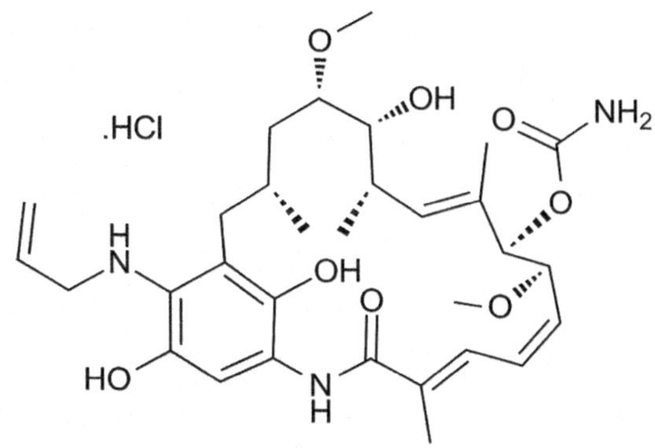

Fig. 3.5 The molecular structure of Retaspimycin hydrochloride (IPI-504)

with unresectable GIST or soft-tissue sarcomas were administered by IPI-504 either twice weekly for 2 weeks on/1 week off or twice weekly for 3 weeks in continuous 3-week cycles. MTD of IPI-504 was determined 400 mg/m^2 twice weekly followed by 1 week off treatment. SD was observed in 70% of patients with GIST and 59% of patients with soft-tissue sarcomas. Metabolic PR occurred in 38% of GIST patients. In addition for patients with GIST, PFS was 10.6 weeks and the median time to tumor progression was 12 weeks (NCT00276302) (Wagner et al. 2013). However a phase III study of IPI-504 showed disappointing results. The RING (Retaspimycin In GIST) trial was a double-blind, placebo-controlled trial of IPI-504 in patients with refractory GIST. The use of imatinib or sunitinib for the treatment of patients was defeated. Unfortunately there was a higher mortality rate in the treatment arm. The difference in the outcome of this study compared with previous studies may be due to the enrolment of patients with the more advanced disease than previous studies (NCT00688766) (Johnston 2009) (Table 3.3).

3.3.2 Resorcinol Derivatives Hsp90 Inhibitors

Radicicol is a resorcyclic acid lactone which achieved from the fungus monosporium bonorden (Fig. 3.6). Its anti-tumor activity was evident in vitro. However, it did not show acceptable anti-tumor activity in vivo probably due to high instability of its reactive epoxide moiety. Radicicol instability has prompted the development of more stable analogues using resorcinol core including ganetespib, NVP-AUY922, AT-13387, and KW-2478 (Garcia-Carbonero et al. 2013). NW457 is a novel radicicol-derivative exerts antitumor activity in vitro and in vivo model of CRC and synergizes irradiation-induced apoptosis. In addition, NW457 reduce hepatotoxicity (Kinzel et al. 2016).

Table 3.3 Clinical studies of Retaspimycin (IPI-504)

Clinical trials identifier	Intervention	Condition	Phase	Main results	References
NCT00606814	IPI-504	16 patients with advanced solid tumors	Ib	4 DLTs was observed:	Riely et al. (2009)
	300, 450, 550 mg/m^2 every 3 W			300 mg/m^2: 1 patient; grade 3 febrile neutropenia	
				450 mg/m^2: 1 patient; grade 3 fatigue	
	Docetaxel			550 mg/m^2: 1 patient; grade 1 asymptomatic sinus bradycardia	
	75 mg/m^2 every 3 W			550 mg/m^2: 1 patient; grade 3 elevated AST with grade 3 acute respiratory distress syndrome	
				All DLTs resolved on trial	
				RP2D: IPI-504 450 mg/m^2 with docetaxel 75 mg/m^2 once every 3 W	
				No pharmacokinetics interactions between docetaxel and IPI-504 have been observed	
NCT00276302	IPI-504	34 patients with unresectable GIST and 17 patients with soft-tissue sarcomas (STS)	I	The MTD was 400 mg/m^2 BIW for 2 W on/1 W off	Wagner et al. (2013)
	90–500 mg/m^2 BIW (2 W on/1 W off)			70% achieved SD; 26/37 of patients with GIST and 10/17 of patients with STS	
				2 patients achieved PR	
				Metabolic PR occurred in 11/29 (38%) of GIST patients	
				AVs: fatigue, headache, and nausea	
NCT00688766	IPI-504	Patients with refractory GIST	III	No improvement in PFS or OS was observed in the treatment arm compared to the placebo arm	Johnston (2009)
	Placebo				

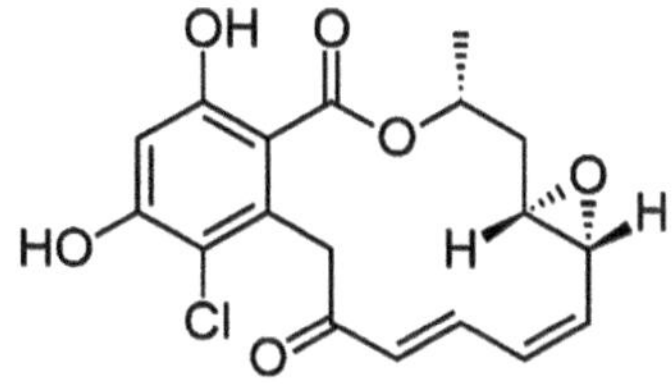

Fig. 3.6 The molecular structure of Radicicol

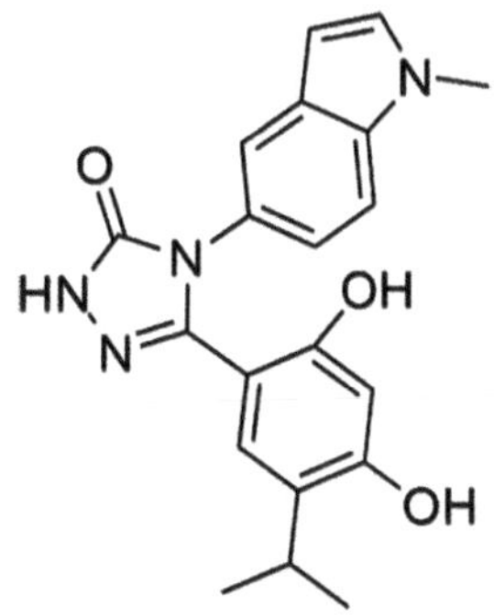

Fig. 3.7 The molecular structure of Ganetespib (STA-9090)

3.3.2.1 Ganetespib (STA-9090)

Ganetespib aother name STA-9090 is a resorcinol-containing triazole with acceptable potency and tumor penetration, and also favorable toxicity in preclinical models (Fig. 3.7). It is reported that ganetespib downregulates thymidylate synthase, and sensitize CRC cell to the effects of 5-FU through reduction of cyclin D1, pRb and inhibition of PI3K/Akt and ERK signaling pathways in vitro and in vivo, suggesting combination therapy with Hsp90 inhibitors and 5-FU is a rational strategy for CRC treatment (Nagaraju et al. 2014; Amerizadeh et al. 2018). Ganetespib showed acceptable tolerability and promising anti-cancer activity in combination with docetaxel in patients with advanced solid tumor malignancies (Kauh et al. 2012). Combination of ganetespib with standard capecitabine and radiation in patients with stage II or III rectal adenocarcinoma revealed some clinical activity and safety (El-Rayes et al. 2015) (Table 3.4).

Ganetespib was well tolerated in a phase I dose-escalation study (7–259 mg/m^2) of 1 h infusion once weekly for 3 weeks of a 28-day cycle. DLTs were observed at the dose of 150 mg/m^2 (amylase elevation) and the dose of 259 mg/m^2 (diarrhea and asthenia). Ganetespib also induced elevations of circulating Hsp70, indicate its biological activity. However, serum levels were variable and did not correlate with the ganetespib dose and tumor response (Goldman et al. 2013). Ganetespib has been also well tolerated in another phase I dose-escalation study (2–25 mg/m^2) 1 h infusion twice weekly for 3 weeks of a 4-week cycle. It shows linear pharmacokinetics, rapid distribution, and half-life of 10–14 h with no accumulation in plasma (Cho et al. 2011). In a phase II trial in patients with metastatic and/or unresectable GIST, who failure to prior therapy received ganetespib by the dose of 200 mg/m^2 IV infusion weekly, 3 week on/1 week off. The best-reported response was SD in 12 patients of 23 evaluable patients. However, no inhibition of activated KIT or its

Table 3.4 Clinical studies of ganetespib (STA-9090)

Clinical trials identifier	Intervention	Condition	Phase	Main results	References
NCT01183364	Ganetespib	27 patients with solid tumor	I	2 DLTs was observed at docetaxel 75 mg/m^2 and ganetespib 200 mg/m^2 including febrile neutropenia and one grade 4 neutropenia	Kauh et al. (2012)
	Docetaxel			1 patient achieved PR	
				12 patients achieved SD for 6 W	
				10 patients achieved SD for 12 W	
				6 patients achieved SD for 18 W	
				RP2D: ganetespib 150 mg/m^2 and docetaxel 75 mg/m^2	
NCT01554969	Ganetespib	16 patients with rectal cancer	I	1 DLT was observed at ganetespib 100 mg/m^2, capecitabine 825 mg/m^2 and RT (50.4 Gy) (grade 3 diarrhea for more than 4 days)	El-Rayes et al. (2015)
	Capecitabin				
	Radiation			12 patients had completed surgical resection	
				Pathologic CR rate was 25%	
				2 patients had residual tumors less than 1 cm	
				67% patients had clearing of lymph nodes disease on the pathologic specimen	
NCT00687934	Ganetespib	53 patients with advanced solid tumors, (8 patients had CRC)	I	4 DLTs was observed	Goldman et al. (2013)
				150 mg/m^2: 1 patient; grade 3 amylase elevation	
				259 mg/m^2: 1 patient grade 3 diarrhea +2 patient grade 3 and grade 4 asthenia	
				MTD: 216 mg/m^2	
				RP2D: 200 mg/m^2 IV at days 1, 8, and 15 every 4 W	
				DCR (objective response and stable disease at $\geq$16 W): 24.4%	

NCT00688116	Ganetespib	29 patients with solid tumors	I	The AEs: more than 20% of patients; mild to moderate of fatigue, nausea, anemia, diarrhea, vomiting, anorexia, and headache	Cho et al. (2011)
				1 DLT was observed at ganetespib 10 mg/m^2 of elevated LFTs	
				MTD: has not been reached	
				1 patient achieved durable PR	
				Several patients had prolonged SD	
NCT01039519	Ganetespib	26 patients with GIST	II	12/23 evaluable patients achieved SD (4 SD ≥16 W, 8 SD ≥8 W)	Demetri et al. (2011)
				No evidence of severe liver, ocular, cardiac or renal toxicity was observed	
				No inhibition of activated KIT or its downstream pathways was observed	
NCT01111838	Ganetespib	17 patients with chemotherapy-refractory, metastatic CRC	II	No objective regression of disease was observed	Cercek et al. (2014)
				2 patients achieved SD for a duration of 6.8 and 5.1 with KRAS-mutant tumors	
				Serious AEs: diarrhea (12%), fatigue (17%), and increased aspartate aminotransferase/alanine aminotransferase (12%) and alkaline phosphatase (6%) levels	

downstream pathways was observed in the analysis of client proteins in paired tumor biopsies (before and after treatment). So they suggested that ganetespib by dosage used in this study is not optimal for inhibition of GISTs with mutant KIT (Demetri et al. 2011). However, ganetespib as a single agent did not demonstrate activity in chemotherapy-refractory metastatic CRC in a phase II trial. They received 200 mg/m^2 IV ganetespib for 3 week on/1 week off until disease progression or unacceptable toxicity. In this study, two patients achieved SD which KRAS-mutant tumors (Cercek et al. 2014).

3.3.2.2 AUY922 or Luminespib

Luminespib, also called NVP-AUY-922; AUY922 or VER-52296 is a resorcinylic isoxazole amide, non-geldanamycin, intravenous (IV) Hsp90 inhibitor with the potential to target various type of cancers in vitro and in vivo (Fig. 3.8). Evidence suggest that AUY-922 could improve the effectiveness of traditional CRC treatment regimens. In this context, a previous study has been reported that AUY922 enhances the apoptosis-inducing potential of chemotherapeutic agents, such as gemcitabine or oxaliplatin in CRC cell lines (Mayor-López et al. 2014). In addition, the combination of AUY922 and trastuzumab represents promising anticancer activity in vitro and in vivo (Wainberg et al. 2013). Furthermore, AUY922 could potentiate the efficacy of radiotherapy by sensitizing CRC cells under hypoxia (Djuzenova et al. 2012). It also sensitized CRC cells to TRAIL and improves its cytotoxicity, without effect on normal colon cells (Lee et al. 2015). However, studies validating the usefulness of AUY-922 in combination with traditional CRC treatment regimens in the clinical setting are scant (Table 3.5).

It has been demonstrated that AUY922 has an anti-proliferative effect that varies in CRC cells with different mutations. CRC cell lines with a mutant KRAS are more sensitive to AUY922-induced apoptosis compared with those carrying wild-type KRAS (Wang et al. 2016a), while mutant BRAF CRC cells are resistant to AUY922 (Wang et al. 2016b) indicating the potential for differential therapeutic selectivity. AUY922 is entering phase I and II clinical trials. In a Japanese clinical trial study, AUY922 demonstrated potential anti-cancer efficacy. In this study 31 patients with

Fig. 3.8 The molecular structure of AUY922 (Luminespib)

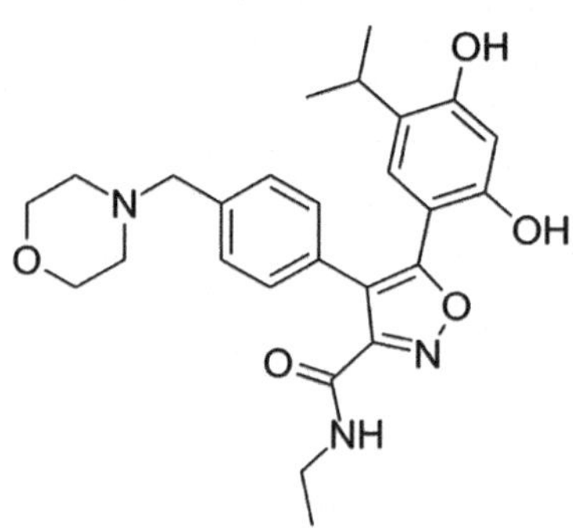

Table 3.5 Clinical studies of AUY922 or Luminespib

Clinical trials identifier	Intervention	Condition	Phase	Results	References
NCT01132625	AUY922	31 patients with advanced solid tumors, (17 patients had CRC)	I	1 patient achieved PR for a duration of >7 months	Doi et al. (2014)
	(IV, QW, 4 W cycle, starting dose 8 mg/m²)			10 patients (32%) achieved a best overall response of SD, lasting ≥2 months	
				No patients achieved a complete response	
				DCR (Disease Control rate; objective response and SD across all dose: 36%)	
				RP2D: 70 mg/m² IV QW	
NCT01226732	AUY 922	23 patients with advanced solid tumors, (16 patients had CRC)	I	4 patients achieved PR	Bendell et al. (2015)
	(22–70 mg/m²). Capecitabine: 1000 mg/m²			8 patients achieved SD, lasting 25.5 W	
NCT01404650	AUY922	25 patients with GIST resistance to imatinib and sunitinib	II	Median PFS was 3.9 months	Bendell et al. (2016)
	(70 mg/m², IV on days 1, 8, and 15 of 3 W cycle)			Median OS was 8.5 months	
				1 patient achieved PR	
				15 patients (60%) achieved the best response of SD	

advanced solid tumors were treated with 7 doses of AUY922 (8–70 mg/m²) once-weekly, IV in a 4-week cycle. PR was observed in one patient with rectal carcinoid tumor with metastatic lung lesions for a duration of >7 months. The best overall response of SD observed in ten patients lasting ≥2 months. The disease control rate (DCR: PR + SD) across all dose levels was 36%. However, a complete response was not observed. Diarrhea, night blindness and nausea were the most common drug-related toxicities (NCT01132625) (Doi et al. 2014). The combination of AUY922 with capecitabine was used by Bendell et al. to treat 23 patients with advanced solid tumors in a phase I dose escalation study (22–70 mg/m²). DLT was observed at the dose of 70 mg/m² of AUY922 and 1250 mg/m² of capecitabine in only one patient and was grade 3 diarrhea. Four patients achieved PR and eight patients achieved SD who previously treated by different systematic chemotherapeutic regimens (NCT01226732) (Bendell et al. 2015). Bendell et al. was also performed phase II trial of AUY922 in patients with GIST who were resistance to imatinib and sunitinib, and observed the promising anti-cancer activity of AUY922. AUY922 was prescribed IV by the dose of 70 mg/mg^2 on days 1, 8, and 15 of 3-week cycles with the median of 2 cycles until disease progression or intolerable toxicity. The median PFS and OS was 3.9 and 8.5 months respectively. In addition, 1 patient achieved PR and 15 patients (60%) achieved the best response of SD. They reported that the most common AEs following AUY922 is diarrhea and ocular toxicities (NCT01404650) (Bendell et al. 2016). Against these encouraging anti-cancer activity of AUY922, another phase I study showed disappointing results. In this study 101 patients with various cancers which the most common were the colon, breast, and ovarian cancer received AUY922 at doses of 2–70 mg/m^2. DLTs was observed in eight patients and included diarrhea, asthenia/fatigue, anorexia, atrial flutter, and visual symptoms. No PR was observed. However, the RP2D was 70 mg/m^2 (Sessa et al. 2013). Therefore, further studies with larger population samples are needed to ensure the efficacy of the drug in the treatment of CRC. A phase II study of AUY922 is under recruiting with the aim to assess DSR in patients with advanced GIST failed to imatinib and sunitinib (NCT01389583). Furthermore, a clinical trial study is under recruiting to analysis pharmacodynamics changes following treatment of patients with advanced solid malignancies with AUY922 (NCT01024283).

3.3.2.3 Onalespib (AT13387)

AT13387 is a second-generation potent, isoindoline derivative and non-ansamycin Hsp90 inhibitor (Kotha and Sreevani 2018) (Fig. 3.9). It has been reported that onalespib delay the emergence of resistance to erlotinib in EGFR-driven xenograft model suggesting their combination as a therapeutic regimen in cancer (Courtin et al. 2015). In a phase I dose-escalation study onalespib was administered infusion twice weekly and then subsequently, weekly of a 4-week cycle in patients with refractory solid tumors. Onalespib was well tolerated with reversible drug-related toxicities and 260 mg/m^2 at DLT (Mahadevan et al. 2012). In another phase I dose-escalation study onalespib in combination with imatinib was well tolerated but exhibited limited antitumor activity in patients with GIST who were TKI-resistant. Regarding to safety the highest dose of onalespib in combination with imatinib was 220 mg/m^2 once weekly for 3 weeks on/1 week off in patients with normal renal function (Wagner et al. 2016). However, Do et al. did not observe the promising effects of onalespib in patients with advanced solid tumors which was administered on a daily for two consecutive days per week schedule. In addition, there was no

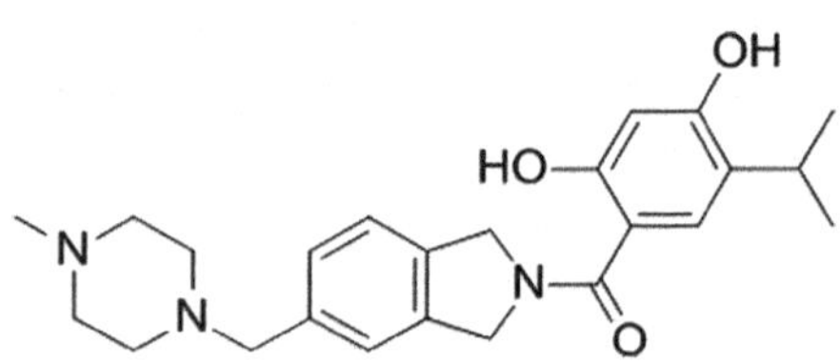

Fig. 3.9 The molecular structure of Onalespib (AT13387)

significant difference between the levels of Hsp70 and Hsp27 in tumor biopsies (Do et al. 2015). A phase I trials of onalespib in combination with CDKI AT7519 is under recruiting to treat patients with metastatic or unresectable solid tumors (NCT02503709). Also another phase I trials of onalespib in combination with dabrafenib and trametinib is under recruiting to treat patients with BRAF V600E/K mutation present, unresectable solid neoplasm or another type of cancer (NCT02097225) (Table 3.6).

Table 3.6 Clinical studies of Onalespib (AT13387) in which patients with colorectal cancer are part of their sample population

Clinical trials identifier	Intervention	Condition	Phase	Main results	References
Not registered	Onalespib; BIW: 10–120 mg/m^2 QW: 150–310 mg/m^2 (3 W on/1 W off)	62 patients with advanced solid tumors	I	One DLT was observed at 120 mg/m^2: visual disturbance RP2D: 260 mg/m^2 for QW and 120 mg/m^2 for BIW One patient with GIST achieved PR 34% achieved SD for a duration of >120 days in 7 patients	Mahadevan et al. (2012)
Not registered	Onalespib: (150–50 mg/m^2 QW IV for 3 W) Imatinib: (400 mg QD for 4 W)	26 patients with TKI-resistant GIST	I	Overall, 81% of patients reported more than one onalespib-related gastrointestinal disorder 35% achieved SD, including two patients who had *KIT* mutations Disease control achieved in (19%) for a duration of 4 months Median PFS: 112 day One patient with PDGFRA-mutant GIST achieved PR	Wagner et al. (2016)
Not registered	Onalespib	31 patients with advanced solid tumors	I	RP2D: 160 mg/m^2 on the QDx2/week schedule Common toxicities: were gastrointestinal, hepatic, and hematologic No responses were observed	Do et al. (2015)

3.3.3 Purine Analogues Derivatives Hsp90 Inhibitors

Purine-based analogues are the first class of fully synthetic developed Hsp90 inhibitors, which their discovery are efforts aimed towards optimization of Hsp90 inhibitors and their development into clinically viable drugs for the treatment of cancer. A number of these inhibitors are including Debio 0932 (CUDC-305), BIIB021 (CNF 2024), MPC-3100 and PU-H71.

3.3.3.1 Debio0932 (CUDC-305)

DEBIO0932, also termed CUDC-305, is a second generation of Hsp90 inhibitor of the imidazopyridine class (Bao et al. 2009) (Fig. 3.10). It exhibits unique pharmacological properties, such as oral bioavailability and blood-brain barrier penetration. Preclinical models show that it exerts sustained tumor growth suppression and prolongs survival. In addition, its ability to cross the blood-brain barrier converts it to a promising agent for treatment of glioblastoma (Canella et al. 2015). A study showed that DEBIO0932 inhibits multiple oncoproteins involved in tumor proliferation including wnt- β-catenin, EGFR, Akt and MAPK (Canella et al. 2015; Moradi-Marjaneh et al. 2018c). In a phase I dose-escalation study, Debio0932 was well tolerated in patients with advanced tumor. However, it showed limited clinical efficiency. It has been suggested as an adjunct treatment of NSCLC (Isambert et al. 2015). It seems that Debio0932 has a potential therapeutic value to treat NSCLC patients. It successfully passed phase I and II trials in patients with NSCLC (van Ingen et al. 2012) (Table 3.7).

3.3.3.2 BIIB021

BIIB021 is an oral purine-based Hsp90 inhibitor that binds competitively with geldanamycin in ATP-binding pocket of Hsp90 (Yan et al. 2017) (Fig. 3.11). It has been reported that BIIB021 may have activity against tumors with acquired multidrug resistance (MDR) (Zhang et al. 2010; Moradi-Marjaneh et al. 2018c). In

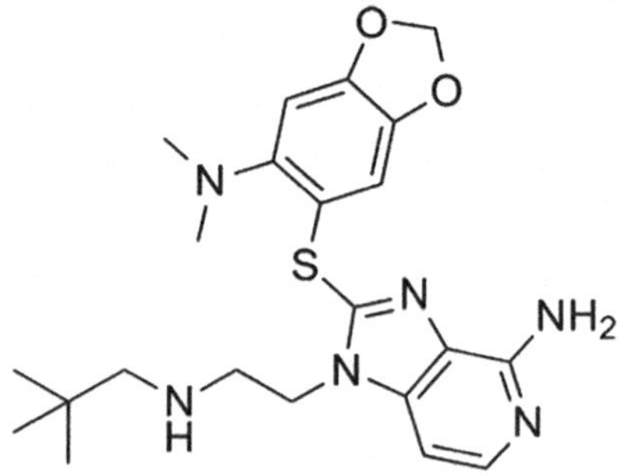

Fig. 3.10 The molecular structure of Debio0932 (CUDC-305)

Table 3.7 Clinical studies of Debio0932

Clinical trials identifier	Intervention	Condition	Phase	Results	References
NCT01168752	Debio0932	50 patients with advanced solid tumors, (14 patients had CRC)	I	3 DLTs was observed at 1600 mg QD; 1 patient, febrile neutropenia and 2 patients, diarrhea and asthenia	Isambert et al. (2015)
	50–1200 mg (QD or QOD)			RP2D: 1000 mg QD	
				2 patient achieved PR	
				12 patients achieved SD	
				Plasma steady state was reached within 9 days	

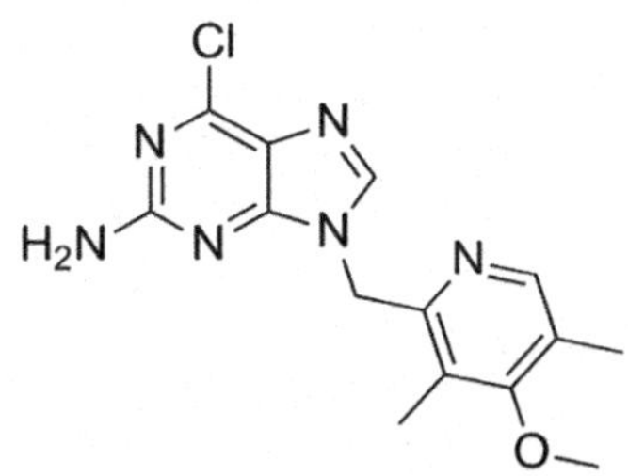

Fig. 3.11 The molecular structure of BIIB021

addition, evidences show that BIIB021 block the constitutive NF-κB activity (Gopalakrishnan et al. 2013) and induces apoptosis in cancer cells (Li et al. 2013). It has also been reported that BIIB021 stimulates autophagy (He et al. 2016) and sensitizes cancer cells to radiation (Wang et al. 2014). The efficacy of this drug is now under investigation in patients with the various type of cancer. Evidence show that BIIB021 exerts antitumor activity in both daily, and intermittent dosing schedules, leading to flexibility for clinical studies (Lundgren et al. 2009). BIIB021 generally well tolerated in patients with advanced solid tumors or chronic lymphocytic leukemia (CLL). The serum levels of Hsp70 increased and serum Her-2 decreased following BIIB021 treatment suggesting its biologic activity. DLT was determined as 800 mg twice weekly (Elfiky et al. 2008). In a phase I dose-escalation study BIIB021 capsules were administered to patients with advanced solid tumors in two schedules; twice weekly with 1 week off or without 1 week off. Both schedules were well tolerated. The Cmax occurred at approximately 90 min and t1/2 was approximately 1 h across dosing cohorts of 25–800 mg BIIB021 twice weekly (Saif et al. 2013). Phase II study of BIIB021 in patients with GIST who were refractory to imatinib and sunitinib showed promising objective responses. Treatment with BIIB021 resulted in a rapid reduction in FDG uptake in many patients, as measured by PET scan (Dickson et al. 2012) (Table 3.8).

Table 3.8 Clinical studies of BIIB021

Clinical trials identifier	Intervention	Condition	Phase	Results	References
NCT00345189	BIIB021	23 patients with advanced solid tumors or chronic lymphocytic leukemia (CLL)	I	2 DLTs was observed	Elfiky et al. (2008)
	QD or BIW (3 W on/1 W off)			800 mg orally BIW; grade 3 syncope, grade 3 dizziness	
				Efficacy: one patient with CLL (25 mg) had a 39% reduction in lymph node size	
				68% of patients with solid tumors achieved SD	
NCT00618735	BIIB021	60 patients advanced solid tumors	I	MTD: 700 mg BIW for 3 W in a 4 W cycle	Saif et al. (2013)
	TIW: (3 W on/1 W off)			MTD: 600 mg BIW for the continuous dosing regimen	
	TIW: continually			AEs were mostly mild or moderate	
				9 patients (20%) achieved SD	
				32 patients (70%) had progressive disease	
NCT00618319	BIIB021	23 patients with GIST	II	5 patient achieved PR for a duration of 25–138 days	Dickson et al. (2012)
	600 mg (BIW) 400 mg (TIW)			10 patients (43%) achieved SD; 4 of 12 subjects who received 600 mg BIW and 6 of 11 subjects (55%) who received 400 mg TIW	
				AEs were mild to moderate	
				3 patients had grade 2 dizziness and there were no grade 3 or 4 dizziness events	
				Cmax was 1.5 μmol and the mean AUC was 2.9 μmol h	

3.3.3.3 MPC3100

MPC-3100 is a fully synthetic potent small-molecule, orally bioavailable, purine-based Hsp90 inhibitor (Spreafico et al. 2015) (Fig. 3.12). Its antitumor activity has been observed in preclinical models of different cancers (Samlowski et al. 2011). In a dose escalation study, 26 patients with various types of cancer who were refractory to available systemic therapy administered by oral MPC-3100 at doses of 50, 100, 165, 245, or 340 mg/m^2 for 3 weeks on/1 week off or total daily doses of 480 mg or 640 mg for 4 weeks continually. SD was observed in 46%. Cmax and AUC (0–12 h) on day 1 increased in an approximately dose-proportional manner. A modest degree of drug accumulation was observed. Hsp70 analysis showed induction in 24 h and during treatment, indicating consistent Hsp90 inhibition by MPC3100 (Samlowski et al. 2011) (Table 3.9).

3.3.3.4 PU-H71

PU-H71 is a purine based analog with remarkable specificity and selectivity toward malignant cells, while exerts minimal toxicity in normal cells, made it an exciting pharmaceutical prospect (Fig. 3.13). It demonstrated potent efficacy in various preclinical models of cancer (Trendowski 2015). Hsp90 inhibition by PU-H71 degrades STK33 (an important enzyme for the viability and proliferation of KRAS mutant cell) in KRAS mutant CRC and induce apoptosis (Azoitei et al. 2012). A study has been shown that PU-H71 exerted efficient and sustained downregulation and inactivation of key oncoproteins and signaling pathway that is imperative for carcinogenesis including Ras/Raf/MAPK, NF-κB, Akt, ERK2, Tyk2, and PKC, both in vitro and in vivo (Caldas-Lopes et al. 2009). However, PU-H71 is only in the initial stages of clinical examination, and its potential for treatment of CRC remains unclear. In a phase I dose-escalation study, PU-H71 was labeled with 124I and intratumoral drug concentration calculated based on PET imaging. Intratumoral

Fig. 3.12 The molecular structure of MPC3100

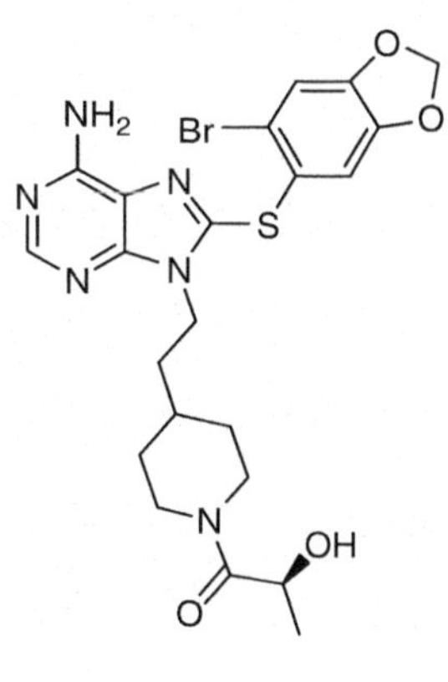

Table 3.9 Clinical studies of MPC-3100

Clinical trials identifier	Intervention	Condition	Phase	Results	References
NCT00920205	MPC-3100	26 patients with various cancer refractory to available systemic therapy	I	2 DLTs was observed:	Samlowski et al. (2011)
	3 W consecutive in a 4 W cycle (50, 100, 165, 245, or 340 mg/m^2)			245 mg/m^2; 2 patients grade 3 supraventricular tachycardia and respiratory failure	
	4 W continuously at total 480 mg or 640 mg/QD			Total daily doses greater than 600 mg were not well tolerated	
				Side-effects were generally manageable or reversible upon discontinuation of MPC-3100	
				The most common AEs was diarrhea, nausea, vomiting and fatigue	
				12 (46%) patients achieved SD	
				Hepatotoxicity was not observed in this study	

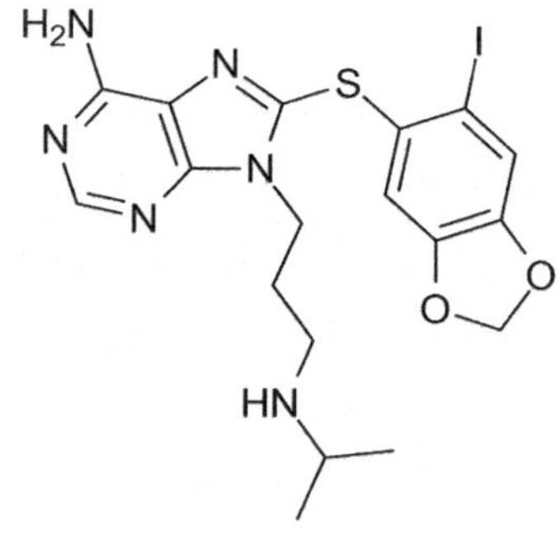

Fig. 3.13 The molecular structure of PU-H71

concentrations were more than the IC50 and achieved 24 h after first dose at doses of 50–300 mg/m^2 which was below DLT-related doses (Gerecitano et al. 2013). In addition, another study of PET imaging of cancer patients using 124I-PUH71 is under recruiting to evaluate the pharmacokinetics, metabolism, biodistribution and radiation dosimetry of PUH71 (NCT01269593) (Table 3.10).

Table 3.10 Clinical studies of PU-H71

Clinical trials identifier	Intervention	Condition	Phase	Results	References
NCT01393509	PU-H71	40 patients with previously treated solid tumors, lymphoma or myeloproliferative neoplasms	I	5 DLTs was observed	Gerecitano et al. (2013)
	(10–400 mg/m^2)			400 mg/m^2: 1 patient; grade 3 mucositis	
	BIW (2 W on/1 W off)			350 mg/m^2: 1 patient; grade 3 AST and ALT, 1 patient;1 patient grade 3 nausea/vomiting + grade 2 intolerable myalgia, 1 patient; grade 3 anemia, 1 patient; grade 2 intolerable headache	
				Intratumoral concentrations of 0.5–8 μM (greater than the IC50 for most cancer cells) achieved 24 h after first dose at doses of 50–300 mg/m^2	
				124I-PU-H71 can be used to visualize PU-H71 uptake	

3.4 Other Hsp90 Inhibitors

3.4.1 XL888

XL888 produced by Exelixis company is a novel, the orally bioavailable, non-benzoquinone, the tropane-derived inhibitor of Hsp90 (Bussenius et al. 2012) (Fig. 3.14), Which showed antitumor activity in a variety of tumor cell types and in vivo models through diverse mechanisms (Paraiso et al. 2012; Haarberg et al. 2013; Vido and Aplin 2015). It has been reported that XL888 overcomes BRAF I inhibitor resistance through overexpression of BIM and the downregulation of Mcl-1. The use of XL888 has been suggested to overcome or delay BRAF inhibitor resistance (Paraiso et al. 2012). This compound is also considered in the treatment of melanoma (Paraiso et al. 2012; Haarberg et al. 2013). XL888 has entered phase I trial in combination with vemurafenib in patients with metastatic BRAF V600

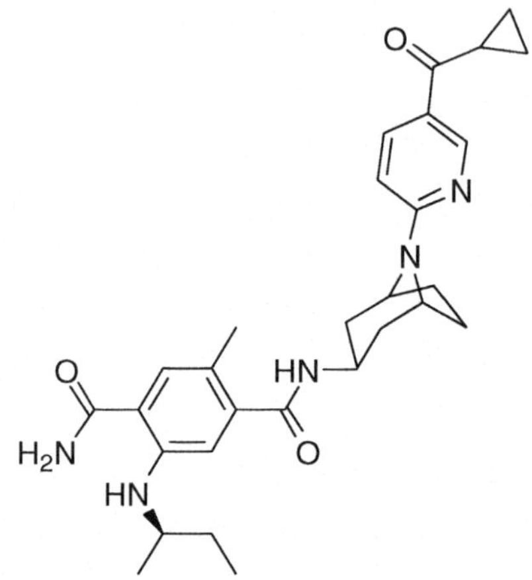

Fig. 3.14 The molecular structure of XL888

mutant melanoma with promising clinical efficiency and acceptable toxicity profile. The RP2D was XL888 90 mg twice per week (BIW) (NCT01657591) (Eroglu et al. 2016). A phase I trial of XL888 in combination with pembrolizumab is under recruiting in previously treated patients with advanced gastrointestinal cancer (NCT03095781) (Akce et al. 2018).

3.4.2 Hsp990 (NVP-Hsp990)

NVP-Hsp990 is a small synthetic molecule, dihydropyridopyrimidinone series, the orally bioavailable inhibitor of Hsp90 (McBride et al. 2014) (Fig. 3.15). Varies studies demonstrated NVP-Hsp990 antitumor activity in vitro and in vivo (Menezes et al. 2012; Stuehmer et al. 2012; Fu et al. 2013). Clinical trials have been initiated to evaluate the potential therapeutic value of NVP-HSP990 in advanced solid tumors. In a dose escalation study, HSP990 alone was well tolerated with DLT of neurological toxicity. Tmax occurred at 3 h and T½ was approximately 20 h. Dose-dependent increasing of Hsp70 was observed. No objective response was observed in patients with advanced solid tumors. However, 39% experience SD. Finally, all patients discontinued treatment due to disease progression (NCT00879905) (Spreafico et al. 2015) (Table 3.11).

3.4.3 SNX-5422 Mesylate (PF-04929113) (PF-113)

SNX-5422 is a prodrug of SNX-2112, water-soluble, orally bioavailable, and potent, non-antibiotic, small molecule inhibitor of Hsp90 (Fig. 3.16). SNX-2112 strongly inhibited tumor growth and angiogenesis in preclinical models of the different type of cancer. Evidence show that SNX-5422 selectively accumulates in tumor tissue

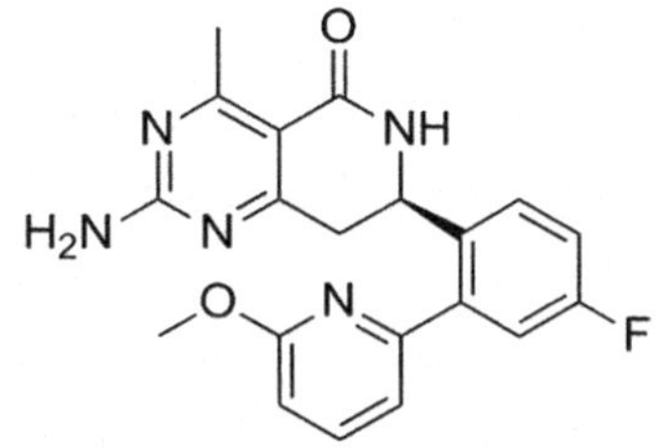

Fig. 3.15 The molecular structure of Hsp990 (NVP-Hsp990)

Table 3.11 Clinical studies of HSP990

Clinical trials identifier	Intervention	Condition	Phase	Results	References
NCT00879905	HSP990 (2.5–60 mg)	64 patients with advanced solid tumors	I	7 DLTs was observed	Spreafico et al. (2015)
				60 mg QW; 2 patients; grade 1 decreased appetite, dysdiadochokinesia, fatigue, hyperreflexia and tremor	
	QW or BIW (4 W cycle)			50 mg QW; 5 patients; grade 3 bilateral myoclonic hand movement, grade 3 diarrhea, grade 3 prolonged QTc, grade 4 ALT and AST increase	
				AEs: diarrhea, fatigue and decreased appetite	
				Pharmacokinetics: rapid absorption, no drug accumulation and large interpatient variability	
				No objective response was observed	
				25 patients (39%) achieved SD	
				RP2D: 50 mg QW	

compared with normal tissue (Chandarlapaty et al. 2008). Despite promising results in preclinical studies, SNX-5422 was withdrawn because of evidence of retinal toxicity. Hence an in vivo study was performed to evaluate the ocular toxicity of SNX-5422. The results showed that both PF-113 and its structurally different back-up, PF-847 exert evidences of irreversible retinal damage with photoreceptor death in animal studies (Infante et al. 2010). However, in another phase I dose-escalation in patients with refractory solid tumors, SNX-5422 showed satisfactory results. MTD for every other day (QOD) and daily (QD) was determined 100 mg/m^2 and 67 mg/m^2 respectively. The observed PR occurred at a wide range of dose levels (5.32 and 56 mg/m^2 QOD) which may be due to genetic sensitivity to Hsp90 inhibition

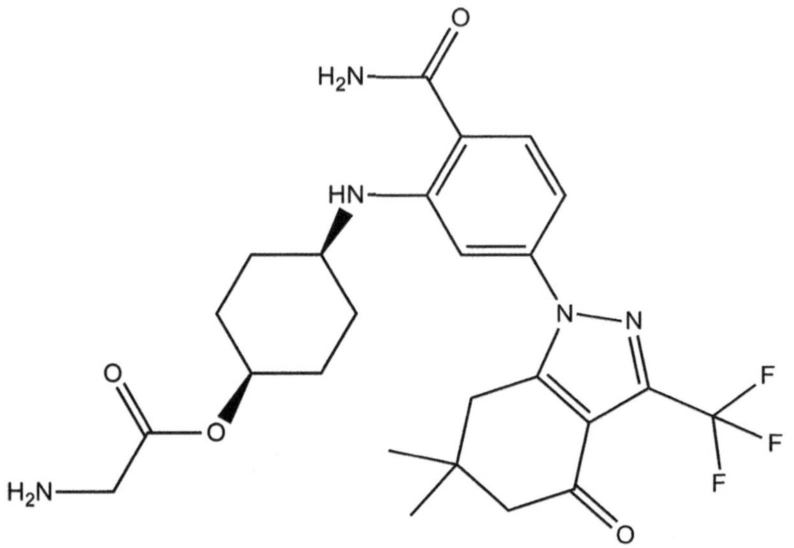

Fig. 3.16 The molecular structure of SNX-5422 mesylate (PF-04929113) (PF-113)

(Infante et al. 2014). SNX-5422 successfully passed phase I B in combination with carboplatin and paclitaxel in patients with advanced non-small-cell lung cancer (NSCLC) (Gutierrez et al. 2017) (Table 3.12).

3.4.4 KW-2478

KW-2478 is a potent, IV, non-ansamycin, non-purine inhibitor of Hsp90, exerts anti-tumor activity in vitro and in vivo models (Fig. 3.17). It has been reported that KW-2478 degrades client proteins such as FGFR3, transcription factor c-Maf and Cyclin D1 in tumor cells (Nakashima et al. 2010). In a phase I single-agent study, KW-2478 was well tolerated and no DLT was observed in any dose. Hence highest dose which was tested determined as RP2D; 176 mg m^{-2} once daily on days 1–5 every 2 weeks (Yong et al. 2016). KW-2478 also passed a phase I/II study in combination with bortezomib in patients with refractory multiple myeloma. Good tolerability of this combination with 39.2% for overall response rate was observed (Cavenagh et al. 2017).

Table 3.12 Clinical studies of PF-113

Clinical trials identifier	Intervention	Condition	Phase	Results	References
NCT00647764	PF-113	44 patients with advanced solid tumor malignancies and lymphoma	I	MTD for QD schedule: 67 mg/m^2	Infante et al. (2010)
	QOD (4–100 mg/m^2)			MTD for QOD schedule was not determined because of retinal toxicity	
	OD (50–89 mg/m^2) on (3 W on/1 W off)			AEs: diarrhea, nausea, and fatigue	
				1 patient achieved PR	
				6 patients achieved SD	
NCT01611623	PF-113	56 patients with advanced solid tumor	I	MTD for QOD: 100 mg/m^2 and for QD:67 mg/m^2	Infante et al. (2014)
	QOD (4–133 mg/m^2)			AEs: diarrhea, nausea, fatigue, and vomiting	
NCT00506805	OD (50–89 mg/m^2) (3 W on/1 W off)			1 patients (QOD schedule) achieved durable complete response	
				1 patients (QOD schedule) achieved PR	
				3 patients (QOD schedule) achieved SD for a duration of 6 months	
				RP2D: 100 mg/m^2 QOD, 3 W on/1 W off	

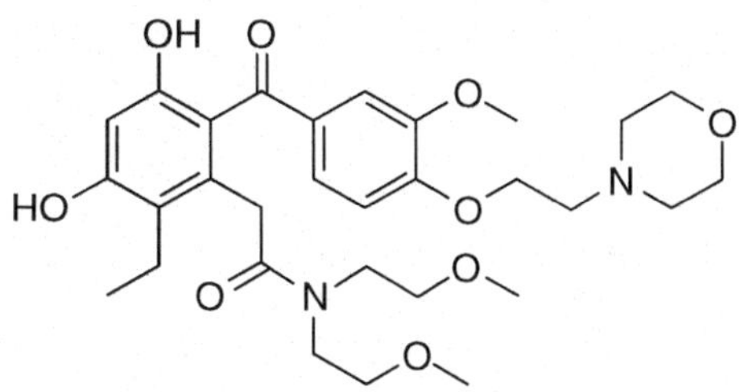

Fig. 3.17 The molecular structure of KW-2478

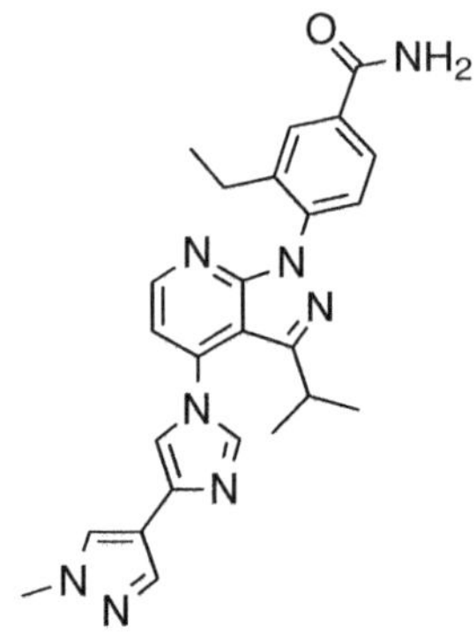

Fig. 3.18 The molecular structure of TAS-116

3.4.5 *Hsp90alpha/beta Inhibitor TAS-116*

The 4-(1H-pyrazolo[3,4-b]pyridine-1-yl)benzamide TAS-116 is orally non-ansamycin, non-purine, non-resorcinol inhibitor of Hsp90 which inhibits cytosolic Hsp90α and β specifically and selectively but interestingly does not inhibit Hsp90 paralogs such as endoplasmic reticulum GRP94 or mitochondrial TRAP1 (Utsugi 2013) (Fig. 3.18). It has potential antineoplastic and radio/chemosensitizing efficiencies and exerts potent antitumor activity in vivo tumor models with minimal ocular toxicity (Ohkubo et al. 2014). In a phase I dose escalation study, TAS-116 showed an acceptable safety profile. The MTD was about 100 and 200 mg/m^2/day with QD and QOD schedule respectively (Yanagitani et al. 2017). It also passed phase II clinical trials in patients with GIST. Although in this study none of the patients achieved a PR, however, 79% experienced SD for a duration of more than 6 weeks (Kurokawa et al. 2017) (Table 3.13).

3.5 Conclusions

Hsp90 proteins play prominent functional roles in cell survival and act as anti-apoptotic proteins. They interact with hundreds of protein clients, of which many are oncoproteins, indicating the high-value of their inhibition as a therapeutic approach in the treatment of CRC. The first Hsp90 inhibitors entering clinical trials were geldanamycin analogues including tanespimycin (17-AAG), alvespimycin (17-DMAG), retaspimycin, IPI-493, etc. In addition, resorcinol derivatives including ganetespib (STA-9090), NVP-AUY922 (VER52296), AT-13387 and KW-2478 entered into phase I and II trials in several types of cancer. Purine analogues including BIIB021, MPC-3100, Debio 0932, PU-H71 were synthetized with the aim of increasing solubility and facilitating consumption in patients. In addition, many other compounds are being developed to improve the anti-cancer activity of Hsp90 inhibitors. Clinical trials have indicated promising anti-cancer effects of Hsp90

Table 3.13 Clinical studies of TAS-116

Clinical trials identifier	Intervention	Condition	Phase	Results	References
NCT02965885	TAS-116	52 patients with advanced solid tumor	I	MTD for QD schedule: 107.5 mg/m^2/day	Yanagitani et al. (2017)
	QD: (4.8–150.5 mg/m^2/day)			MTD for QOD schedule: 210.7 mg/m^2/day	
	QOD: (107.5–295.0 mg/m^2/day)			AEs: gastrointestinal disorders and increased creatinine	
				DLTs:	
				QD schedule, 4 patients (night blindness, visual disorder, AST/ALT/gamma-GTP elevations, and anorexia)	
				QOD schedule, 2 patients (platelet count decreased, febrile neutropenia, pneumonia, respiratory failure, and septic shock)	
				3 patient achieved durable PR	
Not registered	TAS-116	40 patients with GIST	II	MTD for QOD: 100 mg/m^2 and for QD:67 mg/m^2	Kurokawa et al. (2017)
	160 mg/day (5D on/2D off)			AEs: diarrhea, anorexia, increased serum creatinine, and eye disorders	
				There were no treatment-related deaths	
				Diarrhea, nausea, fatigue, and vomiting	
				PFS: was 4.5	
				None of patients had a PR	
				34 patients (79%) achieved SD for a duration of more than 6 weeks	

inhibitors. However, in clinical trials a small percentage of patients had CRC. Therefore, the use of these compounds in the treatment of patients with CRC needs further investigation.

Acknowledgements This research was partly supported by grants awarded by the Mashhad University of Medical Sciences, grant No. 940908 and National Institute for Medical research Development, grant No. 958349.

References

Agnew EB, Wilson RH, Grem JL, Neckers L, Bi D, Takimoto CH (2001) Measurement of the novel antitumor agent 17-(allylamino)-17-demethoxygeldanamycin in human plasma by high-performance liquid chromatography. J Chromatogr B Biomed Sci Appl 755:237–243

Akce M, Alese OB, Shaib WL, Wu CS-Y, Lesinski GB, El-Rayes BF (2018) Phase Ib trial of pembrolizumab and XL888 in patients with advanced gastrointestinal malignancies. J Clin Oncol 36:1–15

Amerizadeh F, Rezaei N, Rahmani F et al (2018) Crocin synergistically enhances the antiproliferative activity of 5-flurouracil through Wnt/PI3K pathway in a mouse model of colitis-associated. J Cell Biochem 119:10250–10261

Archie NT, Klimstra DS, Gonen M et al (2008) A phase 1 dose-escalation study of irinotecan in combination with 17-allylamino-17-demethoxygeldanamycin in patients with solid tumors. Clin Cancer Res 14:6704–6711

Arnold M, Sierra MS, Laversanne M, Soerjomataram I, Jemal A, Bray F (2016) Global patterns and trends in colorectal cancer incidence and mortality. Gut 66:683–691

Azoitei N, Hoffmann CM, Ellegast JM et al (2012) Targeting of KRAS mutant tumors by HSP90 inhibitors involves degradation of STK33. J Exp Med 209:697–711

Bao R, Lai C-J, Qu H et al (2009) CUDC-305, a novel synthetic HSP90 inhibitor with unique pharmacologic properties for cancer therapy. Clin Cancer Res 15:4046–4057

Bedin M, Gaben AM, Saucier C, Mester J (2004) Geldanamycin, an inhibitor of the chaperone activity of HSP90, induces MAPK-independent cell cycle arrest. Int J Cancer 109:643–652

Bedin M, Catelli MG, Cabanie L, Gaben AM, Mester J (2009) Indirect participation of Hsp90 in the regulation of the cyclin E turnover. Biochem Pharmacol 77:151–158

Bendell JC, Jones SF, Hart L et al (2015) A phase I study of the Hsp90 inhibitor AUY922 plus capecitabine for the treatment of patients with advanced solid tumors. Cancer Investig 33:477–482

Bendell JC, Bauer TM, Lamar R et al (2016) A phase 2 study of the Hsp90 inhibitor AUY922 as treatment for patients with refractory gastrointestinal stromal tumors. Cancer Investig 34:265–270

Bussenius J, Blazey CM, Aay N et al (2012) Discovery of XL888: a novel tropane-derived small molecule inhibitor of HSP90. Bioorg Med Chem Lett 22:5396–5404

Caldas-Lopes E, Cerchietti L, Ahn JH et al (2009) Hsp90 inhibitor PU-H71, a multimodal inhibitor of malignancy, induces complete responses in triple-negative breast cancer models. Proc Natl Acad Sci U S A 106:8368–8373

Canella A, Xu J, Meisen WH et al (2015) DEBIO0932, an Hsp90 inhibitor downregulates key signaling pathways and sensitizes glioma cells to temozolomide. AACR 75:1–10

Cavenagh J, Oakervee H, Baetiong-Caguioa P et al (2017) A phase I/II study of KW-2478, an Hsp90 inhibitor, in combination with bortezomib in patients with relapsed/refractory multiple myeloma. Br J Cancer 117:1295–1302

Cercek A, Shia J, Gollub M et al (2014) Ganetespib, a novel Hsp90 inhibitor in patients with KRAS mutated and wild type, refractory metastatic colorectal cancer. Clin Colorectal Cancer 13:207–212

Chandarlapaty S, Sawai A, Ye Q et al (2008) SNX2112, a synthetic heat shock protein 90 inhibitor, has potent antitumor activity against HER kinase–dependent cancers. Clin Cancer Res 14:240–248

Chen JS, Hsu YM, Chen CC, Chen LL, Lee CC, Huang TS (2010) Secreted heat shock protein 90alpha induces colorectal cancer cell invasion through CD91/LRP-1 and NF-kappaB-mediated integrin alphaV expression. J Biol Chem 285:25458–25466
Chen WS, Lee CC, Hsu YM, Chen CC, Huang TS (2011) Identification of heat shock protein 90alpha as an IMH-2 epitope-associated protein and correlation of its mRNA overexpression with colorectal cancer metastasis and poor prognosis. Int J Color Dis 26:1009–1017
Chen WS, Chen CC, Chen LL, Lee CC, Huang TS (2013) Secreted heat shock protein 90alpha (HSP90alpha) induces nuclear factor-kappaB-mediated TCF12 protein expression to down-regulate E-cadherin and to enhance colorectal cancer cell migration and invasion. J Biol Chem 288:9001–9010
Cho D, Heath E, Cleary J et al (2011) A phase I dose-escalation study of the Hsp90 inhibitor ganetespib (STA-9090) administered twice weekly in patients with solid tumors: updated report. J Clin Oncol 29:3051–3051
Courtin A, Smyth T, Hearn K, Lyons J, Thompson N, Wallis NG (2015) The HSP90 inhibitor, onalespib (AT13387), delays the emergence of resistance to erlotinib in an EGFR-driven xenograft model. AACR 75:1–10
de la Mare JA, Jurgens T, Edkins AL (2017) Extracellular Hsp90 and TGFbeta regulate adhesion, migration and anchorage independent growth in a paired colon cancer cell line model. BMC Cancer 17:202
Demetri G, Heinrich M, Chmielowski B et al (2011) An open-label phase II study of the Hsp90 inhibitor ganetespib (STA-9090) in patients (pts) with metastatic and/or unresectable GIST. J Clin Oncol 29:10011–10011
Dickson M, Okuno S, Keohan M et al (2012) Phase II study of the HSP90-inhibitor BIIB021 in gastrointestinal stromal tumors. Ann Oncol 24:252–257
Djuzenova CS, Blassl C, Roloff K et al (2012) Hsp90 inhibitor NVP-AUY922 enhances radiation sensitivity of tumor cell lines under hypoxia. Cancer Biol Ther 13:425–434
Do K, Speranza G, Chang L-C et al (2015) Phase I study of the heat shock protein 90 (Hsp90) inhibitor onalespib (AT13387) administered on a daily for 2 consecutive days per week dosing schedule in patients with advanced solid tumors. Investig New Drugs 33:921–930
Doddareddy MR, Thorat DA, Seo SH et al (2011) Structure based design of heat shock protein 90 inhibitors acting as anticancer agents. Bioorg Med Chem 19:1714–1720
Doi T, Onozawa Y, Fuse N et al (2014) Phase I dose-escalation study of the HSP90 inhibitor AUY922 in Japanese patients with advanced solid tumors. Cancer Chemother Pharmacol 74:629–636
Egorin MJ, Rosen DM, Wolff JH, Callery PS, Musser SM, Eiseman JL (1998) Metabolism of 17-(allylamino)-17-demethoxygeldanamycin (NSC 330507) by murine and human hepatic preparations. Cancer Res 58:2385–2396
Egorin MJ, Zuhowski EG, Rosen DM, Sentz DL, Covey JM, Eiseman JL (2001) Plasma pharmacokinetics and tissue distribution of 17-(allylamino)-17-demethoxygeldanamycin (NSC 330507) in CD2F1 mice1. Cancer Chemother Pharmacol 47:291–302
Elfiky A, Saif M, Beeram M et al (2008) BIIB021, an oral, synthetic non-ansamycin Hsp90 inhibitor: phase I experience. J Clin Oncol 26:2503–2503
El-Rayes BF, Staley CA, Diaz R, Sullivan PS, Shaib WL, Landry JC (2015) Phase I study of ganetespib (G), capecitabine (C), and radiation (RT) in rectal cancer. J Clin Oncol 33:3596–3596
Eroglu Z, Gibney GT, Weber JS et al (2016) Phase I study of vemurafenib and heat shock protein 90 (HSP90) inhibitor XL888 in metastatic BRAF V600 mutant melanoma. J Clin Oncol 34:9544–9544
Fu J, Koul D, Yao J et al (2013) Novel HSP90 inhibitor NVP-HSP990 targets cell cycle regulators to ablate Olig2-positive glioma tumor initiating cells. Cancer Res 15:3062–3074
Garcia-Carbonero R, Carnero A, Paz-Ares L (2013) Inhibition of HSP90 molecular chaperones: moving into the clinic. Lancet Oncol 14:e358–e369
Ge H, Yan Y, Guo L, Tian F, Wu D (2018) Prognostic role of HSPs in human gastrointestinal cancer: a systematic review and meta-analysis. Onco Targets Ther 15:351–359

Gerecitano JF, Modi S, Gajria D et al (2013) Using 124I-PU-H71 PET imaging to predict intratumoral concentration in patients on a phase I trial of PU-H71. J Clin Oncol 31(Suppl): Abstract 11076

Goetz MP, Toft D, Reid J et al (2005) Phase I trial of 17-allylamino-17-demethoxygeldanamycin in patients with advanced cancer. J Clin Oncol 23:1078–1087

Goldman JW, Raju RN, Gordon GA et al (2013) A first in human, safety, pharmacokinetics, and clinical activity phase I study of once weekly administration of the Hsp90 inhibitor ganetespib (STA-9090) in patients with solid malignancies. BMC Cancer 13:152

Gopalakrishnan R, Matta H, Chaudhary PM (2013) A purine scaffold HSP90 inhibitor BIIB021 has selective activity against KSHV-associated primary effusion lymphoma and blocks vFLIP K13-induced NF-Kb. Clin Cancer Res 15:5016–5026

Grammatikakis N, Vultur A, Ramana CV et al (2002) The role of Hsp90N, a new member of the Hsp90 family, in signal transduction and neoplastic transformation. J Biol Chem 277:8312–8320

Gutierrez M, Giaccone G, Liu SV et al (2017) A phase 1B study of SNX-5422 plus carboplatin (C) and paclitaxel (P) in patients with advanced non-small-cell lung cancer (NSCLC). J Clin Oncol 35:e20622

Haarberg HE, Paraiso KH, Wood E et al (2013) Inhibition of Wee1, AKT and CDK4 underlies the efficacy of the HSP9o inhibitor XL888 in an in vivo model of NRAS mutant melanoma. Mol Can Ther 12:901–912

Hamamoto R, Furukawa Y, Morita M et al (2004) SMYD3 encodes a histone methyltransferase involved in the proliferation of cancer cells. Nat Cell Biol 6:731–740

Hanson BE, Vesole DH (2009) Retaspimycin hydrochloride (IPI-504): a novel heat shock protein inhibitor as an anticancer agent. Expert Opin Investig Drugs 18:1375–1383

Hashemzehi M, Behnam-Rassouli R, Hassanian SM et al (2018) Phytosomal-curcumin antagonizes cell growth and migration, induced by thrombin through AMP-kinase in breast cancer. J Cell Biochem 119:5996–6007

He W, Ye X, Huang X et al (2016) Hsp90 inhibitor, BIIB021, induces apoptosis and autophagy by regulating mTOR-Ulk1 pathway in imatinib-sensitive and-resistant chronic myeloid leukemia cells. Int J Oncol 48:1710–1720

Hendrickson AEW, Oberg AL, Glaser G et al (2012) A phase II study of gemcitabine in combination with tanespimycin in advanced epithelial ovarian and primary peritoneal carcinoma. Gynecol Oncol 124:210–215

Hertlein E, Wagner AJ, Jones J et al (2010) 17-DMAG targets the nuclear factor-κB family of proteins to induce apoptosis in chronic lymphocytic leukemia: clinical implications of HSP90 inhibition. Blood 116:45–53

Hubbard J, Erlichman C, Toft DO et al (2011) Phase I study of 17-allylamino-17 demethoxygeldanamycin, gemcitabine and/or cisplatin in patients with refractory solid tumors. Invest New Drugs 29:473–480

Infante J, Weiss G, Jones S et al (2010) 375 A Phase 1 dose escalation study of the oral heat shock protein 90 (Hsp90) inhibitor PF-04929113/SNX-5422 (PF-113) and its associated ocular toxicity. EJC Suppl 7:119

Infante JR, Weiss GJ, Jones S et al (2014) Phase I dose-escalation studies of SNX-5422, an orally bioavailable heat shock protein 90 inhibitor, in patients with refractory solid tumours. Eur J Cancer 50:2897–2904

Isambert N, Delord J-P, Soria J-C et al (2015) Debio0932, a second-generation oral heat shock protein (HSP) inhibitor, in patients with advanced cancer—result of a first-in-man dose-escalation study with a fixed-dose extension phase. Ann Oncol 26:1005–1011

Jhaveri K, Miller K, Rosen L et al (2012) A phase I dose-escalation trial of trastuzumab and alvespimycin hydrochloride (KOS-1022; 17 DMAG) in the treatment of advanced solid tumors. Clin Cancer Res 15:5090–5098

Johnston A (2009) Infinity halts RING trial in advanced gastrointestinal stromal tumors

Kamal A, Thao L, Sensintaffar J et al (2003) A high-affinity conformation of Hsp90 confers tumour selectivity on Hsp90 inhibitors. Nature 425:407–410

Kauh JS, Harvey RD, Owonikoko TK et al (2012) A phase I and pharmacokinetic study of multiple schedules of ganetespib (STA-9090), a heat shock protein 90 inhibitor, in combination with docetaxel for subjects with advanced solid tumor malignancies. Int J Oncol 15:3094–3094
Kinzel L, Ernst A, Orth M et al (2016) A novel HSP90 inhibitor with reduced hepatotoxicity synergizes with radiotherapy to induce apoptosis, abrogate clonogenic survival, and improve tumor control in models of colorectal cancer. Oncotarget 7:43199–43219
Koga F, Kihara K, Neckers L (2009) Inhibition of cancer invasion and metastasis by targeting the molecular chaperone heat-shock protein 90. Anticancer Res 29:797–807
Kotha S, Sreevani G (2018) Cyclotrimerization with propargyl halides as copartners: formal total synthesis of the antitumor Hsp90 inhibitor AT13387. ACS Omega 3:1850–1855
Kurokawa Y, Sawaki A, Komatsu Y et al (2017) 1479PD Phase II study of TAS-116, an oral inhibitor of heat shock protein 90 (HSP90), in metastatic or unresectable gastrointestinal stromal tumor refractory to imatinib, sunitinib and regorafenib. Ann Oncol 28:521–538
Lee DH, Sung KS, Bartlett DL, Kwon YT, Lee YJ (2015) HSP90 inhibitor NVP-AUY922 enhances TRAIL-induced apoptosis by suppressing the JAK2-STAT3-Mcl-1 signal transduction pathway in colorectal cancer cells. Cell Signal 27:293–305
Li M, Zhang X, Zhou W-J et al (2013) Hsp90 inhibitor BIIB021 enhances triptolide-induced apoptosis of human T-cell acute lymphoblastic leukemia cells in vitro mainly by disrupting p53-MDM2 balance. Acta Pharmacol Sin 34:1545–1553
Lundgren K, Zhang H, Brekken J et al (2009) BIIB021, an orally available, fully synthetic small-molecule inhibitor of the heat shock protein Hsp90. Mol Cancer Ther 8:921–929
Mahadevan D, Rensvold DM, Kurtin SE et al (2012) First-in-human phase I study: results of a second-generation non-ansamycin heat shock protein 90 (HSP90) inhibitor AT13387 in refractory solid tumors. J Clin Oncol 30:3028
Margineantu DH, Emerson CB, Diaz D, Hockenbery DM (2007) Hsp90 inhibition decreases mitochondrial protein turnover. PLoS One 2:e1066
Marjaneh RM, Rahmani F, Hassanian SM et al (2018) Phytosomal curcumin inhibits tumor growth in colitis-associated colorectal cancer. J Cell Physiol 233:6785–6798
Matsumiya T, Imaizumi T, Yoshida H, Satoh K, Topham MK, Stafforini DM (2009) The levels of retinoic acid-inducible gene I are regulated by heat shock protein 90-alpha. J Immunol 182:2717–2725
Mayor-López L, Tristante E, Carballo-Santana M et al (2014) Comparative study of 17-AAG and NVP-AUY922 in pancreatic and colorectal cancer cells: are there common determinants of sensitivity? Transl Oncol 7:590–604
McBride CM, Levine B, Xia Y et al (2014) Design, structure–activity relationship, and in vivo characterization of the development candidate NVP-HSP990. J Med Chem 57:9124–9129
Menezes D, Taverna P, Jensen MR et al (2012) The novel oral HSP90 inhibitor NVP-HSP90 exhibits potent and broad-spectrum anti-tumor activities in vitro and in vivo. Mol Cancer Ther: Molcanther 11:730–739
Messaoudi S, Peyrat J, Brion J, Alami M (2008) Recent advances in Hsp90 inhibitors as antitumor agents. Anti Cancer Agents Med Chem 8:761–782
Milicevic Z, Bogojevic D, Mihailovic M, Petrovic M, Krivokapic Z (2008) Molecular characterization of Hsp90 isoforms in colorectal cancer cells and its association with tumour progression. Int J Oncol 32:1169–1178
Mimnaugh EG, Chavany C, Neckers L (1996) Polyubiquitination and proteasomal degradation of the p185c-erbB-2 receptor protein-tyrosine kinase induced by geldanamycin. J Biol Chem 271:22796–22801
Modi S, Stopeck AT, Linden HM et al (2011) HSP90 inhibition is effective in breast cancer: a phase 2 trial of tanespimycin (17AAG) plus trastuzumab in patients with HER2-positive metastatic breast cancer progressing on trastuzumab. Clin Cancer Res 17:5132–5139
Mohammadi A, Yaghoobi M, GholamhoseynianNajar A, Kalantari-Khandani B, Sharifi H, Saravani M (2016a) HSP90 inhibitor enhances anti-proliferative and apoptotic effects of celecoxib on HT-29 colorectal cancer cells via increasing BAX/BCL-2 ratio. Mol Cell Biol 62:62–67

Mohammadi A, Yaghoobi M, Najar AG, Kalantari-Khandani B, Sharifi H, Saravani M (2016b) HSP90 inhibition suppresses PGE2 production via modulating COX-2 and 15-PGDH expression in HT-29 colorectal cancer cells. Inflammation 39:1116–1123

Mohammadian M, Zeynali S, Azarbaijani AF, Ansari MHK, Kheradmand F (2017) Cytotoxic effects of the newly-developed chemotherapeutic agents 17-AAG in combination with oxaliplatin and capecitabine in colorectal cancer cell lines. Res Pharm Sci 12:517–525

Moradi-Marjaneh R, Hassanian SM, Fiuji H et al (2018a) Toll like receptor signaling pathway as a potential therapeutic target in colorectal cancer. J Cell Physiol 233:5613–5622

Moradi-Marjaneh R, Hassanian SM, Ghobadi N et al (2018b) Targeting the death receptor signaling pathway as a potential therapeutic target in the treatment of colorectal cancer. J Cell Physiol 233:6538–6549

Moradi-Marjaneh R, Hassanian SM, Shahidsales S, Avan A, Khazaei M (2018c) Curcumin suppresses colorectal cancer progression through down-regulation of Wnt signaling in cancer stem cells. Basic Clin Cancer Res 10:33–38

Moradi-Marjaneh R, Khazaei M, Seifi S, Hassanian SM, Ferns GA, Avan A (2018d) Pharmacogenetics of anticancer drug sensitivity and toxicity in colorectal cancer. Curr Pharm Des 24:2710–2718

Nagaraju GP, Alese OB, Landry J, Diaz R, El-Rayes BF (2014) HSP90 inhibition downregulates thymidylate synthase and sensitizes colorectal cancer cell lines to the effect of 5FU-based chemotherapy. Oncotarget 5:9980–9991

Nakashima T, Ishii T, Tagaya H et al (2010) New molecular and biological mechanism of antitumor activities of KW-2478, a novel non-ansamycin Hsp90 inhibitor, in multiple myeloma cells. Clin Cancer Res 15:2792–2802

Nowakowski GS, McCollum AK, Ames MM et al (2006) A phase I trial of twice-weekly 17-allylamino-demethoxy-geldanamycin in patients with advanced cancer. Clin Cancer Res 12:6087–6093

Ohkubo S, Kodama Y, Muraoka H et al (2014) TAS-116, a highly selective inhibitor of heat shock protein 90alpha and beta, demonstrates potent antitumor activity and minimal ocular toxicity in preclinical models. Mol Cancer Ther 14:14–22

Pacey S, Wilson RH, Walton MI et al (2011) A phase I study of the heat shock protein 90 inhibitor alvespimycin (17-DMAG) given intravenously to patients with advanced, solid tumors. Clin Cancer Res 17:1561–1570

Paraiso KH, Haarberg HE, Wood E et al (2012) The HSP90 inhibitor XL888 overcomes BRAF inhibitor resistance mediated through diverse mechanisms. Clin Cancer Res 18:2502–2514

Pearl LH, Prodromou C (2001) Structure, function, and mechanism of the Hsp90 molecular chaperone. Adv Protein Chem 59:157–186

Powers MV, Workman P (2007) Inhibitors of the heat shock response. Pharm Biol FEBS Lett 581:3758–3769

Ramalingam SS, Egorin MJ, Ramanathan RK et al (2008) A phase I study of 17-allylamino-17-demethoxygeldanamycin combined with paclitaxel in patients with advanced solid malignancies. Clin Cancer Res 14:3456–3461

Ramanathan RK, Trump DL, Eiseman JL et al (2005) Phase I pharmacokinetic-pharmacodynamic study of 17-(allylamino)-17-demethoxygeldanamycin (17AAG, NSC 330507), a novel inhibitor of heat shock protein 90, in patients with refractory advanced cancers. Clin Cancer Res 11:3385–3391

Riely G, Stoller R, Egorin M et al (2009) A phase Ib trial of IPI-504 (retaspimycin hydrochloride), a novel Hsp90 inhibitor, in combination with docetaxel. J Clin Oncol 27:3547–3547

Saif MW, Takimoto CH, Mita MM et al (2013) A phase 1, dose-escalation, pharmacokinetic and pharmacodynamic study of BIIB021 administered orally in patients with advanced solid tumors. Clin Cancer Res 20:445–455

Samlowski WE, Papadopoulos K, Olszanski AJ et al (2011) Abstract A96: phase 1 study of HSP90 inhibitor MPC-3100 in subjects with refractory or recurrent cancer. AACR 10:12–16

Sessa C, Shapiro GI, Bhalla KN et al (2013) First-in-human phase I dose-escalation study of the HSP90 inhibitor AUY922 in patients with advanced solid tumors. Clin Cancer Res 19:3671–3680

Spreafico A, Delord J, De Mattos-Arruda L et al (2015) A first-in-human phase I, dose-escalation, multicentre study of HSP990 administered orally in adult patients with advanced solid malignancies. Br J Cancer 112:650–659

Stuehmer T, Iskandarov K, Gao Z et al (2012) Preclinical activity of the novel orally bioavailable HSP90 inhibitor NVP-HSP990 against multiple myeloma cells. Anticancer Res 32:453–462

Supko JG, Hickman RL, Grever MR, Malspeis L (1995) Preclinical pharmacologic evaluation of geldanamycin as an antitumor agent. Cancer Chemother Pharmacol 36:305–315

Suzuki S, Kulkarni AB (2010) Extracellular heat shock protein Hsp90β secreted by MG63 osteosarcoma cells inhibits activation of latent TGF-β1. Biochem Biophys Res Commun 398:525–531

Takahashi H, Wang JP, Zheng HC, Masuda S, Takano Y (2011) Overexpression of GRP78 and GRP94 is involved in colorectal carcinogenesis. Histol Histopathol 26:663–671

Taldone T, Sun W, Chiosis G (2009) Discovery and development of heat shock protein 90 inhibitors. Bioorg Med Chem 17:2225–2235

Trendowski M (2015) PU-H71: an improvement on nature's solutions to oncogenic Hsp90 addiction. Pharmacol Res 99:202–216

Utsugi T (2013) New challenges and inspired answers for anticancer drug discovery and development. Jpn J Clin Oncol 43:945–953

Vaishampayan UN, Burger AM, Sausville EA et al (2010) Safety, efficacy, pharmacokinetics and pharmacodynamics of the combination of sorafenib and tanespimycin. Clin Cancer Res 16:3795–3804

van Ingen H, Purcea D, Brienza S, Paz-Ares L (2012) A phase I–II evaluation of the safety and efficacy of the oral HSP90 inhibitor Debio 0932 in combination with SOC in first-and second-line therapy of stage IIIB or IV NSCLC-the halo study (HSP90 inhibition and lung cancer outcomes). Paper presented at the J Thorac Oncol 2012

Vido MJ, Aplin AE (2015) The broad stroke of Hsp90 inhibitors: painting over the RAF inhibitor paradox. J Invest Dermatol 135:2355–2357

Wagner AJ, Chugh R, Rosen LS et al (2013) A phase 1 study of the heat shock protein 90 inhibitor retaspimycin hydrochloride (IPI-504) in patients with gastrointestinal stromal tumors or soft tissue sarcomas. Clin Cancer Res 19:6020–6029

Wagner AJ, Agulnik M, Heinrich MC et al (2016) Dose-escalation study of a second-generation non-ansamycin HSP90 inhibitor, onalespib (AT13387), in combination with imatinib in patients with metastatic gastrointestinal stromal tumour. Eur J Cancer 61:94–101

Wainberg ZA, Anghel A, Rogers AM et al (2013) Inhibition of HSP90 with AUY922 induces synergy in HER2-amplified trastuzumab-resistant breast and gastric cancer. Mol Cancer Ther 12:509–519

Wang X-T, Bao C-H, Jia Y-B et al (2014) BIIB021, a novel Hsp90 inhibitor, sensitizes esophageal squamous cell carcinoma to radiation. Biochem Biophys Res Commun 452:945–950

Wang CY, Guo ST, Wang JY et al (2016a) Inhibition of HSP90 by AUY922 preferentially kills mutant KRAS colon cancer cells by activating Bim through ER stress. Mol Cancer Ther 15:448–459

Wang CY, Guo ST, Wang JY et al (2016b) Reactivation of ERK and Akt confers resistance of mutant BRAF colon cancer cells to the HSP90 inhibitor AUY922. Oncotarget 7:49597–49610

Wrighton KH, Lin X, Feng X-H (2008) Critical regulation of TGFβ signaling by Hsp90. Proc Natl Acad Sci 105:9244–9249

Yan L, Zhang W, Zhang B, Xuan C, Wang D (2017) BIIB021: a novel inhibitor to heat shock protein 90–addicted oncology. Tumour Biol 39:1–5

Yanagitani N, Horiike A, Kitazono S et al (2017) First-in-human phase I study of an oral HSP90 inhibitor, TAS-116, in advanced solid tumors. J Clin Oncol 35:2546–2546

Yong K, Cavet J, Johnson P et al (2016) Phase I study of KW-2478, a novel Hsp90 inhibitor, in patients with B-cell malignancies. Br J Cancer 114:7–13
Zhang WL, Gao XQ, Han JX, Wang GQ, Yue LT (2009) Expressions of heat shock protein (HSP) family HSP 60, 70 and 90alpha in colorectal cancer tissues and their correlations to pathohistological characteristics. Ai Zheng 28:612–618
Zhang H, Neely L, Lundgren K et al (2010) BIIB021, a synthetic Hsp90 inhibitor, has broad application against tumors with acquired multidrug resistance. Int J Cancer 126:1226–1234
Zhou J, Wang W, Deng J et al (2015) Co-inhibition effect of 17-DMAG and oxaliplatin on proliferation and invasion of colorectal cancer cells. Ai Zheng 18:370–375

Chapter 4
Hsp90 in the Migration of Primordial Germ Cells: A Model to Study Long-Distance Cell Migration and Perhaps Cancer?

Marie Lejong, Nathalie Vanmuylder, and Stéphane Louryan

Abstract Primordial Germ Cells (PGC) are the progenitors of the germ-line and differentiate into spermatozoa or oocytes. They are responsible for the transmission of genetic and epigenetic information from one generation to the other. Since the nineteenth century, PGCs have been investigated using different techniques. Regardless of their mode of specification, inheritance or induction, PGCs arise early in development and migrate by a combination of passive and active movements towards the gonadic ridges. The migration of PGCs is very similar to the path taken by metastasis and germ cells are very often proposed as a model for the study of cell migration. Their pathway of migration is regulated by different signals that interact with Hsp90, an ATP-dependent chaperone associated with numerous tumors and used to grade the malignancy. Furthermore, some of the signals regulating PGCs have been proved to have a role in different cancers. This underlines the idea that PGCs could be an interesting tool to study long-distance cell migration and perhaps cancer.

Keywords Cancer · Embryo · Hsp90 · Migration · Primordial germ cell

Abbreviations

actr2	Actin related protein 2 homolog
alk2	Activin receptor-like kinase-2
Bmp	Bone morphogenetic proteins
Bmpr	Bone morphogenetic protein receptor

M. Lejong · N. Vanmuylder · S. Louryan (✉)
Laboratory of Anatomy, Biomechanics and Organogenesis, Faculty of Medicine, Université Libre de Bruxelles, Brussels, Belgium
e-mail: slouryan@ulb.ac.be

A. A. A. Asea, P. Kaur (eds.), *Heat Shock Protein 90 in Human Diseases and Disorders*, Heat Shock Proteins 19,
https://doi.org/10.1007/978-3-030-23158-3_4

EMT	Epithelial-mesenchymal transition
GR	Gonadal ridge
Hsp	Heat shock protein
MOTC	Microtubule organizer center
PGC	Primordial germ cells
ROCK/RhoA	Rho-associated protein kinase
TGFβ	Transforming growth factor β

4.1 Introduction

Primordial Germ Cells (PGCs) strongly express Heat Shock Protein 90 (Hsp90) during their entire life cycle. Understanding the role of Hsp90 during migration and differentiation of PGCs constitutes a major challenge, as well as establishing their putative function during evolution of species. Some relationships between PGCs migration and cancer dissemination could suggest common roles for Hsp90 in both processes. In this contribution, we will discuss the state of the art in the implications of Hsp90 in germs cells migration, and its potential correlation with the metastatic processes.

4.1.1 Identification of the Primordial Germ Cells

Primordial Germ Cells (PGCs) are present in all sexually reproducing animals. They differentiate into spermatozoa or oocytes. In this way, they represent the origin of any new organisms and are responsible for the transmission of genetic and epigenetic information across generations. PGCs are easily recognized thanks to their histological characteristics (Niewkoop and Sutasurya 1979). They are described as large cells (10–20 μm) with a round, oval or pear- shape. The cell nucleus is as well of a substantial size (6–10 μm) with a sparse content rending the nuclear membrane more evident. Depending on the stage of development and the species, the cytoplasm is filled with a various amount of lipid droplets that displace the nucleus in an eccentric position. Glycogen is also identified present and is of primary interest in the PAS-positive identification of these cells. The cytoplasm has a granular appearance with numerous ribosomes and contains a structure called the center sphere, cytoplasmic crescent or attraction sphere. It is made of one or two centrioles, mitochondrias and yolk granules. Eosinophilic granules have been considered as a usefull histological criteria to identify germ cells in fish (Nagai et al. 2001; Gamo 1961; Timmermans 1996; Kazama-Wakabayashi et al. 1999). PGCs also exhibit lamellopodias and fillopodias. This ability is related to their capacity to migrate and will be discussed in greater detail in the description of the migration pathways. Since Weishmann first identified the difference between germinal and somatic cells in the nineteenth century, CGPs were studied by several authors using their morphological

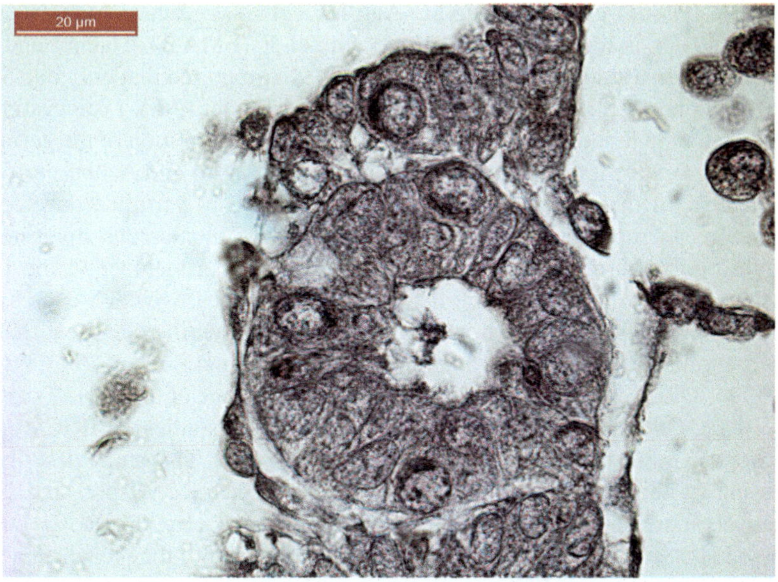

Fig. 4.1 Migrating PGCs in and around the intestinal epithelium, E10 mouse embryo, immunohistochemistry using anti Alkaline Phosphatase antibody. The cytoplasm of germ cells is strongly stained

characteristics (Heys 1931; Everett 1945; Niewkoop and Sutasurya 1979). Different stainings among which we can mention the Periodic acid-Schiff, the Best's Carmine or the Hematoxylin and eosin were used. At this time, the demonstration of alkaline phosphatase activity seemed to be an appropriate way to identify and locate these cells (Fig. 4.1).

Using this technique, Chiquoine (1954) was the first to give a complete description of the origin and migration of the PGCs in the mouse embryo. Nevertheless, this method has been declared unreliable by some because the morphological criteria used to recognize the germ cells could be observed in the entirety of the embryo's tissues (Hargitt 1925; Simkins 1923, 1928). The fact that the selective staining of PGCs with this technique was based on a short incubation in the buffer can explain the raise of criticism. Furthermore, it already underlined the need of a more specific method for the identification of the PGCs. Immunohistochemistry is largely used for the identification of PGCs. Stage-specific embryonic antigens (SSEA) is a glycoprotein found on the PGC's surface and is thus considered as a reliable marker for the study of these cells (Yon and Akbulut 2015; Fox et al. 1981; Shevinsky et al. 1982). Nevertheless, SSEA positivity has been observed in non-germinal tissue depending on the stage of the embryo. Fox et al. (1981) studied the localization of SSEA-1 in mouse embryo and reported positivity on the surface of ectodermal cells, visceral endoderm, primordium of the heart and somites, in the neural crest, the visceral yolk sac and gut endoderm just to mention a few. D'Costa and Petitte (1999) confirmed the observations made by Fox et al. in turkey embryos indicating that SSEA-1 is not germline-specific. Another cell-surface carbohydrate used to recognize PGCs is the embryonic mouse antigen 1 (EMA-1) (Hahnel and Eddy 1986). However, a study made by Urven et al. (1988) reported that only one third to one fifth of the PAS-positive PGCs were labeled by antiEMA-1 suggesting that EMA-1 is not the best marker to identify and follow the migration of the germ cells. The vasa gene was first identified in Drosophila by Lasko and Ashburner (1988). This member of the posterior group genes is required in germline development. Furthermore, the absence of the vasa transcript from somatic cells suggested that this transcript is germline specific. The vasa protein is a member of the DEAD (Asp-Glu-Ala-Asp) family of ATP-dependent RNA helicases which expression is restricted to the germline. Vasa homologues have been identified in Caenorhabditis, mouse, zebrafish, rats, xenopous and humans (Deborah and Bennett 1993; Fujiwara et al. 1994; Olsen et al. 1997; Castrillon et al. 2000; Noce et al. 2001; Tsunekawa et al. 2000). Other markers have been used in various studies on PGCs. Among those, we can quote the work of Vanmuylder et al. (2002). This study described the expression of the two cytosolic isoform of Hsp90 in mice, Hsp86 and Hsp84, respectively corresponding to the human Hsp90α and Hsp90β. PGCs strongly expressed Hsp86 from their appearance on the yolk sac until they reaches the genital ridges. Hsp86-positive PGCs were successively observed at the base of the allantois (Fig. 4.2), in the epithelium lining the hindgut, migrating into the dorsal mesentery (Fig. 4.3) and finally into the genital ridges. Gonadic expression of Hsp86 is limited to the germline. They confirmed the idea of the protective role of the members of the Hsp90 family and that Hsp86 could be the protein that preserve mammals from phenotypic mutations.

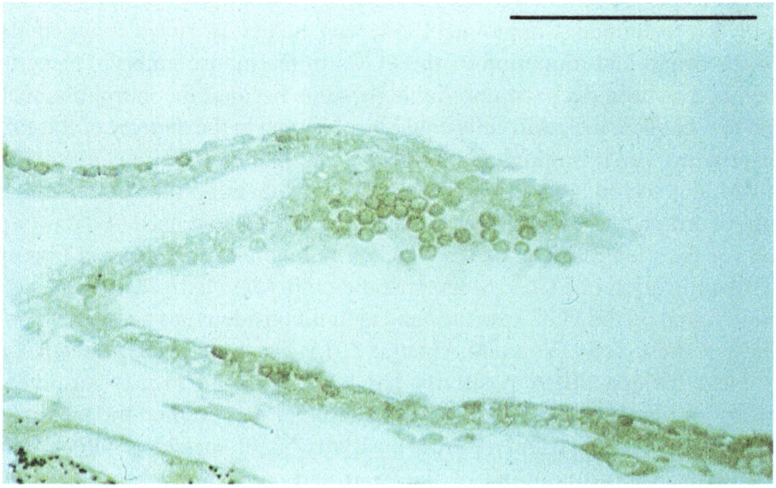

Fig. 4.2 E8.5 mouse embryo. Hsp90 positive PGCs in the allantois, before their intraembryonic migration. Bar: 100 μm

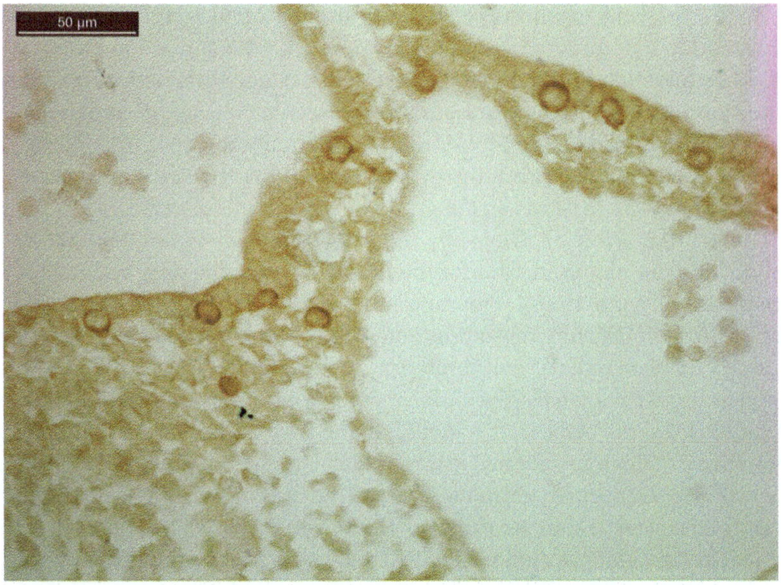

Fig. 4.3 E10 mouse embryo. Hsp90 positive PGCs migrating in the mesentery

4.1.2 Migration Pathways of the Primordial Germ Cells

Primordial germ cells arise early in development and migrate by a combination of passive and active movements towards the gonadic ridges. There are two different mechanisms described for PGCs formation: Inheritance of a germ plasma or induction by signals. The first mode is the one observed in Invertebrates such as *Drosophila* or *Caenorhabidis* and in nonmammalian vertebrates such as frogs and fishes (Kazama-Wakabayashi et al. 1999). The germ plasm, also called pole plasm in flies, consists of maternally derived RNAs, RNA-binding-protein and organelles and is transmitted to the future germ cells during the first divisions of the embryo. On the other side, in mammals, PGCs are specified in the epiblast shortly before or during gastrulation and are under the influence of specific signals secreted by surrounding cells (Ginsburg et al. 1990; McLaren 2003; Molyneaux and Wylie 2004; Saitou and Yamaji 2012; Tam and Zhou 1996). In the mouse, primordial germ cells appear at the base of the allantois at day 7,5. They migrate through the intestinal endoderm between day 8 and day 9 before colonizing the genital ridge through migration in the dorsal mesentery. Fuss and Felix were the first to study the migration of the germ cells in human embryos (Fuss 1911; Felix 1911). The pathway of this migration is very similar to the one observed in mice. Human PGCs arise in the endoderm of the yolk sac at the end of the 3rd week. They migrate through the gut epithelium during the 5th week to reach the dorsal mesentery and finally colonize the gonadic ridges at the beginning of the 6th week (De Felici 2013; McKay et al. 1953). It is of interest to indicate that this pathway of migration is different in other species. In birds, PGCs are first localized in the central zone of the epiblast. They are then moved anteriorly to form the germinal crescent with the apparition of the primitive streak. They enter the forming vascular network at stage Hamburger Hamilton (HH) 10. They are localized in the extra-embryonic vessels at stage 12 and in the embryonic vessels at stage 13. They leave the blood circulation coming out the visceral branches of the aorta and reach the genital ridges (Swift 1914; Ginsburg 1997). In reptiles, the pathway depends on the group of classification. In Anapsids, the migration is interstitial while in Diapsids, the migration will be a combination of the interstitial and circulatory modes for the primitive ones and only circulatory for the superior ones (Dustin 1910). The migration of PGCs is, especially in birds, very similar to the path taken by metastasis and this is why germ cells are often proposed as a model for the study of cell migration. Lejong et al. (2018) reported that geldanamycin, a well-known Hsp90 inhibitor with potent anti-tumor activity, decreases the number of germ cells in the mouse embryo. In 2009, preliminary results of Vanmuylder et al. demonstrated that Geldanamycin administration reduces the number of HSP90-positive germ cells in the mouse embryo. They also found cells with morphological features similar to PGCs in the pelvis as shown in teratoma (Fig. 4.4). In their study, Lejong et al. (2018) compared the immune reactivity of the PGCs for both Hsp90 and Vasa (Fig. 4.5). Those were identical and the authors therefore expressed the idea that Hsp90 could be considered as a reliable marker for PGCs investigation. Furthermore, the cells found in the pelvis suggested that some cells at least got lost. These results suggest that an inhibition of the Hsp90 could decrease the capacity of the germ cells to migrate.

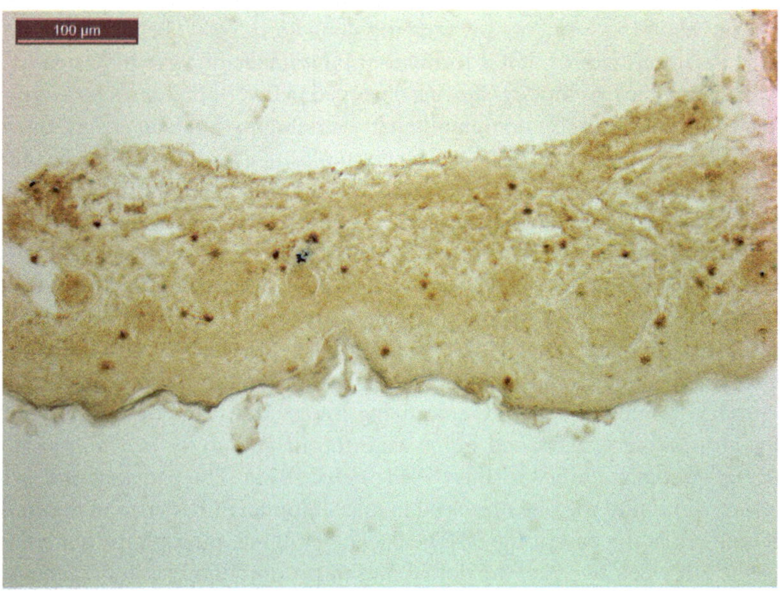

Fig. 4.4 Mouse E18 embryo whose mother received 5 mg of Geldanamycin. Ectopic abnormal HSP90 positive germ cells in perineal area

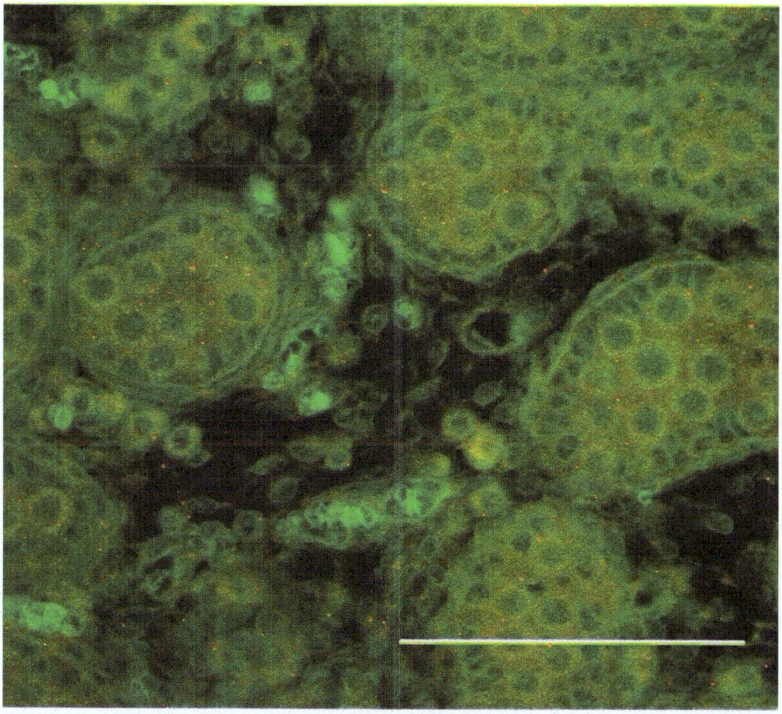

Fig. 4.5 E17 mouse embryo. Double immunostaining with anti-Vasa (*red*) and anti -HSP90 (*green*) antibodies on PGCs demonstrating the same pattern of fixation of both antibodies. Bar: 100 µm

4.1.3 Molecular Control of PGC Migration

PGCs specification and correct migration are under the influence of different signals. In 1996, Tam and Zhou transplanted distal epiblast close to the extraembryonic ectoderm before E6,5 and observed the formation of PGCs even though cells from the distal epiblast are not supposed to generate this cellular type. On the other hand, transplantation of proximal epiblast into the distal region, far away from the extra-embryonic ectoderm, never gave rise to PGCs. They concluded that, in the gastrulating mouse embryo, PGCs specification depends on signals coming from the extraembryonic ectoderm. Bone morphogenetic proteins (Bmps) are members of the transforming growth factor β superfamily of intercellular signaling proteins, TGFβ. At least three Bmps have been demonstrated to be involved in the generation and migration of PGCs: Bmp4, Bmp2 and Bmp8 (Ying et al. 2000, Ying and Zhao 2001; Machev et al. 2004; Lawson et al. 1999; de Sousa Lopes et al. 2004; Dudley et al. 2007). BMP4 is strongly expressed before gastrulation in the extraembryonic ectoderm. Bmp4 homozygous null embryos are depleted of PGCs and lack allantois. Heterozygous embryos are normal regarding the morphological features but contain less PGCs compared to the wild-type embryos. Lawson et al. proposed a model in which two signals are required to generate PGCs and allantois. The first signal consists of a high concentration of Bmp4 which specifies a group of cells that will become either PGCs or allantoic cells depending on a second signal. Bmp2 and Bmp4 have an additive effect in PGC specification. Indeed, Bmp2+/− Bmp4+/− embryos have even less PGCs than the Bmp4 heterozygous mutant embryo. Bmp signaling also regulates PGCs motility. Dudley et al. (2007) studied the role of Bmps on cultured PGCs in migration. They found that Bmp4 has a bi-phasic effect on PGCs number. Treatment with low doses of Bmp4 increases this number whereas high doses decreases it. The BMP-inhibitor Noggin has the opposite effect. Furthermore, they observed that treatment of the slices with Noggin greatly affects PGCs motility. Hsp90 inhibition has been proved to destabilizes TGFβ signaling by decreasing the levels of Bmps receptors such as Alk2, Bmpr1a, Bmpr2 and Actr2 (Haupt et al. 2012). De Sousa Lopes et al. demonstrated that Bmp4 signals through ALK2 for PGCs formation. Furthermore, phosphorylation of the Bmps receptors results in the phosphorylation of Smad-1, Smad-5 and Smad-8. Smad-1 and Smad-5 mutant mouse embryos display a great decrease even a complete lack in PGCs.

Cell migration including the one of the PGCs implies several intracellular modifications including cytoskeleton reorganization. As described earlier, filopodias are one of the morphological features of PGCs in migration. Terayama et al. (2013) made a study on Xenopus primordial germ cells. They results showed that germ cells motility requires actin polymerization and myosin activity and that PGCs migration depends on ROCK/RhoA signaling. Hsp90 binds actin in vitro (Koyasu et al. 1986). Taiyab and Mohan Rao (2011) demonstrated that geldanamycin decreased cell motility and invasion. They showed that Hsp90 inhibition decreases the number of membrane protrusions in F111 cells. They observed a significant decrease in the levels of RhoA and mDia2, members of the Rho family known to be involved in the generation of contractile force and in the formation of lamellipodias.

Closely related to the cytoskeleton is the cadherin superfamily. These glycoproteins are expressed on the surface membrane and play a role in cell adhesion. During their migration, PGCs form a network through cytoplasmic processes or filopodias. Bendel-Stenze et al. (2000) investigated the expression of cadherins during PGCs migration. At E 8,5, PGCs were negative for E-Cadherin. From E9,5, the expression of E-cadherin was in function of the location of the PGCs. The cells still nested in the hindgut remained negative whereas the ones migrating towards the genital ridges started being positive. E-cadherin expression was strongly expressed at the junctions between germ cells and in the genital ridges. The culture of slices of E10,5 embryos with E-cadherin blocking antibodies prevented the condensation of PGCs into the gonadal ridge (GR) and ectopic germ cells were observed. Epithelial mesenchymal transition (EMT) is a process in which cells gain the capacity to migrate. E-cadherin plays a key role in this process. Hsp90 secreted by colorectal cancer cells has been proved to facilitate EMT by decreasing E-cadherin expression (Chen et al. 2013).

Steel factor, is a short-chain helical cytokine that binds to the c-kit receptor, expressed by PGCs throughout their migration. The Steel-c-kit interaction is mandatory for PGCs survival. A mechanism called midline death is set up after E9,5 to get rid of the ectopic cells, mostly in the midline (Gu et al. 2006). PGCs that remain in the midline after E10,5 die by apoptosis. Steel −/− mutant embryos display an important loss of PGCs before E9. Gu et al. (2006) showed that the culture of these embryos in a medium supplemented in Steel factor restored the protection against apoptosis whereas blocking the c-kit receptor removed this protective effect. Hsp90 inhibition by STA-9090 has been proved to downregulate Kit phosphorylation and tumor growth in malignant mast cell tumor (Lin et al. 2008).

The chemokine stromal cell-derived factor 1 (SDF-1) provides directional clues for the migration f PGCs (Ara et al. 2003; Doitsidou et al. 2002; Stebler et al. 2004; Takeuchi et al. 2010). SDF-1 is a chemokine which acts via a G protein-coupled receptor, CXCR4. SDF-1 is involved in lymphocyte and blood vessels development, cardiac septum formation and embryonic viability. Toshiaki et al. demonstrated that PGCs migration in SDF-1−/− mice was unaffected until E9,5. However, only 14% of PGCs were found in the genital ridges of the mutants. The amount of PGCs remaining in the hindgut and the mesentery was also higher. SDF-1 has a key role in post- natal vasculogenesis after ischemic injuries. Ischemia induces an increase in SDF-1 expression which in turn generates the recruitment of endothelium progenitor cells. Wan et al. (2010) demonstrated that Hsp90 inhibition with geldanamycin decreases SDF-1 expression and the neovascularization.

Pfeiffer et al. (2018) also investigated the role of Hsp90 in the migration of PGCS in zebrafish embryos. To do so, they compared the RNA expression profile of motile PGCs with that of PGCS in the gonads and with that of somatic cells. Hsp90 is strongly expressed in PGCs all along their migration. Mutant embryos showed impaired colonization of the gonads due to a significant decrease in the displacement and straightness of the migration path of the PGCs. They observed the presence of two Microtubule Organizing Center (MOTC) in embryos knocked down for Hsp90. This interferes with a correct polarization of the cell and so, prevents a correct migration pathway. In mutant embryos, about twice more PGCs located in ectopic positions were observed compared to the control embryos. Pfeiffer et al. (2018) also noted defects in cell cycle progression with a prolonged time spent in the S/G2/M phases in embryos with decreased or no Hsp90 activity. This again interferes with the polarization of the cell and prevents a correct response to the guidance clues.

4.1.4 Hsp90: An Interesting Tool in Epigenetic Evolution

The first evidence that HSP90 could play a role on biological evolution was brought by Rutherford and Lindquist (1998), in *Drosophila melanogaster*. Heterozygotous HSP83 mutant flies exhibited some homeotic abnormalities. Outcrossing mutant flies with normal strains displayed same kinds of abnormalities. Little changes in temperature increases the amount of abnormalities in heterozygotous, as well as exposition to a HSP90 inhibitor, the geldanamycin. The authors concluded that, with HSP90 buffering inhibition, cryptic variants due to normal variation can phenotypically express. These results were confirmed by Sollars et al. (2003), who observed in drosophila exposed to geldanamycin and trichostatin an abnormal expression of wingless in the eye imaginal discs, associated with abnormal eye phenotypes. Mutation of the trithorax genes were also observed, and the abnormalities were heritable in the offspring.

In the plant *Arabidopsis thaliana*, exposition to two Hsp90 inhibitors (geldanamycin and radicicol) leads to an increase of phenotypical variation (Queitsch et al. 2002). The conclusions were that HSP90 buffers the expression of genetic variations, affects developmental plasticity and increases stability against stochastic process. The cavefish *astyanax mexicanum* was treated by radicicol (Rohner et al. 2013). The surface-dwelling strain (which possess eyes) exhibited morphological variations of both eye and eye-socket, whereas cave-dwelling strains leaded to a reduction of the eyeless orbital cavity. Peuβ et al. (2015) demonstrated in insects that downregulation of Hsp90 can be reduced in cases of genetic variation and that it can be beneficial. In the nematodes *Caenorhabidis elegans* and *c. briggsae*, specific knockdown of Hsp90 demonstrated that HSP90 regulates the mobility of transposable elements, considered as drivers of genomic evolutionary changes (Ryan et al. 2016).

It is also classically considered that the primary role of HSP90 is to prevent aggregation of denatured proteins (Stankiewicz and Mayer 2012). Several examples of refolding of denatured or muted proteins by HSP90 were described (Miyata and Yahara 1992; Buchner 1999; Yahara 1999). Some proteins act as transcriptional regulators. Therefore, failure in refolding of defective proteins by HSP90 can uncover silent mutations. HSP90 can thus be considered by a potential regulator of evolvability (Wagner et al. 1999). The fact that Hsp90 is strongly expressed in the germ-line without any interruption from the first stages of embryonic development to the adult in mammals, including the humans (Figs. 4.6, 4.7, and 4.8) (Vanmuylder

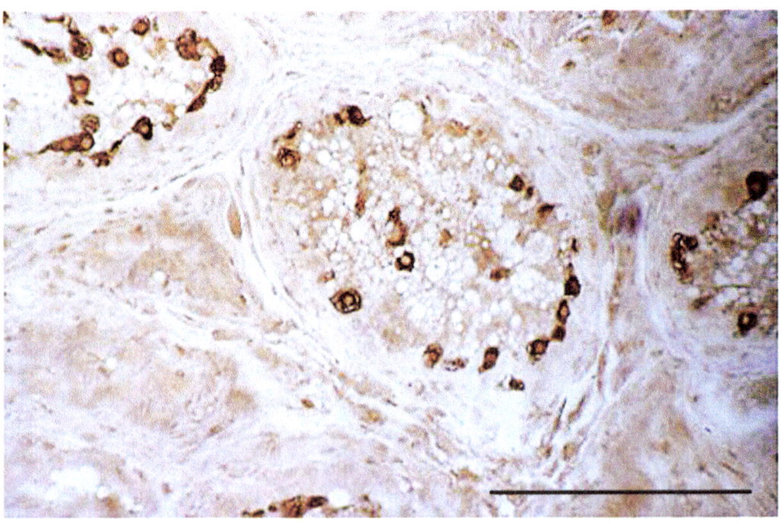

Fig. 4.6 30 weeks male human embryo. HSP90 positive PGC in the testis rudiment. Bar: 100 µm

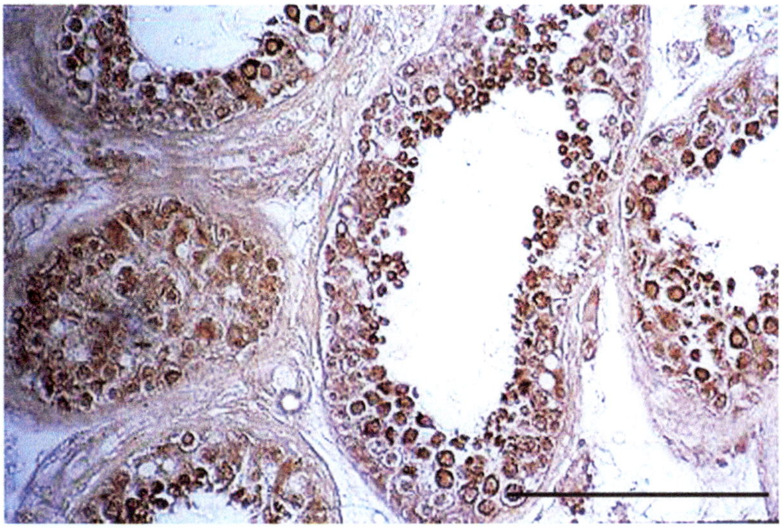

Fig. 4.7 Adult human testis. Hsp90 positivity is restricted to sperm cells. Bar: 100 μm

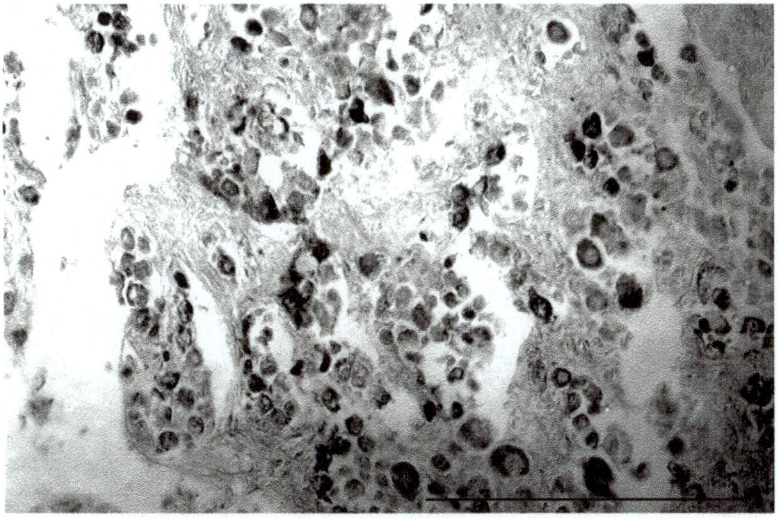

Fig. 4.8 34 weeks female human fetus. HspP90 positive PGC in the ovary rudiment. Bar: 100 μm

et al. 2002; Louryan et al. 2002, 2003) suggest that its protective function could be particularly needful in this cell-line, in order to stabilize the offspring against phenotypical changes due to cryptic genetic variations. Furthermore, the fact that HSP90 function can be reduced by temperature changes suggest a possible role in biological evolution associated with dramatic climatic changes.

4.1.5 Hsp90, Germ Cell and Cancer

The importance of the regulation of cell cycle and migration mechanisms by Hsp90 is well known and characterized as much in germ cells than in cancer cells (Pfeiffer et al. 2018). Hsp90 has a key role in folding/unfolding of proteins but its function extends far beyond. Hsp90 is expressed in the cytosol under normal conditions and is involved in various steps of development, differentiation, apoptosis and oncogenesis (Neckers and Workman 2012). Hsp90 is used as a marker of malignancy for several tumors. The human orthologue HSP90 has been specially studied in the context of cancer spread (reviewed in Wu et al. 2017). For example, Hsp86, corresponding to the human Hsp90α, is overexpressed in pleomorphic adenoma (mixed tumor (Fig. 4.9)) and in some Warthin tumors (cystadenolymphoma) especially in germinative centers (Fig. 4.10) probably due to malignant transformation rates in mixed tumor compared to other adenoma or higher recurrence rate for benign tumors (Vanmuylder et al. 2000; Wang et al. 2013). HSP86 expression in embryonic cells since their appearance at the surface of yolk sac until gonadic mature area is also maintained in yolk sac tumor (Fig. 4.11) (Vanmuylder et al. 2004). Hsp86 inhibition by geldanamycin reduces germ cells migration and increases ectopic germ cell populations, similar to teratomas (Fig. 4.7) (Vanmuylder et al. 2009). Primordial germ cells arise early in development and migrate through different tissues to reach

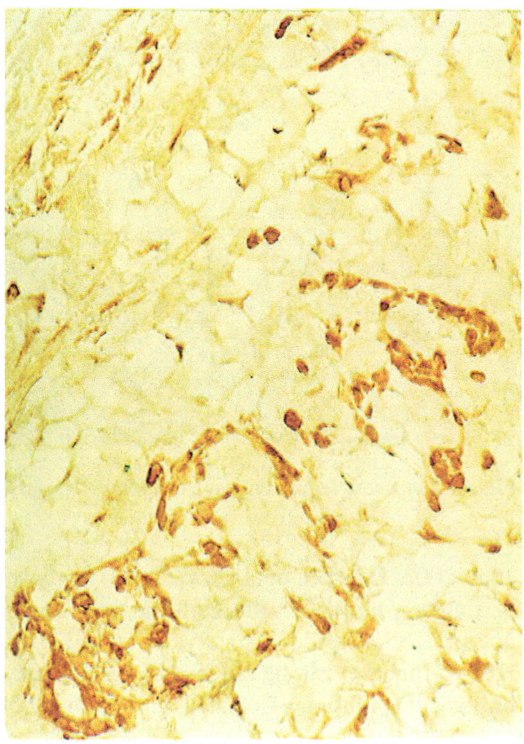

Fig. 4.9 Human salivary pleiomorphic adenoma. HSP90 positive myo-epithelial cells (×220)

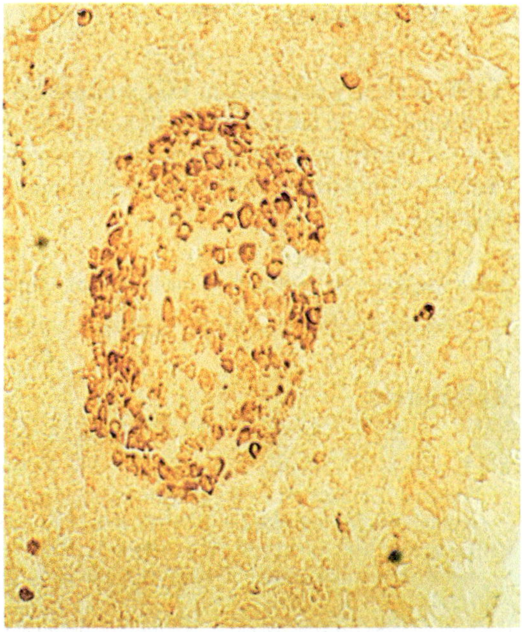

Fig. 4.10 Human salivary Whartin's tumor (cystadenolymphoma). Numerous Hsp90 positive cells are visible in the germinative center (×325)

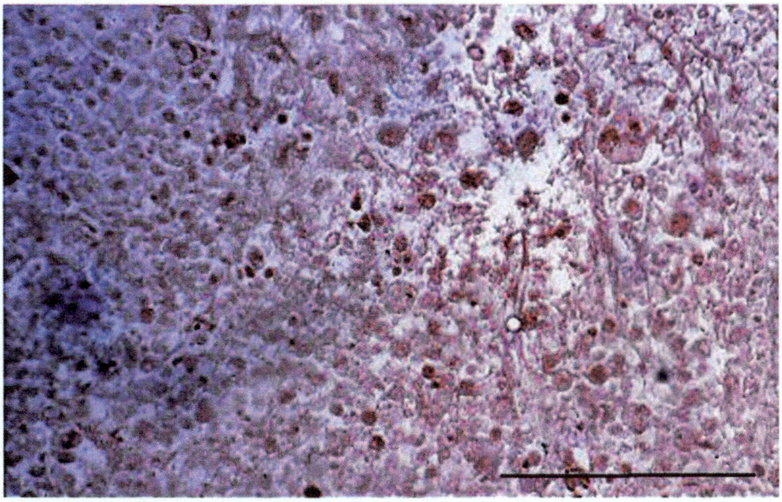

Fig. 4.11 Human yolk sac tumor. Hsp90 positive cells are diffusely visible in the tumor's stroma. Bar: 100 μm

the genital ridges. This mechanism is under the control of several signal pathways some of which have been highlighted in cancer spread. Migration of PGCs seems therefore an interesting model to study long distance cell migration and perhaps cancer.

4.2 Conclusions

Hsp90 is clearly involved in PGCs migration and stabilization, and could play a significant role in the protection of germ lineage against phenotypic variations. The study of Hsp90 in germ cells migration could constitute a now interesting way to establish similarities between embryonic and neoplastic proliferation and migration.

Acknowledgements The authors thank Mrs. M. Choa-Duterre for histological sections.

References

Ara T, Nakamura Y, Egawa T, Sugiyama T, Abe K, Kishimoto T, Matsui Y, Nagasawa T (2003) Impaired colonization of the gonads by primordial gem cells in mice lacking a chemokine, stromal cell-derived factor-1 (SDF-1). Proc Natl Acad Sci 100:5319–5323

Bendel-Stenze MR, Gomperts M, Anderson R, Heasman J, Wylie C (2000) The role of cadherins during primordial germ cell migration and early gonad formation in the mouse. Mech Dev 91:143–152

Buchner J (1999) HSP90 & Co.-a holding for folding. Trends Biochem Sci 24:136–141

Castrillon DH, Quade BJ, Wang TY, Quigley C, Crum CP (2000) The human VASA gene is specifically expressed in the germ cell lineage. Proc Natl Acad Sci U S A 97:9585–9590

Chen WS, Chen CC, Chen LL, Lee CC, Huang TS (2013) Secreted heat shock protein 90α (HSP90α) induces nuclear factor-κB-mediated TCF12 protein expression to down-regulate E-cadherin and to enhance colorectal cancer cell migration and invasion. J Biol Chem 288:9001–9010

Chiquoine DA (1954) The identification, origin, and migration of the primordial germ cells in the mouse embryo. Anat Rec 118:135–146

D'Costa S, Petitte JN (1999) Characterization of stage-specific embryonic antigen-1 (SSEA-1) expression during early development of the turkey embryo. Int J Dev Biol 43:349–356

De Felici M (2013) Origin, migration, and proliferation of human primordial germ cells. In: Coticchio B, De Santis L (eds) Oogenesis. Springer, London, pp 19–37

Deborah LR, Bennett LK (1993) glh-1, a germ-line putative RNA helicase from Caenorhabditis, has four zinc fingers. Proc Natl Acad Sci U S A 90:9300–9304

Doitsidou M, Reichman-Fried M, Stebler J, Koprunner M, Dorries J, Meyer D, Esguerra CV, Leung T, Raz E (2002) Guidance of primordial germ cell migration by the chemokine SDF-1. Cell 111:647–659

Dudley BM, Runyan C, Takeuchi Y, Schaible K, Molyneaux K (2007) BMP signaling regulates PGC numbers and motility in organ culture. Mech Dev 124:68–77

Dustin AP (1910) L'origine et l'évolution des gonocytes chez les reptiles. Arch Biol 25:495–534

Everett NB (1945) The present status of the germ cell problem in vertebrates. Biol Rev Camb Philos Soc 20:45–55

Felix W (1911) Die Entwicklung der Harm- und Geschlechtsorgane. In: Keibel-Mall Handbuch der Entwicklungageschichte des Menschen, vol 2. Hirzel, Leipzig, pp 732–955

Fox N, Damjanov I, Martinez-Hernandez A, Knowel BB, Solter D (1981) Immunohistochemical localization of the early embryonic antigen (SSEA-1) in postimplantation mouse embryos and fetal and adult tissue. Dev Biol 83:391–398

Fujiwara Y, Komiya T, Kawabata H, Sato M, Fujimoto H, Furusawa M, Noce T (1994) Isolation of a DEAD-family protein gene that encodes a murine homolog of Drosophila vasa and its specific expression in germ cell lineage. Proc Natl Acad Sci U S A 91:12258–12262

Fuss A (1911) Uber extraregionare Geschlechtszellen bei einem menschlichen Embryo von 4 Wochen. Anat Am 39:407–409
Gamo H (1961) On the origin of germ cells and formation of gonad primordia in the medaka, *Olyzias latipes*. J Zool 13:101–115
Ginsburg M (1997) Primordial germ cell development in avians. Poult Sci 76:91–95
Ginsburg M, Snow MHL, Mc Laren A (1990) Primordial germ cells in the mouse embryo during gastrulation. Development 110:521–528
Gu Y, Runyan C, Shoemaker A, Surani A, Wylie C (2006) Steel factor controls primordial germ cell survival and motility from the time of their specification in the allantois, and provides a continuous niche throughout their migration. Development 136:1295–1303
Hahnel AC, Eddy EM (1986) Cell surface markers of mouse primordial germ cells defined by two monoclonal antibodies. Gamete Res 15:1235–1244
Hargitt GT (1925) The formation of the sex glands and germ cells of mammals. J Morph Physiol 40:517–557
Haupt A, Joberty G, Bantscheff M, Frohlich H, Stehr H, Schweiger MR, Fischer A, Kerick M, Boerno ST, Dahl A, Lappe M, Lehrach H, Gonzalez C, Drewes G, Lange BMH (2012) Hsp90 inhibition differentially destabilizes MAP linase and TGF-beta signaling components in cancer cells revealed by kinase-targeted chemoproteomics. BMC Cancer 12:38–50
Heys F (1931) The problem of the origin of germ cells. Q Rev Biol 6:1–45
Kazama-Wakabayashi M, Yamaha E, Yamazaki F (1999) The elimination and duplication of lower part of blastoderm effects on the number of primordial germ cells in goldfish. Fish Sci 65:577–582
Koyasu S, Nishida E, Kodawaki T, Matsuzaki F, Lida K, Harada F, Kasuga M, Sakai H, YaharaTwo I (1986) Mammalian heat shock proteins, HSP90 and HSP100, are actin-binding proteins. Proc Natl Acad Sci U S A 83:8054–8058
Lasko PF, Ashburner M (1988) The product of the Drosophila gene vasa is very similar to eukaryotic initiation factor-4A. Nature 335:611–617
Lawson KA, Ray Dunn N, Roelen BAJ, Zeinstra LM, Davis AM, Wright CVE, Korving JPWFM, Hogan BLM (1999) Bmp4 is required for the generation of primordial germ cells in the mouse embryo. Genes Dev 13:424–436
Lejong M, Choa-Duterre M, Vanmuylder N, Louryan S (2018) Geldanamycin administration reduces the amount of primordial germ cells in the mouse embryo. Morphologie 102: 219-224. https://doi.org/10.1016/JMorpho.2018.05.001
Lin TY, Bear M, Du Z, Foley KP, Ying W, Barsoum J, London C (2008) The novel HSP90 inhibitor STA-9090 exhibits activity against Kit-dependent and -independent malignant mast cell tumors. Exp Hematol 36:1266–1277
Louryan S, Evrard L, Glineur R, Vanmuylder N (2002) Protéines de choc thermique, embryogenèse et évolution. Bull Mem Acad R Med Belg 157:293–299
Louryan S, Vanmuylder N, Lambot MA, Rooze M (2003) HSP86: un rôle dans l'évolution humaine ? Anthropol Praehist 114:1–5
Machev N, Fuhrmann G, Viville S (2004) Ontogénèse des cellules germinales primordiales. Med Sci 20:1091–1095
McKay D, Hertig AT, Adams EC, Danziger S (1953) Histochemical observations on the germ cells of human embryos. Anat Rec 17:201–219
McLaren A (2003) Primordial germ cells in the mouse. Dev Biol 262:1–15
Miyata A, Yahara I (1992) The 90-kDa heat shock protein, HSP90, binds and protects casein kinase II from self-aggregation and enhances its kinase activity. J Biol Chem 7:7042–7047
Molyneaux K, Wylie C (2004) Primordial germ cell migration. Int J Dev Biol 48:537–544
Nagai T, Yamaha E, Arai K (2001) Histological differentiation of primordial germ cells in zebrafish. Zool Sci 18:215–223
Neckers L, Workman P (2012) Hsp90 molecular chaperone inhibitors: are we there yet? Clin Cancer Res 18:64–76

Niewkoop D, Sutasurya LA (1979) Primordial germ cells in the chordates. Cambridge University Press, Cambridge

Noce T, Okamoto-Ito S, Tsunekawa N (2001) Vasa homolog genes in mammalian germ cell development. Cell Struct Funct 26:131–136

Olsen LC, Aasland R, Fjose A (1997) A vasa-like gene in zebrafish identifies putative primordial germ cells. Mech Dev 66:95–105

Peuβ R, Eggert H, Armitage SAO, Kurtz J (2015) Downregulation of the evolutionary capacitor Hsp90 is mediated by social cues. Proc R Soc B 282:20152041. https://doi.org/10.1098/rspb.2015.2041

Pfeiffer J, Tarbashevich K, Bandemer J, Palm T, Raz E (2018) Rapid progression through the cell cycle ensures efficient migration of primordial germ cells-the role of Hsp90. Dev Biol 436:84–93

Queitsch C, Sangster TA, Lindquist S (2002) Hsp9O as a capacitor of phenotypic variation. Nature 417:618–624

Rohner N, Jarosz DF, Kowalko JE, Yoshizawa M, Jeffery WR, Borowsky RL, Lindquist S, Tabin CJ (2013) Cryptic variations on morphological evolution: HSP90 as a capacitor for loss of eyes in cavefish. Science 342:1372–1375

Rutherford SL, Lindquist S (1998) Hsp90 as a capacitor for morphological evolution. Nature 396:336–342

Ryan CP, Brownlie JC, Whyard S (2016) Hsp90 and physiological stress are linked to antonomous transposon mobility and heritable genetic change in nematodes. Genome Biol Evol 8:3794–3805

Saitou M, Yamaji M (2012) Primordial germ cells in mice. Cold Spring Harb Perspect Biol 4:1–19

Shevinsky LH, Knowles BB, Damjanov I, Solter D (1982) Monoclonal antibody to murine embryos defines a stage-specific embryonic antigen expressed on mouse embryos and human teratocarcinoma cells. Cell 30:697–705

Simkins CS (1923) Origin and migration of the so-called primordial germ cells in the mouse and rats. Acta Zool 4:241–278

Simkins CS (1928) Origin of sex cells in man. Am J Anat 41:249–272

Sollars V, Lu X, Wang S, Garfinkel MD, Ruden DM (2003) Evidence for an epigenetic mechanism by which Hsp90 acts as a capacitor for morphological evolution. Nat Genet 33:70–74

de Sousa Lopes SM, Roelen BA, Monteiro RM, Emmens R, Lin HY, Li E, Lawson KA, Mummery CL (2004) BMP signaling mediated by ALK2 in the visceral endoderm is necessary for the generation of primordial germ cells in the mouse embryo. Genes Dev 18:1838–1849

Stankiewicz M, Mayer MP (2012) The universe of Hsp90. Biol Mol Concepts 3:79–97

Stebler J, Spieler D, Slanchev K, Molyneaux K, Richter U, Cojocaru V, Tarabykin V, Wylie C, Kessel M, Raz E (2004) Primordial germ cell migration in the chick and mouse embryo: the role of the chemokine SDF-1/CXCL12. Dev Biol 272:351–361

Swift CH (1914) Origin and early history of the primordial germ-cells in the chick. Am J Anat 15:483–516

Taiyab A, Rao CM (2011) Hsp90 modulates actin dynamics: inhibition of Hsp90 leads to decreased cell motility and impairs invasion. Biochim Biophys Acta 1813:213–221

Takeuchi T, Tanigawa Y, Minamide R, Ikenishi K, Komiya T (2010) Analysis of SDF-1/CXCR4 signaling in primordial germ cell migration and survival or differentiation in *Xenopus laevis*. Mech Dev 27:146–158

Tam PP, Zhou SX (1996) The allocation of epiblast cells to ectodermal and germ-line lineages is influenced by the position of the cells in the gastrulating mouse embryo. Dev Biol 178:124–132

Terayama K, Kataoka K, Morichika K, Orii H, Watanabe K, Mochii M (2013) Developmental regulation of locomotive activity in Xenopus primordial germ cells. Develop Growth Differ 55:217–228

Timmermans LPM (1996) Origin and differentiation of primordial germ cells in vertebrates, especially fishes. Neth J Zool 46:147–162

Tsunekawa N, Naito M, Sakai Y, Nishida T, Noce T (2000) Isolation of chicken vasa homolog gene and tracing the origin of primordial germ cells. Development 127:2741–2750
Urven LE, Erickson CA, Abbott UK, McCarrey JR (1988) Analysis of germ line development in the chick embryo using an anti-mouse EC cell antibody. Development 103:299–304
Vanmuylder N, Evrard L, Daelemans P, Dourov N (2000) Chaperones in the parotid gland: localization of heat shock proteins in human salivary glands. Cells Tissues Organs 167:199–205
Vanmuylder N, Werry-Huet A, Rooze M, Louryan S (2002) Heat shock protein HSP86 expression during mouse embryo development, especially in the germ-line. Anat Embryol 205:301–306
Vanmuylder N, Lambot M-A, Rooze M, Noël J-C, Louryan S (2004) HSP86, cellules germinales at yolk sac tumor. Ann Pathol 24:473–475
Vanmuylder N, Larbi H, Choa-Duterre M, Salvia P, Rooze M, Louryan S (2009) Geldanamycin administration reduces the number of HSP86-positive germ cells in the mouse embryo: preliminary results. Rev Med Brux 30:23–27
Wagner GP, Chiu CH, Hansen TF (1999) Is Hsp90 a regulator of evolvability? J Exp Zool 285:116–118
Wan YQ, Zhang XM, Wang XD, Wang BJ, Wang W (2010) 17-AAG, a Hsp90 inhibitor attenuates the hypoxia induced expression of SDF-1α and ILK in mouse RPE cells. Mol Biol Rep 37:1203–1209
Wang G, Gu X, Chen L, Wang Y, Cao B, E Q (2013) Comparison of the expression of 5 heat shock proteins in benign and malignant salivary gland tumor tissues. Oncol Lett 5:1363–1369
Wu J, Liu T, Rios Z, Mei Q, Lin X, Cao S (2017) Heat shock proteins and cancer. Trends Pharmacol Sci 38:226–256
Yahara I (1999) The role of HSP90 in evolution. Gene Cells 4:375–379
Ying Y, Zhao GQ (2001) Cooperation of endoderm-derived BMP2 and extraembryonic ectoderm-derived BMP4 in primordial germ cell generation in the mouse. Dev Biol 232:484–492
Ying Y, Liu XM, Marble A, Lawson KA, Zhao GQ (2000) Requirement of Bmp8b for the generation of primordial germ cells in the mouse. Mol Endocrinol 14:1053–1063
Yon N, Akbulut C (2015) Identification of primordial germ cells: cytological, histological and immunohistochemical aspects. Braz Arch Biol Technol 58:222–228

Chapter 5
Role of Heat Shock Protein 90 in Mammary Tumorigenesis

B. V. Sunil Kumar, Priya K. Gopal, and Ramneek Verma

Abstract Heat shock proteins (HSP) are the proteins with a wide range of functions. They protect the cells from various stressors including heat stress and also act as chaperones thereby aiding in protein folding and maintenance of cellular homeostasis. Various members of HSP family have been associated with cancers. HSP90 is one such member which promotes protein folding and has an important role in breast cancer progression. Reports indicate that mammary tumor cells constitutively secrete HSP90 in order to facilitate tumor invasion and metastasis. Further, neutralisation of extracellular HSP90 or inhibition of its secretion or interruption of its signalling has been found to block cancer invasion. The recent findings that various isoforms of HSP90 are upregulated in breast neoplasia and breast cancer/mammary tumor subjects have elevated serum HSP90 levels suggest an important role of this chaperone in malignant transformation of mammary glands. This chapter briefly reviews recent advances in our understanding of the role of HSP90 in mammary gland tumorigenesis.

Keywords Diagnosis · Expression · HSP90 · HSP90 inhibitors · Mammary gland · Mammary tumor

---

B. V. Sunil Kumar (✉) · R. Verma
Department of Microbial and Environmental Biotechnology, School of Animal Biotechnology, Guru Angad Dev Veterinary and Animal Sciences University, Ludhiana, Punjab, India

P. K. Gopal
Department of Botany, Centre of Advanced Studies, University of Calcutta, Kolkata, West Bengal, India

Department of Microbiology, Muralidhar Girls' College, Kolkata, West Bengal, India

A. A. A. Asea, P. Kaur (eds.), *Heat Shock Protein 90 in Human Diseases and Disorders*, Heat Shock Proteins 19,
https://doi.org/10.1007/978-3-030-23158-3_5

Abbreviations

AKT	Serine-threonine protein kinase
Bcl2	B-cell lymphoma gene 2
Bcl-XL	bcl-2-like 1
Cdk5	Cyclin-dependent kinase-5
Chk	Checkpoint kinase
CSC	Cancer stem cells
Dsh	Phosphoprotein Dishevelled
EGFR2/HER2	Epidermal growth factor receptor-2
FLT3	fms-like tyrosine kinase 3
GRP94	Endoplasmic reticulum Hsp90 homolog
GSK3β	Glycogen synthase kinase 3- beta
HIF-1α	Hypoxia-inducible factor 1-alpha
hsf1	Heat shock transcription factor 1
HSP	Heat shock proteins
IAP	Inhibitor of apoptosis protein
ILK	Integrin-linked kinase
MAPK1	Mitogen-activated protein kinase
MIF	Macrophage migration inhibitory factor
MMP	Matrix metalloproteinases
mTOR	Mammalian target of rapamycin
NFκB	Nuclear factor κ B
PKCα/PKCβ	Protein kinase C-alpha/beta
PTK2	Protein tyrosine kinase 2
Raf	Rapidly accelerated fibrosarcoma
SFRP1	Secreted frizzled related protein
SRC	Steroid receptor co-activator
Src	v-src sarcoma (Schmidt–Ruppin A-2) viral oncogene homolog (avian)
STAT	Signal Transducers and Activators of Transcription
TNBC	Triple negative breast cancer
TRAP	TNF-related activation protein
WNT	Wingless type MMTV integration site family member 1

5.1 Introduction

Homeostasis refers to tendency of an organism to maintain and regulate its internal milieu in a stable state (Martin 2011). Under physiological conditions, cells tend to maintain a state of homeostasis by interacting with the surrounding cells. Conditions such as high temperature, nutrient deficit, infections etc., which try to challenge the cellular homeostasis or simply speaking, "Stress" may signal cells to adjust their metabolism to suit the stressful state (De Maio 2011). In response to the insults, cells releasing some signals in order to combat the stressful circumstances. Heat shock proteins (HSP) are one among those signals that are expressed in response to those cellular insults (De Maio 2011). Initially discovered in drosophila in response to heat shock (Ritossa 1962), these proteins have been related to other types of stress as well. A large number of HSP have been identified till date and have been classified into separate families based on their molecular weights. Six conserved classes of HSP have been identified viz. HSP33, HSP60, HSP70, HSP90, HSP100 and small HSP (sHSP) (Schlesinger 1990). HSP may be expressed constitutively and may aid in vital intracellular processes, while they may also get secreted as extracellular HSP (eHSP) when cells encounter some or other kinds of stress (Giffard et al. 2008; De Maio 2011). As intracellular protein, HSP act as chaperones assisting in proper folding of other host proteins, but few intracellular members also get overexpressed during heat or pathological stress (Bukau et al. 2006; Hartl and Hayer-Hartl 2009; Sunil Kumar et al. 2018).

5.2 HSP in Cancer

Cancer is an outcome of de-regulated cell cycle leading to impairment of normal cell division and cell differentiation (Stivarou et al. 2016; Evan and Vousden 2001). Cancer stem cells (CSC) constitute those cells within a tumor having self-renewing capability and an indeterminate proliferative potential (Dalerba et al. 2007). CSCs usually remain dormant, but upon activation can resist apoptosis, induce angiogenesis and differentiate into heterogeneous populations of cancer cells (Fabian et al. 2009). HSP appear to play important role in carcinogenesis by enabling cells to evade tumour suppression pathways, thereby facilitating tumor progression. HSP also assist cancer metastasis by endowing the neoplastic cells resistance to chemo or hormonal therapy (Calderwood 2010; Zagouri et al. 2012). Levels of HSP also get elevated in different cancers which signals poor disease prognosis (van 't Veer et al. 2002). Activated heat shock transcription factor 1 (hsf1) is required for the transcription of *hsp*. *Hsf*1 itself gets overexpressed at the onset of cancers thereby facilitating tumor progression and metastasis (Wang et al. 2004; Ciocca and Calderwood 2005). Several HSP viz. HSP27, HSP40, HSP60, HSP70, HSP90, HSP110 have been implicated with the transformation of normal cells through disruption of various signalling pathways leading to cancer development and metastasis (Wu et al. 2017). These HSP may be exploited as potential biomarkers of various types of cancers and may be used for cancer diagnosis, assessing tumor progression, or as therapeutic targets for cancer therapy (Wu et al. 2017). In this chapter we shall concentrate on HSP90 and its different isoforms in the light of breast cancer development.

5.2.1 HSP90

The 90 kDa heat shock protein, HSP90, is one of the molecular chaperones, accounting to 1–2% of total eukaryotic cellular proteins under normal physiological conditions (Sreedhar et al. 2004). As a chaperone, it participates in folding of newly synthesized proteins and refolding of denatured ones during stress. It is also involved in signal transduction and degradation of misfolded proteins (Sreedhar et al. 2004). HSP90 binds to a portfolio of client proteins, mostly those involved in signal transduction.

5.2.2 Structure and Isoforms of HSP90

Members of the HSP90 family possess intrinsic ATPase activity that assists in protein folding and protein trafficking. The basic structure of HSP90 includes three domains namely, the N-terminal domain (~25 kDa), the middle domain (~40 kDa), and the C-terminal domain (~12 kDa) (Wu et al. 2017). Drugs such as geldanamycin and tanespimycin bind to ATP binding fold located at the N-terminal domain while, Middle domain is responsible for binding of HSP90 with its client proteins and C-terminal domain aids in HSP90 dimerization. There are four major isoforms of HSP90 namely HSP90A (cytosolic form), HSP90B1 also known as GRP94 (endoplasmic reticulum form), and TRAP (mitochondrial form). Further, HSP90A is sub-classified into HSP90AA (HSP90α) and HSP90AB (HSP90β) isoforms (Sreedhar et al. 2004; Sunil Kumar et al. 2018). HSP90α is also known as inducible/major form while HSP90β is the constitutively expressed minor isoform. Heat shock elements (HSEs) tightly regulate expression of *hsp*90α gene. HSE present in the first intron of *hsp*90β gene acts as an enhancer of constitutive expression as well as for the occasional heat inducibility of the *hsp*90β gene.

Genes encoding mammalian HSP90α and HSP90β isoforms (designated as *hsp*90α and *hsp*90β) contain introns unlike most of the *HSP* and share very less sequence similarity (Sreedhar et al. 2004) and these isoforms are thought to evolve through gene duplication about 500 million years ago (Pepin et al. 2001), while the gene encoding the endoplasmic reticulum isoform of HSP90 (*hsp*90B1) is thought to evolve through paralogous gene duplication very early in eukaryotic evolution (Gupta 1995; Sreedhar et al. 2004). There are five highly conserved regions also known as 'signature sequences' in all the isoforms of HSP90, three are present in the N-terminal domain while two are located in the middle domain (Gupta 1995; Sreedhar et al. 2004). The differential binding of HSP90α and HSP90β isoforms with their respective client proteins is thought to be due to differences in the amino acid sequences of these isoforms. Furthermore, there exists a difference in the dimerization potential between these two isoforms of HSP90 which mostly resides in the 190 amino acid carboxy-terminal domain of the protein. The α form dimerizes more readily as compared to the β isoform (Grammatikakis et al. 2002). Both

HSP90α and HSP90β isoforms interact with the major cytoskeletal proteins, actin and tubulin. However, their degree of binding differs broadly with the β isoform binding more efficiently than α isoform (Cambiazo et al. 1999).

5.2.3 Role of HSP90 in Cell Cycle

Fundamental process such as cell proliferation is highly regulated process with special reference to maintenance and developmental stages of eukaryotic organisms. Heat shock proteins (HSP) are expressed under normal physiological conditions, however their cytosolic expression levels is reported to elevate four to six times than normal under perturbed conditions like heat stress. In order that the cell division cycle can be orchestrated in a synchronized fashion, machinery to stabilize such *ab extra* or *ab intra* perturbations are manifested which involves chaperones- the Heat Shock protein isoforms (De Maio 1999). Petersen and Mitchell (1991) report that irrespective of species, developmental defects may also arise from heat shocks during gastrulation periods. In mammalian embryogenesis, upheaval of thermal stability during gastrulation and subsequent organogenesis such as formation of forebrain, midbrain and hindbrain, central nervous system development, causes invariable prenatal growth retardations (Petersen 1990). Such turbulences can also be fatal to the developing embryo (Edwards 1989). Heat shock induces up-regulated expressions of heat shock proteins which impart "thermotolerance" to growing embryonic cells or otherwise also causing rescue of cells from lethality (Heikkila et al. 1991). Correlation between HSP expression and thermotolerance has been studied over the years. Experimental evidences report influence of specific HSP expressions on the length of cell cycle stages (Milarski et al. 1989; Edwards 1989). At the time of mammalian embryogenesis, arrest of cells in the G1/S & S/G2 phase boundaries rescue cells from heat shock (Hightower and Nover 1991; Walsh and Morris 1989).

Constitutive expressions of HSP71, HSP27, HSP73 and HSP88 at specific cell cycle stages were reported in thermotolerance and heat shock recovery embryo (Walsh et al. 1993). Re-entry to cell cycle can be triggered by altering the levels of HSP27, HSP70 and HSP90. This is ensured by controlling different stages of cell cycle (Sharma et al. 2017). HSP90 expressions is significant to progression of cells through the G0/G1 stage as blocking HSP90 with geldanamycin leads to arrest of cells in the G0 stage by recruitment of HSP90-specific client proteins cdc37 and FKBP52 (Miyata 2005; Kwon et al. 2014). Blocking HSP90 through anti-sense RNA showed retarded rate of cell division in human monoblastoid cells U937, whereas overexpression of HSP90 leads to increased appearance of S-phase cells (Galea-Lauri et al. 1996). Cell cycle regulatory network has heat shock protein chaperons controlling them. They report association, although transiently, with the key factors of cell cycle control factors like Cdk4, Wee-1, pRb, p53, p27/Kip1 and also participate in the nuclear localization of crucial regulatory proteins (Helmbrecht et al. 2000). Overexpression of HSP90 is reported following SV40, polyoma virus and adenovirus infections in eukaryotes (Khandijian and Tuèrler 1983; Simon et al. 1987). In UV irradiated fibroblast cells, recovery of DNA replication was found to associate with a complex structure between Topoisomerase class II, repair enzyme MutL and HSP90 (Stock 1999).

The homologue of yeast cdc37 cell cycle control protein in mammals is a hetero-complex containing Src, Raf and HSP90 (Hutchison et al. 1992; Stancato et al. 1993; Wartmann and Davis 1994; Silverstein et al. 1998). During cell division cycle Cdc37 is coexpressed with cyclin D1 (Stepanova et al. 1996). Disruption of Cdc37/ HSP90 with Cdk4 complex showed decreased half-life of newly synthesized Cdk4, advocating the significance of Cdc37/ HSP90 in stabilizing Cdk4 (Dai et al. 1996; Stepanova et al. 1996; Perdew et al. 1997).

HSP90 interacts with another cell cycle regulatory protein Wee1 (Aligue et al. 1994). Wee1 regulates inhibitory phosphorylation of Cdk1 thereby controlling the length of G2 phase of the cell cycle (Karlseder et al. 1996; Endicott et al. 1999). A temperature sensitive mutant study of HSP90 homologue SWO mutants confirmed interaction of HSP90 with Wee1 is imperative for active Wee1 kinase formation (Helmbrecht et al. 2000).

There are also reports of p53 sequestation in the cytosol by HSP90 via formation of a complex between the two (Blagosklonny et al. 1995). Disruption of this complex leads to prevention of nuclear translocation and proteasomal degradation of p53 thereon (Dasgupta and Momand 1997; Whitesell et al. 1997, 1998). In short the complex formation of HSP90 and other chaperone proteins with p53 (Wild type) is consequential for its ability to manifest or restore transcriptional activity and induce cell cycle arrest. Therefore, HSP remain at the crux of not only counteracting cellular stress levels be it heat shock or oxidative stress through stabilizing transcriptional and translational proteins but also by undoing the effect of stress on regulatory cell cycle proteins as well as on the basic cell cycle sustaining processes.

5.3 HSP90 in Cancer

HSP90 family members are probably the best-studied among all the HSP proteins. HSP90 plays a vital role in cancer development by regulating tumor growth, adhesion, invasion, metastasis, angiogenesis, and apoptosis (Wu et al. 2017). HSP90 is thought to activate certain kinases like protein tyrosine kinase 2 (PTK2), Integrin linked kinase (ILK) and thereby promoting cancer development (Wu et al. 2017). PTK2 is a tyrosine kinase usually localised in the cytosol and involved in cellular adhesion and metastasis, whose level of expression is altered in relation to tumor malignancy (Ochel et al. 1999). Studies have shown that the benzoquinone ansamycin geldanamycin (GA), stimulates the proteolytic degradation of PTK2 in a variety of cancer cells by interfering with the activity of HSP90 suggesting that inhibition of HSP90 leads to degradation of PTK2 thereby decreasing malignancy of cancer cells (Ochel et al. 1999; Wu et al. 2017).

Integrin-linked kinase (ILK) has also been identified as a client protein of HSP90. ILK is involved in integrin-mediated signal transduction and has also been thought to play a vital role in malignant transformation of various cell types. It has been reported that specific HSP90 inhibitors may decrease ILK expression in various cancer cell types and may decrease tumor progression (Aoyagi et al. 2005). It has also been established that CD24 is a client protein of HSP90. CD24 induces angiogenesis in a HSP90-dependent manner and activates STAT3-mediated transcription of VEGF. It is perceived that HSP90 inhibitors decrease angiogenesis by intervening with VEGF receptor expression in human colorectal cancers (Wang et al. 2016). HSP90 inhibitors can also inhibit HIF-1α and NF-κB causing inhibition of epithelial to mesenchymal transition, invasion, and motility of colorectal cancer cells (Nagaraju et al. 2015; Wu et al. 2017).

HSP90 over-expression has been reported in various types of cancers like breast, lung, pancreatic, endometrial, ovarian cancers etc. (Wu et al. 2017). Hence, HSP90 could be used as a potential therapeutic target to impede tumor progression. Many of the HSP90 inhibitors like geldanamycin and tanespimycin offer several advantages over other anti-cancer drugs. A majority of signalling proteins are clients of HSP90. So, HSP90 inhibitors can simultaneously target multiple signalling pathways thereby decreasing the chances of tumor cells' survival. Secondly, HSP90 inhibitors can facilitate specific damage to tumor tissues sparing the normal cells. HSP90 inhibitors also lead to the accumulation of insoluble unfolded proteins which are toxic to cancer cells (Wu et al. 2017).

5.3.1 HSP90 in Breast Cancer

Cancer being the most life threatening disease of the modern world, statistics claim that breast cancer accounts for around 30% female mortality worldwide. Making it only second to lung cancer as a major cause of death in women worldwide (Kakarala and Wicha 2008). Breast cancer is a heterogenous disease, with a high rate of recurrence (Perou et al. 2000). There are several subtypes to the disease based on the expression status of receptors viz. estrogen receptors (ER), epidermal growth factor 2 (HER2 or ErbB2) and progesterone receptor (PR) (Song et al. 2010; Cancer Genome Atlas 2012). Obvious onus is held upon deregulation of cell-survival signalling pathways as the contributing factor towards metastatic competence, atypical proliferation and also therapeutic resistance of breast cancer cells (Pick et al. 2007). In this contest higher expression status of HSP90 has shown to correlate with adverse clinical outcome in breast cancer cases (Eccles et al. 2008). Furthermore, inhibition of HSP90 also leads to modulation of angiogenesis across tumour cells causing loss of their metastatic property (Whitesell and Lindquist 2005). It is a well-known fact that HSP90 is up-regulated in breast cancer cells compared to normal ones and therefore is believed to serve a potential role in tumorigenesis.

During oncogenic progression the normal chaperone function of HSP90 gets debauched to facilitate the massive chaperone support demanded by cancer cells.

Oncogenesis causes constant stress in cancer cells due to aberrant misfolded oncoprotein load, high ROS (Reactive Oxygen species) levels, hypoxia, acidosis and so on which require constant HSP support so that the misfolded proteins do not aggregate inside cells leading to tumor growth arrest (Taipale et al. 2010; Trepel et al. 2010; Owens et al. 2012). Major client proteins of HSP90 include Bcr-Abl oncoprotein, serine-threonine protein kinase (AKT), rapidly accelerated fibrosarcoma protein (RAF-1), steroid receptor co-activator (SRC), hypoxia-inducible factor 1-alpha (HIF-1α)/ signal transducer and activator of transcription 3 (STAT3) in addition to p53 (cellular tumor antigen), BRAF, casein kinase (CK2), checkpoint kinase (Chk4), mutated p53, cyclin-dependent kinase-5 (Cdk5), protein kinase C-alpha (PKCα), PKCβ, mitogen-activated protein kinase (MAPK1)/ extracellular-signal-regulated kinase (ERK2), glycogen synthase kinase 3- beta (GSK3β), Janus kinase (JAK1), nuclear factor κ B (NKκB), c-Raf, epidermal growth factor receptor (ErbB1), ErbB2/HER2, fms-like tyrosine kinase 3 (FLT3) (Slamon et al. 1987; Mimnaugh et al. 1996; Basso et al. 2002a; Chen et al. 2002; Neckers and Ivy 2003; Wagner and Nebreda 2009).

Association of human epidermal growth factor receptor (HER2), a cell receptor tyrosine kinase amplification with aggressive clinical outcome and poor prognosis of about one fourth of breast cancer cases has been reported (Paik et al. 1990; Münster et al. 2001). It is a well documented fact that inhibition of HSP90 degrades HER2 which is a client protein of HSP90 (Trepel et al. 2010). Recently it is reported that cancer cells harbor positive correlation between HER2 and MIF (Macrophage migration inhibitory factor). MIF is a tumor-promoting protein and also a client protein of HSP90. Cancer cells respond to proteotoxic stress by constitutive expression of HSF-1 (heat shock factor-1) (Mendillo et al. 2012). HSF-1 is the major transcriptional regulator of the heat shock proteins. Evidences advocate that in cases of breast cancer with overexpressed HER2 kinase, the HSF1-HSP90 axis is also constitutively activated leading to stabilization of the major cell proliferation protein clients of HSP90 such as mTOR (Mammalian target of rapamycin), MIF, AKT and also HSF-1 itself (Xing et al. 2008; Yarden and Pines 2012). In HER-2 non-expressing cells, Ras-MEK-ERK pathway is activated in concert with the PI3K-AKT-mTOR axis (Pyle et al. 2003).

Stabilized macrophage migration inhibitory factor (MIF) has been observed in breast cancer cells by HSP90 (Schulz et al. 2012a). Stabilization of aberrant MIF is shown to have a pleotropic action in several pathways which is crucial to its role as a tumor promoter. HSP90 inhibition with 17AAG showed a strong inhibition in breast cancer cell growth due to degradation of the MIF protein following HSP90 inhibition (Schulz et al. 2012b). Direct destabilization of HSP90 could therefore be a potential strategy to eliminate HER-2 negative breast cancer population (Vivanco and Sawyers 2002).

Another significant signal transduction component is AKT which resides at the center of cell proliferation and survival signalling circuit. AKT is abnormally or aberrantly activated in several malignancies, particularly mammalian breast carcinomas (Basso et al. 2002a). AKT over expression is also reported to be the cause of chemotherapeutic resistance in breast cancers, apart from being the main reason behind poor prognosis and also the contributing factor behind the high invasive and angiogenic potentiality of breast carcinomas (Basso et al. 2002a). Reports suggest that activating phosphorylations of AKT by upstream PDK1 rests upon the active conformation of HSP90-cdc37 complex (Basso et al. 2002b), as AKT's conformational stability and therefore functional efficacy depends on its interactions with HSP90 (Sato et al. 2000; Soga et al. 2001). This is confirmed by treatment of breast cancer cells with HSP90 inhibitors such as geldanamycin and radicicol. This leads to the abrogation of pan AKT protein expression levels as well as active phosphorylated form of AKT (Soga et al. 2001; Cohen et al. 2003).

In breast tumors, Ras/Raf/MEK pathway is also pivotal to cell survival and proliferation. In most human tumors, especially mammary tumors, B-Raf (isoform of Raf) is mutated (Davies et al. 2002; Wilhelm et al. 2003). Statistically around 7% of total reported cases express this mutation type (Wilhelm et al. 2003). Raf along with its downstream effectors MAPK and MEK (MAPK/Erk kinase) are the clients of HSP90. Raf-1 and mutated B-Raf therefore appear to make the most putative targets for therapeutic approach to breast cancer, however Raf is not the most sensitive client of HSP90. Hence the ideal scenario would be to combine the two drugs to achieve profound repression of Raf and at the same time suppress other activated oncogenic proteins that fall in HSP90 client range.

NF-kB is a crucial cell signalling pathway which regulates several cellular immune response signal depending on the type of stimulus received (Karin and Ben-Neriah 2000). NF-kB mediates cellular inflammatory responses within the cell and also supports oncogenic transformation as well as progression. NF-kB is a nuclear transcriptional regulator which gets activated upon being dissociated from the upstream inhibitory protein IKB (Karin and Delhase 2000). IKB kinase (IKK) phosphorylates the IKB protein leading to its degradation. HSP90 and cdc37 forms complex with IKK, in order to stabilize it and this association is also necessary for proper recruitment of IKK to TNF receptors upon receiving stimulus from the cell exterior (Chen et al. 2002). In triple negative breast cancer (TNBC) cases, aberrant activation of NF-kB is a common and frequent observation (Yamaguchi et al. 2009). A very recent work has identified yet another gene CHORDC1 whose protein Morgana/Chp-1 is reported to be overexpressed in breast cancer particularly the TNBC type (Fusella et al. 2017). Morgana is shown to stabilize and propagate the NF-kB pathway, resist apoptosis and confer oncogenesis progression as well as resistance (Fusella et al. 2014). Morgana binds to HSP90 chaperone protein which is needed for its stability (Gano and Simon 2010; Hong et al. 2013).

JAK-STAT is one of the most important signal transduction pathways, which transcends cellular protein expressions after receiving stimulations from the cytokines and growth factors. Constitutively activated STATs are reported to be primordial for most human malignancies including breast cancers and their overexpression leads to oncogenesis (Bowman et al. 2000). HSP90 plays a crucial role in STAT activation; STAT3 & STAT5 are clients of HSP90 (Epling-Burnette et al. 2001). Treatment of breast cancer cells with geldanamycin, which is a HSP90 inhibitor, showed abrogation of STAT3 protein levels and further induced apoptosis in breast cancer cells (Sato et al. 2003). In another work, TNBC tumorigenesis and chemoresistance is depicted to be an outcome of a concerted involvement of the JAK/STAT signalling pathway and AR (Androgen Receptor) with HSP90 (Agyeman et al. 2016).

WNT pathway is aberrantly activated in more than 60% of breast cancers as evidenced by the cytoplasmic/Nuclear accumulation of β-catenin (Lin et al. 2000; Nakopoulou et al. 2006). Apart from mutations, altercation in the WNT pathway is reported to occur by several other mechanisms such as WNT over-expression, over-expressed phosphorylated Dsh1, decreased SFRP1 levels (Nagahata et al. 2003; Milovanovic et al. 2004; Veeck et al. 2006; Veeck et al. 2008; Collu et al. 2009). MDA-MB-231 cells treated with sFRP1 decreased breast cancer cell mobility, presenting an obvious correlation or role of WNT signalling pathway to breast cancer cell motility (Schlange et al. 2007). β-catenin-TCF association increases cellular transcriptional activity leading to over-expressed Cyclin D1 levels in majority of breast cancer patients (Kelleher et al. 2006). Up-regulation of WNT5A results from STAT3 phosphorylation and activation. The former is the regulator of canonical WNT signalling activation (Darnell et al. 1994; Katoh and Katoh 2007). In breast cancer cases, WNT5A is reported to be highly up-regulated (Iozzo et al. 1995). HSP90 interaction with STAT3 thereby indirectly regulates WNT pathway activation. This is confirmed from the data where blocking of HSP90 resulted in inhibition of WNT5A transcription, an important factor for breast cancer proliferation and tumor growth (Lang et al. 2007; Katoh and Katoh 2007). Also PI3K/AKT signalling pathway is reported to link WNT signalling with HSP90 ATPase activity (Kurashina et al. 2009; Nakayama et al. 2009).

5.3.2 HSP90 in Canine Mammary Tumors

Mammary neoplasia are not only found in human beings, domestic pets are also at an equivalent risk. These neoplasms are most frequent in unsprayed female dogs (Sunil Kumar et al. 2013; Pandey et al. 2015; Sunil Kumar et al. 2017). These tumors share a high degree of similarity in terms of histological patterns with human breast cancer. So, various groups of workers have referred to canine mammary tumors as a model to investigate human breast cancer (Queiroga et al. 2011; Pandey et al. 2015). Various proteins like matrix metalloproteinases (MMP), Mammaglobin, Her2 etc. have been reported to be upregulated at the onset of mammary tumors in dogs (Sunil Kumar et al. 2013; Pandey et al. 2015), which can be used as diagnostic markers of the malady.

Several HSP are also upregulated during malignant transformation of mammary glands in pets. HSP90 expression has been correlated with higher degrees of mammary gland malignancies in dogs and the expression has been found to be highest in solid and complex carcinomas (Badowska-Kozakiewicz and Malicka 2012). HSP90α is located both in intracellular as well as extracellular milieu. The intracellular HSP90α is responsible for chaperoning activity while the extracellular isoform contributes invasiveness to the tumor cells which finally leads to metastasis of primary tumors to distant locations (McCready et al. 2010). In fact, the invasive cancer cells secrete HSP90α to extracellular space via exosomes which is believed to activate MMP2, which facilitates degradation of extracellular matrix leading to metastasis. The extracellular HSP90α also activates plasminogen to plasmin which contributes to cellular migration and invasion (McCready et al. 2010).

Survivin, a member of the inhibitor of apoptosis protein (IAP) family, has also been implicated in both human breast cancer (Jha et al. 2012) and canine mammary tumors (Bongiovanni et al. 2015). This protein negatively regulates cell cycle and opposes apoptosis by inhibiting caspase activation. It is expressed during foetal development and cancers but is absent or expressed at low levels in healthy adult tissues (Altieri 2008). Survivin is one of the client proteins of HSP90 and HSP90 inhibitors target survivin expression by intervening with their interaction which further leads to the latter's misfolding, ubiquitination, and proteasome degradation (Cheung et al. 2010). Although, Bongiovanni et al. (2015) reported no significant correlation between survivin and HSP90 immunolabelling in canine mammary tumors, but the research group has also defended that the lack of correlation might be attributed to the existence of various isoforms of survivin which were not distinguished in their study.

Signal Transducers and Activators of Transcription (STAT) are transcription factors that are usually localised in the cytoplasm but get translocated to the nucleus to regulate gene expression. Activated STAT3 modulates HSP90 promoter activity thereby modulating HSP90 gene expression in response to non-stressful stimuli (Stephanou and Latchman 1999; Petterino et al. 2001). STAT3 expression has been associated with malignant transformation of mammary glands both in women (Garcia et al. 1997) and canines (Petterino et al. 2001) through activation of specific kinase signalling pathways or through involvement in the G1 to S phase cell-cycle transition. The activation and upregulation of STAT3 in canine mammary tumors may also account for the over-expression of HSP90 during malignant transformation of canine mammary glands.

Evasion of apoptosis is one of the trademarks of tumorigenesis. Immortal cells sustain genetic mutations to a greater extent in order to circumvent programmed cell death. Evasion of apoptosis also facilitates metastasis, promotes resistance to immune-based cancer therapy and chemotherapy (Reed 1999; Kumaraguruparan et al. 2006). HSP90 is believed contribute towards mammary gland malignancy by opposing apoptosis. In the past, expression HSP90 has been associated with upregulation of anti-apoptotic proteins, Bcl-2, Bcl-XL, and down-regulation of pro-apoptotic proteins like Bax and caspases in malignant canine mammary tumors (Kumaraguruparan et al. 2006). Bcl2 prevents mitochondrial release of cytochrome c and subsequent activation of caspases thereby arbitrating its anti-apoptotic effects, while, Bcl-XL interacts with Apaf-l and blocks procaspase 9 activation to caspases 9, preventing apoptosis (Petros et al. 2004). It has been concluded that up-regulated HSP90 inhibits crucial steps in apoptotic cascade such as apoptosome formation and Fas signalling and may facilitate development of the malignant phenotype of canine mammary glands (Sreedhar and Csermely 2004; Kumaraguruparan et al. 2006).

There are various reports of HSP90α and HSP90β secretion into circulation during carcinogenesis which has important roles in regulating cancer cell migration and invasion (Brooks et al. 2010; Sims et al. 2011; Rong et al. 2014). We have for the first time reported overexpression and secretion of ER paralog of HSP90 (HSP90B1, GRP94) in the serum of dogs bearing different types of mammary tumors (Sunil Kumar et al. 2018). Stressors like low glucose and hypoxia commonly encountered during malignant transformation of mammary cells induce upregulation of GRP94 which facilitates cell survival by stabilizing and refolding denatured proteins and aiding in their escape from apoptosis (Cawthorn et al. 2012). GRP94 though usually localised in the endoplasmic reticulum of the cells (involved in the maintenance of ER integrity and homeostasis), gets translocated to other cellular compartments and even gets secreted to extracellular spaces during malignancies, and may contribute to cell proliferation, invasion, and evasion of apoptosis (Sunil Kumar et al. 2018; Lee 2014). Reactive oxygen species, hypoxia, injury-released growth factors etc. are the predisposing factors leading to extracellular GRP94 secretion (Li et al. 2012). We have recently reported that a high serum concentration of Grp94 is correlated with malignant transformation of canine mammary glands (Sunil Kumar et al. 2018). We have also observed an overexpression of this isoform of HSP90 in various histotypes of canine mammary tumors by immunohistochemistry (Fig. 5.1a–d). Irrespective of the histological types and grades, Grp94 was overexpressed in all the

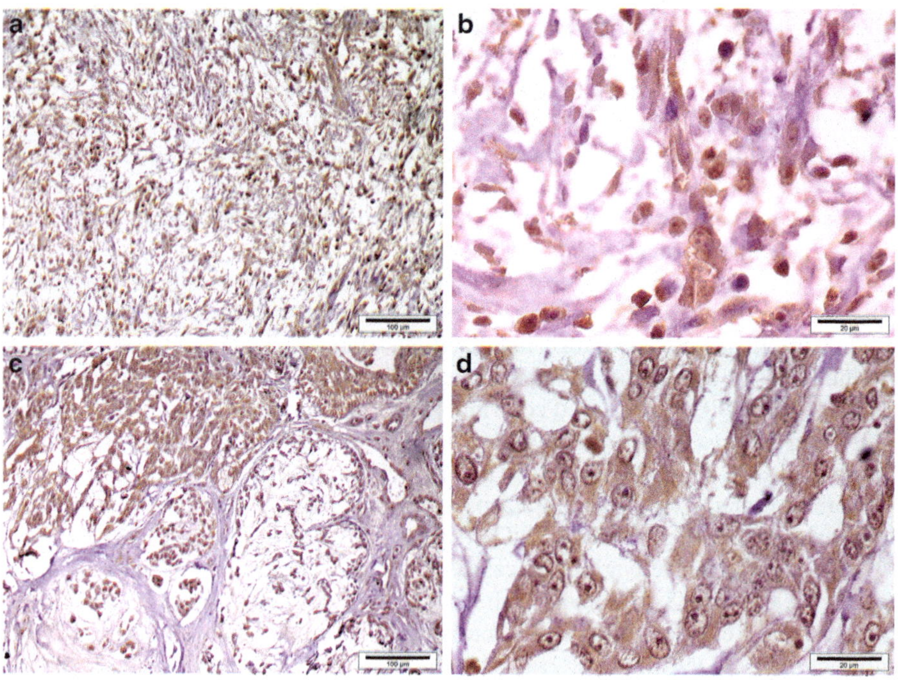

Fig. 5.1 (**a**) Immunohistochemistry of canine mammary tumor (Complex carcinoma showing HSP90B1 immuno-positive cells). Bar = 100 µm. (**b**) Immunohistochemistry of canine mammary tumor (Complex carcinoma showing HSP90B1 immuno-positive cells). Bar = 20 µm. (**c**) Immunohistochemistry of canine mammary tumor (Carcinosarcoma showing HSP90B1 immuno-positive cells). Bar = 100 µm. (**d**) Immunohistochemistry of canine mammary tumor (Carcinosarcoma showing HSP90B1 immuno-positive cells). Bar = 20 µm

types of canine mammary tumors studied. Adding this protein to the current panel of diagnostic biomarkers will certainly improve the sensitivity and specificity of canine mammary tumor diagnosis and prognosis.

5.3.3 HSP90 Inhibitors against Mammary Tumors

Interaction of HSP90 with its client proteins is essential for bringing about a variety of cellular functions like cell cycle regulation, proliferative/anti-apoptotic signalling etc. These client proteins include tyrosine kinases (e.g. Akt and MEK), transcription factors, structural proteins like tubulin, actin etc. (Den and Lu 2012; Goetz et al. 2003). Many of these client proteins and HSP90-client protein complexes are activated in malignancies and lead to aberrant cellular activities which may lead to uncontrolled cell proliferation and resistance to apoptosis. Hence, targeting HSP90 as a part of cancer therapy has recently gained interest among various researchers. Pharmacological inhibition of HSP90 undermines various clients/oncoproteins involved in malignant processes leading to their degradation through the proteasome assisted pathway, thus, HSP90 inhibitors may exert antitumor effects (Den and Lu 2012; Sepp-Lorenzino et al. 1995). Estrogen and progesterone receptors are clients of HSP90, thus HSP90 inhibitors may inhibit signalling of these steroid receptors thereby exhibiting anti-cancer effects (Zagouri et al. 2012; Bagatell et al. 2001).

HSP90 is essential for epidermal growth factor receptor-2 (EGFR2 or HER2) signalling. Inhibition of HSP90 also leads to degradation of HER2 in HER2-positive breast cancer cells (Workman et al. 2007). Triple negative breast cancer (TNBC) subjects lack expression of estrogen receptor (ER), progesterone receptor (PR) and HER2 receptor and have poor prognosis than other types of breast cancer. TNBC subjects have also been found to respond to Hsp90 inhibition through down-regulation of components of the Ras/Raf/MARK pathway (Caldas-Lopes et al. 2009). A number of natural and synthetic inhibitors of HSP90 have been studied till date and have been associated with inhibition of mammary neoplasms, some of them have been discussed in the following paragraphs.

5.3.3.1 Geldanamycin and Radiciol

These are structurally unrelated molecules and target N-terminal domain of HSP90 which is usually a nucleotide binding site. These molecules have higher binding affinity to the nucleotide binding site than either ATP or ADP. These molecules direct client/oncoprotein degradation via proteasome pathway. However, they were not encouraged in clinical trials owing to their poor in-vivo effects, higher toxicity and difficulties in production (Roe et al. 1999).

5.3.3.2 Geldanamycin Analogues

17 allylamino-17 demethoxygeldanamycin (17-AAG; tanespimycin) and 17 dimethylaminoethylamino-17-demethoxygeldanamycin (17-DMAG; alvespimycin) are geldanamycin analogues having less toxicity as compared to natural geldanamycin. 17-AAG specifically inhibits HSP90 in transformed cell with high affinity and is the first HSP90 inhibitor tested in clinical trials (Kamal et al. 2003). This semi-synthetic HSP90 derivative deregulates the client proteins like hsc, keratin 8, keratin 18, akt, c-raf1 and caveolin-1, thereby inhibiting signal transduction (Clarke et al. 2000). 17-DMAG on the other hand, is water soluble and more potent as compared to 17-AAG, However, its toxicity barred further clinical testing (Den and Lu 2012).

5.3.3.3 Synthetic Radicicol Derivatives

These synthetic analogues bear modifications of multiple electrophilic sites which in turn have improved its *in vivo* activity and are effective at minimum doses (Yang et al. 2004). Compounds like STA-9090, NVP-AUY922, KW-2478, and AT-13387, have entered the clinical trials (Lin et al. 2008).

5.3.3.4 Other Synthetic HSP90 Inhibitors

Compounds like retaspimycin HCL, NVP-AUY922, NVP-BEP800, CNF2024/BIIB021, SNX-5422, STA-9090 etc. have shown promising results with better pharmacological and toxicological properties to inhibit HSP90 and treat breast cancer (Zagouri et al. 2012). CNF2024/BIIB021 is the first synthetic HSP90 inhibitor to enter clinical trials. CNF2024/BIIB021 alone or in combination with trastuzumab has shown impressive results to cure HER2 positive breast cancer (Zagouri et al. 2012).

5.4 Conclusions

In conclusion, it could be assumed that HSP90 has a significant role in malignant transformation of mammary glands both in women and pets. It binds to a portfolio of client proteins, mostly those involved in signal transduction. Interaction of HSP90 with its client proteins like tyrosine kinases (e.g. Akt and MEK), transcription factors, structural proteins like tubulin, actin, ErbB etc. is essential for bringing about a variety of cellular functions like cell cycle regulation, proliferative/anti-apoptotic signalling etc. Many of these client proteins and HSP90-client protein complexes are activated in breast malignancies and lead to aberrant cellular activities which may lead to uncontrolled cell proliferation and resistance to apoptosis. Moreover, it is believed that breast cancer cells become resistant to various anti-tumor therapies through HSP90 overexpression. It could be postulated from the above observation that pharmacological HSP90 inhibitors could provide therapeutic modalities in the field of breast cancer treatment. Some of the HSP90 inhibitors have shown impressive results in preclinical studies. However, prospective clinical studies are required to evaluate the role of HSP90 inhibition to treat various subtypes of breast cancer.

Acknowledgements the authors acknowledge Department of Biotechnology, Ministry of Science and Technology (Grant ID: BT/PR6434/GBD/27/414/2012) for carrying out research on diagnostic implication of Hsp90 on mammary tumors.

References

Agyeman AS, Jun WJ, Conzen SD (2016) HSP90 inhibition results in GR, AR and JAK protein degradation, decreased triple-negative breast cancer proliferation and increased paclitaxel sensitivity [abstract]. In: Proceedings of the 107th Annual Meeting of the American Association for Cancer Research, 2016 Apr 16–20, New Orleans, LA. Philadelphia (PA): AACR. Cancer Res 76(14 Suppl):3100. Abstract nr 3100

Aligue R, Akhaven-Niak H, Russell P (1994) A role for HSP90 in cell cycle control: wee 1 tyrosine kinase activity requires interaction with HSP90. EMBO J 13(24):6099–6106

Altieri DC (2008) Survivin, cancer networks and pathway-directed drug discovery. Nat Rev Cancer 8(1):61–70

Aoyagi Y, Fujita N, Tsuruo T (2005) Stabilization of integrin-linked kinase by binding to Hsp90. Biochem Biophys Res Commun 331:1061–1068

Badowska-Kozakiewicz AM, Malicka E (2012) Immunohistochemical evaluation of expression of heat shock proteins HSP70 and HSP90 in mammary gland neoplasms in bitches. Pol J Vet Sci 15(2):209–214

Bagatell R, Khan O, Paine-Murrieta G, Taylor CW, Akinaga S, Whitesell L (2001) Destabilization of steroid receptors by heat shock protein 90-binding drugs: a ligand-independent approach to hormonal therapy of breast cancer. Clin Cancer Res 7:2076–2084

Basso AD, Solit DB, Chiosis G, Giri B, Tsichlis P, Rosen N (2002a) Akt forms an intracellular complex with heat shock protein 90 (HSP90) and Cdc37 and is destabilized by inhibitors of HSP90 function. J Biol Chem 277(42):39858–39866

Basso AD, Solit DB, Munster PN, Rosen N (2002b) Ansamycin antibiotics inhibit Akt activation and cyclin D expression in breast cancer cells that overexpress HER2. Oncogene 21:1159–1166

Blagosklonny MV, Toretsky J, Neckers L (1995) Geldanamycin selectively destabilizes conformationally altered mutated p53. Oncogene 11(5):933–939

Bongiovanni L, Romanucci M, Malatesta D, D'Andrea A, Ciccarelli A, Della Salda L (2015) Survivin and related proteins in canine mammary tumors: immunohistochemical expression. Vet Pathol 52(2):269–275

Bowman T, Garcia R, Turkson J, Jove R (2000) STATs in oncogenesis. Oncogene 19:2474–2488

Brooks SA, Lomax-Browne HJ, Carter TM, Kinch CE, Hall DM (2010) Molecular interactions in cancer cell metastasis. Acta Histochem 112:3–25

Bukau B, Weissman J, Horwich A (2006) Molecular chaperones and protein quality control. Cell 125:443–451

Caldas-Lopes E, Cerchietti L, Ahn JH, Clement CC, Robles AI, Rodina A, Moulick K, Taldone T, Gozman A, Guo Y, Wu N, de Stanchina E, White J, Gross SS, Ma Y, Varticovski L, Melnick A, Chiosis G (2009) Hsp90 inhibitor PU-H71, a multimodal inhibitor of malignancy, induces complete responses in triple-negative breast cancer models. Proc Natl Acad Sci U S A 106:8368–8373

Calderwood SK (2010) Heat shock proteins in breast cancer progression—a suitable case for treatment? Int J Hyperth 26:681–685

Cambiazo V, Gonzalez M, Isamit C, Maccioni RB (1999) The β-isoform of heat shock protein hsp-90 is structurally related with human microtubule-interacting protein Mip-90. FEBS Lett 457:343–347

Cancer Genome Atlas N (2012) Comprehensive molecular portraits of human breast tumours. Nature 490(7418):61–70

Cawthorn TR, Moreno JC, Dharsee M, Tran-Thanh D, Ackloo S, Zhu PH, Sardana G, Chen J, Kupchak P, Jacks LM, Miller NA, Youngson BJ, Iakovlev V, Guidos CJ, Vallis KA, Evans KR, McCready D, Leong WL, Done SJ (2012) Proteomic analyses reveal high expression of decorin and endoplasmin (HSP90B1) are associated with breast cancer metastasis and decreased survival. PLoS One 7(2):e30992

Chen G, Cao P, Goeddel DV (2002) TNF-induced recruitment and activation of the IKK complex require Cdc37 and HSP90. Mol Cell 9:401–410

Cheung CH, Chen HH, Cheng LT, Lyu KW, Kanwar JR, Chang JY (2010) Targeting Hsp90 with small molecule inhibitors induces the over-expression of the anti-apoptotic molecule, survivin, in human A549, HONE-1 and HT-29 cancer cells. Mol Cancer 9:77

Ciocca DR, Calderwood SK (2005) Heat shock proteins in cancer: diagnostic, prognostic, predictive and treatment implications. Cell Stress Chap 10(2):86–103

Clarke PA, Hostein I, Banerji U, Stefano FD, Maloney A, Walton M, Judson I, Workman P (2000) Gene expression profiling of human colon cancer cells following inhibition of signal transduction by 17-allylamino-17-demethoxygeldanamycin, an inhibitor of the hsp90 molecular chaperone. Oncogene 19:4125–4133

Cohen Y, Xing M, Mambo E, Guo Z, Wu G, Trink B, Beller U, Westra WH, Ladenson PW, Sidransky D (2003) BRAF mutation in papillary thyroid carcinoma. J Natl Cancer Inst 95(8):625–627

Collu GM, Meurette M, Brennan K (2009) Is there more to Wnt signaling in breast cancer than stabilisation of β catenin? Breast Cancer Res 11:105

Dai K, Kobayashi R, Beach D (1996) Physical interaction of mammalian CDC37 with CDK4. J Biol Chem 271(36):22030–22034

Dalerba P, Cho RW, Clarke MF (2007) Cancer stem cells: models and concepts. Ann Rev Med 58:267–284

Darnell JE, Kerr IM, Stark GR (1994) JAK-STAT pathways and transcriptional activation in response to IFNs and other extracellular signaling proteins. Science 264:1415–1421

Dasgupta G, Momand J (1997) Geldanamycin prevents nuclear translocation of mutant p53. Exp Cell Res 237(1):29–37

Davies H, Bignell GR, Cox C, Stephens P, Edkins S, Clegg S, Teague J, Woffendin H, Garnett MJ, Bottomley W, Davis N, Dicks E, Ewing R, Floyd Y, Gray K, Hall S, Hawes R, Hughes J, Kosmidou V, Menzies A, Mould C, Parker A, Stevens C, Watt S, Hooper S, Wilson R, Jayatilake H, Gusterson BA, Cooper C, Shipley J, Hargrave D, Pritchard-Jones K, Maitland N, Chenevix-Trench G, Riggins GJ, Bigner DD, Palmieri G, Cossu A, Flanagan A, Nicholson A, Ho JW, Leung SY, Yuen ST, Weber BL, Seigler HF, Darrow TL, Paterson H, Marais R, Marshall CJ, Wooster R, Stratton MR, Futreal PA (2002) Mutations of the BRAF gene in human cancer. Nature 417:949–954

De Maio A (1999) Heat shock proteins: facts, thoughts dreams. Shock 11(1):1–12

De Maio A (2011) Extracellular heat shock proteins, cellular export vesicles, and the stress observation system: a form of communication during injury, infection, and cell damage. Cell Stress Chap 16:235–249

Den RB, Lu B (2012) Heat shock protein 90 inhibition: rationale and clinical potential. Ther Adv Med Oncol 4(4):211–218

Eccles SA, Massey A, Raynaud FI, Sharp SY, Box G, Valenti M, Patterson L, de Haven Brandon A, Gowan S, Boxall F, Aherne W, Rowlands M, Hayes A, Martins V, Urban F et al (2008) NVP-AUY922: a novel heat shock protein 90 inhibitor active against xenograft tumor growth, angiogenesis, and metastasis. Cancer Res 68(8):2850–2860

Edwards MJ (1989) Hyperthermia and the developing central nervous system. Ann Res Inst Environ Med Nagoya Univ 40:355–364

Endicott JA, Noble ME, Tucker JA (1999) Cyclin-dependent kinases: inhibition and substrate recognition. Curr Opin Struct Biol 9:738–744

Epling-Burnette PK, Liu JH, Catlett-Falcone R, Turkson J, Oshiro M, Kothapalli R, Li Y, Wang JM, Yang-Yen HF, Karras J et al (2001) Inhibition of STAT3 signaling leads to apoptosis of leukemic large granular lymphocytes and decreased Mcl-1 expression. J Clin Invest 107:351–362

Evan GI, Vousden KH (2001) Proliferation, cell cycle and apoptosis in cancer. Nature 411:342–348

Fabian A, Barok M, Vereb G, Szollosi J (2009) Die hard: are cancer stem cells the Bruce Willises of tumor biology. Cytometry Part A: J Int Soc Anal Cytol 75:67–74

Fusella F, Ferretti R, Recupero D, Rocca S, Di Savino A, Tornillo G, Silengo L, Turco E, Cabodi S, Provero P, Pandolfi PP, Sapino A, Tarone G, Brancaccio M (2014) Morgana acts as a proto-oncogene through inhibition of a ROCK-PTEN pathway. J Pathol 234:152–163

Fusella F, Seclì L, Busso E, Krepelova A et al (2017) The IKK/NF-κB signaling pathway requires morgana to drive breast cancer metastasis. Nat Commun 8(1):1636–1642

Galea-Lauri J, Latchman DS, Katz DR (1996) The role of the 90-kDa heat shock protein in cell cycle control and differentiation of the monoblastoid cell line U937. Exp Cell Res 226:243–254

Gano JJ, Simon JA (2010) A proteomic investigation of ligand-dependent HSP90 complexes reveals CHORDC1 as a novel ADP-dependent HSP90-interacting protein. Mol Cell Proteomics 9:255–270

Garcia R, Yu CL, Catlett R, Nelson KL, Smithgall T, Fujita DJ, Ethier SP, Jove R (1997) Constitutive activation of Stat3 in fibroblasts transformed by diverse oncoproteins and in breast carcinoma cells. Cell Growth Differ 8:1267–1276

Giffard R, Han R, Emery J, Duan M, Pittet J (2008) Regulation of apoptotic and inflammatory cell signaling in cerebral ischemia: the complex roles of heat shock protein 70. Anesthesiology 109:339–348

Goetz MP, Toft DO, Ames MM, Erlichman C (2003) The Hsp90 chaperone complex as a novel target for cancer therapy. Ann Oncol 14:1169–1176

Grammatikakis N, Vultur A, Ramana CV, Siganou A, Schweinfest CW, Watson DK, Raptis L (2002) The role of Hsp90N, a new member of the Hsp90 family, in signal transduction and neoplastic transformation. J Biol Chem 277:8312–8320

Gupta RS (1995) Phylogenetic analysis of the 90 kD heat shock family of protein sequences and an examination of the relationship among animals, plants, and fungi species. Mol Biol Evol 12:1063–1073

Hartl FU, Hayer-Hartl M (2009) Converging concepts of protein folding in vitro and in vivo. Nat Struct Mol Biol 16:574–581

Heikkila JJ, Krone PH, Ovsenek N (1991) Regulation of heat shock gene expression during Xenopus. Development 17:120–137

Helmbrecht K, Zeise E, Rensing L (2000) Chaperones in cell cycle regulation and mitogenc signal transduction: a review. Cell Prolif 33:341–365

Hightower L, Nover L (1991) Heat shock and development, vol 17. Springer, Berlin, p 229

Hong TJ, Kim S, Wi AR, Lee P, Kang M, Jeong JH, Hahn JS (2013) Dynamic nucleotide-dependent interactions of cysteine- and histidine-rich domain (CHORD)-containing HSP90 cochaperones Chp-1 and melusin with cochaperones PP5 and Sgt1. J Biol Chem 288(1):215–222

Hutchison KA, Brott BK, De Leon JH, Pedrew GH, Jove R, Pratt WB (1992) Reconstitution of the multiprotein complex of pp 60src, HSP90 and p50 in a cell-free system. J Biol Chem 267(5):2902–2908

Iozzo RV, Eichstetter I, Danielson KG (1995) Aberrant expression of the growth factor Wnt-5A in human malignancy. Cancer Res 55:3495–3499

Jha K, Shukla M, Pandey M (2012) Survivin expression and targeting in breast cancer. Surg Oncol 21(2):125–131

Kakarala M, Wicha MS (2008) Implications of the cancer stem-cell hypothesis for breast cancer prevention and therapy. J Clin Oncol 26:2813–2820

Kamal A, Thao L, Sensintaffar J, Zhang L, Boehm MF, Fritz LC, Burrows FJ (2003) A high-affinity conformation of Hsp90 confers tumour selectivity on Hsp90 inhibitors. Nature 425:407–410
Karin M, Ben-Neriah Y (2000) Phosphorylation meets ubiquitination: the control of NF-[kappa]B activity. Annu Rev Immunol 18:621–663
Karin M, Delhase M (2000) The I kappa B kinase (IKK) and NF-kappa B: key elements of proinflammatory signalling. Semin Immunol 12:85–98
Karlseder J, Wissing D, Holzer G, Orel L, Sliutz G, Auer H, Jäättelä M, Simon MM (1996) HSP70 overexpression mediates the escape of a doxorubicin-induced G2 cell cycle arrest. Biochem Biophys Res Comm 220(1):153–159
Katoh M, Katoh M (2007) STAT3 induced WNT5A signaling loop in embryonic stem cells, adult normal tissues, chronic persistent inflammation, rheumatoid arthritis and cancer (review). Int J Mol Med 19:273–278
Kelleher FC, Fennelly D, Rafferty M (2006) Common critical pathways in embryogenesis and cancer. Acta Oncol 45:375–388
Khandijian EW, Tuèrler H (1983) Simian virus 40 and polyoma virus induce synthesis of heat shock proteins in permissive cells. Mol Cell Biol 3(1):1–8
Kumaraguruparan R, Karunagaran D, Balachandran C, Manohar BM, Nagini S (2006) Of humans and canines: a comparative evaluation of heat shock and apoptosis-associated proteins in mammary tumors. Clin Chim Acta 365:168–176
Kurashina R, Ohyashiki JH, Kobayashi C, Hamamura R, Zhang Y, Hirano T, Ohyashiki K (2009) Anti-proliferative activity of heat shock protein (HSP) 90 inhibitors via beta-catenin/TCF7L2 pathway in adult T cell leukemia cells. Cancer Lett 284(1):62–70
Kwon B, Kumar P, Lee HH, Zeng L, Walsh K, Fu Q, Barakat A, Querfurth HW (2014) Aberrant cell cycle re-entry in human and experimental inclusion body myositis and polymyositis. Hum Mol Genet 23:3681–3694
Lang SA, Moser C, Gaumann A, Klein D, Glockzin G, Popp FC, Dahkle MH, Piso P, Schlitt HJ, Geissler EK, Stoeltzing O (2007) Targeting heat shock protein 90 in pancreatic cancer impairs insulin-like growth factor-1 receptor signaling, disrupting an interleukin 6/ signal transducer and activator of transcription 3/ hypoxia inducible factor 1 (alpha) autocrine loop and reduces orthotopic tumour growth. Clin Cancer Res 13:6459–6468
Lee AS (2014) Glucose regulated proteins in cancer: molecular mechanisms and therapeutic potential. Nat Rev Cancer 14(4):263–276
Li W, Sahu D, Tsen F (2012) Secreted heat shock protein-90 (Hsp90) in wound healing and cancer. Biochim Biophys Acta 1823(3):730–741
Lin SY, Xia W, Wang JC, Kwong KY, Spohn B, Wen Y, Pestell RG, Hung MC (2000) β-catenin, a novel prognostic marker for breast cancer: its roles in cyclin D1 expression and cancer progression. Proc Natl Acad Sci U S A 97:4262–4266
Lin TY, Bear M, Du Z, Foley KP, Ying W, Barsoum J, London C (2008) The novel HSP90 inhibitor STA-9090 exhibits activity against kit-dependent and -independent malignant mast cell tumors. Exp Hematol 36:1266–1277
Martin E (2011) A dictionary of biology, 6th edn. Oxford University Press, Oxford, pp 315–316
McCready J, Sims JD, Chan D, Jay DG (2010) Secretion of extracellular hsp90α via exosomes increases cancer cell motility: a role for plasminogen activation. BMC Cancer 10:294–303
Mendillo ML, Santagata S, Koeva M, Bell GW, Hu R, Tamimi RM, Fraenkel E, Ince TA, Whitesell L, Lindquist S (2012) HSF1 drives a transcriptional program distinct from heat shock to support highly malignant human cancers. Cell 150:549–562
Milarski KL, Welch WJ, Morimoto RI (1989) Cell cycle-dependent association of HSP70 with specific cellular proteins. J Cell Biol 108:413–423
Milovanovic T, Planutis K, Nguyen A, Marsh JL, Lin F, Hope C, Holcombe RF (2004) Expression of Wnt genes and frizzled 1 and 2 receptors in normal breast epithelium and infiltrating breast carcinoma. Int J Oncol 25:1337–1342
Mimnaugh EG, Chavany C, Neckers L (1996) Polyubiquitination and proteasomal degradation of the p185c-erbB-2 receptor protein-tyrosine kinase induced by geldanamycin. J Biol Chem 271(37):22796–22801

Miyata Y (2005) Hsp90 inhibitor geldanamycin and its derivatives as novel cancer chemotherapeutic agents. Curr Pharm Des 11(9):1131–1138

Münster PN, Basso A, Solit D, Norton L, Rosen N (2001) Modulation of HSP90 function by ansamycins sensitizes breast cancer cells to chemotherapy-induced apoptosis in an RB- and schedule-dependent manner. See: E. A. Sausville, Combining cytotoxics and 17-allylamino, 17-demethoxygeldanamycin: sequence and tumor biology matters, Clin. Cancer Res., 7: 2155–2158, 2001. Clin Cancer Res 7(8):2228–22236

Nagahata T, Shimada T, Harada A, Nagai H, Onda M, Yokoyama S, Shiba T, Jin E, Kawanami O, Emi M (2003) Amplification, up-regulation and over-expression of DVL-1, the human counterpart of the Drosophila disheveled gene, in primary breast cancers. Cancer Sci 94:515–518

Nagaraju GP, Long TE, Park W, Landry JC, Taliaferro-Smith L, Farris AB, Diaz R, El-Rayes BF (2015) Heat shock protein 90 promotes epithelial to mesenchymal transition, invasion, and migration in colorectal cancer. Mol Carcinog 54:1147–1158

Nakayama M, Hisatsune J, Yamasaki E, Isomoto H, Kurazono H, Hatakeyama M, Azuma T, Yamaoka Y, Yahiro K, Moss J, Hirayama T (2009) Helicobacter pylori VacA-induced inhibition of GSK3 through the PI3K/Akt signaling pathway. J Biol Chem 284(3):1612–1619

Nakopoulou L, Mylona E, Papadaki I, Kavantzas N, Giannopoulou I, Markaki S, Keramopoulos A (2006) Study of phospho-β-catenin subcellular distribution in invasive breast carcinomas in relation to their phenotype and the clinical outcome. Mod Pathol 19:556–563

Neckers L, Ivy SP (2003) Heat shock protein 90. Curr Opin Oncol 15(6):419–424

Ochel HJ, Schulte TW, Nguyen P, Trepel J, Neckers L (1999) The benzoquinone ansamycin geldanamycin stimulates proteolytic degradation of focal adhesion kinase. Mol Genet Metabol 66:24–30

Owens P, Pickup MW, Novitskiy SV, Chytil A, Gorska AE, Aakre ME, West J, Moses HL (2012) Breast Cancer special feature: disruption of bone morphogenetic protein receptor 2 (BMPR2) in mammary tumors promotes metastases through cell autonomous and paracrine mediators. Proc Natl Acad Sci U S A 109(8):2814–2819

Paik S, Hazan R, Fisher ER, Sass RE, Fisher B, Redmond C, Schlessinger J, Lippman ME, King CR (1990) Pathologic findings from the National Surgical Adjuvant Breast and Bowel Project: prognostic significance of erbB-2 protein overexpression in primary breast cancer. J Clin Oncol 8(1):103–112

Pandey M, Sunil Kumar BV, Verma R (2015) Mammaglobin as a diagnostic serum marker of complex canine mammary carcinoma. Res Vet Sci 103:187–192

Pepin K, Momose F, Ishida N, Nagata K (2001) Molecular cloning of horse Hsp90 cDNA and its comparative analysis with other vertebrate Hsp90 sequences. J Vet Med Sci 63:115–124

Perdew GH, Wiegand H, Vanden Heuver JP, Mitchell C, Singh SS (1997) A 50 kilodalton protein associated with Raf and pp 60 (v-Src) protein kinases is a mammalian homolog of the cell cycle control protein Cdc37. Biochemistry 36(12):3600–3607

Perou CM, Sorlie T, Eisen MB, van de Rijn M, Jeffrey SS, Rees CA et al (2000) Molecular portraits of human breast tumours. Nature 406(6797):747–752

Petersen, N.S. (1990) Effects of heat and chemical stress on development. Scandalios: Advances in genetics: responses to environmental stress. San Diego: Academic, 28

Petersen NS, Mitchell HK (1991) Environmentally induced development defects in Drosophila. In: Hightower L, Nover L (eds) Heat shock and development, vol 17. Springer, Berlin, pp 29–43

Petros AM, Olejniczak ET, Fesik SW (2004) Structural biology of the Bcl-2 family of proteins. Biochim Biophys Acta 1644:83–94

Petterino C, Martini M, Castagnaro M (2001) Stat3 expression in canine mammary gland tumors and mammary gland hyperplasia. Eur J Vet Pathol 7(3):117–123

Pick E, Kluger Y, Giltnane JM, Moeder C, Camp RL, Rimm DL, Kluger HM (2007) High HSP90 expression is associated with decreased survival in breast cancer. Cancer Res 67:2932–2937

Pyle ME, Korbonits M, Gueorguiev M, Jordan S, Kola B, Morris DG, Meinhardt A, Powell MP, Claret FX, Zhang Q, Metz C, Bucala R, Grossman AB (2003) Macrophage migration inhibitory factor expression is increased in pituitary adenoma cell nuclei. J Endocrinol 176(1):103–110

Queiroga FL, Raposo T, Carvalho MI, Prada J, Pires I (2011) Canine mammary tumours as a model to study human breast cancer: most recent findings. In Vivo 25:455–466
Reed JC (1999) Mechanisms of apoptosis avoidance in cancer. Curr Opin Oncol 11:68–75
Ritossa F (1962) A new puffing pattern induced by temperature shock and DNP in drosophila. Experientia 18(12):571–573
Roe SM, Prodromou C, O'Brien R, Ladbury JE, Piper PW, Pearl LH (1999) Structural basis for inhibition of the Hsp90 molecular chaperone by the antitumor antibiotics radicicol and geldanamycin. J Med Chem 42:260–266
Rong B, Zhao C, Liu H, Ming Z, Cai X, Gao W, Yang S (2014) Identification and verification of Hsp90-beta as a potential serum biomarker for lung cancer. Am J Cancer Res 4(6):874–885
Sato S, Fujita N, Tsuruo T (2000) Modulation of Akt kinase activity by binding to HSP90. Proc Natl Acad Sci U S A 97:10832–10837
Sato N, Yamamoto T, Sekine Y, Yumioka T, Junicho A, Fuse H, Matsuda T (2003) Involvement of heat-shock protein 90 in the interleukin-6-mediated signaling pathway through STAT3. Biochem Biophys Res Commun 300:847–852
Schlange T, Matsuda Y, Lienhard S, Huber A, Hynes NE (2007) Autocrine WNT signaling contributes to breast cancer cell proliferation via the canonical WNT pathway and EGFR transactivation. Breast Cancer Res 9:R63
Schlesinger MJ (1990) Heat shock proteins. J Biol Chem 265(21):12111–12114
Schulz R, Dobbelstein M, Moll UM (2012a) HSP90 inhibitor antagonizing MIF: the specifics of pleiotropic cancer drug candidates. Oncoimmunology 1:1425–1426
Schulz R, Marchenko N, Holembowski L, Fingerle-Rowson G, Pesic M, Zender L, Dobbelstein M, Moll UM (2012b) Inhibiting the HSP90 chaperone destabilizes macrophage migration inhibitory factor and thereby inhibits breast tumor progression. J Exp Med 209:275–289
Sepp-Lorenzino L, Ma Z, Lebwohl DE, Vinitsky A, Rosen N (1995) Herbimycin A induces the 20 S proteasome- and ubiquitin-dependent degradation of receptor tyrosine kinases. J Biol Chem 270:16580–16587
Sharma R, Kumar D, Jha NK, Jha SK, Ambasta RK, Kumar P (2017) Re-expression of cell cycle markers in aged neurons and muscles: whether cells should divide or die? Biochim Biophys Acta Mol basis Dis 1863(1):324–336
Silverstein AM, Grammatikakis N, Cochran BH, Chinkers M, Pratt WB (1998) p50 (cdc37) binds directly to the catalytic domain of Raf as well as to a site on HSP90 that is topologically adjacent to the tetratricopeptide repeat binding site. J Biol Chem 273(32):20090–20095
Simon MC, Kitchener K, Kao HT, Hickey E, Weber L, Voellmy R, Heintz N, Nevins JR (1987) Selective induction of human heat shock gene transcription by the adenovirus E1A gene products, including the 12S E1A product. Mol Cell Biol 7(8):2884–2890
Sims JD, McCready J, Jay DG (2011) Extracellular Heat Shock Protein (Hsp) 70 and Hsp90α assist in matrix metalloproteinase-2 activation and breast cancer cell migration and invasion. PLoS One 6(4):e18848
Slamon DJ, Clark GM, Wong SG, Levin WJ, Ullrich A, McGuire WL (1987) Human breast cancer: correlation of relapse and survival with amplification of the HER-2/neu oncogenes. Science 235(4785):177–182
Soga S, Sharma SV, Shiotsu Y, Shimizu M, Tahara H, Yamaguchi K, Ikuina Y, Murakata C, Tamaoki T, Kurebayashi J, Schulte TW, Neckers LM, Akinaga S (2001) Stereospecific antitumor activity of radicicol oxime derivatives. Cancer Chemother Pharmacol 48:435–445
Song CH, Park SY, Eom KY, Kim JH, Kim SW, Kim JS, Kim IA (2010) Potential prognostic value of heat-shock protein 90 in the presence of phosphatidylinositol-3-kinase overexpression or loss of PTEN, in invasive breast cancers. Breast Cancer Res 12(2):R20
Sreedhar AS, Csermely P (2004) Heat shock proteins in the regulation of apoptosis: new strategies in tumour therapy: a comprehensive review. Pharmacol Ther 101:227–257
Sreedhar AS, Kalmar E, Csermely P, Shen YF (2004) Hsp90 isoforms: functions, expression and clinical importance. FEBS Lett 562:11–15

Stancato LF, Chow YH, Hutchison KA, Pedrew GH, Jove R, Pratt WB (1993) Raf exists in a native heterocomplex with HSP90 and p50 that can be reconstituted in a cell-free system. J Biol Chem 268(29):21711–21716

Stepanova L, Leng X, Barker SB, Harper JW (1996) Mammalian p50Cdc37 is a protein kinase-targeting subunit of Hsp90 that binds and stabilizes Cdk4. Genes Dev 10(12):1491–1502

Stephanou A, Latchman DS (1999) Transcriptional regulation of the heat shock protein genes by STAT family transcription factors. Gene Expr 7:311–319

Stivarou T, Stellas D, Vartzi G, Thomaidou D, Patsavoudi E (2016) Targeting highly expressed extracellular HSP90 in breast cancer stem cells inhibits tumor growth in vitro and in vivo. Cancer Biol Ther 17(8):799–812

Stock J (1999) Gyrating protein kinases. Curr Biol 9:R364

Sunil Kumar BV, Kumar AK, Padmanath K, Sharma B, Kataria M (2013) Heterologous expression and functional characterization of matrix Metalloproteinase-11 from canine mammary tumor. Anim Biotechnol 24(1):31–43

Sunil Kumar BV, Singh S, Verma R (2017) Anticancer potential of dietary vitamin D and ascorbic acid: a review. Crit Rev Food Sci Nutr 57(12):2623–2635

Sunil Kumar BV, Bhardwaj R, Mahajan K, Kashyap N, Kumar A, Verma R (2018) The overexpression of HSP90B1 is associated with tumorigenesis of canine mammary glands. Mol Cell Biochem 440:23–31

Taipale M, Jarosz DF, Lindquist S (2010) HSP90 at the hub of protein homeostasis: emerging mechanistic insights. Nat Rev Mol Cell Biol 11(7):515–528

Trepel J, Mollapour M, Giaccone G, Neckers L (2010) Targeting the dynamic HSP90 complex in cancer. Nat Rev Cancer 10(8):537–549

van 't Veer LJ, Dai H, van de Vijver MJ, He YD, Hart AA, Mao M, Peterse HL, van der Kooy K, Marton MJ, Witteveen AT, Schreiber GJ, Kerkhoven RM, Roberts C, Linsley PS, Bernards R, Friend SH (2002) Gene expression profiling predicts clinical outcome of breast cancer. Nature 415:530–536

Veeck J, Niederacher D, An H, Klopocki E, Wiesmann F, Betz B, Galm O, Camara O, Dürst M, Kristiansen G, Huszka C, Knüchel R, Dahl E (2006) Aberrant methylation of the Wnt antagonist SFRP1 in breast cancer is associated with unfavourable prognosis. Oncogene 25:3479–3488

Veeck J, Geisler C, Noetzel E, Alkaya S, Hartmann A, Knuchel R, Dahl E (2008) Epigenetic inactivation of the secreted frizzled-related protein-5 (SFRP5) gene in human breast cancer is associated with unfavorable prognosis. Carcinogenesis 29:991–998

Vivanco I, Sawyers CL (2002) The phosphatidylinositol 3-kinase AKT pathway in human cancer. Nat Rev Cancer 2:489–501

Wagner EF, Nebreda AR (2009) Signal integration by JNK and p38 MAPK pathways in cancer development. Nat Rev Cancer 9:537–549

Walsh DA, Morris VB (1989) Heat shock affects cell cycling in the neural plate of cultured rat embryos: a flow cytometric study. Exp Teratol 40:583–592

Walsh D, Li K, Wass J, Dolnikov A, Zeng F, Zhe L, Edwards M (1993) Heat-shock gene expression and cell cycle changes during mammalian embryonic development. Dev Genet 14:127–136

Wang Y, Theriault JR, He H, Gong J, Calderwood SK (2004) Expression of a dominant negative heat shock factor-1 construct inhibits aneuploidy in prostate carcinoma cells. J Biol Chem 279:32651–32659

Wang X, Zhang Y, Zhao Y, Liang Y, Xiang C, Zhou H, Zhang H, Zhang Q, Qing H, Jiang B, Xiong H, Peng L (2016) CD24 promoted cancer cell angiogenesis via Hsp90-mediated STAT3/VEGF signalling pathway in colorectal cancer. Oncotarget 7:55663–55676

Wartmann M, Davis RJ (1994) The native structure of the activated Raf protein kinase is a membrane-bound multi-subunit complex. J Biol Chem 269(9):6695–6701

Whitesell L, Lindquist SL (2005) HSP90 and the chaperoning of cancer. Nat Rev Cancer 5(10):761–772

Whitesell L, Sutphin P, An WG, Schulte T, Blagosklonny MV, Neckers L (1997) Geldanamycin-stimulated destabilization of mutated p53 is mediated by the proteasome in vivo. Oncogene 14(23):2809–2816

Whitesell L, Sutphin PD, Pulcini EJ, Martinez JD, Cook PH (1998) The physical association of multiple molecular chaperone proteins with mutant p53 is altered by geldanamycin, a HSP90-binding agent. Mol Cell Biol 18(3):1517–1524

Wilhelm CCS, Tang L, Rong H, Chen C, Zhang X, McHugh M, Wilkie D, McNabola A, Rowley B, Henderson A, Cao Y, Hofilena G, Housley T, Shujath J, Liu L, Adnane L, Lynch M, Eveleigh D, Gedrich R, Voznesensky A, Riedl B, Wang Q, Post L, Bollag G (2003) BAY 43-9006 exhibits broad spectrum antitumor activity and targets raf/MEK/ERK pathway and receptor tyrosine kinases involved in tumor progression and angiogenesis (abstract A78). Presented at the AACR-NCI-EORTC International Conference Molecular Targets and Cancer Therapeutics, Boston, 17–21 November

Workman P, Burrows F, Neckers L, Rosen N (2007) Drugging the cancer chaperone HSP90: combinatorial therapeutic exploitation of oncogene addiction and tumor stress. Ann N Y Acad Sci 1113:202–216

Wu J, Liu T, Rios Z, Mei Q, Lin X, Cao S (2017) Heat shock proteins and cancer. Trends Pharmacol Sci 38(3):226–256

Xing H, Weng D, Chen G, Tao W, Zhu T, Yang X, Meng L, Wang S, Lu Y, Ma D (2008) Activation of fibronectin/PI-3K/Akt2 leads to chemoresistance to docetaxel by regulating survivin protein expression in ovarian and breast cancer cells. Cancer Lett 261(1):108–119

Yamaguchi N, Ito T, Azuma S, Ito E, Honma R, Yanagjsawa Y, Nishikawa A, Kawamura M, Imai J, Watanabe S, Semba K, Inoue J (2009) Constitutive activation of nuclear factor-kappaB is preferentially involved in the proliferation of basal-like subtype breast cancer cell lines. Cancer Sci 100:1668–1674

Yang ZQ, Geng X, Solit D, Pratilas CA, Rosen N, Danishefsky SJ (2004) New efficient synthesis of resorcinylic macrolides via ynolides: establishment of cycloproparadicicol as synthetically feasible preclinical anticancer agent based on Hsp90 as the target. J Am Chem Soc 126:7881–7889

Yarden Y, Pines G (2012) The ERBB network: at last, cancer therapy meets systems biology. Nat Rev Cancer 12:553–563

Zagouri F, Bournakis E, Koutsoukos K, Papadimitriou CA (2012) Heat shock protein 90 (Hsp90) expression and breast cancer. Pharmaceuticals 5:1008–1020

Chapter 6
Role of HSP90 Inhibitors in the Treatment of Cancer

Geraldine O'Sullivan Coyne, Cecilia Monge, and Alice P. Chen

Abstract The 90-kDa heat shock protein HSP90 is a member of a highly evolutionarily conserved class of molecular chaperone proteins indispensable for the development of cancer; when activated by cellular stress, HSP90 stabilizes oncogenic substrate "client" proteins involved in cellular processes that promote tumorigenesis. HSP90 inhibition attenuates this stabilization of aberrant client proteins in tumor cells, allowing for simultaneous targeting of multiple pathways involved in cancer cell survival. HSP90 inhibitors have been assessed as potential oncologic therapies in several preclinical and clinical studies. Although preclinically promising results have been measured, these results have not translated yet into major clinical efficacy. Combinations of HSP90 inhibitors with approved and investigational oncology drugs may represent further opportunities for the use of these agents in patients with cancer. This chapter reviews some of the important early clinical milestones observed in studies of first- and second-generation HSP90 inhibitors used as single agents and in combination. In the conclusion, possible reasons for the lack of therapeutic benefit in clinical studies are considered.

Keywords Angiogenesis · Heat shock protein 90 · HSP90 inhibitor · Metastasis · Molecular chaperone · Oncogenic driver

Authors Geraldine O'Sullivan Coyne and Cecilia Monge have equally contributed to this chapter.

G. O'Sullivan Coyne · C. Monge · A. P. Chen (✉)
Division of Cancer Treatment and Diagnosis and Center for Cancer Research, National Cancer Institute, National Institutes of Health, Bethesda, MD, USA
e-mail: chenali@mail.nih.gov

A. A. A. Asea, P. Kaur (eds.), *Heat Shock Protein 90 in Human Diseases and Disorders*, Heat Shock Proteins 19,
https://doi.org/10.1007/978-3-030-23158-3_6

Abbreviations

AE	Adverse event
ALK	Anaplastic lymphoma kinase
BRAF	Serine/threonine-protein kinases B-Raf
CDK	Cyclin-dependent kinase
CRAF	Serine/threonine-protein kinases C-Raf
CRC	Colorectal carcinoma
DLT	Dose-limiting toxicity
EGFR	Epidermal growth factor receptor
EMT	Epithelial-to-mesenchymal transition
GDNF	Glial cell line-derived neurotrophic factor
HER2	Human epidermal growth factor receptor 2
HGF	Hepatocyte growth factor
HSP	Heat shock protein
HSP90	Heat shock protein 90
MBC	Metastatic breast cancer
MTD	Maximum tolerated dose
NSCLC	Non-small cell lung cancer
ORR	Overall response rate
TNBC	Triple-negative breast cancer

6.1 Introduction

The development of malignancy in humans is a stepwise process involving both genetic and epigenetic events that drive the transformation of normal cells. Apart from late detection, other factors such as immune evasion and drug resistance also contribute to the lack of effective treatment for many patients with advanced disease, making cancer a leading cause of mortality, with an estimated 9.6 million deaths projected worldwide for 2018 (Bray et al. 2018). Despite significant efforts and advances from the research community to discover and develop new anti-cancer drugs, there continues to be a need for novel, effective oncologic treatments.

6.2 HSP Deregulation and Cancer Development

Heat shock proteins (HSP) are highly conserved, ATP-dependent chaperone molecules that stabilize key proteins (or "client proteins") involved in cellular signal transduction and homeostasis, enabling cells to survive diverse stressors and death stimuli (Soga et al. 2013, Chatterjee and Burns 2017). Mammalian heat shock proteins are classified into six families based on their molecular sizes; well-studied HSP family members include HSP90, HSP70, and HSP27 (Wang et al. 2014). HSP90 modulates hundreds of protein substrates, making it a key regulator of diverse cellular processes such as protein folding, immune regulation, stress response, and DNA repair (Graner 2016; Wu et al. 2017). Approximately 60% of the human kinome is reported to be associated with HSP90, and 30% of E3 ubiquitin ligases and a small fraction of human transcription factors also depend on HSP90 for activity (Taipale et al. 2012). It has been a challenge to precisely define HSP90 client proteins, however, as the binary classification of intracellular proteins as clients or nonclients has been questioned in light of the outcomes of sophisticated high-throughput HSP90 interactome studies demonstrating a continuous distribution for HSP90 binding affinity (Taipale et al. 2012).

For cancer cells within a rapidly proliferating cellular environment, metabolic need and proteolytic stress create a unique dependence on heat shock proteins such as HSP90, as many client proteins are involved in cellular processes (such as cell survival and proliferation) that can promote tumorigenesis if dysregulated (Rappa et al. 2012). HSP90 is overexpressed in many types of cancers (Yufu et al. 1992; McCarthy et al. 2008; Patel et al. 2014) and has been shown to be a negative prognostic factor in some malignancies (Burrows et al. 2004; Pick et al. 2007; Shi et al. 2014). Furthermore, several HSP90 client proteins are known oncoproteins, including BCR-ABL, human epidermal growth factor receptors 1 and 2 (EGFR and HER2, respectively), cyclin-dependent kinase 4 (CDK4), the serine/threonine-protein kinases B-Raf and C-Raf (BRAF and CRAF, respectively), the hepatocyte growth factor (HGF) receptor MET, the glial cell line–derived neurotrophic factor (GDNF) receptor RET, and protein kinase B/AKT (Jhaveri et al. 2014b).

Dysregulation of other heat shock proteins has also been implicated in the development of cancer and other diseases. For example, HSP70 is strongly associated with carcinogenesis, cancer progression, and poor cancer outcomes for various tumor types (Rerole et al. 2011; Wu et al. 2017). Considered a co-chaperone, HSP70 delivers client proteins such as HER2, CRAF, AKT, and CDK4 to HSP90, thereby promoting survival through the inhibition of both cell senescence and apoptotic pathways (Wu et al. 2017). Indeed, the interaction between HSP70 and HSP90 forms the core of the "epichaperome" complex, consisting of numerous chaperone and co-chaperone proteins, that promotes tumor survival and is found in over half of cancer cell lines (Rodina et al. 2016). This interplay between HSP70 and HSP90 also renders HSP70-expressing cancer cells more sensitive to HSP90 inhibition (Ghosh et al. 2008; Rodina et al. 2016). However, though strategies targeting various heat shock proteins have been explored, HSP90 inhibition has demonstrably resulted in ubiquitination and proteasomal degradation of client proteins (Whitesell and Lindquist 2005; Zuehlke and Johnson 2010), supporting the premise that oncogenic transformation and progression is deeply linked to HSP90. Thus, targeted inhibition of HSP90 has emerged as a novel anti-cancer therapeutic strategy.

6.2.1 Development of HSP90 Inhibitors as Anti-cancer Agents

HPS90 and its inhibition have been studied extensively over the past two decades. This has led to the development of several clinically viable inhibitors that disrupt the interaction between HSP90 and its client proteins, which occurs via the N-terminal ATPase domain (Soga et al. 2013). These N-terminal domain–binding molecules are denoted as 'classical inhibitors,' though none of these have received regulatory approval to date. HSP90 inhibitors have several advantageous characteristics as potential oncology therapeutic agents. First, the distribution of HSP90 client proteins across several different cell signaling pathways potentially enables simultaneous modulation of multiple pathways by HSP90 inhibitors. Additionally, tumor cells are reportedly more sensitive to HSP90 inhibition than normal cells (Kamal et al. 2003; Chiosis and Neckers 2006). Finally, the unusual pharmacokinetic profile of these agents—i.e., rapid clearance from plasma and normal tissues, together with prolonged, selective retention in tumor cells (Eiseman et al. 2005, Daozhen et al. 2007)—may contribute to an exploitable therapeutic index (Straume et al. 2012).

Upregulation of other heat shock response proteins, in particular HSP70 and HSP27, occurs in response to HSP90 inhibition (Biamonte et al. 2010) and has been postulated as a mechanism of resistance to these inhibitors. While small-molecule inhibitors of HSP70 have also been investigated preclinically (Radons 2016), as of yet, none have progressed to a clinical trial. However, given that the interplay between HSP70 and HSP90 contributes substantially to the role of these proteins in malignancy, the combination of HSP70 and HSP90 inhibitors has been proposed as a strategy for achieving enhanced antitumor effects.

6.2.1.1 Preclinical Assessments of HSP90 Inhibitors

Molecular studies of HSP90 inhibitors in cancer cell lines have shed light on key mechanisms of action, while subsequent testing in human tumor xenograft models has provided some confirmation of this mechanism of action and has also established the antitumor activity of these agents across several tumor types. The enormous number of HSP90 client proteins implies that the key client proteins driving malignancy may differ across different tumor types, rendering it difficult to determine the precise molecular underpinnings of HSP inhibitor activity in various cancers; indeed, it has been suggested that nearly every oncogenesis-associated protein is an HSP90 client (Chiosis and Neckers 2006; Vartholomaiou et al. 2016).

Despite this complexity, preclinical studies have identified several key oncogenic client proteins for which expression is modulated by HSP90 inhibition, and which therefore may serve as potential selection markers for HSP90 inhibitor therapy. These key client proteins include a number of receptor tyrosine kinases, such as HER2 (Mimnaugh et al. 1996), EGFR (Ahsan et al. 2012), FLT3 (Yao et al. 2003), IGF-1R (Breinig et al. 2011), RET (Alfano et al. 2010), and MET (Miyajima et al.

2013). Several receptor tyrosine kinase fusion proteins that are drivers of oncogenesis are also stabilized by HSP90 and therefore sensitive to HSP90 inhibition, including ALK (anaplastic lymphoma kinase) (Bonvini et al. 2002; Sang et al. 2013) and fibroblast growth factor receptor 2 (FGFR2) fusion proteins (Lamberti et al. 2018). In addition, other signaling kinases, such as RAF-1, BRAF, and AKT, and transcriptional regulators such as NF-κB, HIF-1α, and p53, are also key oncogenic HSP90 client proteins, as is the cell cycle regulator CDK4 (Wu et al. 2017). Just as HSP90 enhances stability of fusion proteins, it can also prolong the half-life of proteins with other types of destabilizing mutations, as is the case for p53 variants with oncogenic point mutations (Whitesell et al. 1998).

In response to HSP90 inhibition, compensatory upregulation of HSP70, and its consequent anti-apoptotic signaling, has been implicated as a mechanism of therapeutic resistance (Garrido et al. 2003; Garrido et al. 2006). HSP70 has been shown to block recruitment of procaspase-9 to the Apaf-1 apoptosome (Beere et al. 2000; Saleh et al. 2000) and to directly antagonize apoptosis-inducing factor (AIF) (Ravagnan et al. 2001). Several in vitro studies have shown that attenuation of HSP70 activity enhances HSP90 inhibitor-mediated cell death (Guo et al. 2005; Powers et al. 2008). Such studies indicate that HSP70 may serve as an important biomarker of HSP90 inhibition, and highlight the potential therapeutic value of HSP70/HSP90 inhibitor combinations. HSP90 inhibitors have demonstrated antitumor activity in several human tumor xenograft models, including those carrying aberrations in the oncogenic client proteins described above. For example, ganetespib and other second-generation HSP90 inhibitors have yielded antitumor activity (ranging from tumor growth inhibition to tumor regression) in multiple breast cancer models, including both HER2-positive and triple-negative breast cancer (TNBC) models (Jensen et al. 2008; Caldas-Lopes et al. 2009; Friedland et al. 2014). HSP90 inhibitors have likewise shown promising single-agent in vivo activity in models of non-small cell lung cancer (NSCLC) driven by ALK fusion proteins or EGFR mutations (Chen et al. 2010; Normant et al. 2011; Graham et al. 2012; Sang et al. 2013) and in melanoma models driven by BRAF mutations (Acquaviva et al. 2014). Such studies have also established in vivo proof-of-concept for HSP90 inhibition, demonstrating decreased levels of various oncogenic HSP90 client proteins (e.g., HER2, BRAF, AKT, EGFR) coincident with antitumor activity in these models. Unfortunately, the impressive preclinical activity of HSP90 inhibitors has not been observed in clinical testing, which has more recently focused on HSP90 inhibitor combination therapies, as described below.

6.2.1.2 Clinical Development of HSP90 Inhibitors

Two natural products prompted the identification and testing of the first generation of N-terminal HSP90 inhibitors: geldanamycin, a benzoquinone ansamycin antibiotic derived from *Streptomyces hygroscopicus*, and radicicol, a macrocyclic lactone antibiotic derived from *Monosporidum bonorden* (Soga et al. 2013; Chatterjee and Burns 2017). However, novel drug development strategies have continued to identify new HSP90-targeting agents, including C-terminal–binding inhibitors. As these agents have progressed through clinical testing, newer agents have been developed to improve upon target binding, efficacy, and toxicity, as discussed below.

6.2.2 First-Generation HSP90 Inhibitors

Despite several encouraging preclinical studies, use of geldanamycin was hampered by unacceptable clinical toxicity and structural volatility, particularly from a quinone ring moiety that contributes to its hepatotoxicity (Supko et al. 1995). The C-17 methoxy group of geldanamycin was eventually replaced by an amine group, resulting in various synthetic analogues created to mitigate these early drug design issues (Table 6.1).

17-N-allylamino-17-demethoxygeldanamycin (17-AAG, tanespimycin; Kosan Biosciences), was the first of the geldanamycin analogs to undergo clinical evaluation in the 1990s. Though it demonstrated a better adverse event (AE) profile than geldanamycin (Yuno et al. 2018), it had poor solubility and bioavailability (Banerji et al. 2005) and did not demonstrate substantial activity as a single agent in multiple phase I and II trials (Goetz et al. 2005; Ronnen et al. 2006; Solit et al. 2008). Tanespimycin did demonstrate modest activity when combined with other antineoplastic agents. A combination with the HER2-targeted antibody trastuzumab in patients with metastatic HER2-positive breast cancer showed an overall response rate (ORR) of 22% and an overall survival (OS) of 17 months (Modi et al. 2011). In a phase I/II study of tanespimycin and bortezomib, an ORR of 27% was reported in

Table 6.1 Classification of HSP90 inhibitors

Inhibitor	Class	Administration Route
First generation		
Geldanamycin	–	Intravenous
Radicicol	–	–
17-AAG (tanespimycin)	GM	Intravenous
17-DMAG (alvespimycin)	GM	Intravenous, oral
IPI-504 (retaspimycin)	GM	Intravenous
Second generation		
AUY922 (luminespib)	RD	Intravenous
STA-9090 (ganetespib)	RD	Intravenous
AT13387 (onalespib)	RD	Intravenous, oral
BIIB021	Purine scaffold	Intravenous
PU-H71	Purine scaffold	Oral
XL888	Aminoterphthalamide	Oral
TAS-116	Pyrazolopyridine	Intravenous
NVP-HSP990	Aminopyrimidine	Oral

GM geldanamycin-based, *RD* radicicol-based

bortezomib-naïve patients with refractory multiple myeloma (Richardson et al. 2011). Though a phase III trial of this combination versus bortezomib alone was initiated in patients with multiple myeloma, this trial was subsequently closed; formulation issues may have been a consideration. Interestingly, Triolimus®—a triplet agent containing tanespimycin together with the microtubule-disrupting chemotherapeutic paclitaxel and the mTOR inhibitor rapamycin—has been formulated using a polymeric micelle drug delivery platform to improve solubility, and is reported to have preclinical activity without the toxicity associated with the previous tanespimycin formulation (Hasenstein et al. 2012).

17-demethoxy-17-N,N-dimethylaminoethylamino-geldanamycin (17-DMAG, alvespimycin; Bristol-Myers Squibb) is another semi-synthetic derivative of geldanamycin, with significant antitumor activity, improved water solubility, and oral bioavailability (Hollingshead et al. 2005; Georgakis et al. 2006). This agent was advanced to clinical trials with both intravenous (IV) and oral (PO) formulations in both solid tumor and hematological malignancies (Hollingshead et al. 2005; Kummar et al. 2010; Lancet et al. 2010). However, significant dose-limiting AEs were reported, including fatigue, nausea, and diarrhea, as well as cardiac, liver, lung, and ocular toxicities (Pacey et al. 2011).

IPI-504 (retaspimycin, Infinity Pharmaceuticals) is a reduced formulation of tanespimycin that showed promise given its improved water solubility (Chatterjee et al. 2016). Retaspimycin hydrochloride has shown single-agent antitumor activity in early-phase trials of patients with soft tissue sarcomas/gastrointestinal stromal tumors, as well as in ALK-rearranged NSCLC; however, significant hepatic toxicity was noted in these studies (Sequist et al. 2010; Wagner et al. 2013). Retaspimycin has also been evaluated together trastuzumab in patients with HER2-positive metastatic breast cancer (MBC) refractory to HER2 targeted therapy; although the combination showed modest activity, it did not meet predefined criteria to expand the trial (Modi et al. 2013).

6.2.3 Second-Generation HSP90 Inhibitors

Second-generation HSP90 inhibitors are improved small molecule classes; the first of these are derivatives of radicicol, as they contain an ATP-binding resorcinol moiety (Neckers and Workman 2012), while molecules in the second class utilize a purine scaffold (Jhaveri et al. 2012). Radicicol itself was deemed unsuitable for clinical development because of little or no activity in animals secondary to chemical instability (Soga et al. 2003).

NVP-AUY922 (luminespib, Novartis) was identified via a high-throughput screen and demonstrated activity in multiple human cancer cell lines and tumor xenograft models (Eccles et al. 2008). Phase II trials have noted activity in both *EGFR*-mutated and *ALK*-rearranged NSCLC as well as in refractory HER2-positive breast cancer, with response rates between 10% and 25% (Schroder et al. 2011; Garon et al. 2012). Recently, promising antitumor activity has also been observed in

NSCLC patients with a rare subtype of *EGFR* exon 20 insertions that are typically refractory to EGFR-specific tyrosine kinase inhibitors (NCT01854034), though results for this study are pending. Dose-limiting toxicities (DLTs) reported in various trials include darkening of vision, atrial flutter, diarrhea, and fatigue. Results for various other ongoing trials are awaited, including combination studies with both chemotherapy and targeted agents (Table 6.2).

AT-13387 (**onalespib**, Astex Therapeutics), is another second-generation inhibitor that has gone through early-phase clinical testing, with a notable difference (relative to other HSP90 inhibitors) of exhibiting prolonged duration of activity in the preclinical setting (Graham et al. 2012). Single-agent trials (Do et al. 2015; Shapiro et al. 2015) as well as combination trials with imatinib (Wagner et al. 2016) or docetaxel (Ramalingam et al. 2015) have been reported. Other combination studies, including with abiraterone (NCT01685268) or crizotinib (NCT01712217), are awaiting results. The agent is currently being studied in a phase I trial in combination with AT7519M, a small molecule inhibitor of CDKs 1, 2, 4, 5, and 9 (NCT02503709; Table 6.2).

STA-9090 (**ganetespib**, Madrigal Pharmaceuticals), a radicicol-derived compound, has been considered one of the most promising HSP90 inhibitors (Chatterjee and Burns 2017), yielding objective responses in phase II studies of MBC (in both patients with HER2-positive tumors and those with TNBC) and NSCLC (in patients with *EML4-ALK* rearrangements) (Socinski et al. 2013; Jhaveri et al. 2014). No objective responses were observed in early-phase studies of ganetespib in patients with hematological malignancies (Lancet et al. 2010b; Padmanabhan et al. 2010). With regard to combination therapies, a phase I study of ganetespib together with paclitaxel and trastuzumab in HER2-positive MBC patients refractory to trastuzumab yielded an ORR of 22% (Jhaveri et al. 2017), and a phase II study of ganetespib in MBC patients was expanded to also examine the combination with paclitaxel following progression on single-agent ganetespib (Cameron et al. 2014). In NSCLC, a significant survival benefit was observed in a phase II study of ganetespib combined with docetaxel in a small subset of patients who had progressed on chemotherapy (Ramalingam et al. 2015); the corresponding phase III study of this combination is ongoing (Ramalingam et al. 2014). However, toxicities have also been problematic for ganetespib combination therapies; in a phase I study of ganetespib combined with the anti-angiogenic agent ziv-aflibercept, the combination resulted in serious adverse events, including small intestinal perforation and one sudden death (a potential gastrointestinal hemorrhage), prompting discontinuation of the trial (Meehan et al. 2018).

PU-H71 (Samus Therapeutics) was developed following implementation of structural biology techniques to identify new HSP90 inhibitors. High-resolution x-ray crystal structures of the HSP90 N-terminal domain, with and without ATP or ATP analogs (Obermann et al. 1998; Li et al. 2012), have enabled the rational design of new HSP90 inhibitors using purine or pyrimidine scaffolds. PU-H71 was the first synthesized second-generation HSP90 inhibitor of this class. A first-in-human study in solid tumors demonstrated that PU-H71 was well tolerated, though no objective responses were reported (Speranza et al. 2018); however, determination of the max-

Table 6.2 Selected ongoing HSP90 inhibitor trials

Tumor type	Phase	Agent tested	Mechanism of action (single agent/combination)	Molecular eligibility criteria	Trial objective	NCT number
Solid tumor; NOS	I	HS-196	ATP binding domain inhibitor linked to near-infrared dye	–	Tumor imaging (contrast agent for tumor detection)	NCT03333031
Solid tumor; NOS	Ia/Ib	TAS-116	Non-resorcinol, non-purine selective inhibitor of cytosolic HSP90α/β isoforms	NSCLC *EGFR*+ or *ALK*+ (only at selected doses)	Safety of different dosing regimes	NCT02965885
Solid tumor; NOS (selected histologies in expansion)	I/IIa	PEN-866	HSP90-targeting ligand linked to SN-38, a topoisomerase 1 inhibitor	–	Safety and MTD	NCT03221400
NSCLC (adenocarcinoma, stage IIIB-IV)	I/II	Luminespib + erlotinib	Radicicol-derived HSP90α/β inhibitor/tyrosine kinase inhibitor targeting *EGFR*	*EGFR*+	Safety, MTD, and RP2D of the combination	NCT01259089
NSCLC (non-squamous, stage IV)	Ib	Luminespib + pemetrexed	Radicicol-derived HSP90α/β inhibitor/antimetabolite chemotherapy	*EGFR*+, *ALK*+, or *KRAS*/*EGFR*/*ALK* WT	Safety and tolerability of the combination	NCT01784640
TNBC	I	Onalespib + paclitaxel	Non-ansamycin HSP90 inhibitor/microtubule inhibitor chemotherapy	–	Safety and RP2D of the combination	NCT02474173
Solid tumor; NOS (expansion: ovarian, fallopian tube, primary peritoneal, and TNBC)	I	Onalespib + olaparib	Non-ansamycin HSP90 inhibitor/PARP inhibitor targeted therapy	–	Safety and MTD	NCT02898207
Solid tumor; NOS	I	Onalespib + AT7519	Non-ansamycin HSP90 inhibitor/CDK 1,2,4,5 and 9 small molecule inhibitor	–	Safety and MTD of the combination	NCT02503709

(continued)

Table 6.2 (continued)

Tumor type	Phase	Agent tested	Mechanism of action (single agent/combination)	Molecular eligibility criteria	Trial objective	NCT number
Solid tumor (gastrointestinal tumors)	Ib	XL888 + pembrolizumab	Selective ATP-competitive inhibitor of HSP90/ anti-PD-1 immune checkpoint inhibitor	–	Safety and RP2D of the combination	NCT03095781
Solid tumor (malignant melanoma, stage III/IV)	I	XL888 + vemurafenib + cobimetinib	Selective ATP-competitive inhibitor of HSP90/BRAF enzyme inhibitor/MEK kinase inhibitor	$BRAF^{V600E}$ mutation–positive	Safety, MTD, and RP2D of the combination	NCT02721459

All trials listed are open only to patients with metastatic disease. Trials are open and/or recruiting patients as of October 2018

NOS not otherwise specified, *MTD* maximum tolerated dose, *NSCLC* non-small cell lung cancer, *EGFR+* epithelial growth factor receptor mutation–positive, *RP2D* recommended phase II dose, *ALK+* anaplastic lymphoma receptor tyrosine kinase rearrangement–positive, *KRAS* Kirsten rat sarcoma viral oncogene, *WT* wild type (non-mutated), *TNBC* triple-negative breast cancer, *PARP* poly ADP-ribose polymerase, *CDK* cyclin-dependent inhibitor, *PD-1* programmed cell death-1 lymphocyte receptor, *MEK* mitogen-activated protein kinase enzyme, *BRAF* serine/threonine kinase B-Raf imum tolerated dose (MTD) was not achieved, and the trial was stopped because of drug supply issues.

CNF2024/BIIB021 (Biogen) has been tested in phase I clinical studies in patients with solid tumors as well as hematologic malignancies; DLTs included hypoglycemia, hyponatremia, liver abnormalities, fatigue, dizziness, and syncope (Saif et al. 2014). Evaluation in phase II trials included patients with gastrointestinal stromal tumors, in which the agent was well tolerated but did not result in significant responses (Shapiro et al. 2015). The combination of BIIB021 with trastuzumab in a phase I study of HER2-positive MBC patients resulted in DLTs including partial seizures, headache, dizziness, fatigue, and rash (Jhaveri et al. 2012).

A number of new HSP90 inhibitors have entered clinical trials in the past few years. **XL888** (Exelixis) is an orally administered small molecule with selective inhibition for both HSP90α and β isoforms (Table 6.1). Both in vitro and in vivo, this agent inhibited growth and induced cell death and tumor regression of melanoma cell lines harboring the oncogenic BRAF V600E mutation and resistant to the BRAF inhibitor vemurafenib, through a variety of mechanisms (Paraiso et al. 2012). In a phase I study of XL888 in patients with refractory solid tumors, the MTD of XL888 was reported to be 135 mg twice weekly, and diarrhea was noted as a DLT (Eroglu et al. 2018). Given these findings, a trial of this agent has been initiated in patients with advanced melanoma harboring the BRAF V600E mutation, and the results are awaited (Table 6.2). TAS-116, another oral agent, has shown good bioavailability and antitumor activity in several xenograft models, as well as reportedly reduced ocular toxicity in various animal species (Ohkubo et al. 2015; Chatterjee and Burns 2017); a phase I study of this agent in pre-defined solid tumor molecular cohorts is ongoing (NCT02965885, Table 6.2). Finally, several radioactive isotopes and clinical imaging agents targeting HSP90 have also been developed (Barrott et al. 2013). One of these, **HS-196**, is an imaging agent containing an HSP90 inhibitor covalently linked to a near-infrared fluorescent dye (Table 6.2). HS-196 selectively and competitively binds to HSP90 in cells following intravenous administration; given that HSP90 is upregulated in a variety of tumor cells, accumulation of this fluorescent dye–tethered agent allows for in vivo detection of these cells due to enhanced uptake of HS-196.

6.2.3.1 Common Toxicities Associated with HSP90 Inhibitors

Clinical development of several HSP90 inhibitors has been halted due to ocular and liver toxicities. Ocular symptoms have been reported as a characteristic toxicity across all classes of HSP90 inhibitors, though, as noted above, not all second-generation inhibitors have resulted in severe ocular toxicity (Jhaveri et al. 2014). Rodent modeling would suggest that prolonged retinal inhibition of HSP90 induces photoreceptor cell death, and that this is dependent on the retina/plasma exposure ratio and retinal elimination rate for each individual inhibitor, rather than their chemical class (Zhou et al. 2013). In the case of elevated liver transaminase toxicities, results from studies of second-generation HSP90 inhibitors suggest that the severe hepatotoxicity that halted development of several first-generation compounds was class-specific (i.e., associated with the quinone ring moiety) rather than target-driven (Jhaveri et al. 2014). Thus, there is evidence that existing and forthcoming second-generation HSP90 inhibitors may avoid both the hepatic and ocular toxicities that have plagued development of previous agents targeting HSP90.

6.2.4 Novel Perspectives and Targets

An entirely new structural class of HSP90 inhibitors has been identified in the past few years. Though they are less well characterized than inhibitors that target the N-terminal ATP binding domain, their interruption of HSP90 chaperone activity is reported to occur in a similar manner to that of the classic inhibitors (Marcu et al. 2000). Compounds that target the C-terminal domain of HSP90 include novobiocin, coumermycin, and 'novologues' such as KU-32 and KU596 (Lancet et al. 2010). These agents were found to have neuroprotective properties when a biarylamide side chain modification was introduced, causing HSP90 client degradation without induction of the heat shock response, prompting clinical evaluation of these agents in neurodegenerative diseases such as diabetic neuropathy (Zhao et al. 2014). Further chemical modifications—specifically, the inclusion of a benzamide side chain—have resulted in compounds with marked anti-proliferative activity against a variety of cancer cell lines (Zhao et al. 2014; Forsberg et al. 2017). These agents have not yet reached the clinic.

Recognition of the importance of T-cells in anti-tumor immunity has led to very promising results in a subset of cancer patients. Multiple clinical trials evaluating several immune checkpoint inhibitor antibodies, cytokines, and engineered T-cell approaches are ongoing, with a view to informing rational combination strategies. HSP90 inhibitors have been identified as compounds that can synergistically potentiate anti-tumor responses when combined with checkpoint immunotherapy (Mbofung et al. 2016), and can sensitize tumors to client protein–specific T-cells (Raveendran et al. 2014). A number of novel clinical combination trials are underway, including XL888 together with the immune checkpoint inhibitor pembrolizumab (NCT03095781, Table 6.2), which will evaluate the hypothesis (informed by preclinical data) that HSP90 has a central role in modulating the tumor microenvironment as well as inflammatory signaling pathways associated with tumor immunosuppression (Akce et al. 2018). Immunotherapy combinations are not only limited to HSP90 inhibitors. Preclinically, HSP70-positive tumors were recognized by natural killer (NK) cells when the NK cells were activated by preincubation with HSP70 peptides and low-dose interleukin-2 (Multhoff et al. 2001); this strategy is currently undergoing testing in a randomized phase II study in patients with stage III/IV NSCLC after chemoradiotherapy (NCT02118415), though no results have been reported to date (Specht et al. 2015).

6.3 Conclusions

HSP90 plays an important role in the biology of human cancer, making it a potential target for therapy that may result in the inhibition of tumor development. Although the combination of HSP90 inhibitors and other oncologic clinical agents may herald exciting future outcomes, the clinical activity of pharmacological inhibition of HSP90 with small molecules (as single agents or in combination with other oncologic agents) has been limited up to now. Even though HSP90 modulates various oncogenic substrate proteins involved in crucial pathways that allow malignant tumors to thrive, the effectiveness of HSP90 inhibitors in the treatment of cancer has not been overwhelmingly successful to date. In current times, when oncology treatment is often decided based on multiomic molecular profiling, the ability to discover a biomarker predictive of response to HSP90 inhibitors is an area worth investigating. Developing the ability to better recognize which tumors are more likely to respond to this type of inhibition may be the door to a more promising future for research and development of novel HSP90 inhibitors in the oncology treatment arena—bringing a more prominent role to HSP90 inhibitors. Over the past 20 years, promising preclinical and clinical results have come from early-phase studies of HSP90 inhibitor agents, but this has not resulted in these agents being incorporated as standard-of-care for any malignancy.

Acknowledgements Special thanks to Sarah Miller, PhD.

References

Acquaviva J, Smith DL, Jimenez J-P, Zhang C, Sequeira M, He S et al (2014) Overcoming acquired BRAF inhibitor resistance in melanoma via targeted inhibition of Hsp90 with Ganetespib. Mol Cancer Ther 13:353–363

Ahsan A, Ramanand SG, Whitehead C, Hiniker SM, Rehemtulla A, Pratt WB et al (2012) Wild-type EGFR is stabilized by direct interaction with HSP90 in cancer cells and tumors. Neoplasia 14:670–677

Akce M, Alese OB, Shaib WL, Wu CS-Y, Lesinski GB, El-Rayes BF (2018) Phase Ib trial of pembrolizumab and XL888 in patients with advanced gastrointestinal malignancies. J Clin Oncol 36:TPS526

Alfano L, Guida T, Provitera L, Vecchio G, Billaud M, Santoro M et al (2010) RET is a Heat Shock Protein 90 (HSP90) client protein and is knocked down upon HSP90 pharmacological block. J Clin Endocrinol 95:3552–3557

Banerji U, O'Donnell A, Scurr M, Pacey S, Stapleton S, Asad Y et al (2005) Phase I pharmacokinetic and pharmacodynamic study of 17-allylamino, 17-demethoxygeldanamycin in patients with advanced malignancies. J Clin Oncol 23:4152–4161

Barrott JJ, Hughes PF, Osada T, Yang XY, Hartman ZC, Loiselle DR et al (2013) Optical and radioiodinated tethered Hsp90 inhibitors reveal selective internalization of ectopic Hsp90 in malignant breast tumor cells. Chem Biol 20:1187–1197

Beere HM, Wolf BB, Cain K, Mosser DD, Mahboubi A, Kuwana T et al (2000) Heat-shock protein 70 inhibits apoptosis by preventing recruitment of procaspase-9 to the Apaf-1 apoptosome. Nat Cell Biol 2:469–475

Biamonte MA, Van de Water R, Arndt JW, Scannevin RH, Perret D, Lee WC (2010) Heat shock protein 90: inhibitors in clinical trials. J Med Chem 53:3–17

Bonvini P, Gastaldi T, Falini B, Rosolen A (2002) Nucleophosmin-anaplastic lymphoma kinase (NPM-ALK), a novel Hsp90-client tyrosine kinase: down-regulation of NPM-ALK expression and tyrosine phosphorylation in ALK(+) CD30(+) lymphoma cells by the Hsp90 antagonist 17-allylamino,17-demethoxygeldanamycin. Cancer Res 62:1559–1566

Bray F, Ferlay J, Soerjomataram I, Siegel RL, Torre LA, Jemal A (2018) Global cancer statistics 2018: GLOBOCAN estimates of incidence and mortality worldwide for 36 cancers in 185 countries. CA Cancer J Clin 68:394–424

Breinig M, Mayer P, Harjung A, Goeppert B, Malz M, Penzel R et al (2011) Heat shock protein 90-sheltered overexpression of insulin-like growth factor 1 receptor contributes to malignancy of thymic epithelial tumors. Clin Cancer Res 17:2237–2249

Burrows F, Zhang H, Kamal A (2004) Hsp90 activation and cell cycle regulation. Cell Cycle 3:1530–1536

Caldas-Lopes E, Cerchietti L, Ahn JH, Clement CC, Robles AI, Rodina A et al (2009) Hsp90 inhibitor PU-H71, a multimodal inhibitor of malignancy, induces complete responses in triple-negative breast cancer models. Proc Natl Acad Sci U S A 106:8368–8373

Cameron DA, Spector N, Cortes J, Mano MS, Canon J-LR, Hickish T et al (2014) Targeting HSP90 in breast cancer: Enchant-1 (NCT01677455) phase 2 proof of concept study of ganetespib in first-line treatment of women with metastatic breast cancer. J Clin Oncol 32:TPS665

Chatterjee S, Burns TF (2017) Targeting heat shock proteins in cancer: a promising therapeutic approach. Int J Mol Sci 18:E1978

Chatterjee S, Bhattacharya S, Socinski MA, Burns TF (2016) HSP90 inhibitors in lung cancer: promise still unfulfilled. Clin Adv Hematol Oncol 14:346–356

Chen Z, Sasaki T, Tan X, Carretero J, Shimamura T, Li D et al (2010) Inhibition of ALK, PI3K/MEK, and HSP90 in murine lung adenocarcinoma induced by EML4-ALK fusion oncogene. Cancer Res 70:9827–9836

Chiosis G, Neckers L (2006) Tumor selectivity of Hsp90 inhibitors: the explanation remains elusive. ACS Chem Biol 1:279–284

Daozhen C, Lu L, Min Y, Xinyu J, Ying H (2007) Synthesis of (131)I-labeled-[(131)I]iodo-17-allylamino-17-demethoxy geldanamycin ([(131)I]iodo-17-AAG) and its biodistribution in mice. Cancer Biother Radiopharm 22:607–612

Do K, Speranza G, Chang L-C, Polley EC, Bishop R, Zhu W et al (2015) Phase I study of the heat shock protein 90 (Hsp90) inhibitor onalespib (AT13387) administered on a daily for 2 consecutive days per week dosing schedule in patients with advanced solid tumors. Investig New Drugs 33:921–930

Eccles SA, Massey A, Raynaud FI, Sharp SY, Box G, Valenti M et al (2008) NVP-AUY922: a novel heat shock protein 90 inhibitor active against xenograft tumor growth, angiogenesis, and metastasis. Cancer Res 68:2850–2860

Eiseman JL, Lan J, Lagattuta TF, Hamburger DR, Joseph E, Covey JM et al (2005) Pharmacokinetics and pharmacodynamics of 17-demethoxy 17-[[(2-dimethylamino)ethyl]amino]geldanamycin (17DMAG, NSC 707545) in C.B-17 SCID mice bearing MDA-MB-231 human breast cancer xenografts. Cancer Chemother Pharmacol 55:21–32

Eroglu Z, Chen YA, Gibney GT, Weber JS, Kudchadkar RR, Khushalani NI et al (2018) Combined BRAF and HSP90 inhibition in patients with unresectable BRAF V600E-mutant melanoma. Clin Cancer Res 24:5516–5524

Forsberg LK, Liu W, Holzbeierlein J, Blagg BSJ (2017) Modified biphenyl Hsp90 C-terminal inhibitors for the treatment of cancer. Bioorg Med Chem Lett 27:4514–4519

Friedland JC, Smith DL, Sang J, Acquaviva J, He S, Zhang C et al (2014) Targeted inhibition of Hsp90 by ganetespib is effective across a broad spectrum of breast cancer subtypes. Investig New Drugs 32:14–24

Garon EB, Moran T, Barlesi F, Gandhi L, Sequist LV, Kim S-W et al (2012) Phase II study of the HSP90 inhibitor AUY922 in patients with previously treated, advanced non-small cell lung cancer (NSCLC). J Clin Oncol 30:7543–7543

Garrido C, Schmitt E, Cande C, Vahsen N, Parcellier A, Kroemer G (2003) HSP27 and HSP70: potentially oncogenic apoptosis inhibitors. Cell Cycle 2:579–584

Garrido C, Brunet M, Didelot C, Zermati Y, Schmitt E, Kroemer G (2006) Heat shock proteins 27 and 70: anti-apoptotic proteins with tumorigenic properties. Cell Cycle 5:2592–2601

Georgakis GV, Li Y, Younes A (2006) The heat shock protein 90 inhibitor 17-AAG induces cell cycle arrest and apoptosis in mantle cell lymphoma cell lines by depleting cyclin D1, Akt, Bid and activating caspase 9. Br J Haematol 135:68–71

Ghosh JC, Dohi T, Kang BH, Altieri DC (2008) Hsp60 regulation of tumor cell apoptosis. J Biol Chem 283:5188–5194

Goetz MP, Toft D, Reid J, Ames M, Stensgard B, Safgren S et al (2005) Phase I trial of 17-allylamino-17-demethoxygeldanamycin in patients with advanced cancer. J Clin Oncol 23:1078–1087

Graham B, Curry J, Smyth T, Fazal L, Feltell R, Harada I et al (2012) The heat shock protein 90 inhibitor, AT13387, displays a long duration of action in vitro and in vivo in non-small cell lung cancer. Cancer Sci 103:522–527

Graner MW (2016) Chapter eight – HSP90 and immune modulation in Cancer. In: Isaacs J, Whitesell L (eds) Advances in cancer research, vol 129. Elsevier Science & Technology, San Diego, pp 191–224

Guo F, Rocha K, Bali P, Pranpat M, Fiskus W, Boyapalle S et al (2005) Abrogation of heat shock protein 70 induction as a strategy to increase antileukemia activity of heat shock protein 90 inhibitor 17-allylamino-demethoxy geldanamycin. Cancer Res 65:10536–10544

Hasenstein JR, Shin HC, Kasmerchak K, Buehler D, Kwon GS, Kozak KR (2012) Antitumor activity of Triolimus: a novel multidrug-loaded micelle containing Paclitaxel, Rapamycin, and 17-AAG. Mol Cancer Ther 11:2233–2242

Hollingshead M, Alley M, Burger AM, Borgel S, Pacula-Cox C, Fiebig HH et al (2005) In vivo antitumor efficacy of 17-DMAG (17-dimethylaminoethylamino-17-demethoxygeldanamycin hydrochloride), a water-soluble geldanamycin derivative. Cancer Chemother Pharmacol 56:115–125

Jensen MR, Schoepfer J, Radimerski T, Massey A, Guy CT, Brueggen J et al (2008) NVP-AUY922: a small molecule HSP90 inhibitor with potent antitumor activity in preclinical breast cancer models. Breast Cancer Res 10:R33

Jhaveri K, Taldone T, Modi S, Chiosis G (2012) Advances in the clinical development of heat shock protein 90 (Hsp90) inhibitors in cancers. Biochim Biophys Acta 1823:742–755

Jhaveri K, Chandarlapaty S, Lake D, Gilewski T, Robson M, Goldfarb S et al (2014a) A phase II open-label study of ganetespib, a novel heat shock protein 90 inhibitor for patients with metastatic breast cancer. Clin Breast Cancer 14:154–160

Jhaveri K, Ochiana SO, Dunphy MP, Gerecitano JF, Corben AD, Peter RI et al (2014b) Heat shock protein 90 inhibitors in the treatment of cancer: current status and future directions. Expert Opin Investig Drugs 23:611–628

Jhaveri K, Wang R, Teplinsky E, Chandarlapaty S, Solit D, Cadoo K et al (2017) A phase I trial of ganetespib in combination with paclitaxel and trastuzumab in patients with human epidermal growth factor receptor-2 (HER2)-positive metastatic breast cancer. Breast Cancer Res 19:89

Kamal A, Thao L, Sensintaffar J, Zhang L, Boehm MF, Fritz LC et al (2003) A high-affinity conformation of Hsp90 confers tumour selectivity on Hsp90 inhibitors. Nature 425:407–410

Kummar S, Gutierrez ME, Gardner ER, Chen X, Figg WD, Zajac-Kaye M et al (2010) Phase I trial of 17-dimethylaminoethylamino-17-demethoxygeldanamycin (17-DMAG), a heat shock protein inhibitor, administered twice weekly in patients with advanced malignancies. Eur J Cancer 46:340–347

Lamberti D, Cristinziano G, Porru M, Leonetti C, Egan JB, Shi CX et al (2018) HSP90 inhibition drives degradation of FGFR2 fusion proteins: implications for treatment of cholangiocarcinoma. Hepatology 69:131–142

Lancet JE, Gojo I, Burton M, Quinn M, Tighe SM, Kersey K et al (2010a) Phase I study of the heat shock protein 90 inhibitor alvespimycin (KOS-1022, 17-DMAG) administered intravenously twice weekly to patients with acute myeloid leukemia. Leukemia 24:699–705

Lancet JE, Smith BD, Bradley R, Komrokji RS, Teofilovici F, Rizzieri DA (2010b) A phase I/II trial of the potent Hsp90 inhibitor STA-9090 administered once weekly in patients with advanced hematologic malignancies. Blood 116:3294–3294

Li J, Sun L, Xu C, Yu F, Zhou H, Zhao Y et al (2012) Structure insights into mechanisms of ATP hydrolysis and the activation of human heat-shock protein 90. Acta Biochim Biophys Sin Shanghai 44:300–306

Marcu MG, Schulte TW, Neckers L (2000) Novobiocin and related coumarins and depletion of heat shock protein 90-dependent signaling proteins. J Natl Cancer Inst 92:242–248

Mbofung RM, McKenzie JA, Malu S, Liu C, Peng W, Kuiatse I et al (2016) Abstract B105: HSP90 inhibitor, ganetespib, enhances responses to cancer immunotherapy through increased expression of interferon response genes. Cancer Immunol Res 4:B105

McCarthy MM, Pick E, Kluger Y, Gould-Rothberg B, Lazova R, Camp RL et al (2008) HSP90 as a marker of progression in melanoma. Ann Oncol 19:590–594

Meehan R, Kummar S, Do K, O'Sullivan Coyne G, Juwara L, Zlott J et al (2018) A phase I study of Ganetespib and Ziv-Aflibercept in patients with advanced carcinomas and sarcomas. Oncologist 23:1269–e1125

Mimnaugh EG, Chavany C, Neckers L (1996) Polyubiquitination and proteasomal degradation of the p185c-erbB-2 receptor protein-tyrosine kinase induced by geldanamycin. J Biol Chem 271:22796–22801

Miyajima N, Tsutsumi S, Sourbier C, Beebe K, Mollapour M, Rivas C et al (2013) The HSP90 inhibitor ganetespib synergizes with the MET kinase inhibitor crizotinib in both crizotinib-sensitive and -resistant MET-driven tumor models. Cancer Res 73:7022–7033

Modi S, Stopeck A, Linden H, Solit D, Chandarlapaty S, Rosen N et al (2011) HSP90 inhibition is effective in breast cancer: a phase II trial of tanespimycin (17-AAG) plus trastuzumab in patients with HER2-positive metastatic breast cancer progressing on trastuzumab. Clin Cancer Res 17:5132–5139

Modi S, Saura C, Henderson C, Lin NU, Mahtani R, Goddard J et al (2013) A multicenter trial evaluating retaspimycin HCL (IPI-504) plus trastuzumab in patients with advanced or metastatic HER2-positive breast cancer. Breast Cancer Res Treat 139:107–113

Multhoff G, Pfister K, Gehrmann M, Hantschel M, Gross C, Hafner M et al (2001) A 14-mer Hsp70 peptide stimulates natural killer (NK) cell activity. Cell Stress Chaperones 6:337–344

Neckers L, Workman P (2012) Hsp90 molecular chaperone inhibitors: are we there yet? Clin Cancer Res 18:64–76

Normant E, Paez G, West KA, Lim AR, Slocum KL, Tunkey C et al (2011) The Hsp90 inhibitor IPI-504 rapidly lowers EML4-ALK levels and induces tumor regression in ALK-driven NSCLC models. Oncogene 30:2581–2586

Obermann WM, Sondermann H, Russo AA, Pavletich NP, Hartl FU (1998) In vivo function of Hsp90 is dependent on ATP binding and ATP hydrolysis. J Cell Biol 143:901–910

Ohkubo S, Kodama Y, Muraoka H, Hitotsumachi H, Yoshimura C, Kitade M et al (2015) TAS-116, a highly selective inhibitor of heat shock protein 90alpha and beta, demonstrates potent antitumor activity and minimal ocular toxicity in preclinical models. Mol Cancer Ther 14:14–22

Pacey S, Wilson RH, Walton M, Eatock MM, Hardcastle A, Zetterlund A et al (2011) A phase I study of the heat shock protein 90 inhibitor alvespimycin (17-DMAG) given intravenously to patients with advanced solid tumors. Clin Cancer Res 17:1561–1570

Padmanabhan S, Kelly KR, Heaney M, Hodges S, Chanel S, Frattini M et al (2010) A phase I study of the potent Hsp90 inhibitor STA-9090 administered twice weekly in subjects with hematologic malignancies. Blood 116:2898–2898

Paraiso KH, Haarberg HE, Wood E, Rebecca VW, Chen YA, Xiang Y et al (2012) The HSP90 inhibitor XL888 overcomes BRAF inhibitor resistance mediated through diverse mechanisms. Clin Cancer Res 18:2502–2514

Patel K, Wen J, Magliocca K, Muller S, Liu Y, Chen ZG et al (2014) Heat shock protein 90 (HSP90) is overexpressed in p16-negative oropharyngeal squamous cell carcinoma, and its inhibition in vitro potentiates the effects of chemoradiation. Cancer Chemother Pharmacol 74:1015–1022

Pick E, Kluger Y, Giltnane JM, Moeder C, Camp RL, Rimm DL et al (2007) High HSP90 expression is associated with decreased survival in breast cancer. Cancer Res 67:2932–2937

Powers MV, Clarke PA, Workman P (2008) Dual targeting of HSC70 and HSP72 inhibits HSP90 function and induces tumor-specific apoptosis. Cancer Cell 14:250–262

Radons J (2016) The human HSP70 family of chaperones: where do we stand? Cell Stress Chaperones 21:379–404

Ramalingam SS, Zaric B, Ceric T, Ciuleanu TE, Spicer JF, Bondarenko I et al (2014) Galaxy-2 trial (NCT01798485): a randomized phase 3 study of ganetespib in combination with docetaxel versus docetaxel alone in patients with advanced lung adenocarcinoma. J Clin Oncol 32:TPS8118

Ramalingam S, Goss G, Rosell R, Schmid-Bindert G, Zaric B, Andric Z et al (2015) A randomized phase II study of ganetespib, a heat shock protein 90 inhibitor, in combination with docetaxel in second-line therapy of advanced non-small cell lung cancer (GALAXY-1). Ann Oncol 26:1741–1748

Rappa F, Farina F, Zummo G, David S, Campanella C, Carini F et al (2012) HSP-molecular chaperones in cancer biogenesis and tumor therapy: an overview. Anticancer Res 32:5139–5150

Ravagnan L, Gurbuxani S, Susin SA, Maisse C, Daugas E, Zamzami N et al (2001) Heat-shock protein 70 antagonizes apoptosis-inducing factor. Nat Cell Biol 3:839–843

Raveendran S, Rao A, Storkus W (2014) Combination immunotherapy of melanoma by inhibiting HSP90 and targeting its client proteins. (TUM7P.934). J Immunol 192:203.216

Rerole AL, Jego G, Garrido C (2011) Hsp70: anti-apoptotic and tumorigenic protein. Methods Mol Biol 787:205–230

Richardson PG, Chanan-Khan AA, Lonial S, Krishnan AY, Carroll MP, Alsina M et al (2011) Tanespimycin and bortezomib combination treatment in patients with relapsed or relapsed and refractory multiple myeloma: results of a phase 1/2 study. Br J Haematol 153:729–740

Rodina A, Wang T, Yan P, Gomes ED, Dunphy MP, Pillarsetty N et al (2016) The epichaperome is an integrated chaperome network that facilitates tumour survival. Nature 538:397–401

Ronnen EA, Kondagunta GV, Ishill N, Sweeney SM, Deluca JK, Schwartz L et al (2006) A phase II trial of 17-(Allylamino)-17-demethoxygeldanamycin in patients with papillary and clear cell renal cell carcinoma. Investig New Drugs 24:543–546

Saif MW, Takimoto C, Mita M, Banerji U, Lamanna N, Castro J et al (2014) A phase 1, dose-escalation, pharmacokinetic and Pharmacodynamic study of BIIB021 administered orally in patients with advanced solid tumors. Clin Cancer Res 20:445–455

Saleh A, Srinivasula SM, Balkir L, Robbins PD, Alnemri ES (2000) Negative regulation of the Apaf-1 apoptosome by Hsp70. Nat Cell Biol 2:476–483

Sang J, Acquaviva J, Friedland JC, Smith DL, Sequeira M, Zhang C et al (2013) Targeted inhibition of the molecular chaperone Hsp90 overcomes ALK inhibitor resistance in non-small cell lung cancer. Cancer Discov 3:430–443

Schroder CP, Pedersen JV, Chua S, Swanton C, Akimov M, Ide S et al (2011) Use of biomarkers and imaging to evaluate the treatment effect of AUY922, an HSP90 inhibitor, in patients with HER2+ or ER+ metastatic breast cancer. J Clin Oncol 29:e11024

Sequist LV, Gettinger S, Senzer NN, Martins RG, Janne PA, Lilenbaum R et al (2010) Activity of IPI-504, a novel heat-shock protein 90 inhibitor, in patients with molecularly defined non-small-cell lung cancer. J Clin Oncol 28:4953–4960

Shapiro GI, Kwak E, Dezube BJ, Yule M, Ayrton J, Lyons J et al (2015) First-in-human phase I dose escalation study of a second-generation non-ansamycin HSP90 inhibitor, AT13387, in patients with advanced solid tumors. Clin Cancer Res 21:87–97

Shi Y, Liu X, Lou J, Han X, Zhang L, Wang Q et al (2014) Plasma levels of heat shock protein 90 alpha associated with lung cancer development and treatment responses. Clin Cancer Res 20:6016–6022

Socinski MA, Goldman J, El-Hariry I, Koczywas M, Vukovic V, Horn L et al (2013) A multicenter phase II study of ganetespib monotherapy in patients with genotypically defined advanced non-small cell lung cancer. Clin Cancer Res 19:3068–3077

Soga S, Shiotsu Y, Akinaga S, Sharma SV (2003) Development of radicicol analogues. Curr Cancer Drug Targets 3:359–369

Soga S, Akinaga S, Shiotsu Y (2013) Hsp90 inhibitors as anti-cancer agents, from basic discoveries to clinical development. Curr Pharm Des 19:366–376

Solit DB, Osman I, Polsky D, Panageas KS, Daud A, Goydos JS et al (2008) Phase II trial of 17-allylamino-17-demethoxygeldanamycin in patients with metastatic melanoma. Clin Cancer Res 14:8302–8307

Specht HM, Ahrens N, Blankenstein C, Duell T, Fietkau R, Gaipl US et al (2015) Heat Shock Protein 70 (Hsp70) peptide activated natural killer (NK) cells for the treatment of patients with non-small cell lung Cancer (NSCLC) after Radiochemotherapy (RCTx) – from preclinical studies to a clinical phase II trial. Front Immunol 6:162

Speranza G, Anderson L, Chen AP, Do K, Eugeni M, Weil M et al (2018) First-in-human study of the epichaperome inhibitor PU-H71: clinical results and metabolic profile. Investig New Drugs 36:230–239

Straume O, Shimamura T, Lampa MJ, Carretero J, Oyan AM, Jia D et al (2012) Suppression of heat shock protein 27 induces long-term dormancy in human breast cancer. Proc Natl Acad Sci U S A 109:8699–8704

Supko JG, Hickman RL, Grever MR, Malspeis L (1995) Preclinical pharmacologic evaluation of geldanamycin as an antitumor agent. Cancer Chemother Pharmacol 36:305–315

Taipale M, Krykbaeva I, Koeva M, Kayatekin C, Westover KD, Karras GI et al (2012) Quantitative analysis of HSP90-client interactions reveals principles of substrate recognition. Cell 150:987–1001

Vartholomaiou E, Echeverría PC, Picard D (2016) Chapter one – unusual suspects in the twilight zone between the Hsp90 interactome and carcinogenesis. In: Isaacs J, Whitesell L (eds) Advances in cancer research, vol 129. Academic, Cambridge, MA, pp 1–30

Wagner AJ, Chugh R, Rosen LS, Morgan JA, George S, Gordon M et al (2013) A phase I study of the HSP90 inhibitor retaspimycin hydrochloride (IPI-504) in patients with gastrointestinal stromal tumors or soft-tissue sarcomas. Clin Cancer Res 19:6020–6029

Wagner AJ, Agulnik M, Heinrich MC, Mahadevan D, Riedel RF, von Mehren M et al (2016) Dose-escalation study of a second-generation non-ansamycin HSP90 inhibitor, onalespib (AT13387), in combination with imatinib in patients with metastatic gastrointestinal stromal tumour. Eur J Cancer 61:94–101

Wang X, Chen M, Zhou J, Zhang X (2014) HSP27, 70 and 90, anti-apoptotic proteins, in clinical cancer therapy (review). Int J Oncol 45:18–30

Whitesell L, Lindquist SL (2005) HSP90 and the chaperoning of cancer. Nat Rev Cancer 5:761–772

Whitesell L, Sutphin PD, Pulcini EJ, Martinez JD, Cook PH (1998) The physical association of multiple molecular chaperone proteins with mutant p53 is altered by geldanamycin, an hsp90-binding agent. Mol Cell Biol 18:1517–1524

Wu J, Liu T, Rios Z, Mei Q, Lin X, Cao S (2017) Heat shock proteins and cancer. Trends Pharmacol Sci 38:226–256

Yao Q, Nishiuchi R, Li Q, Kumar AR, Hudson WA, Kersey JH (2003) FLT3 expressing leukemias are selectively sensitive to inhibitors of the molecular chaperone heat shock protein 90 through destabilization of signal transduction-associated kinases. Clin Cancer Res 9:4483–4493

Yufu Y, Nishimura J, Nawata H (1992) High constitutive expression of heat shock protein 90 alpha in human acute leukemia cells. Leuk Res 16:597–605

Yuno A, Lee MJ, Lee S, Tomita Y, Rekhtman D, Moore B et al (2018) Clinical evaluation and biomarker profiling of Hsp90 inhibitors. Methods Mol Biol 1709:423–441

Zhao H, Anyika M, Girgis A, Blagg BS (2014) Novologues containing a benzamide side chain manifest anti-proliferative activity against two breast cancer cell lines. Bioorg Med Chem Lett 24:3633–3637

Zhou D, Liu Y, Ye J, Ying W, Ogawa LS, Inoue T et al (2013) A rat retinal damage model predicts for potential clinical visual disturbances induced by Hsp90 inhibitors. Toxicol Appl Pharmacol 273:401–409

Zuehlke A, Johnson JL (2010) Hsp90 and co-chaperones twist the functions of diverse client proteins. Biopolymers 93:211–217

Chapter 7
p53-Hsp90 Axis in Human Cancer

Amr Ghaleb and Natalia Marchenko

Abstract The heat shock protein 90 (Hsp90) is a ubiquitous molecular chaperone that is abundantly expressed in cancer cells and plays a pivotal role in correct folding and functions of a variety of oncogenic clients. Hsp90 is up-regulated in response to cellular stresses that cancer cells encounter, such as heat, hypoxia and nutrient deprivation, conditions commonly associated with the tumor microenvironment. P53 is the tumor suppressor gene that is mutated in nearly 50% of all human cancers. When mutated p53 not only lose its tumor suppressive function but also gain novel oncogenic activities via gain-of-function mechanisms leading to increased genomic instability, chemoresistance, and metastasis, which promote tumor progression. In contrast to wild-type p53, mutant p53 is protected from degradation via interaction with Hsp90 leading to marked stabilization of mutant p53 protein in cancer cells. Recent in vivo studies unequivocally have proven that the stabilization of mutant p53 is crucial pre-requisite for its oncogenic functions. The pharmacological targeting the pathways involved in the stabilization of mutant p53, in particular, the Hsp90 chaperone complex, recently attracted a lot of attention as a promising therapeutic approach to treat mutant p53 harboring cancers.

Keywords Cancer · Gain-of-function · Heat shock protein · Hsp90 · Loss of heterozygosity · p53

Abbreviations

DN	dominant negative
GOF	gain-of-function
HDAC	histone deacetylase
HSF1	heat shock factor 1

A. Ghaleb · N. Marchenko (✉)
Department of Pathology, Stony Brook University, Stony Brook, NY, USA
e-mail: natalia.marchenko@stonybrookmedicine.edu

A. A. A. Asea, P. Kaur (eds.), *Heat Shock Protein 90 in Human Diseases and Disorders*, Heat Shock Proteins 19,
https://doi.org/10.1007/978-3-030-23158-3_7

HSP	heat shock protein
LOH	loss of heterozygosity
MDM2	mouse double-minute 2
RTK	receptor tyrosine kinases
SAHA	suberoylanilide hydroxamic acid

7.1 Introduction

The tumor suppressor protein, p53, was initially discovered in the late 1970s, where it was wrongfully identified as the middle antigen of SV40 virus (Kress et al. 1979; Lane and Crawford 1979; Linzer and Levine 1979). Later, p53 was found to be expressed in the cells in response to viral infections and that it accumulates in the nucleus of tumor cells (Kress et al. 1979). Initially, p53 was characterized as an oncoprotein (Eliyahu et al. 1984; Jenkins et al. 1984; Parada et al. 1984). Over the years, starting in the mid-1980s, *TP53* was firmly established as tumor suppressor gene, cell cycle regulator, and cell sensor in response multiple stresses that the cell may encounter (Gatz and Wiesmuller 2006; Menon and Povirk 2014; Soussi 2010).

The tumor suppressive functions of p53 were shown to be inhibited by various mechanisms, such as epigenetic modulations, miRNA, or by a dysregulated CCCTC-binding factor (Saldana-Meyer and Recillas-Targa 2011). However, mutations in TP53 gene are the most frequent inactivating genetic events in most types of cancer ((http://p53.iarc.fr/TP53SomaticMutations.aspx) (Petitjean et al. 2007; Vogelstein et al. 2000)). Unlike other tumor suppressors that are usually inactivated by deletion, 75% of all p53 genetic alterations in TP53 gene are missense mutations within the DNA-binding domain, predominantly located in six most frequent 'hotspots' (http://p53.iarc.fr/TP53SomaticMutations.aspx) (Baugh et al. 2018; Bouaoun et al. 2016). TP53 missense mutations hinder p53 binding to DNA response elements precluding its tumor suppressive transcription program. Clinical data suggest that p53 behaves as a classic "two-hit" tumor suppressor so that a point mutation in one allele of p53 at early stages is followed by loss of heterozygosity (LOH) in the second allele later on during tumor progression (Levine et al. 1991).

Mutant p53 exerts its oncogenic activities either by dominant negative (DN) or by gain-of-function (GOF) mechanisms (Petitjean et al. 2007; Schulz-Heddergott and Moll 2018). In early stages of cancers that are heterozygous for p53, mutant p53 may suppress the expression and/or function of wild-type p53 through DN mechanism (Milner et al. 1991; Willis et al. 2004). Mutant p53 was shown to form heterodimers with wild-type p53 that can cause conformational changes to wild-type p53 or may directly inhibit the DNA-binding activity of wild-type p53 on target genes. Similarly, mutant p53 can interact with other p53 family members- TAp63 and TAp73, to inhibit their tumor suppressive functions (Lang et al. 2004; Muller et al. 2009). Although, in vitro studies greatly support DN effect of mutant p53 (Milner et al. 1991; Willis et al. 2004), in mouse models it was shown to be highly tissue-specific and often is manifested only upon stabilization of mutant p53 after genotoxic stress, e.g., epithelium-specific heterozygous expression of mutant p53 (R270H) exerted DN effect on tumor latency, multiplicity, and progression only after UV exposure but not spontaneous tumors (Wijnhoven et al. 2007). Study on R246S mutant p53 knock-in mouse model has shown that the DN effect on transactivation is detectable after acute p53 activation only (Lee et al. 2012). Also, DNA damage increased the DN activity of mutant p53 in various tissues of heterozygous mutant p53 mice (Lang et al. 2004; Olive et al. 2004).

On the other hand, GOF is commonly defined as "any activity of mutant p53 exerted in the absence of co-expressed wild-type p53" (Oren and Rotter 2010). In accord with this notion, we recently demonstrated that p53 LOH is a necessary prerequisite for mutant p53 GOF in vivo (Alexandrova et al. 2017a). The GOF concept was first introduced by Dittmer et al., where they showed the ability of mutant p53, but not wild-type p53, to transform p53-null cells in vitro and in vivo (Dittmer et al. 1993). Since then, numerous publications have confirmed that missense mutations in the *TP53* gene not only obliterate wild-type p53 tumor suppressive function, but mutant p53 protein also gains novel oncogenic activities. By inducing multiple oncogenic pathways (Adorno et al. 2009; Li et al. 2014; Muller et al. 2009; Yallowitz et al. 2015), microRNA and altering cell metabolism, transcriptional regulation, chromatin structure (Blandino and Di Agostino 2018; Oren and Rotter 2010), mutant p53 promotes oncogenic reprogramming and cancer progression.

7.1.1 Stability of Mutant p53 Protein in Cancer Cells

The stability of a wild-type p53 protein is tightly regulated through degradation by E3 ligase mouse double-minute 2 (MDM2) (Haupt et al. 1997; Lukashchuk and Vousden 2007). As wild-type p53 target gene, MDM2 itself is regulated through a negative feedback loop maintaining a very low and steady level of both proteins under normal condition (Wu et al. 1993). In normal and unstressed cancer cells wild-type p53 protein level is very low due to a rapid turnover by MDM2. This tight control is lost in tumors carrying mutant p53. While initially the stability of mutant p53 was thought to be due to the loss of p53-mediated transactivation of MDM2, it was later shown that mutant p53 is inherently unstable in normal cells, and only tumors but not normal tissues display constitutive stabilization of mutant p53 (Lang et al. 2004; Olive et al. 2004; Terzian et al. 2008). Mutant p53 knock-in models demonstrated that MDM2 is still expressed in a p53-independent mechanism in normal tissues and additional alteration(s) occur upon malignant conversion that stabilizes mutant p53 (Lang et al. 2004; Olive et al. 2004).

Similar to wild-type p53, mutant p53 protein fully preserves its ability to interact with MDM2, and MDM2 protein levels can be easily detectable in mutant p53 expressing cancer cells (Li et al. 2011b). Nevertheless, the profound lack of ubiquitination of mutant p53 indicates a severe degradation defect, due to inactivation of MDM2 enzymatic activity (Li et al. 2011b). Mechanistically, it has been shown that mutant p53 stimulates the formation of triple MDM2-p53-Hsp90 complexes, whereby blocking ubiquitin-protein isopeptide ligase activity of MDM2 and resulting in the stabilization of both mutant p53 and MDM2 (Peng et al. 2001). This evidence indicates that Hsp90 is essential for the stabilization of mutant p53 in cancer cells.

Strikingly, it appears that cancer cells become addicted to high levels of mutant p53 protein, as an acute depletion of mutant p53 markedly reduces cells growth in vitro and in xenografts (Li et al. 2011b), suppresses invasion (Adorno et al. 2009; Li et al. 2011b; Muller et al. 2009; Yan et al. 2008), restores normal mammary architecture in 3D culture in breast cancer cell lines (Freed-Pastor et al. 2012), inhibits metastases in vivo (Morton et al. 2010; Weissmueller et al. 2014), and suppresses mammary stem cells (Yallowitz et al. 2015). Mutant p53 depletion in allotransplanted and autochthonous mouse T/B-lymphoma model curbs tumor growth and extends the survival of mice (Alexandrova et al. 2015). Similarly, the genetic ablation of R248Q mutant p53 inhibits tumor growth and invasion in AOM/DSS-Induced colorectal cancer mouse models. Mechanistically, Schulz-Heddergott et al. demonstrated that mutant p53 protein binds and enhances activating phosphorylation of Stat3, while conditional depletion of mutant p53 from established tumors suppresses Jak2/Stat3 signaling, tumor growth, and invasiveness (Schulz-Heddergott et al. 2018). As other evidence supporting the importance of highly stabilized mutant p53 for the manifestation of GOF activity, we found that in heterozygosity wild-type p53 allele suppresses mutant p53 stabilization in cancer cells in vivo, whereas p53 LOH is associated with stabilization of mutant p53 in cancer cells and, most importantly, with mutant p53 GOF features (Alexandrova et al. 2017a).

Collectively, this proof-of-principal from in vivo studies demonstrate that high levels of mutant p53 protein is essential for the manifestation of mutant p53 GOF and underscore the translational significance of highly stabilized mutant p53 as a potentially attractive therapeutic target. However, targeting mutant p53 is a very challenging task, since mutant p53 is neither a typical enzyme nor localized to the cell surface. On the other hand, the addiction of cancer cells to highly stabilized mutant p53 can be exploited by therapeutic approaches aiming to eliminate mutant p53 from cancer cells. Thus, selectively targeting the mechanisms underlying cancer specific stabilization of mutant p53 protein is the feasible strategy to combat mutant p53 harboring cancers.

7.1.2 Hsp90 in Cancer

A role for Hsp90 in cancer was first implicated in breast tumors where an association of Hsp90 and steroid receptors was observed (Pratt 1987; Shyamala et al. 1989), and in ovarian cancer where increased expression of Hsp90 mRNA was detected (Mileo et al. 1990). To date, it is well known that Hsp90 is ubiquitously expressed in almost all cancer types. The high-level of Hsp90 was shown to be associated with more aggressive phenotypes and poor prognosis in breast cancer patients. Also, the high levels Hsp90 were shown to correlate with more advanced stages of melanomas, leukemia (Chatterjee and Burns 2017) and colon cancer (Milicevic et al. 2008). There are two major isoforms of Hsp90: constitutive Hsp90α and inducible Hsp90β, which together comprise 1–3% of total cellular protein. Although Hsp90 is also expressed in normal tissues, it is activated in cancer cells by various mechanisms: modulation of activity by translational modifications (acetylation, phosphorylation, ubiquitination, S-nitrosylation); activation by interaction with clients; activation of transcription; and subcellular localization (cytoplasm vs. cell surface) (Barrott and Haystead 2013).

As a molecular chaperone, Hsp90 plays an essential role in maintaining proteome homeostasis in cancer cells by facilitating folding and intracellular trafficking of its numerous oncogenic clients. Although Hsp90 is very abundant protein, it gets further upregulated by different stresses experienced by tumor cells such as acidic pH, shortage of nutrients and fluctuating oxygen levels, which commonly occur in cancer (Gabai et al. 1995). It appears that Hsp90 is essential to cancer cells survival by stabilizing and preventing aggregation of many mutated aberrantly folded oncogenic proteins, which drive tumor progression. There are over 400 Hsp90 client proteins (for updated client protein list www.picard.ch/downloads/Hsp90interactors.pdf), with a large number of them playing important roles in different aspects of tumor development such as tumor growth, apoptotic evasion, differentiation and metastasis (Neckers and Workman 2012). The pharmacological inhibition of Hsp90 leads to rapid inhibition of clients activity, with their subsequent degradation leading to depletion of multiple oncoproteins and down-regulation of oncogenic signaling pathways, and ultimately resulting in the modulation of the malignant phenotype (Jackson 2013). The reliance of cancer cells on oncogenic pathways and proteins that are stabilized by Hsp90 is the reason for the sensitivity of these cells to Hsp90 inhibition (Neckers and Workman 2012; Whitesell and Lindquist 2005; Workman et al. 2007). Also, Hsp90 has been found to be secreted by cancer cells, and the role of the secreted Hsp90 has been garnering some attention for its role in cancer cell invasion and metastasis (Eustace and Jay 2004; Eustace et al. 2004; Nolan et al. 2015). Overall, given the clinical importance of numerous oncogenic Hsp90 clients in different types of cancer, targeting Hsp90 and its respective co-chaperones has huge potential for utilization in cancer therapy.

7.1.3 Mutant p53-Hsp90 Axis

Commonly, mutations in TP53 cause an alteration of mutant p53 protein conformation (e.g., residues R249, G245, R282, R175), lower melting temperature, and lead to aberrant protein folding (Schulz-Heddergott and Moll 2018). To prevent protein aggregation and cell death from proteotoxic stress, mutant p53 proteins interact with a number of molecular chaperones, heat shock proteins (HSP) that are essential for stabilization aggregation-prone proteins. Mutant p53 was shown to interact with several HSP: Hsp90 (Blagosklonny et al. 1996), Hsp70 (Sturzbecher et al. 1988), HSC70 (Hinds et al. 1987) and HSP40/DNAJ complex (Parrales et al. 2016).

Recently, mutant p53- Hsp90 interaction was extensively studied, as it was shown that pharmacological Hsp90 inhibition destabilized mutant p53, but not wild-type p53 in cancer cells in vitro (Blagosklonny et al. 1996; Li et al. 2011b; Peng et al. 2001) and in several mouse models (Alexandrova and Marchenko 2015; Schulz-Heddergott et al. 2018). The long-term Hsp90 inhibition by specific inhibitor ganetespib extended the survival of mutant p53 R248Q/− and R172H/R172H knock-in mice by 59% and 48%, respectively, but not p53−/− littermates. Importantly, mutant p53-dependent survival benefits correlated with mutant p53 degradation, apoptosis and prevention of T-cell lymphomagenesis in vivo (Alexandrova et al. 2015). Treating mice with an Hsp90 inhibitor 17AAG suppressed mutant p53 levels and colorectal tumor growth in p53 R248Q/−mice treated with AOM/DSS compared with vehicle controls. Although 17AAG showed some tumor inhibitory effect in p53−/− control mice, it was much less pronounced than in mutant p53 mice (Schulz-Heddergott et al. 2018). Also, ganetespib synergized with a sub-effective dose of the DNA-alkylating cytotoxic agent cyclophosphamide in mutant p53 R248Q lymphoma mouse model by suppressing tumor growth and extending survival. Importantly, ganetespib/cyclophosphamide combinatorial treatment was more beneficial for mutant p53 than p53−/− mice (Alexandrova et al. 2017b). Together, these rigorous in vivo and in vitro studies implicate the potential clinical utility of targeting mutant p53-Hsp90 axis and the mechanisms regulating this interaction to degrade mutant p53 GOF in cancer.

The transcription of inducible HSP is regulated by the master transcription regulator of heat shock response- heat shock factor 1 (HSF1) (Gomez-Pastor et al. 2018). HSF1 induces the transcription of Hsp90α, Hsp70, members of the HSP40/DNAJ family, and numerous co-chaperones (Gomez-Pastor et al. 2018). In normal cells, HSF1 is transiently active in conditions of proteotoxic stress, while tumor cells that under permanent proteotoxic stress due to adverse tumor environment require permanent activation of HSF1. Importantly, activation of HSF1 also appears to be essential for malignant transformation, as shown in several mouse cancer models (Dai et al. 2012; Dai et al. 2007). In addition to regulation of heat shock response, HSF1 drives broad cancer-specific transcriptional programs required for numerous oncogenic processes, such as aberrant cell cycle regulation, signal transduction, translation, metabolism, and invasion (Dai et al. 2007; Mendillo et al. 2012; Min et al. 2007; Zhao et al. 2011). Importantly, recent studies on HSF1 knockout mouse model provide compelling genetic evidence for the existence of crucial cooperation between mutant p53 and HSF1 in malignancy (Dai et al. 2007; Min et al. 2007). Given the importance of both HSF1 and mutant p53 in cancer, we identified the mechanism of interaction between HSF1 and mutant p53 in cancer cells; a novel gain-of-function of mutant p53 as a promoter of HSF1 activity (Li et al. 2014). We found that mutant p53 via stimulation of HSF1 transcriptional activity induces a feed-forward loop to Hsp90, which further stabilizes mutant p53 protein. Mechanistically, mutant p53, through enhanced recycling and/or stability of ErbB2, augments MAPK and PI3K signaling, leading to phospho-activation of HSF1 at Ser326. Furthermore, mutant p53 via direct interaction with phospho-activated HSF1 facilitates the binding of HSF1 to its DNA-binding sites, thereby stimulating the transcription of HSP. In turn, heat shock proteins more potently stabilize their clients ErbB2, EGFR, mutant p53, and HSF1 (and possibly other oncogenes), thus reinforcing tumor progression (Li et al. 2014). This feed-forward loop renders cancer cells resistant to proteotoxic stress, providing a distinct survival advantage. This feed-forward axis was also confirmed in vivo, where we defined its physiological consequences in ErbB2-driven breast cancer mouse model (Yallowitz et al. 2015).

7.1.4 Translational Significance of Mutant p53–Hsp90 Axis in Cancer

Targeting the pathways involved in the stabilization of mutant p53, in particular, the Hsp90 family, has attracted a lot of attention over many years. Hsp90 has proven to be a promising target for cancer treatment and is actively pursued by drug discovery companies (Neckers and Workman 2012; Travers et al. 2012). As discussed above, Hsp90 inhibition was shown to be more efficacious in mutant p53 harboring cells in vitro (Li et al. 2011b) and in vivo (Alexandrova et al. 2015, 2017b; Schulz-Heddergott and Moll 2018) compared to p53null and wild-type p53 counterparts. The first Hsp90 inhibitor identified is benzo-quinone ansamycin geldanamycin (Whitesell et al. 1994). The majority of currently available Hsp90 inhibitors, and all that have been clinically assessed (reviewed in (Butler et al. 2015)), have a common mechanism of action: bind to the nucleotide binding pocket of the N-terminal domain and block the processing of client proteins by preventing ATP binding and hydrolysis (Prodromou et al. 1997; Roe et al. 1999), thus preventing the completion of the Hsp90 chaperone cycle, and clients are subsequently targeted for proteasomal degradation by E3 ubiquitin ligases (Connell et al. 2001; Ehrlich et al. 2009; Xu et al. 2002). As promising as these inhibitors, however variable limitations to their use have risen due to several factors such as insufficient drug dose, the frequency of administration, variable pharmacokinetics, suboptimal formulation and dose-limiting toxicities, including hepatotoxicity (Butler et al. 2015). Additionally, another problem that frequently arises in using N-terminal domain inhibitors is the development of drug resistance, since blocking N-terminal domain of Hsp90 often leads to the activation of HSF1 which in turn induces more Hsp90 and other HSP (Sauvage et al. 2017). However, Hsp90 N-terminal domain inhibitors demonstrated encouraging results in clinical trials in HER-positive breast cancer and non-small cell lung carcinoma, which have high prevalence of p53 mutations- 72% in HER2 positive breast cancer, 81% in squamous cell carcinomas, and 68% in large cell carcinomas (Neckers and Workman 2012; Shepherd et al. 2017).

Further investigating the mechanisms protecting mutant p53 from degradation in cancer cells, we demonstrated that similar to Hsp90 inhibitors, histone deacetylase (HDAC) inhibitor (HDACi) suberoylanilide hydroxamic acid (SAHA) destabilizes mutant p53 at the level of protein degradation. We found that SAHA via inhibition of HDAC6, an essential positive regulator of Hsp90, releases mutant p53 from

Hsp90 inhibitory complex and enables its MDM2- and CHIP-mediated degradation. As a result of mutant p53 degradation, SAHA exhibits preferential cytotoxicity for mutant p53 compared to wild-type and null p53 human cancer cells and strongly chemosensitizes mutant p53 cancer cells for chemotherapy (Li et al. 2011a). As a follow-up of our study, Ingallina et al., demonstrated that HDAC6/Hsp90-dependent mutant p53 accumulation is sustained by RhoA geranylgeranylation downstream of the mevalonate pathway, as well as by RhoA- and actin-dependent transduction providing mechanistic cues to control mutant p53 levels via RhoA/actin cytoskeleton. This study identified statins as a mutant p53 destabilizing compounds that disrupt the SREBP-mevalonate/RhoA signaling pathway, which controls Hsp90-mediated mutant p53 stabilization (Ingallina et al. 2018). As SAHA (vorinostat) and statins are FDA-approved drugs, these mechanistic studies can be instantly translated to the clinic. Indeed, following these publications, a Phase I clinical trial was initiated, which confirmed that simultaneous use of vorinostat (a mutant p53 destabilizing drug) and Pazopanib (an anti-angiogenic VEGFR inhibitor) in metastatic sarcoma and colorectal cancer has preferential response in mutant p53 (45%) vs. wild-type p53 patients (16%) (Fu et al. 2015). With regard to statins, a recent study by Turrell et al. demonstrated that statins sensitivity is dependent on the type of p53 mutation. They identified a p53R270H-specific sensitivity to simvastatin, but not in p53R172H, in murine $Kras^{G12D}$ lung cancer model. Importantly, the "statin-sensitivity" transcriptional signature is also present in human lung tumors suggesting the clinical relevance of this data (Turrell et al. 2017).

As discussed above, we demonstrated that ErbB2 signaling via downstream transduction cascade controls HSF1 activity, HSP expression and as a result, mutant p53 stability (Li et al. 2014; Yallowitz et al. 2015). Following this study, we found that pharmacological interception of this circuit by the dual ErbB2/EGFR kinase inhibitor (lapatinib) downregulates mutant p53 in vitro and in vivo. Inhibition of ErbB2 by lapatinib inhibits transcription factor HSF1, and its target Hsp90, followed by mutant p53 degradation in MDM2 dependent manner. We speculate that mutant p53 sensitizes cancer cells to lapatinib via two complementary mechanisms: mutant p53 mediated amplification of ErbB2 signaling, and simultaneous annihilation of both potent oncogenic drivers, ErbB2 and mutant p53. This conclusion is strongly supported by a meta-analysis of the COSMIC drug sensitivity database of 226 human cancer cell lines (representing breast, as well as other cancers) (http://www.cancerrxgene.org/translation/Drug/119). Specifically, we found that mutant p53 human cell lines were more sensitive to lapatinib than wild-type p53 cells ($p = 0.0408$). Consistently, we found higher sensitivity to lapatinib in R172H/R172H;ErbB2 mammary epithelial cells than their p53null;ErbB2 counterparts (Li and Marchenko 2017). Figure 7.1 is an illustration summarizing the pathways involved in augmenting the mutant p53-Hsp90 axis in cancer and the different druggable targets that have been identified within this pathway.

Adding to the complexity of modulating Hsp90 and its related pathways in treating cancer, a recent follow-up study by our lab (Yallowitz et al. 2018) attempted to identify a mechanism for lapatinib resistance (D'Amato et al. 2015). Several mechanisms of lapatinib resistance have been described in the literature, and they

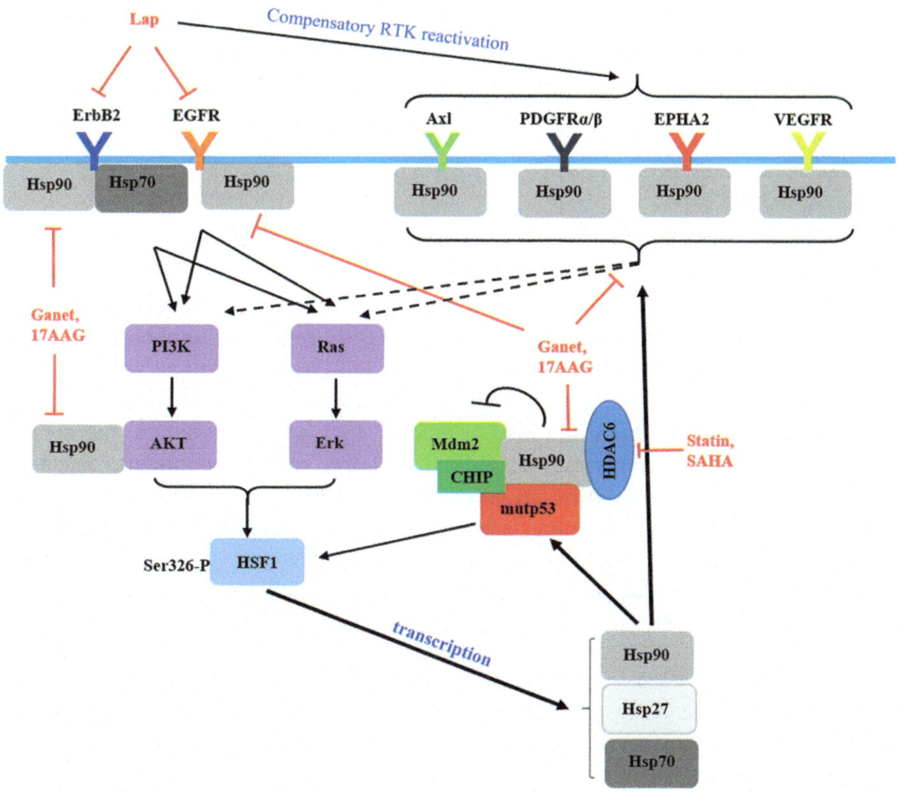

Fig. 7.1 Targeting mutant p53–Hsp90 axis in cancer. Blue line: cell membrane. Above the line represents extracellular material, and below the line represents cell cytoplasm. "Y": different receptor tyrosine kinases. Blocked red lines: drug inhibition. Black arrows: activation. Black broken arrows: compensatory activation pathways. Blocked black line: protein-protein inhibition

primarily involve compensatory activation of several receptor tyrosine kinases (RTKs) (Stuhlmiller et al. 2015). This and the substantial heterogeneity among adaptive RTKs that exists in different cell lines in response to lapatinib represent a major hurdle for the development of successful combinatorial strategies to reverse and/or prevent lapatinib resistance (Stuhlmiller et al. 2015). In our study, we found that lapatinib-resistant cells show chronically activated HSF1 and its transcriptional targets, heat shock proteins, and consequently, higher tolerance to proteotoxic stress. Importantly, lapatinib-resistant tumors and cells retained sensitivity to Hsp90 and HSF1 inhibitors, both in vitro and in vivo. Indeed, HSF1 inhibition simultaneously downregulated ErbB2, adaptive RTKs, and mutant p53, and notably, its combination with lapatinib prevented the development of lapatinib resistance in vitro (Yallowitz et al. 2018). Recently, a novel drug curaxin CBL0137, which has been shown to suppress HSF1 (Neznanov et al. 2011), entered a clinical trial (phase 1) in patients with metastatic or unresectable advanced solid cancers (Incuron) (Burkhart et al. 2014). Hence, pharmacological inhibition of HSF1 opens up a new therapeutic possibility for the clinical application of HSF1 inhibitors to prevent and/or delay the onset of lapatinib resistance in 72% ErbB2 positive breast cancer patients carrying mutant p53.

7.2 Conclusions

Although Hsp90 inhibitors have recently shown some clinical success, it becomes increasingly evident that patient selection and predictive biomarkers seem to be the main hurdle to the successful clinical utilization of Hsp90 inhibitors. Despite nearly 400 proteins interacting with Hsp90 have been identified, the clinical value of these interactions remains unclear. As mutant p53 ablation in cancer cells shows a significant survival benefit in multiple in vivo models, here we discussed the therapeutic potential of targeting mutant p53-Hsp90/HDAC6 axis to achieve mutant p53 depletion in cancer cells. Future retrospective and prospective clinical studies are needed to confirm these preclinical studies and establish mutant p53-Hsp90/HDAC6 axis as an exploitable target.

Acknowledgements This work was supported by the Department of Defense grant W81XWH-16-1-0448 (BC151569) and the Carol Baldwin Breast Cancer Research Fund to N. Marchenko.

References

Adorno M, Cordenonsi M, Montagner M et al (2009) A mutant-p53/Smad complex opposes p63 to empower TGFbeta-induced metastasis. Cell 137:87–98

Alexandrova EM, Marchenko ND (2015) Mutant p53 – heat shock response oncogenic cooperation: a new mechanism of cancer cell survival. Front Endocrinol (Lausanne) 6:53

Alexandrova EM, Yallowitz AR, Li D et al (2015) Improving survival by exploiting tumour dependence on stabilized mutant p53 for treatment. Nature 523:352–356

Alexandrova EM, Mirza SA, Xu S, Schulz-Heddergott R, Marchenko ND, Moll UM (2017a) p53 loss-of-heterozygosity is a necessary prerequisite for mutant p53 stabilization and gain-of-function in vivo. Cell Death Dis 8:e2661

Alexandrova EM, Xu S, Moll UM (2017b) Ganetespib synergizes with cyclophosphamide to improve survival of mice with autochthonous tumors in a mutant p53-dependent manner. Cell Death Dis 8:e2683

Barrott JJ, Haystead TA (2013) Hsp90, an unlikely ally in the war on cancer. FEBS J 280:1381–1396

Baugh EH, Ke H, Levine AJ, Bonneau RA, Chan CS (2018) Why are there hotspot mutations in the TP53 gene in human cancers? Cell Death Differ 25:154–160

Blagosklonny MV, Toretsky J, Bohen S, Neckers L (1996) Mutant conformation of p53 translated in vitro or in vivo requires functional Hsp90. Proc Natl Acad Sci USA 93:8379–8383

Blandino G, Di Agostino S (2018) New therapeutic strategies to treat human cancers expressing mutant p53 proteins. J Exp Clin Cancer Res 37:30

Bouaoun L, Sonkin D, Ardin M et al (2016) TP53 variations in human cancers: new lessons from the IARC TP53 database and genomics data. Hum Mutat 37:865–876

Burkhart C, Fleyshman D, Kohrn R et al (2014) Curaxin CBL0137 eradicates drug resistant cancer stem cells and potentiates efficacy of gemcitabine in preclinical models of pancreatic cancer. Oncotarget 5:11038–11053

Butler LM, Ferraldeschi R, Armstrong HK, Centenera MM, Workman P (2015) Maximizing the therapeutic potential of Hsp90 inhibitors. Mol Cancer Res 13:1445–1451

Chatterjee S, Burns TF (2017) Targeting heat shock proteins in cancer: a promising therapeutic approach. Int J Mol Sci 18:1978

Connell P, Ballinger CA, Jiang J et al (2001) The co-chaperone CHIP regulates protein triage decisions mediated by heat-shock proteins. Nat Cell Biol 3:93–96

D'Amato V, Raimondo L, Formisano L et al (2015) Mechanisms of lapatinib resistance in HER2-driven breast cancer. Cancer Treat Rev 41:877–883

Dai C, Whitesell L, Rogers AB, Lindquist S (2007) Heat shock factor 1 is a powerful multifaceted modifier of carcinogenesis. Cell 130:1005–1018

Dai C, Santagata S, Tang Z et al (2012) Loss of tumor suppressor NF1 activates HSF1 to promote carcinogenesis. J Clin Invest 122:3742–3754

Dittmer D, Pati S, Zambetti G et al (1993) Gain of function mutations in p53. Nat Genet 4:42–46

Ehrlich ES, Wang T, Luo K et al (2009) Regulation of Hsp90 client proteins by a Cullin5-RING E3 ubiquitin ligase. Proc Natl Acad Sci USA 106:20330–20335

Eliyahu D, Raz A, Gruss P, Givol D, Oren M (1984) Participation of p53 cellular tumour antigen in transformation of normal embryonic cells. Nature 312:646–649

Eustace BK, Jay DG (2004) Extracellular roles for the molecular chaperone, hsp90. Cell Cycle 3:1098–1100

Eustace BK, Sakurai T, Stewart JK et al (2004) Functional proteomic screens reveal an essential extracellular role for hsp90 alpha in cancer cell invasiveness. Nat Cell Biol 6:507–514

Freed-Pastor WA, Mizuno H, Zhao X et al (2012) Mutant p53 disrupts mammary tissue architecture via the mevalonate pathway. Cell 148:244–258

Fu S, Hou MM, Naing A et al (2015) Phase I study of pazopanib and vorinostat: a therapeutic approach for inhibiting mutant p53-mediated angiogenesis and facilitating mutant p53 degradation. Ann Oncol 26:1012–1018

Gabai VL, Mosina VA, Budagova KR, Kabakov AE (1995) Spontaneous overexpression of heat-shock proteins in Ehrlich ascites carcinoma cells during in vivo growth. Biochem Mol Biol Int 35:95–102

Gatz SA, Wiesmuller L (2006) p53 in recombination and repair. Cell Death Differ 13:1003–1016

Gomez-Pastor R, Burchfiel ET, Thiele DJ (2018) Regulation of heat shock transcription factors and their roles in physiology and disease. Nat Rev Mol Cell Biol 19:4–19

Haupt Y, Maya R, Kazaz A, Oren M (1997) Mdm2 promotes the rapid degradation of p53. Nature 387:296–299

Hinds PW, Finlay CA, Frey AB, Levine AJ (1987) Immunological evidence for the association of p53 with a heat shock protein, hsc70, in p53-plus-ras-transformed cell lines. Mol Cell Biol 7:2863–2869

Ingallina E, Sorrentino G, Bertolio R et al (2018) Mechanical cues control mutant p53 stability through a mevalonate-RhoA axis. Nat Cell Biol 20:28–35

Jackson SE (2013) Hsp90: structure and function. Top Curr Chem 328:155–240

Jenkins JR, Rudge K, Currie GA (1984) Cellular immortalization by a cDNA clone encoding the transformation-associated phosphoprotein p53. Nature 312:651–654

Kress M, May E, Cassingena R, May P (1979) Simian virus 40-transformed cells express new species of proteins precipitable by anti-simian virus 40 tumor serum. J Virol 31:472–483

Lane DP, Crawford LV (1979) T antigen is bound to a host protein in SV40-transformed cells. Nature 278:261–263

Lang GA, Iwakuma T, Suh YA et al (2004) Gain of function of a p53 hot spot mutation in a mouse model of Li-Fraumeni syndrome. Cell 119:861–872

Lee MK, Teoh WW, Phang BH, Tong WM, Wang ZQ, Sabapathy K (2012) Cell-type, dose, and mutation-type specificity dictate mutant p53 functions in vivo. Cancer Cell 22:751–764

Levine AJ, Momand J, Finlay CA (1991) The p53 tumour suppressor gene. Nature 351:453–456
Li D, Marchenko ND (2017) ErbB2 inhibition by lapatinib promotes degradation of mutant p53 protein in cancer cells. Oncotarget 8:5823–5833
Li D, Marchenko ND, Moll UM (2011a) SAHA shows preferential cytotoxicity in mutant p53 cancer cells by destabilizing mutant p53 through inhibition of the HDAC6-Hsp90 chaperone axis. Cell Death Differ 18:1904–1913
Li D, Marchenko ND, Schulz R et al (2011b) Functional inactivation of endogenous MDM2 and CHIP by Hsp90 causes aberrant stabilization of mutant p53 in human cancer cells. Mol Cancer Res 9:577–588
Li D, Yallowitz A, Ozog L, Marchenko N (2014) A gain-of-function mutant p53-HSF1 feed forward circuit governs adaptation of cancer cells to proteotoxic stress. Cell Death Dis 5:e1194
Linzer DI, Levine AJ (1979) Characterization of a 54K dalton cellular SV40 tumor antigen present in SV40-transformed cells and uninfected embryonal carcinoma cells. Cell 17:43–52
Lukashchuk N, Vousden KH (2007) Ubiquitination and degradation of mutant p53. Mol Cell Biol 27:8284–8295
Mendillo ML, Santagata S, Koeva M et al (2012) HSF1 drives a transcriptional program distinct from heat shock to support highly malignant human cancers. Cell 150:549–562
Menon V, Povirk L (2014) Involvement of p53 in the repair of DNA double strand breaks: multifaceted roles of p53 in homologous recombination repair (HRR) and non-homologous end joining (NHEJ). Subcell Biochem 85:321–336
Mileo AM, Fanuele M, Battaglia F et al (1990) Selective over-expression of mRNA coding for 90 KDa stress-protein in human ovarian cancer. Anticancer Res 10:903–906
Milicevic Z, Bogojevic D, Mihailovic M, Petrovic M, Krivokapic Z (2008) Molecular characterization of hsp90 isoforms in colorectal cancer cells and its association with tumour progression. Int J Oncol 32:1169–1178
Milner J, Medcalf EA, Cook AC (1991) Tumor suppressor p53: analysis of wild-type and mutant p53 complexes. Mol Cell Biol 11:12–19
Min JN, Huang L, Zimonjic DB, Moskophidis D, Mivechi NF (2007) Selective suppression of lymphomas by functional loss of Hsf1 in a p53-deficient mouse model for spontaneous tumors. Oncogene 26:5086–5097
Morton JP, Timpson P, Karim SA et al (2010) Mutant p53 drives metastasis and overcomes growth arrest/senescence in pancreatic cancer. Proc Natl Acad Sci USA 107:246–251
Muller PA, Caswell PT, Doyle B et al (2009) Mutant p53 drives invasion by promoting integrin recycling. Cell 139:1327–1341
Neckers L, Workman P (2012) Hsp90 molecular chaperone inhibitors: are we there yet? Clin Cancer Res 18:64–76
Neznanov N, Komarov AP, Neznanova L, Stanhope-Baker P, Gudkov AV (2011) Proteotoxic stress targeted therapy (PSTT): induction of protein misfolding enhances the antitumor effect of the proteasome inhibitor bortezomib. Oncotarget 2:209–221
Nolan KD, Franco OE, Hance MW, Hayward SW, Isaacs JS (2015) Tumor-secreted Hsp90 subverts polycomb function to drive prostate tumor growth and invasion. J Biol Chem 290:8271–8282
Olive KP, Tuveson DA, Ruhe ZC et al (2004) Mutant p53 gain of function in two mouse models of Li-Fraumeni syndrome. Cell 119:847–860
Oren M, Rotter V (2010) Mutant p53 gain-of-function in cancer. Cold Spring Harb Perspect Biol 2:a001107
Parada LF, Land H, Weinberg RA, Wolf D, Rotter V (1984) Cooperation between gene encoding p53 tumour antigen and ras in cellular transformation. Nature 312:649–651
Parrales A, Ranjan A, Iyer SV et al (2016) DNAJA1 controls the fate of misfolded mutant p53 through the mevalonate pathway. Nat Cell Biol 18:1233–1243
Peng Y, Chen L, Li C, Lu W, Chen J (2001) Inhibition of MDM2 by hsp90 contributes to mutant p53 stabilization. J Biol Chem 276:40583–40590

Petitjean A, Achatz MI, Borresen-Dale AL, Hainaut P, Olivier M (2007) TP53 mutations in human cancers: functional selection and impact on cancer prognosis and outcomes. Oncogene 26:2157–2165

Pratt WB (1987) Transformation of glucocorticoid and progesterone receptors to the DNA-binding state. J Cell Biochem 35:51–68

Prodromou C, Roe SM, O'Brien R, Ladbury JE, Piper PW, Pearl LH (1997) Identification and structural characterization of the ATP/ADP-binding site in the Hsp90 molecular chaperone. Cell 90:65–75

Roe SM, Prodromou C, O'Brien R, Ladbury JE, Piper PW, Pearl LH (1999) Structural basis for inhibition of the Hsp90 molecular chaperone by the antitumor antibiotics radicicol and geldanamycin. J Med Chem 42:260–266

Saldana-Meyer R, Recillas-Targa F (2011) Transcriptional and epigenetic regulation of the p53 tumor suppressor gene. Epigenetics 6:1068–1077

Sauvage F, Messaoudi S, Fattal E, Barratt G, Vergnaud-Gauduchon J (2017) Heat shock proteins and cancer: how can nanomedicine be harnessed? J Control Release 248:133–143

Schulz-Heddergott R, Moll UM (2018) Gain-of-Function (GOF) mutant p53 as actionable therapeutic target. Cancers (Basel) 10:188

Schulz-Heddergott R, Stark N, Edmunds SJ et al (2018) Therapeutic ablation of gain-of-function mutant p53 in colorectal cancer inhibits Stat3-mediated tumor growth and invasion. Cancer Cell 34:298–314 e7

Shepherd FA, Lacas B, Le Teuff G et al (2017) Pooled analysis of the prognostic and predictive effects of TP53 comutation status combined with KRAS or EGFR mutation in early-stage resected non-small-cell lung cancer in four trials of adjuvant chemotherapy. J Clin Oncol 35:2018–2027

Shyamala G, Gauthier Y, Moore SK, Catelli MG, Ullrich SJ (1989) Estrogenic regulation of murine uterine 90-kilodalton heat shock protein gene expression. Mol Cell Biol 9:3567–3570

Soussi T (2010) The history of p53. A perfect example of the drawbacks of scientific paradigms. EMBO Rep 11:822–826

Stuhlmiller TJ, Miller SM, Zawistowski JS et al (2015) Inhibition of Lapatinib-induced kinome reprogramming in ERBB2-positive breast cancer by targeting BET family bromodomains. Cell Rep 11:390–404

Sturzbecher HW, Addison C, Jenkins JR (1988) Characterization of mutant p53-hsp72/73 protein-protein complexes by transient expression in monkey COS cells. Mol Cell Biol 8:3740–3747

Terzian T, Suh YA, Iwakuma T et al (2008) The inherent instability of mutant p53 is alleviated by Mdm2 or p16INK4a loss. Genes Dev 22:1337–1344

Travers J, Sharp S, Workman P (2012) Hsp90 inhibition: two-pronged exploitation of cancer dependencies. Drug Discov Today 17:242–252

Turrell FK, Kerr EM, Gao M et al (2017) Lung tumors with distinct p53 mutations respond similarly to p53 targeted therapy but exhibit genotype-specific statin sensitivity. Genes Dev 31:1339–1353

Vogelstein B, Lane D, Levine AJ (2000) Surfing the p53 network. Nature 408:307–310

Weissmueller S, Manchado E, Saborowski M et al (2014) Mutant p53 drives pancreatic cancer metastasis through cell-autonomous PDGF receptor beta signaling. Cell 157:382–394

Whitesell L, Lindquist SL (2005) Hsp90 and the chaperoning of cancer. Nat Rev Cancer 5:761–772

Whitesell L, Mimnaugh EG, De Costa B, Myers CE, Neckers LM (1994) Inhibition of heat shock protein Hsp90-pp60v-src heteroprotein complex formation by benzoquinone ansamycins: essential role for stress proteins in oncogenic transformation. Proc Natl Acad Sci USA 91:8324–8328

Wijnhoven SW, Speksnijder EN, Liu X et al (2007) Dominant-negative but not gain-of-function effects of a p53.R270H mutation in mouse epithelium tissue after DNA damage. Cancer Res 67:4648–4656

Willis A, Jung EJ, Wakefield T, Chen X (2004) Mutant p53 exerts a dominant negative effect by preventing wild-type p53 from binding to the promoter of its target genes. Oncogene 23:2330–2338

Workman P, Burrows F, Neckers L, Rosen N (2007) Drugging the cancer chaperone Hsp90: combinatorial therapeutic exploitation of oncogene addiction and tumor stress. Ann N Y Acad Sci 1113:202–216

Wu X, Bayle JH, Olson D, Levine AJ (1993) The p53-mdm-2 autoregulatory feedback loop. Genes Dev 7:1126–1132

Xu W, Marcu M, Yuan X, Mimnaugh E, Patterson C, Neckers L (2002) Chaperone-dependent E3 ubiquitin ligase CHIP mediates a degradative pathway for c-ErbB2/Neu. Proc Natl Acad Sci USA 99:12847–12852

Yallowitz AR, Li D, Lobko A, Mott D, Nemajerova A, Marchenko N (2015) Mutant p53 amplifies epidermal growth factor receptor family signaling to promote mammary tumorigenesis. Mol Cancer Res 13:743–754

Yallowitz A, Ghaleb A, Garcia L, Alexandrova EM, Marchenko N (2018) Heat shock factor 1 confers resistance to lapatinib in ERBB2-positive breast cancer cells. Cell Death Dis 9:621

Yan W, Liu G, Scoumanne A, Chen X (2008) Suppression of inhibitor of differentiation 2, a target of mutant p53, is required for gain-of-function mutations. Cancer Res 68:6789–6796

Zhao Y, Liu H, Liu Z et al (2011) Overcoming trastuzumab resistance in breast cancer by targeting dysregulated glucose metabolism. Cancer Res 71:4585–4597

Chapter 8
HSP90 and Its Inhibitors for Cancer Therapy: Use of Nano-delivery System to Improve Its Clinical Application

Prathap Somu and Subhankar Paul

Abstract The molecular chaperone HSP90 (heat shock protein 90) has become a crucial target in cancer therapeutics as its function has been implicated with various types of malignant transformation. Numerous HSP90 inhibitors have been identified so far and many of them have also been clinically tested. Although most of these are natural or their derived inhibitors including geldanamycin and its derivatives 17-AAG and 17-DMAG have shown efficacy, their easy success has been hindered in various stages of the clinical studies due to poor solubility and cytotoxicity. However, recently substantial published documents reported that the systemic targeting of the HSP90 inhibitors using nano-based drug delivery system could provide a possible clinical solution to overcome their limitation. In this chapter, we review the initial development of various HSP90 inhibitors from natural to synthetically derived one and their clinical studies. We also review their limitations and future perspectives as a possible potential agent in the cancer therapeutics by their systemic and control delivery to the target site using the nano-drug delivery system. Also, the application of combined therapy has also been discussed in the current chapter using HSP90 inhibitors and nanocarrier. In addition, we also discuss the therapeutic approaches like photothermal where nano carrier is not only used as a carrier for the systemic delivery of HSP90 inhibitors but also as a therapeutic agent.

Keywords Cancer · Heat shock proteins · Molecular chaperone · Nanomaterials · Targeted delivery · Therapeutic index

P. Somu · S. Paul (✉)
Structural Biology & Nanomedicine Laboratory, Department of Biotechnology & Medical Engineering, National Institute of Technology Rourkela, Rourkela, Odisha, India
e-mail: spaul@nitrkl.ac.in

A. A. A. Asea, P. Kaur (eds.), *Heat Shock Protein 90 in Human Diseases and Disorders*, Heat Shock Proteins 19,
https://doi.org/10.1007/978-3-030-23158-3_8

Abbreviations

17-AAG	17-allylamino-17demethoxygeldanamycin
17-DMAG	17-(2-dimethylaminoethyl)amino-17-demethoxygeldanamycin
ADP	adenosine diphosphate
AhR	aryl hydrocarbon receptor
ATP	adenosine triphosphate
CBR	clinical benefit rate
DLT	dose-limiting toxicities
EGCG	Epigallocatechin 3-gallate
FDG-PET	fluorodeoxyglucose positron emission tomography
GIST	gastrointestinal stromal tumors
GM	Geldanamycin
HOP	HSP90 organizing protein
HSP	heat shock protein
IPI-493	17-desmethoxy-17-amino Geldanamycin
IPI-504	17-allylamino-17-demethoxygeldanamycin Hydroquinone Hydro -chloride
MDR	multi-drug resistance
MTD	maximum tolerated dose
NSCLC	non-small-cell lung carcinoma
ORR	overall response rate
PBMC	peripheral blood mononuclear cells
PCL	poly(ε-caprolactone)
PDGFRA	platelet-derived growth factor receptor alpha
PEG	poly(ethylene glycol)
PET	positron emission tomography
PLGA	poly (lactic-co-glycolic acid)
PR	partial response
RD	radicicol
SD	stable diseases
SUV	standardized uptake value
TNBC	triple-negative breast cancer

8.1 Introduction

According to WHO, cancer is the second most leading cause of death which counts nearly one out of six globally and about ten million new cases are reported every year (Siegel et al. 2017). Among the major therapeutic techniques practiced for the treatment of cancer, chemotherapy and radiation are most common. The genetic vulnerability or instability of cancer cells along with the presence of cellular machinery such as molecular chaperones (heat shock proteins) are responsible to protect them against the initial treatment (Hölzel et al. 2013; Young et al. 2004). For example, when the cancer cells are treated with chemotherapeutic drugs that causes some initial damage but due to the genetic plasticity nature of cancer cells, cancer cells protect themselves by the mutation of the targeted receptor against the chemotherapeutic drugs (Meacham and Morrison 2013) and thus accruing multi-drug resistance (MDR) (Markman et al. 2013; Parveen and Sahoo 2008). Similarly, molecular chaperones, known as heat shock proteins, also help cancer cells from environmental stress due to various therapies such as elevated temperature, oxidative damage caused by photothermal therapy, and hypoxia (caused by low oxygen due to rapid metabolism) (Young et al. 2004). When cancer cells are exposed to environmental stress like elevated temperature and hypoxia, cells induce the synthesis of high amount of heat shock proteins (HSP) that play an important role in regulating and guiding of proteins to their correct structural conformation, which in turn protect cancer cells from various modes of therapy such as photothermal and chemotherapy (Meacham and Morrison 2013). Since most of the client proteins of HSP90 are oncoproteins, such phenomena plays an important role in cancer transformation and progression.

Further, the advancement in the molecular genetics and genomics, more specialized and personalized methods for cancer treatment such as gene therapy, hormone therapy, and immunotherapy have been developed with enormous specificity with reduced systemic toxicity. However, they also have suffered a similar fate due to the genetic instability and vulnerability nature of cancer cells that may cause mutation of the targeted gene or receptor and becoming hormone-independent. So, the two important key factors which play an important role in such adoption are genetic vulnerability nature of cancer cells and molecular chaperones (HSP). Hence, this two factors can be targeted for efficient therapeutic strategy (Audisio et al. 2011; Bagatell and Whitesell 2004). Therefore, one of the recommended way for the treatment of cancer is to target HSP90 that helps cancer cells to adapt such applied environmental stress derived from various therapies like photothermal therapy (Neckers 2002). Further, HSP also function as biochemical buffers to endure numerous genetic alterations present within tumor cells that would otherwise be fatal and also allows mutant proteins to retain or even gain function under such imbalanced signaling condition (Bagatell and Whitesell 2004). Therefore, HSP can be used as therapeutic target for efficient cancer treatment as its expression accounts for the cancer cell's ability to maintain its protein homeostasis even under the hostile conditions such as hypoxia and acidic microenvironment (Pearl and Prodromou 2006; Solit and Chiosis 2008). In mammalian cells, HSP90 is predominant and widespread among other molecular chaperones and also play key role in folding, stability, transportation of various oncoproteins (Neckers 2007; Pearl and Prodromou 2006). Therefore, over the decades, HSP90 has become a main molecular target for the efficient cancer treatment by developing small molecule HSP90 inhibitors, thereby triggering the proteasomal degradation of their oncogenic 'client' proteins for the cancer treatment (Mahalingam et al. 2009).

8.1.1 *HSP 90 as Target Molecule for Cancer Treatment*

Mainly four different types of mammalian HSP have been classified, primarily based on their molecular weight: HSP90, HSP70, HSP60 and small HSP (15–30 kDa). Among them, HSP90 is the most abundant molecular chaperones found in the eukaryotes complexes with huge number of proteins regarded as 'clients' whose folding, stability, transportation is regulated by HSP90. Interestingly, many of its clients are oncoproteins which is why HSP90 plays an important role in the cancer development and has been regarded as a target molecule for the therapy in cancer (Trepel et al. 2010). The Schematic representation of HSP90 cycle and the effects of HSP90 inhibition for its client oncoprotein degradation for cancer therapy were shown in the Fig. 8.1. The synthesis of new client proteins begins with the association of HSP40 and HSP70 through forming HSP70–HSP40 chaperone complex. Then the client proteins are be delivered to the ADP-bound "open" form of HSP90 via the TPR-containing co-chaperone HSP70-HSP90 organizing protein (HOP). The role of HOP in the HSP70–HSP40 chaperone complex provides the reversible link between HSP70 and HSP90 through the MEEVD peptide which allows the transfer of client proteins to HSP90. The binding of client proteins leads to

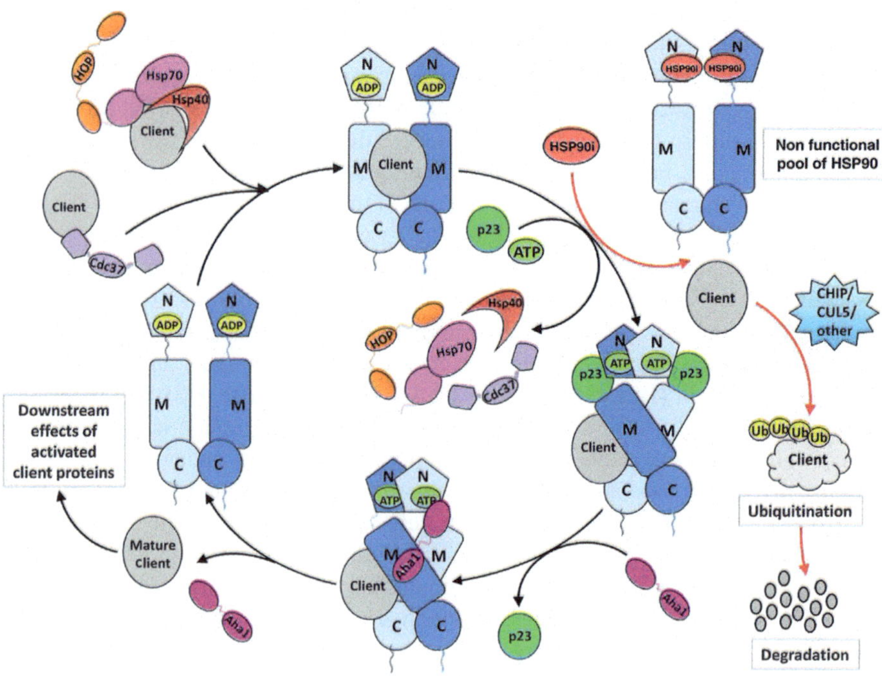

Fig. 8.1 Schematic representation of the role of HSP90 in client protein synthesis and its inhibition by small HSP 90 inhibitors molecules (Butler et al. 2015)

conformational changes conforming the ATP-bound "closed" state of HSP90 with the association of the co-chaperone p23. Then kinases are delivered to HSP90 by an alternative method using a complex with the co-chaperone cell-division cycle 37 homolog (Cdc37). This interaction between HSP90 and other co-chaperone induces conformational changes in client proteins that allow their maturation and activation. The binding of the co-chaperone Activator of HSP90 ATPase 1 (Aha1) to the HSP90 causes hydrolysis of ATP that further leads to the release of mature and active client protein. However, inhibition of HSP90 by small-molecule inhibitors acting at the N-terminal nucleotide site blocks ATP binding and subsequent hydrolysis. The binding of the small-molecule inhibitors at the N-terminal of HSP90 leads to the blocking of ATP binding to HSP90 that causes the termination of the chaperone cycle and degradation of the client proteins by ubiquitin-dependent proteasomal degradation. Thus, there is a gaining interest in the clinical development of HSP90 inhibitors from identification of new HSP90 inhibitors and their modification for the optimization of the pharmaceutical properties including pharmacokinetic, pharmacodynamic and toxicological. Over a decade, considerable progress has happened on many fronts such as introduction of chemically distinct HSP90 inhibitors with improved pharmaceutical properties and the preparation of improved formulations making them potentially a the first-in-class agent for cancer treatment. The development of HSP90 inhibitors from natural to synthetic one and the new nanostructure-based formulations are also discussed further.

8.1.2 First Generation HSP90 Inhibitor: Natural Inhibitor

8.1.2.1 N-Terminal HSP90 Natural Inhibitor

The discovery of geldanamycin (GM) as an HSP90 inhibitor has sparked an interest in both industry and academic sectors for developing small molecule-based HSP90 inhibitors with clinical viability as an effective strategy for the cancer treatment (Patel et al. 2011). More often, natural product/compounds play a vital role during drug discovery. In the case of HSP90, natural products geldanamycin (GA) and radicicol (RD) (Fig. 8.2), mimics the structure which is adopted by ATP in the N-terminal nucleotide-binding pocket of HSP90 (Sidera and Patsavoudi 2014). Thus, the selective binding of both GA and RD to the N-terminal nucleotide-binding pocket of HSP90 eventually leads to inhibition of the intrinsic activities of HSP90 by preventing the ATP binding. This fact increases the recruitment of ubiquitin ligase to HSP90 chaperone complex and induces the depletion of oncogenic HSP90 client proteins via the proteasome degradation of these client proteins (Mimnaugh et al. 1996; Roe et al. 1999; Soga et al. 2013).

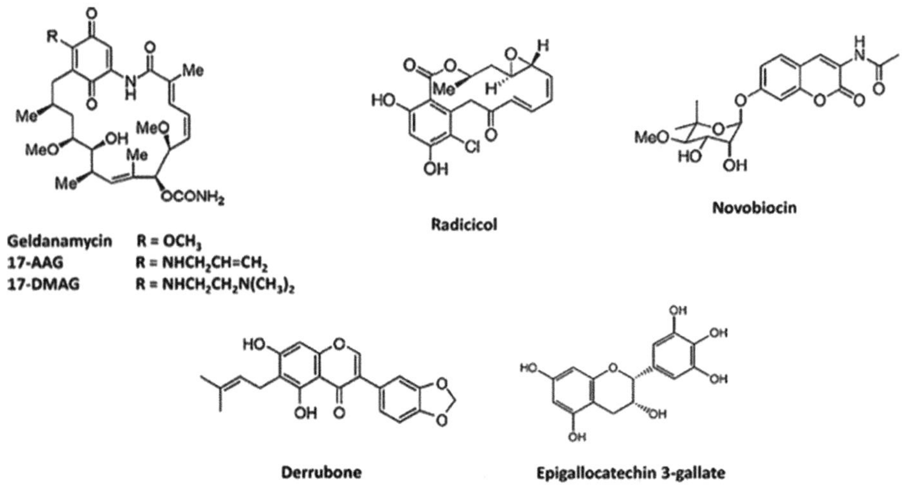

Fig. 8.2 Chemical structures of the natural HSP90 inhibitors (Li et al. 2009)

8.1.2.2 C-Terminal HSP90 Inhibitor

Novobiocin and coumermycin A are two important naturally occurring C-terminal HSP90 inhibitor (Fig. 8.2). Novobiocin is an amino-coumarin containing antibiotic which was isolated from Staphylococcus saprophyticus. It binds to the ATP-binding site present in the C-terminal domain of HSP90 (Byrd et al. 2016). Thus, the binding of novobiocin to HSP90 lead to the disturbance of the interaction of between HSP90 complex and it's both co-chaperones namely p23 and HSP70, thereby the destabilization of HSP90–client protein interactions. Another related coumarin antibiotics as the C-terminal HSP-90 inhibitor, Coumermycin A1 have shown to disrupt C-terminal dimerization and halt the conformational cycle (Allan et al. 2006). Like Novobiocin, Coumermycin A1 also have been shown to disrupt interactions of partner co-chaperones that bind HSP90 at both the C- and N-terminus, *i.e.*, Cdc37, p23, Hsc70, FKBP52, and PP5, thus destabilization of HSP90–client protein complex (Allan et al. 2006; Marcu et al. 2000).

Recently, natural inhibitors like isoflavone derrubone and epigallocatechin 3-gallate (EGCG) have also been isolated and tested (Fig. 8.2). Isoflavone derrubone was isolated from the Indian tree *Derris robusta* and which exhibits its ability to inhibit the interaction of HSP90 and Cdc37 with heme-regulated eIF2α kinase (HRI), an HSP90 client kinase, and also showed anticancer potential in human breast cancer cell lines (Hadden et al. 2007). Similarly, catechin, epigallocatechin 3-gallate (EGCG) was isolated from a green tea polyphenol that also shows its ability to inhibit the transcriptional activity of aryl hydrocarbon receptor (AhR) through binding to the C-terminus of HSP90 (Palermo et al. 2005). The discovery of EGCG has lead to the development of novel HSP90 inhibitors by providing scaffolds for the design of new molecules.

8.1.3 Second Generation HSP90 Inhibitor: Derivatives or Analog of Natural HSP90 Inhibitor

Despite the potential anti-tumor activity of natural HSP90 inhibitors GA and RD, the poor "drug-like properties, limited in vivo stability, poor solubility and significant hepatotoxicity limits their clinical application (Ge et al. 2006; Haque et al. 2016; Wang et al. 2015). Although both GA and RD proved to have poor solubility, they are too toxic and unstable for clinical applications, each of which have a profound effect on the development of new HSP90 inhibitors using them as a lead compound.

8.1.3.1 Analogues of GA

17-Allylamino-17-Demethoxygeldanamycin (17-AAG)

In order to overcome these limitations of GA, new analogues or derivatives were synthesized using the structure of GA by chemical modifications of position 17, which are known as 17-allylamino-17-demethoxygeldanamycin (17-AAG) and 17-(2 dimethyl aminoethyl) amino-17-demethoxygeldanamycin (17-DMAG) (Fig. 8.3) (Hollingshead et al. 2005; Nimmanapalli et al. 2001). 17-AAG is the first HSP90 inhibitor tested for clinical trials as an antineoplastic drug due to its significant anti-cancer potential. However, the approval of 17-AAG was also hindered due to its poor solubility and bioavailability as well as hepatotoxicity and nephrotoxicity effects (Jhaveri et al. 2014). Further, the development of 17-AAG faces several problems like production/formulation issues as well as patent expiry concerns (http://www.myelomabeacon.com/news/2010/07/22/ tanespimycin-development-haulted/).

17-(2-Dimethylaminoethyl)amino-17-Demethoxygeldanamycin (17-DMAG)

Another GA analogue entered into clinical trials was 17-(2-dimethylaminoethyl) amino-17-demethoxygeldanamycin (17-DMAG) in 2005(http://clinicaltrials.gov/ct2/show/NCT00088868). However, 17-DMAG showed very interesting results such as better water solubility, oral bioavailability and greater anticancer activity compared to 17-AAG due to the presence of an ionizable amino group as a result of the substitution of the C-17 methoxy group of GM with N, N-dimethylethylamine (Messaoudi et al. 2008). However, Kosan discontinued 17-DMAG development due to based on several factors, including the strength of intellectual property protection and risk, clinical experience to date, and time to commercialization compared to 17-AAG.

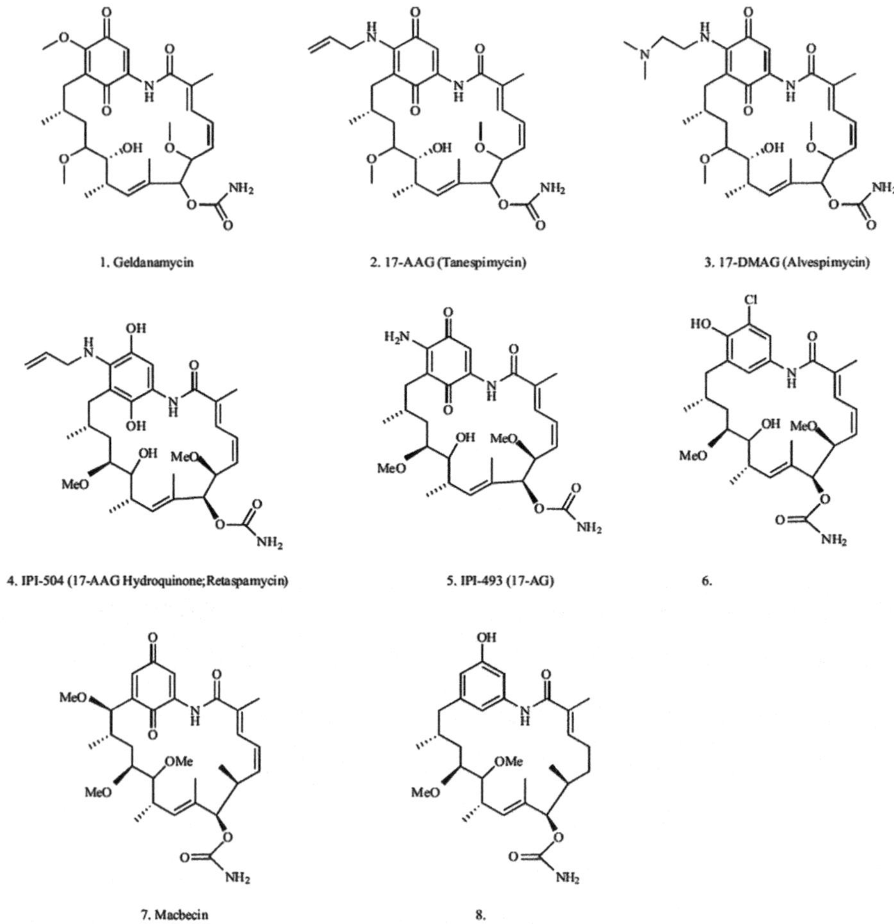

Fig. 8.3 Chemical structures of the N terminal GA analogs HSP90 inhibitors (Sidera and Patsavoudi 2014)

IPI-504 [17-Allylamino-17-Demethoxygeldanamycin Hydroquinone Hydrochloride]

Infinity Pharmaceuticals developed a water-soluble HSP90 inhibitor, IPI-504 (retaspimycin) which is a reduced and hydroquinone hydrochloride salt derivative of 17-AAG (Fig. 8.3) (Sydor et al. 2006). The 17-AAG was reduced using sodium dithionite, followed by conversion to the preparation of IPI-504. Actually, both 17-AAG and IPI-504 co-exist in redox equilibrium in vivo as the hydroquinone forms for IPI-504 and quinine forms for 17-AAG prepared through the oxidoreductases action. Since, the hepatotoxicity of GM is primarily due to the presence of the quinone ring, as in IPI-504, the hydroquinone group is in the reduced form might results in reduced toxicity. The intravenous administration of IPI-504 has reached to

Phase II and III clinical trials (http://clinicaltrials.gov/ct2/show/NCT00817362, and http://clinicaltrials.gov/ct2/show/ NCT00688766). However, since IPI-504 and 17-AAG are interconvertible in vivo *condition* due to the action of the oxidoreductases action, its phase III trial of 17-AAG was terminated (Sydor et al. 2006).

IPI-493 (17-Desmethoxy-17-Amino Geldanamycin)

The Infinity Pharmaceuticals was also developed an orally administrated HSP90 inhibitor17-AG (17-amino-17-demethoxygeldanamycin; IP-493) which is the primary active and long-lived metabolite of IPI-504 and 17-AAG (Fig. 8.3). Although IPI-493 reached Phase I studies, its clinical study was suffered due to its poor pharmaceutical properties such as low solubility and the difficulty of establishing pharmacologically relevant doses of administration (http://clinicaltrials.gov/ct2/show / NCT00724425?term=IPI-493&rank=1). Moreover, due to better clinical relevant properties of IPI-504, the development of IPI-493 agent was stopped and focus was turned exclusively on retaspimycin. Till date, IPI-504 is the most active and clinically feasible GA-derived-HSP90 inhibitor. The analog of ansamycin with devoid of quinone/hydroquinone component was obtained by 'muta-synthesis' and was developed by Kosan Biosciences and displayed comparable activities to the IPI-504 in vitro condition but better efficacy in a mouse xenograft model (Menzella et al. 2009). A genetically engineered microorganism 'Actinosynemma pretiosum Biotica technology was able to secrete the natural HSP90 inhibitor 'macbecin' (Fig. 8.3) (Martin et al. 2008; Tanida et al. 1980). Further, non-quinone analogue of macbecin was developed (Fig. 8.3) with high binding affinity with K_d of 3 nM while the K_d of macbecin was 240 nM) and it was also found to be less toxic than 17- AAG, as it lacks the toxic benzoquinone pharmacophore group with a maximum tolerated dose (MTD) greater than 250 mg/kg.

8.1.3.2 Analogues of RD

The resorcinol core of RD is the critical element for its binding to HSP90. However, In addition to a resorcinol moiety (Fig. 8.4), RD also contains unstable and reactive epoxide and α,β,γ,δ-unsaturated carbonyl groups which makes its unstable in serum and devoid under in vivo condition. However, based on the resorcinol core of RD, a number of molecules have been developed and entered into clinical trials like NVP-AUY922, KW-2478, and AT13387 as well as STA-9090.

NVP-AUY922/VER52296

Workman and colleagues from the Cancer Research UK Center for Cancer Therapeutics, first identified the resorcinol containing pyrazole CCT018159 from an HTS of a library of 56,000 compounds capable of inhibiting the ATPase activity

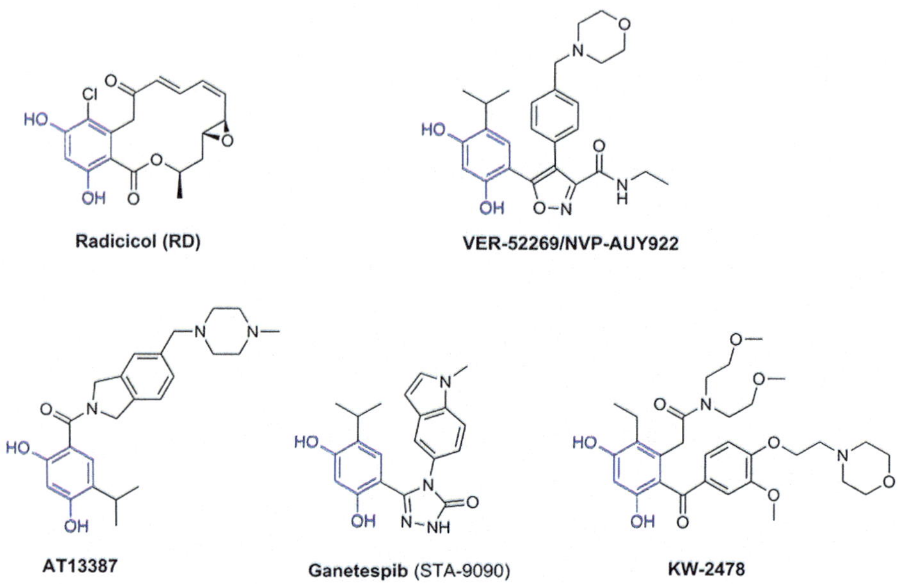

Fig. 8.4 Chemical structures of the N terminal RD analogs HSP90 inhibitors (Jhaveri et al. 2014)

of yeast HSP90 using the malachite green detection of inorganic phosphate (Cheung et al. 2005; Rowlands et al. 2004). They used a structure-based approach and optimized the isoxazole NVP-AUY922/VER52296 as a potent HSP90 inhibitor (Fig. 8.4) (Gaspar et al. 2010; Jensen et al. 2008). Subsequently, this led a candidate, now is being developed by Novartis and, is currently in phase I/II clinical trials for patients with refractory multiple myeloma and HER2 + and ER + metastatic breast cancer as a single agent as well as in combination therapy. From the Phase I trial results in advanced solid tumors, it showed that the maximum tolerated dose (MTD) of weekly was 70 mg/m^2 (Samuel et al. 2010). Further, Dose-limiting toxicities (DLTs) of NVP-AUY922/VER52296 included darkening of vision, atrial flutter, diarrhea, and fatigue. However, around 20% of the patients tested developed night blindness with a grade 3 or higher eye disorders at the MTD. Also, 16 patients have been found to develope SD and 9 had a partial metabolic response fluorodeoxyglucose positron emission tomography (FDG-PET) scans (Samuel et al. 2010). A Phase II expansion trial of NVP-AUY922/VER52296 in HER2-positive and ER + metastatic breast cancer also have been reported two PD responses on FDG-PET (Schroder et al. 2011).

AT13387

AT13387 was discovered by Astex Pharmaceuticals through a fragment-based drug discovery approach for the optimization of a resorcinol-containing group having a high HSP90 binding affinity (Fig. 8.4) (Murray et al. 2010; Woodhead et al. 2010). This fragment screening consisted of a combination of nuclear magnetic resonance and high-throughput X-ray crystallography. One of the fragments have been identified with a phenolic chemotype and selected, which was further optimized via structure-guided design to obtain AT-13387 (Murray et al. 2010; Woodhead et al. 2010).

AT13387 was evaluated in Phase I trials in patients with advanced solid tumors. The various dosing schedules were studied which includes weekly or bi-weekly intravenous infusions for a period of 3 weeks in a 28-day cycle (Do et al. 2012; Jhaveri et al. 2012; Mahadevan et al. 2012). A dosage of 260 mg/m^2 has been identified MTD for the weekly study (Mahadevan et al. 2012). It is currently in three Phase I/II trials under evaluation in either alone or in combination with: (i) abiraterone in the treatment of CRPC which is no longer responding to abiraterone (NCT01685268); (ii) crizotinib in the treatment of NSCLC (NCT01712217); and (iii) imatinib in patients with unresectable and/or metastatic GIST whose tumor has progressed treatment with a maximum of three TKIs (NCT01294202).

Ganetespib (STA-9090)

STA-9090 (Fig. 8.3) is a novel resorcinol HSP90 inhibitor containing triazole developed by Synta Pharmaceuticals (Ying et al. 2011). Goldman et al. have evaluated phase I trial of STA-9090 in 35 patients having advanced solid tumors with an intravenously received the STA-9090 weekly for 3 weeks in a 28-day cycle. It was determined that STA-9090 of MTD was 216 mg/m^2. Dose-limiting toxicity (DLT) of STA-9090 was found as fatigue, diarrhea, and elevated amylase. The other common adverse side effects were anemia, abdominal pain, dyspnea, constipation, and nausea. It was also noticed one partial response (PR) in a patient with rectal cancer and several stable diseases (SD) suffering from gastrointestinal non-small cell lung cancer, stromal tumor, and renal cell carcinoma (Goldman et al. 2010). Another phase I clinical trials were evaluated for ganetespib in advanced solid malignancies using different dosing schedules intravenously such as weekly for 3 weeks or twice-weekly dosing for 3 weeks in a 28-day cycle. DLT of STA-9090 included grade 3 diarrhea, grade 3 amylase elevations as well as grade 3 and 4 asthenia. PR was noticed in one patient with metastatic colorectal cancer whereas SD was observed in 23 patients. Disease control rate was 24.5% of the patients (Goldman et al. 2010). For the twice-weekly dosing trial of STA-9090, DLT of elevated liver enzymes was observed. Ganetespib (STA-9090) was tolerated up to the dosage of 120 mg/m^2; the dose cohort has been expanded to 144 mg/m^2 and dose escalation is ongoing (Cho et al. 2011).

A safety and efficacy of phase I/II study for STA-9090 was conducted in patients suffering from acute myeloid leukemia and other hematological malignancies and evaluated with a dosage schedule of weekly infusions for four consecutive weeks per cycle at three dose levels (120–200 mg/m^2). DLT of elevated liver enzymes was noticed in one patient. There were no adverse responses observed, but one patient with refractory acute myeloid leukemia had developed SD. The recommended phase II dose has not been identified and study is ongoing (Lancet et al. 2010).

Moreover, The combination of ganetespib with chemotherapeutic agents such as doxorubicin and taxanes have shown enhanced cytotoxic effects (Proia et al. 2013). The evaluation of the combination of ganetespib and weekly paclitaxel was planned in the ENCHANT-1 trial. Another trial of ganetespib and docetaxel in Triple-negative breast cancer (TNBC) was planned by Synta Pharmaceuticals. Phase I clinical trial for a combination of ganetespib, trastuzumab and paclitaxel is ongoing in HER2-positive trastuzumab-refractory metastatic breast cancer (MBC) (NCT02060253). Further, Patients with Gastrointestinal Stromal Tumors (GIST) who had failed previous treatment with imatinib and sunitinib, were also treated with single-agent ganetespib with a dosing schedule of weekly for 3 weeks in a 28-day cycle (Demetri et al. 2011). Correlatives included HSP90 client proteins and positron emission tomography (PET) imaging using biopsies both pre- and post-treatment. The primary endpoint was the Clinical Benefit Rate (CBR) (CR + PR + SD >16 weeks). In this study, 12 SD was noticed in 23 patients in this Simons 2-stage model, while 7 out of the 123 patients reported with a >20% decrease in the standardized uptake value (SUV) which measured using PET imaging. These data, therefore, suggested that once-weekly treatment with ganetespib might not be optimal in patients with Gastrointestinal Stromal Tumors (GIST) and therefore accumulation is limited to patients with platelet-derived growth factor receptor alpha (PDGFRA) mutations to allow evaluation of other dosage schedules and combinations.

KW-2478

KW-2478 is one of the important analogs of resorcinol that was discovered by Kyowa Hakko Kirin Pharma using a binding assay where HSP90 molecules were fixed onto a plate and the compounds were added to identify competitive inhibitors of a labeled RD (Fig. 8.4). The endeavors resulted in the identification of competitive inhibitors of RD which was eventually given the clinically fiscal candidature through evaluation of X-ray crystallography, cell-based screening, and in vivo models (Cavenagh et al. 2008; Nakashima et al. 2010). A phase I study of KW-2478, in patients was completed in patients with relapsed/refractory multiple myeloma, chronic lymphocytic leukemia or B-cell non-Hodgkin's lymphoma. KW-2478 (14–99 mg/m^2) was administered with an intravenously dosing schedule of daily on days 1–5 of a 14-day cycle. There were no DLTs observed in the patients up to a dosage of 99 mg/m^2. Drug-related toxicities included grade 1/2 hypertension in one patient and grade 3 QTc prolongation in another. A combination phase I/II trial for KW-2478 and bortezomib was evaluated with administered on days 1, 4, 8, 11 of a 21-day cycle in patients with relapsed and refractory multiple myeloma (Baylon et al. 2013; Jhaveri et al. 2012). In the Phase I trial, a total of 95 patients were involved, while there were 80 patients in the Phase II trial. The common side effects included diarrhea, nausea, vomiting, fatigue and peripheral neuropathy. Also, grade 4 thrombocytopenia was noticed in five patients and grade 4 neutropenia was observed in three patients. The induction of HSP70 was also observed in the all peripheral blood mononuclear cells (PBMC) patients during Phase I trial. Overall response rate (*ORR*) was found to be 39% and progression-free survival (PFS) was 26.4 weeks. ORR was 48% in the bortezomib-naive patients (n = 50) (Baylon et al. 2013).

8.1.3.3 Purine and Purine-Like Analogs

The rational design of HSP90 inhibitors was made possible with the availability of the X-ray crystal structures of ATP/ADP as well as GM and RD bound N-terminal nucleotide-binding domain of HSP90. The GHKL ATPase protein family like HSP90 has a unique and distinguishing Bergerat fold among most other ATPases which results in a distinctive conformation adaptation by a bound nucleotide that enables the selective inhibition of HSP90 by small-molecules over other ATP/ADP binding proteins (Chène 2002). Based on the unique fold of HSP90 as well as a structure-based approach, Chiosis et al., have rationally designed the first reported synthetic HSP90 inhibitor, PU3 (Fig. 8.5, 1 entry), using the co-crystal structures of HSP90 with its ligands such as GM, RD and ATP/ADP (Chiosis et al. 2001). The purine-linker-aryl is an essential motif found in PU3 which was used as a prototype, (Fig. 8.4). By incorporating various functional groups, potent and selective HSP 90 inhibitors candidates with favorable pharmaceutical properties were designed. Among various derivatives of PU3, purines like CNF2024/BIIB021, EC144, MPC-3100, MPC-0767, and PU-H71 as well as the purine-like Debio 0932 (CUDC-305) have advanced to clinical level.

CNF 2024/BIIB021

CNF 2024/BIIB021 was discovered and first developed by Conforma Therapeutics and later by Biogen Idec (Fig. 8.5) (Zhang et al. 2010). CNF2024/BIIB021 is unique amongst other members in this class in that the aryl moiety is attached to the 9-position of the purine (Kasibhatla et al. 2007). Although, CNF 2024/BIIB021 agent was evaluated in phase I and II clinical trials. But the clinical development of CNF 2024/BIIB021 was halted thereafter (Mitchell 2011).

5-(2-Amino-4-Chloro-7-((4-Methoxy-3,5-Dimethylpyridine-2-yl)Methyl)-7H-Pyrrolo[2,3-d]Pyrimidin-5-yl)-2-Methylpent-4-yn-2-ol) [EC 144]

The same company, Biogen Idec also disclosed a series of pyrrolo [2,3-d]pyrimidine (deazapurine) analogs among them, EC144 (Fig. 8.5) proved to have the best combination of pharmacokinetic properties and efficacy in murine cancer models (Shi et al. 2012; Sidera and Patsavoudi 2014). Daiichi Sankyo Inc. also prepared another compound pyrazolo [3,4- d] pyrimidine-based on the tricyclic structure as a potential HSP90 inhibitor (i.e., Fig. 8.5, entry 4) for different cancer cell lines with an IC50 value around the nanomolar range (Ohsuki et al. 2012).

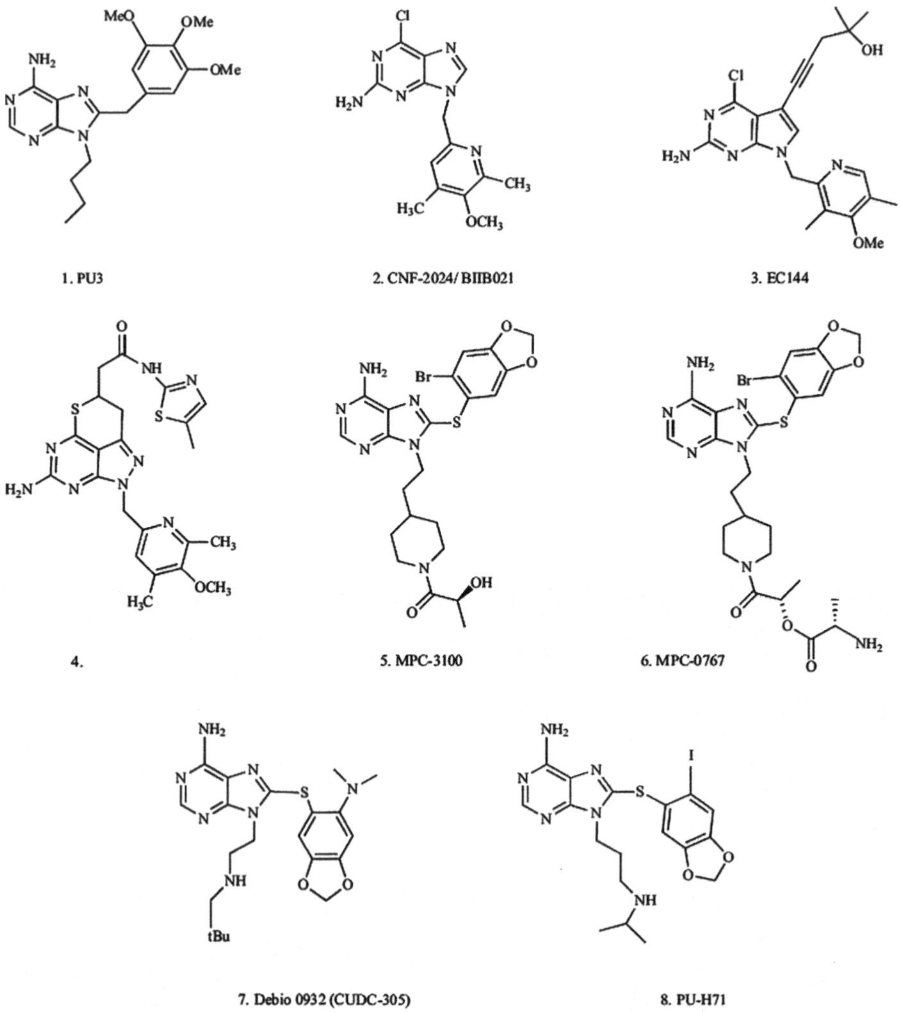

Fig. 8.5 Chemical structures of the purine-scaffold analogs HSP90 inhibitors (Sidera and Patsavoudi 2014)

MPC-3100 and MPC-0767

MPC-3100 (Fig. 8.5) developed for finding HSP90 inhibitor using the purine-scaffold by Myrexis Inc. (Samlowski et al. 2011). MPC-3100 was evaluated in a Phase I trial for 21 days with a week off in a 28-day cycle in 26 patients with recurrent or refractory cancer. Patients received either dosage of 50,100, 165, 245 or 340 mg/m^2 daily for 21 days with a week off in a 28-day cycle or with daily continuously dosage for 28 days of 240 or 320 mg every 12 h. The most common adverse effects developed were nausea, grade 1 or 2 diarrhea, vomiting and fatigue (http://clinicaltrials.gov/show/ NCT00920205). The DLT was supraventricular tachycardia at a dose of 245 mg/m^2. Whereas, patients treated total daily doses >600 mg/day developed Grade 1–3 gastrointestinal and grade 1/2 visual adverse events (Samlowski et al. 2011). Moreover, due to the poor solubility and bioavailability of MPC-3100, Myrexis also developed novel L-alanine ester pro-drug of MPC-3100, MPC-0767 with the planned introduction of a pro-drug, MPC-0767 with improved aqueous solubility compared to MPC-3100 and equivalent tumor regressions efficacy. The Company submitted an IND on MPC-0767 in the first quarter of 2012 (http://www.myriadpharma.com/component/content/article/2-product-pipeline/23-HSP90-inhibitorprogram).

Debio 0932 (CUDC-305)

Scientists at Curis designed the Debio 0932 (CUDC-305) (Fig. 8.5) by replacing the N3 of the purine with a carbon and the iodine by dimethylamine (Bao et al. 2009; Isambert et al. 2012; Taldone et al. 2013). In 2010, Debiopharm initiated the Phase I trial of Debio 0932 in patients with solid tumors or lymphoma (Isambert et al. 2012) (http://clinicaltrials.gov/ct2/show/NCT01168752). Debio 0932 was administered orally with a dosage of 5–100 mg in both daily and every other day regimens with a total of 80 patients. The drug was well tolerated at doses up to 1600 mg every 2 days and 1000 mg/day. The most common adverse side effects of the study were diarrhea, asthenia and decreased appetite with no ocular or cardiotoxicity being reported. One DLT of febrile neutropenia was reported in the every-other-day dosing study where second DLTs of asthenia and diarrhea were observed with the daily dosing study. However, Debio 0932 showed promising activity with partial response (PR) study in two patients, one patient with KRAS-mutant lung cancer and the other patient with breast cancer (Isambert et al. 2012). Further, the daily dosing cohort study has been expanded to a Phase I/II trial of Debio-0932 in combination with standard-of-care in the first- and second-line treatment of non-small cell lung cancer (NCT01714037).

PU-H71

PU-H71 is a purine-based compound discovered at Memorial Sloan-Kettering cancer center, developed by Samus therapeutics and NCI (Fig. 8.5) (Marubayashi et al. 2010; Mitchell 2011; Taldone and Chiosis 2009). A unique feature of PU-H71 is the presence of endogenous iodine atom (^{127}I) that has been replaced with the PET radionuclide ^{124}I which results as an imaging agent ^{124}I-PUH71 (He et al. 2006). Currently, PU-H71 has been evaluated in phase I clinical trial in patients with advanced malignancies and under the license to Samus Therapeutics (http://clinicaltrials.gov/ct2/show/NCT01393509). Recently, Memorial Sloan-Kettering cancer centre has also announced to conduct a pilot study in cancer patients for PET imaging using I-PUH71 (http://clinicaltrials.gov/ct2/show/NCT01269593). We have summerized few important HSP90 inhibitors and their clinical study details in Table 8.1.

Table 8.1 Summary of the important HSP90 inhibitors and details of their clinical study

HSP90 inhibitor class	HSP90 inhibitor	Clinical stage	Combination drug	Disease	Response	References
Geldanamycin derivative	Tanespimycin (17-AAG)	Phase II	No	Melanoma with BRAF mutation	SD was observed in melanoma	Tanida et al. (1980)
	Tanespimycin (17-AAG)	Phase II	Trastuzumab	HER2-positive metastatic breast cancer	Objective responses were observed	Trepel et al. (2010)
	Tanespimycin (17-AAG)	Phase I/ II	Bortezomib	Relapsed or relapsed and refractory multiple myeloma patients	Objective responses were observed	Wang et al. (2015)
	Alvespimycin (17-DMAG)	Phase I	NO	Advanced solid cancers	CR and PR was observed	Won et al. (2011)
	Alvespimycin (17-DMAG)	Phase I	Trastuzumab	Advanced solid tumors	Objective responses and SD	Woodhead et al. (2010)
	Retaspimycin (IPI-504)	Phase II/ III	No	Multiple myeloma patients	SD was observed	Xiong et al. (2008)
	Retaspimycin (IPI-504)	Phase Ib	Docetaxel	NSCLC cells	Objective responses in patients with squamous histology,	Ying et al. (2011)
Resorcinol scaffold	Ganetespib (STA-9090)	Phase I	NO	Solid tumors	SD and PR in 1 colorectal cancer patient, also PR in metastatic melanoma and SD in NSCLC patient	Ohsuki et al. (2012)
	Ganetespib (STA-9090)	Phase II	No	Castrate resistant prostate cancer		NCT01270880
	Ganetespib (STA-9090)	Phase I		Breast cancer	Ongoing	NCT01273896
	Ganetespib (STA-9090)	Phase I	No	Hematologic malignancies	No formal responses	NCT0084872
	Ganetespib (STA-9090)	Phase II	Docetaxel	NSCLC, renal cell cancer, acute myeloid leukemia	Objective responses	Young et al. (2004)

	AUY922	Phase I	No	Advanced solid tumors	SD was observed	Messaoudi et al. (2008)
	AUY922	Phase I	No	NSCLC progressing on chemotherapy	Objective responses were observed, patients with ALK rearrangement, EGFR/ KRAS and also with ALK wild-type and mEGFR	Zhang et al. (2010)
	AUY922	Phase I/ II	Bortezomib	Metastatic melanoma	Ongoing	NCT00708292
	AUY922	Phase Ib/ II	Trastuzumab	HER2+ MBC refractory to trastuzumab	Objective responses	NCT01271920; NCT01402401
	AUY922	Phase II	Erlotinib	mEGFR NSCLC progressing on EGFR inhibitors	Objective responses in EGFRT790M and SD	NCT01259089
	AUY922	Phase I/ II	Cetuximab	KRAS-mutated colorectal cancer	Ongoing	NCT01294826
	AUY922	Phase I/ II	Capecitabine	KRAS-mutated colorectal cancer	Ongoing	NCT01226732
	AT-13387	Phase I	No	Solid tumors	Objective response in GIST and SD in follicular cell thyroid	Nakashima et al. (2010)
	AT-13387	Phase I/ II	No	CRPC progressing on abiraterone	Ongoing	NCT01685268
	AT-13387	Phase I/ II	Abiraterone	CRPC progressing on abiraterone	Ongoing	NCT01685268
	AT-13387	Phase I/ II	Crizotinib	NSCLC	Ongoing	NCT01712217
Other small molecule	SNX-5422	Phase I	No	Patients with refractory solid tumor malignancies and lymphomas	Halted due to excessive ocular toxicity	Zhang et al. (2015)

(continued)

Table 8.1 (continued)

HSP90 inhibitor class	HSP90 inhibitor	Clinical stage	Combination drug	Disease	Response	References
	BIIB021	Phase I	No	Solid tumor and chronic lymphocytic leukemia	Completed	[81]
	DS-2248	Phase I	No	Solid tumors	Ongoing	NCT01288430
	Debio 0932	Phase I	No	Advanced solid tumors/ lymphoma	Ongoing	NCT01168752
	PU-H71	Phase I	No	Advanced solid tumors/ lymphoma	Ongoing	NCT01393509

8.2 Conclusions

Although numerous natural HSP90 inhibitors, as well as their analogues, have been identified with significant anti-tumor activity (as an antineoplastic drug), they failed in clinical trials due to their poor water solubility and poor bioavailability as well as cytotoxic effects such as hepatotoxicity and nephrotoxicity (Petersen et al. 2018). These problems are the main hindrance for the potential application of HSP90 inhibitors as an anti-tumor drug development in cancer. Recently, to overcome such problems, nano-delivery system has also been introduced to allow better drug delivery to the target site by eliminating unwanted toxic effects (Lamotte et al. 2017). In the area of nanomedicine, nano-sized drug carriers can offer several pharmacokinetic advantages compared to conventional methods such as controlled release of drugs leading to an improved bioavailability, biodistribution, therapeutic index, apparent solubility of drugs, prolonged activity and significantly less adverse effects (Lamotte et al. 2017). Recently, the self-assembled biodegradable nanoparticles of recombinant human gelatin (GHG) modified with alpha-tocopheryl succinate (rHG-TOS) have efficiently been encapsulated and delivered 17-AAG, the HSP90 inhibitor in cancer which indeed enhanced the anticancer effect in the tumor model with better biocompatibility (Won et al. 2011). In another study, a lipophilic GA prodrug was prepared and incorporated into methoxy-capped poly(ethylene glycol)-*block*-poly(ε-caprolactone) (mPEG-b-PCL) micelles to reduce the toxicity of GA. This formulation showed very promising results as the distribution volume of lipophilic GA prodrug in mice was decreased significantly and also the area under curve (AUC) was enhanced significantly, compared to the free drug. Moreover, lipophilic GA prodrug- mPEG-b-PCL conjugate systems also exhibited a sustained release of the drug and reduced systemic toxicity with high efficacy against MCF-7 breast cancer cells (Xiong et al. 2008). Sauvage et al. (2016) have successfully applied a strategy for a C-terminal inhibitor of HSP90, 6BrCaQ for its systemic delivery to the target site (Sauvage et al. 2016). As 6BrCaQ is a potential inhibition of HSP90 functions via degradation of its client proteins but its application is limited due to low solubility under in vivo conditions *(*Audisio et al. 2011*)*. Hence, the liposomal nano-encapsulation of 6BrCaQ Sauvage et al. (2016) increased its solubility and induced apoptosis (caspase 3 cleavage) and cell cycle arrest (G2/M) in prostate cancer PC-3 cells. Further, the preliminary in vivo studies of this formulation have shown encouraging anti-cancer potential in the breast cancer nude mice model (unpublished data). The result demonstrates that the encapsulation of insoluble HSP90 inhibitors like 6BrCaQ in suitable nanocarriers like liposome could allow overcoming the limitation associated with insoluble HSP90 inhibitors and their clinical evaluation. In another report, Zhang et al. (2015) prepared mesoporous carbon nanospheres stabilized by phospholipid for systemic delivery of amphiphobic SNX-2112 and demonstrated to enhance its antitumor effect (Zhang et al. 2015). Since SNX-2112 is amphiphobic drug, it is insoluble in both water and oil and its drug delivery was indeed a formidable challenge. The problem was overcome by the systemic delivery of SNX-2112 using carbon nanospheres. Bruni et al. (2017) also formulated poorly water-soluble drugs such as 17-AAG into nano-vectors to enhance its leishmanicidal activity with reduced cytotoxicity (Bruni et al. 2017). There are few reports, where HSP90 inhibitors are used for combinational therapy by loading to a potential nano-carrier having therapeutic properties like photodynamic and photothermal ability. Lin et al. (2016) have recently prepared a nanoformulation of organic and biocompatible nanoporphyrin-based drug delivery for the HSP90 inhibitors like 17AAG or 17DMAG for combination therapy, where the nanocarriers act a carrier as well as a therapeutic agent due to its photodynamic and photothermal potential. The formulation was very effective with an advantage of minimal skin phototoxicity (Lin et al. 2016). In another study, Rochani et al. (2016) have prepared formulation of poly (lactic-*co*-glycolic acid) (PLGA) coated, 17AAG and Fe_3O_4 loaded magnetic nanoparticle, where magnetic hyperthermia of magnetic nanoparticle in combination with the HSP 90 inhibitory effect of 17AAG was used to treat pancreatic cancer treatment and results were promising (Rochani et al. 2016). Although the research of using nanocarrier for HSP90 inhibitors as well as the combination therapy with other drugs is still not sufficient, there are studies which have already entered clinical trials. So we have summarized the literature about nanoparticulate systems carrying an HSP90 inhibitor alone or with other cytotoxic drugs in Table 8.2. Thus, we can conclude that the future perspective of

Table 8.2 The different nanocarriers system used for the delivery of HSP90 inhibitors alone or in combination with other drugs is summarized below

Pharmaceutical agent	Nanocarrier system	Current clinical status	References
Tanespimycin (17-AAG)	Poly(styrene-*co*-maleic-acid) micelles	Preclinical	[92]
Tanespimycin (17-AAG)	PEO-b-PDLLA micelles	Preclinical	[93]
Tanespimycin (17-AAG)	Nanoporphyrins	Preclinical	[90]
Tanespimycin (17-AAG)	Folate-targeted PLGA-PEG-nanoparticles	In vitro	[94]
Tanespimycin (17-AAG)	Albumin nanoparticles	Clinical study stopped after phase I	[95]
Tanespimycin (17-AAG)	Oil-in-water nanoemulsion	The study was aborted after phase I	[96]
17-AAG/rapamycin/paclitaxel (Triolimus)	PEG-b-PLA micelles	Preclinical, combinational therapy using co-delivery of rapamycin and paclitaxel	[97]
17-AAG/docetaxel	HA-PLGA	Preclinical, combinational therapy using co-delivery of rapamycin and paclitaxel	[98]
Lipophilic Geldanamycin prodrugs	mPEG-b-PCL micelles	Preclinical	[85]
Tanespimycin (17-AAG)	Pluronic (P-123/F-127) mixed micelles	In vitro	[99]

HSP90 inhibitor as a potential anti-cancer agent should focus on its systematic delivery to the target site by proper formulation using nano-delivery systems including vesicles, hydrogels. Such approaches might provide the clinical solution by overcoming the above-mentioned limitations such as poor water solubility and cytotoxic effects. Therefore, more work should be accomplished in the line of HSP90 inhibitors using nano-based delivery for higher drug distribution with a lower dose.

Acknowledgements The authors sincerely thank the Science and Engineering Research Board, Government of India, for financial support (Grant No. SERB/F/4290/2016-17) and National Institute of Technology Rourkela, Government of India, for providing the infrastructural facility for the preparation of the chapter.

References

Allan RK, Mok D, Ward BK, Ratajczak T (2006) Modulation of chaperone function and cochaperone interaction by novobiocin in the C-terminal domain of Hsp90 evidence that coumarin antibiotics disrupt Hsp90 dimerization. J Biol Chem 281:7161–7171

Audisio D, Messaoudi S, Cegielkowski L et al (2011) Discovery and biological activity of 6BrCaQ as an inhibitor of the Hsp90 protein folding machinery. ChemMedChem 6:804–815

Bagatell R, Whitesell L (2004) Altered Hsp90 function in cancer: a unique therapeutic opportunity. Mol Cancer Ther 3:1021–1030

Bao R, Lai C-J, Wang D-G et al (2009) Targeting heat shock protein 90 with CUDC-305 overcomes erlotinib resistance in non–small cell lung cancer. Mol Cancer Ther 8(12): 3296–3306

Baylon HG, Caguioa PB, Davies FE et al (2013) A phase 1/2 study of KW-2478, an Hsp 90 inhibitor, in combination with bortezomib (BTZ) in patients (Pts) with relapsed/refractory (R/R) multiple myeloma (MM). Am Soc Hematol 122: 1967

Bruni N, Stella B, Giraudo L, Della Pepa C, Gastaldi D, Dosio F (2017) Nanostructured delivery systems with improved leishmanicidal activity: a critical review. Int J Nanomedicine 12:5289

Butler LM, Ferraldeschi R, Armstrong HK, Centenera MM, Workman P (2015) Maximizing the therapeutic potential of HSP90 inhibitors. Mol Cancer Res 13:1445–1451: molcanres. 0234.2015

Byrd KM, Subramanian C, Sanchez J et al (2016) Synthesis and biological evaluation of novobiocin core analogues as Hsp90 inhibitors. Chem Eur J 22:6921–6931

Cavenagh J, Yong K, Byrne J et al (2008) The safety, pharmacokinetics and pharmacodynamics of KW-2478, a novel HSP90 antagonist, in patients with B-cell malignancies: a first-in-man, phase I, multicentre, open-label, dose escalation study. Am Soc Hematol 112(11): 2777

Chène P (2002) ATPases as drug targets: learning from their structure. Nat Rev Drug Discov 1:665

Cheung K-MJ, Matthews TP, James K et al (2005) The identification, synthesis, protein crystal structure and in vitro biochemical evaluation of a new 3, 4-diarylpyrazole class of Hsp90 inhibitors. Bioorg Med Chem Lett 15:3338–3343

Chiosis G, Timaul MN, Lucas B et al (2001) A small molecule designed to bind to the adenine nucleotide pocket of Hsp90 causes Her2 degradation and the growth arrest and differentiation of breast cancer cells. Chem Biol 8:289–299

Cho D, Heath E, Cleary J et al (2011) A phase I dose-escalation study of the Hsp90 inhibitor ganetespib (STA-9090) administered twice weekly in patients with solid tumors: updated report. J Clin Oncol 29:3051–3051

Demetri G, Heinrich M, Chmielowski B et al (2011) An open-label phase II study of the Hsp90 inhibitor ganetespib (STA-9090) in patients (pts) with metastatic and/or unresectable GIST. J Clin Oncol 29:10011–10011

Do KT, Speranza G, Chen AP et al. (2012) Phase 1 study assessing a two-consecutive-day (QD× 2) dosing schedule of the HSP90 inhibitor, AT13387, in patients with advanced solid tumors. Paper presented at the Proc Am Soc Clin Oncol

Gaspar N, Sharp SY, Eccles SA et al (2010) Mechanistic evaluation of the novel HSP90 inhibitor NVP-AUY922 in adult and pediatric glioblastoma. Mol Cancer Ther 9:1219–1233

Ge J, Normant E, Porter JR et al (2006) Design, synthesis, and biological evaluation of hydroquinone derivatives of 17-amino-17-demethoxygeldanamycin as potent, water-soluble inhibitors of Hsp90. J Med Chem 49:4606–4615

Goldman J, Raju R, Gordon G, Vukovic V, Bradley R, Rosen L (2010) A phase I dose-escalation study of the Hsp90 inhibitor STA-9090 administered once weekly in patients with solid tumors. J Clin Oncol 28:2529–2529

Hadden MK, Galam L, Gestwicki JE, Matts RL, Blagg BS (2007) Derrubone, an inhibitor of the Hsp90 protein folding machinery. J Nat Prod 70:2014–2018

Haque A, Alam Q, Zubair Alam M et al (2016) Current understanding of HSP90 as a novel therapeutic target: an emerging approach for the treatment of cancer. Curr Pharm Des 22:2947–2959

He H, Zatorska D, Kim J et al (2006) Identification of potent water soluble purine-scaffold inhibitors of the heat shock protein 90. J Med Chem 49:381–390

Hollingshead M, Alley M, Burger AM et al (2005) In vivo antitumor efficacy of 17-DMAG (17-dimethylaminoethylamino-17-demethoxygeldanamycin hydrochloride), a water-soluble geldanamycin derivative. Cancer Chemother Pharmacol 56:115–125

Hölzel M, Bovier A, Tüting T (2013) Plasticity of tumour and immune cells: a source of heterogeneity and a cause for therapy resistance? Nat Rev Cancer 13:365

Isambert N, Hollebecque A, Berge Y et al (2012) A phase I study of Debio 0932, an oral HSP90 inhibitor, in patients with solid tumors. Am Soc Clin Oncol 8(12): 3296–3306

Jensen MR, Schoepfer J, Radimerski T et al (2008) NVP-AUY922: a small molecule HSP90 inhibitor with potent antitumor activity in preclinical breast cancer models. Breast Cancer Res 10:R33

Jhaveri K, Taldone T, Modi S, Chiosis G (2012) Advances in the clinical development of heat shock protein 90 (Hsp90) inhibitors in cancers. Biochim Biophys Acta (BBA)-Mol Cell Res 1823:742–755

Jhaveri K, Ochiana SO, Dunphy MP et al (2014) Heat shock protein 90 inhibitors in the treatment of cancer: current status and future directions. Expert Opin Investig Drugs 23:611–628

Kasibhatla SR, Hong K, Biamonte MA et al (2007) Rationally designed high-affinity 2-amino-6-halopurine heat shock protein 90 inhibitors that exhibit potent antitumor activity. J Med Chem 50:2767–2778

Lamotte S, Späth GF, Rachidi N, Prina E (2017) The enemy within: targeting host–parasite interaction for antileishmanial drug discovery. PLoS Negl Trop Dis 11:e0005480

Lancet JE, Smith BD, Bradley R, Komrokji RS, Teofilovici F, Rizzieri DA (2010) A phase I/II trial of the potent Hsp90 inhibitor STA-9090 administered once weekly in patients with advanced hematologic malignancies. Am Soc Hematol 116(21):3294

Li Y, Zhang T, Schwartz SJ, Sun D (2009) New developments in Hsp90 inhibitors as anti-cancer therapeutics: mechanisms, clinical perspective and more potential. Drug Resist Updat 12:17–27

Lin T-y, Guo W, Long Q et al (2016) HSP90 inhibitor encapsulated photo-theranostic nanoparticles for synergistic combination cancer therapy. Theranostics 6:1324

Mahadevan D, Rensvold DM, Kurtin SE et al (2012) First-in-human phase I study: results of a second-generation non-ansamycin heat shock protein 90 (HSP90) inhibitor AT13387 in refractory solid tumors. Am Soc Clin Oncol 3028

Mahalingam D, Swords R, Carew JS, Nawrocki S, Bhalla K, Giles F (2009) Targeting HSP90 for cancer therapy. Br J Cancer 100:1523

Marcu MG, Chadli A, Bouhouche I, Catelli M, Neckers LM (2000) The heat shock protein 90 antagonist novobiocin interacts with a previously unrecognized ATP-binding domain in the carboxyl terminus of the chaperone. J Biol Chem 275:37181–37186

Markman JL, Rekechenetskiy A, Holler E, Ljubimova JY (2013) Nanomedicine therapeutic approaches to overcome cancer drug resistance. Adv Drug Deliv Rev 65:1866–1879
Martin CJ, Gaisser S, Challis IR et al (2008) Molecular characterization of macbecin as an Hsp90 inhibitor. J Med Chem 51:2853–2857
Marubayashi S, Koppikar P, Taldone T et al (2010) HSP90 is a therapeutic target in JAK2-dependent myeloproliferative neoplasms in mice and humans. J Clin Invest 120:3578–3593
Meacham CE, Morrison SJ (2013) Tumour heterogeneity and cancer cell plasticity. Nature 501:328
Menzella HG, Tran T-T, Carney JR et al (2009) Potent non-benzoquinone ansamycin heat shock protein 90 inhibitors from genetic engineering of Streptomyces hygroscopicus. J Med Chem 52:1518–1521
Messaoudi S, Peyrat J, Brion J, Alami M (2008) Recent advances in Hsp90 inhibitors as antitumor agents. Anti-Cancer Agents Med Chem (Formerly Curr Med Chem-Anti-Cancer Agents) 8:761–782
Mimnaugh EG, Chavany C, Neckers L (1996) Polyubiquitination and proteasomal degradation of the p185c-erbB-2 receptor protein-tyrosine kinase induced by geldanamycin. J Biol Chem 271:22796–22801
Mitchell P (2011) Biogen Idec restructures, sharpens neurology focus. Nature Biotechnology 29: 7–8
Murray CW, Carr MG, Callaghan O et al (2010) Fragment-based drug discovery applied to Hsp90. Discovery of two lead series with high ligand efficiency. J Med Chem 53:5942–5955
Nakashima T, Ishii T, Tagaya H et al (2010) New molecular and biological mechanism of antitumor activities of KW-2478, a novel non-ansamycin Hsp90 inhibitor, in multiple myeloma cells. Clin Cancer Res 16:2792–2802: clincanres. 3112. subyr
Neckers L (2002) Hsp90 inhibitors as novel cancer chemotherapeutic agents. Trends Mol Med 8:S55–S61
Neckers L (2007) Heat shock protein 90: the cancer chaperone. In: Heat shock protein cancer. Springer, Dordrecht, pp 231–52
Nimmanapalli R, O'Bryan E, Bhalla K (2001) Geldanamycin and its analogue 17-allylamino-17-demethoxygeldanamycin lowers Bcr-Abl levels and induces apoptosis and differentiation of Bcr-Abl-positive human leukemic blasts. Cancer Res 61:1799–1804
Ohsuki S, Tengeiji A, Ikeda M, Shibata Y, Nagata C, Shimada T (2012) Pyrazolopyrimidine derivative: Google Patents
Palermo CM, Westlake CA, Gasiewicz TA (2005) Epigallocatechin gallate inhibits aryl hydrocarbon receptor gene transcription through an indirect mechanism involving binding to a 90 kDa heat shock protein. Biochemistry 44:5041–5052
Parveen S, Sahoo SK (2008) Polymeric nanoparticles for cancer therapy. J Drug Target 16:108–123
Patel HJ, Modi S, Chiosis G, Taldone T (2011) Advances in the discovery and development of heat-shock protein 90 inhibitors for cancer treatment. Expert Opin Drug Discovery 6:559–587
Pearl LH, Prodromou C (2006) Structure and mechanism of the Hsp90 molecular chaperone machinery. Annu Rev Biochem 75:271–294
Petersen ALOA, Campos TA, Dantas DAS et al (2018) An in vitro assessment on the efficacy of a novel liposomal formulation containing the Hsp-90 inhibitor, 17-AAG. Front Cell Infect Microbiol 8:303
Proia DA, Zhang C, Sequeira M et al (2013) Preclinical activity profile and therapeutic efficacy of the HSP90 inhibitor ganetespib in triple-negative breast cancer. Clin Cancer Res: clincan res 20(2):413–424
Rochani AK, Balasubramanian S, Girija AR et al (2016) Dual mode of cancer cell destruction for pancreatic cancer therapy using Hsp90 inhibitor loaded polymeric nano magnetic formulation. Int J Pharm 511:648–658
Roe SM, Prodromou C, O'Brien R, Ladbury JE, Piper PW, Pearl LH (1999) Structural basis for inhibition of the Hsp90 molecular chaperone by the antitumor antibiotics radicicol and geldanamycin. J Med Chem 42:260–266

Rowlands MG, Newbatt YM, Prodromou C, Pearl LH, Workman P, Aherne W (2004) High-throughput screening assay for inhibitors of heat-shock protein 90 ATPase activity. Anal Biochem 327:176–183

Samlowski W, Papadopoulos K, Olszanski A et al (2011) Phase 1 study of HSP90 inhibitor MPC-3100 in subjects with refractory or recurrent cancer. Mol Targets Cancer Ther 10(11): A96-A96

Samuel T, Sessa C, Britten C et al (2010) AUY922, a novel HSP90 inhibitor: final results of a first-in-human study in patients with advanced solid malignancies. J Clin Oncol 28:2528–2528

Sauvage F, Franzè S, Bruneau A et al (2016) Formulation and in vitro efficacy of liposomes containing the Hsp90 inhibitor 6BrCaQ in prostate cancer cells. Int J Pharm 499:101–109

Schroder C, Pedersen J, Chua S et al (2011) Use of biomarkers and imaging to evaluate the treatment effect of AUY922, an HSP90 inhibitor, in patients with HER2+ or ER+ metastatic breast cancer. J Clin Oncol 29:e11024–e11e24

Shi J, Van de Water R, Hong K et al (2012) EC144 is a potent inhibitor of the heat shock protein 90. J Med Chem 55:7786–7795

Sidera K, Patsavoudi E (2014) HSP90 inhibitors: current development and potential in cancer therapy. Recent Pat Anticancer Drug Discov 9:1–20

Siegel RL, Miller KD, Jemal A (2017) Cancer statistics, 2017. CA Cancer J Clin 67:7–30

Soga S, Akinaga S, Shiotsu Y (2013) Hsp90 inhibitors as anti-cancer agents, from basic discoveries to clinical development. Curr Pharm Des 19:366–376

Solit DB, Chiosis G (2008) Development and application of Hsp90 inhibitors. Drug Discov Today 13:38–43

Sydor JR, Normant E, Pien CS et al (2006) Development of 17-allylamino-17-demethoxygeldanamycin hydroquinone hydrochloride (IPI-504), an anti-cancer agent directed against Hsp90. Proc Natl Acad Sci 103:17408–17413

Taldone T, Chiosis G (2009) Purine-scaffold Hsp90 inhibitors. Curr Top Med Chem 9:1436–1446

Taldone T, Patel PD, Patel M et al (2013) Experimental and structural testing module to analyze paralogue-specificity and affinity in the Hsp90 inhibitors series. J Med Chem 56:6803–6818

Tanida S, Hasegawa T, Higashide E (1980) Macbecins I and II, new antitumor antibiotics. J Antibiot 33:199–204

Trepel J, Mollapour M, Giaccone G, Neckers L (2010) Targeting the dynamic HSP90 complex in cancer. Nat Rev Cancer 10:537

Wang J, Li Z, Lin Z et al (2015) 17-DMCHAG, a new geldanamycin derivative, inhibits prostate cancer cells through Hsp90 inhibition and survivin downregulation. Cancer Lett 362:83–96

Won Y-W, Yoon S-M, Sonn CH, Lee K-M, Kim Y-H (2011) Nano self-assembly of recombinant human gelatin conjugated with α-tocopheryl succinate for Hsp90 inhibitor, 17-AAG, delivery. ACS Nano 5:3839–3848

Woodhead AJ, Angove H, Carr MG et al (2010) Discovery of (2, 4-Dihydroxy-5-isopropylphenyl)-[5-(4-methylpiperazin-1-ylmethyl)-1, 3-dihydroisoindol-2-yl] methanone (AT13387), a novel inhibitor of the molecular chaperone Hsp90 by fragment based drug design. J Med Chem 53:5956–5969

Xiong MP, Yáñez JA, Remsberg CM et al (2008) Formulation of a geldanamycin prodrug in mPEG-b-PCL micelles greatly enhances tolerability and pharmacokinetics in rats. J Control Release 129:33–40

Ying W, Du Z, Sun L et al (2011) Ganetespib, a unique triazolone-containing Hsp90 inhibitor, exhibits potent antitumor activity and a superior safety profile for cancer therapy. Mol Cancer Ther 11(2):475-484

Young JC, Agashe VR, Siegers K, Hartl FU (2004) Pathways of chaperone-mediated protein folding in the cytosol. Nat Rev Mol Cell Biol 5:781

Zhang H, Neely L, Lundgren K et al (2010) BIIB021, a synthetic Hsp90 inhibitor, has broad application against tumors with acquired multidrug resistance. Int J Cancer 126:1226–1234

Zhang X, Zhang T, Ye Y et al (2015) Phospholipid-stabilized mesoporous carbon nanospheres as versatile carriers for systemic delivery of amphiphobic SNX-2112 (a Hsp90 inhibitor) with enhanced antitumor effect. Eur J Pharm Biopharm 94:30–41

Chapter 9
Hsp90 Is a Pivotal Player in Retinal Disease and Cancer

Asmaa Aboelnour, Ahmed E. Noreldin, and Islam M. Saadeldin

Abstract Heat shock proteins (Hsp) are primarily protecting and maintaining cell viability during stressful conditions such as thermal and oxidative challenges through protein refolding and stabilization. Hsp play an essential role to confer eye protection from disease states particularly the diseases affecting the retina. Here, we summarize the Hsp function in normal retina, and their involvement in the pathogenesis of certain retinal diseases such cancer, glaucomatous retina, retinitis pigmentosa, and retinal neurodegeneration, as well as the age-related macular degeneration. This information would provide a better understanding of Hsp function and their involvement in ocular disease pathogenesis that could be a target for therapeutic purposes.

Keywords Cancer · Glaucomatous retina · Hsp90 · Inherited retinal disease · Neurodegenerative disease · Retina

Abbreviations

AMD	Age-related macular degeneration
GFAP	Glial fibrillary acidic protein

A. Aboelnour · A. E. Noreldin
Histology and Cytology Department, Faculty of Veterinary Medicine, Damanhour University, Damanhour, Egypt

I. M. Saadeldin (✉)
Department of Animal Production, College of Food and Agricultural Science, King Saud University, Riyadh, Saudi Arabia

Department of Physiology, Faculty of Veterinary Medicine, Zagazig University, Zagazig, Egypt
e-mail: isaadeldin@ksu.edu.sa

A. A. A. Asea, P. Kaur (eds.), *Heat Shock Protein 90 in Human Diseases and Disorders*, Heat Shock Proteins 19,
https://doi.org/10.1007/978-3-030-23158-3_9

Hsp	Heat shock proteins
IMPDH	Inosine-5′-monophosphate dehydrogenase
RP	Retinitis pigmentosa
RPE	Retinal pigment epithelium

9.1 Introduction

Heat shock proteins (Hsp) are molecules with highly conserved structures, that have been revealed to share in protein quality control in eukaryotes, eubacteria and archaea (Ellis 2007; Lindquist and Craig 1988). Hsp are ubiquitously expressed in various subcellular parts (Jego et al. 2013). Hsp act as stress protein because they are quickly created after exposure to the unfavorable extrinsic factors such as high temperature, cytokine releasing and hypoxia. Italy geneticist Ferruccio Ritossa detected Hsp for the first time in the salivary gland of Drosophila larvae in 1962 (Kiang and Tsokos 1998). He noticed the swelling of the Drosophila's salivary gland after the elevation of the environmental temperature, so he named this phenomenon by heat shock response (HSR). Hsp expression increase to improve the cells resistance to adverse lethal condition(s). The Hsp were nominated according to their molecular weights because their function was obscured at the time of detection. Therefore, the 70 kDa Hsp named as Hsp70. Hsp are six types according to molecular weight variation. They are Hsp40, Hsp60, Hsp70, Hsp90, Hsp110, and small molecule heat shock protein (small Heat Shock Proteins, sHsp) (Lindquist and Craig 1988).

In eukaryotic organisms, heat shock protein 90 represents about 1–2% of cytosolic proteins so it considered the most abundant molecular chaperone (Garcia-Carbonero et al. 2013; Singh et al. 2015; Wainberg et al. 2013). Hsp90 significantly contributes to folding, maturation, stabilization, and activation of Hsp90 client proteins in cancer and normal cells (Waza et al. 2005). Moreover, Hsp90 plays an important role in the maintenance of genome stability, cell signaling and intracellular transport in normal cells (Kaplan and Li 2012). Furthermore, Hsp90 preserves these proteins from proteasomal degradation (Kaplan and Li 2012; Lin et al. 2016). After the exposure of cells to extrinsic stressors, Hsp90 resists against stress *via* two ways. First, Hsp90 correct folding of protein by proper refolding and elevates the rate of the reactivated proteins. Second, Hsp90 destroys non-functional proteins by proteasomal degradation through the polyubiquitination pathway. So, Hsp90 enhances cell survival in stress conditions by maintaining the protein homeostasis (Jolly and Morimoto 2000; Parsell and Lindquist 1993).

Cancer is a critical threat to humanity. In China, it was reported 200 deaths in 2015. The highest mortality rate was recorded in lung cancer cases, gastric cancer, esophageal cancer, and liver cancer respectively (Chen et al. 2016). There is a serious necessity for modern treatments for cancer as the opposition to traditional cytotoxic and cytostatic drugs is now widespread (Sauvage et al. 2017). Cancer cells have activated improper signaling pathways and higher metabolic requirements compared to normal cells thus need a chaperone machinery to survive (Chatterjee and Burns 2017). Mainly, oncoproteins of cancer require increased levels of Hsp chaperonage for their folding, stabilization, aggregation, activation, function and proteolytic degradation (Chatterjee et al. 2016; Neckers and Workman 2012). The expressions/activities of Hsp27, Hsp70, and Hsp90 chaperones are markedly higher in cancer (Garrido et al. 2006; Jego et al. 2013).

Recent studies of the molecular mechanism of tumor incidence have been detected that Hsp90 has a pivotal role in organizing proteins in various signaling pathways which contribute to the tumor carcinogenesis. Therefore, the Hsp90 is considered as a novel important research area in the field of anti-tumor biology (Rui et al. 2018) due to its upregulation in response to oxidative stress conditions in various human tumors (Saif et al. 2013; Tukaj et al. 2017). Moreover, its ATPase activity is boosted 50X in a cancerous microenvironment (Taiyab et al. 2009). Furthermore, Hsp90 stabilizes mutant proteins that manifest throughout cell transformation, thus enabling malignant transformation (Calderwood et al. 2006). Therefore, the suppression of Hsp90 influences processes engaged at the beginning of cancer which can be considered as the “Hallmarks of Cancer” (Miyata et al. 2013; Workman et al. 2007). Consequently, suppression of Hsp90 by proteasomal degradation and inhibitors can be effective in cancer treatment (Proia and Kaufmann 2015; Venkatesan et al. 2016; Wang et al. 2016).

In this chapter, we will first describe Hsp90 in terms of its structure and function inside the cells and then concentrate on the connection of Hsp90 to cancer especially retina cancer and the effects of Hsp90 inhibitors to improve cancer therapy. We will also underline the relationships of Hsp90 to retinal neurodegeneration to understand the role of Hsp90 in the incidence of retinal diseases.

9.1.1 HSP90 in Normal Retina

If metabolic stress affects living cells, they react by transient elevating the expression of HSP. The cause of this denomination is due to the temperature increase was the first-reported stimulator of the Hsp reaction. In mammals, a wide range of detrimental stimuli, like ischemia and hypoxia, and damaging chemical agents, like heavy metals and metabolic poisons are the main Hsp inducers (Kiang and Tsokos 1998). Recently, 90-kDa family (HSP90 family) is one of the HSP families that have drawn the interest of many researchers. Eukaryotic cells have three types of Hsp90s: cytosolic Hsp90 with two isoforms of Hsp90α and Hsp90β, mitochondrial Trap1 (tumor necrosis receptor-associated protein 1) and Grp94 (glucose-regulated protein 94) of the endoplasmic reticulum (ER) (Chiosis 2016; Workman et al. 2007). Hsp90 is a cytosolic protein which is present in either Hsp90β form or Hsp90α form. Although separate genes encode the two isoforms of Hsp90, they are 86% symmetrical at the gene translation level (Moore et al. 1989).

Under normal physiological conditions of eukaryotic cells, Hsp90 represents about 1–2% of total cellular protein so it could be considered one of the most abundant cytoplasmic proteins. However, its expression elevates many folds under stress conditions (Scheibel and Buchner 1998). Although Hsp90 acts as a molecular chaperone-like the other Hsp, it links to a specific subset of cellular proteins, including components of the cytoskeleton, including microfilaments and microtubules (Csermely et al. 1998; Czar et al. 1996); many tyrosine and serine-threonine protein kinases, like cyclin-dependent protein kinase 4 (CDK4), v-Raf/c-Raf and v-Src/c-Src (Csermely et al. 1998; Darimont 1999; Pratt 1998); and transcription factors like steroid hormone receptors (Csermely et al. 1998; Pratt 1998; Scheibel and Buchner 1998). It seems that Hsp90 regulates the signaling capacity of these transcription factors and (Csermely et al. 1998; Pratt 1997, 1998; Ylikomi et al. 1998) and it buffers the destructive changes in the regulatory proteins caused by minor gene mutations (Rutherford and Lindquist 1998).

It has been noticed the strong mRNA expression of murine Hsp90β and Hsp90α through early mammalian ocular development (embryonic days 11.5–14.5). Moreover, Hsp90β mRNA levels still high during later embryogenesis and adulthood but Hsp90α mRNA levels decreased (Tanaka et al. 1995). On the other hand, Kojima et al. (1996) detected that the rat Hsp90α mRNA expressed from embryonic day 17 to adulthood, except around postnatal day 5.

Hsp90 immunoreactivity was detected in most layers of the retina. The strongest immunohistochemical reaction for Hsp90 was detected in the outer limiting membrane (OLM). Lesser intense reactions were observed in the outer plexiform layer (OPL), inner plexiform layer (IPL), and the inner segment (IS). The remainder of the retina staining was noticed in the outer nuclear layer (ONL), inner nuclear layer (INL), and perinuclear regions of cells of the retinal ganglion cells (RGC) and was somewhat higher than the background at the tips of the outer segment (OS), where they are interlocked with processes from the retinal pigment epithelium (RPE). The low reaction was detected in the walls of blood vessels where RGC axons stained for Hsp90 and kept the same reaction as the axons become near the optic disc. On the other hand, a high reaction was observed in the innermost row of ONL nuclei, next to the OPL and near the retina-optic nerve junction. Moreover, RPE cells situated less than 200–250 μm from the optic disc expressed intense Hsp90 reaction, especially inside the cell nuclei. In addition, Hsp90 immunoreactivity was observed in the optic nerve. Furthermore, the Hsp90 immunoreactivity had a clear perinuclear and rare nuclear reaction in the glial cells (Dean and Tytell 2001).

It was suggested that Hsp90 could be involved in the synthesis of an Hsp90-protein substrate heterocomplex (fold some), which contains at least nine accessory proteins and Hsc70 (Pratt 1997). Dean et al. (1999) detected that all retinal layers, from IS to RGC, had a positive immunoreactivity for Hsp90 and Hsc70. These results are harmonious with the functions of Hsp90 and Hsc70 being interdigitated, suggesting both proteins were part of a fold some. The plexiform layers showed the only conflict in the staining patterns. The OPL and IPL had a lower immunoreactivity for Hsc70 than for Hsp90 (Dean et al. 1999). On the other hand, many factors can influence the strength of the immunoreaction in the tissue section, containing many technical aspects of the reactions utilized to detect the immunoreaction and variances in antibody affinity for the antigen. Therefore, other more quantitative analyses must be done to measure the amount of antigen in the retina more accurately.

One functional difference between Hsc70 and Hsp90 that may attribute to their various distributions in the retina is the role of the Hsc70 and Hsp90 in keeping active steroid hormone receptors. As the retina is recognized to be susceptible to corticosteroids and thus it contains receptors (Kobayashi et al. 1998; Mirshahi et al. 1996). The receptor distribution may be related partly to the retinal distribution of Hsp90. Moreover, the plexiform layers include the synaptic connections and cell processes of the retinal cells, so it could be suggested that Hsp90 can play in keeping the structures engaged in forming synaptic junctions. There are some proofs to give a possibility that Hsp may be situated at synaptic connections between retinal cells and that Hsp can preserve synaptic functions in some eukaryotes by altering the properties of their synapses (Karunanithi et al. 1999).

Hsp70 and Hsp90 may play a pivotal part in the preserving of Müller-photoreceptor and Müller–Müller cell junctional specializations localized at the OLM. Karunanithi et al. (1999) have mentioned that the induction of Hsp and especially Hsp70 protects synaptic transmission at the *Drosophila* larval neuromuscular junction. Moreover, his group revealed that the process of Hsp70 induction seems to include pre- and postsynaptic modifications (Karunanithi et al. 1999).

Black et al. (1985) detected that the optic nerve has three distinct morphologic regions near the retina- optic nerve junction, the optic nerve proper (ONP), the retina-optic nerve transition region (ROT) and the optic nerve head (ONH). The pattern of distribution of Hsp90 and Hsc70 in the optic nerve was the same reported by Dean et al. (1999) and was obviously unlike the pattern detected for Hsp25. The majority of the Hsp90 immunostaining was concentrated in the juxtanuclear cytoplasm of the glial cells without any reaction in the nerve fibers.

Dean and Tytell (2001) were reported that there was no sudden elevation in Hsp90 immunoreactivity as the RGC axons entered the optic nerve head and left the retina. They assumed that Hsp25 is the highest distributed Hsp in the axons of the normal optic nerve, while Hsp90 and Hsc70 are localized mainly in the glia of the optic nerve. The functional importance of these variances in distribution stills to be investigated, but it boosts the hypothesis that each of the Hsp has characteristic roles in the different cellular structures of the optic nerve and retina (Dean and Tytell 2001).

It has been supposed that Hap90 is pivotal for neuronal polarization and axonal elongation in the retina. Neuronal function relays on the establishing of various functional and morphological differentiated domains; dendrites, soma, axon initial segment, and axon. The morphological differentiation begins with the specification of one of the initial neurites as the axon and its subsequent elongation. This process needs the dynamics and coordinated reorganization of microtubule and actin cytoskeletons (Bradke and Dotti 1999; Da Silva and Dotti 2002).

Cytoskeleton dynamic behavior is organized by microfilament and microtubule-associated proteins (ABPs or MAPs) and tubulin post-translational modifications as well (Bradke and Dotti 1999; Tapia et al. 2010). Various signaling pathways, like PI3K pathway, are engaged in axon growth and specification (Shi et al. 2003), including proteins like GSK3, Akt (Shi et al. 2004; Yan et al. 2006), Rac, GTPases Rho, Par3/Par6. The axonal specification was suggested to be because of the local accumulation of some specific proteins, that reveals a fine organizing of its synthesis and local degradation mechanisms. There are two suggested mechanisms to concentrate these proteins at the future axon tip. First, kinesins can specifically target proteins to the axon tip. In fact, KIF5 is expressed in several neurites before axon specification but appears absent from the dendrites and selectively accumulated in the emerging axon when the axon is specified (Jacobson et al. 2006). Second, protein degradation in other neurites causes axon selective accumulation of these proteins (Schwamborn et al. 2007).

In non-neuronal systems, Hsp90 proteins assist in the regulation of intracellular protein activity, including protein transport (Johnson et al. 1996). Hsp90 proteins are associated with specific serine and tyrosine kinases (Mimnaugh et al. 1995), calmodulin, actin, and tubulin (Sanchez et al. 1988). Loss of Hsp90 activity suppresses the cellular response to both v-src kinase activity, as well as the nerve growth factor (NGF)- mediated signaling pathway (Jaiswal et al. 1996; Xu and Lindquist 1993), proposing a role for Hsp90 in the organizing of these pathways. Moreover, Hsp90 protein associations can change intracellular protein activation and function (Nathan and Lindquist 1995).

Hsp90 protein and mRNA are basically expressed at high levels in neural tissue, including brain and retina (Tanaka et al. 1995). Therefore, Hsp90 proteins may be supposed to have a selective role in neural function. Bernstein et al. (1996) previously detected that *Hsp90* mRNA is expressed at higher levels at the central region of the primate retina (the fovea) than the retinal periphery (Bernstein et al. 1996). Moreover, *Hsp90* mRNA was detected to be selectively expressed in rodent retinal ganglion cells (RGCs) (Tanaka et al. 1995). Because of the concentration of RGCs in the fovea region of the primate retina (Shapley and Perry 1986), so, it is possible that Hsp90 has an RGC-specific or intensive function.

9.1.2 The Role of HSP90 in Cancer Biology

If the external temperature elevates significantly over normal body temperature, the most cells consequently respond by increasing the production of a limited class of proteins termed heat shock proteins. Inside the cells, the interactions between proteins lead to their proteolytic turnover, intracellular disposition, and folding. Hypoxia and acidosis are common conditions within tumors, and it was detected the elevated chaperone proteins expression in many types of solid tumors (Kimura et al. 1993; Santarosa et al. 1997) which may explain the ability of malignant cells to maintain homeostasis in an antagonistic environment. Chaperone proteins enhance cell resistance against environmental stressors and also consolidate tumor cells to endure the internal alterations, such as mutations of essential signaling molecules that would be fatal (Takayama et al. 2003). Chaperones can act as biochemical buffers for the genetic shakiness that is common in numerous human cancers.

The elevated heat shock protein levels detected in advanced cancers could explain the suitable cellular response to the microenvironment stressors inside the tumor such as nutrient-deprivation, acidosis, hypoxia. On the other hand, Hsp can assist tumor cells to get away from the apoptotic death that caused by the imbalanced signaling accompanied with neoplastic transformation (Takayama et al. 2003). One of the prevalent features of cancer cells is the deterioration of apoptotic signaling, which enhances tumor expansion and survival by separating them from normal regulatory factors so, tumor becomes reluctant to both chemotherapeutic drugs and host defense mechanisms (Jäättelä 1999; Sliutz et al. 1996).

It was mentioned that Hsp90 and its cochaperones modify tumor cell apoptosis. Most of this activity mediated by influences on tumor necrosis factor receptors (Vanden Berghe et al. 2003), AKT (Basso et al. 2002), and nuclear factor-nB function (Chen et al. 2002). On the other hand, Hsp90 may contribute in neoplastic transformation than suppressing apoptosis. Hsp90 is unique among all heat shock proteins as it is not needed for the biogenesis of most polypeptides (Nathan et al. 1997). On the other side, many of its client proteins or cellular substrates are labile signal transducers that have a pivotal role in tissue development, cell survival, and growth control (Pratt 1998).

The more notable Hsp90 client proteins related to cancer are receptor tyrosine kinases (EGFR, HER2, FLT3, and IGF1R), SRC family kinases (FYN, LCK, and SRC), cell cycle G2 checkpoint kinases (POLO-1, MYT1, and WEE1), serine/threonine kinases (CDK4, AKT, and RAF-1), steroid hormone receptors (androgen, glucocorticoid, progesterone, and estrogen), mutant fusion kinases (NPM-ALK and BCR-ABL), transcription factors (HIF-1, HSF-1, and p53) (Karagoz and Rudiger 2015).

9.1.3 Roles of HSP90 in Neurodegeneration

Hsp90 has numerous pivotal preserving roles for the functional viability and stability of cells against the mutating pressure (Chiosis 2006; Whitesell and Lindquist 2005). When protein aggregate in neurodegenerative disorders, the suppression of Hsp90 stimulates heat shock factor-1 (HSF-1) to produce Hsp40 and Hsp70, in addition to other chaperones, which in turn, enhance protein degradation (Klettner 2004; Muchowski and Wacker 2005). On the other hand, Luo et al. (2008) detected the protective effect of Hsp90 on the neuronal proteins of thumping capacity in the neurodegeneration, so enhances the gathering of toxic aggregates.

The heat shock reaction is predominately regulated by HSF-1. When the cells exposed to stressors, they normally react by stimulation of the heat shock response (HSR) associated with elevated induction of cytoprotective Hsp which reduce cytotoxicity resulting from denatured and misfolded proteins (Klettner 2004; Muchowski and Wacker 2005). The most transition events happen on the transcriptional level. The stressors start the destruction of protein by stimulation of HSF-1, which links to upstream regulatory sequences in the promoters of heat

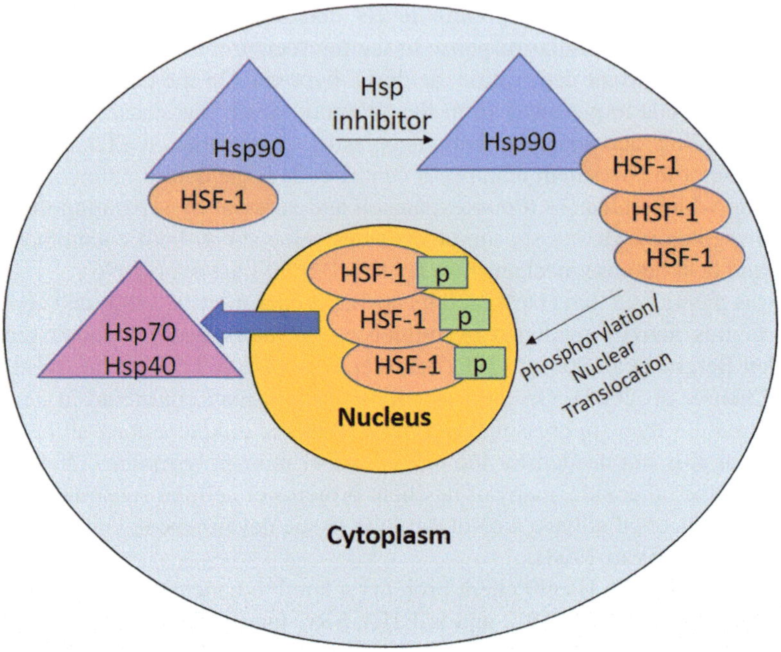

Fig. 9.1 Suppression of Hsp90, heat shock or stress releases HSF-1 from the Hsp90 complex, which leads to its trimerization, activation, phosphorylation and translocation to the nucleus where it initiates a heat shock response, revealed by the production of Hsp like the chaperones Hsp70 and its activator, Hsp40

shock genes (Anckar and Sistonen 2007). The stimulation of HSF-1 begins *via* a multi-step pathway, including a monomer-to trimer transition, nuclear accumulation and extensive post-translational modifications (Fig. 9.1). Hsp90 regulates the function of HSF-1 (Zou et al. 1998). Under normal circumstances, Hsp90 links to HSF-1 and maintains the transcription factor in a monomeric state. Inhibition of Hsp90 or stressor releases HSF-1 from the Hsp90-HSF-1 complex, which leads to its trimerization, activation, and translocation of HSF-1 to the nucleus where it begins a heat shock reaction, revealed by the production of Hsp70 and Hsp40. Neurons have been detected to be reluctant to Hsp stimulation following heat shock (Brown 2007).

Tauopathies are neurodegenerative diseases recognized by tau protein abnormalities which lead to an accumulation of aggregated and hyperphosphorylated tau (Kosik and Shimura 2005). In Alzheimer's disease (AD), Lau et al. (2002) supposed that tau hyperphosphorylation is a pathogenic pathway resulted from thumping stimulation of many kinases, particularly, glycogen synthase kinase-3 beta (GSK3β) and cyclin-dependent protein kinase 5 (CDK5), leading to phosphorylation of tau on pathogenic sites (Lau et al. 2002). Goedert and Jakes (2005) suggested that hyperphosphorylated tau in the AD misfolds microtubules net and synthesis of toxic tau aggregates. In a collection of tauopathies called "frontotemporal dementia and parkinsonism linked to chromosome 17 (FTDP-17)", the mutation results from many transformations in human tau isoforms on chromosome 17, that lead to the accumulation of aggregated tau comparable to that in the AD (Goedert and Jakes 2005). In tauopathies, it has been mentioned that Hsp90 preserves the stability of p25 and p35, neuronal proteins that stimulate CDK5 *via* complex formation resulting in thumping tau phosphorylation, and Hsp90 maintains the mutant but not wild-type tau protein in tauopathies (Luo et al. 2007). Suppression of Hsp90 in mouse and cellular models of tauopathies resulted in a decrease of the pathogenic activity of these proteins and led to a time- and dose-dependent discarding of aggregated tau (Luo et al. 2007).

9.1.4 HSP90 Inhibition Protects Against Inherited Retinal Disease

Hsp90 is a highly conserved and copious molecular chaperone that is engaged in numerous cellular processes, containing the functional maturation of substrate proteins, which are defined as 'clients' (Li and Buchner 2013; Taipale et al. 2010). Many of client proteins are oncogenes, resulting in Hsp90 protruding as a pivotal target in various types of cancer treatment (Pearl et al. 2008). Post-translational modifications and nucleotide binding regulate Hsp90 function (Walton-Diaz et al. 2013). Hsp90 inhibitors connect with a high affinity to the ATP-binding pocket and block the chaperone ATPase cycle resulting in the destruction of client proteins (Li and Buchner 2013; Pearl et al. 2008). Moreover, suppression of Hsp90 function deactivates the chaperone complex with Heat Shock Factor 1 (HSF-1), leading to the activation of HSF-1 and initiation of heat shock protein expression (Zou et al. 1998). Subsequently, Hsp90 suppression can trigger a dual influence, the proteasome-mediated damage of Hsp90 client proteins and activation of HSF-1, which enhances Hsp70 and other chaperones to protect against protein aggregation and lower protein toxicity (Labbadia et al. 2011; Sittler et al. 2001).

9.1.5 Retinitis Pigmentosa

RP is a hereditary photoreceptor degeneration resulted from transformations in the rhodopsin gene which is considered the most prevalent origin of autosomal dominant RP. Mendes and Cheetham (2008) detected protection of the Hsp90 inhibitor 17-N-allylamino-17-demethoxygeldanamycin (17-AAG) against rhodopsin toxicity and P23H, which is deemed to be the most prevalent rhodopsin transformation in the USA. This protection depends on HSF-1 because 17-AAG could not protect mouse embryonic fibroblasts isolated from HSF-1 knock-out mice against P23H rhodopsin aggregation suggesting the dependence of protective influence on the induction of the stress response (Aguila et al. 2014).

Aguila et al. (2014) detected the enhancement of systemic administration of the blood-brain barrier permeable (HSP990), Hsp90 inhibitor, to HSF-1 and the beginning of molecular chaperone expression in vivo in the retina. A single minimum dose of HSP990 was enough to improve photoreceptor survival and visual function of a P23H rhodopsin transgenic rat model which had advanced retinal degeneration. Interestingly, this treatment lowered rhodopsin aggregation and stimulated molecular chaperones but did not influence any phototransduction component, suggesting the efficiency of Hsp90 suppression to induce the proteostasis mechanism which safeguards against misfolded proteins (Aguila et al. 2014). Inosine-5′-monophosphate dehydrogenase (IMPDH) misfolding mutations associated with RP10 is another example of using Hsp90 suppression to target photoreceptor proteostasis imbalances. In this case, Tam et al. (2010) utilized claudin 5 RNAi to temporary lower the permeability of the blood-retinal barrier and let 17-AAG to induce a preservative response in photoreceptors expressing R224P mutant IMPDH accompanied with lowering in mutant aggregated IMPDH and safeguarding the outer nuclear layer (ONL) structure (Tam et al. 2010).

Moreover, suppression of Hsp90 in another class of rhodopsin mutation (R135L) disease model was preventative, but the Hsp90 inhibition machinery was separated from HSF-1. The R135L mutation blocks binding and thumping rhodopsin endocytosis which harmfully influences vesicular traffic and causes rhodopsin hyperphosphorylation (Chuang et al. 2004).

Hsp90 suppression prohibited the mobilization of arrestin to R135L mutant rhodopsin and therefore mitigated thumping endocytosis (Aguila et al. 2014). HSF-1 null cells kept the influence of Hsp90 suppression suggesting that it was independent of HSF-1. Furthermore, a study conducted by Aguila et al. (2014) showed that rhodopsin kinase (GRK1) is a Hsp90 client protein and the influence of Hsp90 suppression on R135L rhodopsin arrest was mediated by lowering of phosphorylation upstream of R135L due to a suitable kinase deficiency (Aguila et al. 2014). This mechanism linked to the decrease of a specific client protein that is interposing an adverse influence of a genetic mutation is distinguished from the enhanced production of protective factors *via* the enhancement of the stress response to combat a mutational consequence. We can conclude the multiple roles of Hsp90 in the retina. Moreover, the usage of Hsp90 inhibitors has potential protective roles against various types of RP *via* various mechanisms.

9.1.6 Age-Related Macular Degeneration (AMD) and RPE Biology

AMD is a multifactorial disease including environmental, genetic, and metabolic factors. Cell death and functional anomalies in the RPE cells have a pivotal role in the AMD progress and are accompanied with elevated oxidative stress (Jarrett and Boulton 2012). Decanini et al. (2007) detected the normal expression of Hsp90 in

RPE cells and its marked elevated expression during the AMD progress. Moreover, Qin et al. (2011) suggested that Hsp90 detected in necrotic RPE cells may act as an inducer for inflammatory reactions in neighboring healthy RPE. Therefore, Hsp90 suppression can prohibit the inflammatory reactions in RPE cells (Wang et al. 2010). In addition, (Wu et al. 2007) detected the ability of the Hsp90 inhibitor geldanamycin to suppress vascular endothelial growth factor expression in hypoxic RPE cells, suggesting that Hsp90 inhibitors can be a potential treatment for both neovascularization and inflammation.

In age-related macular degeneration, damage to the RPE layer leads to the death of foveal cone photoreceptors that are in charge of the sharp central vision (Kaarniranta et al. 2013). This means that RPE cells are essential in the progression of AMD. Recently, inflammasome signaling has been accompanied with the pathogenesis of AMD (Kauppinen et al. 2012; Tarallo et al. 2012). As reviewed by Harris and Rubinsztein (2011), inflammasomes and autophagy control each other, suggesting that autophagy could be in charge of the discarding of unnecessary inflammasome components. Autophagy is the cellular degradation system, particularly when proteasomal degradation has failed. It is established that proteasomal activity reduces during the aging process which is accompanied by the pathogenesis of AMD (Jung et al. 2009; Li et al. 2008). If NLRP3 inflammasome is translated after the priming signal, it is removed from the cell unless safeguarded by a protein complex including Hsp90 (Martinon 2008; Mayor et al. 2007). Mayor et al. revealed that in the obscurity of Hsp90, NLRP3 will be destructed by the proteasome (Mayor et al. 2007).

9.1.7 HSP90 in Glaucomatous Retina

It is well known that Hsp90 is ceaselessly expressed in mammalian cells (Park et al. 2007). Moreover, heat shock stress induces its expression in retinal Müller cells. Immunostaining for Hsp90 was detected to be stronger than Hsp70 in the glaucomatous retinas compared to the control retinas, harmonious with the fact that the apoptotic pathway is one of the key mechanisms included in glaucomatous neuropathy, and that Hsp are upregulated to suppress apoptosis and save neuronal cells. Elevated immunoreactivity for Hsp90 in glaucomatous retinas was detected at the nerve fiber layers and ganglion cell. The GFAP immunolabeling and Hsp90 were similar in both glaucomatous and normal eyes. Importantly, the retinal localization of Hsp90 was similar to GFAP, a glial cell marker, in glaucoma eyes. It is well established that retinal glial cells are stimulated with glaucomatous neuropathy (Tanihara et al. 1997). Moreover, Müller cells have been detected to generate Hsp90 under heat shock stress 32 and to produce neuroprotective cytokines like ciliary neurotrophic factor after retinal injury (Honjo et al. 2000). These cells may hold an inherent neuroprotective capability in the retina. The identical staining pattern of GFAP and Hsp90 indicated that stimulated Müller cells may be a pivotal source of the Hsp (Sakai et al. 2003).

9.1.8 Role of HSP90 in Prevention of Retinal Cancer

Over-expressed Hsp90 were detected in many cancer types like retinoblastoma, glioblastoma (Jiang et al. 2008) and a marked increase has been detected in tumor development (Gyrd-Hansen et al. 2004). Abundant oncogenically activated proteins like FLT3, KIT, CDKs, AKT, and HIF are accompanied with Hsp90 (Li et al. 2009). Among Hsp90 client proteins, many are engaged in the so-called "hallmarks of cancer", first proposed by Hanahan and Weinberg (2011) and subsequently participate in tumorigenic cell stabilization comprises protein kinases. The tyrosine kinase inhibitor resistance might be treated by Hsp90 inhibition, which supposes that the drug opposition in cancer may be dependent on the Hsp90 function due to their abilities as the chaperones to protect the oncoproteins against degradation (He et al. 2013).

Survivin, a suppressor of apoptosis (IAP) gene family member, is one of the Hsp90 client proteins which is over-expressed in abundant cancers containing retinoblastoma (Sudhakar et al. 2013). Survivin shows a multifaceted role in apoptosis suppression, cell division, tumor development and cell cycle organization in different clinical cancers (Kanwar et al. 2013). Inhibition of Survivin causes mitochondrial apoptosis, suppresses cell proliferation and thus elevates the receptivity of tumor cells to chemotherapy (Trabulo et al. 2011).

Hsp90 regulates proteostasis of Survivin by refolding the unfolded/denatured protein to a native state. Hsp90-Survivin can act as a possible aim for RB cancer therapy because the percentage of Survivin and Hsp90 is detected to be high (Jiang et al. 2008). Targeting the ATP-binding pocket of Hsp90 disrupts of Hsp90/Survivin physical interaction and suppresses the downstream signaling cascade of Survivin, resulting in the ubiquitination and inhibition of Survivin signaling pathways leading to tumor growth cease (Meli et al. 2006). These detections suggest the strong possibility of targeting Hsp90-Survivin interaction as an adjuvant therapy in RB control.

9.2 Conclusions

Finally, we can conclude that cell stress especially heat stress is one factor that triggers the onset and progression of Hsp90 in cancer and neurodegenerative diseases where we show the significant impact of Hsp90 on tissues and retina. Thus, the different dynamics and conformational changes of Hsp90 and client proteins could be obviously a landmark for different mechanisms, binding sites and binding ability of Hsp90 for stress response due to the failure of meeting the energy demand which induced by heat stress. Therefore further understanding in the relationship between the client proteins that can be chaperoned by Hsp90 and Hsp90 inhibitors is required for future therapies and the way to control these inhibitors at a physiological level and subsequently removing the factor causing protein unfolding and degradation without affecting the homeostasis of the tissue will be a step forward in treating cancer in general and other aged related diseases of the retina specially AMD.

Acknowledgements We would like to thank the Deanship of Scientific Research and RSSU at King Saud University for their technical support.

References

Aguila M, Bevilacqua D, McCulley C, Schwarz N, Athanasiou D, Kanuga N, Novoselov SS, Lange CA, Ali RR, Bainbridge JW, Gias C, Coffey PJ, Garriga P, Cheetham ME (2014) Hsp90 inhibition protects against inherited retinal degeneration. Hum Mol Genet 23:2164–2175

Anckar J, Sistonen L (2007) Heat shock factor 1 as a coordinator of stress and developmental pathways. Adv Exp Med Biol 594:78–88

Basso AD, Solit DB, Chiosis G, Giri B, Tsichlis P, Rosen N (2002) Akt forms an intracellular complex with heat shock protein 90 (Hsp90) and Cdc37 and is destabilized by inhibitors of Hsp90 function. J Biol Chem 277:39858–39866

Bernstein SL, Borst DE, Neuder ME, Wong P (1996) Characterization of a human fovea cDNA library and regional differential gene expression in the human retina. Genomics 32:301–308

Black JA, Waxman SG, Hildebrand C (1985) Axo-glial relations in the retina-optic nerve junction of the adult rat: freeze-fracture observations on axon membrane structure. J Neurocytol 14:887–907

Bradke F, Dotti CG (1999) The role of local actin instability in axon formation. Science 283:1931–1934

Brown IR (2007) Heat shock proteins and protection of the nervous system. Ann N Y Acad Sci 1113:147–158

Calderwood SK, Khaleque MA, Sawyer DB, Ciocca DR (2006) Heat shock proteins in cancer: chaperones of tumorigenesis. Trends Biochem Sci 31:164–172

Chatterjee S, Burns TF (2017) Targeting heat shock proteins in cancer: a promising therapeutic approach. Int J Mol Sci 18:1978

Chatterjee S, Bhattacharya S, Socinski MA, Burns TF (2016) HSP90 inhibitors in lung cancer: promise still unfulfilled. Clin Adv Hematol Oncol 14:346–356

Chen G, Cao P, Goeddel DV (2002) TNF-induced recruitment and activation of the IKK complex require Cdc37 and Hsp90. Mol Cell 9:401–410

Chen W, Zheng R, Baade PD, Zhang S, Zeng H, Bray F, Jemal A, Yu XQ, He J (2016) Cancer statistics in China, 2015. CA Cancer J Clin 66:115–132

Chiosis G (2006) Targeting chaperones in transformed systems – a focus on Hsp90 and cancer. Expert Opin Ther Targets 10:37–50

Chiosis G (2016) Heat shock proteins in disease–from molecular mechanisms to therapeutics. Curr Top Med Chem 16:2727

Chuang J-Z, Vega C, Jun W, Sung C-H (2004) Structural and functional impairment of endocytic pathways by retinitis pigmentosa mutant rhodopsin-arrestin complexes. J Clin Invest 114:131–140

Csermely P, Schnaider T, Soti C, Prohaszka Z, Nardai G (1998) The 90-kDa molecular chaperone family: structure, function, and clinical applications. A comprehensive review. Pharmacol Ther 79:129–168

Czar MJ, Welsh MJ, Pratt WB (1996) Immunofluorescence localization of the 90-kDa heat-shock protein to cytoskeleton. Eur J Cell Biol 70:322–330

Da Silva JS, Dotti CG (2002) Breaking the neuronal sphere: regulation of the actin cytoskeleton in neuritogenesis. Nat Rev Neurosci 3:694

Darimont BD (1999) The Hsp90 chaperone complex-a potential target for cancer therapy? World J Gastroenterol 5:195–198
Dean DO, Tytell M (2001) Hsp25 and −90 immunoreactivity in the normal rat eye. Invest Ophthalmol Vis Sci 42:3031–3040
Dean DO, Kent CR, Tytell M (1999) Constitutive and inducible heat shock protein 70 immunoreactivity in the normal rat eye. Invest Ophthalmol Vis Sci 40:2952–2962
Decanini A, Nordgaard CL, Feng X, Ferrington DA, Olsen TW (2007) Changes in select redox proteins of the retinal pigment epithelium in age-related macular degeneration. Am J Ophthalmol 143:607–615
Ellis RJ (2007) Protein misassembly. In: Csermely P, Vígh L (eds) Molecular aspects of the stress response: chaperones, membranes and networks. Springer New York, New York, pp 1–13
Garcia-Carbonero R, Carnero A, Paz-Ares L (2013) Inhibition of HSP90 molecular chaperones: moving into the clinic. Lancet Oncol 14:e358–e369
Garrido C, Brunet M, Didelot C, Zermati Y, Schmitt E, Kroemer G (2006) Heat shock proteins 27 and 70: anti-apoptotic proteins with tumorigenic properties. Cell Cycle 5:2592–2601
Goedert M, Jakes R (2005) Mutations causing neurodegenerative tauopathies. Biochim Biophys Acta 1739:240–250
Gyrd-Hansen M, Nylandsted J, Jäättelä M (2004) Heat shock protein 70 promotes cancer cell viability by safeguarding lysosomal integrity. Cell Cycle 3:1484–1485
Hanahan D, Weinberg RA (2011) Hallmarks of cancer: the next generation. Cell 144:646–674
Harris H, Rubinsztein DC (2011) Control of autophagy as a therapy for neurodegenerative disease. Nat Rev Neurol 8:108–117
He S, Zhang C, Shafi AA, Sequeira M, Acquaviva J, Friedland JC, Sang J, Smith DL, Weigel NL, Wada Y (2013) Potent activity of the Hsp90 inhibitor ganetespib in prostate cancer cells irrespective of androgen receptor status or variant receptor expression. Int J Oncol 42:35–43
Honjo M, Tanihara H, Kido N, Inatani M, Okazaki K, Honda Y (2000) Expression of ciliary neurotrophic factor activated by retinal Muller cells in eyes with NMDA- and kainic acid-induced neuronal death. Invest Ophthalmol Vis Sci 41:552–560
Jäättelä M (1999) Escaping cell death: survival proteins in cancer. Exp Cell Res 248:30–43
Jacobson C, Schnapp B, Banker GA (2006) A change in the selective translocation of the Kinesin-1 motor domain marks the initial specification of the axon. Neuron 49:797–804
Jaiswal RK, Weissinger E, Kolch W, Landreth GE (1996) Nerve growth factor-mediated activation of the mitogen-activated protein (MAP) kinase cascade involves a signaling complex containing B-Raf and HSP90. J Biol Chem 271:23626–23629
Jarrett SG, Boulton ME (2012) Consequences of oxidative stress in age-related macular degeneration. Mol Asp Med 33:399–417
Jego G, Hazoumé A, Seigneuric R, Garrido C (2013) Targeting heat shock proteins in cancer. Cancer Lett 332:275–285
Jiang L, Liu X, Li B, He X, Jin Y, Li L, Gao F, Wang N (2008) Heat shock proteins and survivin: relationship and effects on proliferation index of retinoblastoma cells. Histol Histopathol 23:827–832
Johnson J, Corbisier R, Stensgard B, Toft D (1996) The involvement of p23, hsp90, and immunophilins in the assembly of progesterone receptor complexes. J Steroid Biochem Mol Biol 56:31–37
Jolly C, Morimoto RI (2000) Role of the heat shock response and molecular chaperones in oncogenesis and cell death. J Natl Cancer Inst 92:1564–1572
Jung T, Catalgol B, Grune T (2009) The proteasomal system. Mol Asp Med 30:191–296
Kaarniranta K, Sinha D, Blasiak J, Kauppinen A, Vereb Z, Salminen A, Boulton ME, Petrovski G (2013) Autophagy and heterophagy dysregulation leads to retinal pigment epithelium dysfunction and development of age-related macular degeneration. Autophagy 9:973–984
Kanwar JR, Kamalapuram SK, Kanwar RK (2013) Survivin signaling in clinical oncology: a multifaceted dragon. Med Res Rev 33:765–789

Kaplan KB, Li R (2012) A prescription for 'stress'–the role of Hsp90 in genome stability and cellular adaptation. Trends Cell Biol 22:576–583
Karagoz GE, Rudiger SG (2015) Hsp90 interaction with clients. Trends Biochem Sci 40:117–125
Karunanithi S, Barclay JW, Robertson RM, Brown IR, Atwood HL (1999) Neuroprotection at Drosophila synapses conferred by prior heat shock. J Neurosci 19:4360
Kauppinen A, Niskanen H, Suuronen T, Kinnunen K, Salminen A, Kaarniranta K (2012) Oxidative stress activates NLRP3 inflammasomes in ARPE-19 cells – implications for age-related macular degeneration (AMD). Immunol Lett 147:29–33
Kiang JG, Tsokos GC (1998) Heat shock protein 70 kDa: molecular biology, biochemistry, and physiology. Pharmacol Ther 80:183–201
Kimura E, Enns RE, Alcaraz JE, Arboleda J, Slamon DJ, Howell SB (1993) Correlation of the survival of ovarian cancer patients with mRNA expression of the 60-kD heat-shock protein HSP-60. J Clin Oncol Off J Am Soc Clin Oncol 11:891–898
Klettner A (2004) The induction of heat shock proteins as a potential strategy to treat neurodegenerative disorders. Drug News Perspect 17:299–306
Kobayashi K, Kobayashi H, Ueda M, Honda Y (1998) Estrogen receptor expression in bovine and rat retinas. Invest Ophthalmol Vis Sci 39:2105–2110
Kojima M, Hoshimaru M, Aoki T, Takahashi JB, Ohtsuka T, Asahi M, Matsuura N, Kikuchi H (1996) Expression of heat shock proteins in the developing rat retina. Neurosci Lett 205:215–217
Kosik KS, Shimura H (2005) Phosphorylated tau and the neurodegenerative foldopathies. Biochim Biophys Acta 1739:298–310
Labbadia J, Cunliffe H, Weiss A, Katsyuba E, Sathasivam K, Seredenina T, Woodman B, Moussaoui S, Frentzel S, Luthi-Carter R, Paganetti P, Bates GP (2011) Altered chromatin architecture underlies progressive impairment of the heat shock response in mouse models of Huntington disease. J Clin Invest 121:3306–3319
Lau LF, Schachter JB, Seymour PA, Sanner MA (2002) Tau protein phosphorylation as a therapeutic target in Alzheimer's disease. Curr Top Med Chem 2:395–415
Li J, Buchner J (2013) Structure, function and regulation of the hsp90 machinery. Biom J 36:106–117
Li Y, Wang YS, Shen XF, Hui YN, Han J, Zhao W, Zhu J (2008) Alterations of activity and intracellular distribution of the 20S proteasome in ageing retinal pigment epithelial cells. Exp Gerontol 43:1114–1122
Li Y, Zhang T, Schwartz SJ, Sun D (2009) New developments in Hsp90 inhibitors as anti-cancer therapeutics: mechanisms, clinical perspective and more potential. Drug Resist Updat 12:17–27
Lin T-Y, Guo W, Long Q, Ma A, Liu Q, Zhang H, Huang Y, Chandrasekaran S, Pan C, Lam KS (2016) HSP90 inhibitor encapsulated photo-theranostic nanoparticles for synergistic combination cancer therapy. Theranostics 6:1324
Lindquist S, Craig E (1988) The heat-shock proteins. Annu Rev Genet 22:631–677
Luo W, Dou F, Rodina A, Chip S, Kim J, Zhao Q, Moulick K, Aguirre J, Wu N, Greengard P, Chiosis G (2007) Roles of heat-shock protein 90 in maintaining and facilitating the neurodegenerative phenotype in tauopathies. Proc Natl Acad Sci USA 104:9511–9516
Luo W, Rodina A, Chiosis G (2008) Heat shock protein 90: translation from cancer to Alzheimer's disease treatment? BMC Neurosci 9:S7–S7
Martinon F (2008) Detection of immune danger signals by NALP3. J Leukoc Biol 83:507–511
Mayor A, Martinon F, De Smedt T, Petrilli V, Tschopp J (2007) A crucial function of SGT1 and HSP90 in inflammasome activity links mammalian and plant innate immune responses. Nat Immunol 8:497–503
Meli M, Pennati M, Curto M, Daidone MG, Plescia J, Toba S, Altieri DC, Zaffaroni N, Colombo G (2006) Small-molecule targeting of heat shock protein 90 chaperone function: rational identification of a new anticancer lead. J Med Chem 49:7721–7730
Mendes HF, Cheetham ME (2008) Pharmacological manipulation of gain-of-function and dominant-negative mechanisms in rhodopsin retinitis pigmentosa. Hum Mol Genet 17:3043–3054

Mimnaugh EG, Worland PJ, Whitesell L, Neckers LM (1995) Possible role for serine/threonine phosphorylation in the regulation of the heteroprotein complex between the hsp90 stress protein and the pp60v-src tyrosine kinase. J Biol Chem 270:28654–28659
Mirshahi M, Nicolas C, Mirshahi A, Hecquet C, d'Hermies F, Faure JP, Agarwal MK (1996) The mineralocorticoid hormone receptor and action in the eye. Biochem Biophys Res Commun 219:150–156
Miyata Y, Nakamoto H, Neckers L (2013) The therapeutic target Hsp90 and cancer hallmarks. Curr Pharm Des 19:347–365
Moore SK, Kozak C, Robinson EA, Ullrich SJ, Appella E (1989) Murine 86- and 84-kDa heat shock proteins, cDNA sequences, chromosome assignments, and evolutionary origins. J Biol Chem 264:5343–5351
Muchowski PJ, Wacker JL (2005) Modulation of neurodegeneration by molecular chaperones. Nat Rev Neurosci 6:11–22
Nathan DF, Lindquist S (1995) Mutational analysis of Hsp90 function: interactions with a steroid receptor and a protein kinase. Mol Cell Biol 15:3917–3925
Nathan DF, Vos MH, Lindquist S (1997) In vivo functions of the Saccharomyces cerevisiae Hsp90 chaperone. Proc Natl Acad Sci USA 94:12949–12956
Neckers L, Workman P (2012) Hsp90 molecular chaperone inhibitors: are we there yet? Clin Cancer Res 18:64–76
Park JW, Moon C, Yun S, Kim SY, Bae YC, Chun M-H, Moon J-I (2007) Differential expression of heat shock protein mRNAs under in vivo glutathione depletion in the mouse retina. Neurosci Lett 413:260–264
Parsell D, Lindquist S (1993) The function of heat-shock proteins in stress tolerance: degradation and reactivation of damaged proteins. Annu Rev Genet 27:437–496
Pearl LH, Prodromou C, Workman P (2008) The Hsp90 molecular chaperone: an open and shut case for treatment. Biochem J 410:439–453
Pratt WB (1997) The role of the hsp90-based chaperone system in signal transduction by nuclear receptors and receptors signaling via MAP kinase. Annu Rev Pharmacol Toxicol 37:297–326
Pratt WB (1998) The hsp90-based chaperone system: involvement in signal transduction from a variety of hormone and growth factor receptors. Proc Soc Exp Biol Med 217:420–434
Proia DA, Kaufmann GF (2015) Targeting heat-shock protein 90 (HSP90) as a complementary strategy to immune checkpoint blockade for cancer therapy. Cancer Immunol Res 3:583–589
Qin S, Ni M, Wang X, Maurier-Mahe F, Shurland DL, Rodrigues GA (2011) Inhibition of RPE cell sterile inflammatory responses and endotoxin-induced uveitis by a cell-impermeable HSP90 inhibitor. Exp Eye Res 93:889–897
Rui Z, Xiao-Yun G, Xing-Chuang X, You J, Ze-Jian H, Xiang F (2018) Progress in molecular chaperone regulation of heat shock protein 90 and cancer. Chin J Anal Chem 46:301–308
Rutherford SL, Lindquist S (1998) Hsp90 as a capacitor for morphological evolution. Nature 396:336
Saif M, Erlichman C, Dragovich T, Mendelson D, Toft D, Burrows F, Storgard C, Von Hoff D (2013) Open-label, dose-escalation, safety, pharmacokinetic, and pharmacodynamic study of intravenously administered CNF1010 (17-(allylamino)-17-demethoxygeldanamycin [17-AAG]) in patients with solid tumors. Cancer Chemother Pharmacol 71:1345–1355
Sakai M, Sakai H, Nakamura Y, Fukuchi T, Sawaguchi S (2003) Immunolocalization of heat shock proteins in the retina of normal monkey eyes and monkey eyes with laser-induced glaucoma. Jpn J Ophthalmol 47:42–52
Sanchez ER, Redmond T, Scherrer LC, Bresnick EH, Welsh MJ, Pratt WB (1988) Evidence that the 90-kilodalton heat shock protein is associated with tubulin-containing complexes in L cell cytosol and in intact PtK cells. Mol Endocrinol 2:756–760
Santarosa M, Favaro D, Quaia M, Galligioni E (1997) Expression of heat shock protein 72 in renal cell carcinoma: possible role and prognostic implications in cancer patients. Eur J Cancer 33:873–877

Sauvage F, Messaoudi S, Fattal E, Barratt G, Vergnaud-Gauduchon J (2017) Heat shock proteins and cancer: how can nanomedicine be harnessed? J Control Release 248:133–143

Scheibel T, Buchner J (1998) The Hsp90 complex – a super-chaperone machine as a novel drug target. Biochem Pharmacol 56:675–682

Schwamborn JC, Müller M, Becker AH, Püschel AW (2007) Retracted: ubiquitination of the GTPase Rap1B by the ubiquitin ligase Smurf2 is required for the establishment of neuronal polarity. EMBO J 26:1410–1422

Shapley R, Perry VH (1986) Cat and monkey retinal ganglion cells and their visual functional roles. Trends Neurosci 9:229–235

Shi S-H, Jan LY, Jan Y-N (2003) Hippocampal neuronal polarity specified by spatially localized mPar3/mPar6 and PI 3-kinase activity. Cell 112:63–75

Shi S-H, Cheng T, Jan LY, Jan Y-N (2004) APC and GSK-3β are involved in mPar3 targeting to the nascent axon and establishment of neuronal polarity. Curr Biol 14:2025–2032

Singh A, Singh A, Sand JM, Bauer SJ, Hafeez BB, Meske L, Verma AK (2015) Topically applied Hsp90 inhibitor 17AAG inhibits UVR-induced cutaneous squamous cell carcinomas. J Invest Dermatol 135:1098–1107

Sittler A, Lurz R, Lueder G, Priller J, Hayer-Hartl MK, Hartl FU, Lehrach H, Wanker EE (2001) Geldanamycin activates a heat shock response and inhibits huntingtin aggregation in a cell culture model of Huntington's disease. Hum Mol Genet 10:1307–1315

Sliutz G, Karlseder J, Tempfer C, Orel L, Holzer G, Simon MM (1996) Drug resistance against gemcitabine and topotecan mediated by constitutive hsp70 overexpression in vitro: implication of quercetin as sensitiser in chemotherapy. Br J Cancer 74:172–177

Sudhakar J, Venkatesan N, Lakshmanan S, Khetan V, Krishnakumar S, Biswas J (2013) Hypoxic tumor microenvironment in advanced retinoblastoma. Pediatr Blood Cancer 60:1598–1601

Taipale M, Jarosz DF, Lindquist S (2010) HSP90 at the hub of protein homeostasis: emerging mechanistic insights. Nat Rev Mol Cell Biol 11:515–528

Taiyab A, Sreedhar AS, Rao CM (2009) Hsp90 inhibitors, GA and 17AAG, lead to ER stress-induced apoptosis in rat histiocytoma. Biochem Pharmacol 78:142–152

Takayama S, Reed JC, Homma S (2003) Heat-shock proteins as regulators of apoptosis. Oncogene 22:9041–9047

Tam LC, Kiang AS, Campbell M, Keaney J, Farrar GJ, Humphries MM, Kenna PF, Humphries P (2010) Prevention of autosomal dominant retinitis pigmentosa by systemic drug therapy targeting heat shock protein 90 (Hsp90). Hum Mol Genet 19:4421–4436

Tanaka Y, Kobayashi K, Kita M, Kinoshita S, Imanishi J (1995) Messenger RNA expression of heat shock proteins (Hsp) during ocular development. Curr Eye Res 14:1125–1133

Tanihara H, Hangai M, Sawaguchi S, Abe H, Kageyama M, Nakazawa F, Shirasawa E, Honda Y (1997) Up-regulation of glial fibrillary acidic protein in the retina of primate eyes with experimental glaucoma. Arch Ophthalmol 115:752–756

Tapia M, Wandosell F, Garrido JJ (2010) Impaired function of HDAC6 slows down axonal growth and interferes with axon initial segment development. PLoS One 5:e12908

Tarallo V, Hirano Y, Gelfand BD, Dridi S, Kerur N, Kim Y, Cho WG, Kaneko H, Fowler BJ, Bogdanovich S, Albuquerque RJ, Hauswirth WW, Chiodo VA, Kugel JF, Goodrich JA, Ponicsan SL, Chaudhuri G, Murphy MP, Dunaief JL, Ambati BK, Ogura Y, Yoo JW, Lee DK, Provost P, Hinton DR, Nunez G, Baffi JZ, Kleinman ME, Ambati J (2012) DICER1 loss and Alu RNA induce age-related macular degeneration via the NLRP3 inflammasome and MyD88. Cell 149:847–859

Trabulo S, Cardoso A, Santos-Ferreira T, Cardoso A, Simoes S, Pedroso de Lima M (2011) Survivin silencing as a promising strategy to enhance the sensitivity of cancer cells to chemotherapeutic agents. Mol Pharm 8:1120–1131

Tukaj S, Bieber K, Kleszczyński K, Witte M, Cames R, Kalies K, Zillikens D, Ludwig RJ, Fischer TW, Kasperkiewicz M (2017) Topically applied Hsp90 blocker 17AAG inhibits autoantibody-mediated blister-inducing cutaneous inflammation. J Investig Dermatol 137:341–349

Vanden Berghe T, Kalai M, van Loo G, Declercq W, Vandenabeele P (2003) Disruption of HSP90 function reverts tumor necrosis factor-induced necrosis to apoptosis. J Biol Chem 278:5622–5629

Venkatesan N, Kanwar JR, Deepa PR, Navaneethakrishnan S, Joseph C, Krishnakumar S (2016) Targeting HSP90/Survivin using a cell permeable structure based peptido-mimetic shepherdin in retinoblastoma. Chem Biol Interact 252:141–149

Wainberg ZA, Anghel A, Rogers AM, Desai AJ, Kalous O, Conklin D, Ayala R, O'Brien NA, Quadt C, Akimov M (2013) Inhibition of HSP90 with AUY922 induces synergy in HER2-amplified trastuzumab-resistant breast and gastric cancer. Mol Cancer Ther 12(4):509–519

Walton-Diaz A, Khan S, Bourboulia D, Trepel JB, Neckers L, Mollapour M (2013) Contributions of co-chaperones and post-translational modifications towards Hsp90 drug sensitivity. Future Med Chem 5:1059–1071

Wang YQ, Zhang XM, Wang XD, Wang BJ, Wang W (2010) 17-AAG, a Hsp90 inhibitor, attenuates the hypoxia-induced expression of SDF-1alpha and ILK in mouse RPE cells. Mol Biol Rep 37:1203–1209

Wang C, Zhang Y, Guo K, Wang N, Jin H, Liu Y, Qin W (2016) Heat shock proteins in hepatocellular carcinoma: molecular mechanism and therapeutic potential. Int J Cancer 138:1824–1834

Waza M, Adachi H, Katsuno M, Minamiyama M, Sang C, Tanaka F, Inukai A, Doyu M, Sobue G (2005) 17-AAG, an Hsp90 inhibitor, ameliorates polyglutamine-mediated motor neuron degeneration. Nat Med 11:1088

Whitesell L, Lindquist SL (2005) HSP90 and the chaperoning of cancer. Nat Rev Cancer 5:761–772

Workman P, Burrows F, Neckers L, Rosen N (2007) Drugging the cancer chaperone HSP90. Ann N Y Acad Sci 1113:202–216

Wu W-C, Kao Y-H, Hu P-S, Chen J-H (2007) Geldanamycin, a HSP90 inhibitor, attenuates the hypoxia-induced vascular endothelial growth factor expression in retinal pigment epithelium cells in vitro. Exp Eye Res 85:721–731

Xu Y, Lindquist S (1993) Heat-shock protein hsp90 governs the activity of pp60v-src kinase. Proc Natl Acad Sci USA 90:7074–7078

Yan D, Guo L, Wang Y (2006) Requirement of dendritic Akt degradation by the ubiquitin–proteasome system for neuronal polarity. J Cell Biol 174:415–424

Ylikomi T, Wurtz JM, Syvälä H, Passinen S, Pekki A, Haverinen M, Bläuer M, Tuohimaa P, Gronemeyer H (1998) Reappraisal of the role of heat shock proteins as regulators of steroid receptor activity. Crit Rev Biochem Mol Biol 33:437–466

Zou J, Guo Y, Guettouche T, Smith DF, Voellmy R (1998) Repression of heat shock transcription factor HSF1 activation by HSP90 (HSP90 complex) that forms a stress-sensitive complex with HSF1. Cell 94:471–480

Chapter 10
Targeting Hsp-90 Related Disease Entities for Therapeutic Development

Timothy Westlake, Mitchell Sun, Brandon C. Rosenblum, Zhengping Zhuang, and Jared S. Rosenblum

Abstract Heat shock protein 90 (HSP-90) has been identified in many disease processes including cancer, neurodegeneration, autoimmune diseases, and cancers. Great effort has been expended in the development of specific inhibitors of the N-terminal and C-terminal domains. Inhibitors of post-translational modification have also been developed. Herein, we explore the available inhibitors and those in development, discuss the relevant disease processes, and examine the pitfalls and promises of targeting HSP-90 for therapeutic intervention.

Keywords Autoimmune disease · Cancer · Diabetes · Heat shock proteins · Neurodegeneration · Therapeutics

---

T. Westlake · J. S. Rosenblum (✉)
Neuro-Oncology Branch, National Cancer Institute, National Institutes of Health, Bethesda, MD, USA
e-mail: jared.rosenblum@nih.gov

M. Sun
Neuro-Oncology Branch, National Cancer Institute, National Institutes of Health, Bethesda, MD, USA

Liberal Arts and Sciences, University of Illinois at Chicago, Chicago, IL, USA

B. C. Rosenblum
Surgical Neurology Branch, National Institute of Neurologic Disorders and Stroke, Bethesda, MD, USA

Z. Zhuang
Neuro-Oncology Branch, National Cancer Institute, National Institutes of Health, Bethesda, MD, USA

Surgical Neurology Branch, National Institute of Neurologic Disorders and Stroke, Bethesda, MD, USA

A. A. A. Asea, P. Kaur (eds.), *Heat Shock Protein 90 in Human Diseases and Disorders*, Heat Shock Proteins 19,
https://doi.org/10.1007/978-3-030-23158-3_10

Abbreviations

17-AAG	Tanespimycin
ADME	absorption, distribution, metabolism, excretion
AKT	protein kinase B
b-RAF	B-Raf proto-oncogene
c-RAF	RAF proto-oncogene serine/threonine-protein kinase
CTD	C-terminal domain
DMAG	17-dimethylaminoethylamino-17-demethoxygeldanamycin
ER	endoplasmic reticulum
FKBP	FK506 binding protein
FLT	Fms-like tyrosine kinase
GBase	glucocerebrosidase
HDAC	histone deacetylase
HDACI	HDAC inhibitor
HER	human epidermal growth factor receptor
HIF	hypoxia inducible factor
HOP	HSP70-HSP90 organizaing protein
HSF	heat shock factor
HSP	heat shock protein
HTT	huntington protein
IKK	IkB kinase
JAK	Janus kinase
JNK	c-Jun N-terminal kinases
NF-kB	nuclear factor kappa light-chain enhancer of activated B cells
Nrf	nuclear factor erythroid 2-related factor
NTD	N-terminal domain
RAF	rapidly accelerated fibrosarcoma
RASGRP	RAS guanyl-releasing protein
SAHA	suberoyl anilide hydroxamic acid
STAT	signal transducer and activators of transcription
TPR	tetraotricopeptide repeat
VEGFR	vascular endothelial growth factor receptor

10.1 Introduction

Heat-shock protein 90 (HSP-90) regulates the stability, activation, and degradation of a diverse array of proteins associated with growth, proliferation, and survival (Burlison et al. 2006; Neckers and Ivy 2003; Schopf et al. 2017; Schwock et al. 2008). Thus, it is core to regulation of protein stability and protein-degradation pathways and modulating transcription factors, signaling transduction networks, and kinases (Schopf et al. 2017). It facilitates the survival of cells during stress response and exhibits a pronounced anti-apoptotic and stabilization effect. Thus,

HSP-90 has been associated with development and progression of a wide range of pathological conditions including cancers, diabetes, Gaucher disease, neurodegenerative diseases, and autoimmune dysfunction (Hoter et al. 2018; Kasperkiewicz et al. 2011; Lackie et al. 2017; Luo et al. 2010; Rice et al. 2008; Russo et al. 2006; Trepel et al. 2010; Tukaj et al. 2015; Whitesell and Linquist 2005; Yang et al. 2013).

10.1.1 Cancer

A commonality in many human cancers is the overexpression of HSP-90; the drastic two-to-three-fold induction of HSP90 seen in several cancers results in increase stabilization of client proteins (Barrott and Haystead 2013). The pronounced increase of HSP-90 in stress conditions can reach up to 6% of total protein (Prodromou 2016; Taipale et al. 2010). As consequence of increase HSP-90, the stabilization of its client protein results in the protects mutated or up-regulated oncoproteins. The aberrant protection of pro-survival and proliferation-related proteins such as telomerases, B-Raf, Akt, p53, VEGFR, HIF1α, HER-2, tyrosine kinases, steroid hormone receptors contribute to tumorigenesis, metastasis, and invasiveness (Banerji 2009; Beliakoff and Whitesell 2004; Hoter et al. 2018; Jhaveri and Modi 2012; Whitesell and Linquist 2005). As HSP90 acts as a regulator of HSF-1, the major hub of HSF transcriptional expression, HSP90 production causes dysregulation of HSF-1 transcriptional activity which leads to alterations in chaperone expression (Duerfeldt and Blagg 2010).

10.1.2 Neurodegenerative Diseases

HSP-90 works in concert with other chaperone machinery to refold misfolded proteins to prevent toxic accumulation. However, as in the case in various cancers, the stabilization of HSP-90 client proteins can have deleterious consequences outside of the context of normal physiological conditions. The stabilization of proteins associated with diseases results in manifestation of various neurodegenerative disease (Lackie et al. 2017; Luo et al. 2010). The HSP-90 client protein stabilization is a major facilitator for the accumulation of intrinsically disordered proteins (Karagoz et al. 2014; Luo et al. 2010). HSP-90 interacts with and stabilizes Tau, (Dickey et al. 2007; Hoter et al. 2018; Karagoz et al. 2014) a microtubule associated protein that mediates axonal transport. Tau hyperphosphorylation and aggregation is a classical hallmark of Alzheimer's disease; the accumulation is also associated with other neurodegenerative disease such as progressive supranuclear palsy and Pick's disease (Gong and Igbal 2008; Guo et al. 2017; Lee et al. 2001; Shelton et al. 2017). The HSP-90/Tau interface is associated with the neurodegenerative pathologic state and is well explained by enhanced stabilization of hyperphosphorylated Tau, which exacerbates aberrant neural activity seen in tauopathies (Shelton et al. 2017).

The aggregation of intrinsically disordered factors is also associated with Huntington's disease and Parkinson's disease (Lackie et al. 2017; Luo et al. 2010). HSP-90 interacts with Huntington protein (HTT) and leucine-rich repeat kinase 2 (Baldo et al. 2012; Wang et al. 2008). Interestingly, as HSP-90 is an established regulator of HSF-1, evidence suggests that, through the repression of HSP-90, other molecular chaperones systems such as HSP-70 can be enhanced and can facilitate increase neuroprotective function in otherwise pathogenic conditions (Luo et al. 2010).

10.1.3 Gaucher Disease

Gaucher disease is a rare autosomal recessive lysosomal disorder driven by genetic mutations in *GBA* gene encoding the lysosomal enzyme glucocerebrosidase (GCase); the mutations result in metabolic dysfunction and wide-spread organ dysfunction due to effects of the drastic accumulation of GCase substrate (Brady et al. 1966; Hruska et al. 2008; Stirnemann et al. 2017; Yang et al. 2013). The diminished amount of GCase results in the toxic accumulation of the GCase substrate, glucosylceramide (Stirnemann et al. 2017). GBA mutations cause protein misfolding and diminished protein instability resulting in increased retention of GCase in the endoplasmic reticulum (Ron and Horowitz 2005; Stirnemann et al. 2017; Yang et al. 2013). The intrinsic changes in the conformation of mutant GCase results in premature degradation and increased GCase turnover (Yang et al. 2013). HSP-90 is critical for targeting misfolded GCase for proteasomal degradation and directly interacts with GCase to direct the misfolded GCase to cellular ER and proteasomal degradation pathway (Yang et al. 2013). The increase degradation of GCase results in enhanced disease severity. This paradigm is in contrast with other diseases associated with HSP-90, as the interaction does not enhance the accumulation of its client protein but rather directs the client protein for degradation.

10.1.4 Diabetes and Associated Complications

Several of the key players mentioned above constitute a regulatory pathway for insulin sensitivity. Transcription of HSP-70 is regulated by HSF-1, which in turn is activated by inhibition of HSP-90 (Lee et al. 2013). Further, inhibition of HSP-90 leads to inhibition of JNK1 and thus improved insulin signaling; in the mouse model in this study, HSP-90 inhibition reversed hyperglycemia (Lee et al. 2013). In another study of diabetic mice, inhibition of HSP-90 with 17-dimethylaminoethylamino-17-demethoxygeldanamycin (DMAG) lessened renal damage and atherosclerosis incurred by hyperglycemia and hyperlipidemia as evidenced by decrease in albuminuria, renal lesions, and proinflammatory genes (Lazaro et al. 2015). Further study of diabetic atheroprotection with DMAG in a diabetic mouse model found

HSP-90 inhibition to be protective by induction of nuclear factor erythroid-derived 2-like (Nrf2) (Lazaro et al. 2017). Hypercoagulability in diabetes was found to be dependent on a glucose-regulated interaction between HSP-90α and annexin II, which promotes the generation of plasmin (Lei et al. 2004).

10.1.5 Autoimmune Disease

HSP-90 machinery is involved in adaptive and innate immune responses via mediating the activation of immune cells (Srivastava 2002; Taipale et al. 2010). It is instrumental in the function of natural killer cells, macrophages, lymphocytes, dendritic cell maturation, neutrophils (Kasperkiewicz et al. 2011; Srivastava 2002). The chaperone has been implicated in inflammation, antigen presentation, and immune cell activation (Srivastava 2002; Taipale et al. 2010). Client proteins of HSP-90 include inflammation regulating kinases IKK and JAK (Madrigal-Matute et al. 2010; Prodromou 2016; Zhang and Burrows 2004). These kinases modulate transcriptional regulators STAT and NF-kB which in turn dictate the expression of many pro-inflammation factors (Madrigal-Matute et al. 2010; Prodromou 2016). Thus, mounting evidence has demonstrated the importance of HSP-90 in regulating pro-inflammation responses and immune cell leading to the hypothesis that HSP-90 plays a critical function in auto-immune disease (DeBoer et al. 1970; Madrigal-Matute et al. 2010; Ron and Horowitz 2005; Stebbins et al. 1997). In support of this hypothesis, studies have implicated HSP90 in autoimmune diseases such as systemic lupus erythematosus, rheumatoid arthritis, allergic rhinitis, and other autoimmune diseases such as bullous skin diseases (Kasperkiewicz et al. 2011; Rice et al. 2008; Russo et al. 2006; Srivastava 2002).

10.2 Targeting HSP-90

10.2.1 Modulating HSP90 Function by Perturbation of PTMs

The post-translational modifications of HSP-90 alter the chaperone dynamics and perturbs the interaction with co-chaperones, substrates, and can influence enzyme activity (Jackson 2012; Kekapure et al. 2009; Scroggins et al. 2007). Acetylation of HSP-90 at the middle domain results in a marked decrease in its function by impeding the ability to interact with co-chaperones and client proteins, changing the dynamic conformation cycles (Aoyagi and Archer 2005; Kovacs et al. 2005; Mollapour and Neckers 2012; Scroggins et al. 2007). Targeting acetylation presents an avenue to modulate the activity of HSP-90. Reversible protein acetylation regulates a wide range of biochemical processes involving HSP-90 (Kovacs et al. 2005; Yu et al. 2002). The inhibition of histone deacetylase induces hyperacetylation HSP-90; the acetylated form of the chaperone has decreased affinity for ATP and target proteins (Bali et al. 2004). HDAC, while traditionally defined by their role in deacetylation of histones, have been found to act on a larger range of substrates including HSP-90 (Bali et al. 2004; Fiskus et al. 2007). Additionally, HDAC have been found to influence drug resistance to chemotherapeutics and diverse HSP90 inhibitors (Chai et al. 2017; Wang et al. n.d.). Interestingly, HDAC proteins such as HDAC6 also regulate the interactions with HSP90 and HSF master regulator, thereby affecting the transcriptional network of other HSP systems (Boyault et al. 2007; de Zoeten et al. 2011; Prodromou 2016).

Inhibiting deacetylation through HDAC inhibitors (HDACI) presents a promising avenue in which HSP90 chaperone cycling and function can be impeded. Ultimately, HDAC influences the stability of a plethora of downstream targets of HSP-90. HDACI have anti-tumorigenic properties correlating with diminished accumulation of HSP-90 target proteins related to pro-survival and pro-growth (Bali et al. 2004; Ding et al. 2017; Park et al. 2008). The inhibition of HSP-90 chaperone function by HDACIs results in degradation of oncoproteins such as AKT, FLT-3, BCR-ABL, RAF-1, VEGFR1, VEGFR2, JAK2, RASGRP1 and CRAF (Bali et al. 2004; Ding et al. 2017; Park et al. 2008). The resulting degradation of the oncoproteins by HSP-90 stabilization leads to dramatic changes in cell cycle control and proliferation. Modulating HDAC6 and HSP90, through HDACI has been studied in the context of ameliorating autoimmunity by affecting T-regulatory cells (Chiosis et al. 2001). Additionally, NF-κB function is impaired by HDACI inhibition of HDAC6. It is thought that the increase in the acetylation of HSP90 results in reduced stability and degradation of IKK. The reduced stability of IKK in turn causes aberrant NF-κB function (Kovacs et al. 2005; Regna et al. 2015; Trepel et al. 2010). Thus, the HDAC/HSP-90 interface presents a promising target to impede autoimmunity. While initial findings in autoimmunity have shown that targeting HDAC6 show some diminished HSP90 function, further studies are needed (Regna et al. 2015).

HDACI can also promote the stability of HSP90 proteins. In the case of Gaucher's disease, the production of mutated GCase results reduced of accumulation of the enzyme through HSP-90-directed degradation (Wang et al. 2008). HDACI, such as LB-205 and SAHA, results in increase acetylated form of HSP-90 and impairs the binding of HSP-90 to GCase (Yang et al. 2013). The increase in the accumulation of the mutated GCase in part increases the functional activity of the GCase thereby limiting toxic accumulation of the GCase substrate, glucosylceramide (Yang et al. 2013).

10.2.2 Targeting HSP-90 C-terminus

While traditional targeting of HSP-90 for therapeutics have predominantly developed to target the N-terminal domain (NTD) of the protein, novel approaches act to impede the C-terminal function of HSP-90. Inhibition of HSP-90 activity through the NTD perturbs the repressive effect of HSP-90 on HSF-1 which subsequently activates heat shock response. The activation of the heat shock response is thought to facilitate resistance which dampens the effect of HSP90 inhibitors (Yang et al. 2013). Therefore, a major driving force of targeting C-terminus is that prior trials with N-terminal targeting small molecules result in rapid development of resistance to the inhibitory molecules (Donnelly and Blagg 2008; Solárová et al. 2015).

The C-terminal inhibitors are subdivided into two categories; inhibitors that directly target the C-terminus and inhibitors that disrupt the binding of HSP90 to co-chaperones at the C-terminus (Koay et al. 2016). Similar to the NTD, the CTD contains a nucleotide binding site, however, lacks ATPase activity (Schopf et al. 2017). While the nucleotide binding site differs in terms of binding affinity and binding specificity, selective targeting of the nucleotide binding site has shown promising applications in inhibiting chaperone function (Donnelly and Blagg 2008; Solárová et al. 2015). Coumarin-based antibiotics were among the first inhibitors found to target the CTD (Solárová et al. 2015). Initially, this class of small molecules were found to inhibit the ATPase activity of ATP-gyrases; further biochemical classification has shown weak affinity towards the nucleotide binding site of CTD (Burlison et al. 2006; Donnelly and Blagg 2008; Solárová et al. 2015). The binding of novobiocin, coumarin-antibiotic, indeed affects the stability of HSP-90 client proteins and prompted the development of synthetic derivatives of the substrate. These nucleotide binding inhibitors induce conformational changes thought to impede and release protein-interaction by disrupting the dimerization of HSP-90 (Allan et al. 2006; Gormley et al. 1996; Solárová et al. 2015).

Novobiocin exerted anti-tumorigenic properties towards certain cancer lines; however, it lacked the efficacy that N-terminal inhibitors showed. Development of synthetic novobiocin derivatives sought to amend the poor efficacy by improving HSP90 inhibition. KU174, and KU675 analogues of novobiocin have shown strong anti-proliferation activity towards prostate cancer lines (Eskew et al. 2011; Liu et al. 2015; Solárová et al. 2015). Other novobiocin analogues developed have shown potential neuroprotective properties and provide an avenue in which HSP-90 inhibitors can be studied in the context of neurodegeneration (Donnelly and Blagg 2008).

Additional molecules have been tested and developed to target the nucleotide binding site of the CTD. These small molecules include the recent discovery of dihydropyrimidinone and analogs of bisphenol A such as NSC145366 as novel classes of CTD binding compounds (Goode et al. 2017; Terracciano et al. 2018). While, their therapeutic potential has not been fully explored, the continual development of CTD inhibiting agents provides an avenue in which HSP-90 can be inhibited without the potential drawback of driving drug resistance. Current research is focused on the neuroprotective properties of dihydropyridine derivatives and may be candidate therapeutic molecule for Alzheimer's disease (Roe et al. 2018). Recently, a novel C-terminal targeting hexapeptide, amioxyrone, was found to bind to specifically target CTD and inhibit dimerization (Bhatia et al. 2018). The targeting of CTD results in the reduced stabilization, downregulation, and degradation of HSP-90 client oncoproteins without the induction of the heat shock response (Bhatia et al. 2018). The hexapeptide showed effectiveness in leukemic cell lines and leukemia stem cells which demonstrated a novel approach in targeting chronic myeloid leukemia (Bhatia et al. 2018).

The alternative strategy of affecting the C-terminal function is by utilizing small molecules that disrupt the binding of HSP90 to co-chaperones. The C-terminal domain of HSP-90 possess MEEVD residue that regulates the interactions with TPR domain containing co-chaperones (Buchner 1999; Wandinger et al. 2008). TPR co-chaperones have tremendous importance in HSP-90 regulation (Schopf et al. 2017). These chaperones modulate the conformation of HSP-90 and interactions with co-chaperones; therefore the HSP90 chaperone machinery is affected (Schopf et al. 2017). While no natural inhibitors have been discovered to target the MEEVD region, exploration of the TPR-domain binding interface has led to the development of synthetic molecules to target the HSP-90-TPR binding interface (Sidera and Patsavoudi 2014). Recently, C-terminal modulators including modified variants of SM molecules: SM122, SM145, SM253, and SM258, have been developed to interact with Hsp90 and block the binding of TRR-domain containing co-chaperones (Koay et al. 2016). These SM molecules disrupt TRR-containing proteins, FKBP52 and HOP (Koay et al. 2016).

10.2.3 Targeting HSP90 N-Terminus

The inhibition of HSP-90 at the N-terminus can be divided into geldanamycin/ geldanamycin derivatives and purine-based inhibitors. The classic targeting of HSP-90 began with the natural analogs geldanamycin, herbimycin, and macbecin (DeBoer et al. 1970). Out of these three, geldanamycin was the most potent inhibitor due to its ability to more effectively bind to the NTD of HSP-90 and prevent ATP binding to the pocket; it also functioned to inhibit HSP-90 dimerization with heat shock factor 1 (HSF-1) which lead to heat shock response through transcriptional activation of factors such as HSP27, HSP40, HSP70, and HSP90 (Zou et al. 1998). The carbamate group of geldanamycin represents one of its core interacting domains with HSP-90 as it may form a hydrogen bonding network within the pocket and elimination of which abolished geldanamycin function (Stebbins et al. 1997).

The major weaknesses of geldanamycin, however, is its low solubility, difficulty in crossing the blood-brain barrier, and most important of all, its induction of the heat shock response from inhibiting HSP-90. This response is comprised of the cells upregulating transcription of heat-shock proteins to properly compensate for the disruption of protein folding (Sittler et al. 2001). To account for the problems associated with geldanamycin, a semi-synthetic derivative of geldanamycin called tanespimycin (17-AAG) was created that improved the ADME activity while decreasing the toxicity and heat shock response generated by geldanamycin (Goetz et al. 2003). As a result, 17-AAG provided a stronger candidate for HSP-90 inhibition. Despite the improvements made by 17-AAG, induction of the heat shock response and resulting toxicity lead to the computer screening of multiple compounds targeting the NTD to develop new compounds that limited the harmful effects of the previous generation of NTD- targeting inhibitors (Table 10.1).

Table 10.1 HSP90 Inhibitors are categorized by mechanism of action

Main molecular mechanism	Inhibitor	Class of inhibitor	Additional molecular mechanisms	Disease	Findings related to therapeutics
Binds to the ATP-binding pocket of Hsp-90	Geldanamycin	Marcolactams/ benzouinone	Affects virion binding of endoplasmin	Kidney cancer, peripheral nerve damage	Inhibits cell growth and promotes cell death of tumor cells
	Retaspimycin	Macrolactams		Cancer	Inhibits cell growth and promotes cell death of tumor cells
	Tanespimycin (17-AAG)	Macrolactams		Lymphoma, sarcoma, leukemia, prostate cancer	Inhibits tumor cell growth and promotes tumor cell death
	Radicicol	Hydroxygenzoic acid derivatives	Affects virion binding of endoplasmin; affects dihydrolipoyllysine-residue acetyltransferase activity of pyruvate dehydrogenase complex; affects pyruvate dehydrogenase kinase activity of pyruvate dehydrogenase lipoamide kinase isozyme 1; prevents p23 from associating with Hsp90;binds to and inhibits DNA topoisomerase type II proteins and GRP94	Malaria, fungus, viral infection, bacterial infection, cancer	Antibiotic, antimalarial, antifungal, antiangiogenic, antiviral, anti-inflammatory properties; decreased levels of progesterone receptor, Raf-1, p185erb82, and mutant p53; **M2-macrophage phenotype inhibition**
	BIIB021	Purines and purine derivatives		Cancer lymphoma	Inhibits tumor cell growth and promotes tumor cell death
	NVP-BEP800	2-Aminothieno[2,3-*d*] pyrimidine		Breast cancer, cancer	Inhibits tumor cell growth and promotes tumor cell death
	CUDC-305	Imidazopyridine		Cancer	Inhibits tumor cell growth and promotes tumor cell death

(continued)

Table 10.1 (continued)

Main molecular mechanism	Inhibitor	Class of inhibitor	Additional molecular mechanisms	Disease	Findings related to therapeutics
	XL888	Tropane alkaloids		Cancer melanoma	Inhibits tumor cell growth and promotes tumor cell death
	AT13387	Isoindoles		Tumor, gastrointestinal tumor	Inhibits tumor cell growth and promotes tumor cell death
	Tanespimycin (17-AAG)	Alpha amino acid esters		Cancer	Inhibits tumor cell growth and promotes tumor cell death
Antagonizes zinc ion binding of protein S100-A12 and S100-A12	Amlexanox	Chromeno[2,3-b] pyridine-5-ones	Inhibitor of fibroblast growth factor-1 S100 protein binding;antagonizes interleukin-3 receptor binding	Aphthous ulcers	Inhibits slow-reacting substance of anaphylaxis;stabilizes fibroblast growth factor 1
Inhibits voltage-gated potassium channel activity of calcium-activated potassium channel subunit alpha-1	Cromoglicate	Chromones	Antagonizes magnesium ion binding of protein S100-P	Allergic rhinitis	Inhibits mast cell degranulation through preventing release of slow-reacting substance of anaphylaxis and histamine; attenuates inflammatory leukotriene release
Increases binding of HSP70 to HSP90 and p23 dissociation from HSP90 complex	NVP-AUY922	Phenylpropanes		Cancer	Inhibits proliferation of multiple human cancer cell lines in-vitro
Inhibits binding of bRD to HSP90-alpha	KW-2478	Antineoplastics		Multiple myeloma, cancer	

The class of inhibitor and applicable disease are also listed. Several classes of inhibitors have additional mechanisms of action

Purine-based inhibitors were a completely synthetic class of HSP-90 inhibitors designed to target the ATP-binding pocket of the NTD created from the complete crystal structure of HSP90 ATP/ADP complex (Prodromou et al. 1997). These inhibitors have increased potency so that side effects associated with the geldanamycin analogs would be minimized at therapeutic doses. Through screening processes, many derivatives of these purine-based inhibitors were created, the first of which was PU3. This derivative mimicked the binding of ATP in the NTD pocket in its closed conformation (Chiosis et al. 2001). The discovery of PU3 opened the door for a HSP90 NTD inhibitors that could potentially be brought through clinical trials due to the decreased toxicity.

Currently, couple of these purine and purine-like inhibitors are undergoing clinical trials. BIIB021 is a member of the purine inhibitors being used to treat chronic lymphocytic leukemia and in a combination trial to treat HER2 (+) metastatic breast cancer (Table 10.1). Frequent grade 3 and 4 toxicities are associated with its use in chronic lymphocytic leukemia such as fatigue and hyponatremia while diarrhea, partial seizure, and nausea have been associated with its use in metastatic breast cancer (Elfiky et al. 2008). Subsequently, this lead to the use of BIIB021 use in gastrointestinal stromal tumor treatment refractory to imatinib and sunitinib where it was well-tolerated and showed metabolic changes in the patients that primarily lead to the stabilization of the tumors (Dickson et al. 2012).

Another major compound of interest is PU-H71, another purine class inhibitor shown in preclinical studies to be effective against breast cancer, hepatocellular carcinoma, and Bcl-6-dependenet diffuse B-cell lymphoma cell lines. Human studies focused on patients with advanced refractory cancers revealed that PU-H71 was well-tolerated, but its discontinuation of supply did not allow a strong therapeutic index to be determined; Evidence suggested stability of disease as the average response (Speranza et al. 2018).

10.3 Conclusions

Herein, we have examined the role of HSP-90 in several disease entities and the routes that have been examined for therapeutic development. Historically, the C-terminal domain and PTM inhibitors have shown the most promise. Future directions will likely focus on combining these inhibitory steps and perhaps developing conjugated inhibitors to bolster delivery and efficacy.

References

Allan RK, Mok D, Ward BK et al (2006) Modulation of chaperone function and cochaperone interaction by novobiocin in the C-terminal domain of Hsp90: evidence that coumarin antibiotics disrupt Hsp90 dimerization. J Biol Chem 281:7161–7171

Aoyagi S, Archer TK (2005) Modulating molecular chaperone Hsp90 functions through reversible acetylation. Trends Cell Biol 15(11):565–567

Baldo B, Weiss A, Bibel M et al (2012) A screen for enhancers of clearance identifies huntingtin as a heat shock protein 90 (Hsp90) client protein. J Biol Chem 287:1406–1414

Bali P, George P, Cohen P et al (2004) Superior activity of the combination of histone deacetylase inhibitor LAQ824 and the FLT-3 kinase inhibitor PKC412 against human AML cells with mutant FLT-3. Clin Cancer Res 10:4991–4997

Banerji U (2009) Heat shock protein 90 as a drug target: some like it hot. Clin Cancer Res 15:9–14

Barrott JJ, Haystead TA (2013) HSP90, an unlikely ally in the war on cancer. FEBS J 280(6):1381–1396

Beliakoff J, Whitesell L (2004) Hsp90: an emerging target for breast cancer therapy. Anti-Cancer Drugs 15:651–662

Bhatia S, Diedrich D, Frieg B et al (2018) Targeting HSP90 dimerization via the C terminus is effective in imatinib-resistant CML and lacks the heat shock response. Blood 132(3):307–320

Boyault C, Zhang Y, Fritah S et al (2007) HDAC6 controls major cell response pathways to cytotoxic accumulation of protein aggregates. Genes Dev 21:2172–2218

Brady RO, Kanfer JN, Bradley RM et al (1966) Demonstration of a deficiency of glucocerebroside-cleaving enzyme in Gaucher's disease. J Clin Invest 45:1112–1115

Buchner J (1999) Hsp90 & co. – a holding for folding. Trends Biochem Sci 24:136–141

Burlison J, Neckers L, Smith AB et al (2006) Novobiocin: redesigning a DNA gyrase inhibitor for selective inhibition of Hsp90. JACS 128(48):15529–15536

Chai RC, Vieusseux JL, Lang BJ et al (2017) Histone deacteylase activity mediates acquired resistance towards structurally diverse hsp90 inhibitors. Mol Oncol 11(5):567–583

Chiosis G, Timaul MN, Lucas B et al (2001) Small molecule designed to bind to the adenine nucleotide pocket of HSP-90 causes Her2 degradation and the growth arrest and differentiation of breast cancer cells. Chem Biol 8(3):289–299

DeBoer C, Meulman PA, Wnuk RJ et al (1970) Geldanamycin, a new antibiotic. J Antibiot (Tokyo) 23:442–447

Dickey CA, Kamal A, Lundgren K et al (2007) The high-affinity HSP90-CHIP complex recognizes and selectively degrades phosphorylated tau client proteins. J Clin Invest 117:648–658

Dickson MA, Okuno SH, Keohan ML et al (2012) Phase II study of the HSP90-inhibitor BIIB021 in gastrointestinal stromal tumors. Ann Oncol: Off J Eur Soc Med Oncol 24(1):252–257

Ding H, Peterson KL, Correia C et al (2017) Histone deacetylase inhibitors interrupt HSP90-CRAF interactions to upregulate BIM and circumvent drug resistance in lymphoma cells. Leukemia 31(7):1593–1602

Donnelly A, Blagg B (2008) Novobiocin and additional inhibitors of Hsp90 C-terminal nucleotide-binding pocket. Curr Med Chem 15:2702–2717

Duerfeldt A, Blagg B (2010) Hsp90 inhibition: elimination of shock and stress. Bioorg Med Chem Lett 20(17):4983–4987

Elfiky A, Saif MW, Beeram M et al (2008) BIIB021, an oral, synthetic non-ansamycin Hsp90 inhibitor: phase I experience. J Clin Oncol 26:2503

Eskew JD, Sadikot T, Morales P et al (2011) Development and characterization of a novel C-terminal inhibitor of Hsp90 in androgen dependent and independent prostate cancer cells. BMC Cancer 11:468

Fiskus W, Ren Y, Mohapatra A et al (2007) Hydroxamic acid analogue histone deacetylase inhibitors attenuate estrogen receptor-alpha levels and transcriptional activity: a result of hyperacetylation and inhibition of chaperone function of heat shock protein 90. Clin Cancer Res 13:4882–4890

Goetz MP, Toft DO, Ames MM et al (2003) The HSP90 chaperone complex as a novel target for cancer therapy. Ann Oncol 14(8):1169–1176

Gong C, Igbal K (2008) Hyperphosphorylation of microtubule-associated protein tau: a promising therapeutic target for Alzheimer disease. Curr Med Chem 15:2321–2328

Goode KM, Petrov DP, Vickman RE et al (2017) Targeting the Hsp90 C-terminal domain to induce allosteric inhibition and selective client downregulation. Biochim Biophys Acta 1861:1992–2006

Gormley N, Orphanides G, Meyers A et al (1996) The interaction of coumarin antibiotics with dragments of DNA gyrase B. Biochemistry 3:5083–5092

Guo T, Noble W, Hanger DP (2017) Roles of tau protein in health and disease. Acta Neuropathol (Berl) 133:665–704

Hoter A, El-Sabban M, Naim H (2018) The HSP90 family: structure, regulation, function, and implications in health and disease. Int J Mol Sci 19:1–33

Hruska K, LaMarca M, Scott C et al (2008) Gaucher disease: mutation and polymorphism spectrum in the glucocerebrosidase gene (GBA). Hum Mutat 29:567–583

Jackson SE (2012) Hsp90: structure and function. In: Jackson S (ed) Molecular chaperones, Topics in current chemistry, vol 328. Springer, Berlin

Jhaveri K, Modi S (2012) HSP90 inhibitors for cancer therapy and overcoming drug resistance. Adv Pharmacol 65:471–517

Karagoz G, Duarte A, Akoury E et al (2014) Hsp90-tau complex reveals molecular basis for specificity in chaperone action. Cell 156:963–974

Kasperkiewicz M, Müller R, Manz R et al (2011) Heat-shockprotein 90 inhibition in autoimmunity to type VII collagen: evidence that nonmalignant plasma cells are not therapeutic targets. Blood 117(23):6135–6142

Kekapure V, Dannenberg A, Subbaramaiah K (2009) HDAC6 modulates Hsp90 chaperone activity and regulates activation of aryl hydrocarbon receptor signaling. J Biol Chem 284:7436–7445

Koay YC, Wahyudi H, McAlpine SR (2016) Reinventing HSP90 inhibitors: blocking C-terminal binding events to HSP90 by using dimerized inhibitors. Chem Eur J 22:18572–11858

Kovacs J, Murphy P, Gaillard S et al (2005) HDAC6 regulates Hsp90 acetylation and chaperone-dependent activation of glucocorticoid receptor. Mol Cell 18:601–607

Lackie R, Maciejewski A, Ostapchenko V et al (2017) The HSP70/HSP90 chaperone machinery in neurodegenerative diseases. Front Neurosci 11:1–23

Lazaro I, Oguiza A, Recio C et al (2015) Targeting HSP90 ameliorates nephropathy and atherosclerosis through suppression of NF-κB and STAT signaling pathways in diabetic mice. Diabetes 64(10):3600–3613

Lazaro I, Oguiza A, Recio C et al (2017) Interplay between HSP90 and Nrf2 pathways in diabetes-associated atherosclerosis. Clin Investig Arterioscler 29(2):51–59

Lee VM, Goedert M, Trojanowski JQ (2001) Neurodegenerative tauopathies. Annu Rev Neurosci 24:1121–1159

Lee JH, Gao J, Kosinski PA, Elliman SJ et al (2013) Heat shock protein 90 (HSP90) inhibitors activate the heat shock factor 1 (HSF1) stress response pathway and improve glucose regulation in diabetic mice. Biochem Biophys Res Commun 430(3):1109–1113

Lei H, Romeo G, Kazlauskas A (2004) Heat shock protein-90α-dependent translocation of annexin II to the surface of endothelial cells modulates plasmin activity in the diabetic rat aorta. Circ Res 94:902–909

Liu W, Vielhauer GA, Holzbeierlein JM et al (2015) KU675, a concomitant heat-shock protein inhibitor of Hsp90 and Hsc70 that manifests isoform selectivity for Hsp90α in prostate cancer cells. Mol Pharmacol 88(1):121–130

Luo W, Sun W, Taldone T et al (2010) Heat shock protein 90 in neurodegenerative diseases. Mol Neurodegener 5:1–8

Madrigal-Matute J, Lopez-Franco O, Blanco-Colio LM et al (2010) Heat shock protein 90 inhibitors attenuate inflammatory responses in atherosclerosis. Cardiovasc Res 86:330–337

Mollapour M, Neckers L (2012) Post-translational modifications of Hsp90 and their contributions to chaperone regulation. Biochim Biophys Acta 1823:648–655

Neckers L, Ivy SP (2003) Heat shock protein 90. Curr Opin Oncol 15:419–424

Park J, Kim S, Choi M, Lee J et al (2008) Class II histone deacetylases play pivotal roles in heat shock protein 90-mediated proteasomal degradation of vascular endothelial growth factor receptors. Biochem Biophys Res Commun 368:318–322

Prodromou C (2016) Mechanisms of Hsp90 regulation. Biochem J 473:2439–2452

Prodromou C, Rose SM, O'Brien R et al (1997) Identification and structural characterization of the ATP/ADP-binding site in the HSP-90 molecular chaperone. Cell 90(1):65–75

Regna N, Vieson M, Gojmerac A et al (2015) HDAC expression and activity is upregulated in diseased lupus-prone mice. Int Immunopharmacol 29:494–503

Rice JW, Veal JM, Fadden RP et al (2008) Small molecule inhibitors of Hsp90 potently affect inflammatory disease pathways and exhibit activity in models of rheumatoid arthritis. Arthritis Rheum 58(12):3765–3775

Roe M, Wahab B, Torok Z et al (2018) Dihydropyridines allosterically modulate Hsp90 providing a novel mechanism for heat shock protein co-induction and neuroprotection. Front Mol Biosci 5(51):1–14

Ron I, Horowitz M (2005) ER retention and degradation as the molecular basis underlying Gaucher disease heterogeneity. Hum Mol Genet 14(16):2387–2398

Russo CD, Polak PE, Mercado PR et al (2006) The heat-shock protein 90 inhibitor 17-allylamino-17-demethoxygeldanamycin suppresses glial inflammatory responses and ameliorates experimental autoimmune encephalomyelitis. J Neurochem 99(5):1351–1362

Schopf F, Biebl M, Buchner J (2017) The HSP90 chaperone machinery. Nat Rev Mol Cell Biol 18:345–360

Schwock J, Pham NA, Cao MP et al (2008) Efficacy of Hsp90 inhibition for induction of apoptosis and inhibition of growth in cervical carcinoma cells in vitro and in vivo. Cancer Chemother Pharmacol 61:669–681

Scroggins BT, Robzyk K, Wang D et al (2007) An acetylation site in the middle domain of Hsp90 regulates chaperone function. Mol Cell 25:151–159

Shelton L, Koren J, Blair L (2017) Imbalances in the Hsp90 chaperone machinery: implications for Tauopathies. Front Neurosci 11:1–12

Sidera K, Patsavoudi E (2014) HSP90 inhibitors: current development and potential in cancer therapy. Recent Pat Anticancer Drug Discov 9:1–20

Sittler A, Lurz R, Lueder G et al (2001) Geldanamycin activates a heat shock response and inhibits huntingtin aggregation in a cell culture model of Huntington's disease. Hum Mol Genet 10(12):1307–1315

Solárová Z, Mojžiš J, Solár P (2015) Hsp90 inhibitor as a sensitizer of cancer cells to different therapies (review). Int J Oncol 46:907–926

Speranza G, Anderson L, Chen AP et al (2018) First-in-human study of the epichaperome inhibitor PU-H71: clinical results and metabolic profile. Investig New Drugs 36(2):230–239

Srivastava P (2002) Roles of heat-shock proteins in innate and adaptive immunity. Nat Rev Immunol 2(3):185–194

Stebbins CE, Russo AA, Schneider C et al (1997) Crystal structure of an HSP90-geldanamycin complex: targeting of a protein chaperone by an antitumor agent. Cell 18(2):239–250

Stirnemann J, Belmatoug N, Camou F et al (2017) A review of Gaucher disease pathophysiology, clinical presentation and treatments. Int J Mol Sci 18(2):441

Taipale M, Jarosz DF, Lindquist S (2010) HSP90 at the hub of protein homeostasis: emerging mechanistic insights. Nat Rev Mol Cell Biol 11:515–528

Terracciano S, Russo A, Chini MG et al (2018) Discovery of new molecular entities able to strongly interfere with Hsp90 C-terminal domain. Sci Rep 8:1709

Trepel J, Mollapour M, Giaccone G et al (2010) Targeting the dynamic HSP90 complex in cancer. Nat Rev Cancer 10:537–549

Tukaj S, Zillikens D, Kasperkiewicz M (2015) Heat shock protein 90: a pathophysiological factor and novel treatment target in autoimmune bullous skin diseases. Exp Dermatol 24:567–571

Wandinger S, Richter K, Buchner J (2008) The Hsp90 chaperone machinery. J Biol Chem 283(27):18473–18477

Wang L, Xie C, Greggio E et al (2008) The chaperone activity of heat shock protein 90 is critical for maintaining the stability of leucine-rich repeat kinase 2J. Neurosciences 28:3384–3391

Wang Z, Hu P, Tang F et al (n.d.) HDAC6 promotes cell proliferation and confers resistance to temozolomide in glioblastoma. Cancer Lett 379:134–142

Whitesell L, Linquist S (2005) HSP90 and the chaperoning of cancer. Nat Rev Cancer 5:761–772

Yang C, Rahimpour S, Lu J et al (2013) Histone deacetylase inhibitors increase glucocerebrosidase activity in Gaucher disease by modulation of molecular chaperones. Proc Natl Acad Sci U S A 110:966–971

Yu X, Guo ZS, Marcu MG et al (2002) J Natl Cancer Inst 94:504–513

Zhang H, Burrows F (2004) Targeting multiple signal transduction pathways through inhibition of Hsp90. J Mol Med 82:488–499

de Zoeten E, Wang L, Butler K et al (2011) Histone deacetylase 6 and heat shock protein 90 control the functions of foxp3+ T-regulatory cellsMol. Cell Biol 31:2066–2078

Zou JY, Guo YL, Guettouche T et al (1998) Repression of heat shock transcription factor HSF1 activation by HSP90 (HSP90 complex) that forms a stress-sensitive complex with HSF1. Cell 94(4):471–480

Chapter 11
HSP90: A Key Player in Metal-Induced Carcinogenesis?

P. L. Abreu, L. M. R. Ferreira, T. Cunha-Oliveira, M. C. Alpoim, and A. M. Urbano

Abstract Three extensively used metals, cadmium, chromium and nickel, are established human carcinogens. The elucidation of the molecular and cellular mechanisms underlying the carcinogenicity of these metals has centered mostly on the signalling pathways that regulate cellular growth, differentiation and death. Unfortunately, our understanding of the involvement of these pathways in metal-induced carcinogenesis is still very incomplete. More recently, research has extended to include the impact of these metals on mechanisms not traditionally associated with cancer, but that are now increasingly viewed as playing a critical role in carcinogenesis. Among them is the stress response, a highly conserved mechanism employed by all cells for protection against protein damage. Indeed, all three metals induce proteotoxic stress, which warrants following this line of research. The present chapter will critically review published studies on the impact

P. L. Abreu
Instituto de Medicina Molecular João Lobo Antunes, Faculty of Medicine, University of Lisbon, Lisbon, Portugal

L. M. R. Ferreira
Department of Surgery and Diabetes Center, University of California, San Francisco, San Francisco, CA, USA

T. Cunha-Oliveira
CNC – Center for Neuroscience and Cell Biology, University of Coimbra, Coimbra and Cantanhede, Portugal

M. C. Alpoim
CNC – Center for Neuroscience and Cell Biology, University of Coimbra, Coimbra and Cantanhede, Portugal

Centro de Investigação em Meio Ambiente, Genética e Oncobiologia (CIMAGO) and Department of Life Sciences, University of Coimbra, Coimbra, Portugal

A. M. Urbano (✉)
Centro de Investigação em Meio Ambiente, Genética e Oncobiologia (CIMAGO) and Department of Life Sciences, University of Coimbra, Coimbra, Portugal

Molecular Physical Chemistry Research Unit, University of Coimbra, Coimbra, Portugal
e-mail: amurbano@uc.pt

A. A. A. Asea, P. Kaur (eds.), *Heat Shock Protein 90 in Human Diseases and Disorders*, Heat Shock Proteins 19,
https://doi.org/10.1007/978-3-030-23158-3_11

of carcinogenic metals on the expression of the heat shock protein 90 family (HSP90), one of the protein families that mediate the stress response. HSP90 has been consistently found to be overexpressed in many types of cancer and, significantly, HSP90 overexpression has been correlated with increased tumor growth, metastatic potential and resistance to chemotherapy.

Keywords Cadmium · Carcinogenesis · Hexavalent chromium · HSP90 · Nickel · Stress response

Abbreviations

Grp94	glucose regulated protein 94
HIF-1α	hypoxia-inducible factor 1 alpha
HSE	heat shock elements
HSF	heat shock factor(s)
HSP	heat shock protein family
Hsp	heat shock protein(s)
ROS	reactive oxygen species
TRAP1	tumor necrosis factor receptor-associated protein 1

11.1 Introduction

The stress response is a highly conserved mechanism used by all living organisms to recover from proteotoxic stress, i.e., stress that causes damage to proteins (Schlesinger et al. 1982). Ever since the discovery that several of the proteins that mediate this response, the so-called heat shock proteins (Hsp), are consistently overexpressed in many types of cancer, research on the molecular mechanisms of carcinogenesis expanded to include this response. One of the families of Hsp, HSP90, has received particular attention, as its overexpression has been correlated with increased tumor growth, metastatic potential and resistance to chemotherapy (Ciocca et al. 2013; Nahleh et al. 2012; Whitesell and Lindquist 2005).

The exact mechanisms through which the stress response might promote or facilitate carcinogenesis are not known, but it has been hypothesized that the activation of the stress response by mildly cytotoxic exposure to carcinogens may impart the surviving cells with an increased resistance to subsequent stresses, namely those encountered by incipiently transformed cells as they progress to full malignancy. Alternatively, this type of exposure may select for those cells that already possessed that increased resistance. As promoters of proteotoxic stress, carcinogenic metals are potential activators of the stress response. There is also some indication from in vitro studies that pre-incubation with cadmium or hexavalent chromium confers cells thermotolerance, i.e., tolerance to an otherwise lethal thermal stress (Abreu et al. 2018; Goering and Fisher 1995). Whether the observed thermotolerance is due to increased Hsp expression and whether a similar protective effect occurs against stresses more relevant in the context of metal-induced carcinogenicity remains to be investigated.

Here, we critically review literature on the impact on HSP90 expression caused by cadmium, chromium (in the hexavalent state) and nickel, three widely used metals that are established occupational carcinogens (IARC 1990, 1993, 2012; NTP 2014). In addition, we highlight, whenever possible, plausible mechanistic links between HSP90 and metal-induced carcinogenesis. The first section of this chapter is devoted to the stress response and its links to carcinogenesis. Next, we provide information concerning the industrial and commercial applications of these metals, as well as the cancer risks to which several million workers employed in the respective industries are exposed. A brief overview of the molecular mechanisms underpinning their carcinogenicity will follow. Next, our current understanding of the impact of carcinogenic metals on HSP90 expression will be addressed, starting by a brief discussion of the promotion of proteotoxic stress by the three metals. Rather than presenting an exhaustive review of the literature, we aim at providing a critical analysis of selected studies, taking into account the context in which the results reported were generated and assessing their relevance in the field of metal-induced carcinogenesis. This analysis focused mostly on mammalian systems. Finally, the potential of HSP90 as a target for cancer therapy will be discussed.

11.2 Carcinogenesis, the Stress Response and Links Between the Two

Carcinogenesis entails the progression from an initial genetic mutation in a single transformed cell to the development of a fully malignant cancer cell, capable of indefinite growth and metastasis. In this process, abnormal growth and proliferation of cells with an increasing number of mutations eventually leads to the formation of a tumor mass, which will expand and colonize distant sites in the body. This journey is accompanied by a plethora of cellular stresses. At the cell intrinsic level, incipient cancer cells experience redox imbalance, endoplasmic reticulum stress (in part due to the accumulation of high levels of oncoproteins), genotoxic stress (arising from chromosome instability and faulty DNA repair) and deranged metabolism (Acharya et al. 2010; Ferreira 2010; Schar 2001; Vandewynckel et al. 2013). The genesis of a solid tumor is accompanied by yet another set of stresses, cell extrinsic in nature, arising in those areas of the solid tumor microenvironment that are characterized by hypoxia, acidosis and nutrient deprivation (Bertout et al. 2008; Herr and Debatin 2001; Neri and Supuran 2011; Pflaum et al. 2014).

There is growing evidence that tumors harbor a subpopulation of self-renewing cells, responsible for tumor growth, metastasis and resistance to conventional anticancer therapies (Reya et al. 2001). Named cancer stem cells, they were first described in acute leukemia (Lapidot et al. 1994), and later identified in solid tumors (Mani et al. 2008). Cancer stem cells are often found in the most hypoxic and acidic regions of the tumor microenvironment, where they likely experience higher levels of cellular stress (Hjelmeland et al. 2011; Kim et al. 2018).

Several of the above-mentioned stresses are, in turn, strong inducers of proteotoxic stress, i.e., they produce conformational alterations to proteins that may eventually lead to their denaturation and aggregation. To protect themselves against this type of stress, normal cells activate the stress response, a ubiquitous homeostatic system found in all living organisms examined to date (Schlesinger et al. 1982). This response was first observed, in the early 1960s, in the form of a different puffing pattern exhibited by the polytene chromosomes of larval Drosophila tissue submitted to sub-lethal hyperthermia (Ritossa 1962). As chromosome "puffs" are synonymous with intense transcriptional activity, the different puffing pattern revealed that Drosophila cells were responding to heat shock by inducing a set of genes not usually expressed during their stage of larval development.

The products of the small set of genes strongly induced by heat shock belong to a family of highly conserved proteins collectively named heat shock proteins. The designation remains to this day, in spite of the fact that it is now known that these proteins are induced not only by heat shock, but also by any condition capable of inducing proteotoxic stress (Jolly et al. 2000; Morimoto 1993). Hsp mediate the stress response by playing a critical role in protein folding, in the assembly and disassembly of oligomeric protein complexes, in the translocation of proteins to their final subcellular locations and in the regulation of protein degradation (Csermely et al. 1998).

Mammals express many types of Hsp, classified in six families: HSP100, HSP90, HSP70, HSP60, HSP40, and small Hsp (sHSP) (Jolly and Morimoto 2000; Katschinski 2004). Family designations were based on the subunit relative molecular masses of their then known members (or isoforms). New found members are now assigned to a given family based on both subunit relative molecular mass and sequence homology. Of note, initial studies on Hsp did not discriminate between isoforms. For instance, the abbreviations Hsp70 and HSP70 were used to refer to any protein of an approximate molecular mass of 70 kDa strongly induced by heat or other promoters of proteotoxic stress. In an attempt to reduce any confusion that might arise due to this circumstance, the abbreviation HSP will be used throughout this chapter when referring to a family of Hsp (e.g., HSP90, for heat shock protein 90 family), whilst the abbreviation Hsp will be used to discriminate a specific isoform (e.g., Hsp90α, for heat shock protein 90 alpha).

Hsp levels are transcriptionally regulated by transcription factors known as heat shock factors (HSF). HSF are specific to heat shock elements (HSE), regulatory elements within the promoters of Hsp genes (Pirkkala et al. 2001). In mammals, the main regulator of the stress response is heat shock factor 1 (HSF1), one of the several isoforms of HSF present in these organisms (Lindquist 1986; Vihervaara and Sistonen 2014). HSF1 activation under stress conditions involves homotrimerization. In the case of hyperthermia, a "regulatory region" in HSF1 becomes less stable, eventually triggering a separate domain to interact with other HSF1 proteins, forming homotrimers (Hentze et al. 2016). Yet, homotrimerization does not suffice to confer activity to HSF1. In the basal state, HSP90 inhibits HSF1 activity; only when the intracellular levels of unfolded proteins reach a high value does HSP90 get titrated away, allowing HSF1 activation (Voellmy 2004). Interestingly, HSF1 can also sense the redox state of the cell via two cysteine residues that, when oxidized, form a disulfide bond that prompts HSF1 trimerization and concomitant DNA-binding activity (Ahn and Thiele 2003).

Activation of the stress response is likely the key to understanding how tumor cells manage to thrive under the highly adverse circumstances associated with carcinogenesis. The increased HSF1 activity and elevated levels of most Hsp observed in several types of tumors (Ciocca et al. 2013) are in line with this hypothesis. HSF1 activation might not be the only mechanism responsible for the marked upregulation of Hsp in cancer; well-studied oncogenes and tumor suppressor genes have also been implicated. For instance, the activation of the promoters of some Hsp genes by the oncogenic transcription factor c-MYC, or their derepression when the tumor suppressor proteins p53 and p63 lose function, have been put forward as contributors to Hsp upregulation in cancer (Whitesell and Lindquist 2009). HSF1 has been shown to carry out several other functions in cancer cells and is, along with HSP90 and HSP70, one of the most well-studied components of the stress response in the context of carcinogenesis (Vihervaara and Sistonen 2014).

Increased levels of Hsp allow tumor cells to manage the increased burden of damaged and aggregated proteins, which would otherwise trigger programmed cell death. The contribution of the resulting increased pool of Hsp to protein homeostasis is two-fold: it aids the refolding of damaged proteins, preventing their aggregation, and, if the damage is too extensive for refolding to occur, it assists in sequestering and diverting the severely damaged proteins to the proteasome for degradation (Lindquist 1986; Schlesinger 1990; Whitesell and Lindquist 2005). Importantly, several studies have implicated Hsp in various oncogenesis hallmarks, including epithelial cell migration (Piotrowicz et al. 1998), tumor invasiveness and chemotherapy resistance (Oesterreich et al. 1993).

Interestingly, mortalin, a heat shock protein from the Hsp70 family, is overexpressed in embryonic stem cells relative to their differentiated counterparts (Saretzki et al. 2004) and its ectopic expression suffices to induce malignant transformation and inactivation of p53 in cultured cells (Wadhwa et al. 1999). Thus, a provocative view in the field states that the stress response in cancer cells is not merely an adaptive response that enables survival to multiple stresses, but rather an active driver of carcinogenesis (de Billy et al. 2012). In line with this view, the stress response has been implicated in tumor growth, in the acquisition and maintenance of cancer hallmarks and in chemotherapy resistance (Ciocca et al. 2013; Nahleh et al. 2012; Whitesell and Lindquist 2005). In summary, the stress response might be seen as a double-edged sword: developed throughout evolution to protect individual cells and, consequently, the whole organism against stressful conditions, it provides an escape route for incipiently transformed cells, allowing their progression to full malignancy and, ultimately, death of the organism.

11.3 Carcinogenic Metals: General Information

Among the human carcinogens identified by the International Agency for Research on Cancer (IARC), the National Toxicology Program (NTP), and other highly respected regulatory agencies are three metals: cadmium, chromium (in the hexavalent state; Cr(VI)) and nickel (IARC 1990, 1993, 2012; NTP 2014). All three are viewed as occupational carcinogens, as an increased risk of cancer has been unequivocally confirmed only among workers who were exposed, for extended periods of time, to high doses of these carcinogens.

The physical and chemical properties of these three metals make them particularly suitable for a wide variety of commercial and/or industrial applications. Namely, they all are highly resistant to corrosion and, as such, are used to impart this advantage to diverse materials. For instance, both nickel and chromium are components of stainless steel, whose manufacture accounts for ca. 60% and 80% of, respectively, all nickel and chromium produced. Cadmium is mostly used in the manufacture of nickel-cadmium batteries, accounting for ca. 80% of all cadmium produced. All three metals are also used as pigments and in metal finishing, nickel and chromium are used in welding, and each has a wide range of other metal-specific applications. Worldwide, the different cadmium, chromium and nickel industries employ several million workers (IARC 2012).

Although these metals are viewed mostly as occupational carcinogens, there has been, for some time now, concern regarding the general population, as mass production, recycling and disposal of these metals have turned them into widespread environmental pollutants (Jarup 2003). Fossil fuel combustion and tobacco smoke are additional strong contributors to their widespread presence in the environment (IARC 2012). Environmental exposure to cadmium is of particular concern, as this metal has a half-life of approximately 20–30 years in humans, and thus can accumulate to very high levels (Tully et al. 2000). This is in part attributable to the body's poor capacity to metabolically degrade and excrete this metal (Waalkes 2003). In terms of types of cancer, there is now sufficient evidence that cadmium and nickel metal (but not chromium metal), as well as cadmium compounds, hexavalent chromium compounds, and nickel compounds cause lung cancer. Depending on the metal, there is also either sufficient evidence or positive associations between exposure and other respiratory tract cancers, such as cancer of the nasal cavity. In the specific cases of cadmium metal and cadmium compounds, positive associations have been observed between exposure and cancer of the kidney and the prostate (IARC 2012).

11.4 The Generation of Proteotoxic Stress by Cadmium, Hexavalent Chromium and Nickel

Considering that cancer is traditionally viewed as a genetic disease characterized by disorganized growth, research effort in the field of metal carcinogenicity has mostly been devoted to the elucidation of the interactions that carcinogenic metals establish with the genome, the genetic and genomic damage they produce, the impact that they have on signalling pathways and, ultimately, the changes they produce at the phenotypic level (Feng et al. 2018; Holmes et al. 2008; Wang et al. 2018). Understandably, the focus has been mainly on those genes/gene products directly involved in the regulation of cellular proliferation, differentiation and death and whose disruption is thought to dictate malignant growth (Hanahan and Weinberg 2011). In spite of their undeniable importance, a complete discussion of the results obtained in this type of research studies is beyond the scope of this chapter. Instead, this section will briefly describe how cadmium, hexavalent chromium and nickel can generate proteotoxic stress. Nonetheless, the impact of carcinogenic metals on cancer-associated HSP90 client proteins will be very briefly addressed in Sect. 11.6.

Exposure to cadmium, hexavalent chromium and nickel can generate proteotoxic stress by direct and indirect mechanisms. In the case of cadmium, an electrophile (or a soft acid, according to Pearson's principle (Pearson 1968)), proteotoxic stress has been ascribed, to a large extent, to a competition with zinc. Zinc, the second most abundant trace metal found in eukaryotic organisms, is required for essential catalytic functions in hundreds of enzymes, playing also a critical role in the stabilization and folding of protein subdomains (Coleman 1992). By displacing zinc, cadmium compromises the structure and function of affected proteins (Waisberg et al. 2003). In the case of proteins belonging to the cellular antioxidant system, this replacement results in increased levels of reactive oxygen species (ROS) (Dorta et al. 2003; Liu et al. 2009; Martelli et al. 2006; Noel et al. 2004) and, ultimately, additional proteotoxic stress. Moreover, and as already mentioned in a preceding section, an oxidizing intracellular environment can promote the activation of HSF1, through oxidation of two critical cysteine residues of this master regulator of the stress response in vertebrates (Ahn and Thiele 2003; Dai et al. 2007). A similar outcome can result from the binding of cadmium to vicinal sulfhydryl groups in proteins, namely those belonging to the cellular antioxidant defense systems, altering their conformations and impairing their functions (Shimizu et al. 1997; Waalkes 2003; Waisberg et al. 2003). Curiously, back in 1980, when it was not even known that Hsp induction was a generalised defense mechanism against stressful conditions, Levinson and co-workers exposed cells to cadmium and other soft metals with known sulfhydryl binding capacity to test their hypothesis that Hsp induction was somehow related to binding to sulfhydryl-containing (then unknown) targets (Levinson et al. 1980). Substitution of zinc by cadmium has many other important biological consequences, namely in the context of carcinogenesis, but an extended discussion of this topic is out of the scope of this chapter. Let it just be briefly mentioned that, by replacing zinc in proteins and enzymes implicated in gene regulation, DNA repair, redox regulation and in the regulation of signaling pathways, cadmium may strongly contribute to the high degree of genomic instability observed in cadmium-exposed cells, thus contributing to tumor initiation and development (Bishak et al. 2015; Hartwig 2013; Waalkes 2003).

In stark contrast with cadmium and nickel (see below), hexavalent chromium interacts poorly with most biomolecules, so direct damage to proteins is unlikely to be a major contributor to proteotoxic stress (Urbano et al. 2012; Urbano et al. 2008).

However, the intracellular reduction of hexavalent chromium produces a variety of reactive species, some of them strong oxidizers, namely carbon, oxygen and sulphur free radicals, which can act as secondary stressors through the generation of oxidative damage, either directly or through depletion of intracellular antioxidant pools (Abreu et al. 2014). Studies reporting that the addition of *N*-acetylcysteine, a known antioxidant, reversed the impact of cadmium and hexavalent chromium on the expression of, respectively, HSP70 and/or HSP90 give some support to the hypothesis that oxidative stress mediates, at least in part, Hsp induction by these two carcinogens (Han et al. 2007; Xiao et al. 2012). The reactive species generated during the intracellular reduction of hexavalent chromium also cause damage to the DNA and, quite often, the carcinogenicity of this metal ion has been associated with the formation of DNA single- and double-strand breaks due to an abnormal processing of primary lesions by DNA repair systems. As a result of lesion accumulation, particularly double-strand breaks, the cell can either undergo apoptosis or attempt DNA repair by the non-homologous end joining system. Unfortunately, this system generates chromosome rearrangements and, ultimately, genomic instability (Reynolds and Zhitkovich 2007). Genomic instability can also result from uncoupling of centrosome duplication from the cell cycle, as a consequence of prolonged arrest at either S or G2 phases to repair double-strand breaks (Urbano et al. 2008).

Like cadmium and hexavalent chromium, nickel can also generate oxidative stress (Salnikow and Costa 2000). In addition, proteotoxic stress can be generated through binding of nickel to secondary amines in the histidine residues of proteins (Waisberg et al. 2003). Importantly, nickel binds to the His-His-carboxylate motive, present in all dioxygenases, with more affinity than iron, which might explain nickel's effects on the inactivation of histone demethylases, enzymes dependent on the 2-oxoglutarate activity (Chen et al. 2006).

11.5 Investigations Into the Involvement of the Stress Response in Metal Carcinogenesis: A Contextualization

As promoters of proteotoxic stress, carcinogenic metals are potential activators of the stress response. Before starting our discussion of this topic, it must be mentioned that most of the studies that will be discussed in this and the following sections were not aimed at establishing correlations between Hsp induction and carcinogenesis. Thus, the model systems used were not always especially suited for the study of the mechanisms underlying metal-induced carcinogenesis. Studies dating back to the 1970s and 1980s aimed to understand why cells responded to heat shock by a strong induction of a small set of specific proteins (Tissieres et al. 1974). One of these studies, already briefly mentioned in the preceding section, involved several metals, including cadmium and nickel (Levinson et al. 1980). This seminal study showed that exposure of chick embryo cells to 10 μM Cd(II) strongly induced the synthesis of 100, 75, 35 and 25 kDa Hsp. For the 100, 75 and 35 kDa Hsp, enhancement became visible after a 1 h exposure, whereas enhancement of the 25 kDa Hsp required a longer (2 h) exposure. Hsp induction was also observed with copper, zinc and mercury. Nickel, on the contrary, failed to induce these proteins, even at a concentration as high as 500 μM Ni(II). Later, however, HSP70 (but not HSP90) induction by nickel was reported, when primary cultures of rat hepatocytes were exposed, for 4 h, to 2000 μM Ni(II) (Bauman et al. 1993). The same study confirmed cadmium's ability to produce a strong induction of HSP70 in the low micromolar range (4–8 μM Cd(II)). With this exposure regimen, an induction of HSP90 by cadmium could also be detected, albeit smaller than that observed for HSP70. It is important to stress that, in the context of carcinogenesis, the most relevant concentrations are those that induce some, but not overt, cytotoxicity. Thus, although the concentrations of the two metals required for Hsp induction differed by almost three orders of magnitude, their biological significance is likely comparable, as both corresponded to the highest possible toxicity for the corresponding metal, without killing all of the cells. The combined results of these two studies suggested that the impact of metals on Hsp expression depends not only on the metal, but also on the system (e.g., cell type and species), the exposure regimen (duration of the exposure and metal concentration), as well as on the specific Hsp.

Later, the realization that insults that were not overtly cytotoxic promoted Hsp induction led to the suggestion that Hsp levels might be used in molecular toxicology as sensitive biomarkers of early exposure, toxicity and environmental stress. In fact, the first study investigating Hsp induction by hexavalent chromium aimed specifically at verifying the feasibility of using Hsp induction for the rapid detection of low levels of pollutants (Delmas et al. 1998). In this study, rather than relying on protein levels, whose quantification, back then, classically involved the time-consuming and not always straightforward autoradiography or fluorography of electrophoretically separated ^{35}S-methionine-labeled proteins, Delmas and collaborators quantified the transcript levels of a specific Hsp, Hsp72, using an RNase protection method that involved a radiolabeled antisense RNA probe. These authors found that, in HepG2 cells, a 6 h exposure to concentrations as low as 1 μM Cr(VI) significantly increased Hsp72 mRNA. In HT29 cells, Hsp72 induction required a higher Cr(VI) concentration (20 μM). Such higher concentration was, nonetheless, not overtly cytotoxic to this cell line. It is important to note that, in some instances, different study outcomes can be ascribed, at least in part, to the different sensitivities of the methods employed, as is well illustrated in a study by Carroll and Wood (Carroll and Wood 2000). Using immunoblotting, these authors were able to detect HSP90 induction in human keratinocytes after a 1 h exposure to 10 μM Ni(II). Using ^{35}S-methionine-labeling, HSP90 induction only became apparent when the concentration was increased to 1000 μM. The study by Carroll and Wood also confirmed the dependence on the model system employed, as HSP90 induction was much stronger in dermal fibroblasts than in keratinocytes. Finally, as many of the studies did not discriminate which member(s) of a given family was being assessed, comparisons between studies become even more complex.

11.6 The Impact of Cadmium, Hexavalent Chromium, and Nickel on HSP90 Expression

Although activation of the stress response by cadmium, hexavalent chromium and nickel may be mediated by several Hsp families, this chapter will focus specifically on HSP90 induction, with only occasional forays into the induction of other HSP by these metals. HSP90, unlike other HSP, is not required for the correct folding of newly synthesized proteins. Instead, its main role appears to be the stabilization of meta-stable proteins. Significantly, many HSP90 client proteins are involved in the acquisition of cancer hallmarks. Chief among these are several receptor tyrosine kinases and steroid hormone receptors, such as the human epidermal growth factor 2 (HER2), associated with uncontrolled cellular proliferation (Whitesell and Lindquist 2005; Ziemiecki et al. 1986), telomerase, an enzyme required for immortalization and acquisition of cancer stem cell properties (Beck et al. 2011; Holt et al. 1999), AKT, involved in apoptosis (Basso et al. 2002), hypoxia-inducible factor 1 alpha (HIF-1α), essential for angiogenesis (Isaacs et al. 2002) and matrix metalloproteinases (MMP), crucial for successful tissue invasion and metastasis (Eustace et al. 2004). A more complete list of HSP90 client proteins involved in initiating or maintaining cancer hallmarks is provided in Table 11.1. It is important to stress that Table 11.1, which is intended to illustrate the involvement of HSP90 in carcinogenesis, does not constitute an exhaustive list of all reported studies in this field of research.

It has been postulated that stabilization of the above-mentioned proteins by members of the HSP90 family potentiates the metabolic shift and invasiveness observed in tumors (Ferreira 2010; Ferreira et al. 2012; Whitesell and Lindquist 2005). In line with this hypothesis, HSP90 has been consistently found to be overexpressed in many types of cancers and this overexpression correlates with tumor growth, metastatic potential and resistance to chemotherapy (Ciocca et al. 2013; Nahleh et al. 2012; Whitesell and Lindquist 2005). Strikingly, many cancer-associated HSP90 client proteins are modulated by cadmium, hexavalent chromium, or nickel (Table 11.2), with 15 of them being targeted by all three metals (Fig. 11.1).

As is the case with the other HSP of high molecular weight, the chaperoning activity of HSP90 is dependent on ATP. Specifically, ATP must bind to a specific site of their amino terminal domain (N-terminal domain) and suffer hydrolysis. It is believed that ATP binding to HSP90 and its subsequent hydrolysis causes conformational alterations that are essential for substrate binding (Pearl and Prodromou 2006; Wandinger et al. 2008). The ATP-binding pocket of this family of proteins is distinct from the one found in many kinases and in HSP70 (Dutta and Inouye 2000). Although the biological relevance of this feature remains unclear (Whitesell and Lindquist 2005), it has already been exploited for the development of drugs specifically targeting HSP90 (Calderwood et al. 2006), as will be discussed in the last section of this chapter.

Two other domains can be found in HSP90 proteins: a highly charged middle domain (M domain), which is the docking site for the client proteins, and the car-

Table 11.1 HSP90 client proteins involved in initiating or maintaining cancer hallmarks

Cancer hallmark	HSP90 client protein	References
Sustained proliferative signaling	RAF1, HER2, BCR-ABL, Annexins, KIT, PI3K, AKT, EGFR	Ahsan et al. (2012), Citri et al. (2006), Floris et al. (2011), Giulino-Roth et al. (2017), Lei et al. (2004), Sato et al. (2000), Schulte et al. (1995), Solit et al. (2002), and Wu et al. (2008)
Evading growth suppressors	PLK, WEE1, CDK4, CDK6, MYT1, cyclin D, PTEN	Blank et al. (2003), Burrows et al. (2004), Fu et al. (2013), Lokeshwar (2012), Mahony et al. (1998)
Avoiding immune destruction	IKKβ	Lee et al. (2010)
Enabling replicative immortality	Telomerase	Holt et al. (1999)
Tumor-promoting inflammation	IKKβ, NOS, TAK1	Harris et al. (2008), Lee et al. (2010), Presley et al. (2010), Shi et al. (2009), and Trepel et al. (2010)
Invasion and metastasis	MMPs, BRMS1, uPAR, V-ATPase, TAK1, FAK	Asuthkar et al. (2012), Correia et al. (2013), Hartson and Matts (2012), Hurst et al. (2006), Shi et al. (2009), Stellas et al. (2010), and Xiong et al. (2014)
Inducing angiogenesis	FAK, HIF-1α, VEGF-R2, FLT3, SRC, JNK	Bruns et al. (2012), Minet et al. (1999), Nieto-Miguel et al. (2008), Whitesell et al. (1994), Xiong et al. (2014), and Yao et al. (2003)
Genome instability and mutation	BRCA1, BRCA2, CHK1, FANCA, DNA-PKcs, p53, MRE11, RAD50, NBS1	Blagosklonny et al. (1995), Dote et al. (2006), Makhnevych and Houry (2012), Noguchi et al. (2006), Oda et al. (2007), Quanz et al. (2012), and Stecklein et al. (2012)
Resisting cell death	RIPK1, IGF-1R, BCL-2, APAF-1, p53, BID, Survivin	Blagosklonny et al. (1995), Breinig et al. (2011), Dias et al. (2002), Fortugno et al. (2003) Lewis et al. (2000), and Zhao and Wang (2004)
Rewired metabolism	SDH, IDH3G, NDUFS3, MDH2, HIF-1α, SCAP/SREBP, V-ATPase, PKM2, MYC, p53	Blagosklonny et al. (1995), Chae et al. (2013), Hartson and Matts (2012), Kuan et al. (2017), Lee et al. (2018), Minet et al. (1999), and Xu et al. (2017)

boxyl terminal domain (C-terminal domain), which possesses another ATP-binding site and is essential for dimerization (Pearl and Prodromou 2006). It is thought that the chaperoning function of HSP90 requires the assembly of an HSP90 chaperone machine, a dynamic complex comprising, in addition to HSP90 homodimers, HSP70 and proteic co-chaperones (Trepel et al. 2010; Wandinger et al. 2008). Docking of the co-chaperones usually occurs on the M domain, but co-chaperones can also interact with the other two domains and mediate the interaction of HSP90 with its client proteins (Pearl and Prodromou 2006). The fact that HSP90 can chaperone so many distinct client proteins can be explained by the existence of different complexes, assembled from diverse associations of different co-chaperones. So far,

Table 11.2 Cancer-associated HSP90 client proteins modulated by cadmium, hexavalent chromium and/or nickel

Protein/gene modulated by cadmium	References	Protein/gene modulated by hexavalent chromium	References	Protein/gene modulated by nickel	References
Annexin-A3	Lubovac-Pilav et al. (2013)	Annexin-A2	Madureira et al. (2012)	Annexin-A1, -A2, -A8	Wezynfeld et al. (2014)
BCL-2	Fernandez et al. (2003)	BCL-2	Medan et al. (2012)	BCL-2	Lee et al. (2001)
BID	Li et al. (2000)	BID	Hill et al. (2013)	BID	Guo et al. (2016)
Cyclin-D1	Topisirovic et al. (2002)	Cyclin-B1, -D2, -D3, -E2	Stanley et al. (2011)	Cyclin-D1, -E	Ding et al. (2009)
EGFR	Wei et al. (2015)	EGFR	Kim et al. (2015)	EGFR	Chiou et al. (2015)
HIF-1α	Jing et al. (2012)	HIF-1α	Kim et al. (2016)	HIF-1α	Salnikow et al. (2002)
IDH	Kil et al. (2006)	IDH	Shil and Pal (2018)	IDH	O'Leary and Limburg (1977)
IKKβ	Liu et al. (2016)	IKKβ	Chen et al. (2002)	IKKβ	Viemann et al. (2007)
JNK	Chuang et al. (2000)	JNK	O'Hara et al. (2003)	JNK	Wu et al. (2011)
MMP-9	Lian et al. (2015)	MMP-1, -2	Kim et al. (2016) and Lu et al. (2018)	MMP-2, -7, -9, -11, -13	Ota et al. (2018) and Wan et al. (2011)
MYC	Tang and Enger (1991)	MYC	Pratheeshkumar et al. (2016)	MYC	Li et al. (2009)
p53	Tokumoto et al. (2011)	p53	Carlisle et al. (2000)	p53	Maehle et al. (1992)
PI3K/Akt	Tsai et al. (2016)	PI3K/Akt	Kim et al. (2015)	PI3K/Akt	Pan et al. (2011)
SDH	Karthikeyan and Bavani (2009)	SDH	Fernandes et al. (2002)	SDH	Repetto et al. (2001)
CDK-1, -2	Wei et al. (2015)	CDK-1, -2, -4, -6	Stanley et al. (2011)		
CHK1/2	Bork et al. (2010)	CHK1	Ganguly et al. (2018)		
DNA-PKcs	Li et al. (2015)	DNA-PKcs	Hill et al. (2011)		
FAK	Choong et al. (2013)	FAK	Lu et al. (2018)		
IGF-1R	Fujiki et al. (2017)	IGF-1R	He et al. (2013)		
PTEN	Huang et al. (2014)	PTEN	Zhong et al. (2017)		
RAF1	Ju et al. (2017)	RAF1	Bae et al. (2009)		

(continued)

Table 11.2 (continued)

Protein/gene modulated by cadmium	References	Protein/gene modulated by hexavalent chromium	References	Protein/gene modulated by nickel	References
uPAR	Pratheeshkumar et al. (2016)	uPAR	Pratheeshkumar et al. (2016)		
		Plk1	Chun et al. (2010)	Plk3	Li et al. (2017)
NOS	Majumder et al. (2008)			NOS	Gupta et al. (2000)
SRC	Martinez Flores et al. (2013)			SRC	Cabail et al. (2016)
Telomerase	Dai et al. (2010)			Telomerase	Lei et al. (2001)
		MDH2	Shil and Pal (2018)		
		MRE11	Xie et al. (2008)		
		NDUFS3	Zhang et al. (2016)		
		RIPK1	Gavin et al. (2007)		
c-Kit	Weng et al. (2014)				
PKM2	Kim et al. (2014)				
V-ATPase	Herak-Kramberger et al. (1998)				
VEGF-R2	Kim et al. (2012)				
WEE1	Lubovac-Pilav et al. (2013)				
				APAF-1	Guo et al. (2016)
				BRMS1	Ota et al. (2018)

four different proteins have been assigned to HSP90: the classic Hsp90 (i.e., the isoforms Hsp90α and Hsp90β), localizing mostly to the cytosol; glucose regulated protein 94 (Grp94), found in the endoplasmic reticulum; and tumor necrosis factor receptor-associated protein 1 (TRAP1), which localizes to mitochondria (Csermely et al. 1998; Felts et al. 2000). The existence of efficient protein stabilization mechanisms in the mitochondria is amply justified, as these electron- and protein-rich organelles are particularly prone to ROS generation as by-products of respiratory metabolism. Significantly, it was reported that TRAP1 levels are significantly higher in tumor cells than in their normal counterparts (Chae et al. 2013; Kang et al. 2007;

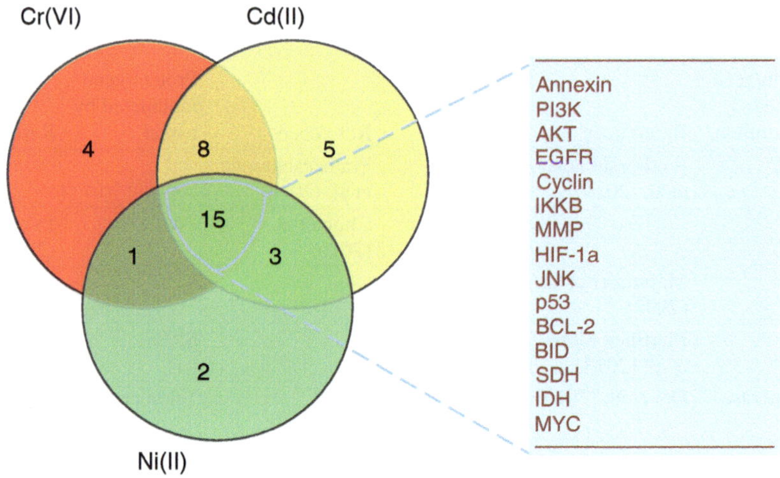

Fig. 11.1 HSP90 clients implicated in cancer are impacted at the protein and/or mRNA level by all three carcinogenic metals: cadmium, hexavalent chromium and nickel. Out of 49 HSP90 client proteins involved in establishing and/or maintaining hallmarks of cancer, 15 are targeted by all three metals. Created using http://genevenn.sourceforge.net/vennresults.php

Leav et al. 2010), likely to ensure mitochondrial genome integrity and safeguard respiratory function and mitochondrial biosynthetic capacity. It must be borne in mind that, although most tumor cells exhibit a higher-than-normal reliance on lactic acid fermentation, they are still dependent on respiration (Abreu and Urbano 2016; Bajzikova et al. 2019; Ferreira 2010; Ferreira et al. 2012; Tan et al. 2015). Interestingly, it has been found that many environmental toxicants, including metals such as cadmium, tend to accumulate in mitochondria (Meyer et al. 2013). As to hexavalent chromium, it was recently reported that, at concentrations in the low micromolar range, it augmented mitochondrial biogenesis in HepG2 cells, suggesting that hexavalent chromium toxicity can be compensated by increasing mitochondrial content (Zhong et al. 2017).

In addition to their different intracellular locations, HSP90 can also be found in the cell surface of many cell types and it can also be secreted to the extracellular environment. Normal cells secrete HSP90 in response to tissue injury, favoring healing by promoting cell motility (Li et al. 2012). HSP90 secretion is mostly regulated by HIF-1α, which is known to be activated by hypoxia and oxidative stress. Considering that many tumors overexpress HIF-1α, which is due, at least in part, to intratumoral hypoxia, this recently identified role of HIF-1α has been proposed as an explanation for the secretion of HSP90 by tumor cells (Li et al. 2012). The consistent observation of HSP90 overexpression in tumors led to the proposal that this feature could be used as a marker of cellular malignancy and to the concept of HSP90 "addiction" (Barrott and Haystead 2013; Trepel et al. 2010). This "addiction" would arise from a constant need of an increased pool of HSP90 to continuously retrieve essential proteins that became misfolded due to extensive proteotoxic stress, as well as to facilitate the function of oncoproteins and mutated tumor suppressor proteins by protecting them from misfolding and degradation.

Our discussion on the impact of carcinogenic metal exposure on HSP90 levels starts with a study that Andrew and collaborators published in the early 2000s, as this is the only study published thus far that involved all three carcinogenic metals (Andrew et al. 2003). In terms of biological relevance, it is noteworthy that this study employed a cell line (BEAS-2B) established from normal human bronchial epithelium, which is the main target of these metals' carcinogenicity, and none of the concentrations tested caused overt cytotoxicity. Aiming at identifying sensitive and specific biomarkers of exposure, Andrew and collaborators used the then relatively recent cDNA microarray technology to assess the impact of these three metals (as well as arsenic) on the expression of 1200 genes. Cells were exposed for 4 h to 3 μM Cd(II), 10 μM Cr(VI) or 3 μg/cm^2 Ni. Each metal modified the expression of a relatively small subset of genes, i.e., the changes observed in gene expression were not the characteristic unspecific response to a highly cytotoxic insult. Specifically, cadmium, hexavalent chromium and nickel altered the expression of, respectively, 25, 44 and 31 genes. Each subset was rather unique, as only 7 of the analyzed genes saw their expression altered by all three metals. Interestingly, HSP90 was one of these 7 genes. The same was not observed for either of the two other HSP assessed in this study, HSP40 and HSP60. For these genes, transcript levels were decreased by Cr(VI) exposure, but remained unaltered upon exposure to Cd(II) and Ni. Whenever affected by these metals, Hsp transcript levels were always decreased.

Decreased Hsp mRNA levels do not, *per se*, support activation of the stress response, as would be expected, considering the capacity of all three metals to produce proteotoxic stress, but this does not necessarily imply otherwise. Indeed, it should be borne in mind that a decrease in Hsp transcript levels does not necessarily translate into a decrease in the corresponding protein levels. There are several reports of decoupling of mRNA and protein steady-state levels (Bauernfeind and Babbitt 2017; Greenbaum et al. 2003), which might have resulted, for instance, from the actions of critical post-transcriptional regulators, such as microRNAs and RNA binding proteins (Glisovic et al. 2008; Janga and Vallabhaneni 2011). Moreover, protein stability and turnover may be affected by post-transcriptional protein modifications (Doherty et al. 2009; Sadoul et al. 2008), further contributing to different profiles of mRNA and protein expression.

Decoupling of mRNA and protein steady state levels was observed in the only study to date that assessed the effects of hexavalent chromium on HSP90 expression at both levels (Abreu et al. 2018). After a 48 h incubation with 1 μM Cr(VI), Hsp90α mRNA levels of BEAS-2B cells, assessed by quantitative reverse-transcription polymerase chain reaction (RT-qPCR), remained the same, whereas Hsp90α protein levels, assessed by enzyme-linked immunosorbent assay (ELISA), were decreased by ca. 60%. Apart from the mechanisms just discussed, different kinetics of the transcriptional and translational programs activated by Cr(VI), as well as different rates of mRNA and protein degradation, might also help explain the observed discrepancy. In fact, previous studies on the kinetics of Hsp expression have shown that Hsp overexpression is a transient event (Diller 2006; Wang et al. 2003) and that, at least under some circumstances, an upregulation observed immediately after shock can be followed by a decline to expression levels lower than the controls during the recovery phase (Liu et al. 2012). In this respect, it is interesting to discuss the results obtained by Rudolf and Cervinka, who investigated the effects, in primary cultures of human skin fibroblasts, of a combined exposure to hexavalent chromium and nickel on, among other parameters, HSP90 expression (Rudolf and Cervinka 2010). First, it was confirmed that cytotoxicity is achieved at much lower concentrations in the case of Cr(VI) (1 μM), than in the case of Ni(II) (250 μM). Secondly, at a non-cytotoxic concentration (25 μM), Ni(II) attenuated significantly the cytotoxicity of Cr(VI). Thirdly, the observed changes in HSP90 expression, which were assessed by immunoblotting, depended on Ni(II) concentration and, for each of the two Ni(II) concentrations tested, were time-dependent. For the non-cytotoxic Ni(II) concentration, a slight induction was observed after 3 h of exposure and this induction became increasingly more pronounced at 6, 12 and 18 h of exposure. For the cytotoxic Ni(II) exposure, a pronounced effect was observed after 3 h of exposure, gradually decreasing afterwards.

In spite of the above considerations concerning decoupling of protein and transcript levels, it might be symptomatic that out of the six studies evaluating the impact of hexavalent chromium on HSP90 protein and/or transcript levels, all employing mammalian cells, four reported a decrease (Abreu et al. 2018; Andrew et al. 2003; Banu et al. 2011; Xiao et al. 2012) and one showed no impact (Izzotti et al. 2002); the only study reporting an increase employed an extremely high concentration of Cr(VI) (600 μM) (Ye and Shi 2001). Altogether, these studies indicate that cells respond to insults that might be viewed as relevant in the context of hexavalent chromium carcinogenicity by decreasing their intracellular HSP90 levels. One hypothesis worth exploring is the promotion of histone deacetylation and/or DNA hypermethylation by hexavalent chromium. Yet, there are other ways through which hexavalent chromium might affect HSP90 activity. Namely, some of the species generated during its intracellular reduction might bind to HSP90, compromising its function. In the case of nickel, binding to HSP90 has been reported in a recent study investigating the inflammatory process caused by nickel eluted from biomedical devices in human monocyte THP-1 cells (nickel is one of the most prevalent contact allergen in the industrialized world (Nielsen et al. 2002)). More specifically, nickel was found to bind to the linker domain on the beta isoform of Hsp90. This binding reduced the interaction of Hsp90β with HIF-1α, and promoted instead the interaction between HIF-1α and HIF-1β, as well as the nuclear localization of HIF-1α (Asakawa et al. 2018).

Nickel has been shown to upregulate HSP90 protein levels in a variety of systems. For instance, using immunoblotting, Hfaiedh and collaborators observed that Gpr94 levels were consistently increased upon exposure of three different mammalian cell lines, A549, COS-7 and HepG2, to 100–400 μM Ni(II) (Hfaiedh et al. 2005). The same study revealed a tissue-specific impact of nickel on Grp94 protein levels: these levels were upregulated in the kidney, yet unaltered in the liver and ovary of Wistar rats. HSP90 induction by nickel in human dermal fibroblasts and keratinocytes was discussed in the previous section (Carroll and Wood 2000). In terms of studies reporting lack of HSP90 induction by nickel, two of them, already mentioned in preceding sections, employed ^{35}S-methionine labeling, a relatively insensitive technique (Bauman et al. 1993; Levinson et al. 1980). Thus, it cannot be excluded that a different outcome might have been obtained using a more sensitive technique. However, in a recent study employing a sensitive technique (two-dimensional polyacrylamide gel electrophoresis (2D-PAGE) in combination with electrospray ionization tandem mass spectrometry (ESI-MS/MS)), exposure of murine bone marrow-derived dendritic cells to Ni(II) did not elicit any changes in HSP90 protein levels at the two concentrations tested, i.e., 250 and 400 μM (the latter described by the authors as "maximal tolerated dose") (Mussotter et al. 2016). Regarding HSP90 mRNA levels, no changes could be observed in cultured human peripheral lung epithelial HPL1D cells when these were exposed for 24 h to non-cytotoxic Ni(II) concentrations (50, 100 and 200 μM), whereas a decrease was observed when the cells were exposed to cytotoxic concentrations (400, 800 and 1600 μM) (Cheng et al. 2003).

As mentioned before, aiming at testing whether Hsp induction was somehow related to binding to sulfhydryl-containing (then unknown) targets, Levinson and co-workers exposed chick embryo cells to cadmium and other soft metals with known sulfhydryl binding capacity (Levinson et al. 1980). Using ^{35}S-methionine labeling, these authors observed that 10 μM Cd(II) strongly induced the synthesis of 100, 75, 35 and 25 kDa Hsp. It is possible, but not certain, that the 75 kDa Hsp band in the autoradiogram included TRAP1. The interpretation of the results of a comprehensive study by Caltabiano and co-workers is also not straightforward in what concerns HSP90 induction by cadmium in human and murine melanoma cell lines (Caltabiano et al. 1986). The authors did report upregulation of HSP90 protein, but neither the magnitude of this upregulation nor the concentration of cadmium that elicited this response are clearly stated, as this aspect was not the major aim of the study. On the contrary, in the study by Bauman and collaborators, it is clearly stated that the two-fold increase in HSP90 protein levels was observed upon exposure of primary cultures of rat hepatocytes to Cd(II) concentration (4–8 μM) that caused some, but not overt cytotoxicity (Bauman et al. 1993).

In spite of these initial results, several studies reported unchanged HSP90 levels upon cadmium exposure. For instance, using RT-PCR and Western immunoblot analysis, Somji and collaborators could not observe any changes in Hsp90, neither at the transcript, nor at the protein levels, when HPT cells (derived from renal cell carcinoma) were exposed acutely (4 h) or chronically (up to 16 days) to Cd(II) (Somji et al. 2002). The concentration used in the acute exposure (53.4 μM) produced some, but not overt, cytotoxicity. Three different concentrations were used for chronic exposure, ranging from a non-cytotoxic concentration (9 μM) to a concentration that produced cell death early in the 16 day time course (45 μM). Whether this lack of response was related to the constitutively high levels of Hsp90 found in these cells remains to be determined.

In another study, Gottschalg and co-workers reported that, under conditions of "phenotypic anchoring" (i.e., in which the cadmium concentration to which cultures of each cell line were exposed were adjusted as to produce similar levels of cytotoxicity), Cd(II) was without effect on HSP90 protein levels in FGC4 cells, HepG2 cells and in rat hepatocytes for concentrations of minimal and mild toxicity (5 and 25%, respectively) (Gottschalg et al. 2006). There is also a report of decreased HSP90 protein levels upon cadmium exposure. This downregulation was observed when cultures of HK-2, a cell line established from normal human kidney, were exposed, for 5 h, to an overtly cytotoxic (100 μM) Cd(II) concentration (Madden et al. 2002). Interestingly, the same insult produced an increase in HSP90 protein levels in cultures of NRK-52, a cell line established from normal rat kidney. Recently, in a study designed to investigate the effects of cadmium on the activation of HSP/HSF1 pathway, Shinkai and collaborators observed that Cd(II), in the low micromolar range, promoted, in bovine aortic endothelial cells, a significant upregulation of Hsp90α and Hsp90β mRNA, but this was not accompanied by a similar increase in Hsp90 protein levels (Shinkai et al. 2017). It was also found that, in these cells, HSF1 interacts with Hsp90, as already observed in other systems (Akerfelt et al. 2010), and that Cd(II) facilitated their dissociation. This disruption might explain, at least in part, the observed upregulation of Hsp90 (as well as other Hsp). Using human recombinant Hsp90β, these authors also showed that, under the conditions used, cadmium modified this isoform at Cys412, located in the M domain, and Cy564, located in the C-terminal domain.

11.7 Targeting HSP90 for Cancer Therapy

Cancer cells are infamous for hijacking physiological processes to sustain uncontrolled growth and metastasis, making them exceptionally resilient. In a vicious cycle, exposure to high levels of stressors selects for higher genetic mutation burdens, which in turn cause even higher levels of cellular stress. Yet, this also results in cancer cells with an excessive reliance in these hijacked processes, a liability that may be exploited in the development of a new generation of cancer therapeutics (Galhardo et al. 2007). Mounting evidence that the stress response is activated in cancer cells has led many investigators to propose several elements of the stress response pathway as new therapeutic targets. As such, much effort has been devoted to the development of Hsp inhibitors (Soo et al. 2008). HSP90 inhibitors, in particular, have been the focus of intense activity in the realm of clinical research, kindled by the prospect of achieving simultaneous inhibition of the chaperoning of several oncogenic proteins crucial for the development and maintenance of cancer hallmarks (Donnelly and Blagg 2008; Hanahan and Weinberg 2011; Soo et al. 2008; Trepel et al. 2010) (Table 11.1). The great majority of HSP90 inhibitors developed so far interact with its N-terminal ATP-binding pocket with greater affinity than ATP. Binding of these compounds thus inhibits the ATPase activity of HSP90, disrupting the chaperone cycle and concomitantly promoting client protein degradation (Trepel et al. 2010; Whitesell and Lindquist 2005). Most tested compounds revealed promising anticancer activities both in vitro and in vivo (Sidera and Patsavoudi 2014). To date, 68 clinical trials investigating HSP90 inhibitors have been registered at clinicaltrials.gov, of which 33 have been completed and 16 are currently ongoing. Yet, no HSP90 inhibitor has been approved for use in the clinic.

Concerns about the safety of this therapeutic strategy have been raised. As HSP90 is ubiquitously expressed and essential for normal cell function, its inhibition may provoke undesirable side effects in normal cells. Interestingly, however, HSP90 has been shown to have ca. 100-fold greater affinity for its inhibitors in cancer cells than in normal cells, leading to an accumulation of the inhibitors within the tumors (Kamal et al. 2003). This differential affinity is likely due to the increased fraction of HSP90 that is involved in multiprotein complexes in cancer cells, possibly as a consequence of the augmented load of misfolded and mutant proteins. In contrast, HSP90 in normal cells is found mostly as free dimers with lower ATPase activity compared with the aforementioned multiprotein complexes (Kamal et al. 2003; Whitesell and Lindquist 2005). These observations indicate the existence of a therapeutic window in which HSP90 inhibitors will efficiently target cancer cells without interfering with neighboring normal cells.

To improve the efficacy of this therapeutic strategy, it will be of utmost importance to choose cancers that are driven by HSP90 client proteins. Indeed, analysis of 15 phase II clinical trials, spanning 10 different types of cancer, using HSP90 inhibitors found several instances of lack of a clinical response due to low expression of HSP90 client proteins in the tumor (Wang et al. 2016). A potential strategy to increase the range of tumors that can be treated with HSP90 inhibitors is to use these molecules synergistically with other therapeutic strategies (Trepel et al. 2010). Currently, there are, for instance, studies investigating the potential of combining HSP90 inhibitors with small molecules targeting co-chaperones essential for HSP90 function, the proteasome machinery or even angiogenesis (Soo et al. 2008; Trepel et al. 2010).

Of note, inhibition of HSP90 by N-terminal domain targeting compounds has been shown to induce the expression of the cytoprotective Hsp70 and Hsp27 proteins (Nahleh et al. 2012). This effect could diminish treatment efficacy, as increased expression of these proteins will protect cancer cells from some of the adverse effects of HSP90 inhibitors. In line with this theory, it was observed that sensitivity of cancer cells to the HSP90 inhibitor geldanamycin can be significantly increased by simultaneous silencing of Hsp70 and/or Hsp27 (Trepel et al. 2010). Mechanistically, HSP90 removes HSF1 trimers from heat shock elements in the genome and sequesters it in unstressed cells, inhibiting HSP70 transcription (Kijima et al. 2018). HSP90 inhibition will thus prolong the duration of gene expression induction by HSF1, in turn promoting the synthesis of Hsp proteins. Moreover, it is known that HSF1 orchestrates a transcriptional program distinct from the heat shock response that promotes carcinogenesis, survival and proliferation of cancer cells (Mendillo et al. 2012), potentially extending the consequences of HSF1 activation well beyond the induction of the stress response.

To overcome the shortcomings inherent to N-terminal domain inhibitors, significant efforts towards the development of HSP90 C-terminal domain inhibitors, which do not seem to have the aforementioned drawbacks, are underway (Ciocca et al. 2013; Donnelly and Blagg 2008; Nahleh et al. 2012). In summary, the development of HSP90 inhibitors has the potential to become a milestone in cancer therapy as a rationally designed approach. Unlike conventional radiotherapy and chemotherapy, which target all dividing cells indiscriminately, HSP90 inhibition should disproportionally affect tumor cells as they acquire an increasingly mutated proteome. As we gain further insights into the molecular mechanisms of cancer and the stress response, HSP90 inhibition is likely to become an important part of our armamentarium against cancer.

11.8 Conclusions

The discovery that, at least under some circumstances, heavy metals induce the expression of Hsp triggered a variety of studies aimed at evaluating whether Hsp levels might be used in molecular toxicology. These studies, which were carried out in a variety of experimental systems, clearly showed that the induction of the different Hsp is dependent on the metal, tissue, cell type and species. Importantly, these studies also unveiled significant differences in induction and recovery kinetics for different stressors and different Hsp. As most of the studies on Hsp induction by metals were not aimed at establishing correlations between Hsp induction and carcinogenesis, the model systems used were not always especially suited for the study of the mechanisms underlying metal-induced carcinogenesis. As a consequence, our understanding of the relationship between HSP90 expression and carcinogenesis is still very incomplete. Nonetheless, these studies provided insight regarding the parameters that might influence HSP90 expression and will definitely prove invaluable for the design of future studies addressing this issue. Future studies should be designed taking into account the relevance of the system in the context of metal-induced carcinogenesis, namely in terms of species, target tissue, metal concentration and kinetics of HSP90 induction and recovery.

Acknowledgements The authors apologize for any studies on the subject that were not mentioned in this review. The authors' original work on the impact of hexavalent chromium on heat shock expression was funded by Centro de Investigação em Meio Ambiente, Genética e Oncobiologia (CIMAGO), Portugal (grant 16/12) and Fundação para a Ciência e a Tecnologia (FCT), Portugal (grants UID/Multi/00070/2019 and SFRH/BPD/101169/2014). The authors were also supported by the following FCT grants: PD/BD/128284/2017 (to PLA), SFRH/BD/91614/2012 (to LMRF), POCI-01-0145-FEDER-029297 (to TCO) and PTDC/MAR-BIO/6149/2014 (to MCA).

References

Abreu PL, Urbano AM (2016) Targeting the Warburg effect for cancer therapy: a long and winding road. In: Rahman A (ed) Frontiers in clinical drug research – anti-cancer agents, vol 3. Bentham Science, Amsterdam, pp 271–324

Abreu PL, Ferreira LM, Alpoim MC, Urbano AM (2014) Impact of hexavalent chromium on mammalian cell bioenergetics: phenotypic changes, molecular basis and potential relevance to chromate-induced lung cancer. Biometals 27:409–443

Abreu PL, Cunha-Oliveira T, Ferreira LMR, Urbano AM (2018) Hexavalent chromium, a lung carcinogen, confers resistance to thermal stress and interferes with heat shock protein expression in human bronchial epithelial cells. Biometals 31:477–487

Acharya A, Das I, Chandhok D, Saha T (2010) Redox regulation in cancer: a double-edged sword with therapeutic potential. Oxidative Med Cell Longev 3:23–34

Ahn SG, Thiele DJ (2003) Redox regulation of mammalian heat shock factor 1 is essential for Hsp gene activation and protection from stress. Genes Dev 17:516–528

Ahsan A, Ramanand SG, Whitehead C et al (2012) Wild-type EGFR is stabilized by direct interaction with HSP90 in cancer cells and tumors. Neoplasia 14:670–677

Akerfelt M, Morimoto RI, Sistonen L (2010) Heat shock factors: integrators of cell stress, development and lifespan. Nat Rev Mol Cell Biol 11:545–555

Andrew AS, Warren AJ, Barchowsky A et al (2003) Genomic and proteomic profiling of responses to toxic metals in human lung cells. Environ Health Perspect 111:825–835

Asakawa S, Onodera R, Kasai K et al (2018) Nickel ions bind to HSP90beta and enhance HIF-1alpha-mediated IL-8 expression. Toxicology 395:45–53

Asuthkar S, Gondi CS, Nalla AK, Velpula KK, Gorantla B, Rao JS (2012) Urokinase-type plasminogen activator receptor (uPAR)-mediated regulation of WNT/beta-catenin signaling is enhanced in irradiated medulloblastoma cells. J Biol Chem 287:20576–20589

Bae D, Camilli TC, Ha NT, Ceryak S (2009) Enhanced clonogenic survival induced by protein tyrosine phosphatase (PTP) inhibition after Cr(VI) exposure is mediated by c-Raf and Ras activity. Cell Signal 21:727–736

Bajzikova M, Kovarova J, Coelho AR et al (2019) Reactivation of dihydroorotate dehydrogenase-driven pyrimidine biosynthesis restores tumor growth of respiration-deficient cancer cells. Cell Metab 29:399–416 e310

Banu SK, Stanley JA, Lee J et al (2011) Hexavalent chromium-induced apoptosis of granulosa cells involves selective sub-cellular translocation of Bcl-2 members, ERK1/2 and p53. Toxicol Appl Pharmacol 251:253–266

Barrott JJ, Haystead TA (2013) Hsp90, an unlikely ally in the war on cancer. FEBS J 280:1381–1396

Basso AD, Solit DB, Chiosis G et al (2002) Akt forms an intracellular complex with heat shock protein 90 (Hsp90) and Cdc37 and is destabilized by inhibitors of Hsp90 function. J Biol Chem 277:39858–39866

Bauernfeind AL, Babbitt CC (2017) The predictive nature of transcript expression levels on protein expression in adult human brain. BMC Genomics 18:322

Bauman JW, Liu J, Klaassen CD (1993) Production of metallothionein and heat-shock proteins in response to metals. Fundam Appl Toxicol 21:15–22

Beck S, Jin X, Sohn YW et al (2011) Telomerase activity-independent function of TERT allows glioma cells to attain cancer stem cell characteristics by inducing EGFR expression. Mol Cells 31:9–15

Bertout JA, Patel SA, Simon MC (2008) The impact of O2 availability on human cancer. Nat Rev Cancer 8:967–975

Bishak YK, Payahoo L, Osatdrahimi A, Nourazarian A (2015) Mechanisms of cadmium carcinogenicity in the gastrointestinal tract. Asian Pac J Cancer Prev 16:9–21

Blagosklonny MV, Toretsky J, Neckers L (1995) Geldanamycin selectively destabilizes and conformationally alters mutated p53. Oncogene 11:933–939

Blank M, Mandel M, Keisari Y, Meruelo D, Lavie G (2003) Enhanced ubiquitinylation of heat shock protein 90 as a potential mechanism for mitotic cell death in cancer cells induced with hypericin. Cancer Res 63:8241–8247

Bork U, Lee WK, Kuchler A, Dittmar T, Thevenod F (2010) Cadmium-induced DNA damage triggers G(2)/M arrest via chk1/2 and cdc2 in p53-deficient kidney proximal tubule cells. Am J Physiol Renal Physiol 298:F255–F265

Breinig M, Mayer P, Harjung A et al (2011) Heat shock protein 90-sheltered overexpression of insulin-like growth factor 1 receptor contributes to malignancy of thymic epithelial tumors. Clin Cancer Res 17:2237–2249
Bruns AF, Yuldasheva N, Latham AM et al (2012) A heat-shock protein axis regulates VEGFR2 proteolysis, blood vessel development and repair. PLoS One 7:e48539
Burrows F, Zhang H, Kamal A (2004) Hsp90 activation and cell cycle regulation. Cell Cycle 3:1530–1536
Cabail MZ, Chen EI, Koller A, Miller WT (2016) Auto-thiophosphorylation activity of Src tyrosine kinase. BMC Biochem 17:13
Calderwood SK, Khaleque MA, Sawyer DB, Ciocca DR (2006) Heat shock proteins in cancer: chaperones of tumorigenesis. Trends Biochem Sci 31:164–172
Caltabiano MM, Koestler TP, Poste G, Greig RG (1986) Induction of 32- and 34-kDa stress proteins by sodium arsenite, heavy metals, and thiol-reactive agents. J Biol Chem 261:13381–13386
Carroll S, Wood EJ (2000) Exposure of human keratinocytes and fibroblasts in vitro to nickel sulphate ions induces synthesis of stress proteins Hsp72 and Hsp90. Acta Derm Venereol 80:94–97
Carlisle DL, Pritchard DE, Singh J, Patierno SR (2000) Chromium(VI) induces p53-dependent apoptosis in diploid human lung and mouse dermal fibroblasts. Mol Carcinog 28:111–118
Chae YC, Angelin A, Lisanti S et al (2013) Landscape of the mitochondrial Hsp90 metabolome in tumours. Nat Commun 4:2139
Chen F, Bower J, Leonard SS et al (2002) Protective roles of NF-kappa B for chromium(VI)-induced cytotoxicity is revealed by expression of Ikappa B kinasebeta mutant. J Biol Chem 277:3342–3349
Chen H, Ke Q, Kluz T, Yan Y, Costa M (2006) Nickel ions increase histone H3 lysine 9 dimethylation and induce transgene silencing. Mol Cell Biol 26:3728–3737
Cheng RY, Zhao A, Alvord WG et al (2003) Gene expression dose-response changes in microarrays after exposure of human peripheral lung epithelial cells to nickel(II). Toxicol Appl Pharmacol 191:22–39
Chiou YH, Liou SH, Wong RH, Chen CY, Lee H (2015) Nickel may contribute to EGFR mutation and synergistically promotes tumor invasion in EGFRmutated lung cancer via nickel-induced microRNA-21 expression. Toxicol Lett 237:46–54
Choong G, Liu Y, Templeton DM (2013) Cadmium affects focal adhesion kinase (FAK) in mesangial cells: involvement of CaMK-II and the actin cytoskeleton. J Cell Biochem 114:1832–1842
Chuang SM, Wang IC, Yang JL (2000) Roles of JNK, p38 and ERK mitogen-activated protein kinases in the growth inhibition and apoptosis induced by cadmium. Carcinogenesis 21:1423–1432
Chun G, Bae D, Nickens K, O'Brien TJ, Patierno SR, Ceryak S (2010) Polo-like kinase 1 enhances survival and mutagenesis after genotoxic stress in normal cells through cell cycle checkpoint bypass. Carcinogenesis 31:785–793
Ciocca DR, Arrigo AP, Calderwood SK (2013) Heat shock proteins and heat shock factor 1 in carcinogenesis and tumor development: an update. Arch Toxicol 87:19–48
Citri A, Harari D, Shohat G et al (2006) Hsp90 recognizes a common surface on client kinases. J Biol Chem 281:14361–14369
Coleman JE (1992) Zinc proteins: enzymes, storage proteins, transcription factors, and replication proteins. Annu Rev Biochem 61:897–946
Correia AL, Mori H, Chen EI, Schmitt FC, Bissell MJ (2013) The hemopexin domain of MMP3 is responsible for mammary epithelial invasion and morphogenesis through extracellular interaction with HSP90beta. Genes Dev 27:805–817
Csermely P, Schnaider T, Soti C, Prohaszka Z, Nardai G (1998) The 90-kDa molecular chaperone family: structure, function, and clinical applications. A comprehensive review. Pharmacol Ther 79:129–168
Dai C, Whitesell L, Rogers AB, Lindquist S (2007) Heat shock factor 1 is a powerful multifaceted modifier of carcinogenesis. Cell 130:1005–1018

Dai W, Chen H, Yu R, He L, Chen B, Chen X (2010) Effects of cadmium on telomerase activity, expressions of TERT, c-myc and P53, and apoptosis of rat hepatocytes. J Huazhong Univ Sci Technolog Med Sci 30:709–713

de Billy E, Travers J, Workman P (2012) Shock about heat shock in cancer. Oncotarget 3:741–743

Delmas F, Schaak S, Gaubin Y et al (1998) Hsp72 mRNA production in cultured human cells submitted to nonlethal aggression by heat, ethanol, or propanol. Application to the detection of low concentrations of chromium(VI) (potassium dichromate). Cell Biol Toxicol 14:39–46

Dias S, Shmelkov SV, Lam G, Rafii S (2002) VEGF(165) promotes survival of leukemic cells by Hsp90-mediated induction of Bcl-2 expression and apoptosis inhibition. Blood 99:2532–2540

Diller KR (2006) Stress protein expression kinetics. Annu Rev Biomed Eng 8:403–424

Ding J, He G, Gong W et al (2009) Effects of nickel on cyclin expression, cell cycle progression and cell proliferation in human pulmonary cells. Cancer Epidemiol Biomark Prev 18:1720–1729

Doherty MK, Hammond DE, Clague MJ, Gaskell SJ, Beynon RJ (2009) Turnover of the human proteome: determination of protein intracellular stability by dynamic SILAC. J Proteome Res 8:104–112

Donnelly A, Blagg BS (2008) Novobiocin and additional inhibitors of the Hsp90 C-terminal nucleotide-binding pocket. Curr Med Chem 15:2702–2717

Dorta DJ, Leite S, DeMarco KC et al (2003) A proposed sequence of events for cadmium-induced mitochondrial impairment. J Inorg Biochem 97:251–257

Dote H, Burgan WE, Camphausen K, Tofilon PJ (2006) Inhibition of hsp90 compromises the DNA damage response to radiation. Cancer Res 66:9211–9220

Dutta R, Inouye M (2000) GHKL, an emergent ATPase/kinase superfamily. Trends Biochem Sci 25:24–28

Eustace BK, Sakurai T, Stewart JK et al (2004) Functional proteomic screens reveal an essential extracellular role for hsp90 alpha in cancer cell invasiveness. Nat Cell Biol 6:507–514

Felts SJ, Owen BA, Nguyen P et al (2000) The hsp90-related protein TRAP1 is a mitochondrial protein with distinct functional properties. J Biol Chem 275:3305–3312

Feng H, Liu J, Hu G, Jia G (2018) The role of epigenetics in the toxic effects induced by hexavalent chromium. React Oxygen Spec 5:107–117

Fernandes MA, Santos MS, Alpoim MC, Madeira VM, Vicente JA (2002) Chromium(VI) interaction with plant and animal mitochondrial bioenergetics: a comparative study. J Biochem Mol Toxicol 16:53–63

Fernandez EL, Gustafson AL, Andersson M, Hellman B, Dencker L (2003) Cadmium-induced changes in apoptotic gene expression levels and DNA damage in mouse embryos are blocked by zinc. Toxicol Sci 76:162–170

Ferreira LM (2010) Cancer metabolism: the Warburg effect today. Exp Mol Pathol 89:372–380

Ferreira LM, Hebrant A, Dumont JE (2012) Metabolic reprogramming of the tumor. Oncogene 31:3999–4011

Floris G, Sciot R, Wozniak A et al (2011) The novel HSP90 inhibitor, IPI-493, is highly effective in human gastrostrointestinal stromal tumor xenografts carrying heterogeneous KIT mutations. Clin Cancer Res 17:5604–5614

Fortugno P, Beltrami E, Plescia J et al (2003) Regulation of survivin function by Hsp90. Proc Natl Acad Sci U S A 100:13791–13796

Fu J, Koul D, Yao J et al (2013) Novel HSP90 inhibitor NVP-HSP990 targets cell-cycle regulators to ablate Olig2-positive glioma tumor-initiating cells. Cancer Res 73:3062–3074

Fujiki K, Inamura H, Miyayama T, Matsuoka M (2017) Involvement of Notch1 signaling in malignant progression of A549 cells subjected to prolonged cadmium exposure. J Biol Chem 292:7942–7953

Galhardo RS, Hastings PJ, Rosenberg SM (2007) Mutation as a stress response and the regulation of evolvability. Crit Rev Biochem Mol Biol 42:399–435

Ganguly A, Guo L, Sun L, Suo F, Du LL, Russell P (2018) Tdp1 processes chromate-induced single-strand DNA breaks that collapse replication forks. PLoS Genet 14:e1007595

Gavin IM, Gillis B, Arbieva Z, Prabhakar BS (2007) Identification of human cell responses to hexavalent chromium. Environ Mol Mutagen 48:650–657

Giulino-Roth L, van Besien HJ, Dalton T et al (2017) Inhibition of Hsp90 suppresses PI3K/AKT/mTOR signaling and has antitumor activity in Burkitt lymphoma. Mol Cancer Ther 16:1779–1790

Glisovic T, Bachorik JL, Yong J, Dreyfuss G (2008) RNA-binding proteins and post-transcriptional gene regulation. FEBS Lett 582:1977–1986

Goering PL, Fisher BR (1995) Metals and stress proteins. In: Goyer R, Cherian MG (eds) Toxicology of metals, vol 115. Springer, Berlin, pp 229–266

Gottschalg E, Moore NE, Ryan AK et al (2006) Phenotypic anchoring of arsenic and cadmium toxicity in three hepatic-related cell systems reveals compound- and cell-specific selective up-regulation of stress protein expression: implications for fingerprint profiling of cytotoxicity. Chem Biol Interact 161:251–261

Greenbaum D, Colangelo C, Williams K, Gerstein M (2003) Comparing protein abundance and mRNA expression levels on a genomic scale. Genome Biol 4:117

Guo H, Cui H, Fang J et al (2016) Nickel chloride-induced apoptosis via mitochondria- and Fas-mediated caspase-dependent pathways in broiler chickens. Oncotarget 7:79747–79760

Gupta S, Ahmad N, Husain MM, Srivastava RC (2000) Involvement of nitric oxide in nickel-induced hyperglycemia in rats. Nitric Oxide 4:129–138

Han SG, Castranova V, Vallyathan V (2007) Comparative cytotoxicity of cadmium and mercury in a human bronchial epithelial cell line (BEAS-2B) and its role in oxidative stress and induction of heat shock protein 70. J Toxicol Environ Health 70:852–860

Hanahan D, Weinberg RA (2011) Hallmarks of cancer: the next generation. Cell 144:646–674

Hartson SD, Matts RL (2012) Approaches for defining the Hsp90-dependent proteome. Biochim Biophys Acta 1823:656–667

Hartwig A (2013) Cadmium and cancer. Met Ions Life Sci 11:491–507

Harris MB, Mitchell BM, Sood SG, Webb RC, Venema RC (2008) Increased nitric oxide synthase activity and Hsp90 association in skeletal muscle following chronic exercise. Eur J Appl Physiol 104:795–802

He J, Qian X, Carpenter R et al (2013) Repression of miR-143 mediates Cr (VI)-induced tumor angiogenesis via IGF-IR/IRS1/ERK/IL-8 pathway. Toxicol Sci 134:26–38

Hentze N, Le Breton L, Wiesner J, Kempf G, Mayer MP (2016) Molecular mechanism of thermosensory function of human heat shock transcription factor Hsf1. Elife:5

Herak-Kramberger CM, Brown D, Sabolic I (1998) Cadmium inhibits vacuolar H(+)-ATPase and endocytosis in rat kidney cortex. Kidney Int 53:1713–1726

Herr I, Debatin KM (2001) Cellular stress response and apoptosis in cancer therapy. Blood 98:2603–2614

Hfaiedh N, Allagui MS, El Feki A et al (2005) Effects of nickel poisoning on expression pattern of the 72/73 and 94 kDa stress proteins in rat organs and in the COS-7, HepG2, and A549 cell lines. J Biochem Mol Toxicol 19:12–18

Hill R, Madureira PA, Waisman DM, Lee PW (2011) DNA-PKCS binding to p53 on the p21WAF1/CIP1 promoter blocks transcription resulting in cell death. Oncotarget 2:1094–1108

Hill R, Rabb M, Madureira PA et al (2013) Gemcitabine-mediated tumour regression and p53-dependent gene expression: implications for colon and pancreatic cancer therapy. Cell Death Dis 4:e791

Hjelmeland AB, Wu Q, Heddleston JM et al (2011) Acidic stress promotes a glioma stem cell phenotype. Cell Death Differ 18:829–840

Holmes AL, Wise SS, Wise JP Sr (2008) Carcinogenicity of hexavalent chromium. Indian J Med Res 128:353–372

Holt SE, Aisner DL, Baur J et al (1999) Functional requirement of p23 and Hsp90 in telomerase complexes. Genes Dev 13:817–826

Huang YY, Xia MZ, Wang H et al (2014) Cadmium selectively induces MIP-2 and COX-2 through PTEN-mediated Akt activation in RAW264.7 cells. Toxicol Sci 138:310–321

Hurst DR, Mehta A, Moore BP et al (2006) Breast cancer metastasis suppressor 1 (BRMS1) is stabilized by the Hsp90 chaperone. Biochem Biophys Res Commun 348:1429–1435
IARC (1990) Chromium, nickel and welding. IARC Monogr Eval Carcinog Risks Hum 49:1–648
IARC (1993) Beryllium, cadmium, mercury, and exposures in the glass manufacturing industry. IARC Monogr Eval Carcinog Risks Hum 58:1–415
IARC (2012) Arsenic, metals, fibres and dusts. IARC Monogr Eval Carcinog Risks Hum 100 C:1–465
Isaacs JS, Jung YJ, Mimnaugh EG et al (2002) Hsp90 regulates a von Hippel Lindau-independent hypoxia-inducible factor-1 alpha-degradative pathway. J Biol Chem 277:29936–29944
Izzotti A, Cartiglia C, Balansky R et al (2002) Selective induction of gene expression in rat lung by hexavalent chromium. Mol Carcinog 35:75–84
Janga SC, Vallabhaneni S (2011) MicroRNAs as post-transcriptional machines and their interplay with cellular networks. Adv Exp Med Biol 722:59–74
Jarup L (2003) Hazards of heavy metal contamination. Br Med Bull 68:167–182
Jing Y, Liu LZ, Jiang Y et al (2012) Cadmium increases HIF-1 and VEGF expression through ROS, ERK, and AKT signaling pathways and induces malignant transformation of human bronchial epithelial cells. Toxicol Sci 125:10–19
Jolly C, Morimoto RI (2000) Role of the heat shock response and molecular chaperones in oncogenesis and cell death. J Natl Cancer Inst 92:1564–1572
Ju H, Arumugam P, Lee J, Song JM (2017) Impact of environmental pollutant cadmium on the establishment of a Cancer stem cell population in breast and hepatic Cancer. ACS Omega 2:563–572
Kamal A, Thao L, Sensintaffar J et al (2003) A high-affinity conformation of Hsp90 confers tumour selectivity on Hsp90 inhibitors. Nature 425:407–410
Kang BH, Plescia J, Dohi T et al (2007) Regulation of tumor cell mitochondrial homeostasis by an organelle-specific Hsp90 chaperone network. Cell 131:257–270
Karthikeyan J, Bavani G (2009) Effect of cadmium on lactate dehyrogenase isoenzyme, succinate dehydrogenase and Na(+)-K(+)-ATPase in liver tissue of rat. J Environ Biol 30:895–898
Katschinski DM (2004) On heat and cells and proteins. News Physiol Sci 19:11–15
Kijima T, Prince TL, Tigue ML et al (2018) HSP90 inhibitors disrupt a transient HSP90-HSF1 interaction and identify a noncanonical model of HSP90-mediated HSF1 regulation. Sci Rep 8:6976
Kil IS, Shin SW, Yeo HS, Lee YS, Park JW (2006) Mitochondrial NADP+-dependent isocitrate dehydrogenase protects cadmium-induced apoptosis. Mol Pharmacol 70:1053–1061
Kim D, Dai J, Fai LY et al (2015) Constitutive activation of epidermal growth factor receptor promotes tumorigenesis of Cr(VI)-transformed cells through decreased reactive oxygen species and apoptosis resistance development. J Biol Chem 290:2213–2224
Kim D, Dai J, Park YH et al (2016) Activation of epidermal growth factor receptor/p38/hypoxia-inducible factor-1alpha is pivotal for angiogenesis and tumorigenesis of malignantly transformed cells induced by hexavalent chromium. J Biol Chem 291:16271–16281
Kim H, Lin Q, Glazer PM, Yun Z (2018) The hypoxic tumor microenvironment in vivo selects the cancer stem cell fate of breast cancer cells. Breast Cancer Res 20:16
Kim J, Lim W, Ko Y et al (2012) The effects of cadmium on VEGF-mediated angiogenesis in HUVECs. J Appl Toxicol 32:342–349
Kim SY, Sohn SJ, Won AJ, Kim HS, Moon A (2014) Identification of noninvasive biomarkers for nephrotoxicity using HK-2 human kidney epithelial cells. Toxicol Sci 140:247–258
Kuan YC, Hashidume T, Shibata T et al (2017) Heat shock protein 90 modulates lipid homeostasis by regulating the stability and function of sterol regulatory element-binding protein (SREBP) and SREBP cleavage-activating protein. J Biol Chem 292:3016–3028
Lapidot T, Sirard C, Vormoor J et al (1994) A cell initiating human acute myeloid leukaemia after transplantation into SCID mice. Nature 367:645–648
Leav I, Plescia J, Goel HL et al (2010) Cytoprotective mitochondrial chaperone TRAP-1 as a novel molecular target in localized and metastatic prostate cancer. Am J Pathol 176:393–401

Lee J, Zhang LL, Wu W et al (2018) Activation of MYC, a bona fide client of HSP90, contributes to intrinsic ibrutinib resistance in mantle cell lymphoma. Blood Adv 2:2039–2051

Lee KH, Jang Y, Chung JH (2010) Heat shock protein 90 regulates IkappaB kinase complex and NF-kappaB activation in angiotensin II-induced cardiac cell hypertrophy. Exp Mol Med 42:703–711

Lee SH, Choi JG, Cho MH (2001) Apoptosis, bcl2 expression, and cell cycle analyses in nickel(II)-treated normal rat kidney cells. J Korean Med Sci 16:165–168

Lei H, Romeo G, Kazlauskas A (2004) Heat shock protein 90alpha-dependent translocation of annexin II to the surface of endothelial cells modulates plasmin activity in the diabetic rat aorta. Circ Res 94:902–909

Lei YX, Chen JK, Wu ZL (2001) Detection of DNA strand breaks, DNA-protein crosslinks, and telomerase activity in nickel-transformed BALB/c-3T3 cells. Teratog Carcinog Mutagen 21:463–471

Levinson W, Oppermann H, Jackson J (1980) Transition series metals and sulfhydryl reagents induce the synthesis of four proteins in eukaryotic cells. Biochim Biophys Acta 606:170–180

Lewis J, Devin A, Miller A et al (2000) Disruption of hsp90 function results in degradation of the death domain kinase, receptor-interacting protein (RIP), and blockage of tumor necrosis factor-induced nuclear factor-kappaB activation. J Biol Chem 275:10519–10526

Li C, Park S, Zhang X, Dai W, Xu D (2017) Mutual regulation between polo-like kinase 3 and SIAH2 E3 ubiquitin ligase defines a regulatory network that fine-tunes the cellular response to hypoxia and nickel. J Biol Chem 292:11431–11444

Li M, Kondo T, Zhao QL et al (2000) Apoptosis induced by cadmium in human lymphoma U937 cells through Ca2+–calpain and caspase-mitochondriadependent pathways. J Biol Chem 275:39702–39709

Li Q, Kluz T, Sun H, Costa M (2009) Mechanisms of c-myc degradation by nickel compounds and hypoxia. PLoS One 4:e8531

Li W, Gu X, Zhang X et al (2015) Cadmium delays non-homologous end joining (NHEJ) repair via inhibition of DNA-PKcs phosphorylation and downregulation of XRCC4 and ligase IV. Mutat Res 779:112–123

Li W, Sahu D, Tsen F (2012) Secreted heat shock protein-90 (Hsp90) in wound healing and cancer. Biochim Biophys Acta 1823:730–741

Lian S, Xia Y, Khoi PN et al (2015) Cadmium induces matrix metalloproteinase-9 expression via ROS-dependent EGFR, NF-small ka, CyrillicB, and AP-1 pathways in human endothelial cells. Toxicology 338:104–116

Lindquist S (1986) The heat-shock response. Annu Rev Biochem 55:1151–1191

Liu J, Qu W, Kadiiska MB (2009) Role of oxidative stress in cadmium toxicity and carcinogenesis. Toxicol Appl Pharmacol 238:209–214

Liu GT, Wang JF, Cramer G et al (2012) Transcriptomic analysis of grape (*Vitis vinifera* L.) leaves during and after recovery from heat stress. BMC Plant Biol 12:174

Liu M, Chen S, Yueh MF et al (2016) Cadmium and arsenic override NF-kappaB developmental regulation of the intestinal UGT1A1 gene and control of hyperbilirubinemia. Biochem Pharmacol 110-111:37–46

Lokeshwar VB (2012) Wee1-Hsp90 inhibitor combination treatment: molecular therapy with potentially broad applicability. Cell Cycle 11:3722–3723

Lu J, Zhou Z, Tang M et al (2018) Role of LKB1 in migration and invasion of Cr(VI)-transformed human bronchial epithelial Beas-2B cells. Anti-Cancer Drugs 29:660–673

Lubovac-Pilav Z, Borras DM, Ponce E, Louie MC (2013) Using expression profiling to understand the effects of chronic cadmium exposure on MCF-7 breast cancer cells. PLoS One 8:e84646

Madden EF, Akkerman M, Fowler BA (2002) A comparison of 60, 70, and 90 kDa stress protein expression in normal rat NRK-52 and human HK-2 kidney cell lines following in vitro exposure to arsenite and cadmium alone or in combination. J Biochem Mol Toxicol 16:24–32

Madureira PA, Hill R, Lee PW, Waisman DM (2012) Genotoxic agents promote the nuclear accumulation of annexin A2: role of annexin A2 in mitigating DNA damage. PLoS One 7:e50591

Maehle L, Metcalf RA, Ryberg D, Bennett WP, Harris CC, Haugen A (1992) Altered p53 gene structure and expression in human epithelial cells after exposure to nickel. Cancer Res 52:218–221

Mahony D, Parry DA, Lees E (1998) Active cdk6 complexes are predominantly nuclear and represent only a minority of the cdk6 in T cells. Oncogene 16:603–611

Majumder S, Muley A, Kolluru GK et al (2008) Cadmium reduces nitric oxide production by impairing phosphorylation of endothelial nitric oxide synthase. Biochem Cell Biol 86:1–10

Makhnevych T, Houry WA (2012) The role of Hsp90 in protein complex assembly. Biochim Biophys Acta 1823:674–682

Mani SA, Guo W, Liao MJ et al (2008) The epithelial-mesenchymal transition generates cells with properties of stem cells. Cell 133:704–715

Martelli A, Rousselet E, Dycke C, Bouron A, Moulis JM (2006) Cadmium toxicity in animal cells by interference with essential metals. Biochimie 88:1807–1814

Martinez Flores K, Uribe Marin BC, Souza Arroyo V et al (2013) Hepatocytes display a compensatory survival response against cadmium toxicity by a mechanism mediated by EGFR and Src. Toxicol In Vitro 27:1031–1042

Medan D, Luanpitpong S, Azad N et al (2012) Multifunctional role of Bcl-2 in malignant transformation and tumorigenesis of Cr(VI)-transformed lung cells. PLoS One 7:e37045

Mendillo ML, Santagata S, Koeva M et al (2012) HSF1 drives a transcriptional program distinct from heat shock to support highly malignant human cancers. Cell 150:549–562

Meyer JN, Leung MC, Rooney JP et al (2013) Mitochondria as a target of environmental toxicants. Toxicol Sci 134:1–17

Minet E, Mottet D, Michel G et al (1999) Hypoxia-induced activation of HIF-1: role of HIF-1alpha-Hsp90 interaction. FEBS Lett 460:251–256

Morimoto RI (1993) Cells in stress: transcriptional activation of heat shock genes. Science 259:1409–1410

Mussotter F, Tomm JM, El Ali Z et al (2016) Proteomics analysis of dendritic cell activation by contact allergens reveals possible biomarkers regulated by Nrf2. Toxicol Appl Pharmacol 313:170–179

Nahleh Z, Tfayli A, Najm A, El Sayed A, Nahle Z (2012) Heat shock proteins in cancer: targeting the ‘chaperones’. Future Med Chem 4:927–935

Neri D, Supuran CT (2011) Interfering with pH regulation in tumours as a therapeutic strategy. Nat Rev Drug Discov 10:767–777

Nielsen NH, Linneberg A, Menne T et al (2002) Incidence of allergic contact sensitization in Danish adults between 1990 and 1998; the Copenhagen allergy study, Denmark. Br J Dermatol 147:487–492

Nieto-Miguel T, Gajate C, Gonzalez-Camacho F, Mollinedo F (2008) Proapoptotic role of Hsp90 by its interaction with c-Jun N-terminal kinase in lipid rafts in edelfosine-mediated antileukemic therapy. Oncogene 27:1779–1787

Noguchi M, Yu D, Hirayama R et al (2006) Inhibition of homologous recombination repair in irradiated tumor cells pretreated with Hsp90 inhibitor 17-allylamino-17-demethoxygeldanamycin. Biochem Biophys Res Commun 351:658–663

Noel L, Guerin T, Kolf-Clauw M (2004) Subchronic dietary exposure of rats to cadmium alters the metabolism of metals essential to bone health. Food Chem Toxicol 42:1203–1210

NTP (2014) Report on carcinogens, Thirteenth Edition. In U.S. Department of Health and Human Services, PHSs (ed)

O’Hara KA, Klei LR, Barchowsky A (2003) Selective activation of Src family kinases and JNK by low levels of chromium(VI). Toxicol Appl Pharmacol 190:214–223

O’Leary MH, Limburg JA (1977) Isotope effect studies of the role of metal ions in isocitrate dehydrogenase. Biochemistry 16:1129–1135

Oda T, Hayano T, Miyaso H, Takahashi N, Yamashita T (2007) Hsp90 regulates the Fanconi anemia DNA damage response pathway. Blood 109:5016–5026

Oesterreich S, Weng CN, Qiu M et al (1993) The small heat shock protein hsp27 is correlated with growth and drug resistance in human breast cancer cell lines. Cancer Res 53:4443–4448

Ota H, Shionome T, Suguro H et al (2018) Nickel chloride administration prevents the growth of oral squamous cell carcinoma. Oncotarget 9:24109–24121

Pan JJ, Chang QS, Wang X et al (2011) Activation of Akt/GSK3beta and Akt/Bcl-2 signaling pathways in nickel-transformed BEAS-2B cells. Int J Oncol 39:1285–1294

Pearl LH, Prodromou C (2006) Structure and mechanism of the Hsp90 molecular chaperone machinery. Annu Rev Biochem 75:271–294

Pearson RG (1968) Hard and soft acids and bases, HSAB, part 1: fundamental principles. J Chem Educ 45:581

Pflaum J, Schlosser S, Muller M (2014) p53 family and cellular stress responses in cancer. Front Oncol 4:285

Piotrowicz RS, Hickey E, Levin EG (1998) Heat shock protein 27 kDa expression and phosphorylation regulates endothelial cell migration. FASEB J 12:1481–1490

Pirkkala L, Nykanen P, Sistonen L (2001) Roles of the heat shock transcription factors in regulation of the heat shock response and beyond. FASEB J 15:1118–1131

Pratheeshkumar P, Son YO, Divya SP et al (2016) Hexavalent chromium induces malignant transformation of human lung bronchial epithelial cells via ROSdependent activation of miR-21-PDCD4 signaling. Oncotarget 7:51193–51210

Presley T, Vedam K, Druhan LJ, Ilangovan G (2010) Hyperthermia-induced Hsp90.eNOS preserves mitochondrial respiration in hyperglycemic endothelial cells by down-regulating Glut-1 and up-regulating G6PD activity. J Biol Chem 285:38194–38203

Quanz M, Herbette A, Sayarath M et al (2012) Heat shock protein 90alpha (Hsp90alpha) is phosphorylated in response to DNA damage and accumulates in repair foci. J Biol Chem 287:8803–8815

Repetto G, del Peso A, Sanz P, Repetto M (2001) In vitro effects of lithium and nickel at different levels on neuro-2a mouse neuroblastoma cells. Toxicol In Vitro 15:363–368

Reya T, Morrison SJ, Clarke MF, Weissman IL (2001) Stem cells, cancer, and cancer stem cells. Nature 414:105–111

Reynolds M, Zhitkovich A (2007) Cellular vitamin C increases chromate toxicity via a death program requiring mismatch repair but not p53. Carcinogenesis 28:1613–1620

Ritossa F (1962) A new puffing pattern induced by temperature shock and DNP in Drosophila. Exp Dermatol 18:571–573

Rudolf E, Cervinka M (2010) Nickel modifies the cytotoxicity of hexavalent chromium in human dermal fibroblasts. Toxicol Lett 197:143–150

Sadoul K, Boyault C, Pabion M, Khochbin S (2008) Regulation of protein turnover by acetyltransferases and deacetylases. Biochimie 90:306–312

Salnikow K, Costa M (2000) Epigenetic mechanisms of nickel carcinogenesis. J Environ Pathol Toxicol Oncol 19:307–318

Salnikow K, Davidson T, Costa M (2002) The role of hypoxia-inducible signaling pathway in nickel carcinogenesis. Environ Health Perspect 110(Suppl 5):831–834

Saretzki G, Armstrong L, Leake A, Lako M, von Zglinicki T (2004) Stress defense in murine embryonic stem cells is superior to that of various differentiated murine cells. Stem Cells 22:962–971

Sato S, Fujita N, Tsuruo T (2000) Modulation of Akt kinase activity by binding to Hsp90. Proc Natl Acad Sci U S A 97:10832–10837

Schar P (2001) Spontaneous DNA damage, genome instability, and cancer – when DNA replication escapes control. Cell 104:329–332

Schlesinger MJ (1990) Heat shock proteins. J Biol Chem 265:12111–12114

Schlesinger MJ, Ashburner M, Tissières A (eds) (1982) Heat shock, from bacteria to man. Cold Spring Harbor Laboratory, Cold Spring Harbor

Schulte TW, Blagosklonny MV, Ingui C, Neckers L (1995) Disruption of the Raf-1-Hsp90 molecular complex results in destabilization of Raf-1 and loss of Raf-1-Ras association. J Biol Chem 270:24585–24588
Shi L, Zhang Z, Fang S et al (2009) Heat shock protein 90 (Hsp90) regulates the stability of transforming growth factor beta-activated kinase 1 (TAK1) in interleukin-1beta-induced cell signaling. Mol Immunol 46:541–550
Shil K, Pal S (2018) Metabolic adaptability in hexavalent chromium-treated renal tissue: an in vivo study. Clin Kidney J 11:222–229
Shimizu M, Hochadel JF, Waalkes MP (1997) Effects of glutathione depletion on cadmium-induced metallothionein synthesis, cytotoxicity, and proto-oncogene expression in cultured rat myoblasts. J Toxicol Environ Health 51:609–621
Shinkai Y, Masuda A, Akiyama M, Xian M, Kumagai Y (2017) Cadmium-mediated activation of the HSP90/HSF1 pathway regulated by reactive persulfides/polysulfides. Toxicol Sci 156:412–421
Sidera K, Patsavoudi E (2014) HSP90 inhibitors: current development and potential in cancer therapy. Recent Pat Anticancer Drug Discov 9:1–20
Somji S, Ann Sens M, Garrett SH et al (2002) Expression of hsp 90 in the human kidney and in proximal tubule cells exposed to heat, sodium arsenite and cadmium chloride. Toxicol Lett 133:241–254
Solit DB, Zheng FF, Drobnjak M et al (2002) 17-Allylamino-17-demethoxygeldanamycin induces the degradation of androgen receptor and HER-2/neu and inhibits the growth of prostate cancer xenografts. Clin Cancer Res 8:986–993
Soo ET, Yip GW, Lwin ZM, Kumar SD, Bay BH (2008) Heat shock proteins as novel therapeutic targets in cancer. In Vivo 22:311–315
Stanley JA, Lee J, Nithy TK, Arosh JA, Burghardt RC, Banu SK (2011) Chromium-VI arrests cell cycle and decreases granulosa cell proliferation by down-regulating cyclin-dependent kinases (CDK) and cyclins and up-regulating CDK-inhibitors. Reprod Toxicol 32:112–123
Stecklein SR, Kumaraswamy E, Behbod F et al (2012) BRCA1 and HSP90 cooperate in homologous and non-homologous DNA double-strand-break repair and G2/M checkpoint activation. Proc Natl Acad Sci U S A 109:13650–13655
Stellas D, El Hamidieh A, Patsavoudi E (2010) Monoclonal antibody 4C5 prevents activation of MMP2 and MMP9 by disrupting their interaction with extracellular HSP90 and inhibits formation of metastatic breast cancer cell deposits. BMC Cell Biol 11:51
Tan AS, Baty JW, Dong LF et al (2015) Mitochondrial genome acquisition restores respiratory function and tumorigenic potential of cancer cells without mitochondrial DNA. Cell Metab 21:81–94
Tang N, Enger MD (1991) Cadmium induces hypertrophy accompanied by increased myc mRNA accumulation in NRK-49F cells. Cell Biol Toxicol 7:401–411
Tissieres A, Mitchell HK, Tracy UM (1974) Protein synthesis in salivary glands of *Drosophila melanogaster*: relation to chromosome puffs. J Mol Biol 84:389–398
Tokumoto M, Fujiwara Y, Shimada A et al (2011) Cadmium toxicity is caused by accumulation of p53 through the down-regulation of Ube2d family genes in vitro and in vivo. J Toxicol Sci 36:191–200
Topisirovic I, Capili AD, Borden KL (2002) Gamma interferon and cadmium treatments modulate eukaryotic initiation factor 4E-dependent mRNA transport of cyclin D1 in a PML-dependent manner. Mol Cell Biol 22:6183–6198
Trepel J, Mollapour M, Giaccone G, Neckers L (2010) Targeting the dynamic HSP90 complex in cancer. Nat Rev Cancer 10:537–549
Tsai JS, Chao CH, Lin LY (2016) Cadmium activates multiple signaling pathways that coordinately stimulate Akt activity to enhance c-Myc mRNA stability. PLoS One 11:e0147011
Tully DB, Collins BJ, Overstreet JD et al (2000) Effects of arsenic, cadmium, chromium, and lead on gene expression regulated by a battery of 13 different promoters in recombinant HepG2 cells. Toxicol Appl Pharmacol 168:79–90

Urbano AM, Rodrigues CFD, Alpoim MC (2008) Hexavalent chromium exposure, genomic instability and lung cancer. Gene Ther Mol Biol 12B:219–238
Urbano AM, Ferreira LM, Alpoim MC (2012) Molecular and cellular mechanisms of hexavalent chromium-induced lung cancer: an updated perspective. Curr Drug Metab 13:284–305
Vandewynckel YP, Laukens D, Geerts A et al (2013) The paradox of the unfolded protein response in cancer. Anticancer Res 33:4683–4694
Viemann D, Schmidt M, Tenbrock K et al (2007) The contact allergen nickel triggers a unique inflammatory and proangiogenic gene expression pattern via activation of NF-kappaB and hypoxia-inducible factor-1alpha. J Immunol 178:3198–3207
Vihervaara A, Sistonen L (2014) HSF1 at a glance. J Cell Sci 127:261–266
Voellmy R (2004) On mechanisms that control heat shock transcription factor activity in metazoan cells. Cell Stress Chaperones 9:122–133
Waalkes MP (2003) Cadmium carcinogenesis. Mutat Res 533:107–120
Wadhwa R, Takano S, Mitsui Y, Kaul SC (1999) NIH 3T3 cells malignantly transformed by mot-2 show inactivation and cytoplasmic sequestration of the p53 protein. Cell Res 9:261–269
Waisberg M, Joseph P, Hale B, Beyersmann D (2003) Molecular and cellular mechanisms of cadmium carcinogenesis. Toxicology 192:95–117
Wan R, Mo Y, Chien S et al (2011) The role of hypoxia inducible factor-1alpha in the increased MMP-2 and MMP-9 production by human monocytes exposed to nickel nanoparticles. Nanotoxicology 5:568–582
Wandinger SK, Richter K, Buchner J (2008) The Hsp90 chaperone machinery. J Biol Chem 283:18473–18477
Wang S, Diller KR, Aggarwal SJ (2003) Kinetics study of endogenous heat shock protein 70 expression. J Biomech Eng 125:794–797
Wang H, Lu M, Yao M, Zhu W (2016) Effects of treatment with an Hsp90 inhibitor in tumors based on 15 phase II clinical trials. Mol Clin Oncol 5:326–334
Wang Z, Wu J, Humphries B et al (2018) Upregulation of histone-lysine methyltransferases plays a causal role in hexavalent chromium-induced cancer stem cell-like property and cell transformation. Toxicol Appl Pharmacol 342:22–30
Wei Z, Song X, Shaikh ZA (2015) Cadmium promotes the proliferation of triple-negative breast cancer cells through EGFR-mediated cell cycle regulation. Toxicol Appl Pharmacol 289:98–108
Weng S, Wang W, Li Y et al (2014) Continuous cadmium exposure from weaning to maturity induces downregulation of ovarian follicle development-related SCF/c-kit gene expression and the corresponding changes of DNA methylation/microRNA pattern. Toxicol Lett 225:367–377
Wezynfeld NE, Bossak K, Goch W, Bonna A, Bal W, Fraczyk T (2014) Human annexins A1, A2, and A8 as potential molecular targets for Ni(II) ions. Chem Res Toxicol 27:1996–2009
Whitesell L, Mimnaugh EG, De Costa B, Myers CE, Neckers LM (1994) Inhibition of heat shock protein HSP90-pp60v-src heteroprotein complex formation by benzoquinone ansamycins: essential role for stress proteins in oncogenic transformation. Proc Natl Acad Sci U S A 91:8324–8328
Whitesell L, Lindquist SL (2005) HSP90 and the chaperoning of cancer. Nat Rev Cancer 5:761–772
Whitesell L, Lindquist S (2009) Inhibiting the transcription factor HSF1 as an anticancer strategy. Expert Opin Ther Targets 13:469–478
Wu HC, Yang CY, Hung DZ et al (2011) Nickel(II) induced JNK activation-regulated mitochondria-dependent apoptotic pathway leading to cultured rat pancreatic beta-cell death. Toxicology 289:103–111
Wu LX, Xu JH, Zhang KZ et al (2008) Disruption of the Bcr-Abl/Hsp90 protein complex: a possible mechanism to inhibit Bcr-Abl-positive human leukemic blasts by novobiocin. Leukemia 22:1402–1409
Xiao F, Li Y, Dai L et al (2012) Hexavalent chromium targets mitochondrial respiratory chain complex I to induce reactive oxygen species-dependent caspase-3 activation in L-02 hepatocytes. Int J Mol Med 30:629–635

Xie H, Wise SS, Wise JP Sr (2008) Deficient repair of particulate hexavalent chromium-induced DNA double strand breaks leads to neoplastic transformation. Mutat Res 649:230–238

Xiong X, Wang Y, Liu C et al (2014) Heat shock protein 90beta stabilizes focal adhesion kinase and enhances cell migration and invasion in breast cancer cells. Exp Cell Res 326:78–89

Xu Q, Tu J, Dou C et al (2017) HSP90 promotes cell glycolysis, proliferation and inhibits apoptosis by regulating PKM2 abundance via Thr-328 phosphorylation in hepatocellular carcinoma. Mol Cancer 16:178

Yao Q, Nishiuchi R, Li Q, Kumar AR, Hudson WA, Kersey JH (2003) FLT3 expressing leukemias are selectively sensitive to inhibitors of the molecular chaperone heat shock protein 90 through destabilization of signal transduction-associated kinases. Clin Cancer Res 9:4483–4493

Ye J, Shi X (2001) Gene expression profile in response to chromium-induced cell stress in A549 cells. Mol Cell Biochem 222:189–197

Zhang Y, Zhang Y, Zhong C, Xiao F (2016) Cr(VI) induces premature senescence through ROS-mediated p53 pathway in L-02 hepatocytes. Sci Rep 6:34578

Zhao C, Wang E (2004) Heat shock protein 90 suppresses tumor necrosis factor alpha induced apoptosis by preventing the cleavage of bid in NIH3T3 fibroblasts. Cell Signal 16:313–321

Zhong X, de Cassia da Silveira ESR, Zhong C (2017) Mitochondrial biogenesis in response to chromium (VI) toxicity in human liver cells. Int J Mol Sci 18

Ziemiecki A, Catelli MG, Joab I, Moncharmont B (1986) Association of the heat shock protein hsp90 with steroid hormone receptors and tyrosine kinase oncogene products. Biochem Biophys Res Commun 138:1298–1307

Part II
Biomolecular Aspects of HSP90

Chapter 12
Hsp90 and Its Role in Heme-Maturation of Client Proteins: Implications for Human Diseases

Arnab Ghosh and Dennis J. Stuehr

Abstract Hemeproteins are essential for life and heme insertion is an essential step in their maturation. Maturation of hemeprotein requires that they incorporate heme and become active, but knowledge of this essential cellular process remains incomplete. However recent studies on chaperon Hsp90 has revealed that it drives functional heme insertion in vital hemeproteins like inducible nitric oxide synthase (iNOS), soluble guanylate cyclase (sGC) and hemoglobin (Hb). In all three cases Hsp90 interacts with the heme-free or apo-protein and then drives the heme insertion by an ATP dependent process before dissociating from the heme-replete proteins. Given the diverse role of chaperon Hsp90, and in particular to it being a major therapeutic target in drug discovery programs these findings add up to Hsp90's repertoire of being a druggable target and opens up more avenues in regulating growth of diseased cells in those pathologic conditions where these hemeproteins are dysfunctional.

Keywords Angiogenesis · Bronchodilation · Erythropoiesis · Heme · Hemeprotein · Metastasis

Abbreviations

AHSP	Alpha hemoglobin stabilizing protein
AIF	Apoptosis-inducing factor
Apaf-1	Apoptotic protease activating factor 1
Ask-1	Apoptosis signal-regulating kinase 1
ATP	Adenosine triphosphate
cGMP	Cyclic guanosine monophosphate

A. Ghosh (✉) · D. J. Stuehr
Department of Inflammation and Immunity, Lerner Research Institute, The Cleveland Clinic, Cleveland, OH, USA
e-mail: ghosha3@ccf.org

A. A. A. Asea, P. Kaur (eds.), *Heat Shock Protein 90 in Human Diseases and Disorders*, Heat Shock Proteins 19,
https://doi.org/10.1007/978-3-030-23158-3_12

ECM	Extracellular matrix
FLVCRb	Feline leukemia virus subgroup c receptor b
Hb	Hemoglobin
HDM	House dust mite
Her-2	Human epidermal growth factor receptor 2
Hip	hsp70-interacting protein
H-NOX	Heme nitric oxide/oxygen
Hop	hsp70/hsp90 organizing protein
Hsp90	Heat shock protein 90
iNOS	Inducible nitric oxide synthase
IPF	Idiopathic pulmonary fibrosis
MMPs	Matrix metalloproteinases
NO	Nitric oxide
NOS	Nitric oxide synthase
NOX	NADPH oxidase
per2	Period circadian regulator 2
ROS	Reactive oxygen species
sGC	Soluble guanylate cyclase
TGF-β1	Transforming growth factor-β1
VEGF	Vascular endothelial growth factor

12.1 Introduction

Hsp90 is a ubiquitously-expressed, ATP-dependent chaperone that helps to fold, stabilize, or modify the functions of select client proteins (Li and Buchner 2013; Makhnevych and Houry 2012). Hsp90 functions through its subdomain molecular motions and its inherent ATPase activity to help control client protein maturation, trafficking, and lifetime in cells (Faou and Hoogenraad 2012; Prodromou 2012; Taipale et al. 2010). The molecular-level impacts of hsp90 on various client proteins are just beginning to be elucidated (Krukenberg et al. 2011; Park et al. 2011; Ratzke et al. 2012; Southworth and Agard 2011). Hsp90 primarily exist as homodimers whose activity is regulated by ATP. It belongs to the HSP90 family of proteins which represents a well-conserved group of proteins with an average molecular mass of 90 kDa. There are two major Hsp90 isoforms, the Hsp90α1 (HspC1) and the Hsp90β (HspC3), commonly known as Hsp90 (Csermely et al. 1998; Chen et al. 2005). These two proteins in humans are expressed by genes *HSP90AA1 and HSP90AB1* respectively. While Hsp90α1 (HspC1) represents the stress-inducible isoform, Hsp90β (HspC3) is constitutively expressed.

Like many other chaperones, Hsp90 is a hydrophobic protein whose hydrophobicity further increases after heat shock (Yamamoto et al. 1991; Iwasaki et al. 1989). Hsp90 is the major soluble protein of the cell and most commonly located in the cytoplasm. It is also reported to be present in the extracellular milieu where they act as potent stimulators of immune responses (Pockley and Multhoff 2008; Pockley et al. 2008). Hsp90 activity is regulated by post-translational modifications and its association with numerous co-chaperones and client proteins involved in signal transduction and transcriptional regulation. Hsp90 is known to associate with several heme proteins, and the ascribed outcomes include assisting in protein heme-maturation, stabilization, function, or activity, and in shaping enzyme product distribution (McClellan et al. 2007). Herein we present the current knowledge of Hsp90 function with regard to heme-maturation of client proteins, role of hsp90 in tumorigenesis and pulmonary diseases and how Hsp90 function in client hemeprotein maturation may relate to human diseases where Hsp90 is a therapeutic target.

12.1.1 Heme Proteins in Biology

Heme proteins are the most ubiquitously distributed metalloprotein groups in nature. These proteins usually contain heme or an iron (Fe) protoporphyin IX (or a derivative of this macrocycle) in the active site. Depending on the nature of the side chains present in the macrocycle and the type of bond involved in the interaction with the protein, the hemes are classified in three major classes; a, b and c (Severance and Hamza 2009). Hemes a and c are synthesized from heme b, which is the most abundant heme found in hemeproteins, with chemical modification occurring in the side chains as shown in Fig. 12.1. This metallated protoporphyrin is corordinated to the protein through histidine, methionine, tyrosine or cysteine axial ligands (Paoli et al. 2002). The representative classes of heme proteins is summarized in Fig. 12.2.

Heme proteins are involved in a significant array of biological functions including cell energetics (cytochromes), oxygen transport and storage (hemoglobin and myoglobin), oxygenation (monooxygenases), hydrogen peroxide degradation (peroxidases, catalases), oxygen reduction (heme-copper oxidases), heme or small molecule sensing (FixL, HemeAT sensors), transcription regulation (Per2, RevErb

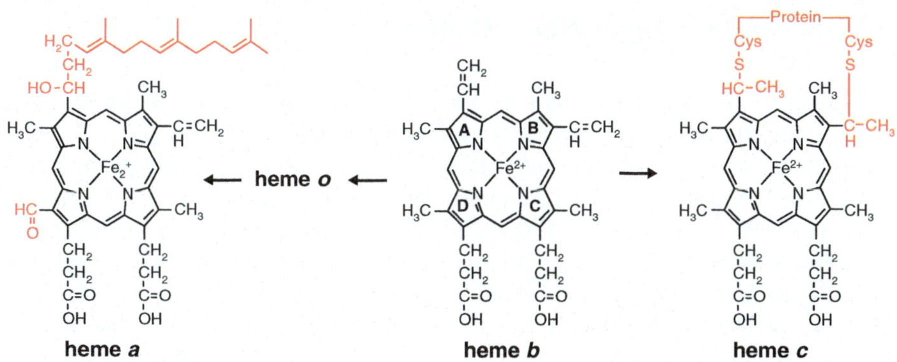

Fig. 12.1 Structures for heme a, b and c. Heme a and c are synthesized from heme b via side chain modifications shown in red. Pyrrole rings nomenclature (A, B, C and D) are depicted using the Hans Fischer system. (Adapted from Severance and Hamza 2009)

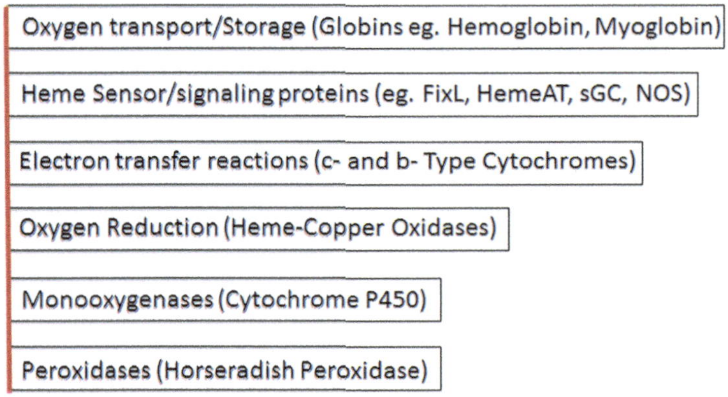

Fig. 12.2 Representative classes of heme proteins

αβ, Bach1), cell signaling (sGC, NOSs), numerous enzymatic transformations and host defense (Ponka 1999; Smith and Veitch 1998; Taylor and Zhulin 1999). It is remarkable that heme proteins use the same heme cofactor, Fe-protoporphyrin IX, to confer such diverse functions. This is attributed to the variety of interactions between the heme cofactor and the protein polypeptide chain, such as the coordination of heme iron by various axial ligands (His, Met, Cys, water, etc.) and the hydrogen-bonding interactions in the heme distal pocket (Lu et al. 2009; Spiro et al. 2013; Poulos 2014). Thus the reactive versatility of the heme iron is exploited via the structural nature of the protein matrix. In addition to these interactions, post-translational modifications (such as nitration, glycosylation, amino acid cross links such as Tyr-His cross links in heme-copper oxidase etc.) also play key roles in tuning the structure and function of heme proteins (Yamakura and Ikeda 2006; Liu et al. 2012; Yoshikawa et al. 1998).

12.1.2 Maturation of Heme Proteins

In a majority of cases, the heme cofactor of hemeproteins is essential for function, but because free heme is reactive its production is tightly regulated (Taketani 2005). The specific steps of heme biosynthesis are well-documented and take place in both the cytosolic and mitochondrial compartments of a cell (Dumont et al. 1991; Steiner et al. 1996; Wang et al. 1996), with the final three steps occurring in the mitochondria. However, with the exception of cytochrome c biogenesis (Richard-Fogal et al. 2009) and some aspects of heme acquisition and catabolism (West and Oates 2008; Yi et al. 2009), relatively little is known about how heme is transported out of mitochondria in eukaryotic cells and how it becomes inserted into soluble proteins in the cytosol, or how these processes might be regulated (Fleming and Hamza 2012; Severance and Hamza 2009). These steps are critical, given that free heme is potentially cytotoxic (Tsiftsoglou et al. 2006) and is normally kept at low intracellular levels (Wijayanti et al. 2004). New studies have identified a mitochondrial heme exporter FLVCRb, which is essential for erythropoiesis (Byon et al. 2013). How this heme exporter pertains to heme transport into soluble hemeproteins in the cytosol remains to be investigated. Heme insertion into catalase was shown to possibly occur within the peroxisome (Lazarow and de Duve 1973), and myeloperoxidase trafficking in cells was found to depend on its heme content (Nauseef et al. 1992). More importantly recent studies from our group (Ghosh et al. 2011; Ghosh and Stuehr 2012) as well as previous studies from Osawa's group (Billecke et al. 2004) have shown specific involvement of chaperon Hsp90 in cytosolic heme insertion into soluble proteins. Inducible nitric oxide synthase (iNOS), soluble guanylate cyclase (sGC) and hemoglobin (Hb) are our new finds of hsp90 client proteins which show remarkable hsp90 dependence on their heme maturation.

12.1.3 HSP90 and Its Role in Client Hemeprotein Maturation

Osawa and colleagues showed for the first time that hsp90 is required for cellular heme insertion into neuronal nitric oxide synthase (nNOS) (Billecke et al. 2004). Later we uncovered a role for Hsp90 in iNOS heme insertion (Ghosh et al. 2011). In the latter case, Hsp90 was shown to primarily associate with an apo-iNOS monomer in cells, and then found to drive heme insertion into the apo-enzyme by an ATP-dependent process, after which the Hsp90 interaction with the heme-replete, mature iNOS fell apart. We then found that heme insertion into the β subunit of soluble guanylate cyclase (sGC) is also Hsp90-dependent (Ghosh and Stuehr 2012). Given that sGC structural homologs (H-NOX) and NOS have markedly different protein structures and heme environments (Crane et al. 1998; Martin et al. 2010), these findings hint that Hsp90 may play a broader role in heme protein maturation than was previously realized. In the sGC studies, Hsp90 associated primarily with the heme-free form of sGC-β1 in cells, but the association fell apart once heme became inserted. Although the association did not require hsp90 to have an intact ATPase activity, as judged from results using Hsp90 inhibitors radicicol or novobiocin, or by using an ATPase defective Hsp90 mutant (D88N), an intact ATPase activity was essential to actually drive heme insertion into the apo-sGCβ1. Thus, the model suggested for Hsp90 function in sGC maturation mimics the model proposed for driving heme insertion into apo-iNOS (Fig. 12.3). These similarities imply Hsp90 may operate through a common mechanism to target and stabilize heme-free forms of client heme proteins, and then enable their maturation by driving heme insertion in an ATP-dependent process. Other studies on transmembrane heme proteins like NADPH oxidases (NOX) have shown preliminary evidence of a Hsp90 regulation (Chen et al. 2011). NOX enzymes utilize NADPH to synthesize superoxide. This NOX functionality is dependent on its heme and is a major source of cellular reactive oxygen species (ROS) (Cai and Harrison 2000; Griendling et al. 2000). Because of the intrinsic toxicity of ROS, the regulation of

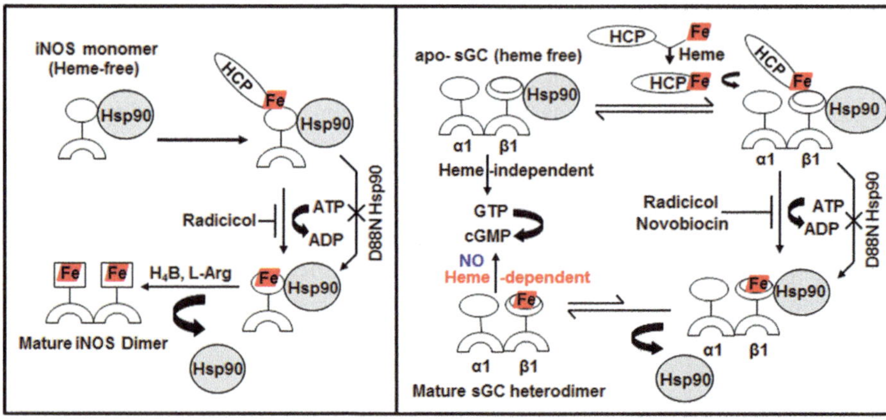

Fig. 12.3 Similarities in Hsp90-apo-protein interactions with respect to heme deplete/replete states of sGC and iNOS. HCP indicates an unknown heme carrier protein which may eventually be Hsp90

NOX enzymes has developed a significant degree of complexity. Hsp90 has been shown to bind to the C-terminal domain of NOX5 and regulate its superoxide production, suggesting that Hsp90 may regulate certain parameters of cellular redox. However a more in depth study or direct study relating Hsp90 to heme-maturation of NOX-5 is lacking and warrants further studies.

Recently we found that Hsp90 chaperones hemoglobin (Hb) maturation in erythroid and non-erythroid cells (RAW, A549 cells) following a similar mechanism and further reinforcing this concept (Ghosh et al. 2018) (Fig. 12.4). While in erythroid cells, Hb-α and Hb-β/γ (-$\beta_{adult}/\gamma_{fetal}$ Hb) are independently chaperoned by AHSP (alpha Hb stabilizing protein) and Hsp90 respectively, Hsp90 promotes heme-maturation of both Hb-αβ subunits in non-erythroid cells which lack AHSP. Although a role for AHSP in Hb-α maturation had been established (Kihm et al. 2002; Feng et al. 2005; Yu et al. 2007), what chaperones might aid partner globin maturation (Hb-γ and Hb-β) was unclear. Our study revealed that Hsp90 enabled maturation of both Hb-γ and Hb-β by associating with and stabilizing their immature, heme-free forms and by driving their heme insertion reactions in an ATP-dependent manner. This was demonstrated in the erythroid-like K562 cell line and in two human erythroid progenitor cell types (HiDEP-1 and HUDEP-2) during their in vitro differentiation to mature erythroid cells, implying that Hsp90's role is relevant for globin maturation during erythropoiesis. Thus, Hsp90 appeared to act as a counterpart to AHSP by chaperoning Hb-γ and Hb-β for their heme insertion reactions, as would be required to form functional fetal (α2γ2) and adult (α2β2) tetramers during erythropoiesis. Under all circumstances, the globin associations of Hsp90 or AHSP remained specific. Together, our findings support the view that Hb-α and Hb-β maturation are independently chaperoned in erythroid cells (Weiss and dos Santos 2009) and confirm that Hb-α maturation in erythroid cells is independent of Hsp90. However, when Hb-α was coexpressed with Hb-β in nonerythroid cells either

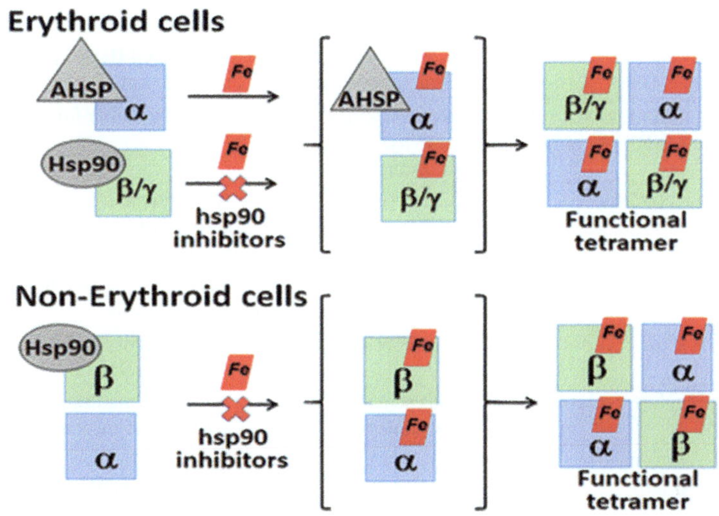

Fig. 12.4 Hsp90 chaperons Hb maturation

naturally or as a consequence of transfection, we saw that its heme insertion then became Hsp90-dependent. Thus, Hsp90 is needed for heme insertion into at least two (Hb-β and Hb-γ) and as many as three (Hb-α) globins, depending on the circumstances under which they are expressed (i.e., with or without AHSP), and therefore Hsp90 plays an unexpectedly broad role in globin maturation. This concept may help explain how Hb maturation can succeed in nonerythroid cells, which all lack AHSP expression. A model for Hb maturation that incorporates our current findings is presented in Fig. 12.4. Thus Hsp90-assisted heme insertion into Hb-β and Hb-γ appeared to follow the tenets established for maturation of other heme proteins like NOS and sGC (Ghosh et al. 2011; Ghosh and Stuehr 2012).

Our findings also revealed that low doses of nitric oxide (NO) can contribute to sGC maturation by triggering a rapid Hsp90-dependent heme insertion into the apo-sGCβ1 (heme-free) population, ultimately resulting in a mature sGC heterodimer (Ghosh et al. 2014). Our study showed that at resting or steady state the majority of apo-sGCβ1 is bound to Hsp90 and activation by NO caused a transient swapping of binding partners of sGC-β1, i.e. from Hsp90 to sGC-α1 subunit, which forms the active sGCα1β1 heterodimer. This partner protein swap resulting in elevated active sGC-α1β1 heterodimer showed for the first time that NO can trigger a change in protein interactions by inserting heme into apo-sGCβ1. This finding filled a void in an earlier work done by Ignarro and colleagues in the 80s which showed that NO-heme moiety could be transferred into heme-free or apo-sGCβ1, through an exchange reaction with NO-hemeproteins to activate the enzyme (Ignarro et al. 1986). This suggests that a NO-Hsp90 synergy may be essential for maturation and activation of certain hemeproteins.

Table 12.1 Hsp90 chaperon hemeprotein interactions in eukaryotes

Protein	Function of heme	Associated Hsp90 chaperon, possible function
Eukaryotic initiation factor 2-alpha (eIF2) kinase HRI (heme regulated inhibitor)	Repress kinase activity, signal transduction	Amplify/enable function
NO synthases (NOSs)	Catalytic O_2 activation	Aid heme maturation, boost activity of heme replete enzyme, stabilize protein
Cytochrome P450 2B1	Catalytic O_2 activation	GRP94 (?), enable heme reconstitution
NADPH oxidase 1, 2, and 5	Catalytic O_2 reduction	Boost activity and influence enzyme products
Soluble guanylate cyclase (sGC)	Gas sensing signal transduction	Aid heme maturation, boost activity and increase enzyme lifetime
Hemoglobin (Hb)	O_2 transport	Aid heme maturation

Based on our recent finds of client hemeproteins of Hsp90 and previous literature on Hsp90 regulation of hemeproteins, we present a summary of currently known Hsp90 hemeprotein interactions in Table 12.1.

12.1.4 HSP90 and Human Diseases

Hsp90 is the epicenter to a broad spectrum of human diseases ranging from various form of cancers (Trepel et al. 2010; Miyata et al. 2013; Tóth et al. 2015), neurodegenerative diseases (Luo et al. 2010), asthma and pulmonary diseases (Ghosh et al. 2016; Bonniaud et al. 2017), autoimmune and endocrine disorders to many forms of retinal dysfunction (Ratajczak et al. 2015; Aguilà and Cheetham 2016). The ability of Hsp90 to protect unfolded proteins from aggregation, assist in proteosomal degradation and modulate several growth and signaling pathways simultaneously, makes it an attractive target for therapeutic intervention (Miyata et al. 2013). Elevated levels of chaperon Hsp90 has been found in a wide spectrum of cancers suggesting a central role in survival and growth of malignant cells (Calderwood et al. 2006; Whitesell and Lindquist 2005; Ciocca and Calderwood 2005). Such enhanced levels of Hsp90 is postulated as a protective effect from various stress parameters (e.g. hypoxia, ischemia, heavy metals etc.) which cells encounter under pathologic conditions. Since several oncoproteins are clients of Hsp90, targeting Hsp90 represents a useful anti-cancer approach. Over the years numerous oncoproteins have been identified as putative targets and Hsp90 inhibition is a promising approach in anti-cancer strategies (Xu and Neckers 2007; Kamal et al. 2003). These approaches are strengthened by the fact that inhibition of Hsp90 contributes to degradation of many oncoproteins thereby intensifying these strategies (Ali et al. 2006; Jackson et al. 2004; Chen et al. 2010).

12.1.5 Role of HSP90 in Tumorogenesis

Hsp90 and its chaperon proteins are involved in multiple cellular signaling pathways which regulate apoptosis and promote cell survival (Trepel et al. 2010). In this context, three acquired capabilities of cancer cells growth and sustenance are, evasion of apoptosis, sustained angiogenesis and tissue invasion and metastasis (Miyata et al. 2013). A great body of literature indicates critical inhibitory contribution of Hsp90 to apoptosis which is key to normal cell growth or adverse tumor progression. Hsp90 is known to inhibit apoptosis by directly binding to Apaf-1, blocking cytochrome c-mediated oligomerization of Apaf-1 and activation of pro-caspase 9 (Pandey et al. 2000). It also inhibits apoptosis by forming a ternary complex with pro-apoptotic kinase Ask-1 and Akt (Zhang et al. 2005). Moreover Hsp90 exerts anti-apoptotic activity by blocking the mitochondrial-cytosolic transition of apoptosis-inducing factor (AIF) and endonuclease G (Fulda et al. 2010). With regard to angiogenesis Hsp90 is known to promote angiogenesis and metastasis by chaperoning certain client proteins including VEGF, NOS, and MMP-2 (Whitesell and Lindquist 2005). Together these functions of chaperon Hsp90 make it clearly anti-apoptotic and pro-angiogenic (Miyata et al. 2013). Additionally Hsp90 also plays a role in the modulation of the extracellular matrix (ECM), since Hsp90β was identified as interaction partner of MMP-3 in the extracellular matrix accountable for mammary epithelial invasion and morphogenesis (Correia et al. 2013). In combination with the co-chaperones Hsp70, Hsp40, Hip, Hop, and p23, extracellular Hsp90α is known to interact with and promote the proteolytic activity of MMP-2 in an ATP-independent manner (Sims et al. 2011; Song et al. 2010). It is known that degradation of ECM is a signal for the beginning of invasion and metastasis, and MMPs are important molecules involved in this process (Nelson et al. 2000). Hsp90 and MMP-9 is known to constitute as a complex in anaplastic large cell lymphomas, and MMP-9 could be activated by Hsp90 to promote cell invasion (Lagarrigue et al. 2010). These functions of Hsp90 highlight its critical role in fostering tumour growth and metastasis.

Considering the above facts it is not surprising that Hsp90 is central to tumorigenesis and the evidences are evergrowing. A study showed that Hsp90 derived from tumor cells (breast and colon tumors) displayed high ATPase activity relative to normal cells (Kamal et al. 2003). The tumor Hsp90 displayed very high affinity for Hsp90 inhibitor 17-AAG even at nanomolar levels. On this basis it was inferred that Hsp90 chaperons mutant proteins within these tumors resulting in chaperoning and stabilization of such oncoproteins, saving them from degradation and explaining Hsp90-dependent malignant progression which was unresolved before. Likewise Hsp90 was found to stabilize the conformation of mutant proteins that arise during transformation including v-Src, Bcr-Abl, and p53 (Nimmanapalli et al. 2001; Neckers 2002), replicating multiple effect of these mutants on the cancer cell phenotypes. An increased activity of Hsp90 has been described in lung cancer expressing a mutated form of the epidermal growth factor receptor which belongs to the group of Hsp90 client proteins (Shimamura and Shapiro 2008). In melanoma,

Hsp90 showed high expression in the tumors than nevi and was associated with disease progression, suggesting it to be a valuable drug target in melanoma as well as a useful diagnostic marker (McCarthy et al. 2008). Moreover in bladder and epithelial ovarian carcinomas a high expression of Hsp90 was observed at advanced stages, suggesting that it maybe an indicator of aggressiveness (Lebret et al. 2003; Elpek et al. 2003). Studies by Wang et al., showed that the level of plasma Hsp90α was significantly enhanced in patients with malignant tumours of breast, lung, pancreas and liver in comparison with normal people, and patients with benign tumours (Wang et al. 2009), highlighting the importance of cancer cell invasions.

The tumor selectivity of Hsp90 inhibitors makes Hsp90 a unique therapeutic target. What stands out here is the high ATPase of tumor Hsp90 which maybe the driving force as we see in our heme-maturation of client hemeproteins under normal conditions (Ghosh et al. 2011; Ghosh and Stuehr 2012). The ability of cancer cells within tumors to expand requires angiogenic ability and neovascularization. Since such angiogenic pathways are turned on during tumor progression (Hanahan and Folkman 1996) and the vascularization machinery becomes active enough to sprout new blood vessels and sustain neoplastic growth. As new blood vessel growth requires functional Hsp90 (Miyata et al. 2013), active heme-maturation of Hsp90 clients such as iNOS (Ghosh et al. 2011) Hb and sGC (Ghosh et al. 2018; Ghosh and Stuehr 2012) may also occur during angiogenesis within these malignant tumors to support these processes. These events can now be envisioned based on our recent finds.

12.1.6 Role of HSP90 in Asthma and Pulmonary Diseases

The ability of Hsp90 to modulate cell fate might have vital repercussions not only for cancer but also for other progressive human diseases, such as those of the pulmonary system e.g. asthma (Ghosh and Erzurum 2011) and pulmonary fibrosis (Mora et al. 2017). Asthma is defined by airway inflammation and hyper responsiveness, and contributes to morbidity and mortality world wide (Ghosh and Erzurum 2011). sGC is a key enzyme of the NO signaling pathway, and is activated by NO produced from NOS enzymes (Ghosh et al. 2014; Stasch et al. 2011), thereby activating synthesis of the second messenger cGMP, which produces vascular smooth muscle relaxation or vasodilation as a downstream effect (Murad 2006; Bryan et al. 2009). This constitutes the NO-sGC-cGMP signal pathway and is a well known dilation pathway in the vasculature. In a recent study (Ghosh et al. 2016) we found that the NO-sGC-cGMP pathway plays a significant role in lung bronchodilation, and that lung sGC becomes dysfunctional in asthma due to high levels of NO generated from iNOS induction during inflammation in the airway epithelium (Ghosh et al. 2016; Guo et al. 1995), which desensitizes the sGC resident in the airway smooth muscles below (Fig. 12.5). This dysfunctional sGC is heme-free, does not respond to its natural activator NO, but can be activated by sGC agonists like BAY 60-2770 (Pankey et al. 2011), which can activate the enzyme

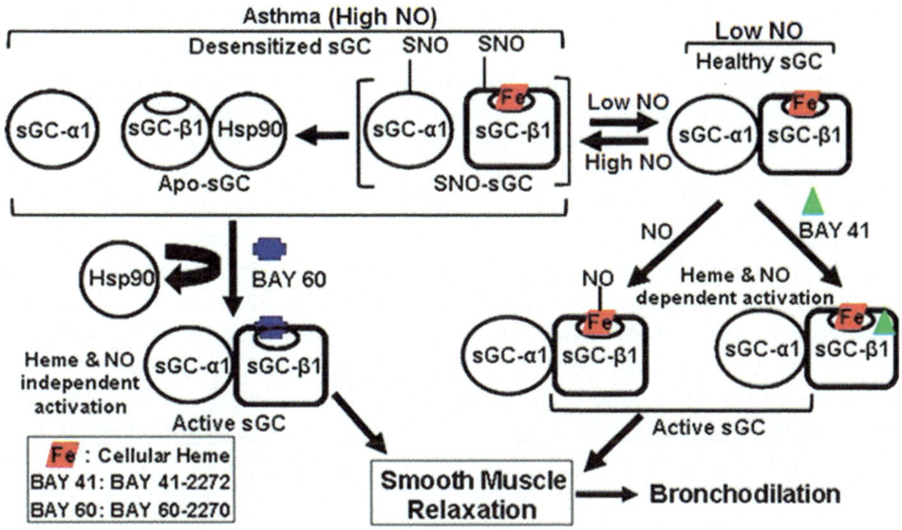

Fig. 12.5 Schematic representation depicting ways by which different populations of sGC can be activated in healthy and diseased conditions which leads to smooth muscle relaxation and ultimate bronchodilation

independent of NO to produce bronchodilation and such agonists are future drugs for asthma. In this study Hsp90 was associated with heme-free sGC, and seemed to stabilize it rather than priming it for degradation, which makes this pathological sGC more drug receptive. In other studies on asthma, relating to airway epithelium and inflammation, Hsp90 is more directly implicated. Extracellular Hsp90α (eHSp90α) is shown to mediate HDM-induced human bronchial epithelial dysfunction, suggesting that eHsp90α is a potential therapeutic target for treatment of asthma (Dong et al. 2017). Another report (Pezzulo et al. 2018) suggests that Hsp90 inhibition reverts IL-13, IL-17-induced goblet cell metaplasia in the human airway epithelia. Hsp90 inhibitors have been also been shown to reduce airway inflammation in mouse models of allergic asthma (Dimitropolou et al. 2010), and another study showed that Hsp90 inhibitors hampered airway relaxation (Intapad et al. 2012). Given our current understanding of iNOS heme-maturation most of these processes of Hsp90 inhibition may involve inhibition of iNOS heme-insertion, suppressing iNOS induction in the airway epithelium and reducing inflammation (Ghosh et al. 2016).

Idiopathic pulmonary fibrosis (IPF) is a progressive disease of the lung parenchyma, causing significant morbidity and mortality (Hopkins et al. 2016; Overgaard et al. 2016). Transforming growth factor (TGF)-β1 is a key cytokine involved in the process of fibrogenesis (Froese et al. 2016). It causes myofibroblast proliferation and differentiation and increases the synthesis of collagen, fibronection and other extracellular matrix components (ECM). A recent study (Sibinska et al. 2017) demonstrated that Hsp90 has a direct role in TGF-β1 signaling pathway and Hsp90 inhibition reduced lung fibrogenesis and fibrosis progression in mice. The study also showed that Hsp90 is overexpressed in IPF lungs and fibrosis can be inhibited using a water soluble hsp90 inhibitor, 17-DMAG, which targets the Hsp90 ATPase similar to 17-AAG (Sibinska et al. 2017; Kamal et al. 2003). Another study (Sontake et al. 2017) found that Hsp90 was elevated in expression and in its ATPase activity in lung biopsies of patients with IPF. These findings bear great semblance with regard to Hsp90 expression and ATPase to what we see in a majority of cancers (Miyata et al. 2013) and suggests that targeting Hsp90 is an effective strategy for treating fibrotic lung disease. More recently sGC agonists (Sander et al. 2017) are also being tried to treat age-related fibrosis. Since Hsp90 and other chaperon machinery slows down with age (Verbeke et al. 2001), causing low NO levels which may reduce sGC activation thereby making such sGC agonists as the drugs of choice. Moreover a slow Hsp90 chaperon machinery may cause reduced heme-maturation causing buildup of heme-free sGC, which can be activated by BAY 60-2770 like drugs, further encouraging the use of these agonists to treat age-related fibrosis.

12.1.7 Therapeutic Aspects of HSP90 Inhibition

Hsp90 has emerged as an important molecule in anti-tumour therapy, and several drug classes have been found to target its ATP-binding domain resulting in inactivation of the chaperone. The first hsp90 inhibitor, 17-AAG (tanespimycin), entered clinical trials in 1999, while a second inhibitor 17-DMAG entered the first-in-human study in 2004 (Trepel et al. 2010). From this period onwards, many other Hsp90 inhibitors are currently undergoing clinical evaluation in cancer patients, owing to extensive efforts in rational drug design and discovery (Eccles et al. 2008; Chiosis and Tao 2006). Various other newer class of hsp90 inhibitors are also on the rise (Neckers and Workman 2012; Fuhrmann-Stroissnigg et al. 2017).

A wealth of evidence now indicates that, after successive dosing of Hsp90 inhibitors to animals bearing human tumours, blocks tumour growth efficiently. However, tumours often start to re-grow after withdrawal of the inhibitor (Fadden et al. 2010). A similar observation is made in patients with solid tumours treated with several structurally unrelated Hsp90 inhibitors, thus rendering hsp90 inhibitors an unsuitable agent in monotherapy. However in certain tumour entities, single-agent inhibitor therapy was found to suppress tumour growth (Socinski et al. 2013). Likewise Hsp90 inhibitors in combination with cutting-edge targeted therapies is the pathway to move forward. It has been shown that a combination of the Hsp90 inhibitor SNX-5422 and trastuzumab (herceptin), a monoclonal antibody that blocks the Her-2 receptor, led to a synergistic regression of tumour growth in a xenograft model of human breast cancer (Fadden et al. 2010). As another example, a phase II trial, in which the combinatorial administration of 17-AAG (tanespimycin) plus trastuzumab displayed significant anti-cancer activity in patients with Her-2-positive metastatic breast cancer, which was previously seen to progress on trastuzumab (Modi et al. 2011). Thus the synergistic effects in tumour regression observed in

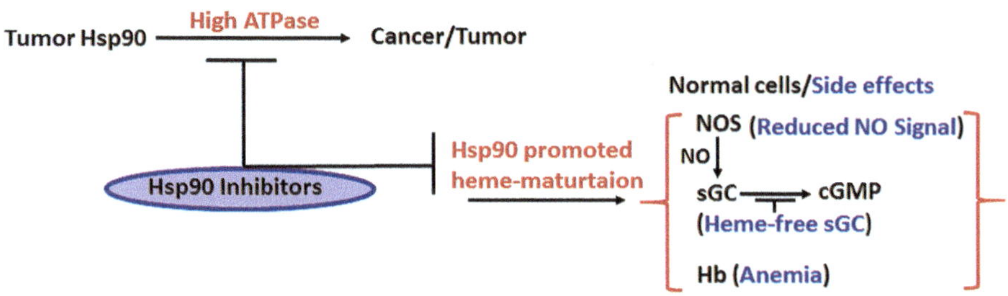

Fig. 12.6 Good effect of Hsp90 inhibitors on tumor cells while adverse effect on normal cells

animal studies after combinatorial administration of hsp90 inhibitors and potent anti-cancer drugs hold true in human trials, hence targeted therapies hold potential for future drug designing. Based on our current knowledge of hemeprotein maturation and the role played by Hsp90, use of Hsp90 inhibitors which target the ATPase function of the chaperon can also give rise to certain side effects (Fig. 12.6). These inhibitors can adversely effect normal cells by blocking the heme-maturation of iNOS, sGC or Hb (Fig. 12.6). For example, the Hsp90 inhibitors that are being developed for cancer treatment might unintentionally block Hb maturation in the recipient. Indeed, anemia has been commonly reported as a side effect during the clinical trials of Hsp90 inhibitor drug candidates (Pillai and Ramalingam 2012; Do et al. 2015). Likewise the Hsp90 inhibitors being used for asthma and IPF may also block heme-maturation of sGC, creating pathologic heme-free sGC and obstructing bronchodilation cascades. Hence widespread use of these Hsp90 inhibitors may require a more cautious approach or a combination therapy with sGC agonist like drugs may help in future therapeutics in such cases.

12.2 Conclusions

There are at least 13 Hsp90 inhibitors (Kim 2009) currently undergoing clincal trials from a broad range of tumors including breast, prostrate, gastrointestinal, melanoma, and hematological malignancies, and in vitro studies continue to identify newer small molecule compounds that selectively target Hsp90 chaperon in its various paralog forms (Taldone et al. 2014; Gomez-Monterrey et al. 2012). Thus this evergrowing progress of Hsp90 inhibitor drugs and their relevance to specific diseases holds great promise for further exciting developments in the future. Given the novel role of Hsp90 in hemeprotein maturation the effects of an overactive or a downregulated Hsp90 can both be deleterious to cellular homeostasis. This is now more evident from the fact that Hsp90 regulates heme maturation of three key hemeproteins (iNOS, sGC and Hb) and intertwines Hb maturation to the NO-sGC-cGMP dilation cascade. Together these realizations provide a platform to explore these concepts which may help in future drug designing.

Acknowledgements This work is supported by National Institute of Health Grant HL081064 (to D.J.S and A.G.) and a Research Centre for Excellence Grant from the Cleveland Clinic (to A.G. and D.J.S.).

References

Aguilà M, Cheetham ME (2016) Hsp90 as a potential therapeutic target in retinal disease. Adv Exp Med Biol 854:161–167

Ali MM, Roe SM, Vaughan CK et al (2006) Crystal structure of an Hsp90-nucleotide-p23/Sba1 closed chaperone complex. Nature 440:1013–1017

Billecke SS, Draganov DI, Morishima Y et al (2004) The role of hsp90 in heme-dependent activation of apo-neuronal nitric-oxide synthase. J Biol Chem 279:30252–30258

Bonniaud P, Bellaye PS, Burgy O, Kolb M (2017) Heat shock protein: a hot topic in idiopathic pulmonary fibrosis. Eur Respir J 49:1602152

Bryan NS, Bian K, Murad F (2009) Discovery of the nitric oxide signaling pathway and targets for drug development. Front Biosci 14:1–18

Byon JC, Chen J, Doty RT, Abkowitz JL (2013) FLVCR is necessary for erythroid maturation, may contribute to platelet maturation, but is dispensable for normal hematopoietic stem cell function. Blood 122:2903–2910

Cai H, Harrison DG (2000) Endothelial dysfunction in cardiovascular diseases—the role of oxidant stress. Circ Res 87:840–844

Calderwood SK, Khaleque MA, Sawyer DB, Ciocca DR (2006) Heat shock proteins in cancer: chaperones of tumorigenesis. Trends Biochem Sci 31:164–172

Chen B, Piel WH, Gui L, Bruford E, Monteiro A (2005) The HSP90 family of genes in the human genome: insights into their divergence and evolution. Genomics 86:627–637

Chen Z et al (2010) Inhibition of ALK, PI3K/MEK, and HSP90 in murine lung adenocarcinoma induced by EML4-ALK fusion oncogene. Cancer Res 70:9827–9836

Chen F, Pandey D, Chadli A, Catravas JD, Chen T, Fulton DJ (2011) Hsp90 regulates NADPH oxidase activity and is necessary for superoxide but not hydrogen peroxide production. Antioxid Redox Signal 14:2107–2119

Chiosis G, Tao H (2006) Purine-scaffold Hsp90 inhibitors. IDrugs 9:778–782

Ciocca DR, Calderwood SK (2005) Heat shock proteins in cancer: diagnostic, prognostic, predictive, and treatment implications. Cell Stress Chaperones 10:86–103

Correia AL, Mori H, Chen EI, Schmitt FC, Bissell MJ (2013) The hemopexin domain of MMP3 is responsible for mammary epithelial invasion and morphogenesis through extracellular interaction with HSP90beta. Genes Dev 27:805–817

Crane BR, Arvai AS, Ghosh DK et al (1998) Structure of nitric oxide synthase oxygenase dimer with pterin and substrate. Science 279:2121–2126

Csermely P, Schnaider T, Soti C, Prohaszka Z, Nardai G (1998) The 90-kDa molecular chaperone family: structure, function, and clinical applications. A comprehensive review. Pharmacol Ther 79:129–168

Dimitropoulou C, Snead C, Gallops J, Catravas JD (2010) Heat shock protein 90 inhibitors reduce airway inflammation in a mouse model of allergic asthma. FASEB J 24(Suppl 1)

Do K, Speranza G, Chang LC et al (2015) Phase I study of the heat shock protein 90 (Hsp90) inhibitor onalespib (AT13387) administered on a daily for 2 consecutive days per week dosing schedule in patients with advanced solid tumors. Investig New Drugs 33:921–930

Dong HM, Le YQ, Wang YH et al (2017) Extracellular heat shock protein 90α mediates HDM-induced bronchial epithelial barrier dysfunction by activating RhoA/MLC signaling. Respir Res 18:111

Dumont ME, Cardillo TS, Hayes MK, Sherman F (1991) Role of cytochrome c heme lyase in mitochondrial import and accumulation of cytochrome c in *Saccharomyces cerevisiae*. Mol Cell Biol 11:5487–5496

Eccles SA, Massey A, Raynaud FI et al (2008) NVP-AUY922: a novel heat shock protein 90 inhibitor active against xenograft tumor growth, angiogenesis, and metastasis. Cancer Res. 2008 68:2850–2860

Elpek GO, Karaveli S, Simsek T, Keles N, Aksoy NH (2003) Expression of heat-shock proteins hsp27, hsp70 and hsp90 in malignant epithelial tumour of the ovaries. APMIS 111:523–530

Fadden P, Huang KH, Veal JM et al (2010) Application of chemoproteomics to drug discovery: identification of a clinical candidate targeting hsp90. Chem Biol 17:686–694

Faou P, Hoogenraad NJ (2012) Tom34: a cytosolic cochaperone of the Hsp90/Hsp70 protein complex involved in mitochondrial protein import. Biochim Biophys Acta 1823:348–357

Feng L, Zhou S, Gu L et al (2005) Structure of oxidized alpha-haemoglobin bound to AHSP reveals a protective mechanism for haem. Nature 435:697–701

Fleming MD, Hamza I (2012) Mitochondrial heme: an exit strategy at last. J Clin Invest 122:4328–4330

Froese AR, Shimbori C, Bellaye PS et al (2016) Stretch-induced activation of transforming growth factor-β1 in pulmonary fibrosis. Am J Respir Crit Care Med 194:84–96

Fuhrmann-Stroissnigg H, Ling YY, Zhao J et al (2017) Identification of HSP90 inhibitors as a novel class of senolytics. Nat Commun 8:422

Fulda S, Galluzzi L, Kroemer G (2010) Targeting mitochondria for cancer therapy. Nat Rev Drug Discov 9:447–464

Ghosh S, Erzurum SC (2011) Nitric oxide metabolism in asthma pathophysiology. Biochim Biophys Acta 1810:1008–1016

Ghosh A, Stuehr DJ (2012) Soluble guanylyl cyclase requires heat shock protein 90 for heme insertion during maturation of the NO-active enzyme. Proc Natl Acad Sci U S A 109:12998–13003

Ghosh A, Chawla-Sarkar M, Stuehr DJ (2011) Hsp90 interacts with inducible NO synthase client protein in its heme-free state and then drives heme insertion by an ATP-dependent process. FASEB J 25:2049–2060

Ghosh A, Stasch JP, Papapetropoulos A, Stuehr DJ (2014) Nitric oxide and heat shock protein 90 activate soluble guanylate cyclase by driving rapid change in its subunit interactions and heme content. J Biol Chem 289:15259–15271

Ghosh A, Koziol-White CJ, Asosingh K et al (2016) Soluble guanylate cyclase as an alternative target for bronchodilator therapy in asthma. Proc Natl Acad Sci U S A 113:E2355–E2362

Ghosh A, Garee G, Sweeny EA, Nakamura Y, Stuehr DJ (2018) Hsp90 chaperones hemoglobin maturation in erythroid and non-erythroid cells. Proc Natl Acad Sci U S A 115:E1117–E1126

Gomez-Monterrey I, Sala M, Musella S, Campiglia P (2012) Heat shock protein 90 inhibitors as therapeutic agents. Recent Pat Anticancer Drug Discov 7:313–336

Griendling KK, Sorescu D, Ushio-Fukai M (2000) NAD(P)H oxidase: role in cardiovascular biology and disease. Circ Res 86:494–501

Guo FH, De Raeve HR, Rice TW, Stuehr DJ, Thunnissen FB, Erzurum SC (1995) Continuous nitric oxide synthesis by inducible nitric oxide synthase in normal human airway epithelium in vivo. Proc Natl Acad Sci U S A 92:7809–7813

Hanahan D, Folkman J (1996) Patterns and emerging mechanisms of the angiogenic switch during tumorigenesis. Cell 86:353–364

Hopkins RB, Burke N, Fell C, Dion G, Kolb M (2016) Epidemiology and survival of idiopathic pulmonary fibrosis from national data in Canada. Eur Respir J 48:187–195

Ignarro LJ, Adams JB, Horwitz PM, Wood KS (1986) Activation of soluble guanylate cyclase by NO-hemoproteins involves NO-heme exchange. Comparison of heme-containing and heme-deficient enzyme forms. J Biol Chem 261:4997–5002

Intapad S, Dimitropoulou C, Snead C, Piyachaturawat P, Catravas JD (2012) Regulation of asthmatic airway relaxation by estrogen and heat shock protein 90. J Cell Physiol 227:3036–3043

Iwasaki M, Saito H, Yamamoto M, Korach KS, Hirogome T, Sugano H (1989) Purification of heat shock protein 90 from calf uterus and rat liver and characterization of the highly hydrophobic region. Biochim Biophys Acta 992:1–8
Jackson SE, Queitsch C, Toft D (2004) Hsp90: from structure to phenotype. Nat Struct Mol Biol 11:1152–1155
Kamal A, Thao L, Sensintaffar J et al (2003) A high-affinity conformation of Hsp90 confers tumour selectivity on Hsp90 inhibitors. Nature 425:407–410
Kihm AJ, Kong Y, Hong W et al (2002) An abundant erythroid protein that stabilizes free alpha-haemoglobin. Nature 417:758–763
Kim YS (2009) Update on Hsp90 inhibitors in clinical trial. Curr Top Med Chem 9:1479–1492
Krukenberg KA, Street TO, Lavery LA, Agard DA (2011) Conformational dynamics of the molecular chaperone Hsp90. Q Rev Biophys 44:229–255
Lagarrigue F, Dupuis-Coronas S, Ramel D et al (2010) Matrix metalloproteinase-9 is upregulated in nucleophosmin-anaplastic lymphoma kinase-positive anaplastic lymphomas and activated at the cell surface by the chaperone heat shock protein 90 to promote cell invasion. Cancer Res 70:6978–6987
Lazarow PB, de Duve C (1973) The synthesis and turnover of rat liver peroxisomes. V. Intracellular pathway of catalase synthesis. J Cell Biol 59:507–524
Lebret T, Watson RW, Molinié V et al (2003) Heat shock proteins HSP27, HSP60, HSP70, and HSP90: expression in bladder carcinoma. Cancer 98:970–977
Li J, Buchner J (2013) Structure, function and regulation of the hsp90 machinery. Biom J 36:106–117
Liu X, Yu Y, Hu C, Zhang W, Lu Y, Wang J (2012) Significant increase of oxidase activity through the genetic incorporation of a tyrosine-histidine cross-link in a myoglobin model of heme-copper oxidase. Angew Chem Int Ed Eng 51:4312–4316
Lu Y, Yeung N, Sieracki N, Marshall NM (2009) Design of functional metalloproteins. Nature 460:855–862
Luo W, Sun W, Taldone T, Rodina A, Chiosis G (2010) Heat shock protein 90 in neurodegenerative diseases. Mol Neurodegener 5:24
Makhnevych T, Houry WA (2012) The role of Hsp90 in protein complex assembly. Biochim Biophys Acta 1823:674–682
Martin F, Baskaran P, Ma X et al (2010) Structure of cinaciguat (BAY 58-2667) bound to Nostoc H-NOX domain reveals insights into heme-mimetic activation of the soluble guanylyl cyclase. J Biol Chem 285:22651–22657
McCarthy MM, Pick E, Kluger Y et al (2008) HSP90 as a marker of progression in melanoma. Ann Oncol 19:590–594
McClellan AJ, Xia Y, Deutschbauer AM, Davis RW, Gerstein M, Frydman J (2007) Diverse cellular functions of the Hsp90 molecular chaperone uncovered using systems approaches. Cell 131:121–135
Miyata Y, Nakamoto H, Neckers L (2013) The therapeutic target Hsp90 and cancer hallmarks. Curr Pharm Des 19:347–365
Modi S, Stopeck A, Linden H et al (2011) HSP90 inhibition is effective in breast cancer: a phase II trial of tanespimycin (17-AAG) plus trastuzumab in patients with HER2-positive metastatic breast cancer progressing on trastuzumab. Clin Cancer Res 17:5132–5139
Mora AL, Rojas M, Pardo A, Selman M (2017) Emerging therapies for idiopathic pulmonary fibrosis, a progressive age-related disease. Nat Rev Drug Discov 16:755–772
Murad F (2006) Shattuck Lecture. Nitric oxide and cyclic GMP in cell signaling and drug development. N Engl J Med 355:2003–2011
Nauseef WM, McCormick S, Yi H (1992) Roles of heme insertion and the mannose-6-phosphate receptor in processing of the human myeloid lysosomal enzyme, myeloperoxidase. Blood 80:2622–2633
Neckers L (2002) Hsp90 inhibitors as novel cancer chemotherapeutic agents. Trends Mol Med 8:S55–S61

Neckers L, Workman P (2012) Hsp90 molecular chaperone inhibitors: are we there yet? Clin Cancer Res 18:64–76

Nelson AR, Fingleton B, Rothenberg ML, Matrisian LM (2000) Matrix metalloproteinases: biologic activity and clinical implications. J Clin Oncol 18:1135–1149

Nimmanapalli R, O'Bryan E, Bhalla K (2001) Geldanamycin and its analogue 17-allylamino-17-demethoxygeldanamycin lowers Bcr-Abl levels and induces apoptosis and differentiation of Bcr-Abl-positive human leukemic blasts. Cancer Res 61:1799–1804

Overgaard D, Kaldan G, Marsaa K, Nielsen TL, Shaker SB, Egerod I (2016) The lived experience with idiopathic pulmonary fibrosis: a qualitative study. Eur Respir J 47:1472–1480

Pandey P, Saleh A, Nakazawa A et al (2000) Negative regulation of cytochrome c-mediated oligomerization of Apaf-1 and activation of procaspase-9 by heat shock protein 90. EMBO J 19:4310–4322

Pankey EA, Bhartiya M, Badejo AM Jr (2011) Pulmonary and systemic vasodilator responses to the soluble guanylyl cyclase activator, BAY 60-2770, are not dependent on endogenous nitric oxide or reduced heme. Am J Physiol Heart Circ Physiol 300:H792–H802

Paoli M, Marles-Wright J, Smith A (2002) Structure-function relationships in heme-proteins. DNA Cell Biol. 2002 21:271–280

Park SJ, Kostic M, Dyson HJ (2011) Dynamic interaction of Hsp90 with its client protein p53. J Mol Biol 411:158–173

Pezzulo A, Tudas RA, Stewart CG et al (2018) HSP90 inhibition reverts IL13- and IL17-induced goblet cell metaplasia in human airway epithelia. Am J Respir Crit Care Med 197:A1296

Pillai RN, Ramalingam SS (2012) Hsp90 inhibitors. J Thorac Oncol 7:S407–S408

Pockley AG, Multhoff G (2008) Cell stress proteins in extracellular fluids: friend or foe? Novartis Found Symp 291:86–95

Pockley AG, Muthana M, Calderwood SK (2008) The dual immunoregulatory roles of stress proteins. Trends Biochem Sci 33:71–79

Ponka P (1999) Cell biology of heme. Am J Med Sci 318:241–256

Poulos TL (2014) Heme enzyme structure and function. Chem Rev 114:3919–3962

Prodromou C (2012) The 'active life' of Hsp90 complexes. Biochim Biophys Acta 1823:614–623

Ratajczak T, Ward BK, Walsh JP, Cluning C (2015) Hsp90 as a therapeutic target in endocrinology: current evidence. Res Rep Endocr Disord 5:141–155

Ratzke C, Berkemeier F, Hugel T (2012) Heat shock protein 90's mechanochemical cycle is dominated by thermal fluctuations. Proc Natl Acad Sci U S A 109:161–166

Richard-Fogal CL, Frawley ER, Bonner ER, Zhu H, San Francisco B, Kranz RG (2009) A conserved haem redox and trafficking pathway for cofactor attachment. EMBO J 28:2349–2359

Sandner P, Berger P, Zenzmaier C (2017) The potential of sGC modulators for the treatment of age-related fibrosis: a mini-review. Gerontology 63:216–227

Severance S, Hamza I (2009) Trafficking of heme and porphyrins in metazoa. Chem Rev 109:4596–4616

Shimamura T, Shapiro GI (2008) Heat shock protein 90 inhibition in lung cancer. J Thorac Oncol 3:S152–S159

Sibinska Z, Tian X, Korfei M et al (2017) Amplified canonical transforming growth factor-β signalling *via* heat shock protein 90 in pulmonary fibrosis. Eur Respir J 49:1501941

Sims JD, McCready J, Jay DG (2011) Extracellular heat shock protein (Hsp)70 and Hsp90α assist in matrix metalloproteinase-2 activation and breast cancer cell migration and invasion. PLoS One 6:e18848

Smith AT, Veitch NC (1998) Substrate binding and catalysis in heme peroxidases. Curr Opin Chem Biol 2:269–278

Socinski MA, Goldman J, El-Hariry I et al (2013) A multicenter phase II study of ganetespib monotherapy in patients with genotypically defined advanced non-small cell lung cancer. Clin Cancer Res 19:3068–3077

Song X, Wang X, Zhuo W et al (2010) The regulatory mechanism of extracellular Hsp90{alpha} on matrix metalloproteinase-2 processing and tumor angiogenesis. J Biol Chem 285:40039–40049

Sontake V, Wang Y, Kasam RK et al (2017) Hsp90 regulation of fibroblast activation in pulmonary fibrosis. JCI Insight 2:e91454
Southworth DR, Agard DA (2011) Client-loading conformation of the Hsp90 molecular chaperone revealed in the cryo-EM structure of the human Hsp90: Hop complex. Mol Cell 42:771–781
Spiro TG, Soldatova AV, Balakrishnan G (2013) CO, NO and O_2 as vibrational probes of heme protein interactions. Coord Chem Rev 257:511–527
Stasch JP, Pacher P, Evgenov OV (2011) Soluble guanylate cyclase as an emerging therapeutic target in cardiopulmonary disease. Circulation 123:2263–2273
Steiner H, Kispal G, Zollner A, Haid A, Neupert W, Lill R (1996) Heme binding to a conserved Cys-Pro-Val motif is crucial for the catalytic function of mitochondrial heme lyases. J Biol Chem 271:32605–32611
Taipale M, Jarosz DF, Lindquist S (2010) HSP90 at the hub of protein homeostasis: emerging mechanistic insights. Nat Rev Mol Cell Biol 11:515–528
Taketani S (2005) Aquisition, mobilization and utilization of cellular iron and heme: endless findings and growing evidence of tight regulation. Tohoku J Exp Med 205:297–318
Taldone T, Ochiana SO, Patel PD, Chiosis G (2014) Selective targeting of the stress chaperome as a therapeutic strategy. Trends Pharmacol Sci 35:592–603
Taylor BL, Zhulin IB (1999) PAS domains: internal sensors of oxygen, redox potential, and light. Microbiol Mol Biol Rev 63:479–506
Tóth ME, Gombos I, Sántha M (2015) Heat shock proteins and their role in human diseases. Acta Biol Szeged 59:121–141
Trepel J, Mollapour M, Giaccone G, Neckers L (2010) Targeting the dynamic HSP90 complex in cancer. Nat Rev Cancer 10:537–549
Tsiftsoglou AS, Tsamadou AI, Papadopoulou LC (2006) Heme as key regulator of major mammalian cellular functions: molecular, cellular, and pharmacological aspects. Pharmacol Ther 111:327–345
Verbeke P, Fonager J, Clark BF, Rattan SI (2001) Heat shock response and ageing: mechanisms and applications. Cell Biol Int 25:845–857
Wang X, Dumont ME, Sherman F (1996) Sequence requirements for mitochondrial import of yeast cytochrome c. J Biol Chem 271:6594–6604
Wang X, Song X, Zhuo W et al (2009) The regulatory mechanism of Hsp90alpha secretion and its function in tumor malignancy. Proc Natl Acad Sci U S A 106:21288–21293
Weiss MJ, dos Santos CO (2009) Chaperoning erythropoiesis. Blood 113:2136–2144
West AR, Oates PS (2008) Mechanisms of heme iron absorption: current questions and controversies. World J Gastroenterol 14:4101–4110
Whitesell L, Lindquist SL (2005) HSP90 and the chaperoning of cancer. Nat Rev Cancer 5:761–772
Wijayanti N, Katz N, Immenschuh S (2004) Biology of heme in health and disease. Curr Med Chem 11:981–986
Xu W, Neckers L (2007) Targeting the molecular chaperone heat shock protein 90 provides a multifaceted effect on diverse cell signaling pathways of cancer cells. Clin Cancer Res 13:1625–1629
Yamakura F, Ikeda K (2006) Modification of tryptophan and tryptophan residues in proteins by reactive nitrogen species. Nitric Oxide 14:152–161
Yamamoto M, Takahashi Y, Inano K, Horigome T, Sugano H (1991) Characterization of the hydrophobic region of heat shock protein 90. J Biochem 110:141–145
Yi L, Jenkins PM, Leichert LI, Jakob U, Martens JR, Ragsdale SW (2009) Heme regulatory motifs in heme oxygenase-2 form a thiol/disulfide redox switch that responds to the cellular redox state. J Biol Chem 284:20556–20561
Yoshikawa S, Shinzawa-Itoh K, Nakashima R et al (1998) Redox-coupled crystal structural changes in bovine heart cytochrome c oxidase. Science 280:1723–1729
Yu X, Kong Y, Dore LC et al (2007) An erythroid chaperone that facilitates folding of alpha-globin subunits for hemoglobin synthesis. J Clin Invest 117:1856–1865
Zhang R, Luo D, Miao R et al (2005) Hsp90-Akt phosphorylates ASK1 and inhibits ASK1-mediated apoptosis. Oncogene 24:3954–3963

Chapter 13
Moonlighting Functions of Heat Shock Protein 90

Chang Chen and Constance Jeffery

Abstract Hsp90 is a highly expressed and ubiquitous chaperone in eukaryotes and bacteria. It works with hundreds of client proteins and is regulated by dozens of co-chaperones. Its functions in folding, stabilizing, assembling and disassembling proteins and complexes that are involved in many key processes in the cell, including antigen cross-presentation, stabilization of the cytoskeleton, signaling pathways, stabilization of steroid receptors and other transcription factors, assembly and disassembly of transcription machinery, DNA repair, and the cell cycle. This ubiquitous and versatile intracellular protein is found to have even more functions outside the cell. In this review we discuss the idea that Hsp90 is a moonlighting protein with roles as a secreted cytokine and as a cell surface apoptotic signal and receptor for bacterial cells and lipopolysaccharide.

Keywords Chaperone · Cytokine · HSP90 · Moonlighting protein · Receptor · Secretion

Abbreviations

A2MR	Alpha 2 macroglobulin receptor
BMDC	Bone marrow derived dendritic cells
CD11	Integrin subunit
CD18	Integrin subunit

C. Chen
Department of Bioengineering, University of Illinois at Chicago, Chicago, IL, USA

C. Jeffery (✉)
Department of Bioengineering, University of Illinois at Chicago, Chicago, IL, USA

Department of Biological Sciences, University of Illinois at Chicago, Chicago, IL, USA
e-mail: cjeffery@uic.edu

A. A. A. Asea, P. Kaur (eds.), *Heat Shock Protein 90 in Human Diseases and Disorders*, Heat Shock Proteins 19,
https://doi.org/10.1007/978-3-030-23158-3_13

CXCR4	Chemokine receptor 4
ER	Endoplasmic reticulum
GDF5	Growth differentiation factor 5
HEp-2	Human epithelial type 2
Hsc70	Heat shock cognate 71 kDa protein
hsp75	Heat shock protein 75 kDa mitochondrial
Hsp90	Heat shock protein Hsp 90
HtpG	High temperature protein G/C62.5
LBP	Lipopolysaccharide-binding protein
LOX-1	Lectin-like oxidized LDL receptor-1
LPS	Lipopolysaccharide
LRP-1	LDL receptor-related protein 1/ CD91
MoonProt	Moonlighting proteins database
PTMs	Post-translational modifications
TLR4	Toll-like receptor 4
TRAP1	Heat shock protein 75 kDa mitochondrial

13.1 Introduction

Heat shock protein 90 (Hsp90) is a highly-expressed protein that acts as an ATP-dependent molecular chaperone (Csermely and Kahn 1991; Nadeau et al. 1993; Panaretou et al. 1998; Obermann et al. 1998). It is important for maintaining cellular proteostasis in physiological and stress conditions (reviewed in Schopf et al. 2017; Pearl 2016) through its function in general protein folding and stabilization, including assembly of complexes and aiding client proteins to interact with ligands. By interacting with a diverse set of client proteins, including kinases, nuclear receptors and hundreds of other proteins, it plays key roles in antigen cross-presentation, stabilization of the cytoskeleton, signaling pathways, stabilization of steroid receptors and transcription factors, assembly and disassembly of transcription machinery, DNA repair, the cell cycle and many other cell processes. These numerous and diverse interactions are enabled and regulated by a large number of co-chaperones that modulate its ATP binding and hydrolysis, conformational changes, and interactions with client proteins (reviewed in Schopf et al. 2017).

Hsp90 is highly conserved in eukaryotes and bacteria, but it is not found in archaea (Chen et al. 2006). Eukaryotes have two Hsp90 proteins in the cytoplasm, Hsc82 and Hsp82 in the yeast *S. cerevisiae*, and Hsp90α and Hsp90β in *Homo sapiens* (Rebbe et al. 1987). There are also versions in the endoplasmic reticulum (endoplasmin/Grp94/Hsp90B1), mitochondria (hsp75/TRAP1) (Song et al. 1995) and chloroplasts (Hsp90c). Bacteria usually contain only one Hsp90 gene. HtpG (High temperature protein G/C62.5) is the homologue in *E. coli* (Bardwell and Craig 1987). Most of the discussion below about the moonlighting functions of Hsp90 is about mammalian Hsp90α unless indicated otherwise.

In addition to its many roles inside the cell, Hsp90 is a moonlighting protein with several more functions outside the cell. Moonlighting proteins comprise a group of multifunctional proteins that have multiple biochemical or biophysical functions performed by a single polypeptide chain (Jeffery 1999). They do not include proteins that are considered multifunctional due to gene fusions, different versions of the protein due to multiple RNA splice variants, or pleiotropic effects. Taxon specific crystallins in the lens of the eye of several species were among the first proteins to be identified as moonlighting proteins. Zeta crystallin is the same protein as a quinone oxidoreductase, which catalyzes the conversion between quinone and semiquinone (Rao et al. 1992). Delta-2 crystallin is an arginosuccinate lyase, which cleaves arginosuccinate to produce arginine and fumarate in the urea cycle (Wistow and Piatigorsky 1990). Lambda crystallin is also an enzyme, L-gulonate 3-dehydrogenase, and catalyzes NAD-linked dehydrogenation in the urinate cycle (Ishikura et al. 2005). As another type of example, aconitase, an enzyme in the citric acid cycle that uses an iron-sulfur cluster in its active site, has a second function in which it binds to mRNA and regulates translation of proteins involved in iron uptake (Philpott et al. 1994; Banerjee et al. 2007; Kennedy et al. 1992).

Additional examples of protein moonlighting are found throughout the evolutionary tree – in plants, bacteria, archaea, insects, mammals, including several proteins in humans. Moonlighting proteins have been found that perform different functions when in different locations within a cell, expressed in different cell types, as part of different multimers, or upon binding of substrates, products, cofactors or other small molecule ligands (Jeffery 1999, 2004, Jeffery 2009). The Moonlighting Proteins Database (MoonProt, www.moonlightingproteins.org), launched online in 2014 (Mani et al. 2014), contains more than 300 moonlighting proteins. They cover diverse types of proteins including, but not limited to, enzymes, transcription factors, chaperones, receptors, and ribosomal proteins. Some perform their distinct functions at different times, but others perform multiple functions simultaneously. The ability of a single protein to participate in multiple cellular activities can be valuable to the cell, for example, in coordinating multiple biochemical processes or metabolic pathways. In addition to Hsp90, the known moonlighting proteins include many other heat shock proteins and chaperones, including DnaK, Hsp60/GroEL, Hsp70, DegP, FtsH, calreticulin, DegQ, peroxiredoxin, and protein deglycase 1 (reviewed in Jeffery 2018; Pockley and Henderson 2017; Chen et al. 2017).

13.2 Moonlighting Functions of Hsp90

13.2.1 Hsp90 on the Cell Surface

Some of the moonlighting intracellular chaperones, as well as dozens of other intracellular proteins, have been found to perform a second function on the cell surface. Most often, they act as adhesins that interact with other cell types, soluble host proteins, or extracellular matrix (Kainulainen and Korhonen 2014; Jeffery 2018), including DnaK from *Bifidobacterium* (Ruiz et al., 2011), *Lactococcus lactis* (Yuan and Wong, 1995), *Neisseria meningititis* (Tzeng et al., 2008), and *Mycobacterium tuberculosis* (Kennaway et al., 2005), Hsp60 from *Legionella pneumophila* (Garduño et al., 1998) and *Listeria* (Kim et al., 2006), and human calreticulin (Saito et al. 1999). How or why these particular proteins are secreted and attached to the cell surface is not known; their physiochemical features are similar to other cytosolic proteins (Amblee and Jeffery 2015). Proteomics studies of proteins on the cell surface of dozens of species have identified many additional intracellular proteins on the cell surface (reviewed in Wang and Jeffery 2016), but it is not yet clear if those proteins also have a second function on the cell surface or the same function as inside the cell. In some cases, the proteins might have been identified in the surface proteomics studies due to experimental challenges such as being closely associated with cytoplasmic domains of transmembrane proteins. One such study identified Hsp90 on the surface of *E. coli* strain BL21 (Thein et al. 2010), although its function on the bacterial surface is unknown.

Hsp90 has been found to have additional functions when displayed on the surface of mammalian cells. The amount of Hsp90 secreted was shown to increase due to a number of factors, including activation of endothelial cells and interaction of cells with extracellular matrix components like fibronectin, with much more of the Hsp90α isoform than the Hsp90β secreted (Song and Luo 2010). One extracellular function of Hsp90 is as a signal on the surface of human apoptotic cells to trigger engulfment by dendritic cells. Apoptosis is important for development, differentiation and as a response to stresses that damage cells, but it's important to remove these dying cells to maintain tissue homeostasis. Zhu and coworkers found that the display of Hsp90, along with Hsp60 and Hsp70, is an early response in apoptotic process of EL4, E.G7, and HL60 cells to a variety of cell stresses including uv light and cisplatin (Zhu et al. 2016). The Hsp proteins appeared on the cell surface even before phosphatidylserine, an early apoptotic signal, became exposed on the surface. Once on the cell surface, the Hsp serve as an "eat me" signal to prompt bone marrow derived dendritic cells (BMDC) to phagocytose the dying cells (Fig. 13.1). The phagocytic cells also engulf latex beads covered with Hsp60, 70 or 90, but the internalization was prevented by the addition of soluble Hsp60, 70, or 90, presumably because they block the receptor for the Hsp on the dendritic cells. The receptor for Hsp90 on the dendritic cells was found to be the lectin-like oxidized LDL receptor-1 (LOX-1). Expression of LOX-1 on the surface of CHO cells enabled Hsp90 also to bind to those cells.

Hsp90 on mammalian cell surfaces is also involved in sensing bacterial proteins and initiating an immune response. The cell surface protein JlpA on *Campylobacter jejuni*, a common cause of food poisoning, was found to interact directly and specifically with cell surface-exposed Hsp90α on human epithelial type 2 (HEp-2) cells (Jin et al. 2003). Binding resulted in the initiation of signaling pathways involving NF-κB and p38 MAP kinase that lead to activation of proinflammatory immune responses. The authors noted that because Hsp90α does not contain a transmembrane domain, there must be at least one additional cell surface protein involved that can transduce the signal into the cell. The cell surface protein NadA from *Neisseria*

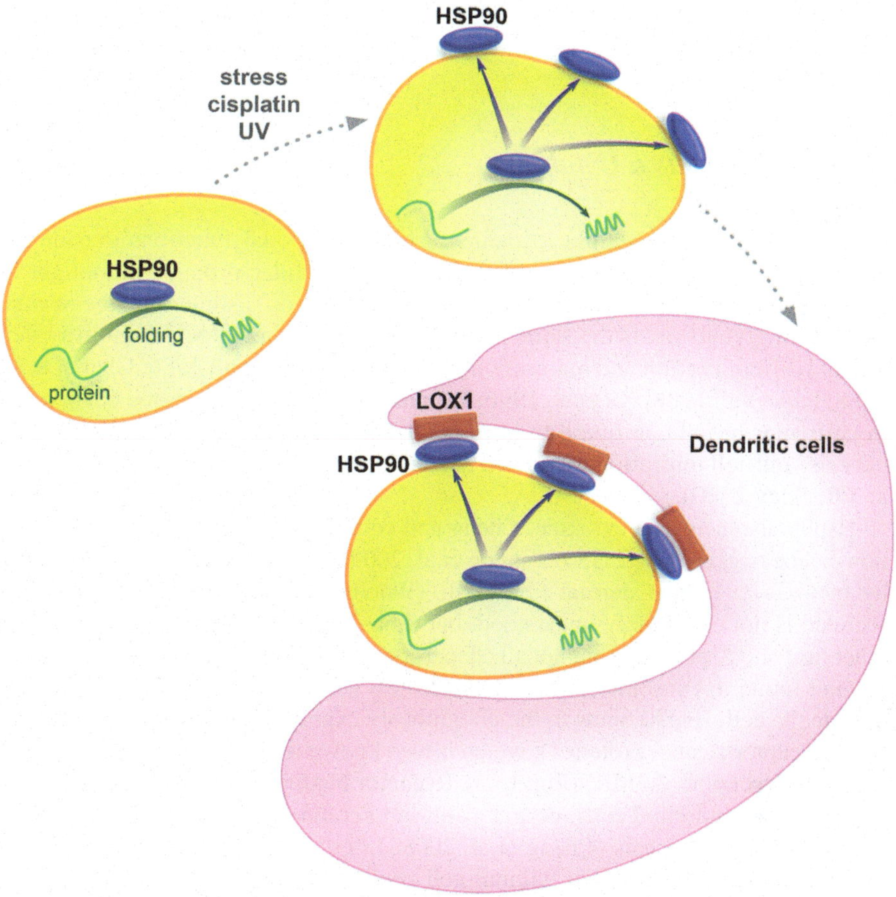

Fig. 13.1 Surface expression of Hsp90 triggers engulfment by dendritic cells. Hsp90 functions as a chaperone inside the cell under nonstress conditions. It becomes expressed on the cell surface as an "eat me" signal in response to stresses that trigger apoptosis, including UV light and cisplatin. The secreted Hsp90 binds to the LOX-1 receptor on dendritic cells (pink), and the dendritic cells engulf and destroy the cells displaying Hsp90

meningitidis also binds to Hsp90α on the surface of monocytes (Cecchini et al. 2011). Binding results in cell activation and the induction of cytokine and chemokine secretion.

In addition to detecting bacterial proteins, Hsp90 also acts as part of a complex that binds to bacterial lipopolysaccharide (LPS) and triggers an immune response (Triantafilou and Triantafilou 2002). After lipopolysaccharide-binding protein (LBP) binds and transfers the LPS to membrane-bound CD14, the LPS is passed into the membrane where it binds to a complex of receptors (Triantafilou et al. 2001a, b). The complex includes chemokine receptor 4 (CXCR4), Hsc70, Hsp90α, and growth differentiation factor 5 (GDF5). This complex recruits Toll-like receptor

4 (TLR4) and/or integrins (CD11 or CD18) to act as transmembrane signaling proteins to activate cytoplasmic signaling pathways (Triantafilou et al. 2001a, b).

13.2.2 *Hsp90 as a Cytokine*

Another way for intracellular proteins to have additional functions outside the cell is as a soluble signaling protein. Many intracellular proteins moonlight as secreted cytokines and growth factors. Phosphoglucose isomerase, the second enzyme in glycolysis, is the same protein as neuroleukin, autocrine motility factor, and differentiation and maturation mediator (Gurney et al. 1986; Chaput et al. 1988; Faik et al. 1988; Xu et al. 1996; Watanabe et al. 1996). Secreted chaperones in particular have been found to have many roles in immunomodulation, angiogenesis, and cell migration (reviewed in Henderson and Pockley 2005; Henderson and Pockley 2010).

Hsp90 also acts as a cytokine. Cheng and coworkers showed that Hsp90 secretion can be triggered by TGF-α (Cheng et al. 2008). Secretion can also be triggered by hypoxia in human dermal fibroblasts (Woodley et al. 2009). How Hsp90 is secreted is not completely understood, but it has been found together with several other heat shock proteins (but not all chaperones) in the lumen of exosomes, and the amount found in exosomes increases under heat stress (Clayton et al. 2005). Once outside the cell, Hsp90 binds to the ubiquitously expressed surface receptor LRP-1 (LDL receptor-related protein 1), which is also known as CD91 or the alpha 2 macroglobulin receptor (A2MR) (Fig. 13.2). Receptor binding causes an increase in cell migration by epidermal cells, dermal cells and keratinocytes and promotes wound healing and angiogenesis (Cheng et al. 2008; Li et al. 2007; Song and Luo 2010; Jayaprakash et al. 2015), but without cell proliferation. The interaction with the alpha2 macroglobulin receptor CD91 is also important in another pathway. Hsp90, along with heat shock proteins gp96 and Hsp70, can form a complex with antigens and bind to CD91 for aiding uptake of the antigen by antigen-presenting cells (Basu et al. 2001).

An understanding of how Hsp90 performs its immunomodulatory, angiogenesis, and wound healing activities may aid in the development of novel treatments for promoting wound healing, which is especially needed for the treatment of the difficult to treat skin wounds and ulcers common in diabetic patients. It was shown that topical Hsp90α speeds up wound healing in mice (Li et al. 2007). The Cheng group also demonstrated that application of a purified 115 amino acid fragment of Hsp90α accelerated both acute and diabetic wound healing in mice (Chen et al. 2011). Interestingly, only the Hsp90α isoform promotes wound healing, and the other major isoform, Hsp90β, does not. As mentioned above, the Hsp90β isoform is not secreted at the same level as Hsp90α, and it does not have significant effects on cell migration (Jayaprakash et al. 2015). Hsp90's role as a cytokine is also often co-opted by cancer cells, as reviewed by Hance and co-workers (Hance et al. 2014). A

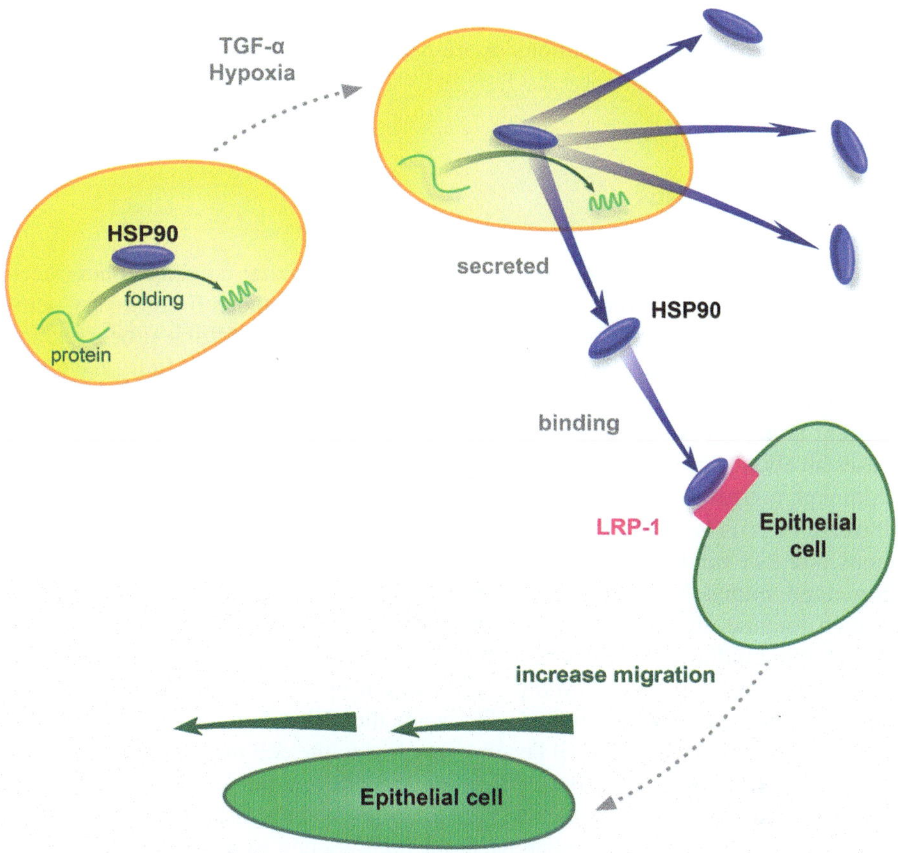

Fig. 13.2 Secreted Hsp90 interacts with the LRP-1 receptor on epithelial cells. Hsp90 secretion is triggered by TGF-α or HIF-1-α. Secreted Hsp90 binds to the LRP-1 receptor on epithelial cells (light green). Receptor binding causes an increase in epithelial cell migration (dark green cells and arrows)

better understanding of its roles in cancer cell migration could lead to novel therapeutics for metastasis.

13.2.3 Potential Kinase and Autophosphorylating Activity

It's interesting to note that in addition to using ATP during the chaperone function, several studies have shown that Hsp90 homologues can undergo autophosphorylation or even phosphorylate other proteins. This includes Hsp90 homologues from *Brassica napus* (rapeseed plant) (Park et al. 1998) and rat (Csermely and Kahn 1991; Langer et al. 2002). The chaperone function and the extracellular functions do not involve autophosphorylation of Hsp90 or phosphorylation of client proteins or co-chaperones by Hsp90, so whether or not these activities have an as yet unknown role in any of the known functions or are part of an additional moonlighting function, is not clear.

13.3 Conclusions

Hsp90's many activities and interactions with client proteins, co-chaperones, receptors, and other macromolecules both inside and outside the cell are a very active area of research. Because of its key roles in many central cellular processes, it is important in both health and disease, and inhibitors of Hsp90 activity have been found to be effective anticancer therapeutics (Hance et al. 2014). Hsp90 is also a current target for developing treatments for other diseases as well, including diseases involving immune pathways (Neckers and Workman 2012; Verma et al. 2016). Although a great deal has been learned about its many roles as a chaperone inside the cell, there is less information about how Hsp90 performs its extracellular functions. Are ATP binding and hydrolysis needed? Are any additional proteins needed? Structural studies have identified multiple domains and conformations of Hsp90. Which domains are involved in the moonlighting function? Which amino acids are involved in binding to receptor? In what conformation, "closed", "open", extended, does it bind to the receptor? Many additional questions are common to the dozens of intracellular proteins that moonlight outside the cell and on the cell surface. How is Hsp90 secreted? How does it become attached to the cell membrane? Why is the Hsp90α isoform secreted from cells when the other major cytoplasmic isoform, Hsp90β, is not? Are the mitochondrial, chloroplast, and ER Hsp90 homologues also moonlighting proteins? Do other bacterial homologues have moonlighting functions outside the cell? What is the mechanism to increase secretion of Hsp90 under some cellular conditions? It undergoes phosphorylation by several kinases as well as other post-translational modifications (PTMs) (Mollapour and Neckers 2012), and PTMs have been found to play a role in switching functions in many other moonlighting proteins (Jeffery 2016). Do any of the PTMs affect secretion of Hsp90? Although much has been learned about the Hsp90 protein family in recent decades, there is still much to be learned. Understanding the molecular mechanisms of Hsp90's extracellular functions is needed to help clarify its roles in health and disease and could provide information about novel cellular pathways including how moonlighting proteins like Hsp90 are secreted. In addition, elucidating the connections between sequence, structure, and function in moonlighting proteins like members of the Hsp90 family is also needed to increase our general understanding of proteins, for example, improving our ability to predict all of the functions of a protein, understanding how protein functions evolve, and designing novel proteins.

Acknowledgements Research on this project in the Jeffery lab is supported by an award from the University of Illinois Cancer Center.

References

Amblee V, Jeffery CJ (2015) Physical features of intracellular proteins that moonlight on the cell surface. PLoS One 10:e0130575

Banerjee S, Nandyala AK, Raviprasad P, Ahmed N, Hasnain SE (2007) Iron-dependent RNA-binding activity of *Mycobacterium tuberculosis* aconitase. J Bacteriol 189:4046–4052

Bardwell JC, Craig EA (1987) Eukaryotic Mr 83,000 heat shock protein has a homologue in *Escherichia coli*. Proc Natl Acad Sci USA 84:5177–5181

Basu S, Binder RJ, Ramalingam T, Srivastava PK (2001) CD91 is a common receptor for heat shock proteins gp96, hsp90, hsp70, and calreticulin. Immunity 14:303–313

Cecchini P, Tavano R, de Laureto PP, Franzoso S, Mazzon C, Montanari P, Papini E (2011) The soluble recombinant *Neisseria meningitidis* adhesin NadAΔ351–405 stimulates human monocytes by binding to extracellular Hsp90. PLoS One 6:e25089

Chaput M, Claes V, Portetelle D, Cludts I, Cravador A, Burny A, Gras H, Tartar A (1988) The neurotrophic factor neuroleukin is 90% homologous with phosphohexose isomerase. Nature 332:454–455

Chen B, Zhong D, Monteiro A (2006) Comparative genomics and evolution of the HSP90 family of genes across all kingdoms of organisms. BMC Genomics 7:156

Chen C, Zabad S, Liu H, Wang W, Jeffery C (2017) MoonProt 2.0: an expansion and update of the moonlighting proteins database. Nucleic Acids Res 46:D640–D644

Cheng CF, Fan J, Fedesco M, Guan S, Li Y, Bandyopadhyay B, Bright AM, Yerushalmi D, Liang M, Chen M, Han YP, Woodley DT, Li W (2008) Transforming growth factor α (TGFα)-stimulated secretion of HSP90α: using the receptor LRP-1/CD91 to promote human skin cell migration against a TGFβ-rich environment during wound healing. Mol Cell Biol 28:3344–3358

Cheng CF, Sahu D, Tsen F, Zhao Z, Fan J, Kim R, Wang X, O'Brien K, Li Y, Kuang Y, Chen M, Woodley DT, Li W (2011) A fragment of secreted Hsp90α carries properties that enable it to accelerate effectively both acute and diabetic wound healing in mice. J Clin Invest 121:4348–4361

Clayton A, Turkes A, Navabi H, Mason MD, Tabi Z (2005) Induction of heat shock proteins in B-cell exosomes. J Cell Sci 118:3631–3638

Csermely P, Kahn CR (1991) The 90-kDa heat shock protein (hsp-90) possesses an ATP binding site and autophosphorylating activity. J Biol Chem 266:4943–4950

Faik P, Walker JI, Redmill AA, Morgan MJ (1988) Mouse glucose-6-phosphate isomerase and neuroleukin have identical 3′ sequences. Nature 332:455–457

Garduño RA, Garduño E, Hoffman PS (1998) Surface-associated hsp60 chaperonin of *Legionella pneumophila* mediates invasion in a HeLa cell model. Infect Immun 66:4602–4610

Gurney ME, Heinrich SP, Lee MR, Yin HS (1986) Molecular cloning and expression of neuroleukin, a neurotrophic factor for spinal and sensory neurons. Science 234:566–574

Hance M, Nolan K, Isaacs J (2014) The double-edged sword: conserved functions of extracellular hsp90 in wound healing and cancer. Cancers 6:1065–1097

Henderson B, Pockley AG (eds) (2005) Molecular chaperones and cell signaling. Cambridge University Press, Cambridge

Henderson B, Pockley AG (2010) Molecular chaperones and protein-folding catalysts as intercellular signaling regulators in immunity and inflammation. J Leukoc Biol 88:445–462

Ishikura S, Usami N, Araki M, Hara A (2005) Structural and functional characterization of rabbit and human L-gulonate 3-dehydrogenase. J Biochem 137:303–314

Jayaprakash P, Dong H, Zou M, Bhatia A, O'Brien K, Chen M, Woodley DT, Li W (2015) HSP90α and HSP90β co-operate a stress-response mechanism to cope with hypoxia and nutrient paucity during wound healing. J Cell Sci 128:1475–1480

Jeffery CJ (1999) Moonlighting proteins. Trends Biochem 24:8–11

Jeffery CJ (2004) Molecular mechanisms for multitasking: recent crystal structures of moonlighting proteins. Curr Opin Struct Biol 14:663–668

Jeffery CJ (2009) Moonlighting proteins—an update. Mol BioSyst 5:345–350

Jeffery CJ (2016) Protein species and moonlighting proteins: very small changes in a protein's covalent structure can change its biochemical function. J Proteome 134:19–24

Jeffery C (2018) Intracellular proteins moonlighting as bacterial adhesion factors. AIMS Microbiol 4:362–376

Jin S, Song YC, Emili A, Sherman PM, Chan VL (2003) JlpA of campylobacter jejuni interacts with surface-exposed heat shock protein 90α and triggers signalling pathways leading to the activation of NF-κB and p38 MAP kinase in epithelial cells. Cell Microbiol 5:165–174

Kainulainen V, Korhonen T (2014) Dancing to another tune—adhesive moonlighting proteins in bacteria. Biology 3:178–204

Kennaway CK, Benesch JL, Gohlke U, Wang L, Robinson CV, Orlova EV, Saibil HR, Keep NH (2005) Dodecameric structure of the small heat shock protein Acr1 from *Mycobacterium tuberculosis*. J Biol Chem 280:33419–33425

Kennedy MC, Mende-Mueller L, Blondin GA, Beinert H (1992) Purification and characterization of cytosolic aconitase from beef liver and its relationship to the iron-responsive element binding protein. Proc Natl Acad Sci USA 89:11730–11734

Kim KP, Jagadeesan B, Burkholder KM, Jaradat ZW, Wampler JL, Lathrop AA, Morgan MT, Bhunia AK (2006) Adhesion characteristics of Listeria adhesion protein (LAP)-expressing *Escherichia coli* to Caco-2 cells and of recombinant LAP to eukaryotic receptor Hsp60 as examined in a surface plasmon resonance sensor. FEMS Microbiol Lett 256:324–332

Langer T, Schlatter H, Fasold H (2002) Evidence that the novobiocin-sensitive ATP-binding site of the heat shock protein 90 (hsp90) is necessary for its autophosphorylation. Cell Biol Int 26:653–657

Li W, Li Y, Guan S, Fan J, Cheng CF, Bright AM, Chinn C, Chen M, Woodley DT (2007) Extracellular heat shock protein-90α: linking hypoxia to skin cell motility and wound healing. EMBO J 26:1221–1233

Mani M, Chen C, Amblee V, Liu H, Mathur T, Zwicke G, Zabad S, Patel B, Thakkar J, Jeffery CJ (2014) MoonProt: a database for proteins that are known to moonlight. Nucleic Acids Res 43:D277–D282

Mollapour M, Neckers L (2012) Post-translational modifications of Hsp90 and their contributions to chaperone regulation. Biochim Biophys Acta 1823:648–655

Nadeau K, Das A, Walsh CT (1993) Hsp90 chaperonins possess ATPase activity and bind heat shock transcription factors and peptidyl prolyl isomerases. J Biol Chem 268:1479–1487

Neckers L, Workman P (2012) Hsp90 molecular chaperone inhibitors: are we there yet? Clin Cancer Res 18:64–76

Obermann WM, Sondermann H, Russo AA, Pavletich NP, Hartl FU (1998) In vivo function of Hsp90 is dependent on ATP binding and ATP hydrolysis. J Cell Biol 143:901–910

Panaretou B, Prodromou C, Roe SM, O'Brien R, Ladbury JE, Piper PW, Pearl LH (1998) ATP binding and hydrolysis are essential to the function of the Hsp90 molecular chaperone in vivo. EMBO J 17:4829–4836

Park M, Kang CY, Krishna P (1998) *Brassica napus* hsp90 can autophosphorylate and phosphorylate other protein substrates. Mol Cell Biochem 185:33–38

Pearl LH (2016) The HSP90 molecular chaperone—an enigmatic ATPase. Biopolymers 105:594–607

Philpott CC, Klausner RD, Rouault TA (1994) The bifunctional iron-responsive element binding protein/cytosolic aconitase: the role of active-site residues in ligand binding and regulation. Proc Natl Acad Sci USA 91:7321–7325

Pockley AG, Henderson B (2017) Extracellular cell stress (heat shock) proteins—immune responses and disease: an overview. Philos Trans R Soc B Biol Sci 373:20160522

Rao PV, Krishna CM, Zigler JS (1992) Identification and characterization of the enzymatic activity of zeta-crystallin from Guinea pig lens. A novel NADPH: quinone oxidoreductase. J Biol Chem 267:96–102

Rebbe NF, Ware J, Bertina RM, Modrich P, Stafford DW (1987) Nucleotide sequence of a cDNA for a member of the human 90-kDa heat-shock protein family. Gene 53:235–245

Ruiz L, Ruas-Madiedo P, Gueimonde M, Clara GDLR, Margolles A, Sánchez B (2011) How do bifidobacteria counteract environmental challenges? Mechanisms involved and physiological consequences. Genes Nutr 6:307–318

Saito Y, Ihara Y, Leach MR, Cohen-Doyle MF, Williams DB (1999) Calreticulin functions in vitro as a molecular chaperone for both glycosylated and non-glycosylated proteins. EMBO J 18:6718–6729

Schopf FH, Biebl MM, Buchner J (2017) The HSP90 chaperone machinery. Nat Rev Mol Cell Biol 18:345–360

Song X, Luo Y (2010) The regulatory mechanism of Hsp90α secretion from endothelial cells and its role in angiogenesis during wound healing. Biochem Biophys Res Commun 398:111–117

Song HY, Dunbar JD, Zhang YX, Guo D, Donner DB (1995) Identification of a protein with homology to hsp90 that binds the type 1 tumor necrosis factor receptor. J Biol Chem 270:3574–3581

Thein M, Sauer G, Paramasivam N, Grin I, Linke D (2010) Efficient subfractionation of gram-negative bacteria for proteomics studies. J Proteome Res 9:6135–6147

Triantafilou M, Triantafilou K (2002) Lipopolysaccharide recognition: CD14, TLRs and the LPS-activation cluster. Trends Immunol 23:301–304

Triantafilou K, Triantafilou M, Dedrick RL (2001a) A CD14-independent LPS receptor cluster. Nat Immunol 2:338 345

Triantafilou K, Triantafilou M, Ladha S, Mackie A, Dedrick RL, Fernandez N, Cherry R (2001b) Fluorescence recovery after photobleaching reveals that LPS rapidly transfers from CD14 to hsp70 and hsp90 on the cell membrane. J Cell Sci 114:2535–2545

Tzeng YL, Kahler CM, Zhang X, Stephens DS (2008) MisR/MisS two-component regulon in *Neisseria meningitidis*. Infect Immun 76:704–716

Verma S, Goyal S, Jamal S, Singh A, Grover A (2016) Hsp90: friends, clients and natural foes. Biochimie 127:227–240

Wang W, Jeffery CJ (2016) An analysis of surface proteomics results reveals novel candidates for intracellular/surface moonlighting proteins in bacteria. Mol BioSyst 12:1420–1431

Watanabe H, Takehana K, Date M, Shinozaki T, Raz A (1996) Tumor cell autocrine motility factor is the neuroleukin/phosphohexose isomerase polypeptide. Cancer Res 56:2960–2963

Wistow GJ, Piatigorsky J (1990) Gene conversion and splice-site slippage in the argininosuccinate lyases/δ-crystallins of the duck lens: members of an enzyme superfamily. Gene 96:263–270

Woodley DT, Fan J, Cheng CF, Li Y, Chen M, Bu G, Li W (2009) Participation of the lipoprotein receptor LRP1 in hypoxia-HSP90α autocrine signaling to promote keratinocyte migration. J Cell Sci 122:1495–1498

Xu W, Seiter K, Feldman E, Ahmed T, Chiao JW (1996) The differentiation and maturation mediator for human myeloid leukemia cells shares homology with neuroleukin or phosphoglucose isomerase. Blood 87:4502–4506

Yuan G, Wong SL (1995) Isolation and characterization of *Bacillus subtilis* groE regulatory mutants: evidence for orf39 in the dnaK operon as a repressor gene in regulating the expression of both groE and dnaK. J Bacteriol 177:6462–6468

Zhu H, Fang X, Zhang D, Wu W, Shao M, Wang L, Gu J (2016) Membrane-bound heat shock proteins facilitate the uptake of dying cells and cross-presentation of cellular antigen. Apoptosis 21:96–109

Chapter 14
Hsp90 as a Member of Dicarboxylate Clamp TPR Protein Interaction Network: Implication in Human Diseases and Prospect as a Drug Target

Rajnish Kumar, Bengt Winblad, and Pavel F. Pavlov

Abstract Heat shock protein (Hsp) 90 kDa is a widely expressed molecular chaperone and is involved in folding of broad range of client proteins, intracellular transport and degradation of damaged and misfolded proteins. The function of Hsp90 is mediated through its partner co-chaperones, which either affects the ATPase activity or directly helps Hsp90 to interact with its specific client proteins. Tetratricopeptide repeat (TPR) domain containing proteins represent a major class of co-chaperones which interact with the extreme C-terminus of Hsp90 through a dicarboxylate clamp mechanism. We have recently suggested that Hsp90 and Hsp70 molecular chaperones belong to dicarboxylate clamp protein interaction network where proteins containing similar C-terminus as that of Hsp90/Hsp70 interact with TPR motif containing proteins through dicarboxylate clamp mechanism. Recent findings suggest that several of TPR co-chaperones have been involved in variety of human diseases such as tauopathy and amyloidopathy in Alzheimer's disease, cancer, metabolic disorders, inflammation and others. In this chapter, we discuss the potential of Hsp90 TPR containing co-chaperones as drug targets in human disorders.

Keywords Alzheimer's disease · Dicarboxylate clamp · Molecular chaperones · Protein interaction network · Tetratricopeptide repeat motif · TPR proteins

---

R. Kumar
Centre for Alzheimer Research, Department of Neurobiology, Care Sciences and Society, Division of Neurogeriatrics, Karolinska Institute, Solna, Sweden

B. Winblad · P. F. Pavlov (✉)
Centre for Alzheimer Research, Department of Neurobiology, Care Sciences and Society, Division of Neurogeriatrics, Karolinska Institute, Solna, Sweden

Memory Clinic, Theme Aging, Karolinska University Hospital, Huddinge, Sweden
e-mail: Pavel.pavlov@ki.se

A. A. A. Asea, P. Kaur (eds.), *Heat Shock Protein 90 in Human Diseases and Disorders*, Heat Shock Proteins 19,
https://doi.org/10.1007/978-3-030-23158-3_14

Abbreviations

Aha1	Activator of 90 kDa heat shock protein ATPase homolog 1
AIP	Arylhydrocarbon receptor-interacting protein
CFTR	Cystic fibrosis transmembrane conductance regulator
CHIP	C terminus of HSC70-interacting protein
CK1, CK2	Casein kinase 1 and casein kinase 2
Cyp40	Cyclophilin 40 kDa
dcTPR	Dicarboxylate clamp tetratricopeptide repeat
FKBP	FK506-binding protein
GO	Gene ontology
GR	Glucocorticoid receptor
GSK3β	Glycogen synthase kinase 3 beta
HCV	Hepatitis C virus
HOP	Hsp70-Hsp90 organizing protein
Hsp	Heat shock protein
p23	Prostaglandin E synthase 3
PP5	Protein phosphatase 5
PPIase	Peptidyl-prolyl cis-trans isomerase
PPIs	Protein-protein interactions
SGTA	Small, glutamine-rich, tetratricopeptide repeat protein alpha
Tom70	Translocase of the outer mitochondrial membrane
TPR	Tetratricopeptide repeat

14.1 Introduction

Heat shock protein (Hsp) 90 is an abundantly expressed ATP-dependent molecular chaperone which is involved in a variety of cellular functions such as regulation of proteolysis, supervision of protein folding, activation, aggregation and degradation. The function of Hsp90 is dependent on ATP hydrolysis and complexation with its co-chaperones (Ferraro et al. 2018). Hsp90α which is induced under heat response and Hsp90β which is constitutively expressed are the two isoforms present in mammalian cells. Impairment in Hsp90 cellular functions is associated with development of several diseases namely neurodegeneration, cancer, infections and inflammation (Jolly and Morimoto 2000; Luo et al. 2010).

Structurally, Hsp90 is present as a homodimer (Verba et al. 2016) containing three different functional domains. In the open, ATP free state Hsp90 dimerization is mediated only by C-terminal domain. The N-terminal domain has an ATP binding pocket and it upon binding of the ATP leads to formation of a closed state where the N-terminal domains also forms the dimer. The middle domain of the Hsp90 is responsible for binding to the client proteins whereas the C-terminal domain which ends in MEEVD peptide is responsible for binding to the tetratricopeptide repeat (TPR) containing proteins, which upon binding also forms the Hsp90 dimer (Fig. 14.1).

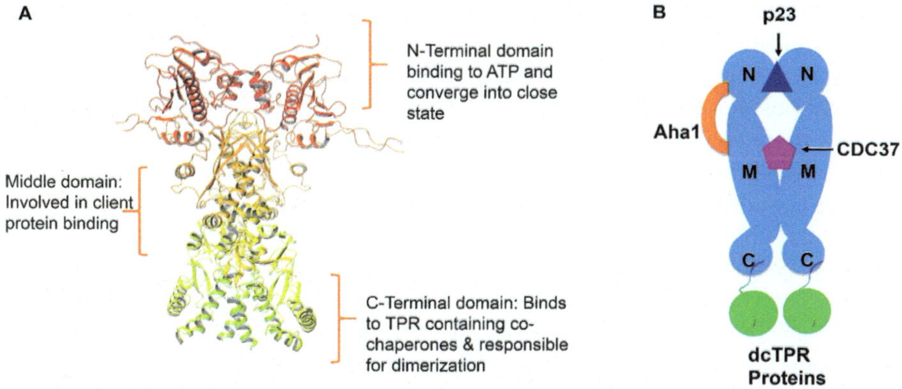

Fig. 14.1 Hsp90 and its co-chaperones. (**a**) The 3D ribbon representation of cryo-EM structure of Hsp90 dimer (Verba et al. 2016). The ribbons are rainbow colored from N terminus to C terminus and the image is made by PyMol (Schrodinger 2015). (**b**) The schematic representation of binding of different co-chaperones to the Hsp90 dimer

Hsp90 is known to bind to several diverse co-chaperones which regulate its functional activity by binding to its different domains. Structural and functional data of the Hsp90 super-complex with its co-chaperones has been nicely reviewed by Buchner et al., (Schopf et al. 2017). Figure 14.1b depicts the schematics of binding of prostaglandin E synthase 3 (p23), activator of 90 kDa heat shock protein ATPase homolog 1 (Aha1), CDC37 and dicarboxylate clamp tetratricopeptide repeat (dcTPR) containing co-chaperones to the Hsp90 dimer. In an effort to explore the vast network of Hsp90 interactors, a thorough protein-protein interactions (PPIs) database of Hsp90 chaperone machinery is maintained by Didier Picard (https://www.picard.ch/Hsp90Int/index.php) based on parallel mining of all the major PPIs databases (Echeverría et al. 2011). The Hsp90 co-chaperones can be largely classified into TPR containing and non-TPR co-chaperones. In this chapter, we discuss the TPR containing co-chaperones and Hsp90 as a part of dcTPR interaction network, including their involvement in the disease and their prospects as a target for drug design and development.

14.2 The dcTPR Domain Containing Co-chaperones of Hsp90

The dcTPR domain containing co-chaperones is among the major class of Hsp90/Hsp70 co-chaperones and their interaction mechanism is well understood through structural biology. The TPR motif represents a repeat of 34 amino acid residues in the primary structure of the protein and consists of 7 antiparallel alpha helices. Out of the several hundred TPR motif-containing proteins annotated in the human database ~20 different proteins are known to interact with Hsp70 and Hsp90 via the dicarboxylate clamp mechanism, (Scheufler et al. 2000). In a systematic study Richter et al. screened the TPR domain containing proteins of *C. elegans* with potential to interact with Hsp90/70 and ranked them on the basis of their similarity with the known dcTPR proteins (Haslbeck et al. 2013). The study identified 49 dcTPR proteins in humans on the basis of a score above the cutoff value. Out of these, many proteins are well characterized and known to interact with either Hsp90 or Hsp70 while several of them are still needed to be characterized.

The dcTPR containing co-chaperones are involved in several functions such as regulation of conformational dynamics of the Hsp90 super-complex with client proteins, selection and recruitment of clients, and client maturation. For example, Hsp70-Hsp90 Organizing Protein (HOP)/Sti1 is known to function as an adaptor for Hsp90 and Hsp70 and also inhibits the ATPase activity of Hsp90 (Johnson et al. 1998; Prodromou et al. 1999). Another dcTPR co-chaperone, protein phosphatase 5 (PP5)/Ppt1 is involved in dephosphorylation of Hsp90 and thereby modulating its conformational dynamics and client maturation (Soroka et al. 2012). Several members of the dcTPR containing co-chaperones of Hsp90/70 contains a peptidyl-prolyl cis-trans isomerase (PPIase) domain and belongs to group of FK506 binding protein (FKBP) such as FKBP51, FKBP52, FKBP38, FKBP37 also known as arylhydrocarbon receptor-interacting protein (AIP), FKBP36, and FKBPL proteins. These PPIase containing dcTPR co-chaperones are involved in selection, recruitment and folding of client proteins (Bose et al. 1996; Guy et al. 2015). Other functions of dcTPR co-chaperones involves protein ubiquitination (C terminus of HSC70-interacting protein; CHIP), mitochondrial targeting (TOM70, TOM34), folding of specific substrates (UNC45) etc.

14.3 Dicarboxylate Clamp Mechanism

The TPR domain containing molecular co-chaperones interact with Hsp90/Hsp70 through formation of so called dicarboxylate clamp which represent the conserved electrostatically driven type of interaction between residues present in the TPR domain and two carboxylic groups of C-terminal aspartate or to lesser extent glutamate residues. Other residues within Hsp90 C-terminal peptide MEEVD form additional electrostatic and hydrophobic contacts within the binding pocket on TPR proteins. The dicarboxylate clamp mechanism is conserved in most of the TPR-Hsp90/Hsp70 interactions. In a recent study, we reported the crystal structure of FKBP51 in complex with the C-terminal MEEVD peptide of Hsp90 in order to elucidate the mechanism of their interaction additionally using the conservational profile of several of the dcTPR proteins. In case of FKBP51, the K352 and K272 which formed hydrogen bond and salt bridge respectively with the Hsp90 C-terminal MEEVD peptide were conserved. Additionally, K329, N322 and E273 involved in formation of contacts with the Hsp90 C-terminal were also shown to be conserved (Kumar et al. 2017). A representative dicarboxylate clamp formed by the interaction between TPR domain of FKBP38 and MEEVD peptide (PDB ID: 5MGX) is shown in Fig. 14.2. The MEEVD peptide sits very firmly into the TPR domain and the it

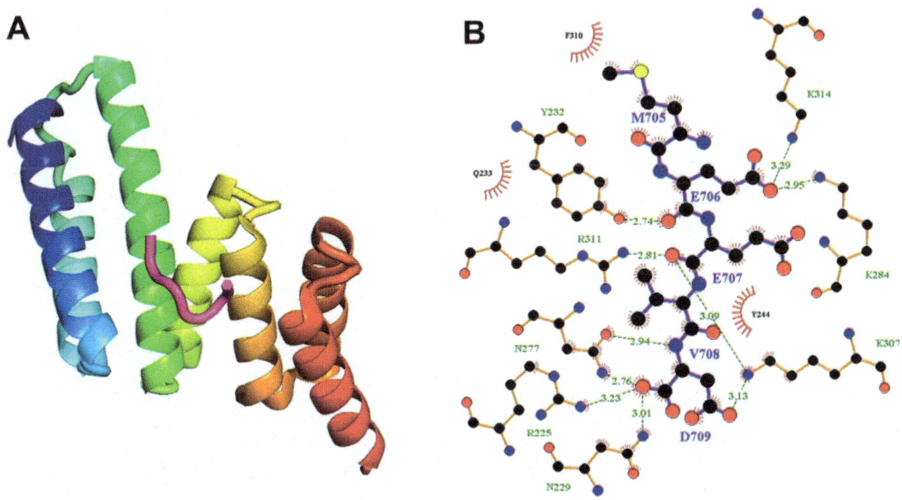

Fig. 14.2 A representative dicarboxylate clamp between FKBP38 TPR domain and MEEVD peptide (PDB ID: 5MGX). (**a**) MEEVD peptide bound to the TPR domain of FKBP38 depicted as cartoon using PyMol (Schrodinger 2015). (**b**) 2D interaction diagram created by LigPlot+ showing the important amino acid residue interactions in the formation of dicarboxylate clamp

can be seen clearly that the two carboxylic functional groups of the terminal aspartic acid (D709) forms a very strong clamp involving R225, N229, N277 and K307 residues with a distance ranging from 2.76 to 3.23 Å. The other residues such as V708, E707 and E706 also forms hydrogen bonds and salt bridges with N277, R311, K307, K284, Y232 and K314 with varying bond distance ranging from 2.74 to 3.29 Å.

14.4 Other Potential dcTPR Interacting Proteins

The TPR interaction network is not fully understood and it motivated us to explore other set of proteins except Hsp90/Hsp70 which can interact with the dcTPR containing proteins. In a recent study, we performed a systematic search of the global human protein database for the proteins having similar C-terminal residues as that of Hsp90/Hsp70 (MEEVD/IIEVD). The search resulted in identification of 112 proteins which may interact with the dcTPR containing proteins. We have confirmed interactions of selected proteins containing Hsp70/Hsp90-like C-termini with dcTPR proteins both in vitro and in situ. Interestingly, the, Gene Ontology (GO) functional analysis of identified proteins revealed enrichment of a few functional classes namely molecular chaperones and proteins of ubiquitin-proteasomal system highlighting the possibility of functional specialization of proteins containing Hsp70/Hsp90-like carboxy-terminal sequences. Therefore, Hsp90, Hsp70 and other proteins containing C-termini with Hsp90/Hsp70-like signature are engaged into interactions with dcTPR proteins via dicarboxylate clamp mechanism forming dicarboxylate clamp protein interaction network (Fig. 14.3).

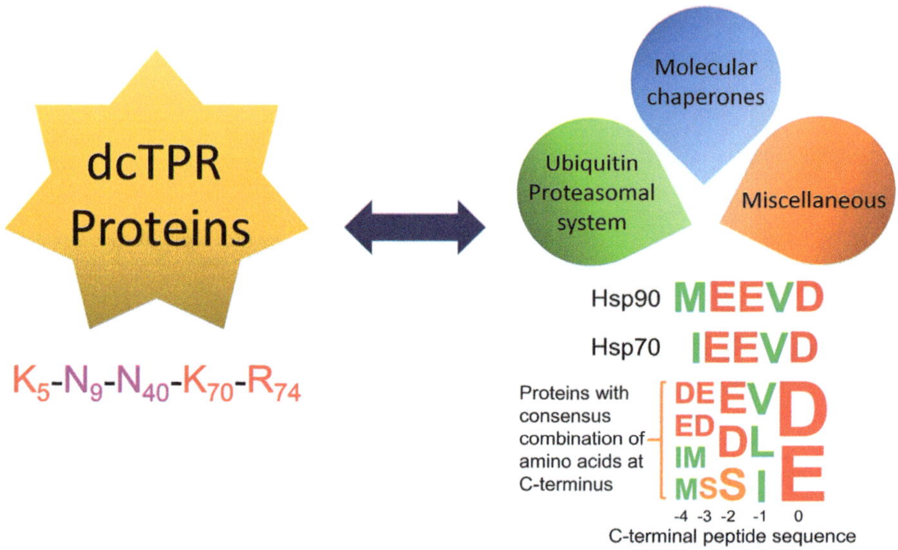

Fig. 14.3 Dicarboxylate clamp TPR interacting protein network. The dcTPR proteins contain characteristic and conserved amino acid signature of K_5-N_9-N_{40}-K_{70}-R_{74} which binds not only to the C-terminal peptide MEEVD and IEEVD of Hsp90 and Hsp70, respectively, but to other protein containing a similar C-terminal sequence resulting in dicarboxylate clamp TPR interacting protein network. Functionally, these dcTPR interacting proteins can be classified as molecular chaperones, proteins involved in ubiquitin proteasomal system and proteins with miscellaneous functions as indicated by our recent study (Bernadotte et al. 2018)

14.5 Regulation of Dicarboxylate Clamp Protein Interaction Network

Taking into account the high complexity of dcTPR protein interaction network, it is not surprising that such interactions would be tightly regulated within living cells. Several factors or mechanisms of regulation have been described in the scientific literature:

14.5.1 Spatial Regulation of dcTPR Protein Interaction Network

Close proximity interactions of dcTPR proteins with proteins containing C-termini similar to that of Hsp70/Hsp90 would require co-localization of interacting partners in the same sub-cellular compartment or organelle. Direct contacts between interacting proteins facilitated by their co-localization would be particularly important for the low-affinity dcTPR interactions (Fan et al. 2011).

14.5.2 Regulation of dcTPR Protein Interaction Network by Small Molecules

Ligand-induced switch between FKBP51 and FKBP52 bound to Hsp90 has been described for cytosolic steroid hormone complexes (Davies et al. 2002). It has been found that in the glucocorticoid receptor (GR) complex, binding of ligand to the GR induced switch from the Hsp90-GR-FKBP51 complex to the Hsp90-GR-FKBP52-dynein complex which was subsequently translocated to the nucleus (Davies et al. 2002). It is believed that hormone binding induces conformational change within the complex in favor of FKBP52 binding however, the exact molecular mechanism of the switch remains elusive.

14.5.3 Regulation of dcTPR Protein Interaction Network by Phosphorylation of Serine/Threonine Residues Within Hsp70/Hsp90-Like C-Terminal Peptides

Hsp90, Hsp70 as well as many of proteins containing Hsp70/Hsp90-like C-terminal peptides contain serine or threonine residues within last 10 amino acids of their sequence. Muller et al. found that these residues can be phosphorylated in vitro and in situ by several protein kinases including CK1, CK2 and GSK3β (Muller et al. 2013). Phosphorylation inhibits binding of CHIP co-chaperone to Hsp90 and promotes binding of HOP protein. Peptide phosphorylation was suggested to play a role in regulation of folding/degradation balance and much higher levels of phosphorylated Hsp90 C-terminal peptide were found in highly proliferative cancer cells (Muller et al. 2013).

14.5.4 Regulation of dcTPR Protein Interaction Network by Phosphorylation of dcTPR Proteins

Phosphorylation of particular dcTPR proteins can affect their ability to interact with their respective partners within dcTPR interaction network. FKBP52 phosphorylation by CK2 at threonine 141 was found to be inhibitory for its interaction with Hsp90 (Miyata et al. 1997). Thr-141 is positioned outside of TPR domain of FKBP52 at interface between FK1 and FK2 domains (Miyata et al. 1997). These results indicate existence of potential phosphorylation sites on any of dcTPR proteins that would regulate their binding to the partner proteins within dcTPR protein interaction network.

14.5.5 Regulation of dcTPR Protein Interaction Network by Calcium/Calmodulin Binding

FKBP38 and FKBP52 are two dcTPR proteins that contain calmodulin binding sequences in proximity of TPR domains (Davies and Sanchez 2005; Edlich et al. 2007). Interaction of these immunophilins with Hsp90 is calmodulin-dependent since depletion of calcium/calmodulin or removal of calmodulin binding sequences from these dcTPR proteins inhibited Hsp90 binding and affects downstream protein function (Cheung-Flynn et al. 2003; Edlich et al. 2007).

14.5.6 S100 Proteins as Endogenous Inhibitors of dcTPR-Hsp90 Interactions

S100 proteins is a family of conserved proteins capable to interact with several targets in calcium-dependent manner. In has been found that several members of S100 protein family, namely S100A1, S100A2, S100A6, S100B and S100P associate with the TPR domains of dcTPR proteins FKBP38 (Shimamoto et al. 2014), FKBP52 and Cyp40 (Shimamoto et al. 2010), CHIP (Shimamoto et al. 2013), HOP and Tom70 (Shimamoto et al. 2008) and PP5 (Yamaguchi et al. 2012) in calcium-regulated manner. S100 proteins compete with Hsp90 for the binding to dcTPR proteins affecting chaperone-dependent functions (Shimamoto et al. 2008, 2010, 2013, 2014; Yamaguchi et al. 2012). Further work is required to elucidate target specificity for S100-dcTPR protein interactions and possibility of regulation of these interactions in vivo.

14.6 Role of the Hsp90 dcTPR Co-chaperones in Disease Pathologies

By controlling intracellular protein homeostasis, Hsp90 chaperone machinery regulates multitude of vital processes in the living organisms. Imbalances in its functioning could potentiate progression of numerous human diseases and vice versa, the disease course can be corrected via modulation of the Hsp90 chaperone machinery. The dcTPR co-chaperones of Hsp90 have been implicated in a number of human diseases. The evidences would come from human genetic studies, knockout animal models, in vitro experiments as well as drug development programs. Good example of this would be FK506-binding protein 51 kDa, FKBP51, component of Hsp90-steroid hormone receptor complexes, negative regulator of glucocorticoid receptor (GR) signaling and positive regulator of androgen response.

14.6.1 FKBP51 as Potential Drug Target

FKBP51 consists of two domains: FK506-binding domain possessing PPIase activity and dcTPR domain mediating FKBP51 interaction with Hsp90. High levels of FKBP51 are linked to GR resistance and reduced stress coping behavior. In humans common allelic variants in the *FKBP5* gene are associated with increased risk of developing affective disorders like anxiety, depression and post-traumatic stress disorder (for recent reviews see (Criado-Marrero et al. 2018; Matosin et al. 2018). *FKBP5* knockout mice are viable, fertile, have normal lifespan yet, they have profound phenotype alterations related to GR signaling. Studies in *FKBP5* knockout mice revealed an increased resistance to stress stimuli, significantly lower levels of stress hormones in blood after the stress and anti-depressive behavior (O'Leary et al. 2011). Moreover, these mice showed reduced hypersensitivity in several persistent pain models in rodents (Maiaru et al. 2016). Also *FKBP5* null mice exhibit resistance to high fat diet-induced obesity, hepatic steatosis and elevated glucose and insulin tolerance (Stechschulte et al. 2016). In vitro experiments revealed that FKBP51 is highly expressed in several types of cancer cells, in particular, prostate cancer (Ni et al. 2010), malignant melanoma (Romano et al. 2013; Srivastava et al. 2015), glioma (D'Arrigo et al. 2017), oral squamous cell carcinoma (Russo et al. 2017). Recently, first specific inhibitors of PPIase activity of FKBP51 have been developed (Gaali et al. 2015). These molecules (SAFit1 and SAFit2) showed their potential as anti-depressants (Gaali et al. 2015), anxiolytic properties (Hartmann et al. 2015), reduced chronic pain (Maiaru et al. 2016, 2018) and exhibit anti-obesity properties (Balsevich et al. 2017) in rodent preclinical models.

14.6.2 The dcTPR Proteins as Potential Targets Against Neurodegenerative Disorders

Recently, TPR co-chaperones have been implicated into development of Alzheimer's disease that is characterized by abnormal accumulation of aggregated proteins – amyloid β (Aβ) in senile plaques and hyperphosphorylated tau (*p*-tau) in the paired helical filaments. FKBP51 have been found to play an important role in metabolism of tau. Tau levels were decreased in FKBP51 knockout mice and overexpression of FKBP51 preserved tau in mouse model of tauopathy (Blair et al. 2013). It has been suggested that strategies aimed at attenuating FKBP51 levels or its interaction with Hsp90 have the potential to be therapeutically relevant for AD and other tauopathies (Blair et al. 2013). PP5 is another dcTPR protein implicated in metabolism of tau. PP5 resides in cytoplasm in inactive form and is activated upon binding to Hsp90 (Yang et al. 2005). PP5 can dephosphorylate tau and its activity is decreased in AD patients (Liu et al. 2005). Molecules preventing interaction of PP5 with Hsp90 increase phosphatase activity (Zeke et al. 2005). Several drug development programs have been focused on inhibition of protein kinases that phosphorylate tau.

Activation of PP5 could serve as a vivid alternative to reduce levels of p-tau. Moreover, activated PP5 was also shown to protect cells against Aβ toxicity (Sanchez-Ortiz et al. 2009). Small glutamine-rich tetratricopeptide repeat protein alpha (SGTA) is another dcTPR co-chaperone implicated into Aβ metabolism. Inactivation of SGTA homologue in *C. elegans* protected transgenic nematode against Aβ toxicity (Fonte et al. 2002). Therefore, inhibitors of Hsp70/Hsp90 interactions with several dcTPR motif partners potentially have neuroprotective activity associated with Aβ and tau metabolism. Involvement of dcTPR proteins into other neurodegenerative disorders has not yet been reported. However, disease progression of many neurodegenerative disorders including Parkinson's disease, Huntington's disease, amyotrophic lateral sclerosis is driven by the protein aggregation, therefore, modulation of molecular chaperone machinery can have therapeutic potential for the treatment of these disorders.

14.6.3 Importance of Dicarboxylate Clamp Mediated Hsp90-dcTPR Protein Interactions for Various Disease-Associated Conditions

As mentioned above many dcTPR proteins have been associated with different human diseases including AD, cancer, stress-related, metabolic disorders and several other conditions. The evidence was based on the alteration of dcTPR gene expression associated with the disease, its protein levels, or affected function but the importance of dicarboxylate clamp mechanism was not investigated in many studies. Several reports mechanistically implicate dicarboxylate clamp mediated interactions into pathological conditions. In cystic fibrosis, FKBP38 facilitates folding of cystic fibrosis transmembrane conductance regulator (CFTR) (Banasavadi-Siddegowda et al. 2011). Mutations in CFTR often result in inability of CFTR to be correctly folded in the endoplasmic reticulum and in its subsequent degradation. It has been found that Hsp90 negatively regulates FKBP38 ability to facilitate CFTR folding confirmed by mutagenesis studies of amino acids responsible for dicarboxylate clamp interactions (Banasavadi-Siddegowda et al. 2011). Dicarboxylate clamp mediated FKBP38-Hsp90 interaction was also shown to be necessary for hepatitis C virus (HCV) replication in host cells (Okamoto et al. 2006). Expression of FKBP38 lacking residues involved in dicarboxylate clamp interactions with Hsp90 acts as dominant negative mutant inhibiting HCV replication (Okamoto et al. 2006). In prostatic cancer cells, point mutations in FKBP51 that abolish its interaction with Hsp90 reduce its ability to enhance androgen receptor dependent transcription and chaperone complex assembly (Ni et al. 2010).

14.7 Small Molecules as Inhibitors of Hsp90 Interaction with dcTPR Proteins

In general, the PPIs are difficult to target because of the physical nature of interactions between two proteins (Fry 2015). The protein interaction interface is usually too shallow and is not well-defined in most of the cases. Due to relatively small interaction area between the dcTPR domain of co-chaperones and Hsp70/Hsp90 C-terminal peptides and well-defined binding pocket (Fig. 14.2) it is possible to identify small molecules inhibiting such interaction.

14.7.1 Hsp90 as a Drug Target

A large number of inhibitors of Hsp90 N-terminal (ATP binding domain) have been reported and reviewed nicely (Yin et al. 2009). The C-terminal half of Hsp90 represents a second site for pharmacologic intervention of chaperone function. Numerous novobiocin analogues interacting with C-terminal domain of Hsp90 have been synthesized and investigated in various in vitro systems (for review see (Dlugosz and Janecka 2017)) showing strong anti-proliferative effect in cancer cells. One of Hsp90 C-terminal domain binding molecules, (−)-epigallocatechin-3-gallate has been reported as a modifier of Hsp90 interaction with several dcTPR co-chaperones including AIP (Yin et al. 2009). Despite obvious potential of Hsp90 inhibitors for the treatment of disorders ranging from neurodegeneration to cancer efficacy of this class of therapeutic agents has been relatively limited to date, with promising responses mainly observed in breast and lung cancer (Butler et al. 2015). In addition, adverse events and some significant toxicities have been documented in clinical trials. It has to be stressed that molecules binding Hsp90 and interfering with its interaction with dcTPR co-chaperones would be unspecific towards particular chaperone co-chaperone interactions. Uncoupling Hsp90 from its co-chaperones would have profound systemic effect on cell functions restricting potential therapeutic areas of these drugs to anti-cancer therapies.

14.7.2 DcTPR Proteins Targeting with Small Molecules

Several reports describe binding of low molecular weight compounds to the TPR domain of dcTPR co-chaperones. Poly-unsaturated fatty acids as well as long-chain fatty acyl-CoA esters bind to the TPR domain of PP5 stimulating its phosphatase activity at low micromomolar concentrations (Kang et al. 2001; Ramsey and Chinkers 2002). Moreover, they as well as microtubule-dissociating drug nocodazole dissociate PP5 from its complex with Hsp90 and Hsp70 (Zeke et al. 2005). Despite the fact that these molecules cannot be considered as good drug candidates due to their pharmacophore structure or toxicity, these data illustrate feasibility of dcTPR domain targeting approach by small molecules. These observations were further corroborated by the study of Yi and Regan (Yi and Regan 2008) who have developed primary high throughput screen probing interactions between the Hsp90 C-terminal peptide and the TPR motif protein HOP. By high throughput screening of a commercially available compounds libraries they have identified 6 compounds containing a 7-azapteridine ring system (pyrimido[5,4-e][1,2,4]triazine-5,7-dione). These compounds inhibited Hsp90-HOP interactions in sub-micromolar concentrations and inhibited growth of several cancer cell lines (Yi et al. 2009). Recently, we have identified a novel compound GMP-1 that disrupts interactions between Hsp90/70, molecular chaperone and protein import receptor Tom70. Treatment with GMP-1 demonstrated neuroprotective effects in the mouse and drosophila model of AD pathology (Pavlov et al. 2018).

14.7.3 *Problem of Specific Targeting of TPR Domain by Small Molecules*

Disruption of Hsp90 interactions with particular dcTPR co-chaperone apparently will not inhibit general chaperone function. It represents an attractive strategy to specifically modulate various chaperone-mediated cellular activities if sufficient discrimination between different dcTPR proteins would be achieved. Binding site for the MEEVD peptide has similar tertiary structure and conserved amino acids responsible for the interaction, but the overall homology between different dcTPR domains of co-chaperones is below 50% (Scheufler et al. 2000). Theoretically, this provides an opportunity to develop small molecules (modulators) competing for MEEVD peptide binding and specifically blocking the interactions between Hsp70/Hsp90 and particular TPR motif co-chaperones. These molecules would comprise functional groups for anchoring to dicarboxylate clamp binding pocket and groups providing specificity for each of the target dcTPR protein. Availability of the crystal structures of dcTPR co-chaperones with Hsp70/Hsp90 C-terminal peptides allows implementation of in silico docking and computational drug design to identify potent and selective modulators of Hsp90 function.

14.8 Conclusions

Hsp90 is a major chaperone functioning in close coordination with several of its co-chaperones including dcTPR containing proteins. These dcTPR containing proteins bind to the C-terminal of Hsp90/70 through a dicarboxylate mechanism. Functionally, the dcTPR interaction network have not been fully characterized and in our studies, we have identified and characterized several proteins having a similar

C-terminal sequence as that of Hsp90/70 and have a potential to interact with the dcTPR proteins through a dicarboxylate mechanism. This chapter explores the basic mechanism of the dcTPR interaction describes the dcTPR interaction protein network and discuss perspectives of therapeutic modulation of molecular chaperone network.

Acknowledements The work was supported by research grants from the Swedish Research Council (2015-02774, 2018-002843), Stiftelsen Olle Enqvist Byggmästare, Margareta af Ugglas Foundation, Foundation for Geriatric Diseases at Karolinska Institutet, Loo & Hans Osterman Foundation, KI Foundations, Lindhés Advokatbyrå AB Foundation, Gunvor and Josef Anérs Foundation, the Swedish Brain Foundation, Magnus Bergvalls Foundation, Gun and Bertil Stohnes Foundation, Tore Nilssons Foundation for medical research, and the Foundation for Old Servants.

References

Balsevich G, Hausl AS, Meyer CW et al (2017) Stress-responsive FKBP51 regulates AKT2-AS160 signaling and metabolic function. Nat Commun 8:1725

Banasavadi-Siddegowda YK, Mai J, Fan Y et al (2011) FKBP38 peptidylprolyl isomerase promotes the folding of cystic fibrosis transmembrane conductance regulator in the endoplasmic reticulum. J Biol Chem 286:43071–43080

Bernadotte A, Kumar R, Winblad B, Pavlov PF (2018) In silico identification and biochemical characterization of the human dicarboxylate clamp TPR protein interaction network. FEBS Open Biol 8:1830–1843

Blair LJ, Nordhues BA, Hill SE et al (2013) Accelerated neurodegeneration through chaperone-mediated oligomerization of tau. J Clin Invest 123:4158–4169

Bose S, Weikl T, Bugl H, Buchner J (1996) Chaperone function of Hsp90-associated proteins. Science 274:1715–1717

Butler LM, Ferraldeschi R, Armstrong HK, Centenera MM, Workman P (2015) Maximizing the therapeutic potential of HSP90 inhibitors. Mol Cancer Res 13:1445–1451

Cheung-Flynn J, Roberts PJ, Riggs DL, Smith DF (2003) C-terminal sequences outside the tetratricopeptide repeat domain of FKBP51 and FKBP52 cause differential binding to Hsp90. J Biol Chem 278:17388–17394

Criado-Marrero M, Rein T, Binder EB, Porter JT, Koren J, 3rd, Blair LJ (2018) Hsp90 and FKBP51: complex regulators of psychiatric diseases. Philos Trans R Soc Lond Ser B Biol Sci 373

D'Arrigo P, Russo M, Rea A et al (2017) A regulatory role for the co-chaperone FKBP51s in PD-L1 expression in glioma. Oncotarget 8:68291–68304

Davies TH, Sanchez ER (2005) FKBP52. Int J Biochem Cell Biol 37:42–47

Davies TH, Ning YM, Sanchez ER (2002) A new first step in activation of steroid receptors: hormone-induced switching of FKBP51 and FKBP52 immunophilins. J Biol Chem 277:4597–4600

Dlugosz A, Janecka A (2017) Novobiocin analogs as potential anticancer agents. Mini-Rev Med Chem 17:728–733

Echeverría PC, Bernthaler A, Dupuis P, Mayer B, Picard D (2011) An interaction network predicted from public data as a discovery tool: application to the Hsp90 molecular chaperone machine. PLoS One 6:e26044

Edlich F, Erdmann F, Jarczowski F, Moutty MC, Weiwad M, Fischer G (2007) The Bcl-2 regulator FKBP38-calmodulin-Ca2+ is inhibited by Hsp90. J Biol Chem 282:15341–15348

Fan AC, Kozlov G, Hoegl A et al (2011) Interaction between the human mitochondrial import receptors Tom20 and Tom70 in vitro suggests a chaperone displacement mechanism. J Biol Chem 286:32208–32219
Ferraro M, D'Annessa I, Moroni E et al (2018) Allosteric modulators of HSP90 and HSP70: dynamics meets function through structure-based drug design. J Med Chem 62:60–87
Fonte V, Kapulkin WJ, Taft A, Fluet A, Friedman D, Link CD (2002) Interaction of intracellular beta amyloid peptide with chaperone proteins. Proc Natl Acad Sci USA 99:9439–9444
Fry DC (2015) Targeting protein-protein interactions for drug discovery. Methods Mol Biol 1278:93–106
Gaali S, Kirschner A, Cuboni S et al (2015) Selective inhibitors of the FK506-binding protein 51 by induced fit. Nat Chem Biol 11:33–37
Guy NC, Garcia YA, Sivils JC, Galigniana MD, Cox MB (2015) Functions of the Hsp90-binding FKBP immunophilins. Subcell Biochem 78:35–68
Hartmann J, Wagner KV, Gaali S et al (2015) Pharmacological inhibition of the psychiatric risk factor FKBP51 has anxiolytic properties. J Neurosci 35:9007–9016
Haslbeck V, Eckl JM, Kaiser CJO, Papsdorf K, Hessling M, Richter K (2013) Chaperone-interacting TPR proteins in *Caenorhabditis elegans*. J Mol Biol 425:2922–2939
Johnson BD, Schumacher RJ, Ross ED, Toft DO (1998) Hop modulates Hsp70/Hsp90 interactions in protein folding. J Biol Chem 273:3679–3686
Jolly C, Morimoto RI (2000) Role of the heat shock response and molecular chaperones in oncogenesis and cell death. J Natl Cancer Inst 92:1564–1572
Kang H, Sayner SL, Gross KL, Russell LC, Chinkers M (2001) Identification of amino acids in the tetratricopeptide repeat and C-terminal domains of protein phosphatase 5 involved in autoinhibition and lipid activation. Biochemistry 40:10485–10490
Kumar R, Moche M, Winblad B, Pavlov PF (2017) Combined x-ray crystallography and computational modeling approach to investigate the Hsp90 C-terminal peptide binding to FKBP51. Sci Rep 7:14288
Liu F, Iqbal K, Grundke-Iqbal I, Rossie S, Gong CX (2005) Dephosphorylation of tau by protein phosphatase 5: impairment in Alzheimer's disease. J Biol Chem 280:1790–1796
Luo W, Sun W, Taldone T, Rodina A, Chiosis G (2010) Heat shock protein 90 in neurodegenerative diseases. Mol Neurodegener 5:24–24
Maiaru M, Tochiki KK, Cox MB et al (2016) The stress regulator FKBP51 drives chronic pain by modulating spinal glucocorticoid signaling. Sci Transl Med 8:325ra19
Maiaru M, Morgan OB, Mao T et al (2018) The stress regulator FKBP51: a novel and promising druggable target for the treatment of persistent pain states across sexes. Pain 159:1224–1234
Matosin N, Halldorsdottir T, Binder EB (2018) Understanding the molecular mechanisms underpinning gene by environment interactions in psychiatric disorders: the FKBP5 model. Biol Psychiatry 83:821–830
Miyata Y, Chambraud B, Radanyi C et al (1997) Phosphorylation of the immunosuppressant FK506-binding protein FKBP52 by casein kinase II: regulation of HSP90-binding activity of FKBP52. Proc Natl Acad Sci USA 94:14500–14505
Muller P, Ruckova E, Halada P et al (2013) C-terminal phosphorylation of Hsp70 and Hsp90 regulates alternate binding to co-chaperones CHIP and HOP to determine cellular protein folding/degradation balances. Oncogene 32:3101–3110
Ni L, Yang CS, Gioeli D, Frierson H, Toft DO, Paschal BM (2010) FKBP51 promotes assembly of the Hsp90 chaperone complex and regulates androgen receptor signaling in prostate cancer cells. Mol Cell Biol 30:1243–1253
Okamoto T, Nishimura Y, Ichimura T et al (2006) Hepatitis C virus RNA replication is regulated by FKBP8 and Hsp90. EMBO J 25:5015–5025
O'Leary JC III, Dharia S, Blair LJ et al (2011) A new anti-depressive strategy for the elderly: ablation of FKBP5/FKBP51. PLoS One 6:e24840

Pavlov PF, Hutter-Paier B, Havas D, Windisch M, Winblad B (2018) Development of GMP-1 a molecular chaperone network modulator protecting mitochondrial function and its assessment in fly and mice models of Alzheimer's disease. J Cell Mol Med 22:3464–3474

Prodromou C, Siligardi G, O'Brien R et al (1999) Regulation of Hsp90 ATPase activity by tetratricopeptide repeat (TPR)-domain co-chaperones. EMBO J 18:754–762

Ramsey AJ, Chinkers M (2002) Identification of potential physiological activators of protein phosphatase 5. Biochemistry 41:5625–5632

Romano S, Staibano S, Greco A et al (2013) FK506 binding protein 51 positively regulates melanoma stemness and metastatic potential. Cell Death Dis 4:e578

Russo D, Merolla F, Mascolo M et al (2017) FKBP51 immunohistochemical expression: a new prognostic biomarker for OSCC? Int J Mol Sci 18:443

Sanchez-Ortiz E, Hahm BK, Armstrong DL, Rossie S (2009) Protein phosphatase 5 protects neurons against amyloid-beta toxicity. J Neurochem 111:391–402

Scheufler C, Brinker A, Bourenkov G et al (2000) Structure of TPR domain-peptide complexes: critical elements in the assembly of the Hsp70-Hsp90 multichaperone machine. Cell 101:199–210

Schopf FH, Biebl MM, Buchner J (2017) The HSP90 chaperone machinery. Nat Rev Mol Cell Biol 18:345–360

Schrodinger LLC (2015) The PyMOL molecular graphics system, Version 1.8

Shimamoto S, Takata M, Tokuda M, Oohira F, Tokumitsu H, Kobayashi R (2008) Interactions of S100A2 and S100A6 with the tetratricopeptide repeat proteins, Hsp90/Hsp70-organizing protein and kinesin light chain. J Biol Chem 283:28246–28258

Shimamoto S, Kubota Y, Tokumitsu H, Kobayashi R (2010) S100 proteins regulate the interaction of Hsp90 with Cyclophilin 40 and FKBP52 through their tetratricopeptide repeats. FEBS Lett 584:1119–1125

Shimamoto S, Kubota Y, Yamaguchi F, Tokumitsu H, Kobayashi R (2013) Ca2+/S100 proteins act as upstream regulators of the chaperone-associated ubiquitin ligase CHIP (C terminus of Hsc70-interacting protein). J Biol Chem 288:7158–7168

Shimamoto S, Tsuchiya M, Yamaguchi F, Kubota Y, Tokumitsu H, Kobayashi R (2014) Ca2+/S100 proteins inhibit the interaction of FKBP38 with Bcl-2 and Hsp90. Biochem J 458:141–152

Soroka J, Wandinger SK, Mausbacher N et al (2012) Conformational switching of the molecular chaperone Hsp90 via regulated phosphorylation. Mol Cell 45:517–528

Srivastava SK, Bhardwaj A, Arora S et al (2015) Interleukin-8 is a key mediator of FKBP51-induced melanoma growth, angiogenesis and metastasis. Br J Cancer 112:1772–1781

Stechschulte LA, Qiu B, Warrier M et al (2016) FKBP51 null mice are resistant to diet-induced obesity and the PPARgamma agonist rosiglitazone. Endocrinology 157:3888–3900

Verba KA, Wang RY, Arakawa A et al (2016) Atomic structure of Hsp90-Cdc37-Cdk4 reveals that Hsp90 traps and stabilizes an unfolded kinase. Science 352:1542–1547

Yamaguchi F, Umeda Y, Shimamoto S et al (2012) S100 proteins modulate protein phosphatase 5 function: a link between CA2+ signal transduction and protein dephosphorylation. J Biol Chem 287:13787–13798

Yang J, Roe SM, Cliff MJ et al (2005) Molecular basis for TPR domain-mediated regulation of protein phosphatase 5. EMBO J 24:1–10

Yi F, Regan L (2008) A novel class of small molecule inhibitors of Hsp90. ACS Chem Biol 3:645–654

Yi F, Zhu P, Southall N et al (2009) An AlphaScreen-based high-throughput screen to identify inhibitors of Hsp90-cochaperone interaction. J Biomol Screen 14:273–281

Yin Z, Henry EC, Gasiewicz TA (2009) (-)-Epigallocatechin-3-gallate is a novel Hsp90 inhibitor. Biochemistry 48:336–345

Zeke T, Morrice N, Vazquez-Martin C, Cohen PT (2005) Human protein phosphatase 5 dissociates from heat-shock proteins and is proteolytically activated in response to arachidonic acid and the microtubule-depolymerizing drug nocodazole. Biochem J 385:45–56

Chapter 15
The 'Complex World' of the Hsp90 Co-chaperone R2TP

Chrisostomos Prodromou

Abstract The Hsp90 co-chaperone R2TP consists of the AAA+ ATPases, RUVBL1 (Rvb1p in yeast) and RUVBL2 (Rvb2p in yeast), which together make up a hetero-hexameric ring, in complex with PIH1D1 (Pih1p in yeast) and RPAP3 (Tah1p in yeast). R2TP is involved in the activation of client proteins, such as phosphatidylinositol 3 kinase-related kinases, including mTORC1, ATM, DNA-PK, SMG and ATR/ATRIP, or in the assembly of protein complexes including those of RNA polymerase and snoRNPs, amongst others. In other cases, the role of the TP component (RPAP3-PIH1D1) of R2TP, and consequently Hsp90, is controversial. None-the-less, the extensive role of RUVBL1-RUVBL2 complex in cells, either with or without Hsp90, means that dysfunction of these AAA$^+$ ATPases, Hsp90 or components of the complexes they assemble leads to diseases such as cancer, ciliary dyskinesia and in the case of defects in ATM to ataxia telangiectasia-like syndrome. Recent advances in determining the structure of the R2TP complex has led to an increased understanding of the assembly and function of the R2TP complex. In this review we discuss the current structural advances in determining the architecture of the R2TP complex and the advances made in understanding its active state.

Keywords Axonemal dynein · Carcinogenesis · Co-chaperone · Hsp90 · R2TP · RUVBL1-RUVBL2

Abbreviations

EM	Electron microscopy
Hsp90	Heat shock protein 90
R2	RUVBL1-RUVBL2 complex

C. Prodromou (✉)
Biochemistry and Biomedicine, University of Sussex, Falmer, Brighton, UK
e-mail: chris.prodromou@sussex.ac.uk

A. A. A. Asea, P. Kaur (eds.), *Heat Shock Protein 90 in Human Diseases and Disorders*, Heat Shock Proteins 19,
https://doi.org/10.1007/978-3-030-23158-3_15

R2TP RUVBL1 RUVBL2-RPAP3-PIH1D1 or Rvb1p-Rvb2p-Pih1p-Tah1p complex

15.1 Introduction

R2TP complex consists of RUVBL1 (also known as Pontin, TIP49, TIP49a and in yeast as Rvb1p), RUVBL2 (also known as Reptin, TIP48, TIP49b and in yeast as Rvb2p), RPAP3 (Tah1p in yeast) and PIH1D1 (Pih1p in yeast). The RUVBL1-RUVBL2 (R2) proteins are AAA^+ ATPases and together with RPAP3 and PIH1D1 form the specialized Hsp90 co-chaperone, R2TP. This co-chaperone complex has been implicated in a number of biological processes including the activation of phosphatidylinositol 3 kinase-related kinases, including mTORC1, ATM, DNA-PK, SMG and ATR/ATRIP, the assembly of RNA polymerase and snoRNPs, including telomerase, the ribosome and the spliceosome, for the loading of DNA remodelling complexes such as INO80 and SWI and in DNA damage and repair systems, such as the Fanconi anaemia core complex, and in the assembly of axonemal dyneins (von Morgen et al. 2015). Because R2TP is involved in such diverse biological systems it comes as no surprise that dysfunction of the complex and its components can have drastic effects on human health.

The overexpression of R2 complex was initially reported in hepatocellular carcinoma and colorectal cancer, but later in other tissue including breast, lung, gastric, esophageal, pancreatic, kidney, bladder as well as in lymphatic, and leukemic cancers (Mao and Houry 2017). They have also been implicated in a variety of cellular processes highly relevant to cancer. For example, they interact with the oncogenes c-myc and β-catenin and are able to modulate their transcriptional activities (reviewed in (Grigoletto et al. 2011)). snoRNA biogenesis was shown to be deregulated in tumorigenesis (reviewed in (Mannoor et al. 2012)), while defects in PIH1D3, an orthologue of PIH1D1, or in DNAAF1 (LRRC50), a dynein assembly factor, can cause ciliary dyskinesia (Hartill et al. 2018; Miao et al. 2016; Olcese et al. 2017), and defects in ATM can cause ataxia telangiectasia-like syndrome (Nallar and Kalvakolanu 2013; Olcese et al. 2017; Taylor et al. 2004). The role of R2 in DNA damage, repair and modelling is also highly significant (reviewed in (Grigoletto et al. 2011; Huber et al. 2008; Jha and Dutta 2009; Mao and Houry 2017)). R2 has also been reported to localize to the cytoplasm, where it promotes cancer cell invasion and metastasis (Zhang et al. 2015). Clearly, it is not surprising that due to the various roles that R2 plays in carcinogenesis, that it is proving to be an important biomarker for the diagnosis and prognosis of various cancers (Mikesch et al. 2018)). Silencing experiments in xenografts indicate that they may be suitable targets for cancer therapy (reviewed in (Grigoletto et al. 2011). Consequently, R2 is being recognized as a current target for the development of novel therapeutic anticancer drugs.

The R2TP complex acts as a specialized co-chaperone able to recruit Hsp90 and client proteins. It appears that Hsp90 does not recognize client proteins directly and that specialized co-chaperones are required to allow recognition of the client. Often these are more than mere scaffold proteins and are themselves regulators of Hsp90 itself. This is exemplified by the co-chaperones Cdc37 and Sti1, which recruit kinases and steroid hormone receptors, respectively, for maturation by Hsp90, but simultaneously regulate the ATPase activity of the chaperone (Prodromou 2012; Prodromou and Morgan 2016; Prodromou et al. 1999; Siligardi et al. 2002). Hsp90 is also under many levels of regulation that likely can impact on the functioning of the R2TP complex (Prodromou 2016). Within recent years, details of the R2TP structure have been elucidated and our understanding of the mechanistic action of this co-chaperone has increased (Martino et al. 2018; Maurizy et al. 2018; Rivera-Calzada et al. 2017), although there are many details that remain to be established. Initial structural studies using human RUVBL1 indicated a hexameric ring structure with marginal ATPase activity (Matias et al. 2006). Support for an R2 hexameric ring structure was provided initially by transmission electron microscopy (EM) studies using the yeast Ruv1p-Ruv2p complex (Gribun et al. 2008). The yeast Rvb1p-Rvb2p complex showed enhanced ATPase activity, and nucleotide could induce conformational changes as observed by EM. However, a serious of dodecameric structures conflicted with the hexameric ring structure as being the active state of the complex (Puri et al. 2007). This review aims to consolidate the structural and biochemical data that has led to our increased understanding of the core R2TP complex. Finally, we look at emerging roles of R2 in axonemal dynein assembly and draw parallels with equivalent components found in the R2TP complex.

15.1.1 Common Features Among the Co-chaperones Recruiting Client Proteins

In yeast, Sti1p was the first co-chaperone shown to be able to regulate the ATPase activity of Hsp90 (Prodromou et al. 1999). In humans the homologue HOP (also known as p60) is responsible, together with Hsp70, for recruiting steroid-hormone receptors to the Hsp90 complex. It has been proposed that the ability of Hop/Sti1p to inhibit the Hsp90 ATPase activity is a prerequisite for client protein loading by providing a stable platform that allows binding to Hsp90 (Prodromou 2012; Prodromou and Morgan 2016). Furthermore, for transfer of a client to Hsp90 the low affinity client-protein binding state of Hsp70 is required, which is promoted by Hsp70 bound ATP. In support of this, rabbit reticulocyte HOP was shown to promote the binding of ATP to Hsp70, and simultaneously increase the dissociation of ADP (Gross and Hessefort 1996). In contrast, Sti1p has been reported to activate the ATPase activity of Ssa1p (a yeast Hsp70) (Wegele et al. 2003). Consequently, other factors such as the Ssa1p nucleotide exchange factor Fes1p, that favours ADP release and exchange with ATP, might be involved in client protein release form Ssa1p (Kabani et al. 2002). The situation is further complicated because human Hop protein appears not to be an effecter of the ATPase activity of Hsp70 and does not act as a nucleotide exchange factor, but simply prefers to bind Hsp70 in its ADP state (Johnson et al. 1998). Once again other nucleotide exchange factors may be important to allow client protein release from the human Hsp70 to Hsp90. However, whatever the effect of Hop and Sti1p on the ATPase activity of Hsp70, the release of the client protein must require that Hsp70 be converted to an ATP-like state so the client protein can be released to Hsp90. Clearly, the precise molecular details of how this might be achieved requires further study.

Similarly to Hop/Sti1p, Cdc37 (also known as p50) has been shown to be able to inhibit the ATPase activity of Hsp90 (Siligardi et al. 2002). Structural details on how this is achieved have been documented (Roe et al. 2004). It appears that human Cdc37, by binding between the N-terminal domains of Hsp90, prevents N-terminal dimerization. Thus, Cdc37 interacts with the 'ATP lids', a mobile segment of the N-terminal domains of Hsp90, and physically blocks dimerization by preventing closer of the 'ATP lids' (Roe et al. 2004). Furthermore, Arg 167 of Cdc37 is able to hydrogen bond with the catalytic Glu 33 residue of Hsp90 and thus prevents it from carrying out ATP hydrolysis (Roe et al. 2004). Cdc37, has also been shown to inhibit the ATPase activity of kinase client-protein that it delivers to Hsp90 (Polier et al. 2013). Thus, a number of common features can be established for the Sti1p/HOP and Cdc37 systems that deliver client proteins to Hsp90. Firstly, they don't simply act as scaffold proteins, but as regulators of Hsp90 activity. And secondly, they may also act to modulate the activity of additional components, such as Hsp70 by HOP/Sti1p and client kinases by Cdc37. The Tah1p-Pih1p component of the R2TP complex seems to serve a similar role to those of Cdc37 and HOP. Firstly, the PIH1D1/Pih1p component is known to be able to recruit PIKKs, via the TTT complex (Tel2-Tti1-Tti2), while Tah1p/RPAP3 scaffolds Hsp90 (Hurov et al. 2010; Millson et al. 2008). Secondly, Tah1p-Pih1p in yeast has been shown to be able to inhibit the ATPase activity of Hsp90 (Eckert et al. 2010), while it appears to be able to activate the ATPase activity of the Rvb1-Rvb2 complex (Rivera-Calzada et al. 2017). However, how these co-chaperones achieve their regulatory roles is still largely unknown and it is likely that further structural studies will be required to elucidate the mechanisms involved.

15.1.2 Structure of Tah1p- and RPAP3- MEEVD And Pih1p Complex

Biochemical and structural studies have shown that Tah1p consists of a TPR domain that interacts with the conserved MEEVD peptide sequence of the HSP90 family of proteins (HSP) and an unstructured C-terminal region that binds Pih1p (Millson et al. 2008; Pal et al. 2014) (Figs. 15.1 and 15.2a). The TPR domain was shown to bind specifically to both yeast and human HSP (K_d = 0.32–0.78 μM), showing only weak affinity towards the yeast Hsp70, Ssa1p (K_d = 16 μM). Tah1p dimerization has been observed and it has been suggested that this may prevent mixed yeast Hsp90

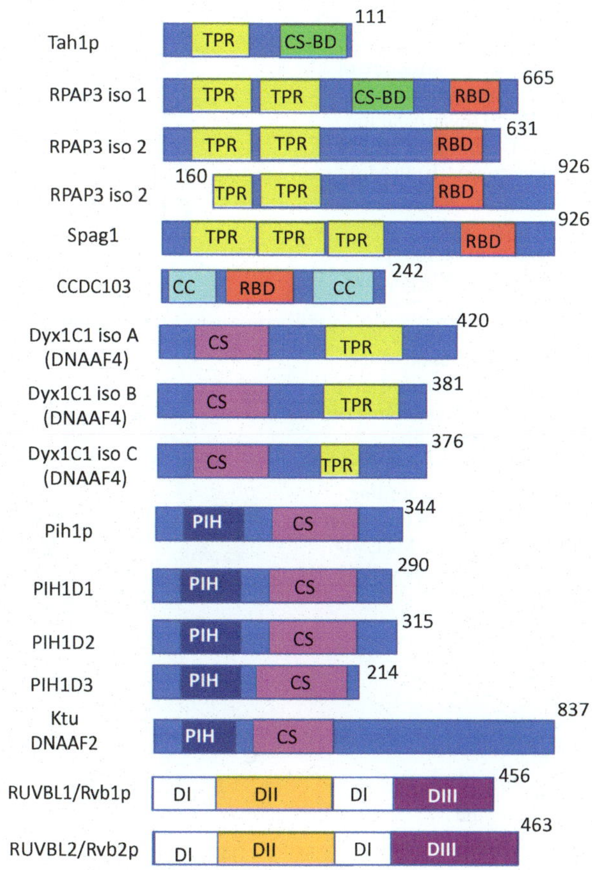

Fig. 15.1 Domain structure of R2TP and co-chaperones showing common domains. Domains are indicated by colour. Yellow, tetratricopeptide domain; red, domains similar to the RPAP3 RUVBL-binding domain; green, domain binding the CS domain of PIH1D1/Pih1p; cyan, coiled coil domain; magenta, CS domain; blue, PIH domain; and for the R2 proteins, white, DI domains; orange DII domains and purple, DIII domains. Protein lengths are indicated but relative sizes are approximate for clarity

co-chaperone complexes from forming by ensuring that both MEEVD sites of Hsp90 are simultaneously occupied by Tah1p (Morgan et al. 2015) (Fig. 15.2b). In this regard it may be significant that RPAP3, the human homologue of Tah1, contains two TPR modules. However, specificity for Hsp90 is determined by a requirement for a methionine amino-acid residue, conserved in the C-terminal MEEVD sequence of HSP, in contrast to the valine found in the conserved VEEVD peptide sequence of HSP70 proteins, which was positively discriminated against (Millson et al. 2008). In yeast, the methionine sits in a hydrophobic pocket lined by Ser 78 from both protomers of the Tah1p dimer, and by Val74, Lys 79 and Tyr 82 from protomer 1 and Gln 81 from the second Tah1p protomer (Fig. 15.2c). Finally, Tah1p was shown to have a mild activating effect on the ATPase activity of yeast Hsp90 (Millson et al. 2008).

The molecular details showing the interaction between the unstructured C-terminal region of Tah1p with the N-terminal CS domain of Pih1p has been determined (Pal et al. 2014; Quinternet et al. 2015) (Figs. 15.1 and 15.3a). The complex formed between Tah1p and Pih1p was mainly due to bridging interactions between

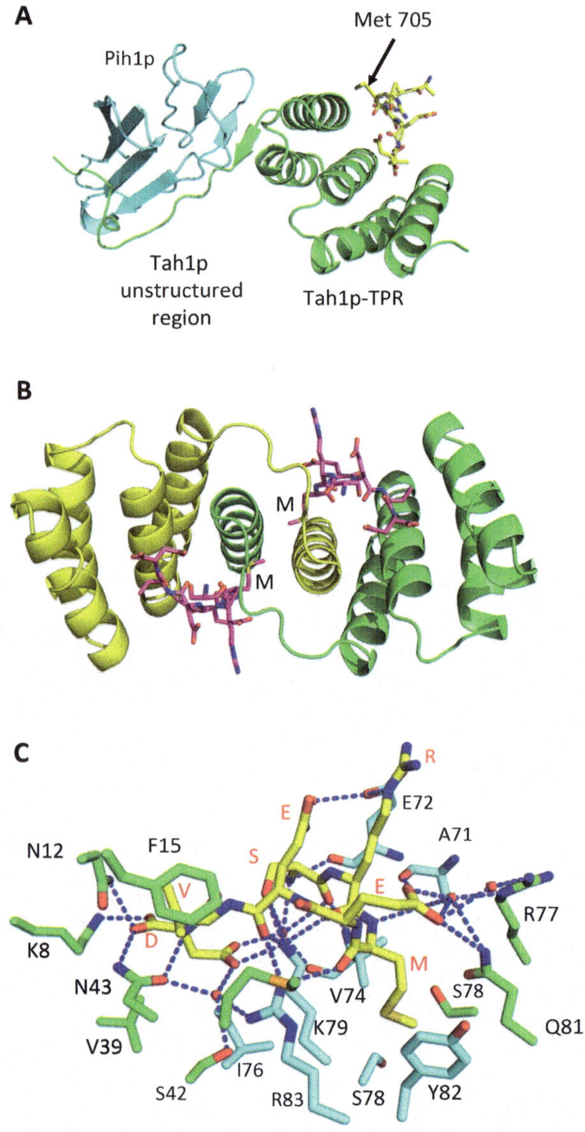

Fig. 15.2 Structure of Tah1p and Pih1p. (**a**) PyMol cartoon structure of Tah1p (green) in complex with Pih1p (cyan) and the MEEVD peptide of Hsp90 (yellow stick). (**b**) PyMol cartoon showing the helix-swap-helix dimerization of Tah1p, with the MEEVD peptide shown in stick representation. (**c**) PyMol stick representation of the interactions between Tah1p and the MEEVD peptide of Hsp90. Amino acid residues are coloured (green or cyan) according to the protomer of the Tah1p dimer. The MEEVD peptide is shown as yellow sticks

the edges of the two β-sheets of the β-sandwich Pih1p-CS domain and a third set of associations with the extreme distal part of the unstructured C-terminal segment of Tah1p, by forming a mixed hydrophobic and polar interface with the face of the four-stranded β-sheet of Pih1p (Fig. 15.3a). Additionally, there is a set of interactions that spatially fix the relative orientation of the Tah1p TPR domain to the CS domain of Pih1p (Fig. 15.3b). This could play a significant role in the docking of Tah1p-Pih1p complex on to the DII domain side of the Rvb1p-Rvb2p complex. However, work using NMR spectroscopy did not support these specific stabilizing interactions but suggested that others could be present (Quinternet et al. 2015). It

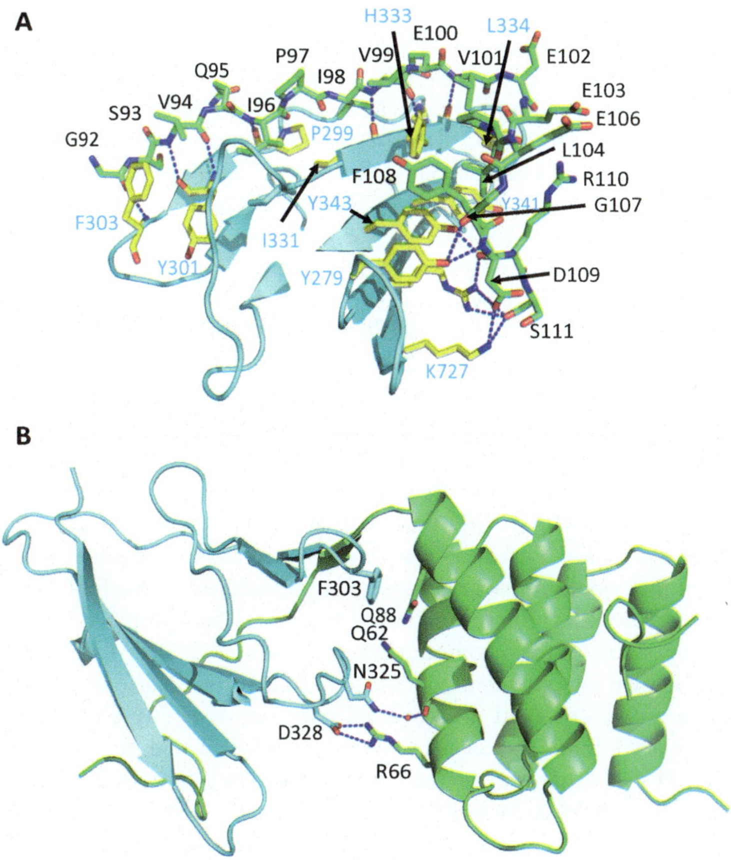

Fig. 15.3 PyMol cartoon of the Tah1p interaction with Pih1p. (**a**) Interactions between Tah1p (green) and Pih1p (yellow). (**b**) Interactions between the TPR domain of Tah1p (green) with the CS domain of Pih1p (cyan)

has been noted that Pih1p in complex with Tah1p shows additional contacts to either the middle- or C-terminal domain of HSP90 (Quinternet et al. 2015). This indicates that Hsp90 is not simply tethered to Tah1p by its flexible conserved MEEVD sequences, but is constrained into a single bound conformation.

It has been reported that Pih1p can also interact with the snoRNA U3 assembly factor Rsa1p and perhaps in a mutually exclusively way with the core protein Nop58p in the presence of Rsa1p (Quinternet et al. 2015). Thus, potentially Pih1p could act as a selection factor for the recruitment of different client protein specific components required by the client protein in question. Two further orthologues of PIH1D1 have also been identified. PIH1D2 appears to associate with SPAG1, an RPAP3-like protein, and PIH1D3 with DYX1C1 isoform C (Maurizy et al. 2018). Clearly, this represents modified R2TP like complexes that are able to recruit different client proteins and suggests that the R2 components can be recruited into multiple cellular process by additional co-factors that make up the specific co-chaperone in question. In humans, the 12.5 kD protein Tah1p is represented by a much larger protein known as RPAP3 or Spagh (75.7 kD) (Fig.15.1). The structure of the two TPR domains of RPAP3 in complex with MEEVD peptides and the C-terminal RUVBL binding domain (RBD) has been determined (Martino et al. 2018; Maurizy et al. 2018; Pal et al. 2014). Hsp90 appears to bind both TPR domain modules of RPAP3 and mutagenesis data suggest that Hsp90 binds with stronger affinity towards TPR1 (K_d = 0.94 μM) than to TPR2 (K_d = 9.7 μM). The human TPR-MEEVD structures of RPAP3 are essentially similar to those of the yeast Tah1p-MEEVD complex, except that Tah1p requires homodimerization too form a full complement of TPR α-helices (Morgan et al. 2015). The RBD C-terminal domain of RPAP3 appears to be specific for the ATPase domain of the RUVBL2 component of R2TP (Martino et al. 2018; Maurizy et al. 2018) (Fig. 15.3a), which suggests that the recruitment of RPAP3 into the R2TP complex is fundamentally different to that of the yeast Tah1p protein. Three isoforms of RPAP3 have been identified and it appears that isoform 1 interacts with PIH1D1 (Maurizy et al. 2018; Yoshida et al. 2013) (Fig. 15.1). This interaction is mediated through amino acid residues 400–420 of RPAP3 (Martino et al. 2018), which appears to be lacking in RPAP3 isoform 2. For isoform 3 the first 159 amino acid residues appear to be missing and this would impact on the structural integrity of the first TPR module. The significance of this remains unknown, but it may be significant that Tah1p can be either monomeric or form homodimers (Millson et al. 2008; Morgan et al. 2015). CCDC103, known as coiled coil domain containing protein, also contains an RBD-like domain and is involved in axonemal complex assembly and when mutated leads to primary ciliary dyskinesia. Using the LUMIER and proteomic data three types of R2TP-like complex have been identified. The first is R2T, which contains RPAP3 isoform 2 and thus lacks PIH1D1. The second is R2SP and contains SPAG1 and PIH1D2. And he third was named R2SD and contains SPAG1 and DNAAF2, although this complex was only detected in the LUMIER assay (Maurizy et al. 2018). DNAAF2 was also detected in complex with DYX1C1 isoform A (DYX1C1 is also known as DNAAF4) and PIH1D3 with DYX1C1 isoform C (Maurizy et al. 2018; Olcese et al. 2017; Tarkar et al. 2013) (Fig. 15.1). Interestingly, mutation of PIH1D3 results in primary ciliary dyskinesia (Olcese et al. 2017). Clearly, these components are client specific and help to recruit such specific clients to the R2 complex.

15.1.3 Structure of the TEL2-PIH1D1 Complex

The recruitment of phosphatidylinositol 3-kinase-related kinases depends not only on the PIH1D1 component of the R2TP complex but also on another complex known as the Triple T complex (TTT), which consists of TTI1, TTI2 and TEL2 of which the structure of TEL2 has been determined (Takai et al. 2010). The TTT

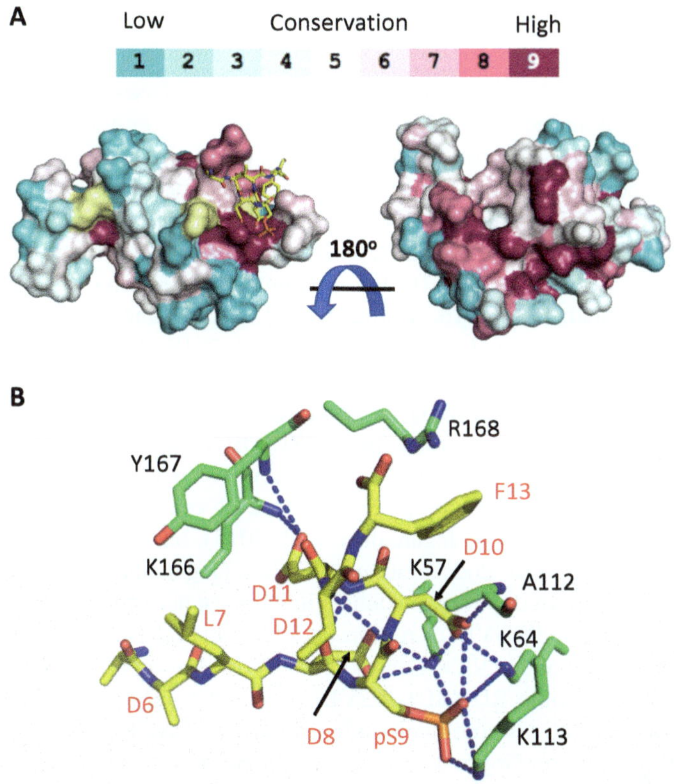

Fig. 15.4 Structure of the PIH domain of PIH1D1. (**a**) PyMol space fill showing conservation on both faces of the PIH domain. Left panel, the conserved TEL2 DpSDD motif is bound to a highly conserved pocket of the PIH domain. Right panel, conservation of the R2 binding face. (**b**) PyMol stick representation of the interactions of the conserved DpSDD motif of TEL2 with the Pih domain of Pih1p

complex is recruited by a phosphorylation dependent interaction with PIH1D1 (Horejsi et al. 2010; Pal et al. 2014) (Fig. 15.4a). It has been proposed that such phosphorylation-dependent PIH1D1 interactions define the substrate specificity of the R2TP complex (Horejsi et al. 2014). Examples include interactions with MRE11, ECD and UBR5 amongst others (Horejsi et al. 2014). The phosphorylation motif bound by PIH1D1 consists of a conserved DpS(T)DD peptide sequence (Fig. 15.4b), which when mutated leads to an instability of PIKKs that is not due to a lack of association with the TTT complex, but due to a lack of recruitment to the R2TP complex (Horejsi et al. 2010). The phosphoserine of TEL2 is bound by a number of basic side chains including Lys 57, Lys 64 and Lys 113 among a number of other interactions that anchor the DpSDD sequence to PIH1D1 (Fig. 15.4b).

15.1.4 Structure of the R2 Complex

RUVBL1 and RUVBL2 are highly conserved AAA$^+$ ATPases (ATPases associated with diverse cellular activities) and play important roles in c-Myc and Wnt pathways, processes such as chromatin remodelling, developmental and transcriptional regulation, apoptosis, chromatin remodelling, axonemal dynein, telomerase and RNA polymerase II assembly, snRNA assembly of snRNPs, snoRNA assembly in the biogenesis of RNA (rRNA, tRNA and snRNA) and PIKK maturation amongst other processes (Boulon et al. 2010; Jha and Dutta 2009; Kakihara and Houry 2012; Lakshminarasimhan et al. 2016; Malinova et al. 2017; Nano and Houry 2013; Pal et al. 2014; von Morgen et al. 2015; Zur Lage et al. 2018). RUVBL1 and RUVBL2 and their yeast equivalents, Rvb1p and Rvb2p respectively, are split ATPases, where the ATPase domains (DI and DIII) are separated by the insertion of the DII domain (Fig. 15.1). Structural studies using human RUVBL1 showed it to assemble into a hexameric ring structure in its ADP bound form, with marginal ATPase activity (Matias et al. 2006). A hexameric ring structure of yeast Rvb1p-Rvb2p was also reported (Gribun et al. 2008; Jeganathan et al. 2015), but with around 10% of the protein assembling into dodecameric complex (Jeganathan et al. 2015). For RUVBL2 it was reported to form hexamers in the presence of both ADP and ATP. In contrast, RUVBL1 and RUVBL2 mixtures could form both hexamers and dodecamers with either nucleotide (Niewiarowski et al. 2010; Puri et al. 2007). A dodecameric structure of human RUVBL1-RUVBL2 was also reported and although the complex was purified without nucleotide a mixture of ADP and ATP were found bound to the complex (Gorynia et al. 2011). In addition to these numerous structures reported so far, a stretched side-view dodecameric structure was seen that was attributed to differences in the conformation of the DII domains (Lopez-Perrote et al. 2012). In this stretched conformation the DII domains are pulled upright and slightly rotated relative to the compact dodecameric state. Interestingly, a truncated mutant of RUVBL1-RUVBL2 that lacked the external fragment of the DII domains lead to an increased ATPase activity (Gorynia et al. 2011). However, the significance of this remains unknown.

The first crystal x-ray structure of intact Rvb1-Rvb2 complex from *Chaetomium thermophilum* provided a 3.6 Å dodecameric structure of the complex (Lakomek et al. 2015) (Fig. 15.5a). The DII domains of Rvb1 and Rvb2 were found to occupy different positions; the Rvb1 DII domains form contacts between the hexameric rings while the Rvb2 DII domains are rotated 100° outward and thus occupy a lateral position (Fig. 15.5b). Interestingly, ATP was found bound to Rvb1, even though ATP was not added during the purification, while the Rvb2 ATPase site was unoccupied by nucleotide. This suggests that ATP hydrolysis and nucleotide binding may be a nonconcerted process. In addition, a higher resolution structure with added ADP-BeF_3 was obtained at 2.9 Å and in this structure ADP was bound to both Rvb1 and Rvb2 molecules, but density for BeF_3 was absent. The overall architecture of each hexameric ring is 118 Å with a central hole of 24–25 Å in diameter. The nucleotide binding sites open towards the interface of adjacent Rvb1 and Rvb2 molecules

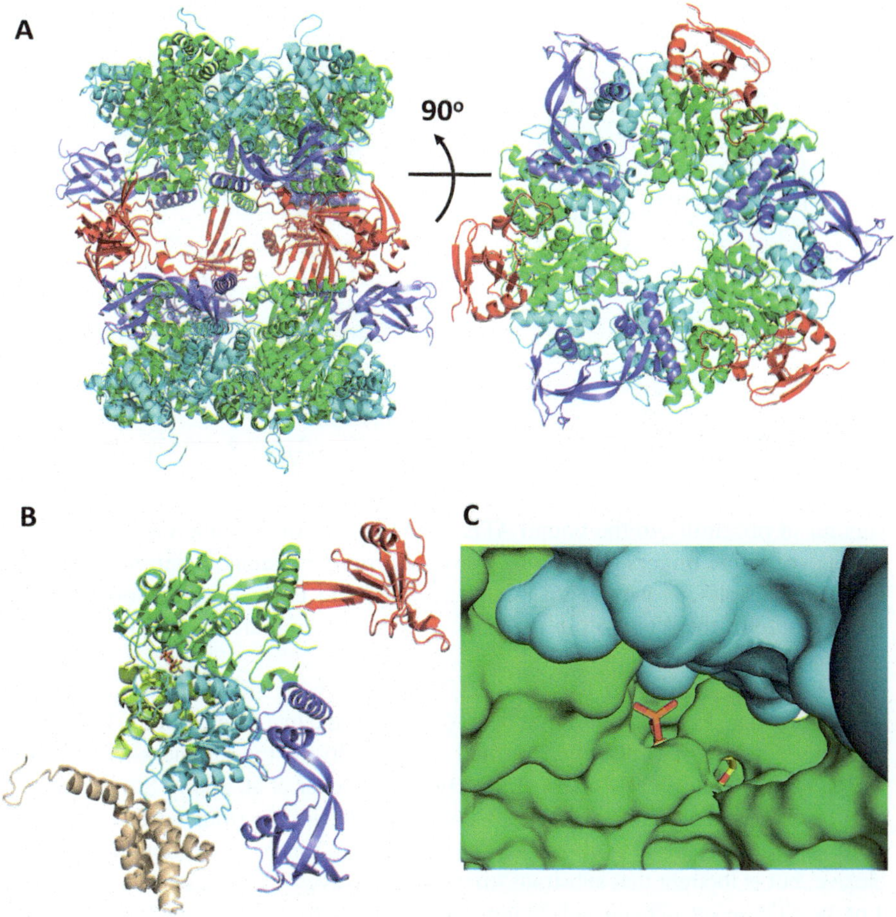

Fig. 15.5 The dodecameric structure of *Chaetomium thermophilum*. (**a**) Dodecameric structure. Left panel, side view and right panel top view. Rvb1, green and red and Rvb2 blue and cyan. The DII domains (blue and red) adopt very different conformations. (**b**) conformations adopted by the DI, DII and DIII domains of Rvb1 and Rvb2. Rvb1, coloured red, green and olive and Rvb2, coloured wheat, cyan and blue, respectively. (**c**) Rvb1 (green) and Rvb2 (cyan) shown in PyMol surface representation. ATP is shown in stick representation and is bound within the ATP binding site of Rvb1. Exchange with ADP is clearly restricted

(Fig. 15.5c). Access in and out of the ATPase site is restricted and likely limits the ATPase activity of the R2 complex. The structure of the ATPase site itself conforms to the walker type and ATP is similarly bound in both the *Chaetomium thermophilum* and human structures (Gorynia et al. 2011; Walker et al. 1982). The Walker-A (P-loop) and -B motifs are involved in co-ordinating the positions of the β- and γ-phosphates (Fig. 15.6), and the invariant lysine 77 (*Chaetomium thermophilum*), is the most commonly mutated residue of the P-loop. Another feature of Walker ATPases is the presence of one or more arginine residues from the neighbouring

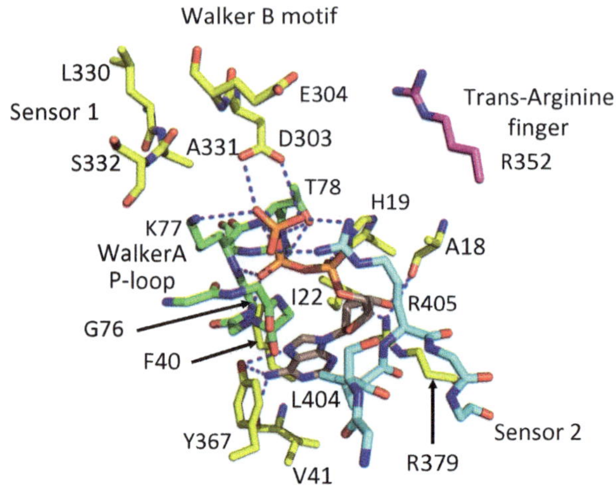

Fig. 15.6 Rvb1 and Rvb2 interactions with ATP. ATP is bound within the Rvb1 ATP binding site of *Chaetomium thermophilum*. Amino acid residues are coloured according to the conserved walker motifs. The trans-arginine finger (R352) is from Rvb2

subunit in proximity to the bound ATP, which are normally called arginine fingers or trans-arginine fingers (Fig. 15.6). Another arginine residue, positioned at the tip of an alpha helix, is part of the sensor 2 motif and is involved in contacting the γ-phosphate of the ATP within the same protomer (Ammelburg et al. 2006; Botos et al. 2004).

The ATPase activity of RUVBL1-RUVBL2 has been shown to be required for in vivo function. For example, it has been shown that the inactivation of the RUVBL1/2 activity decreases mTOR activation and the stability of the telomerase component TERC (Jonsson et al. 2004; Kim et al. 2013; Venteicher et al. 2008). Stimulation has also been reported for the yeast RVB proteins with double-stranded DNA with a 5 prime overhang (Gribun et al. 2008). The mechanism for this activation has remained elusive, but is thought that substrate stimulation involves a glutamate switch in sensor 1 of the ATPase site (Zhang and Wigley 2008) (Fig. 15.6). It is thought that suppression of ATPase activity in AAA$^+$ ATPases serves to allow directionality and the correct assembly of complex systems rather than acting in a simple catalytic role (Zhang and Wigley 2008). It therefore appears that ATP hydrolysis may simply serve as a control switch signaling the end of a process so that components can be recycled. Clearly the arrangement of subunits and protomers in a hexameric ring limits the ability of ADP and ATP to passively exchange with the ATP binding sites of the R2 complex (Fig. 15.5c). In fact, RUVBL1 has been shown to have one the lowest surface accessible areas between the DI and DIII domains with respect to a number of AAA$^+$ ATPases, indicating that nucleotide cannot be easily exchanged (Matias et al. 2006). It may be of significance then that a complex of Tah1p-Pih1p could stimulate the ATPase activity of the yeast Rvb1p-Rvb2p complex (Rivera-Calzada et al. 2017), although the precise mechanism was not established. Significantly, substrates interacting with R2 can also stimulate the ATPase activity of the complex (Zaarur et al. 2015). However, with observations that both substrate and the co-chaperones Pih1p-Tah1p showing activation, it is likely that both components are required for full activation. Studies using the dodecameric yeast Rvb1p-Rvb2p complex suggest that the nucleotide state determines the precise conformation of the DII domains and leads to changes in the hexamer to hexamer interaction interface and a rotation of 29° of one of the hexameric rings relative to the other (Ewens et al. 2016). However, the significance of this remains unknown, especially in the light of the R2-INO80 complex in which INO80 is in complex with a single hexameric R2 complex (Aramayo et al. 2018). However, from the array of R2 monomeric and hexameric structures seen it is clear that the precise active state of R2 remained controversial and elusive.

15.1.5 Structure of the R2TP Complex

The structure of the yeast and human R2TP complex was recently determined by EM (Martino et al. 2018; Rivera-Calzada et al. 2017). The yeast structure shows that the Pih1p-Tah1p complex is cradled within the DII domains (Fig. 15.7). In the human structure, PIH1D1 binds in a similar position to the equivalent yeast protein. However, RPAP3 interacts through its RBD domain with the ATPase side of the R2 hexameric ring and then interacts with PIH1D1 through a central region consisting of amino acid residues 400–420. Recent advances in understanding the active state of the R2 complex has come from such structural studies; namely R2 in complex with Tah1p/RPAP3 and Pih1p/PIH1D1 and with components of the INO80 chromatin remodelling complex (Aramayo et al. 2018; Eustermann et al. 2018; Martino et al. 2018; Rivera-Calzada et al. 2017). These studies show that a single hexameric R2 ring (Fig. 15.7) is probably the active state, rather than the dodecameric structures seen in earlier studies (Jeganathan et al. 2015).

In a similar way to PIH1D1/Pih1p binding, the ATPase component of the INO80 complex (INO80A, amino acid residues 267–1556) is cradled by the DII domains,

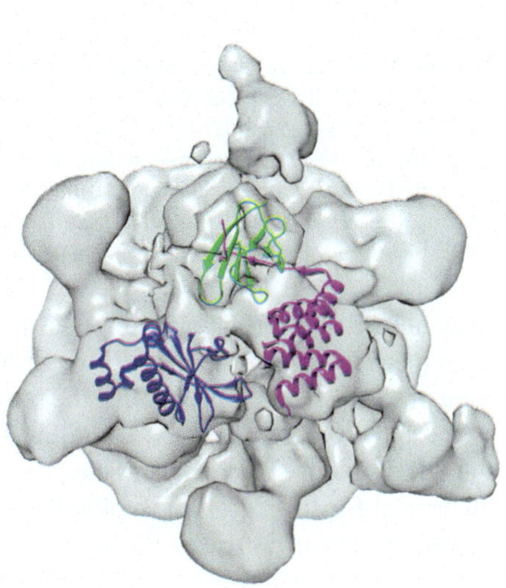

Fig. 15.7 The structure of yeast R2TP. The R2 component is shown in grey and Tah1p and Pih1p are fitted into density not represented by the DII domains

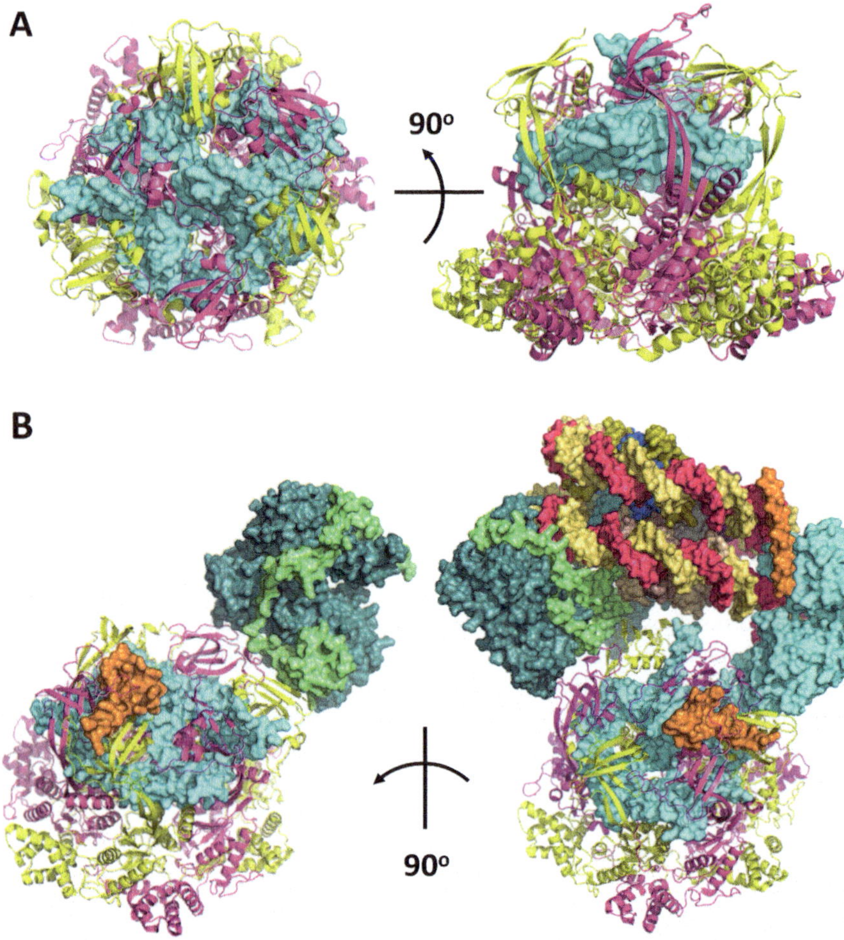

Fig. 15.8 PyMol structure representation of the R2TP-INO80 nucleosome complex. (**a**) Two views of the R2-INO80 complex. INO80 (space-filled cyan) is cradled between the DII domains of RUVBL1 (magenta) and RUUVBL2 (yellow). The DII domains form an intricate interaction with INO80A. (**b**) The intricate interaction of INO80A (cyan) with the DII domains of RUVBL1 (magenta) and RUVBL2 (yellow) positions these domains to allow further interactions with ARP5 (green), IES2 (gold) and IES6 (olive). Red and yellow, represent the DNA of the histone complex (PDB 6FML)

in which each DII domain adopts a different conformation, but is also in contact with the INO80A component (Fig. 15.8a). Remarkably, this asymmetry in the DII domains may well help establish the position of a number of additional components of the INO80 core complex, including IES2, IES6 and ARP5, all of which contact the DII domains (Fig. 15.8b). Thus, the DII domains and their positioning appear to be critical for formation of the INO80 complex. These interactions are further supported by structures of the INO80 chromatin remodelling complex interacting with a nucleosome (Aramayo et al. 2018; Eustermann et al. 2018), and those of the ATP-dependent chromatin remodelling complex SWR-C (Nguyen et al. 2013). These structures suggest that interaction with the PIH1D1/Pih1p and RPAP3/Tah1p components is probably impossible. Thus, reports suggesting that Hsp90 may play a role in INO80 and SWR-C assembly may be either due to a requirement by one of the components of these complexes for Hsp90, such as IES1 of the INO80 complex, or a requirement for the remodelling of the R2 (or Rvb1-Rvb2) complex (Zhao et al. 2005). Further work is therefore required to understand the role of HSP proteins in such complexes.

15.1.6 R2TP-Prefoldin Complex and Additional Factors

In addition to Hsp90, the R2TP complex has also been seen to be associated with Prefoldin-like complex (PFDL) and a number of additional factors that in some cases appear to be client specific (Fig. 15.9). The PFDL consists of two subunits in common with the canonical Prefoldin complex (PFD), PFDN2 and PFDN6, as well as three other specific subunits, URI1, PDGR1 and UXT. The canonical PFD assists the folding of actin and α and β-tubulin together with the chaperone, CCT (Siegers et al. 1999; Vainberg et al. 1998). In contrast, the R2TP-PFDL complex has been implicated in snoRNPs, nuclear RNA polymerases, PIKK-containing complexes, TSC1-TSC2, TSC1-TBCD17 and the U5 snRNP complexes (Cloutier et al. 2009; Cloutier and Coulombe 2010; Cloutier et al. 2017; Izumi et al. 2012). The PFDL

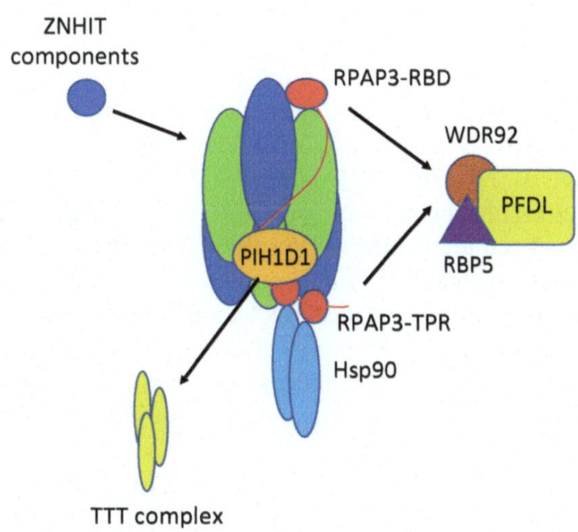

Fig. 15.9 Schematic showing the Hsp90-R2TP-PFDL interactions. Additionally, the interaction of the TTT complex and Prefoldin complex (PFDL-WDR92 and RBP5) is shown together with possible ZNHIT interactions that may modify the activity of the complex or help select specific clients. Other possible components include DNAAF4 (DYX1C1)-DNAAF2 (Ktu) and DNAAF4-PIH1D3 complex, which are not shown

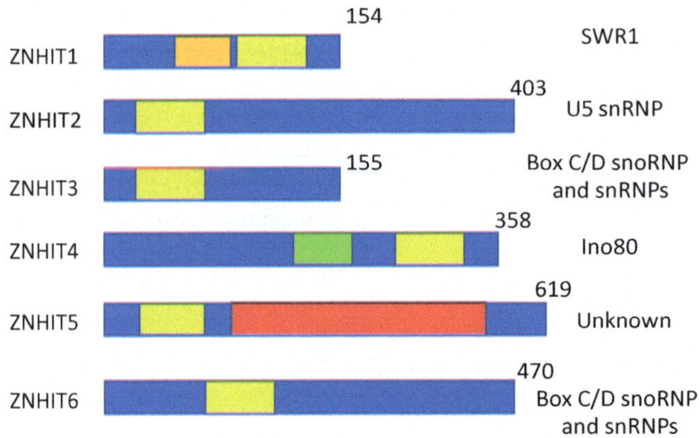

Fig. 15.10 Additional co-chaperones of the R2TP complex. ZnHIT proteins with their role in a variety of R2TP complexes indicated to the right. The length of each protein is indicated, but lengths are approximate for clarity. Similar domains are represented by colour. Yellow, HIT domain, orange, NR1D2 interacting domain, green, IES2 interacting domain, red, dead box helicase

complex appears to consist of a complex that also contains WDR92 (also known as monad) and RBP5. It has been suggested that WDR92 can interact with RPAP3 (Itsuki et al. 2008; Sardiu et al. 2008) and that RBP5 interacts with URI (Mita et al. 2013). However, it should be noted that in *S. cerevisiae* the WDR2 and PFDL modules are completely absent. While this is surprising it might explain the difference between RPAP3 and the smaller Tah1p component of yeast, since RPAP3 has been implicated in the recruitment of WDR92.

In addition to the PFDL complex a number of additional factors have also been identified that are required for client protein maturation and assembly of complexes. For example, box C/D assembly requires additional zinc finger containing proteins (Fig. 15.10) such as NUFIP1, ZNHIT6/BCD1, ZNHIT3 (Bizarro et al. 2014; McKeegan et al. 2007; Rothe et al. 2014) and ZNHIT2, the latter of which interacts with RUVBL2, has been implicated in U5 snRNP, U4 snRNP, snoRNP and telomerase assembly together with R2TP-PFDL complex (Cloutier et al. 2017). In contrast, ZNHIT1 is required for assembly of the SWR1 complex, while others such as ZNHIT5 have yet to be characterised (Verheggen et al. 2015). Clearly, such additional factors are key partners of the R2TP complex and may act as bridging components or alter the activity of the R2TP complex in order to allow the assembly of specific clients.

15.2 Conclusions

Recent structural advances have allowed some mechanistic insight into the function of R2 complex. Most significantly, it appears that the dodecameric ring may be an inactive state and that it is converted to an active form upon interaction with substrate, as seen in the R2 -INO80 and -SWR-C structures, or with co-chaperones, such as Tah1p/RPAP3 and Pih1p/PIH1D1. Biochemical studies support this idea since the ATPase activity of R2 can be stimulated by interaction with substrate, via a glutamate switch, while the yeast Tah1p-Pih1p complex can stimulate the ATPase activity of the yeast R2 complex. Furthermore, these studies suggest that the DII domain side of the hexameric ring is the prime location for substrate and co-chaperone interaction. However, the ATPase side of the hexameric ring has also shown to allow binding of the co-chaperone RPAP3 via its RBD domain, which in stark contrast is not found with yeast Tah1p protein. Further work is now required in order to understand the role of Hsp90 and associated co-chaperones in the maturation of client protein into protein complexes. For example, does Hsp90 act directly on client proteins or does it help modulate the structure of the R2 complex? Can both client protein and Hsp90 exist within the R2 complex or is the apparent requirement for Hsp90 dependent on a single component of those complexes? These are just some of the questions that have been raised from the current structural and biochemical studies and further work is now needed in order to answer these. What is clear though is that we need to understand the role of the many of the components that make up the R2TP complex, including that of PFDL, Hsp90 and many of the newly identified co-chaperones such as ZNHIT2, if we are going to fully understand the role of these proteins in human disease. Structural and biochemical experiments will be required in order to advance our understanding of what appears to be a large and versatile multi-chaperone system. However, what seems to be emerging is that this large chaperone system could be acting partly as a scaffold to bring components that need to be assembled into large complexes possibly by utilizing the ATPase activity of both Hsp90 and the R2 complex.

References

Ammelburg M, Frickey T, Lupas AN (2006) Classification of AAA+ proteins. J Struct Biol 156:2–11

Aramayo RJ, Willhoft O, Ayala R, Bythell-Douglas R, Wigley DB, Zhang X (2018) Cryo-EM structures of the human INO80 chromatin-remodeling complex. Nat Struct Mol Biol 25:37–44

Bizarro J, Charron C, Boulon S et al (2014) Proteomic and 3D structure analyses highlight the C/D box snoRNP assembly mechanism and its control. J Cell Biol 207:463–480

Botos I, Melnikov EE, Cherry S et al (2004) Crystal structure of the AAA+ alpha domain of *E. coli* Lon protease at 1.9A resolution. J Struct Biol 146:113–122

Boulon S, Pradet-Balade B, Verheggen C et al (2010) HSP90 and its R2TP/Prefoldin-like cochaperone are involved in the cytoplasmic assembly of RNA polymerase II. Mol Cell 39:912–924

Cloutier P, Coulombe B (2010) New insights into the biogenesis of nuclear RNA polymerases? Biochem Cell Biol 88:211–221

Cloutier P, Al-Khoury R, Lavallee-Adam M et al (2009) High-resolution mapping of the protein interaction network for the human transcription machinery and affinity purification of RNA polymerase II-associated complexes. Methods 48:381–386

Cloutier P, Poitras C, Durand M et al (2017) R2TP/Prefoldin-like component RUVBL1/RUVBL2 directly interacts with ZNHIT2 to regulate assembly of U5 small nuclear ribonucleoprotein. Nat Commun 8:15615

Eckert K, Saliou JM, Monlezun L et al (2010) The Pih1-Tah1 cochaperone complex inhibits Hsp90 molecular chaperone ATPase activity. J Biol Chem 285:31304–31312
Eustermann S, Schall K, Kostrewa D et al (2018) Structural basis for ATP-dependent chromatin remodelling by the INO80 complex. Nature 556:386–390
Ewens CA, Su M, Zhao L, Nano N, Houry WA, Southworth DR (2016) Architecture and nucleotide-dependent conformational changes of the Rvb1-Rvb2 AAA+ complex revealed by cryoelectron microscopy. Structure 24:657–666
Gorynia S, Bandeiras TM, Pinho FG et al (2011) Structural and functional insights into a dodecameric molecular machine – the RuvBL1/RuvBL2 complex. J Struct Biol 176:279–291
Gribun A, Cheung KL, Huen J, Ortega J, Houry WA (2008) Yeast Rvb1 and Rvb2 are ATP-dependent DNA helicases that form a heterohexameric complex. J Mol Biol 376:1320–1333
Grigoletto A, Lestienne P, Rosenbaum J (2011) The multifaceted proteins Reptin and Pontin as major players in cancer. Biochim Biophys Acta 1815:147–157
Gross M, Hessefort S (1996) Purification and characterisation of a 66-kDa protein from rabbit reticulocyte lysate which promotes the recycling of hsp70. J Biol Chem 271:16833–16841
Hartill VL, van de Hoek G, Patel MP et al (2018) DNAAF1 links heart laterality with the AAA+ ATPase RUVBL1 and ciliary intraflagellar transport. Hum Mol Genet 27:529–545
Horejsi Z, Takai H, Adelman CA et al (2010) CK2 phospho-dependent binding of R2TP complex to TEL2 is essential for mTOR and SMG1 stability. Mol Cell 39:839–850
Horejsi Z, Stach L, Flower TG et al (2014) Phosphorylation-dependent PIH1D1 interactions define substrate specificity of the R2TP cochaperone complex. Cell Rep 7:19–26
Huber O, Menard L, Haurie V, Nicou A, Taras D, Rosenbaum J (2008) Pontin and reptin, two related ATPases with multiple roles in cancer. Cancer Res 68:6873–6876
Hurov KE, Cotta-Ramusino C, Elledge SJ (2010) A genetic screen identifies the triple T complex required for DNA damage signaling and ATM and ATR stability. Genes Dev 24:1939–1950
Itsuki Y, Saeki M, Nakahara H et al (2008) Molecular cloning of novel Monad binding protein containing tetratricopeptide repeat domains. FEBS Lett 582:2365–2370
Izumi N, Yamashita A, Hirano H, Ohno S (2012) Heat shock protein 90 regulates phosphatidylinositol 3-kinase-related protein kinase family proteins together with the RUVBL1/2 and Tel2-containing co-factor complex. Cancer Sci 103:50–57
Jeganathan A, Leong V, Zhao L et al (2015) Yeast rvb1 and rvb2 proteins oligomerize as a conformationally variable dodecamer with low frequency. J Mol Biol 427:1875–1886
Jha S, Dutta A (2009) RVB1/RVB2: running rings around molecular biology. Mol Cell 34:521–533
Johnson BD, Schumacher RJ, Ross ED, Toft DO (1998) Hop modulates hsp70/hsp90 interactions in protein folding. J Biol Chem 273:3679–3686
Jonsson ZO, Jha S, Wohlschlegel JA, Dutta A (2004) Rvb1p/Rvb2p recruit Arp5p and assemble a functional Ino80 chromatin remodeling complex. Mol Cell 16:465–477
Kabani M, Beckerich JM, Brodsky JL (2002) Nucleotide exchange factor for the yeast Hsp70 molecular chaperone Ssa1p. Mol Cell Biol 22:4677–4689
Kakihara Y, Houry WA (2012) The R2TP complex: discovery and functions. Biochim Biophys Acta 1823:101–107
Kim SG, Hoffman GR, Poulogiannis G et al (2013) Metabolic stress controls mTORC1 lysosomal localization and dimerization by regulating the TTT-RUVBL1/2 complex. Mol Cell 49:172–185
Lakomek K, Stoehr G, Tosi A, Schmailzl M, Hopfner KP (2015) Structural basis for dodecameric assembly states and conformational plasticity of the full-length AAA+ ATPases Rvb1 . Rvb2. Structure 23:483–495
Lakshminarasimhan M, Boanca G, Banks CA et al (2016) Proteomic and genomic analyses of the Rvb1 and Rvb2 interaction network upon deletion of R2TP complex components. Mol Cell Proteomics 15:960–974
Lopez-Perrote A, Munoz-Hernandez H, Gil D, Llorca O (2012) Conformational transitions regulate the exposure of a DNA-binding domain in the RuvBL1-RuvBL2 complex. Nucleic Acids Res 40:11086–11099

Malinova A, Cvackova Z, Mateju D et al (2017) Assembly of the U5 snRNP component PRPF8 is controlled by the HSP90/R2TP chaperones. J Cell Biol 216:1579–1596
Mannoor K, Liao J, Jiang F (2012) Small nucleolar RNAs in cancer. Biochim Biophys Acta 1826:121–128
Mao YQ, Houry WA (2017) The role of Pontin and Reptin in cellular physiology and cancer etiology. Front Mol Biosci 4:58
Martino F, Pal M, Munoz-Hernandez H et al (2018) RPAP3 provides a flexible scaffold for coupling HSP90 to the human R2TP co-chaperone complex. Nat Commun 9:1501
Matias PM, Gorynia S, Donner P, Carrondo MA (2006) Crystal structure of the human AAA+ protein RuvBL1. J Biol Chem 281:38918–38929
Maurizy C, Quinternet M, Abel Y et al (2018) The RPAP3-Cterminal domain identifies R2TP-like quaternary chaperones. Nat Commun 9:2093
McKeegan KS, Debieux CM, Boulon S, Bertrand E, Watkins NJ (2007) A dynamic scaffold of pre-snoRNP factors facilitates human box C/D snoRNP assembly. Mol Cell Biol 27:6782–6793
Miao C, Jiang Q, Li H et al (2016) Mutations in the motile cilia gene DNAAF1 are associated with neural tube defects in humans. G3 (Bethesda) 6:3307–3316
Mikesch JH, Hartmann W, Angenendt L et al (2018) AAA+ ATPases Reptin and Pontin as potential diagnostic and prognostic biomarkers in salivary gland cancer – a short report. Cell Oncol (Dordr) 41:455–462
Millson SH, Vaughan CK, Zhai C et al (2008) Chaperone ligand-discrimination by the TPR-domain protein Tah1. Biochem J 413:261–268
Mita P, Savas JN, Ha S, Djouder N, Yates JR 3rd, Logan SK (2013) Analysis of URI nuclear interaction with RPB5 and components of the R2TP/prefoldin-like complex. PLoS One 8:e63879
Morgan RM, Pal M, Roe SM, Pearl LH, Prodromou C (2015) Tah1 helix-swap dimerization prevents mixed Hsp90 co-chaperone complexes. Acta Crystallogr D Biol Crystallogr 71:1197–1206
Nallar SC, Kalvakolanu DV (2013) Regulation of snoRNAs in cancer: close encounters with interferon. J Interf Cytokine Res 33:189–198
Nano N, Houry WA (2013) Chaperone-like activity of the AAA+ proteins Rvb1 and Rvb2 in the assembly of various complexes. Philos Trans R Soc Lond Ser B Biol Sci 368:20110399
Nguyen VQ, Ranjan A, Stengel F et al (2013) Molecular architecture of the ATP-dependent chromatin-remodeling complex SWR1. Cell 154:1220–1231
Niewiarowski A, Bradley AS, Gor J, McKay AR, Perkins SJ, Tsaneva IR (2010) Oligomeric assembly and interactions within the human RuvB-like RuvBL1 and RuvBL2 complexes. Biochem J 429:113–125
Olcese C, Patel MP, Shoemark A et al (2017) X-linked primary ciliary dyskinesia due to mutations in the cytoplasmic axonemal dynein assembly factor PIH1D3. Nat Commun 8:14279
Pal M, Morgan M, Phelps SE et al (2014) Structural basis for phosphorylation-dependent recruitment of Tel2 to Hsp90 by Pih1. Structure 22:805–818
Polier S, Samant RS, Clarke PA, Workman P, Prodromou C, Pearl LH (2013) ATP-competitive inhibitors block protein kinase recruitment to the Hsp90-Cdc37 system. Nat Chem Biol 9:307–312
Prodromou C (2012) The 'active life' of Hsp90 complexes. Biochim Biophys Acta 1823:614–623
Prodromou C (2016) Mechanisms of Hsp90 regulation. Biochem J 473:2439–2452
Prodromou C, Morgan RML (2016) "Tuning" the ATPase activity of Hsp90. Adv Biochem Health D 14:469–490
Prodromou C, Siligardi G, O'Brien R et al (1999) Regulation of Hsp90 ATPase activity by tetratricopeptide repeat (TPR)-domain co-chaperones. EMBO J 18:754–762
Puri T, Wendler P, Sigala B, Saibil H, Tsaneva IR (2007) Dodecameric structure and ATPase activity of the human TIP48/TIP49 complex. J Mol Biol 366:179–192
Quinternet M, Rothe B, Barbier M et al (2015) Structure/function analysis of protein-protein interactions developed by the yeast Pih1 platform protein and its Partners in box C/D snoRNP assembly. J Mol Biol 427:2816–2839

Rivera-Calzada A, Pal M, Munoz-Hernandez H et al (2017) The structure of the R2TP complex defines a platform for recruiting diverse client proteins to the HSP90 molecular chaperone system. Structure 25:1145–52 e4

Roe SM, Ali MM, Meyer P et al (2004) The mechanism of Hsp90 regulation by the protein kinase-specific cochaperone p50(cdc37). Cell 116:87–98

Rothe B, Saliou JM, Quinternet M et al (2014) Protein Hit1, a novel box C/D snoRNP assembly factor, controls cellular concentration of the scaffolding protein Rsa1 by direct interaction. Nucleic Acids Res 42:10731–10747

Sardiu ME, Cai Y, Jin J et al (2008) Probabilistic assembly of human protein interaction networks from label-free quantitative proteomics. Proc Natl Acad Sci U S A 105:1454–1459

Siegers K, Waldmann T, Leroux MR et al (1999) Compartmentation of protein folding in vivo: sequestration of non-native polypeptide by the chaperonin-GimC system. EMBO J 18:75–84

Siligardi G, Panaretou B, Meyer P et al (2002) Regulation of Hsp90 ATPase activity by the co-chaperone Cdc37p/p50^{cdc37}. J Biol Chem 277:20151–20159

Takai H, Xie Y, de Lange T, Pavletich NP (2010) Tel2 structure and function in the Hsp90-dependent maturation of mTOR and ATR complexes. Genes Dev 24:2019–2030

Tarkar A, Loges NT, Slagle CE et al (2013) DYX1C1 is required for axonemal dynein assembly and ciliary motility. Nat Genet 45:995–1003

Taylor AM, Groom A, Byrd PJ (2004) Ataxia-telangiectasia-like disorder (ATLD)-its clinical presentation and molecular basis. DNA Repair (Amst) 3:1219–1225

Vainberg IE, Lewis SA, Rommelaere H et al (1998) Prefoldin, a chaperone that delivers unfolded proteins to cytosolic chaperonin. Cell 93:863–873

Venteicher AS, Meng Z, Mason PJ, Veenstra TD, Artandi SE (2008) Identification of ATPases pontin and reptin as telomerase components essential for holoenzyme assembly. Cell 132:945–957

Verheggen C, Pradet-Balade B, Bertrand E (2015) SnoRNPs, ZNHIT proteins and the R2TP pathway. Oncotarget 6:41399–41400

von Morgen P, Horejsi Z, Macurek L (2015) Substrate recognition and function of the R2TP complex in response to cellular stress. Front Genet 6:69

Walker JE, Saraste M, Runswick MJ, Gay NJ (1982) Distantly related sequences in the alpha-subunits and beta-subunits of ATP synthase, myosin, kinases and other ATP-requiring enzymes and a common nucleotide binding fold. EMBO J 1:945–951

Wegele H, Haslbeck M, Reinstein J, Buchner J (2003) Sti1 is a novel activator of the Ssa proteins. J Biol Chem 278:25970–25976

Yoshida M, Saeki M, Egusa H et al (2013) RPAP3 splicing variant isoform 1 interacts with PIH1D1 to compose R2TP complex for cell survival. Biochem Biophys Res Commun 430:320–324

Zaarur N, Xu X, Lestienne P et al (2015) RuvbL1 and RuvbL2 enhance aggresome formation and disaggregate amyloid fibrils. EMBO J 34:2363–2382

Zhang X, Wigley DB (2008) The 'glutamate switch' provides a link between ATPase activity and ligand binding in AAA+ proteins. Nat Struct Mol Biol 15:1223–1227

Zhang X, Ren J, Yan L et al (2015) Cytoplasmic expression of pontin in renal cell carcinoma correlates with tumor invasion, metastasis and patients' survival. PLoS One 10:e0118659

Zhao R, Davey M, Hsu YC et al (2005) Navigating the chaperone network: an integrative map of physical and genetic interactions mediated by the hsp90 chaperone. Cell 120:715–727

Zur Lage P, Stefanopoulou P, Styczynska-Soczka K et al (2018) Ciliary dynein motor preassembly is regulated by Wdr92 in association with HSP90 co-chaperone, R2TP. J Cell Biol 217:2583–2598

Chapter 16
Functions of SGT1, a Co-chaperone

Yohei Niikura and Katsumi Kitagawa

Abstract SGT1/SUGT1, a co-chaperone of HSP90, is involved in multiple cellular activities including cullin E3 ubiquitin ligase activity. In various species, SGT1 homologs have been identified, indicating that SGT1 genes and proteins are highly conserved. SGT1 functions in multiple distinct biological processes, such as plant and mammal innate immune systems, yeast CBF3 kinetochore assembly, human CENP-A deposition and kinetochore assembly, SCF function and cyclic AMP signaling, centrosome organization and cytokinesis, and brain and heart diseases. Domain-specific functions, interactors, and roles in subcellular localization of SGT1 are described in this chapter. Importantly, SGT1 contributes to cancer development by stabilizing oncoproteins, and the SGT1-HSP90 complex is a potential target for therapies aimed at cancer, brain and heart disease. Recent advances in our understanding of the physiologic role of SGT1 are briefly reviewed in this chapter.

Keywords Brain and heart diseases · Cancer · E3 ubiquitin ligase · Kinetochore · Post-translational modification (PTM) · SGT1/SUGT1

Abbreviations

17-AAG	17-allylaminogeldanamycin
AC	Adenylyl cyclase
ALS	Amyotrophic lateral sclerosis

Y. Niikura (✉)
MOE Key Laboratory of Model Animal for Disease Study, Model Animal Research Center, Nanjing University, Nanjing, Jiangsu Province, China
e-mail: niikura@nicemice.cn

K. Kitagawa (✉)
Greehey Children's Cancer Research Institute, Department of Molecular Medicine, UT Health San Antonio, San Antonio, TX, USA
e-mail: kitagawaK@uthscsa.edu

A. A. A. Asea, P. Kaur (eds.), *Heat Shock Protein 90 in Human Diseases and Disorders*, Heat Shock Proteins 19,
https://doi.org/10.1007/978-3-030-23158-3_16

CBF3	Centromere binding factor 3
CDPK	Calcium-dependent protein kinase
CENP-A	Centromere protein-A
CENP-B	Centromere protein-B
CHORD	Cysteine- and histidine-rich domain
CompPASS	Comparative Proteomic Analysis Software Suite
CS	CHORD protein and SGT1-specific
CUL	Cullin
CWP	Cell wall protein
DCM	Dilated cardiomyopathy
eEF1	Eukaryotic elongation factor 1
GA	Geldanamycin
HR	Hypersensitive response
HSP	Heat shock protein
IL	Interleukin
IL-1	Interleukin 1
LRR	Leucine rich repeats
MSU	Monosodium urate
NF-κB	Nuclear factor kappa-light-chain-enhancer of activated B cells
NLR	Nucleotide binding leucine-rich repeat receptor
Nod	Nucleotide-binding oligomerization domain
PAMP	Pathogen-associated molecular pattern
PCD	Programmed cell death
PCP	Phencyclidine
PGN	Peptidoglycan
PLK	Polo-like kinase
PQC	Protein quality control
PTM	Post-translational modification
R proteins	Resistance proteins
RAR1	Required for Mla12 resistance
S. cerevisiae	*Saccharomyces cerevisiae*
SCF	SKP1, Cullin, F-box
SGS	SGT1-specific
SGT1/SUGT1	Suppressor of G2 allele of SKP1
TNF-α	Tumor necrosis factor alpha
TPR	Tetratricopeptide repeat
UBE3A	Ubiquitin-protein ligase E3A
UPS	Ubiquitin-dependent 26S proteasome system

16.1 Introduction

SGT1 was originally isolated as a dosage suppressor of *skp1-4* mutant in yeast *S. cerevisiae*, in which *SGT1* is an essential gene (Kitagawa et al. 1999). The *skp1-4* mutation causes defects in the kinetochore function in budding yeast (Connelly and Hieter 1996). Sgt1 physically interacts with Skp1 in vivo and in vitro, and different *sgt1* conditional mutants arrest with either a G1 or a G2 DNA content. In various species, SGT1 homologs were identified, indicating that SGT1 genes and proteins are highly conserved. In humans, a single copy of *SGT1* encodes two isoforms, SGT1A and its splice variant SGT1B, which has 33 extra amino acids instead of Ser^{110} in the TPR domain (Niikura and Kitagawa 2003) (Fig. 16.1). In

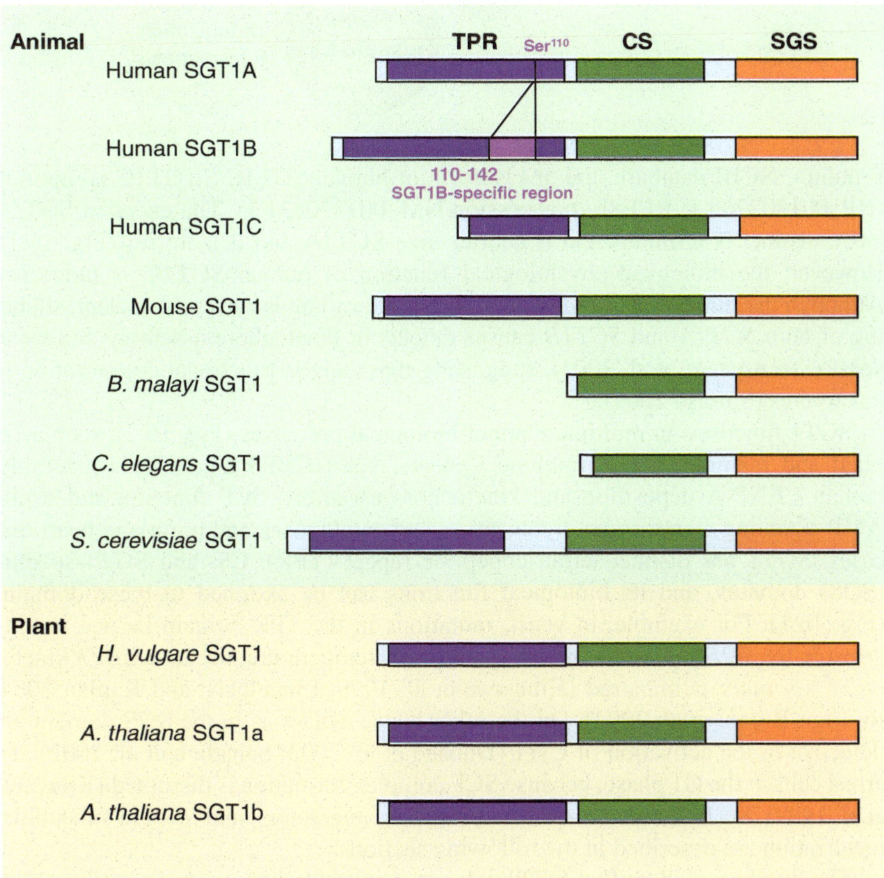

Fig. 16.1 SGT1 proteins among different species in eukaryotes. SGT1 proteins in animals (top) and plants (bottom). TPR, tetratricopeptide repeat-containing domain (InterPro: IPR013026); CS, CHORD protein and SGT1-specific domain (InterPro: IPR007052); SGS, SGT1-specific domain (InterPro: IPR007699). The human SGT1B/SUGT1B-specific region that contains a 33-amino acid insertion in the TPR domain and lacks Ser110 of SGT1A is shown

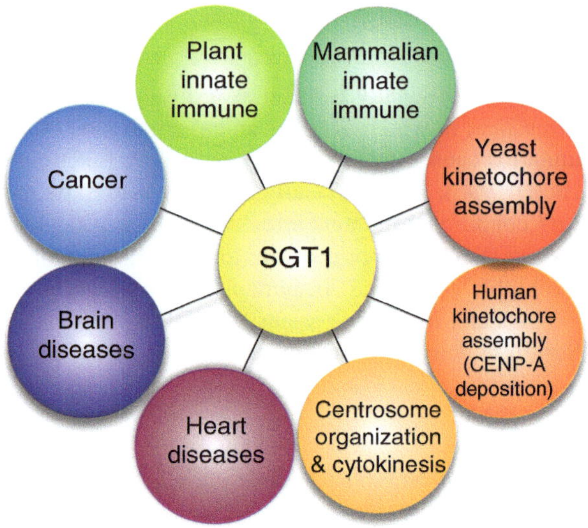

Fig. 16.2 Multiple functions of SGT1. Currently known SGT1 functions in multiple distinct biological processes are summarized

Protein – NCBI database, the 3rd isoform of human SGT1C/SUGT1C is reported (NP_001307760.1; REFSEQ: accession NM_001320831.1). The encoded SGT1C has a distinct N-terminus and is shorter than SGT1A and B isoforms (Fig. 16.1). However, the biological-physiological function of human SGT1C is unknown. Although the functional importance of the splice variant is currently unclear, silencing of both *SGT1A* and *SGT1B* causes defects in kinetochore assembly similar to yeast (Steensgaard et al. 2004), suggesting that there is functional conservation in eukaryotes (Shirasu 2009).

SGT1 functions in multiple distinct biological processes (Fig. 16.2), such as in plant and mammal innate immune systems, yeast CBF3 kinetochore assembly, human CENP-A deposition and kinetochore assembly, SCF function and cyclic AMP signaling, centrosome organization and cytokinesis, and brain and heart diseases. SGT1 has distinct tetratricopeptide repeat (TPR), CS, and SGT1-specific (SGS) domains, and its biological functions can be assigned to these domains (Fig. 16.1). For example, in yeast, mutations in the TPR domain induce mitotic arrest in the G2/M phase at the non-permissive temperature, because CBF3 kinetochore assembly is impaired (Kitagawa et al. 1999; Lingelbach and Kaplan 2004; Rodrigo-Brenni et al. 2004). On the other hand, mutations in the SGS domain are defective in the activation of Cyr1 (Dubacq et al. 2002; Schadick et al. 2002) and arrest cells at the G1 phase, because SCF complex formation is disrupted (Kitagawa et al. 1999). SGT1's domain-specific functions, interactors, and roles in subcellular localization are described in the following sections.

The number of identified SGT1-intractors is expanding every year (Fig. 16.3; Table 16.1) and major interactors of SGT1 currently identified includes, but is not limited to heat shock proteins (e.g., HSP90 and HSP70), CHORD domain-containing proteins (e.g., RAR1/melusin and Chp-1/CHORDC1), plant and mammal defense/resistance signaling proteins including NLR proteins, kinetochore

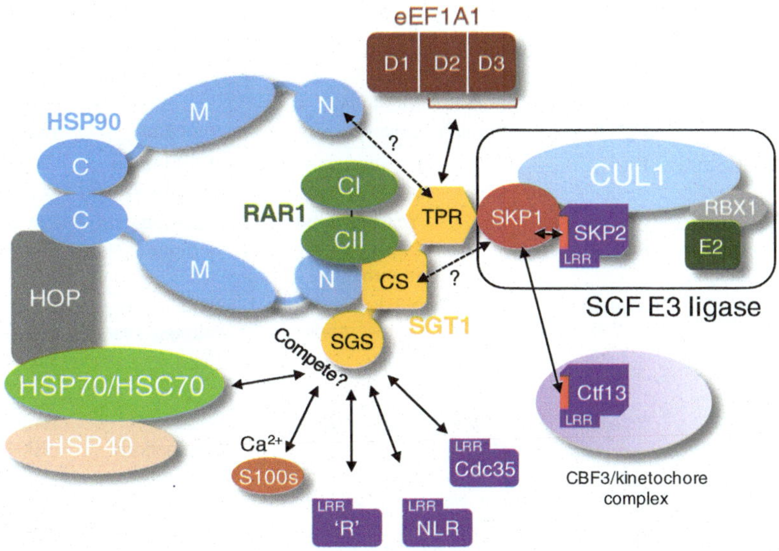

Fig. 16.3 SGT1 protein interactions through specific domains. The TPR domain interacts with the SKP1 component of SCF E3 ligase and with the CBF3 yeast kinetochore complex. The TPR domain and the D2 and D3 domains of eEF1A1 are involved in complex formation (Novosylna et al. 2015). The CS domain interacts with the N-terminal domain (ND) of HSP90. Simultaneous coassociation of RAR1-CHORD2 and HSP90-ND with the CS domain was suggested (Kadota et al. 2010; Shirasu 2009). The SGS domain interacts with various LRR proteins (plant R proteins, yeast Cyr1/Cdc35 adenylate cyclase, and NLR proteins). In plants, RAR1 is essential for R-protein activity, and bridges interactions between SGT1 and HSP90. Coexpression of the SGS binding protein S100A6 reduces SGT1-HSP70 interaction through a Ca^{2+}- dependent binding to the SGS domain in human cells (Spiechowicz et al. 2007) ("Compete?"). The double-headed arrow indicates the interaction. The dotted double-headed arrow indicates the possibility of interaction. Red boxes in SKP2 and Ctf13 indicate F-box motifs. Other client proteins of HSP90 and SGT1-interactors are omitted for simplicity (see also Table 16.1). This figure was partly adapted from Pearl et al. (2008). CI, CHORD I; CII, CHORD II

proteins (e.g., yeast Ctf13 and human MIS12 complex), components and interactors of ubiquitin E3 ligases (e.g., SKP1, Cullins, CSNs, UBE3A, and E3 substrates), kinases (e.g., PIK1 and PLK1), and disease-specific components (e.g., UBE3A, eEF1A, and melusin) at specific organs. Because SGT1 has a co-chaperone activity involved in multiple functions through heat shock chaperone proteins (e.g., HSP90 and HSP70), and such chaperone machinery is often tightly associated with the ubiquitin-dependent 26S proteasome system (UPS) (Kleiger and Mayor 2014; Mimnaugh et al. 1996), it is logical to conclude that the SGT1 co-chaperone and its interactors selectively collaborate or compete for binding to subpopulations of cellular HSP90. Thus, changes in the relative levels of SGT1 co-chaperone and its interactors may have marked effects on HSP90 function. The molecular mechanism of SGT1 with specific SGT1 interactors as currently understood is described below, including the relevance to post-translational modifications (PTMs) of SGT1.

Table 16.1 SGT1 interactors

Protain pair		Organism (of interactor with SGT1/Sgt1)	SGT1 domain	Method of detection	Function	References
SGT1[a]	HSP90	Barley	TPR + CS	Co-immunoprecipitation (co-IP)	Plant innate immune system	Takahashi et al. (2003)
		N. benthamiana	–	Yeast two-hybrid (Y2H), co-IP, in vitro	Plant innate immune system	Liu et al. (2004)
		Arabidopsis	CS	Y2H	Plant innate immune system	Boter et al. (2007)
		Arabidopsis	–	Co-IP	Plant innate immune system	Hubert et al. (2003)
		Yeast (*S. cerevisiae*)	TPR + CS	Y2H, co-IP,	Yeast kinetochore assembly	Bansal et al. (2004)
		Yeast (*S. cerevisiae*)	CS	In vitro		Catlett and Kaplan (2006)
		Human	CS	Co-IP, in vitro, NMR	Human CENP-A deposition and kinetochore assembly	Lee et al. (2004)
		Human	–	Co-IP	Human CENP-A deposition and kinetochore assembly	Niikura et al. (2006)
	HSP70	*Arabidopsis*	SGS	Co-IP	Plant innate immune system	Noel et al. (2007)
		Yeast (*S. cerevisiae*)	–	IP-mass spectrometry		Bansal et al. (2004)
		Human	–	Co-IP, in vitro		Spiechowicz et al. (2007)
	HSC70	Human	–	Co-IP, IP-mass spectrometry		Niikura et al. (2006)
	RAR1	Barley	CS	Y2H, co-IP	Plant innate immune system	Azevedo et al. (2002)
		N. benthamiana	(Not TPR)	Y2H, co-IP, in vitro	Plant innate immune system	Liu et al. (2002)
		Rice (*O. sativa*)	CS	Y2H, biomolecular fluorescence complementation (BiFC)	Plant innate immune system	Wang et al. (2008)

(continued)

Table 16.1 (continued)

Protain pair		Organism (of interactor with SGT1/Sgt1)	SGT1 domain	Method of detection	Function	References
		Arabidopsis	–	Y2H, fluorescence resonance energy transfer–fluorescence lifetime imaging microscopy (FRET-FLIM)	Plant innate immune system	Bayle et al. (2008)
		Wheat (*Triticum aestivum L*)	–	Y2H	Plant innate immune system	Tai et al. (2008)
		Wheat (*Triticum aestivum L*)	(Not TPR in Y2H)	Y2H, BiFC assay	Plant innate immune system	Wang et al. (2015)
	Melusin	Human	–	Co-IP	Heart	Sbroggio et al. (2008)
	Chp-1	Mouse (with human SGT1)	–	In vitro	Heart	Hong et al. (2013)
	Bs2	*N. benthamiana*	–	Co-IP	Plant innate immune system	Leister et al. (2005)
	Prf	*N. benthamiana* (with all *N. benthamiana* SGT1 and tomato SlSGT1-1 and SlSGT1-2)	–	Co-IP	Plant innate immune system	Kud et al. (2013)
	SRC2-1	Co-IP, BiFC assay in *N. benthamiana*	–	Y2H, co-IP, BiFC assay	Plant innate immune system	Liu et al. (2016)
	Rad6	Rice (*O. sativa*)	–	In vitro	Plant innate immune system	Yamamoto et al. (2004)
	MLA1	Barley	[b]	Y2H	Plant innate immune system	Bieri et al. (2004)
	PIK1	Pepper (*Capsicum annuum*)	–	Y2H, co-IP, BiFC assay	Plant innate immune system	Kim et al. (2014)

(continued)

Table 16.1 (continued)

Protain pair		Organism (of interactor with SGT1/ Sgt1)	SGT1 domain	Method of detection	Function	References
	AvrBsT	*Xanthomonas campestris* pv *vesicatoria* (*Xcv*) (with pepper SGT1)	–	Y2H, co-IP, BiFC assay	Plant innate immune system	Kim et al. (2014)
	NLRP3	Human	CS + SGS	Y2H, co-IP	Mammalian innate immune system	Mayor et al. (2007)
	NLRP2	Human	–	Co-IP	Mammalian innate immune system	da Silva Correia et al. (2007) and Mayor et al. (2007)
	NLRP4	Human	–	Co-IP	Mammalian innate immune system	da Silva Correia et al. (2007) and Mayor et al. (2007)
	NLR12	Human	–	Co-IP	Mammalian innate immune system	Mayor et al. (2007)
	Nod1	Human	–	Co-IP	Mammalian innate immune system	Mayor et al. (2007)
		Human	CS + SGS	Co-IP	Mammalian innate immune system	da Silva Correia et al. (2007)
	Nod2	Human	–	Co-IP	Mammalian innate immune system	da Silva Correia et al. (2007) and Mayor et al. (2007)
	NLRC4/ IPAF	Human	–	Co-IP	Mammalian innate immune system	Mayor et al. (2007)
	SKP1	Barley	–	Co-IP		Azevedo et al. (2002)
		N. benthamiana	–	Y2H, co-IP, in vitro	Plant innate immune system	Liu et al. (2002)
		Yeast (*S. cerevisiae*)	–	Co-IP, in vitro	SCF funtion (G1 function)	Kitagawa et al. (1999)

(continued)

Table 16.1 (continued)

Protain pair		Organism (of interactor with SGT1/ Sgt1)	SGT1 domain	Method of detection	Function	References
		Yeast (*S. cerevisiae*)	TPR + CS	Y2H, co-IP, in vitro	Yeast kinetochore assembly	Bansal et al. (2004)
		Yeast (*S. cerevisiae*)	TPR	In vitro		Catlett and Kaplan (2006)
		Human	–	Co-IP		Niikura et al. (2006)
	Ctf13	Yeast (*S. cerevisiae*)	TPR + CS	Co-IP, in vitro	Yeast kinetochore assembly	Lingelbach and Kaplan (2004)
	MIS12	Human			Human kinetochore assembly	Davies and Kaplan (2010)
	DSN1	Human			Human kinetochore assembly	Davies and Kaplan (2010)
	NSL1	Human			Human kinetochore assembly	Davies and Kaplan (2010)
	NNF1	Human			Human kinetochore assembly	Davies and Kaplan (2010)
	Cyr1/ Cdc35	Yeast (*S. cerevisiae*)	SGS	Co-IP	Cyclic AMP signaling (G1 function)	Dubacq et al. (2002)
	CUL1	*N. benthamiana*	–	Co-IP	Plant innate immune system, SCF function	Liu et al. (2002)
		Yeast	–	Co-IP	Yeast kinetochore assembly	Kitagawa et al. (1999)
	CUL4	Human	–	Co-IP	Human CENP-A deposition and kinetochore assembly	Niikura et al. (2017)
	CSN4	*N. benthamiana*	–	Co-IP	Plant innate immune system, SCF function	Liu et al. (2002)
		Barley	–	Co-IP	Plant innate immune system, SCF function	Azevedo et al. (2002)

(continued)

Table 16.1 (continued)

Protain pair		Organism (of interactor with SGT1/Sgt1)	SGT1 domain	Method of detection	Function	References
	CSN5	*Barley*	–	Co-IP	Plant innate immune system, SCF function	Azevedo et al. (2002)
	S100A6	Human	SGS	Co-IP, in vitro	SGT1 phosphorylation	Nowotny et al. (2003)
	S100B	Human	–	In vitro		Nowotny et al. (2003)
	S100P	Human	–	In vitro		Nowotny et al. (2003)
	PLK1	Human	–	Co-IP, in vitro (PLK1 in HeLa cell lysates)	Mammalian kinetochore assembly	Liu et al. (2012)
	PHLPP1	Human	–	Co-IP, in vitro	SCF function	Gao et al. (2013)
		Human	–	Co-IP	Human kinetochore assembly	Gangula and Maddika 2017[c]
	RNF41	Human	–	Co-IP, IP-mass spectrometry	Human kinetochore assembly, SCF function	Gangula and Maddika 2017[c]
	eEF1A1	Human	TPR	Co-IP, in vitro	Brain	Novosylna et al. (2015)
	UBE3A	Human	–	Co-IP, IP-mass spectrometry	Brain	Martinez-Noel et al. (2018)

[a]Arabidopsis and humans contain two copies of SGT1: SGT1a/SUGT1A and SGT1b/SUGT1B
[b]In yeast-two hybrid screening, the identified prey clones encoded either the complete coding sequence of SGT1 or three different truncations in the TPR domain, as well as one truncation immediately N-terminal to the SGS domain. No prey clones were recovered encoding the CS plus SGS domains alone
[c]Interactors identified only by mass spectrometric analysis are not shown in this table

The last three sections summarize three types of disorders (i.e., cancer, brain disease, and heart diseases) in which mammalian SGT1 is involved. Overexpression of SGT1 may play a role in tumorigenesis, and SGT1 protein was highly expressed in tumor tissues and xenografts (Ogi et al. 2015). Past and current data from our group and others suggest that Sgt1 is involved in cancer development by stabilizing oncoproteins (Gao et al. 2013; Ogi et al. 2015; Iwatsuki et al. 2010), and highlight SGT1 as a potential therapeutic target. Localization of SGT1 in mammalian brain is consistently observed, indicating a strong association between SGT1 protein levels and brain diseases and/or disorders (Pickering et al. 2013; Spiechowicz et al. 2006). Genetic studies and interactors of Sgt1 identified in *Drosophila* suggest that Sgt1

Table 16.2 Excluded protein/genes referred to as "SGT1/Sgt1" in this review

Name in used reference (abbreviation)	Full name	Species	References
hSgt1	Human suppressor of Gcr two	Human	Kainou et al. (2006)
SpSgt1	Sz. pombe suppressor of Gcr two	Yeast (Schizosaccharomyces pombe)	Sato et al. (1999) and Kainou et al. (2006)
AtSGT1	*Arabidopsis thaliana* UDP-glucose: SA glucosyltransferase 1	Plant (Arabidopsis thaliana)	Sato et al. (1999)
SGTA/hSGT	Small glutamine-rich tetratricopeptide repeat-containing protein alpha	Human	Kordes et al. (1998)
SGT1 (gene)	*Solanum melongena* L. solasodine galactosyltransferase	Eggplant (Solanum melongena L.)	Moehs et al. (1997)

has an essential role in generating neuroblasts. In humans, SGT1 physically interacts with some proteins where mutation or deletion is significantly associated with mental and neurologic disorders (Martinez-Noel et al. 2018; McLachlan et al. 2018; Novosylna et al. 2015). Gene and protein expression of SGT1 is specifically upregulated in heart failure, and lack of melusin (animal RAR1 homolog and SGT1 interactor) directly leads to cardiac defects in mice (Brancaccio et al. 2003a). These varied effects of SGT1 suggest the importance of addressing current gaps to better understand the organ-specific and disease-specific roles of SGT1. Lastly, current problems and limits including future directions and ideas of where research might go next are briefly lined up and summarized in Conclusion. To avoid confusion of nomenclature, we listed proteins/genes that we exclude the content in the present review (Table 16.2).

16.1.1 SGT1 Structure, Interactors and Protein Profiles (Stability and Expression Profiles)

According to previous analyses of amino acid sequences, in humans SGT1 contains five conserved domains: a TPR (tetratricopeptide repeat) domain, two variable regions (VR1 and VR2), a CS (CHORD protein and SGT1-specific) domain, and an SGS (SGT1-specific) domain (Azevedo et al. 2002). Characterization of each domain, specific interactors through these domains (Fig. 16.3; Table 16.1) (partly adapted from (Shirasu 2009) and (Spiechowicz and Filipek 2005)), and protein stability and expression profiles are described below.

The TPR domain consists of three or more TPR motifs, which are degenerate, 34-amino acid repeats often arranged in tandem arrays (Blatch and Lassle 1999). The TPR motif mediates protein – protein interactions and possibly protein – lipid interactions (Lapouge et al. 2000) involved in functionally different proteins and biological systems. Most TPR-containing proteins associate with multiprotein complexes, and previous reports suggest that TPR motifs are required for function of chaperone, cell-cycle, transcription, and protein transport complexes (Blatch and Lassle 1999). For example, many TPR motif-containing proteins might function as scaffolds for the assembly of multiprotein complexes such as the anaphase promoting complex (APC) or the peroxisomal import receptor complex (Lapouge et al. 2000). Interactions between the SGT1-TPR domain and the D2 and D3 domains of eEF1A1 have been reported (Fig. 16.3; Table 16.1), suggesting that SGT1 competes with viral RNA for binding to eEF1A. eEF1A improves virus performance via interaction with viral genomic RNA and viral RNA-dependent RNA polymerase (Novosylna et al. 2015). Mutations in eEF1 complex subunits are strongly associated with neurodevelopmental abnormalities (McLachlan et al. 2018) (see also following Sect. 16.1.10: **Mammalian SGT1 in brain diseases**). Previous 3D structural data revealed that the TPR motif consists of two antiparallel α-helices, such that tandem arrays of TPR motifs harbor a right-handed helical structure with an amphipathic channel that might hold the complementary structure of a partner protein (Blatch and Lassle 1999). Thus, the TPR motif might well represent an ancient protein-protein interaction module that has been recruited by various proteins (Blatch and Lassle 1999).

The CS domain is characterized as being similar to that in the human p23 protein. Examples of CS-containing proteins are SGT1, Rar1 (required for Mla12 resistance), melusin (an animal RAR1 homolog), B5 + B5R flavo-hemo cytochrome NAD(P)H oxidoreductase type B in humans and mice, and NudC, which is involved in migration to the nucleus during mitosis in Aspergillus nidulans (Garcia-Ranea et al. 2002). The phylogenetic link between p23 and HSP20/a-crystallin and their similar functions in chaperone activity suggest a common function, probably as stabilization folding structures and/or protein–protein interactions (Garcia-Ranea et al. 2002). Two CS domains are found at the N-terminus of Ubiquitin carboxy-terminal hydrolase 19 (USP19), these domains might function in the interaction of USP19 with cellular inhibitors of apoptosis (IAPs) (Mei et al. 2011). The cytoplasmic isoform USP19_b up-regulates protein levels of the polyglutamine (polyQ)-containing proteins ataxin-3 (Atx3) and huntingtin (Htt), and promotes aggregation of their polyQ-expanded content though the HSP90 species in cell models (He et al. 2016) (see also following Sect. 16.1.10: **Mammalian SGT1 in brain diseases**).

Our group and others have found that both TPR and CS domains are important for SGT1-HSP90 interactions (Azevedo et al. 2002; Bansal et al. 2004; Bansal et al. 2009a; Bansal et al. 2009b; Dubacq et al. 2002; Lingelbach and Kaplan 2004; Shirasu 2009) (Fig. 16.3; Table 16.1). These interactions are also required for assembly of the core kinetochore complex (see also following Sect. 16.1.5: **Yeast Sgt1 in kinetochore assembly**). Our group also demonstrated that both TPR and CS domains are important for Sgt1-Skp1 interactions (Fig. 16.3; Table 16.1) (Bansal et al. 2004), and identified point mutations (L31P) that disrupt dimerization and Skp1 binding in our in vitro translation system using reticulocyte lysates (presumably including abundant Hsp90) (Bansal et al. 2009b) (see also following Sect. 16.1.5: **Yeast Sgt1 in kinetochore assembly**).

Willhoft et al. determined the crystal structure of the interacting domains of *S. cerevisiae* Sgt1 and Skp1 (Willhoft et al. 2017). Sgt1 associates with the Skp1-BTB/POZ domain through Sgt1-TPR domain using residues that are conserved in humans. Their data indicate that the G2/M phenotype of Sgt1 ts-mutants is a consequence of disruption of Skp1 association with Sgt1, suggesting that the interaction with Skp1 is an essential function of Sgt1 in yeast. The TPR domain of SGT1 has functional and structural similarity to that of protein phosphatase 5, which binds to the C-terminal pentapeptide MEEVD of HSP90 (Das et al. 1998; Russell et al. 1999; Scheufler et al. 2000). The results of deletion analyses suggest that the CS domain of plant SGT1 is required and sufficient for HSP90 interactions, but the TPR domain is not (Takahashi et al. 2003). The TPR domain is not present in SGT1 homologs of several *Caenorhabditis* species and *Brugia malayi*, further suggesting that it is not essential for conserved SGT1 function *per se* (Shirasu 2009).

Moreover, Lee et al. reported that the CS domain is a binding module of human SGT1 for HSP90, suggesting that CS-containing proteins are involved in recruiting heat shock proteins to multiprotein assemblies (Lee et al. 2004). In this study, unlike p23, the interaction of SGT1 with HSP90 can occur in the absence of ATP, suggesting that the CS domain is a binding module for HSP90 distinct from p23-like domains. Further structural differences between the SGT1-CS domain and p23 appeared in how they bind to HSP90 (Shirasu 2009; Kadota et al. 2010). The HSP90 binding side of the SGT1-CS domain consists of a four-stranded β-sheet, which is analogous to the HSP90-binding side of p23, but does not contain the C-terminal strand that is important for much p23-HSP90 binding (Boter et al. 2007; Ali et al. 2006). Unlike p23, the SGT1-CS domain binds to the ADP-bound state of the N-terminal domain (ND) of Hsp90 with the opened lid segment in the Hsp90-ND–Sgt1-CS complex (Zhang et al. 2008). The SGT1-CS binds to Hsp90-ND away from the ATP binding pocket and lid segment, whereas p23 interacts with residues in the segment that are available only in the ATP-bound closed conformation of the chaperone (Kadota et al. 2008; Zhang et al. 2008; Ali et al. 2006).

Importantly, the SGT1-CS domain also associates with residues in the HSP90-ND, which moves largely during the ATPase cycle (Kadota et al. 2008; Zhang et al. 2008). Therefore, the ATPase-driven conformational change within the HSP90 dimer would strongly induce dissociation of SGT1. SGT1 and p23 do not interact with HSP90 at the same time, although SGT1 could also bind to the ATP-bound (closed) state of HSP90 and p23 binds to a different site (Kadota et al. 2008). Indeed, HSP90 cochaperones Chp-1 and melusin interact with cochaperones PP5 and SGT1 in the presence of ATP (Hong et al. 2013) (Table 16.1). Collectively, these findings suggest that SGT1 and p23 have distinct functions to regulate HSP90 activity. The in vitro co-chaperone activity of SGT1 via this different form of interaction (Zabka et al. 2008) suggests that HSP90 may bind with different co-chaperones, each of which may provide tailored chaperone activity. In *S. cerevisiae*, mutation of *SGT1* or *CSN1* renders yeast hypersensitive to overexpression of other co-chaperones, suggesting that co-chaperones selectively compete for binding to subpopulations of cellular Hsp90 and changes in the relative levels of co-chaperones may have marked effects on Hsp90 function (Johnson et al. 2014). The study of large-scale mutagenesis and NMR analysis of plant SGT1 also verified that the CS domain is required and sufficient for interactions with HSP90 (Boter et al. 2007). In rice, the salt bridges and hydrogen bonds in the OsSgt1-CS and OsHsp90-ND domains played important roles in the stability of the OsSgt1-OsHsp90 complex (Yan et al. 2012).

However, in vitro studies by demonstrated that deleting the TPR domain of yeast SGT1 greatly reduces its interaction with HSP90 (Hsc82), but the CS domain retains weak binding activity (Catlett and Kaplan 2006). The TPR domain of yeast SGT1 is required for SGT1-HSP90 interaction in yeast two-hybrid analyses. The TPR domain of yeast SGT1 may bind to the C-terminal end of HSP90 in yeast, similar to other TPR-containing co-chaperones. Because HSP90 functions as a dimer (Pearl and Prodromou 2006), and SGT1-TPR mediates dimerization of SGT1 in plants (Noel et al. 2007; Nyarko et al. 2007; Taube et al. 2014) and yeast (Bansal et al. 2009a; Bansal et al. 2009b), HSP90-SGT1 interaction may be stabilized by SGT1 dimerization (Shirasu 2009). Indeed, our in vitro analysis indicated that MBP-Hsp82 (a yeast Hsp90 protein) moderately stimulates (but does not inhibit) the association between His_6-Sgt1 and GST-Sgt1 (Bansal et al. 2009b) (see also following Sect. 16.1.5: **Yeast Sgt1 in kinetochore assembly**). However, in another in vitro experiment using reticulocyte lysates (presumably including abundant Hsp90) and in in vivo experiments, we found that GA, an Hsp90 inhibitor, substantially inhibits Sgt1-Skp1 association but not Sgt1 dimerization. These results suggest that Hsp82 (a yeast Hsp90 protein) is required for Sgt1 binding to Skp1 but not for generating Sgt1 dimers. Because the SGT1-HSP90 complex is involved in multiple functions, effects of Hsp90 on Sgt1 dimerization may vary depending on the specific function and local condition where the complex is involved. Further studies of how Hsp90 affects Sgt1 dimerization in specific "tailored" function(s) are needed.

In addition, the CS domain is required for Arabidopsis Sgt1 (AtSgt1) to interact with AtRar1 and barley Rar1 (HvRar1) in the yeast two-hybrid system (Azevedo et al. 2002). HvRar1 is a convergence point in signaling triggered by multiple *R* genes that confer resistance to the pathogenic powdery mildew (*Blumeria graminis f. sp. hordei*) (Shirasu et al. 1999). HvRar1 contains two highly similar but distinct zinc-binding domains called CHORD-I and -II (cysteine- and histidine-rich domain). Both domains are conserved and organized in tandem in protozoa, plants, and metazoa (Shirasu et al. 1999). CHORD is a novel class among eukaryotic zinc-binding domains and is also required for development in *C. elegans* (Shirasu et al. 1999). The CS and CHORD domains are fused into a single polypeptide chain in metazoans but are found in separate proteins in plants; this is thought to indicate interactions between CS and CHORD following the Rosetta Stone principle (Azevedo et al. 2002; Marcotte et al. 1999; Shirasu et al. 1999). Indeed, the CS motif of the AtSgt1 interacts with the CHORD-II domain of the AtRar1 in the yeast two-hybrid system (Azevedo et al. 2002). RAR1-SGT1 interactions also have been detected in bread wheat (*Triticum aestivum* L.) (Tai 2008) (Fig. 16.3; Table 16.1). In various species, CHORD and CS domains are associated, either within the same protein (e.g., mammalian melusin and Chp1), or in separate but interacting proteins (e.g., plant Rar1 and Sgt1) (Zhang et al. 2010). Both RAR1-CHORD and SGT1-CS domains independently interact with HSP90 (Zhang et al. 2010), and the N-terminal domain (ND) of HSP90 and the RAR1-CHORD2 domain bind to opposite sides of the CS domain (Boter et al. 2007; Kadota et al. 2010; Shirasu 2009), suggesting that RAR1-CHORD2 and HSP90 simultaneously co-associate with SGT1 (Kadota et al. 2010; Shirasu 2009) (Fig. 16.3). SGT1-CS and CS domains of CHORD proteins bind in a nucleotide (ATP)-independent manner to HSP90 (Catlett and Kaplan 2006; Lee et al. 2004).

Zhang et al. have determined the structure of an Hsp90-CS-CHORD ternary complex, indicating the dynamic nature of Hsp90-Rar1-Sgt1 complexes (Zhang et al. 2010). They proposed a new model for dynamic interactions of the SGT1-HSP90-RAR1 complex, in which CHORD-containing proteins are a novel family of HSP90 co-chaperones (e.g., mammalian melusin) that act in conjunction with SGT1 (Shirasu 2009; Boter et al. 2007). Rar1 and Sgt1 are essential for functioning of nucleotide binding leucine-rich repeat receptors (NLRs) (Ting et al. 2008), encoded by disease resistance (*R*) genes that mediate an innate immunity mechanism in plants (Shirasu and Schulze-Lefert 2003). Siligardi et al. determined stoichiometric interactions of the Hsp90-Sgt1-Rar1 complex by CD and SRCD spectroscopy in the presence of excess ADP. They found that a symmetric $Sgt1_2$-$Hsp90_2$-$Rar1_1$complex formed, which could allow two NLR molecules to simultaneously bind (Siligardi et al. 2017). In mammals, there are two CHORD proteins, the muscle-specific protein melusin, which is involved in the cardiomyocyte hypertrophic response (Brancaccio et al. 2003b) (see also following Sect. 16.1.11: **Mammalian SGT1 in heart diseases**), and Chp-1 (CHORD-containing protein 1). Unlike Rar1, Chp-1 contains a C-terminal CS domain, which is similar to that of SGT1, is necessary but not sufficient for interaction with the N-terminal domain of HSP90 (Wu et al. 2005). Chp-1 has not been detected in the active Nod1 complex (da Silva Correia et al. 2007) and its interaction with SGT1 has not been demonstrated, although Chp-1 and RAR proteins appear similar (Pearl et al. 2008).

Moreover, HSP90-mediated function of NLR receptors in plant and animal innate immunity depends on the SGT1 and, at least in plants, on the Rar1 (Zhang et al. 2010). Mammalian SGT1 is similarly essential for activation of NLRs, and mediates an innate immune response in animals (da Silva Correia et al. 2007; Mayor et al. 2007) (see also following Sect. 16.1.4: **SGT1 in Mammalian Innate Immune System**). Both the CS and SGS domains of SGT1 were required for interaction with Nod1, the LRR (leucine rich repeats) domain of NALP3, and endogenous HSP90 (da Silva Correia et al. 2007; Mayor et al. 2007). Interestingly, fusion of human SGT1 (hSGT1) to NOD1 LRR significantly enhanced protein solubility, and the fusion protein was stabilized by co-expression of mouse Hsp90a in *E.coli* (Hong and Hahn 2016). Biochemical analyses suggest most NLR proteins require RAR1 for stabilization, although genetic analyses suggest *RAR1* specificity for particular

NLR pathways. As shown in barley MLA alleles, in which RAR1 dependency is located in the LRR domain, stability of NLR is likely to be determined by the LRR domain (Halterman and Wise 2004; Shen et al. 2003). SGT1 is also required for the steady-state accumulation of certain NLRs, such as Rx (Azevedo et al. 2002; Boter et al. 2007) and N (Mestre and Baulcombe 2006), as in the case of RAR1. The CS and SGS domains of SGT1 are essential for steady-state accumulation of certain NLRs (Boter et al. 2007). In the same manner, mutation, silencing, or inhibition of HSP90 reduces the levels of Rx (Lu et al. 2003), RPM1 (Hubert et al. 2003; Holt et al. 2005), and RPS5 (Holt et al. 2005). Therefore, a seemingly large number of NLR proteins require HSP90 for their stability, possibly to maintain their sensory signal-competent state (Kadota et al. 2010; Shirasu 2009). Stuttmann et al. discussed the possible contribution of an SGT1-chaperone complex in the folding and maturation of LRR-containing proteins and its evolutionary consequences for emergence of novel LRR interaction surfaces (Stuttmann et al. 2008).

The transcriptional regulation of RAR1 and SGT1 encoding genes is different, perhaps reflecting their distinct functions, although these proteins interact with each other (Shirasu 2009). *SGT1a* and *SGT1b* mRNA are strongly induced upon *Ha* inoculation or under various stress conditions, but public microarray data indicate that *RAR1* mRNA levels are very low and do not respond to pathogenic infection (Noel et al. 2007; Azevedo et al. 2006). However, SGT1 protein levels do not significantly change upon infection (Austin et al. 2002). Similarly, *HSP90.1* mRNA is highly expressed upon *Pst* inoculation and stress conditions, but total protein levels are not significantly changed (Takahashi et al. 2003). It is possible that newly synthesized SGT1 and HSP90 deal with stress conditions (Shirasu 2009) or such new protein synthesis could be induced upon stress. On the other hand, melusin is co-expressed with HSP90 and HSP70 in animals in response to mechanical stresses, suggesting a strong functional link to the SGT1-HSP90 chaperone machinery under cellular stress (Sbroggio et al. 2008).

Our group previously reported protein expression levels of human SGT1/SUGT1 in different human tissue lysates when a splice variant SGT1B/SUGT1B was identified: humans contain at least two copies of SGT1: SGT1A/SUGT1A and SGT1B/SUGT1B (Niikura and Kitagawa 2003; Zou et al. 2004). We performed immunoblot analyses of different tissue lysates from donors with unknown diseases (Blot Human Tissues; IMGENEX, San Diego, CA), using a mouse monoclonal antibody (BD Transduction Laboratories, Franklin Lake, NJ) against a C-terminal region of SGT1A (amino acids 191–306) identical to a region in SGT1B (amino acids 223–338). Protein levels of SGT1A and SGT1B were higher in brain, liver, lung, and testis than in other tissues. Levels of SGT1B were slightly lower than those of SGT1A in human liver and lung. These results suggest that SGT1A and SGT1B have tissue-specific roles and that their roles may differ in the liver and lung. Because of the limited information for the tissue lysates we used, further tissue-specific studies comparing normal samples and samples from specific diseases could be of clinical value. In Protein – NCBI database, the 3rd isoform of human SGT1C/SUGT1C is reported (NP_001307760.1; REFSEQ: accession NM_001320831.1). The human SGT1C lacks an alternate exon (ca. 62 amino acids of the SGT1A/B-TPR domain) and contains another short alternate exon in the 5′ region corresponding to the amino acids sequence: MLSQKE initiating translation at an alternate start codon, compared to human SGT1A. The encoded SGT1C has a distinct N-terminus and is shorter than the SGT1A and B isoforms (Fig. 16.1). However, the biological-physiological function is unknown. Spiechowicz et al. examined Sgt1 protein expression in rat tissues using Northern and Western blots; they found that the Sgt1 protein is expressed in various tissues and that brain, skeletal muscle, and spleen contain high levels of this protein and its mRNA (Spiechowicz and Filipek 2005). These data might encourage study of SGT1 function in brain, although it is difficult to speculate about its roles at present (see also following Sect. 16.1.10: **Mammalian SGT1 in brain diseases**).

The SGS domain has ~90 amino acids, unique to the SGT1-like proteins and SGT1 orthologs among various species. In *S. cerevisiae*, Dubacq et al. observed coimmunoprecipitation of Sgt1-13myc and 3HA-Cyr1/Cdc35, but not Skp1 with 3HA-Cyr1/Cdc35, under conditions in which they readily observed the coimmunoprecipitation of Sgt1–13myc (Dubacq et al. 2002). These data suggest that Sgt1 forms exclusive complexes with Cyr1/Cdc35 and Skp1 (Fig. 16.3; Table 16.1). Their results suggest that SGT1 contributes to adenylyl cyclase activity and the *sgt1* allele with S371N mutation in the SGS domain is deficient for this function. These authors termed a highly conserved C-terminal region of the Sgt1 as AC (adenylyl cyclase) domain. Sgt1 coprecipitated with Cdc35-1 that has a L901H mutation in LRR domain, as it did with wild-type Cdc35/Cyr1, but Sgt1-S371N mutated in the SGS/AC domain did not coprecipitate with Cdc35-1. The faulty physical interaction between Cdc35-1 and Sgt1-S371N correlates with thermosensitivity of strains containing both mutations, and other LRR-containing proteins bound with Sgt1 in a yeast two-hybrid screen. Overall, these results support the idea that the C terminus of Sgt1 may also interact with the LRR domain of Cdc35. Moreover, one of the two point mutations found in the *sgt1-5* mutant (E364K) affects another highly conserved residue that maps near the S371N substitution, and the *sgt1-5* mutant has a thermosensitive G1 arrest phenotype that resembles that of *cdc35/cyr1* mutants (Kitagawa et al. 1999). Therefore, these results suggest that the *sgt1-5* mutant might also have an adenylyl cyclase deficiency (see also following Sect. 16.1.6: **Yeast Sgt1 in G1 function**). It is also unclear if the SGS domain of SGT1 is required for *Candida* Cyr1 recognition of bacterial peptidoglycan (PGN) (Shirasu 2009; Xu et al. 2008), as it is for SCF ubiquitin-ligase activity in Arabidopsis (Gray et al. 2003).

In plants, SGT1 interacts with the LRR domains of Bs2 and MLA1 (Leister et al. 2005; Boter et al. 2007). The SGS domain of SGT1 is sufficient for interaction with the LRR for MLA1. Human SGT1 likewise interacted with the LRR domain of NLRP3 as bait in yeast two-hybrid screening (Mayor et al. 2007). Moreover, SGT1 interacted with NLRP2, NLRP4, NLRP12, Nod1, Nod2, and NLRC4 (da Silva Correia et al. 2007; Mayor et al. 2007). Notably, HSP90 was always present along with the different NLRs and the different NALP3 domains in the immunoprecipitates (Mayor et al. 2007). HSP90 can bind to the NACHT domain of NLRP3 without SGT1, but for binding to the LRR domain, SGT1 is always present with

HSP90 in the immunoprecipitates (Mayor et al. 2007). Both CS and SGS domains of SGT1 are required for interactions with Nod1, the LRR domain of NALP3, and endogenous HSP90, suggesting that SGT1 can bind to HSP90 and an NLR protein simultaneously (da Silva Correia et al. 2007; Mayor et al. 2007). In plants, HSP90 associates with RPM1 (Hubert et al. 2003), N (Liu et al. 2004), and I2 (de la Fuente van Bentem et al. 2005). In the same manner as plant NLRs, human NLR proteins require HSP90 chaperone activity to maintain steady-state levels (da Silva Correia et al. 2007; Mayor et al. 2007; Hahn 2005). Using biochemical fractionation, Noel et al. demonstrated that the SGT1-SGS domain, which interacts with HSP70 and NLR proteins, is required for the nuclear localization of SGT1, suggesting that SGT1 may shuttle into the nucleus with its substrate (Noel et al. 2007; Wang et al. 2008) (see also following Sect. 16.1.2: **SGT1 Protein Subcellular Localization**). In humans, calcyclin S100A6 and other S100 proteins interact with SGT1 through its 71 C-terminal residues (amino acids 263–333) in the SGS domain, like its homolog CacyBP/SIP, in a calcium-dependent manner (Nowotny et al. 2003). In this study, S100A6 did not influence SKP1-SGT1 binding, suggesting that separate SGT1 domains are responsible for interactions with S100A6 and SKP1. Phosphorylation of the S100A6-binding domain of SGT1 by protein kinase CK2 (casein kinase 2) was inhibited by S100A6, suggesting that S100A6 binding is required to regulate the phosphorylation of SGT1 (see also following Sect. 16.1.5: **Yeast Sgt1 in kinetochore assembly** for Sgt1 phosphorylation by CK2). Breen et al. reported that calcyclin S100A6 regulates pulmonary fibroblast proliferation, morphology, and cytoskeletal organization, suggesting a role for calcyclin in modulating calcium-dependent signaling events that regulate progression through the cell cycle (Breen and Tang 2003). Coexpression of the SGS binding protein S100A6 reduces SGT1-HSP70 interaction through a Ca^{2+}- dependent binding to the SGS domain in human cells (Spiechowicz et al. 2007). It is yet unclear if S100 proteins and HSP70 compete to bind to SGT1-SGS domain, and if they do, how this competition is regulated for the SGT1 nuclear translocation and the SGT1-NLR interaction (Fig. 16.3: see also following sections).

16.1.2 SGT1 Protein Subcellular Localization

In yeast, the major pool of HSP90 is presumably in the cytoplasm, but HSP90 can be shuttled into the nucleus when it binds to substrates such as the glucocorticoid receptor (Pratt and Toft 2003). In plants, fluorescently tagged RAR1 and SGT1 were found both in the cytoplasm and nucleus (Noel et al. 2007; Wang et al. 2008). These results were verified by biochemical fractionation detection with anti-SGT1 antibody (Noel et al. 2007). The results also demonstrated that the SGT1-SGS domain that interacts with HSP70 and NLR proteins is required for the nuclear localization of SGT1, suggesting that SGT1 may shuttle into the nucleus with its substrate (Noel et al. 2007). A study using a bimolecular fluorescence complement assay (BiFC) revealed that RAR1 and SGT1 can associate in both the cytoplasm and the nucleus (Wang et al. 2008), but a study using fluorescence resonance energy transfer–fluorescence lifetime imaging microscopy (FRET-FLIM) captured the RAR1-SGT1 interaction only in the cytoplasm (9). Limitations of these studies are that the proper functionality of the fluorescently tagged proteins was not validated, and these proteins were overexpressed (Shirasu 2009). Forced nuclear localization of SGT1 causes tobacco (*Nicotiana tabacum*) N protein to be confined to nuclei, suggesting that nucleocytoplasmic partitioning of R protein including the tobacco N receptor is modulated by SGT1 phosphorylation catalyzed by plant MAPK (Hoser et al. 2013, 2014).

In human HEp-2 cells, Prus et al. showed that SGT1 translocates to the nucleus due to heat shock, and increased intracellular concentration of dependent binding to the SGS domain in human transduced by S100A6 is necessary for nuclear translocation of the SGT1 protein (Prus and Filipek 2011). Further, their results suggest that S100A6- Ca^{2+}-mediated SGT1 dephosphorylation on Ser249 and Ser299 promotes its nuclear translocation, most likely due to disruption of the SGT1-HSP90 interaction (Prus et al. 2011). SGT1-SGS domain, which interacts with HSP70, NLR, and S100 proteins, is required for the nuclear localization of SGT1, suggesting that SGT1 may shuttle into the nucleus with its substrate (Noel et al. 2007; Wang et al. 2008). However, coexpression of the SGS binding protein S100A6 reduces SGT1-HSP70 interaction through a Ca^{2+}- dependent binding to the SGS domain in human cells (Spiechowicz et al. 2007). It is yet unclear if S100 proteins and HSP70 compete to bind to SGT1-SGS domain, and if they do, how this competition is regulated for the SGT1 nuclear translocation and the SGT1-NLR interaction (Fig. 16.3: see also previous and following sections).

In human cells, Liu et al. reported that SGT1 dynamically localizes at the kinetochores, which lack microtubule attachments during prometaphase (Liu et al. 2012). In their study, SGT1 was present at kinetochores after nocodazole treatment, but not after paclitaxel treatment, suggesting that SGT1 is recruited to the kinetochores that lack microtubule attachments. PLK1 was required for the prometaphase-specific localization of SGT1 to the kinetochore, and PLK1-mediated phosphorylation of SGT1 serine 331 (S331) at the kinetochore enhanced the association of HSP90-SGT1 chaperone with the MIS12 complex, coordinating the recruitment of the NDC80 complex to form efficient microtubule-binding sites. Further, we found that human SGT1 is required for CENP-A deposition and kinetochore assembly (Niikura et al. 2017). Based on these data, we proposed a hypothetical pathway depicting the SGT1 phosphorylation and dephosphorylation and its contribution to CENP-A deposition (see also following Sect. 16.1.7: **Human SGT1 is required for CENP-A deposition and kinetochore assembly**).

16.1.3 Plant SGT1 in Plant Innate Immune System

SGT1, initially identified as an interacting partner of RAR1 in yeast two-hybrid screens (Azevedo et al. 2002; Liu et al. 2002), plays a central role in plant disease resistance triggered by a number of resistance (R) proteins. Several groups identified SGT1 as an essential component of *R* gene-mediated disease resistance in plants (Muskett and Parker 2003). Microorganisms continually attack eukaryotes with sophisticated strategies to colonize their hosts. Plants respond to pathogenic infection via a surveillance system consisting of R proteins that catch corresponding pathogen determinants (Dangl and Jones 2001). In breeding agriculturally important plants, *R* genes have been widely used and have considerably contributed to the genetic value of modern crop species (Shirasu 2009). Many *R* genes have been isolated and characterized in detail over the past 15 years from a wide range of plant species, and most encode NLR proteins (Dangl and Jones 2001; Kadota et al. 2010; Shirasu 2009). Various R proteins activate resistance responses that often include rapid ion fluxes, generation of reactive oxygen species (ROS), and production of antimicrobial compounds. These responses often coincide with localized programmed cell death, the hypersensitive response (HR), at the site of pathogen invasion (Shirasu and Schulze-Lefert 2000). Common signal transduction pathways are likely used upon pathogen recognition, because many *R* gene products share universal response patterns and structural similarities and modules such as a nucleotide-binding site and a leucine-rich repeat domain. Some genes required for *R* gene function have been identified and represent such downstream signaling components. For instance, *EDS1* encodes a lipase-like protein, and *NDR1* encodes a putative membrane-associated protein. However, the biochemical function(s) of these proteins is still unclear. In addition, because R proteins share many similarities with components of the animal innate immune system, it has been suggested that some downstream signaling components may be common to both plants and animals (Staskawicz et al. 2001).

RAR1 (required for Mla12 resistance) is an essential component of resistance generated by many *R* genes (Shirasu et al. 1999; Tornero et al. 2002; Muskett et al. 2002). The importance of *RAR1* in NLR- dependent resistance pathways was also demonstrated by gene silencing methods in several plants (Kadota et al. 2010; Shirasu 2009). However, the precise biochemical function of RAR1 remains obscure. RAR1 consists of two zinc-binding modules termed cysteine- and histidine-rich domain (CHORD)-I and CHORD-II, which are characterized by six conserved cysteines and three conserved histidines (Kadota et al. 2010; Shirasu 2009). RAR1 homologs are absent in yeast but are present in plants, protozoa, metazoa, and fungi (Azevedo et al. 2002). In all tested eukaryotes except yeast, CHORD-containing proteins have been identified. Because of the conserved primary structure of CHORD domains and their tandem organization, these proteins may serve important cellular functions. For example, silencing of the *C. elegans* CHORD-containing gene, chp, results in semisterility and embryo lethality, suggesting an essential function of the wild-type gene in nematode development, demonstrating an involvement of RAR1 homologs in animal development (Shirasu et al. 1999). An animal RAR1 homolog, melusin, originally identified as a β1-integrin interacting protein in the yeast two-hybrid system, is expressed in heart muscle, where it is required for sensing stress to prevent cardiac failure in response to chronic pressure overload (Brancaccio et al. 2003a) (see also following Sect. 16.1.11: **Mammalian SGT1 in Heart Diseases**). Therefore, CHORD proteins are involved in various biological systems, suggesting that RAR1 may be a component of conserved fundamental processes to many cellular activities (Takahashi et al. 2003).

RAR1 interacts with SGT1, and mutation analysis in *Arabidopsis* and gene silencing experiments in barley and *Nicotiana benthamiana* confirmed that SGT1 is involved in disease resistance mediated by diverse R proteins (Azevedo et al. 2002; Austin et al. 2002; Peart et al. 2002; Liu et al. 2002; Tor et al. 2002). SGT1 is required for immune responses triggered by both NLR- and non-NLR-type sensors, and also by overexpression of a truncated form of calcium-dependent protein kinase (CDPK) (Shirasu 2009). In *Arabidopsis*, two SGT1 isoforms, SGT1a and SGT1b, were identified. The loss of *Ha* resistance in the *sgt1b* mutant that would otherwise be provided by RPP5 (4) or RPP7 (122) suggests the importance of SGT1b (Shirasu 2009). While RAR1 appears to function only in immunity, SGT1 is involved in various and essential functions such as auxin and jasmonate responses, heat shock tolerance, growth, and development (Shirasu 2009). *Sgt1*, but not *Rar1*, is essential for the *RB*-mediated resistance to late blight in potatoes, marking a difference between RAR1 and SGT1, and providing additional evidence for the universal role of the *Sgt1* gene in various *R* gene-mediated plant defense responses (Bhaskar et al. 2008). In *Nicotiana benthamiana*, Rpi-blb2 triggered a hypersensitive response through SGT1-mediated, but not RAR-mediated or HSP90-mediated pathways (Oh et al. 2014a). It is suggested that the biochemical activity of SGT1 is highly conserved in eukaryotes, because plant and human *SGT1* genes can complement cell-cycle defects in yeast *sgt1* mutants (Azevedo et al. 2002; Kitagawa et al. 1999). Consistent with this suggestion, Arabidopsis SGT1b is required for the SCF (TIR1)-mediated auxin response (Gray et al. 2003). In plants, SGT1 interacts with the SCF complex subunits SKP1, CUL1, and components of the COP9 signalosome that regulates the SCF complex by removing the small ubiquitin-like protein RUB1 from CUL1 (Liu et al. 2012; Azevedo et al. 2002). Silencing SCF and components of the COP9 signalosome in plants compromises virus resistance, indicating the importance of ubiquitylation in disease resistance (Liu et al. 2002).

Takahashi et al. identified HSP90 as another RAR1-interacting protein by yeast two-hybrid screening. RAR1 interacts with the N-terminal half of HSP90 that contains the ATPase domain (Takahashi et al. 2003). HSP90 also specifically interacts with SGT1 that contains a tetratricopeptide repeat (TPR) motif and CS (CHORD protein and SGT1-specific) motifs with similarity to the co-chaperone p23. They demonstrated that the specific HSP90 inhibitor geldanamycin (GDA) inhibits the hypersensitive response (HR) and subsequent cell death and resistance mediated by the *Arabidopsis* R protein RPS2 against *Pseudomonas syringae pv. tomato* DC3000 (*avrRpt2*). Mutations in an *Arabidopsis* HSP90 isoform attenuated *RPS2*-mediated resistance and HR cell death, demonstrating that AtHSP90.1 is required for full

RPS2 resistance and is rapidly induced upon pathogen challenge. Further, multiple groups have reported the involvement of SGT1 in plant disease resistance. AvrRps4-independent HR cell death resulting from RPS4 expression depends on the three plant signaling components EDS1, SGT1 and HSP90 (Zhang et al. 2004). HSP90 and SGT1 are required for MI-1-mediated pest resistance (Bhattarai et al. 2007). In soybean *Glycine max*, *GmRAR1* and *GmSGT1* are required for basal, *R* gene–mediated, and systemic acquired resistance (Fu et al. 2009), and GmSGT1 is an essential component for soybean in resisting pathogens (Yan et al. 2014). The cell wall protein fraction (CWP)-induced defense system appears to be regulated by jasmonic acid (JA)-mediated and SGT1-, RAR1- and NPR1-dependent signaling pathways (Kawamura et al. 2009).

TaRAR1 and TaSGT1 interact with TaHsp90 to function in bread wheat (*Triticum aestivum L.*) seedling growth and stripe rust resistance (Wang et al. 2015). Thus, RAR1 and SGT1 might function as co-chaperones of HSP90 in fundamental processes for plant disease resistance. Further, many other studies showed that SGT1 is involved in plant disease resistance or stress response/resistance, although characterization of the involvement of HSP90 and RAR1 remains unclear in some studies (Bhavsar et al. 2013; Ishiga et al. 2011; Kumar and Kirti 2015; Meldau et al. 2011; Oh et al. 2014b; Shanmugam et al. 2016; Uppalapati et al. 2011; Wang et al. 2010; Xing et al. 2013; Yusof et al. 2016; Park et al. 2017). SGT1 interacts with the Prf resistance protein and contributes to Prf-mediated defense responses by stabilizing Prf protein via its co-chaperone activity (Kud et al. 2013) (Table 16.1). Pepper SGT1 interacts with PIK1 and the bacterial effector protein AvrBsT and promotes the hypersensitive cell death associated with PIK1-mediated phosphorylation in plants (Kim et al. 2014) (Table 16.1). SGT1 interacts with PcINF1/SRC2-1 and its nucleocytoplasmic partitioning, is involved in pepper's immunity against *P. capsici*, providing a molecular link between Ca(2+) signaling associated SRC2-1 and SGT1-mediated defense signaling (Liu et al. 2016) (Table 16.1).

In plants, several cases have been reported in which pathogens require RAR1, SGT1, and HSP90 for virulence (Kadota et al. 2010; Shirasu 2009). The first example is that the *Pst* effector AvrB requires RAR1 to induce chlorosis in *Arabidopsis* (Shang et al. 2006). Coimmunoprecipitation of RAR1 and AvrB indicated that RAR1 is a virulence target of AvrB (Shang et al. 2006). Moreover, Eitas et al. reported that AvrB-triggered chlorosis is caused by activation of TAO1, an NLR protein (Eitas et al. 2008). Therefore, a simple and plausible explanation is that AvrB weakly activates TAO1, possibly via RIN4, and that TAO1 requires RAR1 for its stabilization to trigger AvrB-dependent chlorosis. A second example is that the bacterial effector AvrPtoB from *Pst*. AvrPtoB suppresses immunity responses induced by a bacterial pathogen-associated molecular pattern (PAMP), flg22, a component of bacterial flagellin (Hann and Rathjen 2007; He et al. 2006); SGT1 or RAR1 are required for AvrPtoB-dependent suppression (Hann and Rathjen 2007). Analogous to AvrB, AvrPtoB may weakly activate an NLR that suppresses PAMP-dependent immunity responses. The third example is a necrotrophic pathogen, *Botrytis cinerea,* the fungus that requires host SGT1 for its virulence (El Oirdi and Bouarab 2007). Necrotrophs such as *B. cinerea* may activate NLR-dependent responses to induce cell death and subsequently obtain nutrients from the dead cells. It remains still unclear whether *B. cinerea* directly targets and activates the chaperone machinery (Shirasu 2009). Cuzick et al. inoculated five different defense mutants and revealed that *sgt1b* buds and flowers were more resistant to *F. culmorum* infection, although the *sgt1b* mutation does not affect *F. culmorum*-induced accumulation of reactive oxygen species and did not induce programmed cell death (PCD) (Cuzick et al. 2009). They suggested testing an Arabidopsis line expressing an anti-apoptotic protein and evaluating outcomes on both Fusarium disease and DON mycotoxin levels or evaluating specific cell types using a more sensitive method to detect apoptosis such as terminal deoxynucleotidyl transferase-mediated dUTP nick end-labelling (TUNEL) assays. In *Nicotiana bethamiana*, NbSGT1 expression is induced by potato virus X (PVX) and enhances virus accumulation and infection. However, it remains unclear whether HSP chaperones (e.g., HSP70, HSP90) and RAR1 proteins form a complex necessary for PVX infection or if other viral or cellular protein partners exist for SGT1. Meanwhile, NbHsp90, NbRAR1, and NbSGT1 negatively regulate bacterial wilt disease caused by Ralstonia solanacearum (Ito et al. 2015).

16.1.4 SGT1 in Mammalian Innate Immune System

The family of mammalian Nod-like receptors (or NLRs) (Cam et al. 2008; Ting et al. 2008) is composed of critical intracellular immune proteins structurally related to plant resistance proteins. NLRs are functionally related to the plant R proteins, and the largest class of R proteins contain a conserved Toll-IL-1 receptor domain or a coiled-coil domain, followed by a nucleotide-binding domain and a LRR (leucine-rich repeat) at the C-terminus (Inohara et al. 2005; Pearl et al. 2008). For example, NLRs, NALP3, and IPAF can each form the 'inflammasome' (i.e. multiprotein pro-inflammatory complexes). Further, mutations in the gene encoding Nod2, another NLR, are positively associated with Crohn's disease (Mayor et al. 2007). Mayor et al. showed that many NLRs interacted with the SGT1 and HSP90, both of which have plant orthologs essential for R-protein responses (Mayor et al. 2007). The assembly of multiprotein proinflammatory complexes called 'inflammasomes' drives activation of the proinflammatory proteins caspase-1 and/or caspase-5 (Ogura et al. 2006; Martinon and Tschopp 2007). Members of the nucleotide-binding oligomerization domain (Nod)–like receptor protein family are central components of inflammasomes. Proinflammatory caspases are recruited by the NLR proteins (Martinon et al. 2002) so that they can be activated to process the proinflammatory cytokines pro-interleukin-1b (pro-IL-1b) and pro-IL-18 in their cleaved and active forms (Martinon and Tschopp 2007). Mutations in the gene encoding the NLR NALP3 (also called 'cryopyrin') are associated with several autoinflammatory disorders, suggesting that inflammasome complexes are essentially involved in inflammatory responses (Hoffman et al. 2001; Hawkins et al. 2004). This association with disease indicates that tight control of inflammasome activity is required to avoid pathologic consequences. However, the regulation and mechanisms of the activators and the modulators of NLR- inflammasome activity are still unclear.

Bacterial components are detected by the leucine-rich repeat (LRR) domains of NALP3 and IPAF, the two best-characterized NLR members with inflammasome activity. NALP3 detects peptidoglycan (PGN) and RNA, whereas IPAF detects flagellin (Martinon et al. 2004; Kanneganti et al. 2006; Franchi et al. 2006; Miao et al. 2006). Endogenous microbe-independent stress signals also induce activation of the NALP3-containing inflammasome. For instance, activation of caspase-1, which depends on the NALP3-containing inflammasome, is induced by exposure of macrophages to ATP or crystals of monosodium urate (MSU). However, several cytoplasmic proteins, including pyrin, a protein associated with the autoinflammatory disease familial Mediterranean fever, dampen some inflammasome activity (Chae et al. 2003).

Mammalian NLRs are structurally related to plant R proteins that detect, through host-pathogen, gene-for-gene interaction (Chisholm et al. 2006), specific 'avirulence' gene products produced by pathogens. Members of the NLR and R-protein families incorporate amino-terminal 'effector domains' (such as the death-fold, coiled-coil or TIR domains), Nods (such as NACHT or NB-ARC), and a variable number of LRRs. Inclusion of similar domains by both NLR and R proteins is highly suggestive of a similar mechanism of activation (Inohara et al. 2005; Meylan et al. 2006; Ting and Davis 2005). Mayor et al. showed that various NLRs interacted with SGT1 and the HSP90 (Mayor et al. 2007). Inflammatory response to bacterial PGN or MSU mediated by NALP3 was abrogated by SGT1 depletion with siRNA or HSP90 inhibitor geldanamycin. Geldanamycin treatment also inhibited Nod2-mediated activation of the transcription factor NF-κB and reduced NALP3-mediated gout-like inflammation in mice. SGT1 is required for both NOD1 and NOD2 to respond to PGN derivatives (da Silva Correia et al. 2007). They provided evidence that, analogous to findings in plants, mammalian innate immune responses involving NLRs are regulated by HSP90 and SGT1. Recently, Novosylna et al. suggested that SGT1 competes with viral RNA for binding to eEF1A, which improves virus performance via interaction with viral genomic RNA and viral RNA-dependent RNA polymerase (Novosylna et al. 2015); however, involvement of NLRs is unknown. Further studies are needed to elucidate how SGT1 and HSP90 cooperate with other components to regulate NLR- inflammasome activity.

Another type of HSP70 (stress-inducible form) and HSC70 (heat shock cognate 70, constitutively expressed form), present as Ssa1 and Ssb1 in yeast, cooperate with Hsp90 and interact with SGT1 (Bansal et al. 2004; Niikura et al. 2006; Noel et al. 2007; Spiechowicz et al. 2007). In contrast to HSP90, HSP70 captures newly synthesized proteins or unfolded polypeptides under stress conditions (Wegele et al. 2004). Some substrates of HSP70 are transferred to HSP90 with the help of co-chaperones Hop (Sti1 in yeast) and/or HSP40 (Ydj1 in yeast), which mediates the last step of protein maturation (Wegele et al. 2006). Hop (HSP70-HSP90 Organizing Protein) links together HSP90 and HSP70 by forming a multi-chaperone complex (Wegele et al. 2004) (Fig. 16.3; Table 16.1). SGT1 interacts with HSP90 and HSP70, as in the case of Hop (Bansal et al. 2004; Noel et al. 2007; Spiechowicz et al. 2007).

SGT1-HSP70 interaction can be mediated by a HSP90-Hop complex, because Hop and SGT1 can bind to HSP90 simultaneously (Catlett and Kaplan 2006).

The specificity of SGT1-HSP70 interaction was verified by several studies. First, the interaction was confirmed as mediated at the SGS domain of SGT1 in both plants and humans (Niikura et al. 2006; Noel et al. 2007; Spiechowicz et al. 2007). Second, coexpression of the SGS binding protein S100A6 (calcyclin) (Nowotny et al. 2003) was reported to reduce SGT1-HSP70 interaction through Ca^{2+}- dependent binding to the C-terminal part of SGT1 in human cells (Spiechowicz et al. 2007). However, it is yet unclear if S100 proteins and HSP70 compete to bind to SGT1-SGS domain, and if they do, how this competition is regulated for the SGT1 nuclear translocation and the SGT1-NLR interaction (Fig. 16.3: see also previous sections). The SGT1-NLR interaction may be initially mediated by HSP70, then subsequently transferred to HSP90 (Shirasu 2009). Overexpression of HSP70 decreases NLR-dependent immunity in *Arabidopsis* but does not increase instability of R proteins (Noel et al. 2007). In humans, some studies postulated that extracellular localized Hsp70 may play a key role in inducing the cellular immune response (Milani et al. 2002; Multhoff 2002). Asea et al. suggested that HSP70 function as a chaperone and a cytokine, demonstrating that it stimulates cytokine production through a CD14-dependent pathway (Asea et al. 2000). They found that HSP70 bound with high affinity to the plasma membrane, elicited a rapid intracellular calcium flux, activated nuclear factor (NF)-κB, and upregulated expression of TNF-α, interleukin (IL)-1β, and IL-6 in human monocytes. Future studies will likely focus on specific functions of the cochaperone SGT1 and its contribution to the roles of HSP70 in immunity.

16.1.5 Yeast SGT1 in Kinetochore Assembly

The centromere, a single chromosomal region that spindle microtubules must attach to, is essential for chromosome segregation and inheritance, and requires structural and regulatory proteins for its activity and coordination within the cell cycle. The structural protein complex located at the centromere is called the kinetochore. The kinetochore is thought to generate a signal that arrests cells during mitosis when it is not properly attached to microtubules, thereby preventing aberrant chromosome transmission to daughter cells, which can lead to aneuploidy and tumorigenesis (3, 4). In many eukaryotes, the centromere consists of large arrays of repetitive DNA up to mega base pairs of DNA. On such "regional centromeres," kinetochore assembly is mainly defined by epigenetic regulation (Verdaasdonk and Bloom 2011) (see also following Sect. 16.1.7: **Human SGT1 is required for CENP-A deposition and kinetochore assembly**). By contrast, a clade of budding yeasts (Saccharomycetaceae) has a "point centromere" consisting of 120–200 base pairs of DNA, on which kinetochore assembly is defined by the consensus DNA sequence (Dujon 2010; Pluta et al. 1995). Recently it was suggested that even point centromeres could undergo rapid evolutionary changes (Kobayashi et al. 2015). The kinetochore of the budding yeast *Saccharomyces cerevisiae* (*S. cerevisiae*), has been characterized thoroughly, both genetically and biochemically; thus, its molecular structure is the most well detailed to date. In *S. cerevisiae*, the kinetochore consists of over 70 different proteins, many of which are conserved in mammals (Kitagawa and Hieter 2001). In *S. cerevisiae*, the centromere DNA is a 125-bp region that contains three conserved regions, CDEI, CDEII, and CDEIII (Fitzgerald-Hayes et al. 1982; Hieter et al. 1985). CDEIII (25 bp) is essential for centromere function (Hegemann et al. 1988) and is the site where CBF3 binds to centromeric DNA. CBF3 contains four proteins: Ndc10, Cep3, Ctf13 (Lechner and Carbon 1991; Doheny et al. 1993; Goh and Kilmartin 1993; Jiang et al. 1993; Lechner 1994; Strunnikov et al. 1995; Connelly and Hieter 1996; Stemmann and Lechner 1996), and Skp1 (Connelly and Hieter 1996; Stemmann and Lechner 1996). All of the four proteins are essential for viability and mutations in any one abolish the CDEIII-binding activity of CBF3 (Sorger et al. 1995; Kaplan et al. 1997). All the described kinetochore proteins, except the CDEI-binding Cbf1, localize to kinetochores in the CBF3 complex-dependent manner (Westermann et al. 2007). Therefore, the CBF3 complex is the essential structure of the kinetochore and studying the mechanism of CBF3 assembly is of great interest in yeast kinetochore biology.

SGT1 was isolated as a dosage suppressor of *skp1-4*, a kinetochore-defective mutant (Kitagawa et al. 1999) as described above. The *sgt1-3* mutant exhibits a number of characteristics consistent with a role for Sgt1 in kinetochore function. Those include the chromosome mis-segregation phenotype and G2/M accumulation at non-permissive temperature in a *MAD2*-dependent manner. The genetic interactions among *SGT1*, *SKP1*, and *CTF13* further support a role for *SGT1* in kinetochore function. Sgt1, although not a component of CBF3, can physically interact with Skp1 and is required for CDEIII-binding activity. Sgt1 and Skp1 are required for Ctf13 activation and subsequent assembly of the CBF3 complex (Kitagawa et al. 1999). However, the mechanisms of its activation and CBF3 assembly were not characterized in detail.

Formation of the active Ctf13-Skp1 complex may require Hsp90 (Stemmann et al. 2002). Sgt1 has highly conserved tetratricopeptide repeat (TPR) (Kitagawa et al. 1999) and CS (CHORD protein and SGT1-specific) motifs. In other proteins these motifs are required for interaction with Hsp90 (Dubacq et al. 2002; Azevedo et al. 2002). Our group reported that molecular chaperones from the Hsp90 and Hsp70 families form a complex with Sgt1 in yeast (Bansal et al. 2004). In this study, we determined that Skp1 and Hsc82 (a yeast Hsp90 protein) bind to the N-terminal region of Sgt1 that contains TPR motifs. When we performed sequence and phenotypic analyses of *sgt1* mutants, our results strongly suggested that the N-terminal region containing the Hsc82-binding and Skp1-binding domains of Sgt1 is important for the kinetochore function of Sgt1. Results of binding analyses revealed that Hsp90's binding to Sgt1 stimulates the binding of Sgt1 to Skp1 and that Sgt1 and Hsp90 stimulate the binding of Skp1 to Ctf13, the F-box core kinetochore protein. These data indicate that Sgt1 and Hsp90 function in assembling CBF3 by activating Skp1 and Ctf13 in yeast. Thus, we proposed a model in which Sgt1 and its interaction with Hsp90 are required for assembly of the core kinetochore complex, an initial step in kinetochore activation.

In in vitro and in vivo immunoprecipitation and analytical ultracentrifugation analyses, we showed that Sgt1 forms homodimers (Bansal et al. 2009a). We used Sgt1 deletion proteins to identify the dimerization domain, and reported that the Skp1-binding domain (amino acids 1–211) contains the Sgt1 homodimerization domain. The Sgt1 mutant proteins that dimerized neither bind to Skp1, suggesting that Sgt1 dimerization is important for Sgt1-Skp1 binding. The temperature and benomyl sensitivity and the chromosome missegregation phenotype of *sgt1-L31P* were suppressed by restoring dimerization activity of a dimerization-deficient *sgt1* mutant (*sgt1-L31P*) by using the CENP-B (centromere protein-B) dimerization domain. This study strongly suggests that Sgt1 dimerization is required for kinetochore assembly. Further, our group reported that protein kinase CK2 (casein kinase 2) phosphorylates Sgt1 at Ser^{361}, which inhibits Sgt1 dimerization (Bansal et al. 2009a). Therefore, CK2 appears to regulate Sgt1 dimerization and subsequent kinetochore assembly negatively in budding yeast. Nevertheless, Nyarko et al. have shown that human SGT1A does not form dimers efficiently performing gel-filtration chromatography (Nyarko et al. 2007), although the kinetochore function of Sgt1 is evolutionarily conserved between humans and yeast (Bansal et al. 2009a) (see also following Sect. 16.1.7: **Human SGT1 is required for CENP-A deposition and kinetochore assembly**). In analytical ultracentrifuge experiments, we also have shown that human SGT1A exists mainly as a monomer (Bansal et al. 2008). Future experiments will determine whether human SGT1 is phosphorylated by CK2 in vivo. If so, it will be interesting to study the functional role of phosphorylation of human SGT1.

Willhoft et al. determined the crystal structure of the interacting domains of *Saccharomyces cerevisiae* Sgt1 and Skp1 (Willhoft et al. 2017). Dimerization of yeast Sgt1 occurs via an insertion that is absent from monomeric human Sgt1. They identified point mutations (H59A and D61R) that disrupt dimerization and Skp1 binding in vitro and found that the interaction with Skp1 is an essential function of Sgt1 in yeast. Their data indicate that the G2/M phenotype of Sgt1 ts-mutants results from disruption of Skp1's association with Sgt1, however, mutation (H59A or D61R) of the dimerization interface that they identified in vitro does not result in temperature sensitivity. Discrepancies between our results and those of Willhoft et al. could be attributed to the different sites of dimerization-deficient point mutations (Willhoft et al., H59A and D61R; Our group, L31P) used for temperature sensitivity assays, although all reside in the Sgt1-TPR domain. As Willhoft et al. noted, since derivatives of G2/M mutant alleles can be rescued by attaching an artificial dimerization domain (Bansal et al. 2009b), and as phosphorylation of the Sgt1-SGS domain negatively regulates both dimerization and the interaction with Skp1 in vivo (Bansal et al. 2009a), we cannot exclude the possibility that additional or different dimerization interfaces exist in vivo.

16.1.6 SGT1 in SCF Function and Cyclic AMP Signaling: Yeast SGT1 in G1 Function

When the *SGT1* gene was identified in budding yeast, it was reported that *SGT1* is required for cell cycle progression at both the G1/S and G2/M transitions (Kitagawa et al. 1999). Among 5 *sgt1* mutant strains that could suppress the temperature sensitivity of the *skp1-4* mutation, strains containing *sgt1-1*, *sgt1-2*, *sgt1-4*, and *sgt1-5* alleles accumulated with G1 DNA content after being shifted to 37 °C for 3 h. In contrast to the *sgt1-3* mutant strain that arrested with largely G2 DNA content and markedly increased chromosome missegregation, *sgt1-5* cells arrested before the α factor point, which is indicative of the early G1 phase. Therefore, in this study, *sgt1-5* and *sgt1-3* are termed as a G1 and G2 allele mutant, respectively. Sgt1 showed association with Skp1 as described before, and *sgt1-5* mutants showed defects of SCF substrates (i.e., defects in Sic1p destruction in vivo and Cln1 ubiquitylation in vitro). Although the defect responsible for the G1 arrest is unknown, *sgt1-5* mutants may be defective in their ability to destroy early G1 targets of SCF. However, there are three known SCF targets operating in G1-Cdk inhibitor: Far1p and the bud site emergence regulators Gic1p and Gic2p (Brown et al. 1997; Jaquenoud et al. 1998). But *sgt1-5* arrest was independent of these potential targets, since *sgt1-5 far1Δ1* and *sgt1-5gic1Δgic2Δ* mutants still exhibited the G1 arrest phenotype at a nonpermissive temperature. Yamamoto et al. reported interactions between a ubiquitin-conjugating enzyme, Rad6, from rice, *Oryza sativa* L. cv. Nipponbare (OsRad6), and Sgt1 (OsSgt1), suggesting that the SCF complex is involved in functions requiring the Rad6-mediated pathway, such as ubiquitination and DNA repair (Yamamoto et al. 2004). Further studies are required to identify new SCF substrates operating in G1 through Sgt1-Skp1.

In plants and humans, the SGT1-HSP90 complex is involved in multiple functions (Fig. 16.2; see also previous and following sections), and the HSP90 chaperone machinery is often tightly associated with the ubiquitin-dependent 26S proteasome system (UPS) (Kleiger and Mayor 2014; Mimnaugh et al. 1996). This association is probably a part of a protein quality control mechanism that ensures prompt degradation of unfolded or misfolded sensors to avoid inappropriate activation of signal pathways in multiple regulations (Kadota et al. 2010; Shirasu 2009). A member of the RAR1-SGT1-HSP90 chaperone complex associates with several components involved in protein degradation pathways (Kadota et al. 2010; Shirasu 2009) (Fig. 16.3; Table 16.1). In yeast, Sgt1 directly binds to Skp1, a component of the SCF (SKP1, Cullin, F-box protein) ubiquitin ligase complex (see also previous sections), acting as an adaptor to link HSP90 and SCF (Catlett and Kaplan 2006; Kitagawa et al. 1999) (Fig. 16.3, Table 16.1). SKP1 and its interactor CULLIN1 (CUL1) were also found in an SGT1 complex in plants (Azevedo et al. 2002; Liu et al. 2002), and the SKP1-SGT1 interaction was reported as direct (Liu et al. 2002) (Fig. 16.3; Table 16.1). *Arabidopsis* SGT1b is consistently required for the SCF (TIR1)-mediated degradation of Aux/IAA proteins as auxin hormone responses (Gray et al. 2003). HSP90 can possibly connect protein folding and degradation pathways, because it can interact at the same time with SGT1 and CHIP (carboxy terminus of the Hsc70-interacting protein), which contains a ubiquitin ligase domain (Zhang et al. 2008). In plants, RAR1 and SGT1 also associate with the COP9 signalosome (CSN) (Azevedo et al. 2002; Liu et al. 2002), which deneddylates CUL1, thereby inactivating the SCF complex in the CRL (Cullin-RING E3 ubiquitin-Ligases) assembling cycle (Merlet et al. 2009; Lyapina et al. 2001).

The importance of ubiquitin-dependent degradation in signaling of innate immunity was demonstrated by the experimental result that silencing SKP1 or the CSN components impairs *N* gene resistance against tobacco mosaic virus (Liu et al. 2002). However, another NLR protein, Rx, does not require an SGT1 TPR domain (Boter et al. 2007), which is the interaction domain for SKP1 in yeast (Catlett and Kaplan 2006). Thus, the SKP1-SGT1 interaction may not be critical for innate immunity. The SKP1/CSN-dependent ubiquitin pathway may function downstream of the immune sensors, and SGT1's function in the ubiquitylation machinery through SKP1 may be to mediate degradation of misfolded NLR proteins (Shirasu 2009).

While the SGT1-HSP90 complex is associated with the polyubiquitylation degradation pathway through the SCF complex, the SGT1-HSP90 complex is also required for signalling ubiquitylation (mono- or diubiquitylation) of CENP-A and CENP-A deposition at centromeres (Niikura et al. 2017) (see also following Sect. 16.1.7: **Human SGT1 is required for CENP-A deposition and kinetochore assembly**). In contrast to SCF function, our study showed that the SGT1-HSP90 complex contributes the E3 ligase activity of the CUL4A complex (CUL4A-RBX1-COPS8) in CENP-A ubiquitylation (Niikura et al. 2017). We previously reported that the CUL4A complex uses adaptor COPS8/CSN8 to target CENP-A, suggesting that COPS8 itself is the specific adaptor of the CUL4A-RBX1 complex that is required to recognize and ubiquitylate the substrate CENP-A in a non-canonical CRL4 machinery (Niikura et al. 2015). In our study, SKP1 siRNA treatment did not cause any signal reduction of CENP-A at centromeres (Niikura et al. 2017). Therefore, we concluded that the SGT1-HSP90 complex participates in CENP-A deposition at centromeres in a SKP1-independent and/or SCF-independent manner. This conclusion is consistent with our previous report that the CUL4A-RBX1 complex, which does not require SKP1 to function, is required for CENP-A deposition at centromeres (Niikura et al. 2015). Further study is needed to clarify the role of human SGT1 in G1 and/or SCF function, including how G1/SCF function correlates to signalling ubiquitylation of CENP-A and how these different functions are distinctively regulated.

Both Git7 (the Sgt1 homolog) in fission yeast and Sgt1 in budding yeast are required for the cyclic AMP signaling pathway (Dubacq et al. 2002; Schadick et al. 2002). To avoid confusion of nomenclature, we note that the hSgt1 (human suppressor of Gcr two; **not** suppressor of G2 allele of *skp1*) orthologue of *Sz. pombe* (SpSgt1) (Kainou et al. 2006) is not Git7, neither is the *S. cerevisiae* Sgt1 homolog in fission yeast. Therefore, we excluded the content about SpSgt*1* in the present review (Table 16.1).

In *S. pombe*, transcription of the *fbp1* gene encoding fructose-1,6-bisphosphatase is repressed by glucose (Hoffman and Winston 1989; Vassarotti and Friesen 1985). Schadick et al. described cloning and characterization of the *S. pombe git7* gene that conferred constitutive transcription of *fbp1* gene (Schadick et al. 2002). The *git* (glucose-insensitive transcription) genes encode components of a cyclic AMP (cAMP) signaling pathway and are required for adenylate cyclase activation. Because *S. pombe* cells lack both adenylate cyclase or protein kinase A (the major effector of cAMP in eukaryotes), Git7 is essential, but this requirement appears to be due to functions in septation and cell walls (see also following Sect. 16.1.8: **Sgt1 in centrosome organization and cytokinesis**). The *git7* mutants tested by Schadick et al. showed sensitivity to the microtubule-destabilizing drug benomyl, although no chromosome instability. Two alleles of *git7* (*git7-93 and git7-GFP*) encoding proteins altered in the highly conserved C-terminus are functional for cell growth and septation, but defective in glucose-triggered cAMP signaling. *S. cerevisiae* and human *SGT1* genes suppress *git7-93* but not *git-235* for glucose repression of *fbp1* transcription and benomyl sensitivity, suggesting that the Git7/Sgt1 proteins may act as multimers, such that Git7-93p but not Git7-235 can deliver the orthologous proteins to species-specific targets.

In *S. cerevisiae*, Dubacq et al. also identified *SGT1* as an allele-specific suppressor of cdc35-1, a thermosensitive mutation in the leucine-rich repeat domain of the adenylyl cyclase Cyr1/Cdc35. They demonstrated that Sgt1 and Cyr1/Cdc35 physically interact and that activity of the cAMP pathway is affected in an *sgt1* conditional mutant. Increasing intracellular levels of cAMP via deleting the PDE2 gene, which encodes a high-affinity cAMP phosphodiesterase, modified the *sgt1-5* (G1 allele) phenotype and the cAMP pathway was affected in a conditional null mutant for *SGT1*. Sequence analysis revealed that the CS domain of Sgt1 adopts a fold similar to that of the Hsp90 cochaperone p23, suggesting that Sgt1 has features of a cochaperone. Thus, these results suggest that Sgt1 is an activator of adenylyl cyclase in *S. cerevisiae* and may function in assembly or conformational activation of specific multiprotein complexes. These studies of fission and budding yeasts suggest that members of the Git7/Sgt1 protein family might play a conserved role in the regulation of adenylate cyclase activation in *S. pombe*, *S. cerevisiae*, and humans.

However, in *Candida albicans*, genetic depletion of Sgt1 phenocopies depletion of Hsp90, inducing yeast to filament morphogenesis and invasive growth, and reduce kidney fungal burden in a murine model of systemic infection (Shapiro et al. 2012). In this study, Sgt1 physically interacted with Cyr1/Cdc35, and depletion of either Sgt1 or Hsp90 activated cAMP-PKA signaling. This result suggests that Sgt1 interacts with Cyr1 and represses cAMP-dependent PKA activity in *C. albicans*, whereas in *S. cerevisiae* Sgt1 activates Cyr1-mediated signaling (Dubacq et al. 2002), thus reflecting functional divergence in the Sgt1-dependent circuitry. Further investigation is expected to study the functional divergence of the cAMP-PKA signaling through Sgt1-Hsp90 and Cyr1/Cdc35.

16.1.7 Human SGT1 Is Required for CENP-A Deposition and Kinetochore Assembly

Studies of Sgtl in budding yeast suggest that Sgtl is required for kinetochore assembly. Further questions were raised about the function of mammalian SGT1. Steensgaard et al. reported that RNA interference-mediated depletion of SGT1 from HeLa cells markedly alters the mitotic spindle and causes problems in chromosome alignment (Steensgaard et al. 2004). SGT1-depleted cells arrest at mitosis due to activation of the spindle checkpoint (also known as spindle assembly checkpoint). However, the checkpoint response is significantly weakened in SGT1-depleted cells, correlated with a large reduction in kinetochore levels of spindle checkpoint proteins (MAD1, MAD2, and BUBR1). Presumably loss of complete kinetochore-microtubule attachment activates the spindle checkpoint response of reduced potency, with cells exiting mitosis aberrantly, inducing tetraploid G1 cells and eventually cell death. These phenotypes are due to mislocalization of Hec1, CENP-E, CENP-F, CENP-I, but not CENP-C, to mitotic kinetochores. Reduced potency of the spindle checkpoint correlates with extensive kinetochore damage caused by loss of SGT1, and the residual spindle checkpoint response may be due to traces of MAD1 and MAD2 at kinetochores of SGT1-depleted cells. These studies suggest that SGT1 is an essential protein and a critical assembly factor for the mammalian kinetochore, corroborating the hypothesis of a kinetochore assembly pathway that is conserved from yeast to humans.

In the budding yeast *Saccharomyces cerevisiae*, Hsp90 is required for the assembly of the kinetochore's core complex, CBF3 (Stemmann et al. 2002), and the interaction between Hsp90 and Sgtl is also required for kinetochore assembly (Bansal et al. 2004; Lingelbach and Kaplan 2004). Human centromere DNA consists of repetitive alpha-satellite sequences that are several kilobases to megabases long. The overall structure of the human regional centromere is very different from that in yeast, and yeast CBF3 is not conserved in humans (Kitagawa and Hieter 2001; Biggins and Walczak 2003). Nevertheless, some inner (e.g., CENP-C), central (e.g., CENP-H and CENP-I), and outer kinetochore proteins (e.g., HEC1) and the spindle checkpoint components (e.g., BUB1 and MAD2) are conserved (Kitagawa and Hieter 2001; Biggins and Walczak 2003). Human SGT1 (SUGT1) can rescue the sgt1-null mutant (Kitagawa et al. 1999). As described in previous sections, studies in SGT1-depleted human cells showed that CENP-I, CENP-F, and HEC, but not CENP-C, are absent from kinetochores (Steensgaard et al. 2004). These results suggest that the function of SGT1 is conserved in humans and yeast. However, the role of HSP90 at human kinetochores had not been explored before our study using the HSP90 inhibitor 17-allylaminogeldanamycin (17-AAG), which is currently in clinical trials. 17-AAG is thought to exert antitumor activity by simultaneously targeting several oncogenic signaling pathways.

Our group reported a novel mechanism by which 17-AAG inhibits cell proliferation, and we provide the first evidence that HSP90 is required for the assembly of kinetochore protein complexes in humans (Niikura et al. 2006). HeLa cells treated with 17-AAG showed delocalization of several kinetochore proteins, including CENP-I and CENP-H but excluding CENP-B and CENP-C. Consistently, 17-AAG treatment induced a mitotic arrest that depends on the spindle checkpoint protein MAD2, and induced misalignment of chromosomes and aneuploidy. HSP90 interacts with SGT1 in human cells and depletion of SGT1 sensitizes HeLa cells to 17-AAG. Overexpression of SGT1 rescued the proper localization of specific kinetochore proteins and chromosome alignment in cells treated with 17-AAG. Inhibition of HSP90 by siRNA prevented the kinetochore localization of several kinetochore proteins, and 17-AAG inhibited the HSP90-SGT1 association. These biochemical and genetic results suggest that HSP90, through its interaction with SGT1, is required for kinetochore assembly. Therefore, the HSP90-SGT1 chaperone complex is important for formation of functional kinetochores required for accurate assembly of many protein complexes. However, its targets and exact contribution to kinetochore assembly still remain to be studied.

Polo-like kinase 1 (PLK1) is essential for this attachment process as a key regulator of mitosis, but its molecular mechanism is still unclear. Liu et al. reported that SGT1 is a PLK1 substrate during mitosis (Liu et al. 2012). They showed that SGT1 localizes at the kinetochores in cells treated with nocodazole but not paclitaxel (Taxol), which lack microtubule attachments during prometaphase. PLK1 is required for the kinetochore localization of SGT1 and phosphorylates serine 331 (S331) of SGT1 at the kinetochores. This phosphorylation event enhances the interaction of the HSP90-SGT1 chaperone with the MIS12 complex to stabilize this complex at the kinetochores. Therefore, SGT1 S331 phosphorylation coordinates the recruitment of the NDC80 complex to form proper kinetochore-microtubule attachments. Abrogation of SGT1 phosphorylation destabilizes the MIS12 and NDC80 complexes at the kinetochores, impairs stable microtubule attachment, and eventually results in chromosome misalignment to delay the anaphase onset. These results demonstrated a mechanism for PLK1 in promoting kinetochore-microtubule attachment to ensure chromosome stability.

Integrating the model proposed by Liu et al. (2012), we proposed that one possible pathway of the functional contribution of the SGT1-HSP90 complex is initiated by PLK1-mediated phosphorylation of SGT1 S331 (in humans) after CDK1-mediated priming phosphorylation of SGT1 S249 (this priming phosphorylation generates a docking site for the Polo-box domain of PLK1) (Niikura et al. 2017) (Fig. 16.4, G2/M and prometaphase at CEN). The timing of SGT1 delocalization at the kinetochore (i.e., metaphase) and dephosphorylation of S331 (i.e., anaphase) reported by Liu et al. is slightly earlier than the timing of CENP-A deposition at the centromere (i.e., telophase/early G1) reported previously (Jansen et al. 2007). These results suggest that SGT1 phosphorylation is inhibitory at the activation time (i.e., telophase/early G1) of the CUL4A complex. Proper dephosphorylation could be predicted (Fig. 16.4, "Dephosphorylation?"), although murine Sgt1-S302 (equivalent to human SGT1-S331) may be required for the Hsp90-Sgt1 chaperone to interact with the MIS12 complex and properly assembly of the MIS12 complex at the kinetochores, but not for the kinetochore localization of murine Sgt1 itself (Liu et al. 2012). Eventually each type of SGT1 phosphorylation may contribute to pro-

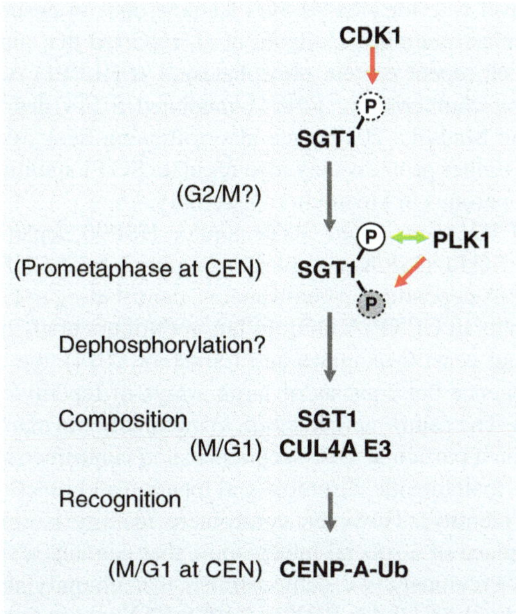

Fig. 16.4 A hypothetical pathway depicting the functional contribution of the SGT1-HSP90 complex to CENP-A deposition. In humans, CDK1-mediated priming phosphorylation occurs on SGT1 serine 249 (S249, "P" in white), and this priming phosphorylation generates a docking site for the Polo-box domain of PLK1 (Liu et al. 2012). Subsequently, PLK1-mediated phosphorylation on SGT1 serine 331 (S331, "P" in gray) is followed. These two phosphorylations may lead to the prometaphase localization of SGT1 at the kinetochorve, activation of the CUL4A complex, and subsequent ubiquitylation of CENP-A. Dephosphorylation of human SGT1 S331 from anaphase (Liu et al. 2012) could contribute to composition of the CUL4A complex and recognition of CENP-A by COPS8 ("Composition" and "Recognition"), and subsequent CENP-A deposition. The silver arrow indicates the direction of the pathway, the red arrow indicates phosphorylation, and the green double-headed arrow indicates the interaction

metaphase localization of SGT1 to the kinetochore, activation of the CUL4A complex (Fig. 16.4, "Composition" and "Recognition", see also below), and subsequent ubiquitylation of CENP-A (Fig. 16.4, M/G1 and M/G1 at CEN).

Davies et al. reported that because of its targeting of the MIS12 complex (Table 16.1), the SGT1-HSP90 complex together with SKP1 is important for the accurate assembly of kinetochore protein complexes (Davies and Kaplan 2010). Inhibition of HSP90 or SGT1 destabilizes the MIS12 complex and delays proper chromosome alignment due to inefficient kinetochore-microtubule attachment. Interestingly, coinhibition of SGT1 and the SCF subunit, SKP1, increases MIS12 complexes (DSN1 and $NDC80^{HEC1}$) at kinetochores and restores timely chromosome alignment but forms less robust kinetochore-microtubule attachments. They proposed that a balance between MIS12 complex assembly and turnover is required for the efficient and accurate assembly of kinetochore–microtubule-binding sites, and that this balance depends on the SCF ubiquitin ligase subunit SKP1. These findings support a novel role for HSP90-SGT1 chaperones in ensuring the fidelity of multiprotein complex assembly. Gangula et al. reported that pleckstrin homology domain leucine-rich repeat protein phosphatase 1 (PHLPP1) functions in kinetochore assembly by counteracting RNF41-mediated SGT1 degradation in human cells (Gangula and Maddika 2017) (see also following Sect. 16.1.9: **Mammalian SGT1 in cancer**). Other proteins may also regulate SGT1 stability and functions of HSP90-SGT1 chaperones in kinetochore assembly.

Our studies of RNAi-mediated SGT1 and/or HSP90 depletion in HeLa cells revealed that the SGT1-HSP90 complex is required for CENP-A ubiquitylation in vivo and CENP-A deposition at centromeres, contributing to E3 ligase activity of the CUL4A complex in CENP-A ubiquitylation (Niikura et al. 2017). In most species except budding yeast (Fukagawa and Earnshaw 2014), the centromere has no defined DNA sequence but consists of large arrays of repetitive DNA composing centromeric DNA. The centromere location in mammals and many other eukaryotes does not depend on a particular DNA sequence, and centromeric DNA is not sufficient or necessary to define the chromosomal location and function of a centromere (i.e., centromere identity). However, centromere identity is epigenetically determined by the presence of a special nucleosome that contains a specific histone H3 variant CENP-A. Previously, we demonstrated that ubiquitylation on lysine 124 (K124) mediated by the CUL4A-RBX1-COPS8 E3 ligase is required for CENP-A deposition at the centromere (Niikura et al. 2015). In contrast to SCF function, our study showed that the SGT1-HSP90 complex contributes the E3 ligase activity of the CUL4A complex (CUL4A-RBX1-COPS8) in CENP-A ubiquitylation (Niikura et al. 2017). Our data support the concept that the CUL4A complex uses adaptor COPS8/CSN8 to target CENP-A, suggesting that COPS8 itself is the specific adaptor of the CUL4A-RBX1 complex required to recognize and ubiquitylate the substrate CENP-A in a non-canonical CRL4 machinery (Niikura et al. 2015). However, the mechanism that controls the E3 ligase activity of the CUL4A-RBX1-COPS8 complex remained unclear.

We found that the SGT1-HSP90 complex is required for CENP-A ubiquitylation in vivo and CENP-A deposition at centromeres applying RNA interference (RNAi)-mediated SGT1 and/or HSP90 depletion in HeLa cells (Niikura et al. 2017). Our group and other group previously demonstrated in vivo interactions of HSP90-CUL4 (Dias et al. 2002) and SGT1A-CUL4A (Table 16.1). We reported that the CUL4A complex targets CENP-A through the adaptor COPS8/CSN8 (Niikura et al. 2015). Thus, we hypothesized that depletion of SGT1 or HSP90 protein leads to loss of interaction among components of the CUL4A complex. Indeed, SGT1 or HSP90 siRNA disrupted an association between COPS8 and CENP-A and an interaction between COPS8 and CUL4A. These results suggest that the SGT1-HSP90 complex is required for composition of the CUL4A complex and recognition of CENP-A by COPS8 (Fig. 16.4, “Composition” and “Recognition”), clarifying how the SGT1-HSP90 complex contributes to E3 ligase activity of the CUL4A complex in CENP-A ubiquitylation. In this study, SKP1 siRNA treatment did not cause any signal reduction of CENP-A at centromeres. Therefore, we concluded that the SGT1-HSP90 complex participates in CENP-A deposition at centromeres in a

SKP1-independent and/or SCF-independent manner. This conclusion is consistent with our previous report that the CUL4A-RBX1 complex, which does not require SKP1 to function, is required for CENP-A deposition at centromeres (Niikura et al. 2015).

It is unlikely that SKP1 turnover affects the CENP-A loading pathway, because our results suggest that SKP1 is not required for recruitment of CENP-A to centromeres. Destabilization of the MIS12 complex at the kinetochore observed by Davies et al. could be, at least in part, due to the defect in CENP-A recruitment, because CENP-A is at the top of a hierarchy of the pathway that determines the assembly of kinetochore components (Liu et al. 2006). This idea is supported by our results showing that SGT1 siRNA treatment did not significantly change the recruitment of endogenous MIS12, HEC1, and SKA1 proteins at LacO arrays, after ectopic loci were forcibly determined through LacO-LacI-CENP-A interaction. Taken together, these data suggest that the SGT1 defect–induced reduction of immunofluorescence signals of the central-outer kinetochore proteins at the kinetochore, including ones reported previously (Davies and Kaplan 2010; Niikura et al. 2006; Steensgaard et al. 2004), are explained by CENP-A mislocalization caused by SGT1 defects.

16.1.8 SGT1 in Centrosome Organization and Cytokinesis

Lange et al. detected abnormal centrosome separation and maturation, aberrant spindles, and impaired chromosome segregation in *Drosophila* Hsp90 mutant cells or mammalian cells treated with the Hsp90 inhibitor geldanamycin, and these phenotypes are correlated with decreased Polo protein levels (Lange et al. 2000). These results suggest that Hsp90 is required to ensure proper centrosome function in *Drosophila* and vertebrate cells. Moreover, de Carcer et al. reported that stabilization of Polo is a mechanism by which Hsp90 contributes to the maintenance of functional centrosomes (de Carcer et al. 2001). Martins et al. reported that mutations in sgt1 do not affect overall kinetochore assembly or spindle assembly checkpoint in *Drosophila*; however, they observed similar phenotypes to Hsp90 mutant cells reported previously (Martins et al. 2009). In *sgt1* mutant cells, centrosomes fail to mature, and pericentriolar material components do not localize normally, resulting in highly abnormal spindles (Martins et al. 2009). In *sgt1* mutant neuroblasts, they observed lower overall levels of Polo, and overexpression of the Polo kinase that resulted in a substantial rescue of the centrosome defects, suggesting that Sgt1 is involved in the stabilization of Polo allowing normal centrosome maturation, entry and progression though mitosis. Further study is needed to elucidate how the SGT1-HSP90 chaperone complex is regulated at opposite polar ends (kinetochore and centrosome) of microtubules, and how these functions are synchronized and communicated for proper function.

In fission yeast, Git7 (the Sgt1 homolog) seems to be required for both cell wall integrity and septation (Schadick et al. 2002). The *git7-27* and *git7-235* mutant cells fail to form septa, and form defective septa that lead to either cell lysis during cytokinesis or a failure to undergo cytokinesis. Rodrigo-Brenni et al. suggest that Skp1-Sgt1 interaction is critical for assembly of CBF3 complexes in budding yeast (Rodrigo-Brenni et al. 2004). They have shown that Skp1 and Sgt1 contribute to a final rate-limiting step in assembly, the binding of the core CBF3 subunit Ctf13 to Ndc10, using in vitro binding assays and a unique protocol for synchronizing the assembly of kinetochores in cells (Rodrigo-Brenni et al. 2004). Further, they reported that inhibiting the assembly of CBF3 causes disorganized septins and defects in cell polarity that cause cytokinesis failures (Gillis et al. 2005). Further studies are needed to elucidate how the SGT1-HSP90 chaperone complex is regulated at different stages in mitosis and how these stages are interconnected to achieve proper chromosome segregation.

16.1.9 Mammalian SGT1 in Cancer: The Oncogenic Role of the Cochaperone SGT1

Defective orchestration of oncogenes and tumor suppressor genes is a common propensity of various types of cancer. Because dysregulated oncoproteins are unstable, cancer cells use HSP90 as a chaperone to promote folding and function of mutated or overexpressed oncoproteins (Trepel et al. 2010). The inhibition of one or more oncoproteins can often suppress the proliferation and survival of cancer cells (Weinstein and Joe 2008). Thus, HSP90 inhibitors continue to be developed as cancer therapies and tested in clinical trials (Trepel et al. 2010). SGT1 is a highly conserved cochaperone of HSP90 (Bansal et al. 2004; Hubert et al. 2003; Lingelbach and Kaplan 2004; Takahashi et al. 2003; Lee et al. 2004). Cochaperones associate with HSP90, regulate the ATPase activity of HSP90, and recruit client proteins to HSP90. Therefore, co-chaperones are required for HSP90 function and have been proposed as potential targets for cancer therapy (Trepel et al. 2010). SGT1 is involved in multiple specific cellular functions (Fig. 16.2) including ubiquitination (Kitagawa et al. 1999), the cyclic AMP pathway (Dubacq et al. 2002; Schadick et al. 2002), centrosome maturation (Martins et al. 2009), immune response (Austin et al. 2002; Azevedo et al. 2002; da Silva Correia et al. 2007; Mayor et al. 2007), and CENP-A deposition and subsequent kinetochore assembly (see previous sections) (Niikura et al. 2017; Niikura et al. 2006; Bansal et al. 2004; Steensgaard et al. 2004) presumably as a co-chaperone. Thus, depletion of SGT1 causes misalignment of chromosomes, activation of a weakened spindle checkpoint, and, potentially, aneuploidy.

SGT1 is reportedly overexpressed in tumor tissues. *SGT1* mRNA levels are increased in colorectal cancer, and these levels are highly associated with rate of recurrence and poorer prognosis (Iwatsuki et al. 2010). SGT1 interacts with PHLPP1 (Table 16.1), and overexpression of SGT1 upregulates Akt signaling by promoting beta-TrCP-dependent PHLPP1 degradation in gastric cancer cells (Gao et al. 2013). PHLPP1 functions in kinetochore assembly by counteracting RNF41-mediated

SGT1 degradation in human cells (Gangula and Maddika 2017) (see also previous Sect. 16.1.7: **Human SGT1 is required for CENP-A deposition and kinetochore assembly**). These findings suggest that overexpression of SGT1 play a role in tumorigenesis. Our group investigated SGT1 protein levels in tumor tissues and pediatric tumor xenografts (Ogi et al. 2015). SGT1 was highly expressed in colon, breast and lung tumor tissues and in Ewing sarcoma and rhabdomyosarcoma xenografts. We also found that *Sgt1* heterozygous knockout resulted in suppressed Hras-mediated transformation in vitro and tumor formation in $p53^{-/-}$ mouse embryonic fibroblast cells, and significantly increased survival of $p53^{-/-}$ mice. These effects appeared to be independent of mitotic defects: kinetochores are assembled properly and the spindle checkpoint is normal in *Sgt1*$^{+/-}$ MEF cells, and there is no difference in ploidy between *Sgt1*$^{+/+}$ MEF cells and *Sgt1*$^{+/-}$ MEF cells. Therefore, our group concluded that aneuploidy does not seem to be the mechanism of the tumor suppression in *Sgt1*$^{+/-}$ mice. Depletion of Sgt1 expression inhibited the proliferation of cancer cells (Ewing sarcoma and rhabdomyosarcoma cells) and destabilized EWS-FLI1 and PAX3-FOXO1 oncogenic fusion proteins, respectively, which are required for cellular growth. These results suggest that SGT1 is involved in cancer development by stabilizing oncoproteins and point to SGT1 as a potential therapeutic target.

16.1.10 Mammalian SGT1 in Brain Diseases

Spiechowicz et al. examined the expression of Sgt1 protein in rat tissues using Northern and Western blots (Spiechowicz and Filipek 2005). They found that Sgt1 protein is expressed in various tissues and that brain, skeletal muscle, and spleen contain high levels of this protein and its mRNA. They also examined localization of Sgt1 in areas of the rat brain such as cortex, cerebellum and hippocampus by Western blots; the cerebellum and cortex contained the highest amounts of Sgt1. They found Sgt1 protein in some populations of neurons, like the Purkinje cells of the cerebellum and in the glial cells of white matter, using immunohistochemistry. These data might suggest the study of Sgt1 function in brain, although it is difficult to speculate what are its roles in this tissue at this stage. Spiechowicz et al. also analyzed SGT1 localization in human brain by immunohistochemistry (Spiechowicz et al. 2006) by comparing density of SGT1-immunopositive neurons in cortical layers of brain sections from healthy older adults and those with Alzheimer's disease (AD). They found a significant decrease in SGT1-immunopositive neurons in the temporal (up to 25-fold), angular (up to 11-fold), and posterior cingulate cortex (up to fivefold), suggesting that the presence of SGT1 in post-mitotic neurons indicates the involvement of this protein in a process different from that required for activity of the kinetochore. Therefore, SGT1 could be a possible marker of neuronal degeneration in AD.

Oligodendrocytes are glial cells that generate and support the myelin sheath around the axons in the central nervous system. The myelin sheath is an electrical insulating proteolipid layer that transmits nervous impulses to and between the cell bodies through the axon. Damaging or losing this layer may affect axonal function, which is observed in diseases such as multiple sclerosis, amyotrophic lateral sclerosis (ALS), and schizophrenia (Nonneman et al. 2014). Tavares et al. identified SGT1 among 39 splice variants using an oligodendrocyte proteome dataset deposited in ProteomeXchange to determine the level of oligodendrocyte-associated proteins (Tavares et al. 2017). These findings may encourage other research groups working with central nervous system disorders to investigate the biological significance of SGT1 that could be targeted in novel treatment strategies and diagnostic methods. Pickering et al. performed proteomic analysis of the medial prefrontal cortex extracted from rats treated with phencyclidine (PCP), which mimics many aspects of schizophrenia, yet the underlying mechanism of neurochemical adaptation for PCP is unknown (Pickering et al. 2013). They found that Sgt1/Sugt1 protein was downregulated among 11 downregulated proteins they identified. He et al. demonstrated that cytoplasmic isoform USP19_b upregulates protein levels of the polyglutamine (polyQ)-containing proteins, ataxin-3 (Atx3) and huntingtin (Htt), and promotes aggregation of their polyQ-expanded through the HSP90 chaperone system in cell models (He et al. 2016). Wang et al. discuss the mechanisms behind impaired heat shock protein function in brain aging, including transcriptional regulation, posttranslational modification, and communication across cells and organs (Wang et al. 2018). Further investigation to study the involvement of SGT1 in this HSP90 chaperone system is expected.

These studies consistently indicated a strong association between SGT1 protein levels and its brain-cellular localization and brain diseases and/or disorders; however, the regulation of protein interaction and signaling is still unclear. Anderson et al. performed a genetic screen to identify regulators of neuroblast self-renewal in *Drosophila*, and identified a mutation in *sgt1* that had fewer neuroblasts. In this study, *sgt1* neuroblasts have two polarity phenotypes: failure to establish apical cortical polarity at prophase, and lack of cortical Scribble localization throughout the cell cycle (Andersen et al. 2012). Double mutants lacking Sgt1 and Pins (a microtubule-induced polarity pathway component) resulted in neuroblasts without detectable cortical polarity and formation of "neuroblast tumors." Mutants in hsp83 (encoding the Hsp90), LKB1, or AMPKα all show similar prophase apical cortical polarity defects (but no Scribble phenotype), and activated AMPKα rescued the *sgt1* mutant phenotype. Therefore, they proposed that an Sgt1/Hsp90–LKB1–AMPK pathway acts redundantly with a microtubule-induced polarity pathway to generate neuroblast cortical polarity, and the absence of neuroblast cortical polarity can produce neuroblast tumors.

Martinez-Noel et al. performed multiple immunoprecipitation using UBE3A as bait in SH-SY5Y cells, and detected SGT1 as one of the "high confidence interacting proteins" by Comparative Proteomic Analysis Software Suite (CompPASS) (Martinez-Noel et al. 2018) (Table 16.1). Perturbations in activity and dosage of the UBE3A ubiquitin-ligase have been linked to Angelman syndrome and autism spectrum disorders (Martinez-Noel et al. 2018). SGT1 interactors are shown in Table 16.1, however, more detailed studies concerning interactions of SGT1 with its targets are required to better understand the function of this protein in various cells and tissues, including the cells of the central nervous system.

Mammalian translation elongation factor eEF1A is involved in ribosomal polypeptide synthesis and many additional roles including the regulation of cytoskeleton dynamics, proteasome-mediated protein degradation, cell signaling related to apoptosis and cancer, centrosomal activation and formation of the complexes with α-, β-, and γ-tubulins, and chaperone-like activity in eukaryotes. Notably, eEF1A is engaged in the viral replication apparatus, in which it interacts with both protein and RNA components. Novosylna et al. demonstrated the interaction between SGT1 and eEF1A1 using both purified proteins and cell lysates, identifying that the TPR domain of SGT1 and the D2 and D3 domains of eEF1A1 are involved in complex formation (Novosylna et al. 2015) (Fig. 16.3; Table 16.1). Modeling of the SGT1-eEF1A1 complex suggested both shape and charge complementarities of the eEF1A1-SGT1 interface stabilized by a number of predicted salt bridges. Because such an interaction mode is more typical for protein-nucleic acid interaction, they suggested that SGT1 competes with viral RNA for binding to eEF1A.

Mutations in eEF1 complex subunits are strongly associated with neurodevelopmental abnormalities (McLachlan et al. 2018). The eEF1A2 subunit is most strongly associated with disorders of neurodevelopment. eEF1A2 was first implicated in neurological disorders when a spontaneous 15.8 kb deletion including the promoter and first exon of the gene was identified in mice displaying a severe neurodegenerative phenotype (McLachlan et al. 2018). Mice homozygous for the deletion undergo muscular and neuronal degeneration, the onset of which coincides with the down-regulation of eEF1A1 in these cell types (Khalyfa et al. 2001; Newbery et al. 2007). Unlike eEF1A1, eEF1A2 is not expressed in tissues other than neurons and muscle, so it is reasonable to assume that mutations would be better tolerated than those in ubiquitously expressed eEF1 subunits. Pathogenic mutations in eEF1A1 would presumably show a lethal phenotype (McLachlan et al. 2018). Studies of mutations in *EEF1A2* in humans suggest that haploinsufficiency for *EEF1A2* is not lethal and that the missense mutations may result in some form of cellular toxicity. This toxicity could result from a dominant negative effect, possibly on eEF1A1 in developing neurons, or in altered protein-protein interaction profiles in which different mutations lead to different edgotypes (Sahni et al. 2015). It will be interesting to assess the effects of each mutation on binding to other eEF1 complex subunits in future.

For *EEF1A2, EEF1B2, EEF1D,* and *VARS,* the phenotype associated with mutations is predominantly one of abnormal neurodevelopment (McLachlan et al. 2018). While transcripts encoding eEF1 subunits eEF1A1, eEF1B2, eEF1D eEF1G and VARS are expressed throughout brain development, generally declining after birth, eEF1A2 expression increases throughout embryonic development, peaking after birth in a pattern consistent with that seen for synapse development and maturation. This pattern is consistent with data from mice and rats (McLachlan et al. 2018). Mice that carry biallelic mutations in the gene encoding eEF1A2, either p.(G70S) missense or null, die before 4 weeks of age of seizures, sudden death, and/or neurodegeneration (Davies et al. 2017). These results suggest further studies of SGT1 interactions with other eEF1 subunits, especially eEF1A2 mutants, and the involvement of SGT1 in neurodevelopment through these subunits. Because there is significant evidence to support the hypothesis that mutations to key components of the translational machinery affect neurons more severely than other cell types (McLachlan et al. 2018), and perturbations in protein synthesis can lead to neurodegeneration (Lee et al. 2006), future studies may focus the role of SGT1 in neuronal protein translation and synthesis via effects of mutations in elongation factors.

16.1.11 Mammalian SGT1 in Heart Diseases

Expression of SGT1 gene and protein is specifically up-regulated in heart failure (Sharma et al. 2004; Kapustian et al. 2013). Moreover, the possible involvement of molecular chaperones has been reported in cardioprotection (Latchman 2001). The SGT1-interactor, Hsp70, and the small chaperone Hsp27 might exert their cardioprotective role in heat shock, ischemia, and experimental infarction (Kapustian et al. 2013). Overexpression of Hsp70 in animals improved performance of the cardiovascular system after ischemia/reperfusion (Radford et al. 1996; Marber et al. 1995). These cardioprotective properties may derive from the potential antiapoptotic action of Hsp proteins, as shown, for example, for the Hsp60 protein (Gupta and Knowlton 2005). All these data suggest that Hsp proteins are implicated in many molecular processes involved in the etiology of cardiac diseases, but their role and mechanism still remain to be studied. Some heat shock proteins, such as Hsp70 and Hsp60, are involved in autoimmune diseases, including cardiospecific ones (Kapustian et al. 2013). Expression and cellular localization of HSP70 and HSP60 is altered in dilated cardiomyopathy (DCM)-affected hearts (Kapustian et al. 2013), and autoantibodies against HSP proteins, including HSP60 and HSP70, were present in sera of patients with DCM (Portig et al. 1997). Kapustian et al. found that the levels of autoantibodies against HSP90 and SGT1 proteins were significantly higher in patients with DCM and ischemic heart disease than in sera of healthy donors. They found changes in subcellular localization of HSP90 in the left ventricle of DCM hearts while the total level of this protein remained unchanged. SGT1 protein levels were increased in DCM, with no changes in subcellular localization. Taken together, their data suggested that HSP90 and SGT1 might be involved in the progression of heart failure and thus might serve as markers for cardiomyopathies of different origin.

While Rar1 is clearly involved in plant innate immunity, the involvement of mammalian CHORD proteins in NLR function is controversial, with conflicting observations (da Silva Correia et al.2007; Hahn 2005) (see also previous Sect. 16.1.3: **Plant SGT1 in Plant Innate Immune System**). However, melusin (animal RAR1 homolog), which is specifically expressed in the skeletal and cardiac muscles of vertebrates (Brancaccio et al. 1999), is directly implicated in the response to cardiac mechanical stress during pressure overload (Brancaccio et al. 2003a; De Acetis et al. 2005). Brancaccio et al. and De Acetis et al. demonstrated that melusin is required to generate cardiomyocyte hypertrophy in response to stress stimuli, such as mechanical overload using loss and gain of function genetically modified mouse models (Brancaccio et al. 2003b; De Acetis et al. 2005). Lack of melusin leads to reduced left ventricle hypertrophy and promotes the evolution toward heart dilation in response to pressure overload (Brancaccio et al. 2003b). On the other hand, forced melusin expression in heart leads to the development of sustained concentric hypertrophy and prevents the evolution toward heart failure (De Acetis et al. 2005).

Melusin controls the phosphorylation state of the kinases ERKs, AKT and GSK3b; these pathways are required to activate the hypertrophy response in cardiomyocytes (De Acetis et al. 2005; Brancaccio et al. 2003a). However, functional partners that mediate melusin signal transduction are still unknown, since melusin does not possess a kinase or phosphatase domain. Sbroggio et al. demonstrated that melusin interacts with Hsp90 and Sgt1 in mouse and monkey cell, respectively, and behaves like a chaperone using gene expression profiling and co-immunoprecipitation analysis (Sbroggio et al. 2008). In this study, after C2C12 differentiated myotubes with radicicol (an inhibitor of the Hsp90 ATPase activity) for 24 hours, melusin protein levels did not change, but they were enhanced at the higher radicicol dose. Therefore, melusin is not a client protein, but possibly a chaperone or a co-chaperone. While the expression of Hsp90 and Hsp70 is regulated by different stress stimuli, melusin expression is regulated only by mechanical stretch but not by heat shock, consistent with the data showing that melusin is required in triggering cardiomyocyte hypertrophy in response to mechanical load, but not to neurohumoral stimuli (Brancaccio et al. 2003a). In summary, the study of Sbroggio et al. confirmed that the interaction profile and the molecular machinery of the CHORD-containing protein are evolutionary conserved, suggesting that melusin's ability to affect signal transduction in heart hypertrophy can be exerted by its own chaperone activity on crucial signaling components, or by driving Hsp90 on particular transduction pathways. Further study of the involvement and the contribution of SGT1 in these pathways is required, and how aberrant mitotic activity of SGT1-HSP90 is engaged in heart failure (e.g., DCM and/or cardiomyocyte hypertrophy).

16.2 Conclusions

SGT1 functions in multiple distinct biological processes as a co-chaperone of HSP90, and the numbers of identified SGT1 interactors are growing every year. Structural analysis of conformational regulations of its functional domains through the specific interactors would enhance our understanding of SGT1. Subcellular localization and expression levels of SGT1 protein are dynamically but tightly controlled by many factors: environmental stimuli, protein-protein interactions, and dimerization and PTM(s) of SGT1 itself. Elucidation of the chaperone machinery of SGT1-HSP90 complex (including but not limited to ubiquitin-proteasome machinery) in specific cells, tissues, and organs and in certain pathologic environments is an interesting research direction for the future. In plants, the SGT1-HSP90 complex is regulated by RAR1, a CHORD-containing protein, resulting in the stabilization of NLR proteins. Still unclear are the precise biochemical functions of plant RAR1 through the SGT1-HSP90 complex for stabilization of NLR proteins and downstream signaling components that require *R* gene function. In mammals, the SGT1-HSP90 complex targets NLR-type sensors as the substrates. Future research will elucidate the regulation and mechanisms of the activators and the modulators of NLR- inflammasome activity through the SGT1-HSP90 complex. Because SGT1 is required for cell cycle progression and functions at both the G1/S and G2/M and SGT1 protein is widely expressed in different tissues, future work will investigate how SGT1 changes its functions in cell cycle in different tissues and organs. In tissues such as the brain, heart, and muscle where mitosis is infrequent and spontaneous regeneration does not occur, errors of G1(G0)/S functions (e.g., SCF function and cyclic AMP signaling) would be a serious problem. On the other hand, in tissues such as the blood, skin, bone, and gut, where mitosis is frequent and cells have high turnover rates of regeneration, error of G2/M functions (e.g., CENP-A deposition, kinetochore assembly, centrosome organization, and cytokinesis) would be a serious problem. Thus, cell cycle-specific SGT1 function involved in different types of cancer in different organs is a likely focus for future research, as is the involvement of SGT1 function in development and regeneration of brain neurons and the heart. The SGT1- HSP90 complex is a potential target for cancer, brain and heart disease therapy. Combinational therapies directed at these targets including other distinctive interactor(s) could be more effective and specific than current approaches.

Acknowledgements We thank past and current researchers at Greehey Children's Cancer Research Institute at UT Health Science Center San Antonio School of Medicine, The Research Institute at Nationwide Children's Hospital, and St. Jude Children's Research Hospital for their helpful discussions. This work was supported by NCI R21CA205659 and institutional support from Greehey Children's Cancer Research Institute and MOE Key Laboratory of Model Animal for Disease Study, Model Animal Research Center, Nanjing University.

References

Ali MM, Roe SM, Vaughan CK, Meyer P, Panaretou B, Piper PW, Prodromou C, Pearl LH (2006) Crystal structure of an Hsp90-nucleotide-p23/Sba1 closed chaperone complex. Nature 440(7087):1013–1017. https://doi.org/10.1038/nature04716

Andersen RO, Turnbull DW, Johnson EA, Doe CQ (2012) Sgt1 acts via an LKB1/AMPK pathway to establish cortical polarity in larval neuroblasts. Dev Biol 363(1):258–265. https://doi.org/10.1016/j.ydbio.2011.12.047

Asea A, Kraeft SK, Kurt-Jones EA, Stevenson MA, Chen LB, Finberg RW, Koo GC, Calderwood SK (2000) HSP70 stimulates cytokine production through a CD14-dependant pathway, demonstrating its dual role as a chaperone and cytokine. Nat Med 6(4):435–442. https://doi.org/10.1038/74697

Austin MJ, Muskett P, Kahn K, Feys BJ, Jones JD, Parker JE (2002) Regulatory role of SGT1 in early R gene-mediated plant defenses. Science 295(5562):2077–2080. https://doi.org/10.1126/science.1067747

Azevedo C, Sadanandom A, Kitagawa K, Freialdenhoven A, Shirasu K, Schulze-Lefert P (2002) The RAR1 interactor SGT1, an essential component of R gene-triggered disease resistance. Science 295(5562):2073–2076. https://doi.org/10.1126/science.1067554

Azevedo C, Betsuyaku S, Peart J, Takahashi A, Noel L, Sadanandom A, Casais C, Parker J, Shirasu K (2006) Role of SGT1 in resistance protein accumulation in plant immunity. EMBO J 25(9):2007–2016. https://doi.org/10.1038/sj.emboj.7601084

Bansal PK, Abdulle R, Kitagawa K (2004) Sgt1 associates with Hsp90: an initial step of assembly of the core kinetochore complex. Mol Cell Biol 24(18):8069–8079. https://doi.org/10.1128/MCB.24.18.8069-8079.2004

Bansal PK, Nourse A, Abdulle R, Kitagawa K (2008) SGT1 dimerization is required for yeast kinetochore assembly. J Biol Chem. doi:M806281200 [pii]. https://doi.org/10.1074/jbc.M806281200

Bansal PK, Mishra A, High AA, Abdulle R, Kitagawa K (2009a) Sgt1 dimerization is negatively regulated by protein kinase CK2-mediated phosphorylation at Ser361. J Biol Chem 284(28):18692–18698. https://doi.org/10.1074/jbc.M109.012732

Bansal PK, Nourse A, Abdulle R, Kitagawa K (2009b) Sgt1 dimerization is required for yeast kinetochore assembly. J Biol Chem 284(6):3586–3592. https://doi.org/10.1074/jbc.M806281200

Bayle V, Nussaume L, Bhat RA (2008) Combination of novel green fluorescent protein mutant TSapphire and DsRed variant mOrange to set up a versatile in planta FRET-FLIM assay. Plant Physiol 148(1):51–60. Epub 2008/07/16. https://doi.org/10.1104/pp.108.117358. PubMed PMID: 18621983; PMCID: PMC2528103

Bhaskar PB, Raasch JA, Kramer LC, Neumann P, Wielgus SM, Austin-Phillips S, Jiang J (2008) Sgt1, but not Rar1, is essential for the RB-mediated broad-spectrum resistance to potato late blight. BMC Plant Biol 8:8. https://doi.org/10.1186/1471-2229-8-8

Bhattarai KK, Li Q, Liu Y, Dinesh-Kumar SP, Kaloshian I (2007) The MI-1-mediated pest resistance requires Hsp90 and Sgt1. Plant Physiol 144(1):312–323. https://doi.org/10.1104/pp.107.097246

Bhavsar AP, Brown NF, Stoepel J, Wiermer M, Martin DD, Hsu KJ, Imami K, Ross CJ, Hayden MR, Foster LJ, Li X, Hieter P, Finlay BB (2013) The Salmonella type III effector SspH2 specifically exploits the NLR co-chaperone activity of SGT1 to subvert immunity. PLoS Pathog 9(7):e1003518. https://doi.org/10.1371/journal.ppat.1003518

Bieri S, Mauch S, Shen QH, Peart J, Devoto A, Casais C, Ceron F, Schulze S, Steinbiss HH, Shirasu K (2004) Schulze-Lefert P. RAR1 positively controls steady state levels of barley MLA resistance proteins and enables sufficient MLA6 accumulation for effective resistance. Plant cell 16(12):3480–3495. Epub 2004/11/19. https://doi.org/10.1105/tpc.104.026682. PubMed PMID: 15548741; PMCID: PMC535887

Biggins S, Walczak CE (2003) Captivating capture: how microtubules attach to kinetochores. Curr Biol 13(11):R449–R460

Blatch GL, Lassle M (1999) The tetratricopeptide repeat: a structural motif mediating protein-protein interactions. BioEssays 21(11):932–939. https://doi.org/10.1002/(SICI)1521-1878(199911)21:11<932::AID-BIES5>3.0.CO;2-N

Boter M, Amigues B, Peart J, Breuer C, Kadota Y, Casais C, Moore G, Kleanthous C, Ochsenbein F, Shirasu K, Guerois R (2007) Structural and functional analysis of SGT1 reveals that its interaction with HSP90 is required for the accumulation of Rx, an R protein involved in plant immunity. Plant Cell 19(11):3791–3804. https://doi.org/10.1105/tpc.107.050427

Brancaccio M, Guazzone S, Menini N, Sibona E, Hirsch E, De Andrea M, Rocchi M, Altruda F, Tarone G, Silengo L (1999) Melusin is a new muscle-specific interactor for beta(1) integrin cytoplasmic domain. J Biol Chem 274(41):29282–29288. Epub 1999/10/03. https://doi.org/10.1074/jbc.274.41.29282

Brancaccio M, Fratta L, Notte A, Hirsch E, Poulet R, Guazzone S, De Acetis M, Vecchione C, Marino G, Altruda F, Silengo L, Tarone G, Lembo G (2003a) Melusin, a muscle-specific integrin beta1-interacting protein, is required to prevent cardiac failure in response to chronic pressure overload. Nat Med 9(1):68–75. https://doi.org/10.1038/nm805

Brancaccio M, Menini N, Bongioanni D, Ferretti R, De Acetis M, Silengo L, Tarone G (2003b) Chp-1 and melusin, two CHORD containing proteins in vertebrates. FEBS Lett 551(1–3):47–52
Breen EC, Tang K (2003) Calcyclin (S100A6) regulates pulmonary fibroblast proliferation, morphology, and cytoskeletal organization in vitro. J Cell Biochem 88(4):848–854. https://doi.org/10.1002/jcb.10398
Brown JL, Jaquenoud M, Gulli MP, Chant J, Peter M (1997) Novel Cdc42-binding proteins Gic1 and Gic2 control cell polarity in yeast. Genes Dev 11(22):2972–2982
Cam HP, Noma K, Ebina H, Levin HL, Grewal SI (2008) Host genome surveillance for retrotransposons by transposon-derived proteins. Nature 451(7177):431–436. doi:nature06499 [pii]. https://doi.org/10.1038/nature06499
Catlett MG, Kaplan KB (2006) Sgt1p is a unique co-chaperone that acts as a client adaptor to link Hsp90 to Skp1p. J Biol Chem 281(44):33739–33748. https://doi.org/10.1074/jbc.M603847200
Chae JJ, Komarow HD, Cheng J, Wood G, Raben N, Liu PP, Kastner DL (2003) Targeted disruption of pyrin, the FMF protein, causes heightened sensitivity to endotoxin and a defect in macrophage apoptosis. Mol Cell 11(3):591–604
Chisholm ST, Coaker G, Day B, Staskawicz BJ (2006) Host-microbe interactions: shaping the evolution of the plant immune response. Cell 124(4):803–814. https://doi.org/10.1016/j.cell.2006.02.008
Connelly C, Hieter P (1996) Budding yeast SKP1 encodes an evolutionarily conserved kinetochore protein required for cell cycle progression. Cell 86(2):275–285
Cuzick A, Maguire K, Hammond-Kosack KE (2009) Lack of the plant signalling component SGT1b enhances disease resistance to Fusarium culmorum in Arabidopsis buds and flowers. New Phytol 181(4):901–912. https://doi.org/10.1111/j.1469-8137.2008.02712.x
da Silva CJ, Miranda Y, Leonard N, Ulevitch R (2007) SGT1 is essential for Nod1 activation. Proc Natl Acad Sci U S A 104(16):6764–6769. https://doi.org/10.1073/pnas.0610926104
Dangl JL, Jones JD (2001) Plant pathogens and integrated defence responses to infection. Nature 411(6839):826–833. https://doi.org/10.1038/35081161
Das AK, Cohen PW, Barford D (1998) The structure of the tetratricopeptide repeats of protein phosphatase 5: implications for TPR-mediated protein-protein interactions. EMBO J 17(5):1192–1199. https://doi.org/10.1093/emboj/17.5.1192
Davies AE, Kaplan KB (2010) Hsp90-Sgt1 and Skp1 target human Mis12 complexes to ensure efficient formation of kinetochore-microtubule binding sites. J Cell Biol 189(2):261–274. Epub 2010/04/21. https://doi.org/10.1083/jcb.200910036. PubMed PMID: 20404110; PMCID: PMC2856898
Davies FC, Hope JE, McLachlan F, Nunez F, Doig J, Bengani H, Smith C, Abbott CM (2017) Biallelic mutations in the gene encoding eEF1A2 cause seizures and sudden death in F0 mice. Sci Rep 7:46019. Epub 2017/04/06. https://doi.org/10.1038/srep46019. PubMed PMID: 28378778; PMCID: PMC5380952
De Acetis M, Notte A, Accornero F, Selvetella G, Brancaccio M, Vecchione C, Sbroggio M, Collino F, Pacchioni B, Lanfranchi G, Aretini A, Ferretti R, Maffei A, Altruda F, Silengo L, Tarone G, Lembo G (2005) Cardiac overexpression of melusin protects from dilated cardiomyopathy due to long-standing pressure overload. Circ Res 96(10):1087–1094. https://doi.org/10.1161/01.RES.0000168028.36081.e0
de Carcer G, do Carmo Avides M, Lallena MJ, Glover DM, Gonzalez C (2001) Requirement of Hsp90 for centrosomal function reflects its regulation of Polo kinase stability. EMBO J 20(11):2878–2884. https://doi.org/10.1093/emboj/20.11.2878
de la Fuente van Bentem S, Vossen JH, de Vries KJ, van Wees S, Tameling WI, Dekker HL, de Koster CG, Haring MA, Takken FL, Cornelissen BJ (2005) Heat shock protein 90 and its co-chaperone protein phosphatase 5 interact with distinct regions of the tomato I-2 disease resistance protein. Plant J 43(2):284–298. https://doi.org/10.1111/j.1365-313X.2005.02450.x
Dias DC, Dolios G, Wang R, Pan ZQ (2002) CUL7: a DOC domain-containing cullin selectively binds Skp1.Fbx29 to form an SCF-like complex. Proc Natl Acad Sci U S A 99(26):16601–16606. https://doi.org/10.1073/pnas.252646399

Doheny KF, Sorger PK, Hyman AA, Tugendreich S, Spencer F, Hieter P (1993) Identification of essential components of the *S. cerevisiae* kinetochore. Cell 73(4):761–774
Dubacq C, Guerois R, Courbeyrette R, Kitagawa K, Mann C (2002) Sgt1p contributes to cyclic AMP pathway activity and physically interacts with the adenylyl cyclase Cyr1p/Cdc35p in budding yeast. Eukaryot Cell 1(4):568–582
Dujon B (2010) Yeast evolutionary genomics. Nat Rev Genet 11(7):512–524. https://doi.org/10.1038/nrg2811
Eitas TK, Nimchuk ZL, Dangl JL (2008) Arabidopsis TAO1 is a TIR-NB-LRR protein that contributes to disease resistance induced by the Pseudomonas syringae effector AvrB. Proc Natl Acad Sci U S A 105(17):6475–6480. https://doi.org/10.1073/pnas.0802157105
El Oirdi M, Bouarab K (2007) Plant signalling components EDS1 and SGT1 enhance disease caused by the necrotrophic pathogen Botrytis cinerea. New Phytol 175(1):131–139. https://doi.org/10.1111/j.1469-8137.2007.02086.x
Fitzgerald-Hayes M, Clarke L, Carbon J (1982) Nucleotide sequence comparisons and functional analysis of yeast centromere DNAs. Cell 29(1):235–244
Franchi L, Amer A, Body-Malapel M, Kanneganti TD, Ozoren N, Jagirdar R, Inohara N, Vandenabeele P, Bertin J, Coyle A, Grant EP, Nunez G (2006) Cytosolic flagellin requires Ipaf for activation of caspase-1 and interleukin 1beta in salmonella-infected macrophages. Nat Immunol 7(6):576–582. https://doi.org/10.1038/ni1346
Fu DQ, Ghabrial S, Kachroo A (2009) GmRAR1 and GmSGT1 are required for basal, R gene-mediated and systemic acquired resistance in soybean. Mol Plant-Microbe Interact 22(1):86–95. https://doi.org/10.1094/MPMI-22-1-0086
Fukagawa T, Earnshaw WC (2014) The centromere: chromatin foundation for the kinetochore machinery. Dev Cell 30(5):496–508. https://doi.org/10.1016/j.devcel.2014.08.016
Gangula NR, Maddika S (2017) Interplay between the phosphatase PHLPP1 and E3 ligase RNF41 stimulates proper kinetochore assembly via the outer-kinetochore protein SGT1. J Biol Chem 292(34):13947–13958. Epub 2017/07/12. https://doi.org/10.1074/jbc.M117.782896. PubMed PMID: 28696259; PMCID: PMC5572923
Gao G, Kun T, Sheng Y, Qian M, Kong F, Liu X, Yu Z, Zhang H, Zhang Q, Gu J, Zhang X (2013) SGT1 regulates Akt signaling by promoting beta-TrCP-dependent PHLPP1 degradation in gastric cancer cells. Mol Biol Rep 40(4):2947–2953. https://doi.org/10.1007/s11033-012-2363-8
Garcia-Ranea JA, Mirey G, Camonis J, Valencia A (2002) p23 and HSP20/alpha-crystallin proteins define a conserved sequence domain present in other eukaryotic protein families. FEBS Lett 529(2–3):162–167. Epub 2002/10/10. https://doi.org/10.1016/s0014-5793(02)03321-5
Gillis AN, Thomas S, Hansen SD, Kaplan KB (2005) A novel role for the CBF3 kinetochore-scaffold complex in regulating septin dynamics and cytokinesis. J Cell Biol 171(5):773–784. https://doi.org/10.1083/jcb.200507017
Goh PY, Kilmartin JV (1993) NDC10: a gene involved in chromosome segregation in *Saccharomyces cerevisiae*. J Cell Biol 121(3):503–512
Gray WM, Muskett PR, Chuang HW, Parker JE (2003) Arabidopsis SGT1b is required for SCF(TIR1)-mediated auxin response. Plant Cell 15(6):1310–1319
Gupta S, Knowlton AA (2005) HSP60, Bax, apoptosis and the heart. J Cell Mol Med 9(1):51–58
Hahn JS (2005) Regulation of Nod1 by Hsp90 chaperone complex. FEBS Lett 579(20):4513–4519. https://doi.org/10.1016/j.febslet.2005.07.024
Halterman DA, Wise RP (2004) A single-amino acid substitution in the sixth leucine-rich repeat of barley MLA6 and MLA13 alleviates dependence on RAR1 for disease resistance signaling. Plant J 38(2):215–226. https://doi.org/10.1111/j.1365-313X.2004.02032.x
Hann DR, Rathjen JP (2007) Early events in the pathogenicity of Pseudomonas syringae on Nicotiana benthamiana. Plant J 49(4):607–618. https://doi.org/10.1111/j.1365-313X.2006.02981.x
Hawkins PN, Lachmann HJ, Aganna E, McDermott MF (2004) Spectrum of clinical features in Muckle-Wells syndrome and response to anakinra. Arthritis Rheum 50(2):607–612. https://doi.org/10.1002/art.20033

He P, Shan L, Lin NC, Martin GB, Kemmerling B, Nurnberger T, Sheen J (2006) Specific bacterial suppressors of MAMP signaling upstream of MAPKKK in Arabidopsis innate immunity. Cell 125(3):563–575. https://doi.org/10.1016/j.cell.2006.02.047

He WT, Zheng XM, Zhang YH, Gao YG, Song AX, van der Goot FG, Hu HY (2016) Cytoplasmic ubiquitin-specific protease 19 (USP19) modulates aggregation of polyglutamine-expanded Ataxin-3 and huntingtin through the HSP90 chaperone. PLoS One 11(1):e0147515. https://doi.org/10.1371/journal.pone.0147515

Hegemann JH, Shero JH, Cottarel G, Philippsen P, Hieter P (1988) Mutational analysis of centromere DNA from chromosome VI of *Saccharomyces cerevisiae*. Mol Cell Biol 8(6):2523–2535

Hieter P, Pridmore D, Hegemann JH, Thomas M, Davis RW, Philippsen P (1985) Functional selection and analysis of yeast centromeric DNA. Cell 42(3):913–921

Hoffman CS, Winston F (1989) A transcriptionally regulated expression vector for the fission yeast Schizosaccharomyces pombe. Gene 84(2):473–479

Hoffman HM, Mueller JL, Broide DH, Wanderer AA, Kolodner RD (2001) Mutation of a new gene encoding a putative pyrin-like protein causes familial cold autoinflammatory syndrome and Muckle-Wells syndrome. Nat Genet 29(3):301–305. https://doi.org/10.1038/ng756

Holt BF 3rd, Belkhadir Y, Dangl JL (2005) Antagonistic control of disease resistance protein stability in the plant immune system. Science 309(5736):929–932. https://doi.org/10.1126/science.1109977

Hong TJ, Hahn JS (2016) Application of SGT1-Hsp90 chaperone complex for soluble expression of NOD1 LRR domain in *E. coli*. Biochem Biophys Res Commun 478(4):1647–1652. https://doi.org/10.1016/j.bbrc.2016.08.174

Hong TJ, Kim S, Wi AR, Lee P, Kang M, Jeong JH, Hahn JS (2013) Dynamic nucleotide-dependent interactions of cysteine- and histidine-rich domain (CHORD)-containing Hsp90 cochaperones Chp-1 and melusin with cochaperones PP5 and Sgt1. J Biol Chem 288(1):215–222. https://doi.org/10.1074/jbc.M112.398636

Hoser R, Zurczak M, Lichocka M, Zuzga S, Dadlez M, Samuel MA, Ellis BE, Stuttmann J, Parker JE, Hennig J, Krzymowska M (2013) Nucleocytoplasmic partitioning of tobacco N receptor is modulated by SGT1. New Phytol 200(1):158–171. https://doi.org/10.1111/nph.12347

Hoser R, Lichocka M, Zurczak M, Hennig J, Krzymowska M (2014) Emerging role of SGT1 as a regulator of NB-LRR-receptor nucleocytoplasmic partitioning. Plant Signal Behav 9(4)

Hubert DA, Tornero P, Belkhadir Y, Krishna P, Takahashi A, Shirasu K, Dangl JL (2003) Cytosolic HSP90 associates with and modulates the Arabidopsis RPM1 disease resistance protein. EMBO J 22(21):5679–5689. https://doi.org/10.1093/emboj/cdg547

Inohara, Chamaillard, McDonald C, Nunez G (2005) NOD-LRR proteins: role in host-microbial interactions and inflammatory disease. Annu Rev Biochem 74:355–383. https://doi.org/10.1146/annurev.biochem.74.082803.133347

Ishiga Y, Uppalapati SR, Ishiga T, Mysore KS (2011) Involvement of SGT1 in COR-mediated signal transduction pathway leading to disease symptom development. Plant Signal Behav 6(7):1072–1073

Ito M, Ohnishi K, Hikichi Y, Kiba A (2015) Molecular chaperons and co-chaperons, Hsp90, RAR1, and SGT1 negatively regulate bacterial wilt disease caused by Ralstonia solanacearum in Nicotiana benthamiana. Plant Signal Behav 10(6):e970410. https://doi.org/10.4161/15592316.2014.970410

Iwatsuki M, Mimori K, Sato T, Toh H, Yokobori T, Tanaka F, Ishikawa K, Baba H, Mori M (2010) Overexpression of SUGT1 in human colorectal cancer and its clinicopathological significance. Int J Oncol 36(3):569–575

Jansen LE, Black BE, Foltz DR, Cleveland DW (2007) Propagation of centromeric chromatin requires exit from mitosis. J Cell Biol 176(6):795–805. https://doi.org/10.1083/jcb.200701066

Jaquenoud M, Gulli MP, Peter K, Peter M (1998) The Cdc42p effector Gic2p is targeted for ubiquitin-dependent degradation by the SCFGrr1 complex. EMBO J 17(18):5360–5373. https://doi.org/10.1093/emboj/17.18.5360

Jiang W, Lechner J, Carbon J (1993) Isolation and characterization of a gene (CBF2) specifying a protein component of the budding yeast kinetochore. J Cell Biol 121(3):513–519

Johnson JL, Zuehlke AD, Tenge VR, Langworthy JC (2014) Mutation of essential Hsp90 co-chaperones SGT1 or CNS1 renders yeast hypersensitive to overexpression of other co-chaperones. Curr Genet 60(4):265–276. https://doi.org/10.1007/s00294-014-0432-3

Kadota Y, Amigues B, Ducassou L, Madaoui H, Ochsenbein F, Guerois R, Shirasu K (2008) Structural and functional analysis of SGT1-HSP90 core complex required for innate immunity in plants. EMBO Rep 9(12):1209–1215. https://doi.org/10.1038/embor.2008.185

Kadota Y, Shirasu K, Guerois R (2010) NLR sensors meet at the SGT1-HSP90 crossroad. Trends Biochem Sci 35(4):199–207. https://doi.org/10.1016/j.tibs.2009.12.005

Kainou T, Shinzato T, Sasaki K, Mitsui Y, Giga-Hama Y, Kumagai H, Uemura H (2006) Spsgt1, a new essential gene of Schizosaccharomyces pombe, is involved in carbohydrate metabolism. Yeast 23(1):35–53. https://doi.org/10.1002/yea.1336

Kanneganti TD, Ozoren N, Body-Malapel M, Amer A, Park JH, Franchi L, Whitfield J, Barchet W, Colonna M, Vandenabeele P, Bertin J, Coyle A, Grant EP, Akira S, Nunez G (2006) Bacterial RNA and small antiviral compounds activate caspase-1 through cryopyrin/Nalp3. Nature 440(7081):233–236. https://doi.org/10.1038/nature04517

Kaplan KB, Hyman AA, Sorger PK (1997) Regulating the yeast kinetochore by ubiquitin-dependent degradation and Skp1p-mediated phosphorylation. Cell 91(4):491–500

Kapustian LL, Vigontina OA, Rozhko OT, Ryabenko DV, Michowski W, Lesniak W, Filipek A, Kroupskaya IV, Sidorik LL (2013) Hsp90 and its co-chaperone, Sgt1, as autoantigens in dilated cardiomyopathy. Heart Vessel 28(1):114–119. https://doi.org/10.1007/s00380-011-0226-1

Kawamura Y, Takenaka S, Hase S, Kubota M, Ichinose Y, Kanayama Y, Nakaho K, Klessig DF, Takahashi H (2009) Enhanced defense responses in Arabidopsis induced by the cell wall protein fractions from Pythium oligandrum require SGT1, RAR1, NPR1 and JAR1. Plant Cell Physiol 50(5):924–934. https://doi.org/10.1093/pcp/pcp044

Khalyfa A, Bourbeau D, Chen E, Petroulakis E, Pan J, Xu S, Wang E (2001) Characterization of elongation factor-1A (eEF1A-1) and eEF1A-2/S1 protein expression in normal and wasted mice. J Biol Chem 276(25):22915–22922. Epub 2001/04/11. https://doi.org/10.1074/jbc.M101011200

Kim NH, Kim DS, Chung EH, Hwang BK (2014) Pepper suppressor of the G2 allele of skp1 interacts with the receptor-like cytoplasmic kinase1 and type III effector AvrBsT and promotes the hypersensitive cell death response in a phosphorylation-dependent manner. Plant Physiol 165(1):76–91. https://doi.org/10.1104/pp.114.238840

Kitagawa K, Hieter P (2001) Evolutionary conservation between budding yeast and human kinetochores. Nat Rev Mol Cell Biol 2(9):678–687. https://doi.org/10.1038/35089568

Kitagawa K, Skowyra D, Elledge SJ, Harper JW, Hieter P (1999) SGT1 encodes an essential component of the yeast kinetochore assembly pathway and a novel subunit of the SCF ubiquitin ligase complex. Mol Cell 4(1):21–33. doi:S1097-2765(00)80184-7 [pii]

Kleiger G, Mayor T (2014) Perilous journey: a tour of the ubiquitin-proteasome system. Trends Cell Biol 24(6):352–359. https://doi.org/10.1016/j.tcb.2013.12.003

Kobayashi N, Suzuki Y, Schoenfeld LW, Muller CA, Nieduszynski C, Wolfe KH, Tanaka TU (2015) Discovery of an unconventional centromere in budding yeast redefines evolution of point centromeres. Curr Biol 25(15):2026–2033. https://doi.org/10.1016/j.cub.2015.06.023

Kordes E, Savelyeva L, Schwab M, Rommelaere J, Jauniaux JC, Cziepluch C (1998) Isolation and characterization of human SGT and identification of homologues in Saccharomyces cerevisiae and *Caenorhabditis elegans*. Genomics 52(1):90–94. https://doi.org/10.1006/geno.1998.5385

Kud J, Zhao Z, Du X, Liu Y, Zhao Y, Xiao F (2013) SGT1 interacts with the Prf resistance protein and is required for Prf accumulation and Prf-mediated defense signaling. Biochem Biophys Res Commun 431(3):501–505. https://doi.org/10.1016/j.bbrc.2013.01.028

Kumar D, Kirti PB (2015) Pathogen-induced SGT1 of Arachis diogoi induces cell death and enhanced disease resistance in tobacco and peanut. Plant Biotechnol J 13(1):73–84. https://doi.org/10.1111/pbi.12237

Lange BM, Bachi A, Wilm M, Gonzalez C (2000) Hsp90 is a core centrosomal component and is required at different stages of the centrosome cycle in Drosophila and vertebrates. EMBO J 19(6):1252–1262. https://doi.org/10.1093/emboj/19.6.1252
Lapouge K, Smith SJ, Walker PA, Gamblin SJ, Smerdon SJ, Rittinger K (2000) Structure of the TPR domain of p67phox in complex with Rac.GTP. Mol Cell 6(4):899–907
Latchman DS (2001) Heat shock proteins and cardiac protection. Cardiovasc Res 51(4):637–646
Lechner J (1994) A zinc finger protein, essential for chromosome segregation, constitutes a putative DNA binding subunit of the *Saccharomyces cerevisiae* kinetochore complex, Cbf3. EMBO J 13(21):5203–5211
Lechner J, Carbon J (1991) A 240 kd multisubunit protein complex, CBF3, is a major component of the budding yeast centromere. Cell 64(4):717–725
Lee YT, Jacob J, Michowski W, Nowotny M, Kuznicki J, Chazin WJ (2004) Human Sgt1 binds HSP90 through the CHORD-Sgt1 domain and not the tetratricopeptide repeat domain. J Biol Chem 279(16):16511–16517. https://doi.org/10.1074/jbc.M400215200
Lee JW, Beebe K, Nangle LA, Jang J, Longo-Guess CM, Cook SA, Davisson MT, Sundberg JP, Schimmel P, Ackerman SL (2006) Editing-defective tRNA synthetase causes protein misfolding and neurodegeneration. Nature 443(7107):50–55. Epub 2006/08/15. https://doi.org/10.1038/nature05096
Leister RT, Dahlbeck D, Day B, Li Y, Chesnokova O, Staskawicz BJ (2005) Molecular genetic evidence for the role of SGT1 in the intramolecular complementation of Bs2 protein activity in Nicotiana benthamiana. Plant Cell 17(4):1268–1278. https://doi.org/10.1105/tpc.104.029637
Lingelbach LB, Kaplan KB (2004) The interaction between Sgt1p and Skp1p is regulated by HSP90 chaperones and is required for proper CBF3 assembly. Mol Cell Biol 24(20):8938–8950. https://doi.org/10.1128/MCB.24.20.8938-8950.2004
Liu Y, Schiff M, Serino G, Deng XW, Dinesh-Kumar SP (2002) Role of SCF ubiquitin-ligase and the COP9 signalosome in the N gene-mediated resistance response to tobacco mosaic virus. Plant Cell 14(7):1483–1496
Liu Y, Burch-Smith T, Schiff M, Feng S, Dinesh-Kumar SP (2004) Molecular chaperone Hsp90 associates with resistance protein N and its signaling proteins SGT1 and Rar1 to modulate an innate immune response in plants. J Biol Chem 279(3):2101–2108. https://doi.org/10.1074/jbc.M310029200
Liu ST, Rattner JB, Jablonski SA, Yen TJ (2006) Mapping the assembly pathways that specify formation of the trilaminar kinetochore plates in human cells. J Cell Biol 175(1):41–53. doi:jcb.200606020 [pii]. https://doi.org/10.1083/jcb.200606020
Liu XS, Song B, Tang J, Liu W, Kuang S, Liu X (2012) Plk1 phosphorylates Sgt1 at the kinetochores to promote timely kinetochore-microtubule attachment. Mol Cell Biol 32(19):4053–4067. https://doi.org/10.1128/MCB.00516-12
Liu ZQ, Liu YY, Shi LP, Yang S, Shen L, Yu HX, Wang RZ, Wen JY, Tang Q, Hussain A, Khan MI, Hu J, Liu CL, Zhang YW, Cheng W, He SL (2016) SGT1 is required in PcINF1/SRC2-1 induced pepper defense response by interacting with SRC2-1. Sci Rep 6:21651. https://doi.org/10.1038/srep21651
Lu R, Malcuit I, Moffett P, Ruiz MT, Peart J, Wu AJ, Rathjen JP, Bendahmane A, Day L, Baulcombe DC (2003) High throughput virus-induced gene silencing implicates heat shock protein 90 in plant disease resistance. EMBO J 22(21):5690–5699. https://doi.org/10.1093/emboj/cdg546
Lyapina S, Cope G, Shevchenko A, Serino G, Tsuge T, Zhou C, Wolf DA, Wei N, Shevchenko A, Deshaies RJ (2001) Promotion of NEDD-CUL1 conjugate cleavage by COP9 signalosome. Science 292(5520):1382–1385. https://doi.org/10.1126/science.1059780
Marber MS, Mestril R, Chi SH, Sayen MR, Yellon DM, Dillmann WH (1995) Overexpression of the rat inducible 70-kD heat stress protein in a transgenic mouse increases the resistance of the heart to ischemic injury. J Clin Invest 95(4):1446–1456. https://doi.org/10.1172/JCI117815
Marcotte EM, Pellegrini M, Thompson MJ, Yeates TO, Eisenberg D (1999) A combined algorithm for genome-wide prediction of protein function. Nature 402(6757):83–86. https://doi.org/10.1038/47048

Martinez-Noel G, Luck K, Kuhnle S, Desbuleux A, Szajner P, Galligan JT, Rodriguez D, Zheng L, Boyland K, Leclere F, Zhong Q, Hill DE, Vidal M, Howley PM (2018) Network analysis of UBE3A/E6AP-associated proteins provides connections to several distinct cellular processes. J Mol Biol 430(7):1024–1050. http://dx.doi.org/10.1016/j.jmb.2018.01.021. PubMed PMID: 29426014; PMCID: PMC5866790

Martinon F, Tschopp J (2007) Inflammatory caspases and inflammasomes: master switches of inflammation. Cell Death Differ 14(1):10–22. https://doi.org/10.1038/sj.cdd.4402038

Martinon F, Burns K, Tschopp J (2002) The inflammasome: a molecular platform triggering activation of inflammatory caspases and processing of proIL-beta. Mol Cell 10(2):417–426

Martinon F, Agostini L, Meylan E, Tschopp J (2004) Identification of bacterial muramyl dipeptide as activator of the NALP3/cryopyrin inflammasome. Curr Biol 14(21):1929–1934. https://doi.org/10.1016/j.cub.2004.10.027

Martins T, Maia AF, Steffensen S, Sunkel CE (2009) Sgt1, a co-chaperone of Hsp90 stabilizes Polo and is required for centrosome organization. EMBO J 28(3):234–247. https://doi.org/10.1038/emboj.2008.283

Mayor A, Martinon F, De Smedt T, Petrilli V, Tschopp J (2007) A crucial function of SGT1 and HSP90 in inflammasome activity links mammalian and plant innate immune responses. Nat Immunol 8(5):497–503. https://doi.org/10.1038/ni1459

McLachlan F, Sires AM, Abbott CM (2018) The role of translation elongation factor eEF1 subunits in neurodevelopmental disorders. Hum Mutat. https://doi.org/10.1002/humu.23677

Mei Y, Hahn AA, Hu S, Yang X (2011) The USP19 deubiquitinase regulates the stability of c-IAP1 and c-IAP2. J Biol Chem 286(41):35380–35387. https://doi.org/10.1074/jbc.M111.282020

Meldau S, Baldwin IT, Wu J (2011) SGT1 regulates wounding- and herbivory-induced jasmonic acid accumulation and Nicotiana attenuata's resistance to the specialist lepidopteran herbivore Manduca sexta. New Phytol 189(4):1143–1156. https://doi.org/10.1111/j.1469-8137.2010.03558.x

Merlet J, Burger J, Gomes JE, Pintard L (2009) Regulation of cullin-RING E3 ubiquitin-ligases by neddylation and dimerization. Cell Mol Life Sci 66(11–12):1924–1938. https://doi.org/10.1007/s00018-009-8712-7

Mestre P, Baulcombe DC (2006) Elicitor-mediated oligomerization of the tobacco N disease resistance protein. Plant Cell 18(2):491–501. https://doi.org/10.1105/tpc.105.037234

Meylan E, Tschopp J, Karin M (2006) Intracellular pattern recognition receptors in the host response. Nature 442(7098):39–44. https://doi.org/10.1038/nature04946

Miao EA, Alpuche-Aranda CM, Dors M, Clark AE, Bader MW, Miller SI, Aderem A (2006) Cytoplasmic flagellin activates caspase-1 and secretion of interleukin 1beta via Ipaf. Nat Immunol 7(6):569–575. https://doi.org/10.1038/ni1344

Milani V, Noessner E, Ghose S, Kuppner M, Ahrens B, Scharner A, Gastpar R, Issels RD (2002) Heat shock protein 70: role in antigen presentation and immune stimulation. Int J Hyperth 18(6):563–575. https://doi.org/10.1080/02656730210166140

Mimnaugh EG, Chavany C, Neckers L (1996) Polyubiquitination and proteasomal degradation of the p185c-erbB-2 receptor protein-tyrosine kinase induced by geldanamycin. J Biol Chem 271(37):22796–22801

Moehs CP, Allen PV, Friedman M, Belknap WR (1997) Cloning and expression of solanidine UDP-glucose glucosyltransferase from potato. Plant J 11(2):227–236

Multhoff G (2002) Activation of natural killer cells by heat shock protein 70. Int J Hyperth 18(6):576–585. https://doi.org/10.1080/0265673021000017109

Muskett P, Parker J (2003) Role of SGT1 in the regulation of plant R gene signalling. Microbes Infect 5(11):969–976

Muskett PR, Kahn K, Austin MJ, Moisan LJ, Sadanandom A, Shirasu K, Jones JD, Parker JE (2002) Arabidopsis RAR1 exerts rate-limiting control of R gene-mediated defenses against multiple pathogens. Plant Cell 14(5):979–992

Newbery HJ, Loh DH, O'Donoghue JE, Tomlinson VA, Chau YY, Boyd JA, Bergmann JH, Brownstein D, Abbott CM (2007) Translation elongation factor eEF1A2 is essential for post-weaning survival in mice. J Biol Chem 282(39):28951–28959. Epub 2007/07/21. https://doi.org/10.1074/jbc.M703962200

Niikura Y, Kitagawa K (2003) Identification of a novel splice variant: human SGT1B (SUGT1B). DNA Seq 14(6):436–441

Niikura Y, Ohta S, Vandenbeldt KJ, Abdulle R, McEwen BF, Kitagawa K (2006) 17-AAG, an Hsp90 inhibitor, causes kinetochore defects: a novel mechanism by which 17-AAG inhibits cell proliferation. Oncogene 25(30):4133–4146. https://doi.org/10.1038/sj.onc.1209461

Niikura Y, Kitagawa R, Ogi H, Abdulle R, Pagala V, Kitagawa K (2015) CENP-A K124 ubiquitylation is required for CENP-A deposition at the centromere. Dev Cell 32(5):589–603. https://doi.org/10.1016/j.devcel.2015.01.024

Niikura Y, Kitagawa R, Ogi H, Kitagawa K (2017) SGT1-HSP90 complex is required for CENP-A deposition at centromeres. Cell Cycle 16(18):1683–1694. https://doi.org/10.1080/15384101.2017.1325039

Noel LD, Cagna G, Stuttmann J, Wirthmuller L, Betsuyaku S, Witte CP, Bhat R, Pochon N, Colby T, Parker JE (2007) Interaction between SGT1 and cytosolic/nuclear HSC70 chaperones regulates Arabidopsis immune responses. Plant Cell 19(12):4061–4076. https://doi.org/10.1105/tpc.107.051896

Nonneman A, Robberecht W, Van Den Bosch L (2014) The role of oligodendroglial dysfunction in amyotrophic lateral sclerosis. Neurodegener Dis Manag 4(3):223–239. https://doi.org/10.2217/nmt.14.21

Novosylna O, Jurewicz E, Pydiura N, Goral A, Filipek A, Negrutskii B, El'skaya A (2015) Translation elongation factor eEF1A1 is a novel partner of a multifunctional protein Sgt1. Biochimie 119:137–145. https://doi.org/10.1016/j.biochi.2015.10.026

Nowotny M, Spiechowicz M, Jastrzebska B, Filipek A, Kitagawa K, Kuznicki J (2003) Calcium-regulated interaction of Sgt1 with S100A6 (calcyclin) and other S100 proteins. J Biol Chem 278(29):26923–26928. https://doi.org/10.1074/jbc.M211518200

Nyarko A, Mosbahi K, Rowe AJ, Leech A, Boter M, Shirasu K, Kleanthous C (2007) TPR-mediated self-association of plant SGT1. Biochemistry 46(40):11331–11341. https://doi.org/10.1021/bi700735t

Ogi H, Sakuraba Y, Kitagawa R, Xiao L, Shen C, Cynthia MA, Ohta S, Arnold MA, Ramirez N, Houghton PJ, Kitagawa K (2015) The oncogenic role of the cochaperone Sgt1. Oncogene 4:e149. https://doi.org/10.1038/oncsis.2015.12

Ogura Y, Sutterwala FS, Flavell RA (2006) The inflammasome: first line of the immune response to cell stress. Cell 126(4):659–662. https://doi.org/10.1016/j.cell.2006.08.002

Oh SK, Kim H, Choi D (2014a) Rpi-blb2-mediated late blight resistance in Nicotiana benthamiana requires SGT1 and salicylic acid-mediated signaling but not RAR1 or HSP90. FEBS Lett 588(7):1109–1115. https://doi.org/10.1016/j.febslet.2014.02.028

Oh SK, Kwon SY, Choi D (2014b) Rpi-blb2-mediated hypersensitive cell death caused by *Phytophthora infestans* AVRblb2 requires SGT1, but not EDS1, NDR1, salicylic acid-, jasmonic acid-, or ethylene-mediated signaling. Plant Pathol J 30(3):254–260. https://doi.org/10.5423/PPJ.OA.03.2014.0027

Park SC, Cheong MS, Kim EJ, Kim JH, Chi YH, Jang MK (2017) Antifungal effect of Arabidopsis SGT1 proteins via mitochondrial reactive oxygen species. J Agric Food Chem 65(38):8340–8347. https://doi.org/10.1021/acs.jafc.7b02808

Pearl LH, Prodromou C (2006) Structure and mechanism of the Hsp90 molecular chaperone machinery. Annu Rev Biochem 75:271–294. https://doi.org/10.1146/annurev.biochem.75.103004.142738

Pearl LH, Prodromou C, Workman P (2008) The Hsp90 molecular chaperone: an open and shut case for treatment. Biochem J 410(3):439–453. https://doi.org/10.1042/BJ20071640

Peart JR, Lu R, Sadanandom A, Malcuit I, Moffett P, Brice DC, Schauser L, Jaggard DA, Xiao S, Coleman MJ, Dow M, Jones JD, Shirasu K, Baulcombe DC (2002) Ubiquitin ligase-associated protein SGT1 is required for host and nonhost disease resistance in plants. Proc Natl Acad Sci U S A 99(16):10865–10869. https://doi.org/10.1073/pnas.152330599

Pickering C, Ericson M, Soderpalm B (2013) Chronic phencyclidine increases synapsin-1 and synaptic adaptation proteins in the medial prefrontal cortex. ISRN Psychiatry 2013:620361. https://doi.org/10.1155/2013/620361

Pluta AF, Mackay AM, Ainsztein AM, Goldberg IG, Earnshaw WC (1995) The centromere: hub of chromosomal activities. Science 270(5242):1591–1594

Portig I, Pankuweit S, Maisch B (1997) Antibodies against stress proteins in sera of patients with dilated cardiomyopathy. J Mol Cell Cardiol 29(8):2245–2251. https://doi.org/10.1006/jmcc.1997.0463

Pratt WB, Toft DO (2003) Regulation of signaling protein function and trafficking by the hsp90/hsp70-based chaperone machinery. Exp Biol Med (Maywood) 228(2):111–133

Prus W, Filipek A (2011) S100A6 mediates nuclear translocation of Sgt1: a heat shock-regulated protein. Amino Acids 41(4):781–787. https://doi.org/10.1007/s00726-010-0526-2

Prus W, Zabka M, Bieganowski P, Filipek A (2011) Nuclear translocation of Sgt1 depends on its phosphorylation state. Int J Biochem Cell Biol 43(12):1747–1753. https://doi.org/10.1016/j.biocel.2011.08.010

Radford NB, Fina M, Benjamin IJ, Moreadith RW, Graves KH, Zhao P, Gavva S, Wiethoff A, Sherry AD, Malloy CR, Williams RS (1996) Cardioprotective effects of 70-kDa heat shock protein in transgenic mice. Proc Natl Acad Sci U S A 93(6):2339–2342

Rodrigo-Brenni MC, Thomas S, Bouck DC, Kaplan KB (2004) Sgt1p and Skp1p modulate the assembly and turnover of CBF3 complexes required for proper kinetochore function. Mol Biol Cell 15(7):3366–3378. https://doi.org/10.1091/mbc.e03-12-0887

Russell LC, Whitt SR, Chen MS, Chinkers M (1999) Identification of conserved residues required for the binding of a tetratricopeptide repeat domain to heat shock protein 90. J Biol Chem 274(29):20060–20063

Sahni N, Yi S, Taipale M, Fuxman Bass JI, Coulombe-Huntington J, Yang F, Peng J, Weile J, Karras GI, Wang Y, Kovacs IA, Kamburov A, Krykbaeva I, Lam MH, Tucker G, Khurana V, Sharma A, Liu YY, Yachie N, Zhong Q, Shen Y, Palagi A, San-Miguel A, Fan C, Balcha D, Dricot A, Jordan DM, Walsh JM, Shah AA, Yang X, Stoyanova AK, Leighton A, Calderwood MA, Jacob Y, Cusick ME, Salehi-Ashtiani K, Whitesell LJ, Sunyaev S, Berger B, Barabasi AL, Charloteaux B, Hill DE, Hao T, Roth FP, Xia Y, Walhout AJM, Lindquist S, Vidal M (2015) Widespread macromolecular interaction perturbations in human genetic disorders. Cell 161(3):647–660. Epub 2015/04/25. https://doi.org/10.1016/j.cell.2015.04.013. PubMed PMID: 25910212; PMCID: PMC4441215

Sato T, Jigami Y, Suzuki T, Uemura H (1999) A human gene, hSGT1, can substitute for GCR2, which encodes a general regulatory factor of glycolytic gene expression in *Saccharomyces cerevisiae*. Mol Gen Genet 260(6):535–540

Sbroggio M, Ferretti R, Percivalle E, Gutkowska M, Zylicz A, Michowski W, Kuznicki J, Accornero F, Pacchioni B, Lanfranchi G, Hamm J, Turco E, Silengo L, Tarone G, Brancaccio M (2008) The mammalian CHORD-containing protein melusin is a stress response protein interacting with Hsp90 and Sgt1. FEBS Lett 582(13):1788–1794. https://doi.org/10.1016/j.febslet.2008.04.058

Schadick K, Fourcade HM, Boumenot P, Seitz JJ, Morrell JL, Chang L, Gould KL, Partridge JF, Allshire RC, Kitagawa K, Hieter P, Hoffman CS (2002) Schizosaccharomyces pombe Git7p, a member of the *Saccharomyces cerevisiae* Sgtlp family, is required for glucose and cyclic AMP signaling, cell wall integrity, and septation. Eukaryot Cell 1(4):558–567

Scheufler C, Brinker A, Bourenkov G, Pegoraro S, Moroder L, Bartunik H, Hartl FU, Moarefi I (2000) Structure of TPR domain-peptide complexes: critical elements in the assembly of the Hsp70-Hsp90 multichaperone machine. Cell 101(2):199–210. https://doi.org/10.1016/S0092-8674(00)80830-2

Shang Y, Li X, Cui H, He P, Thilmony R, Chintamanani S, Zwiesler-Vollick J, Gopalan S, Tang X, Zhou JM (2006) RAR1, a central player in plant immunity, is targeted by Pseudomonas syringae effector AvrB. Proc Natl Acad Sci U S A 103(50):19200–19205. https://doi.org/10.1073/pnas.0607279103

Shanmugam A, Thamilarasan SK, Park JI, Jung MY, Nou IS (2016) Characterization and abiotic stress-responsive expression analysis of SGT1 genes in *Brassica oleracea*. Genome 59(4):243–251. https://doi.org/10.1139/gen-2015-0128

Shapiro RS, Zaas AK, Betancourt-Quiroz M, Perfect JR, Cowen LE (2012) The Hsp90 co-chaperone Sgt1 governs *Candida albicans* morphogenesis and drug resistance. PLoS One 7(9):e44734. https://doi.org/10.1371/journal.pone.0044734

Sharma UC, Barenbrug P, Pokharel S, Dassen WR, Pinto YM, Maessen JG (2004) Systematic review of the outcome of aortic valve replacement in patients with aortic stenosis. Ann Thorac Surg 78(1):90–95. https://doi.org/10.1016/j.athoracsur.2004.02.020

Shen QH, Zhou F, Bieri S, Haizel T, Shirasu K, Schulze-Lefert P (2003) Recognition specificity and RAR1/SGT1 dependence in barley Mla disease resistance genes to the powdery mildew fungus. Plant Cell 15(3):732–744

Shirasu K (2009) The HSP90-SGT1 chaperone complex for NLR immune sensors. Annu Rev Plant Biol 60:139–164. https://doi.org/10.1146/annurev.arplant.59.032607.092906

Shirasu K, Schulze-Lefert P (2000) Regulators of cell death in disease resistance. Plant Mol Biol 44(3):371–385

Shirasu K, Schulze-Lefert P (2003) Complex formation, promiscuity and multi-functionality: protein interactions in disease-resistance pathways. Trends Plant Sci 8(6):252–258. https://doi.org/10.1016/S1360-1385(03)00104-3

Shirasu K, Lahaye T, Tan MW, Zhou F, Azevedo C, Schulze-Lefert P (1999) A novel class of eukaryotic zinc-binding proteins is required for disease resistance signaling in barley and development in *C. elegans*. Cell 99(4):355–366

Siligardi G, Zhang M, Prodromou C (2017) The stoichiometric interaction of the Hsp90-Sgt1-Rar1 complex by CD and SRCD spectroscopy. Front Mol Biosci 4:95. https://doi.org/10.3389/fmolb.2017.00095

Sorger PK, Doheny KF, Hieter P, Kopski KM, Huffaker TC, Hyman AA (1995) Two genes required for the binding of an essential *Saccharomyces cerevisiae* kinetochore complex to DNA. Proc Natl Acad Sci U S A 92(26):12026–12030

Spiechowicz M, Filipek A (2005) The expression and function of Sgt1 protein in eukaryotic cells. Acta Neurobiol Exp (Wars) 65(2):161–165

Spiechowicz M, Bernstein HG, Dobrowolny H, Lesniak W, Mawrin C, Bogerts B, Kuznicki J, Filipek A (2006) Density of Sgt1-immunopositive neurons is decreased in the cerebral cortex of Alzheimer's disease brain. Neurochem Int 49(5):487–493. https://doi.org/10.1016/j.neuint.2006.03.009

Spiechowicz M, Zylicz A, Bieganowski P, Kuznicki J, Filipek A (2007) Hsp70 is a new target of Sgt1--an interaction modulated by S100A6. Biochem Biophys Res Commun 357(4):1148–1153. https://doi.org/10.1016/j.bbrc.2007.04.073

Staskawicz BJ, Mudgett MB, Dangl JL, Galan JE (2001) Common and contrasting themes of plant and animal diseases. Science 292(5525):2285–2289. https://doi.org/10.1126/science.1062013

Steensgaard P, Garre M, Muradore I, Transidico P, Nigg EA, Kitagawa K, Earnshaw WC, Faretta M, Musacchio A (2004) Sgt1 is required for human kinetochore assembly. EMBO Rep 5(6):626–631. https://doi.org/10.1038/sj.embor.7400154

Stemmann O, Lechner J (1996) The *Saccharomyces cerevisiae* kinetochore contains a cyclin-CDK complexing homologue, as identified by in vitro reconstitution. EMBO J 15(14):3611–3620

Stemmann O, Neidig A, Kocher T, Wilm M, Lechner J (2002) Hsp90 enables Ctf13p/Skp1p to nucleate the budding yeast kinetochore. Proc Natl Acad Sci U S A 99(13):8585–8590. https://doi.org/10.1073/pnas.082223899

Strunnikov AV, Kingsbury J, Koshland D (1995) CEP3 encodes a centromere protein of *Saccharomyces cerevisiae*. J Cell Biol 128(5):749–760

Stuttmann J, Parker JE, Noel LD (2008) Staying in the fold: the SGT1/chaperone machinery in maintenance and evolution of leucine-rich repeat proteins. Plant Signal Behav 3(5):283–285

Tai YS (2008) Interactome of signaling networks in wheat: the protein-protein interaction between TaRAR1 and TaSGT1. Mol Biol Rep 35(3):337–343.. Epub 2007/06/15. https://doi.org/10.1007/s11033-007-9091-5

Takahashi A, Casais C, Ichimura K, Shirasu K (2003) HSP90 interacts with RAR1 and SGT1 and is essential for RPS2-mediated disease resistance in Arabidopsis. Proc Natl Acad Sci U S A 100(20):11777–11782. https://doi.org/10.1073/pnas.2033934100

Taube M, Pienkowska JR, Jarmolowski A, Kozak M (2014) Low-resolution structure of the full-length barley (*Hordeum vulgare*) SGT1 protein in solution, obtained using small-angle X-ray scattering. PLoS One 9(4):e93313. https://doi.org/10.1371/journal.pone.0093313

Tavares R, Wajnberg G, Scherer NM, Pauletti BA, Cassoli JS, Ferreira CG, Paes Leme AF, de Araujo-Souza PS, Martins-de-Souza D, Passetti F (2017) Unveiling alterative splice diversity from human oligodendrocyte proteome data. J Proteomics. 151:293–301. https://doi.org/10.1016/j.jprot.2016.05.023. Epub 2016/05/26. PubMed PMID: 27222040

Ting JP, Davis BK (2005) CATERPILLER: a novel gene family important in immunity, cell death, and diseases. Annu Rev Immunol 23:387–414. https://doi.org/10.1146/annurev.immunol.23.021704.115616

Ting JP, Lovering RC, Alnemri ES, Bertin J, Boss JM, Davis BK, Flavell RA, Girardin SE, Godzik A, Harton JA, Hoffman HM, Hugot JP, Inohara N, Mackenzie A, Maltais LJ, Nunez G, Ogura Y, Otten LA, Philpott D, Reed JC, Reith W, Schreiber S, Steimle V, Ward PA (2008) The NLR gene family: a standard nomenclature. Immunity 28(3):285–287. https://doi.org/10.1016/j.immuni.2008.02.005

Tor M, Gordon P, Cuzick A, Eulgem T, Sinapidou E, Mert-Turk F, Can C, Dangl JL, Holub EB (2002) Arabidopsis SGT1b is required for defense signaling conferred by several downy mildew resistance genes. Plant Cell 14(5):993–1003

Tornero P, Merritt P, Sadanandom A, Shirasu K, Innes RW, Dangl JL (2002) RAR1 and NDR1 contribute quantitatively to disease resistance in Arabidopsis, and their relative contributions are dependent on the R gene assayed. Plant Cell 14(5):1005–1015

Trepel J, Mollapour M, Giaccone G, Neckers L (2010) Targeting the dynamic HSP90 complex in cancer. Nat Rev Cancer 10(8):537–549. https://doi.org/10.1038/nrc2887

Uppalapati SR, Ishiga Y, Ryu CM, Ishiga T, Wang K, Noel LD, Parker JE, Mysore KS (2011) SGT1 contributes to coronatine signaling and Pseudomonas syringae pv. tomato disease symptom development in tomato and Arabidopsis. New Phytol 189(1):83–93. https://doi.org/10.1111/j.1469-8137.2010.03470.x

Vassarotti A, Friesen JD (1985) Isolation of the fructose-1,6-bisphosphatase gene of the yeast Schizosaccharomyces pombe. Evidence for transcriptional regulation. J Biol Chem 260(10):6348–6353

Verdaasdonk JS, Bloom K (2011) Centromeres: unique chromatin structures that drive chromosome segregation. Nat Rev Mol Cell Biol 12(5):320–332. https://doi.org/10.1038/nrm3107

Wang Y, Gao M, Li Q, Wang L, Wang J, Jeon JS, Qu N, Zhang Y, He Z (2008) OsRAR1 and OsSGT1 physically interact and function in rice basal disease resistance. Mol Plant-Microbe Interact 21(3):294–303. https://doi.org/10.1094/MPMI-21-3-0294

Wang K, Uppalapati SR, Zhu X, Dinesh-Kumar SP, Mysore KS (2010) SGT1 positively regulates the process of plant cell death during both compatible and incompatible plant-pathogen interactions. Mol Plant Pathol 11(5):597–611. https://doi.org/10.1111/j.1364-3703.2010.00631.x

Wang GF, Fan R, Wang X, Wang D, Zhang X (2015) TaRAR1 and TaSGT1 associate with TaHsp90 to function in bread wheat (*Triticum aestivum* L.) seedling growth and stripe rust resistance. Plant Mol Biol 87(6):577–589. https://doi.org/10.1007/s11103-015-0298-x

Wang K, Shang Y, Dou F (2018) Brain aging: Hsp90 and neurodegenerative diseases. Adv Exp Med Biol 1086:93–103. https://doi.org/10.1007/978-981-13-1117-8_6

Wegele H, Muller L, Buchner J (2004) Hsp70 and Hsp90 – a relay team for protein folding. Rev Physiol Biochem Pharmacol 151:1–44. https://doi.org/10.1007/s10254-003-0021-1

Wegele H, Wandinger SK, Schmid AB, Reinstein J, Buchner J (2006) Substrate transfer from the chaperone Hsp70 to Hsp90. J Mol Biol 356(3):802–811. https://doi.org/10.1016/j.jmb.2005.12.008

Weinstein IB, Joe A (2008) Oncogene addiction. Cancer Res 68(9):3077–3080.; discussion 3080. https://doi.org/10.1158/0008-5472.CAN-07-3293

Westermann S, Drubin DG, Barnes G (2007) Structures and functions of yeast kinetochore complexes. Annu Rev Biochem 76:563–591

Willhoft O, Kerr R, Patel D, Zhang W, Al-Jassar C, Daviter T, Millson SH, Thalassinos K, Vaughan CK (2017) The crystal structure of the Sgt1-Skp1 complex: the link between Hsp90 and both SCF E3 ubiquitin ligases and kinetochores. Sci Rep 7:41626. https://doi.org/10.1038/srep41626

Wu J, Luo S, Jiang H, Li H (2005) Mammalian CHORD-containing protein 1 is a novel heat shock protein 90-interacting protein. FEBS Lett 579(2):421–426. https://doi.org/10.1016/j.febslet.2004.12.005

Xing L, Qian C, Cao A, Li Y, Jiang Z, Li M, Jin X, Hu J, Zhang Y, Wang X, Chen P (2013) The Hv-SGT1 gene from *Haynaldia villosa* contributes to resistances towards both biotrophic and hemi-biotrophic pathogens in common wheat (*Triticum aestivum* L.). PLoS One 8(9):e72571. https://doi.org/10.1371/journal.pone.0072571

Xu XL, Lee RT, Fang HM, Wang YM, Li R, Zou H, Zhu Y, Wang Y (2008) Bacterial peptidoglycan triggers *Candida albicans* hyphal growth by directly activating the adenylyl cyclase Cyr1p. Cell Host Microbe 4(1):28–39. https://doi.org/10.1016/j.chom.2008.05.014

Yamamoto T, Mori Y, Ishibashi T, Uchiyama Y, Sakaguchi N, Furukawa T, Hashimoto J, Kimura S, Sakaguchi K (2004) Characterization of Rad6 from a higher plant, rice (*Oryza sativa* L.) and its interaction with Sgt1, a subunit of the SCF ubiquitin ligase complex. Biochem Biophys Res Commun 314(2):434–439

Yan JJ, Zhang YB, Ding Y (2012) Binding mechanism between Hsp90 and Sgt1 explored by homology modeling and molecular dynamics simulations in rice. J Mol Model 18(10):4665–4673. https://doi.org/10.1007/s00894-012-1464-6

Yan Q, Cui X, Su L, Xu N, Guo N, Xing H, Dou D (2014) GmSGT1 is differently required for soybean Rps genes-mediated and basal resistance to Phytophthora sojae. Plant Cell Rep 33(8):1275–1288. https://doi.org/10.1007/s00299-014-1615-6

Yusof NA, Hashim NH, Beddoe T, Mahadi NM, Illias RM, Bakar FD, Murad AM (2016) Thermotolerance and molecular chaperone function of an SGT1-like protein from the psychrophilic yeast, Glaciozyma Antarctica. Cell Stress Chaperones 21(4):707–715. https://doi.org/10.1007/s12192-016-0696-2

Zabka M, Lesniak W, Prus W, Kuznicki J, Filipek A (2008) Sgt1 has co-chaperone properties and is up-regulated by heat shock. Biochem Biophys Res Commun 370(1):179–183. https://doi.org/10.1016/j.bbrc.2008.03.055

Zhang Y, Dorey S, Swiderski M, Jones JD (2004) Expression of RPS4 in tobacco induces an AvrRps4-independent HR that requires EDS1, SGT1 and HSP90. Plant J 40(2):213–224. https://doi.org/10.1111/j.1365-313X.2004.02201.x

Zhang M, Boter M, Li K, Kadota Y, Panaretou B, Prodromou C, Shirasu K, Pearl LH (2008) Structural and functional coupling of Hsp90- and Sgt1-centred multi-protein complexes. EMBO J 27(20):2789–2798. https://doi.org/10.1038/emboj.2008.190

Zhang M, Kadota Y, Prodromou C, Shirasu K, Pearl LH (2010) Structural basis for assembly of Hsp90-Sgt1-CHORD protein complexes: implications for chaperoning of NLR innate immunity receptors. Mol Cell 39(2):269–281. https://doi.org/10.1016/j.molcel.2010.05.010

Zou X, Ji C, Wang L, Wu M, Zheng H, Xu J, Jin F, Gu S, Ying K, Xie Y, Mao Y (2004) Molecular cloning and characterization of SGT1.2, a novel splice variant of *Homo sapiens* SGT1. DNA Seq 15(2):140–143

Chapter 17
Sti1/Hop Plays a Pivotal Role in Hsp90 Regulation Beyond Bridging Hsp70

Michael Reidy

Abstract Since its initial characterization, Hop (Hsp90/Hsp70 organizing protein), known as Sti1 in yeast (stress inducible) is mostly understood to serve as a bridge that facilitates transfer of substrate "client" proteins from Hsp70 to Hsp90. Recent work has shown that Sti1 regulates Hsp90 in a manner distinct from its role as a bridge to Hsp70. This second function of Sti1 seems to be to position Hsp90 for subsequent steps of the client maturation cycle, after the client has been transferred from Hsp70. Thus, Sti1/Hop occupies a central gatekeeper role in the Hsp90 reaction cycle, by first facilitating client access to Hsp90 and then promoting the next steps of the cycle.

Keywords Chaperone · Co-chaperone · Hop · Hsp · Hsp70 · Hsp90 · Sti1

Abbreviations

3D	Three dimensional
AR	Androgen receptor
ATP	Adenosine triphosphate
cryoEM	Cryoelectron microscopy
DP	Aspartate/proline-rich motif
EM	Electron microscopy
FOA	5′Fluoro-orotic acid
GR	Glucocorticoid receptor
Hop	Hsp90/Hsp70 organizing protein
Hsp	Heat shock protein
MAP	Mitogen-activated protein

M. Reidy (✉)
Laboratory of Biochemistry and Genetics, National Institute of Diabetes and Digestive and Kidney Diseases, National Institutes of Health, Bethesda, MD, USA
e-mail: michael.reidy@nih.gov

A. A. A. Asea, P. Kaur (eds.), *Heat Shock Protein 90 in Human Diseases and Disorders*, Heat Shock Proteins 19,
https://doi.org/10.1007/978-3-030-23158-3_17

SdC	Sti1-dependent carboxy-terminal proximal
SdN	Sti1-dependent amino-terminal proximal
TPR	Tetratricopeptide repeat

17.1 Introduction

In order to discuss the role of Sti1 in Hsp90 regulation, it is necessary to first discuss Hsp90. Hsp90 is an evolutionarily conserved, essential (in eukaryotes) homodimer that regulates a diverse and expanding list of substrate proteins, known as clients (Karagoz and Rudiger 2015; Prodromou 2016). Through an open-close client binding mechanism, Fig. 17.1, Hsp90 regulates the activities and stabilities of client proteins such as kinases, transcription factors and metabolic enzymes as well as glucocorticoid and androgen receptors (GR and AR, respectively) in mammals. A large and diverse cohort of Hsp90 co-chaperones assist in the regulation of the eukaryotic Hsp90 reaction cycle. Table 17.1 lists the major co-chaperones, their names in yeast and mammalian systems and their known role or function in Hsp90 regulation.

Hsp90 is a modular protein with an amino-terminal (N) ATPase domain connected to a middle (M) domain via a charged linker, and a carboxy-terminal (C) dimerization domain immediately adjacent to the M-domain. Hsp90 exists as a homodimer joined at the C-terminal domain, and in the apo state the N-terminal domains are positioned the furthest possible from each other, called the "open

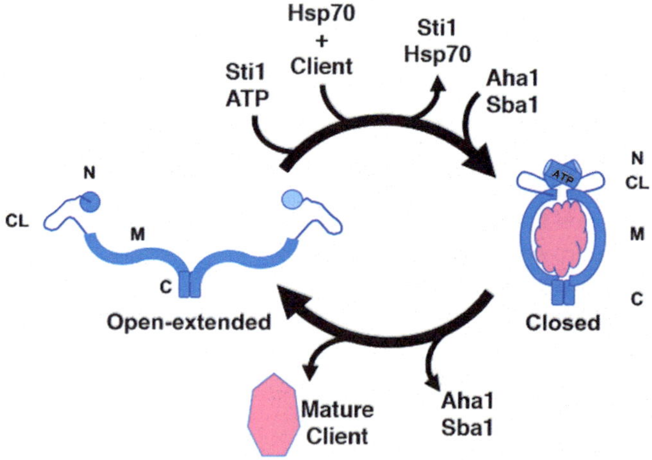

Fig. 17.1 A simplified view of the Hsp90 chaperone cycle. Hsp90 (blue) regulates the activity and stability of client proteins (pink) via an open-close mechanism that is assisted by various co-chaperones such as Hsp70, Sti1, Aha1 and Sba1 (see Table 17.1). The relative location of the N-terminal (N), charged linker (CL), middle (M) and C-terminal (C) domains are shown in the open (left) and closed (right) conformations. (See text for more details)

Table 17.1 The human and yeast names and functions of the major Hsp90 co-chaperones

Human	Yeast	Role in Hsp90 regulation
Hsc70, Hsp70	Ssa1–4	Delivery of clients
Hop	Sti1	Bridging Hsp90 and Hsp70, formation of client loading conformation, has three TPR domains
Aha	Aha1	Stimulator of Hsp90 ATP hydrolysis
p23	Sba1	Stabilizes ATP-bound, N-terminally dimerized conformation
Cdc37	Cdc37	Required for kinase clients
FKBP51, FKBP52, CyP-40	Cpr6, 7	Cyclophilin (peptidyl-prolyl cis-trans isomerase), has TPR domains

extended" conformation (Krukenberg et al. 2009; Schopf et al. 2017). In the ATP state, the N-terminal domains of both protomers make physical contact. In order to do so they must first rotate inwards and dock to the N-terminal portion of the M-domain. Hydrolysis only occurs when the N-terminal domains come into contact with each other in the "closed" conformation. Clients are bound in the interior space formed by the two M-domains in the closed conformation (Verba et al. 2016). Interaction with Hsp90 modifies the activity of the client in a specific manner. For example, GR and AR can only bind to their hormone ligands when in complex with Hsp90 and other co-chaperones (Pratt et al. 2006), while in the case of MAP kinases Hsp90 interaction is required for the phosphorylation and activation of target transcription factors (Pratt et al. 2006). After ATP hydrolysis the N-terminal domains dissociate and Hsp90 returns to the open extended conformation, releasing the mature client and resetting the cycle.

Different co-chaperones bind to distinct conformations of Hsp90. Since the position of the N-terminal domains relative to each other changes dramatically depending on the nucleotide state, several possible conformations theoretically exist between open-extended and closed. For example, Sba1 (p23 in humans) binds only the ATP-bound, fully closed conformation of Hsp90 (Ali et al. 2006). Presence of Sba1 in Hsp90 pull-downs from cell lysates treated with non-hydrolysable ATP analogs is generally regarded as a useful sensor for the closed conformation of Hsp90 (Zuehlke and Johnson 2012). In addition to the conformation of Hsp90 influencing the interaction of co-chaperones, the binding of some co-chaperones directly influences the conformation of Hsp90. For example, using cryoEM to construct 3D structures of complexes formed from purified proteins, the Agard lab showed that Hsp90 adopted a "client loading" conformation when bound to Hop (Southworth and Agard 2011). This conformation was partially closed relative to the open-extended and Sba1-bound conformations. Hydrophobic residues in the interior of the Hsp90 dimer M-domains aligned into patches when bound to Hop. These patches were postulated to be the sites of client interaction. The degree of conformational requirements for co-chaperone interaction reflect the tight regulation these co-chaperones exert on the Hsp90 reaction cycle.

17.1.1 STI1/Hop Structure/Function

Sti1/Hop is a 66 kilodalton protein containing three tetratricopeptide repeat (TPR) domains, termed 1, 2A and 2B, and two aspartate/proline motif (DP) domains, see Fig. 17.2. While Sti1 is not essential for *Saccharomyces cerevisiae*, cells lacking Sti1 are hypersensitive to Hsp90 specific inhibitors such as radicicol and fail to grow under sub-optimal conditions (Chang et al. 1997). Early co-crystallographic studies demonstrated that carboxylate clamps in Hop TPR1 and TPR2A specifically interacted with the C-terminal EEVD motifs present in Hsp70 (GPTIVEEVD) and Hsp90 (MEEVD), respectively (Scheufler et al. 2000). From these studies a straightforward model of Sti1/Hop bridging Hsp90 and Hsp70 was developed: TPR1 bound to Hsp70 and TPR2A bound to Hsp90, forming a tripartite complex that allowed the client bound to Hsp70 to be transferred to Hsp90. However, subsequent studies that explored the interaction between Hsp70 and Hop suggested that this model was inadequate (Carrigan et al. 2004), since Hsp70 proteins lacking the C-terminal EEVD motif bound Hop like wild type. Also, the authors found that mutations in TPR2A, TPR2B and DP2 negatively impacted Hsp70 binding, which was not predicted based on the contemporary model.

It was then shown using yeast genetics that Sti1 regulated Hsp70 and Hsp90 independently on some level. By studying the effects of mutations in Hsp70 on the stability of self-templating amyloids, called prions, that propagated in some strains of yeast in a chaperone-dependent manner, it was found that destabilization of the [*PSI*$^+$] prion by the dominant *SSA1-21* allele required the C-terminal GPTIVEED motif of Hsp70 that had been shown previously to be important for TPR interaction. The authors then showed that Sti1 was required for the *SSA1-21* effect on prion stability (Jones et al. 2004). In a follow-up study it was shown that mutations in Sti1 that mediated the *SSA1-21* effect did not affect the cells' sensitivity to radicicol. In this way the regulation of Hsp70 by Sti1 could be differentiated from that of Hsp90, since radicicol treatment had no effect on [*PSI*$^+$] prion stability (Song and Masison 2005). This suggested Sti1 could regulate Hsp70 separately from Hsp90. However, mutations in Sti1 that affected either prion stability or radicicol sensitivity negatively impacted maturation of exogenously expressed mammalian GR in yeast, a pathway that was known to require both Hsp90 and Hsp70. These findings supported the bridging model. Thus it was possible that Sti1 performed an important role in Hsp90 regulation in addition to bridging Hsp90 and Hsp70. Another notable finding from this study was the implication that Hsp90 was a ligand for Sti1 TPR2B. The mystery surrounding the ligand for TPR2B deepened when it was

Fig. 17.2 Domain structure of Hop/Sti1. Hop/Sti1 contains three tetratricopeptide (TPR) domains and two aspartate/proline motif domains. TPR1 and DP1 are separated from the rest of the molecule by a flexible linker. TPR2A, 2B and DP2 comprise a rigid module that makes extensive contacts with Hsp90. Not to scale

shown that mutations in both TPR1 and TPR2B were necessary to abolish physical and genetic interactions with Hsp70 (Flom et al. 2006, 2007), contradicting some results from the studies described above. Importantly, these reports also found that the Sti1 TPR2A domain alone was not sufficient to interact with Hsp90 but required the presence of the TPR2B domain as well, which did agree with findings from those same earlier studies (Song and Masison 2005).

Using negative staining EM and cryo-EM of stabilized Hop:Hsp90 complexes to construct high-resolution structures, the Agard group established a model for client transfer from Hsp70 to Hsp90 bridged by Hop (Southworth and Agard 2011). In their reconstructed models, two molecules of Hop bound on opposite sides of one V-shaped Hsp90 dimer. The orientation of the M-domain relative to the C-domain in each protomer resulted in a more closed Hsp90 dimer compared to Hsp90 without Hop, under similar conditions. The only structures of Hsp90 N-terminal domain that fit the density they observed were from the ATP-bound, fully closed Hsp90 crystal structure. This was remarkable because in the closed structure, the N-terminal domains make extensive contacts whereas in the Hop-bound reconstructions they were measured to be on average 80 Å apart. Thus, the conformation that Hsp90 adopted when bound to Hop represented an unobserved conformation they termed the "client-loading" conformation.

In the cryo-EM structures of Hop:Hsp90, the authors noted that only two of Hop's three TPR domains were clearly represented as electron densities that spanned across and out from the Hsp90 dimer (Southworth and Agard 2011). The missing TPR domain in the structures was probably due its flexibility. Some density was observed below the C-terminal domain that could represent the position of the third TPR domain. Unfortunately, this meant that the authors could not confidently predict the orientation of Hop relative to the Hsp90 dimer, because it was unclear whether the observed densities were TPR1 and 2A or TPR2A and 2B. To solve this the authors used gold particle labeling of the His6-tag on the N-terminus of Hop. From these images, the authors concluded that they were observing TPR1 and TPR2A, however they could not rule out the possibility of the alternative (TPR2A and 2B). This orientation, with TPR1 positioned adjacent to the client binding region of Hsp90, fit well with the idea that Hop facilitated client transfer from Hsp70 to Hsp90.

When the authors added Hsp70 to the reaction mixtures, they observed very little change in the Hop:Hsp90 complexes themselves, indicating that the client loading conformation is independent of Hsp70 (Southworth and Agard 2011). Interestingly, a single Hop:Hsp90 complex bound a single molecule of Hsp70, in agreement with the bridging model of Hop/Sti1 function. The authors concluded that binding of Hop/Sti1 to Hsp90 was probably performing a more active role in Hsp90 regulation than merely serving as a bridge between Hsp90 and Hsp70.

Two reports published the next year 1 month apart, one from the Mayer lab and the other from the Buchner lab demonstrated that Sti1 TPR2B interacted physically with Hsp90 using biochemical, biophysical and mass-spectrometry methods (Lee et al. 2012; Schmid et al. 2012). Both studies reaffirmed that TPR2A bound the Hsp90 C-terminal MEEVD motif and identified the N-terminal portion of the

M-domain of Hsp90 as the site of interaction with TPR2B, in agreement with the Agard model. Also, both the Mayer and Buchner models agreed that the TPR1 domain of Sti1 swings freely via flexible linker to allow interaction with Hsp70. Thus, both models positioned Hop/Sti1 in an orientation opposite to that suggested by the Agard model. The Mayer and Buchner models differed, however, in two important ways. First, the model put forth by the Mayer lab had the TPR2A of Sti1 bound to the C-terminal MEEVD of one Hsp90 protomer while the TPR2A and 2B domains of the same Sti1 molecule made contacts with the M-domain of the other Hsp90 protomer. The Buchner model also identified contacts between Sti1 TPR2A and 2B and the M-domain of Hsp90, but on the same protomer that bound TPR2A via the MEEVD motif.

The second big difference in the two studies was that the Buchner lab went one step further and found that both TPR1 and TPR2B bound to Hsp70. The finding that Sti1 TPR2B made contacts with both Hsp90 and Hsp70 clarified confusion arising from differing results from the earlier studies (Flom et al. 2007; Song and Masison 2005). In the Buchner model, Sti1 TPR2A binds the Hsp90 MEEVD motif while TPR2B makes contacts in the M-domain in such a way as to make the carboxylate clamp accessible. TPR1 binds to Hsp70 and through the flexibility provided by the linker positions Hsp70 to make contact with and transfer to TPR2B. Thus, the client bound to Hsp70 is oriented in such a way to facilitate its transfer to Hsp90. The major difference in the Agard and Buchner models, the orientation of Hop/Sti1 with respect to the Hsp90 dimer, may actually be resolved by allowing that the densities that were assigned to TPR1 are actually TPR2B, and it is TPR1 that is not present in their reconstructions because of the flexible linker. The subsequent finding that TPR2A and 2B form a rigid module that is separated from TPR1 by a long flexible linker reinforces this notion (Rohl et al. 2015). In a further refinement to the Buchner model, it was shown that Hsp90 influences the affinity of Sti1 for Hsp70. In the absence of Hsp90, Sti1 binds Hsp70 via the TPR2B domain, but when Sti1 is bound to Hsp90, TPR1 is the preferred binding site for Hsp70 (Rohl et al. 2015). These findings provide a possible mechanism for the separate regulation of Hsp70 and Hsp90 observed previously (Jones et al. 2004; Song and Masison 2005).

17.1.2 Hsp90 and Hsp70 Interact Independently of STI1/HOP

As we have seen, a host of biochemical, biophysical and structural studies have established a reasonable model for the molecular mechanism by which Sti1/Hop physically bridges Hsp70 and Hsp90. Yet, as mentioned above, loss of Sti1/Hop is not lethal under optimal growth conditions in yeast (Chang et al. 1997) or *C. elegans* (Gaiser et al. 2009; Song et al. 2009). On the other hand, Sti1/Hop does become essential when yeast cells are stressed by temperature changes or inhibition of Hsp90 ATPase (Chang et al. 1997). In *sti1*Δ yeast cells exogenously expressed mammalian glucocorticoid receptor fails to mature properly when exposed to hormone (Chang et al. 1997). While *C. elegans* lacking Sti1/Hop appear to have little phenotype under optimal conditions, they do have a reduced lifespan, developmental defects and weakened resistance to stress (Gaiser et al. 2009; Song et al. 2009). And in mice, loss of Sti1/Hop is embryonically lethal (Beraldo et al. 2013). Thus the role Sti1/Hop plays must be important for Hsp90-regulated processes that are more complex than merely maintaining viability under optimal conditions. Indeed, Hsp70 and Hsp90 interact directly (Genest et al. 2013; Kravats et al. 2018), and this Sti1/Hop independent binding is conserved from prokaryotes to humans.

The observation that yeast cells lacking Sti1 were viable under optimal conditions but temperature sensitive was reported over 20 years ago (Chang et al. 1997). Ten years later a report from the Johnson lab identified mutations in yeast Hsp90 that were synthetically lethal with *sti1*Δ (Flom et al. 2007). In cells expressing Sti1, these Hsp90 mutants co-purified with less Hsp70 than wild type Hsp90. They also observed that Hsp70 co-purified with Hsp90 in cells lacking Sti1. The authors noted that one of their mutants, G309S (in the *HSC82* isoform of yeast Hsp90), relied on Sti1 to interact with Hsp70. Taken together with results of earlier in vitro studies that found Hsp90 and Hsp70 could only co-purify in the presence of Hop, findings that were refuted by later studies (see below) (Chen et al. 1996; Johnson et al. 1996), the authors concluded that there may be another unknown factor that mediated the interaction of Hsp70 and Hsp90 in the absence of Sti1. Later studies showed that the interaction between Hsp70 and Hsp90 was in fact direct (Genest et al. 2013; Kravats et al. 2018).

Several studies from the laboratory of Sue Wickner have demonstrated unequivocally that Hsp70 and Hsp90 interact directly. Working primarily with the *E. coli* Hsp90 paralog HtpG, Genest and co-workers showed that prokaryotic Hsp90 and Hsp70 directly interacted with each other, which was expected since prokaryotes lack a Hop/Sti1 paralog (as well other Hsp90 co-chaperones) (Genest et al. 2013). Remarkably, this study also identified a putative Hsp70 binding site on *E. coli* Hsp90. In a follow-up study in collaboration with the Johnson and Masison labs, the Wickner group reported that the direct interaction between Hsp90 and Hsp70 was evolutionarily conserved in yeast (Kravats et al. 2018). The mutations in bacterial Hsp90 that affected direct interaction with Hsp70 had an identical affect when the analogous mutations were made in yeast Hsp90, and the most severe of these mutants were synthetically lethal with loss of Sti1. Remarkably, these residues (K394 and K399 in *HSP82*) were very close on the 3D structure of Hsp90 to G309, the residue identified as important for Hsp70 interaction reported earlier (Flom et al. 2007) (see above). Using purified proteins, the authors went on to show that yeast Hsp90 and Hsp70 physically interacted in vitro, although weakly, and the association of the mutant Hsp90s and Hsp70 was strengthened by Sti1. Together, these findings provided strong biochemical and genetic evidence that the Hsp70 binding site on Hsp90 was conserved from bacteria to yeast.

Our recent paper investigated the nature of mutations in Hsp90 that were lethal only in cells lacking Sti1 (discussed in more detail below), it was shown that the human paralogs of Hsp90 and Hsp70 directly interacted, demonstrating conservation of this interaction to humans (Reidy et al. 2018). Human Hsp90β is able to support viability in yeast cells as the only source of Hsp90, but only when Sti1 is present (Scheufler et al. 2000). We showed through genetics and biochemistry that a forward mutation that enhanced direct interaction with yeast Hsp70 (see below) or co-expression of a human Hsp70 isoform overcame the dependence of human Hsp90β on Sti1. These results provided strong evidence for the idea that the direct Hsp90-Hsp70 interaction was conserved in humans. Taken together, these studies support the idea that basic cellular functions are maintained by a direct Hsp90-Hsp70 interaction, and provide strong evidence that this interaction is, in eukaryotes, essential.

17.1.3 Beyond the Bridge: A Second Function for STI1/HOP?

Through various collaborations studying different mutations in yeast Hsp90 that affected aspects such as Hsp70 or client interaction, as well as through reports by other labs, we observed that point mutations in Hsp90 that were synthetically lethal with deletion of *STI1* clustered into two distinct regions on the 3D structure of Hsp90, see Fig. 17.3. Mutations in the first cluster such as G309S (Flom et al. 2007), V391E (originally described as important for Aha1 interaction in vitro, it was observed to be Sti1 dependent in vivo) (Retzlaff et al. 2010), and K399C (Kravats et al. 2018) were located in or near the Hsp70 binding site identified by Sue

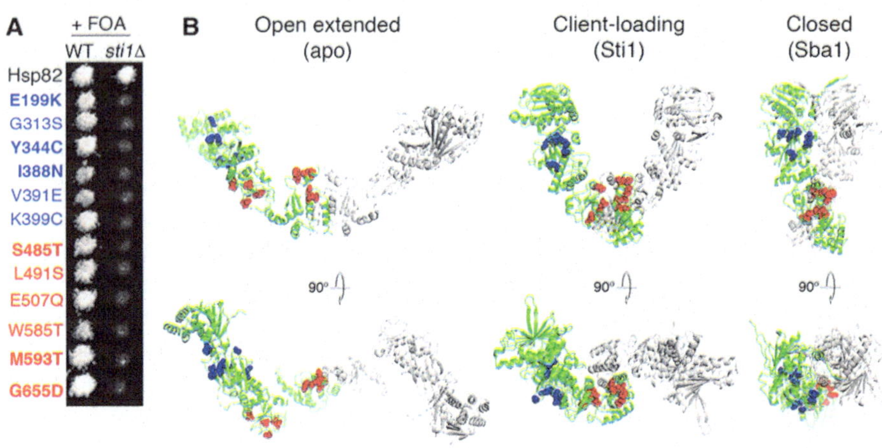

Fig. 17.3 Sti1-dependent mutants in Hsp90 lie in two distinct regions on the Hsp90 crystal structure. (**a**) Wild type (*STI1*$^+$) and *sti1*Δ cells expressing the indicated point mutations in *HSP82* (an isoform of yeast Hsp90) were grown on media lacking FOA then replica-plated to media containing FOA (shown here). The absence of growth of the mutants in *sti1*Δ cells (right column) demonstrates the inability of the indicated mutation to support viability as the only source of Hsp90. All mutants supported viability in *STI1*$^+$ cells (left column). (**b**) The mutants from A indicated as spheres on various conformations of Hsp90. Blue are SdN mutants and red are SdC mutants (see text for more detail). (Adapted from Reidy et al. 2018. *Genetics* 2018. Used by permission of the authors)

Wickner's lab. The dependence on Sti1 conferred by mutations in this area, which we labeled "SdN" (Sti1-dependent N-terminal proximal) made sense, since they were deficient in binding Hsp70 directly and thus should need Sti1 to serve as a bridge between Hsp90 and Hsp70. Less clear, however, was the functional requirement provided by Sti1 in the case of the second cluster of Sti1 dependent mutations, such as S481Y, L487S and M589A in the *HSC82* isoform (Flom et al. 2007) and E507Q and W585T in the *HSP82* isoform (Genest et al. 2013) (the Hsp82 mutants' Sti1 dependence was at the time unpublished). This second cluster of Sti1-dependent mutations was located at the junction of the M- and C-terminal domains, in a region identified as important for client interaction (Genest et al. 2013; Zuehlke et al. 2017) and was termed "SdC" (Sti1-dependent C-terminal proximal).

In order to convince ourselves that point mutations in Hsp90 that were Sti1 dependent were confined to two distinct regions on the Hsp90 structure, we conducted an unbiased screen to find new Sti1 dependent mutations. A pool of a *LEU2*-marked plasmid containing randomly mutagenized *hsp82* alleles was introduced via transformation into a strain that was deleted for *STI1* and both chromosomal copies of yeast Hsp90. In this parent strain viability is maintained by a copy of *HSP82* on a *URA3*-marked plasmid. Leu$^+$ transformants expressing mutant *hsp82* alleles that function like wild type under optimal conditions allowed for loss of the parental *URA3* plasmid when grown non-selectively and were thus able to grow on media containing 5′-fluoro-orotic acid (FOA), which is toxic to cells expressing *URA3*. Alternately, mutations that inactivated Hsp90 function were not able to lose the parental *URA3* plasmid and were sensitive to FOA. The FOA sensitive colonies from the first transformation that became FOA resistant upon introduction of a *STI1*-expressing *TRP1* plasmid in a second round of transformations were scored as Sti1 dependent. The mutant *hsp82* allele on the *LEU2* plasmid was then recovered and sequenced. After screening ~10,000 initial Leu$^+$ transformants in this way, we identified 6 new Sti1 dependent Hsp90 mutations. Remarkably, three of them (E199K, Y344C and I388N) mapped to the SdN region, and the other three (S485T, M593T and G655D) mapped to the SdC region (Reidy et al. 2018), see Fig. 17.3. Two of the SdC mutants identified were different substitutions in the same residue identified by the Johnson lab, but in the other isoform of yeast Hsp90. In total 12 Sti1-dependent mutants were identified, six from each class, Fig. 17.3. Two from each class were chosen to study further. One of the two chosen from each class had a more severe phenotype in *STI1* cells than the other, displaying radicicol sensitivity and a sorbitol-suppressed high temperature growth defect. This latter phenotype likely is due to a cell wall defect downstream of Hsp90, regulated by Hsp90 client MAP kinases such as Slt2 (Millson et al. 2005; Piper et al. 2006).

Any combination of individual SdN and SdC mutations in the same allele resulted in loss of viability in *STI1* containing cells under optimal conditions, strong genetic evidence that the different mutations affected different parts of a vital pathway. As expected, purified versions of the SdN mutant proteins bound much less Hsp70 compared to wild type Hsp82. However, the SdC mutations had no effect on Hsp70 binding, suggesting these mutations rely on Sti1 for something other than mediating interaction with Hsp70. One of the advantages of the yeast system is the ability to rapidly perform forward genetics such as second-site suppressor screens. In this type of analysis, the phenotype conferred by the mutation-of-interest is suppressed by additional mutations, either targeted or generated randomly, in the same allele. We used this approach combined with other yeast genetic techniques and biochemistry to help understand why the SdN and SdC mutations depended on Sti1.

In the first screen we identified a second site suppressor of the SdN mutation K399C's Sti1 dependency. The isolated suppressor, E402K, was located right next to K399C, and we found that restoring positive charge in this region was responsible for the suppressive effect. Because E402R worked slightly better, it was chosen for further study. The SdN suppressor E402R dramatically increased direct binding to Hsp70 alone and when combined with K399C restored direct Hsp70 interaction to wild type levels. This result was exactly what could be predicted based on the knowledge of Sti1/Hop functions in Hsp90-Hsp70 bridging. However, combining the Hsp70-binding-enhancing mutation E402R with either SdC mutation failed to relieve Sti1 dependence. Furthermore, when expressed in cells containing Sti1, E402R made the SdC radicicol sensitivity and high temperature growth defects worse. Thus, increasing Hsp70 interaction in the context of the client-binding defective SdC mutation was deleterious to proper functioning of Hsp90. These results were strong evidence that Hsp90 SdC mutants required a function of Sti1 that was not its ability to mediate interaction with Hsp70. What was this second function of Sti1? The answer to this question was also obtained using a second-site suppressor screen. Several suppressor mutations in residue A107 (P, G, and T) in the N-terminal domain, and A577V on the other end of the molecule in the SdC region, were identified through their ability to rescue SdC mutant growth defects in *STI1* cells. The mutation A107N was already known to promote N-terminal dimerization (Millson et al. 2010; Prodromou et al. 2000; Vaughan et al. 2009). As a result A107N increased the intrinsic ATPase rate compared to wild type. The Neckers lab reported that A107N suppressed the growth defects of W585T (a severe SdC mutant) in *STI1*-containing cells by promoting closure of the dimer to compensate for the weakened client interaction of W585T (Zuehlke et al. 2017). We found that A107N also rescued SdC mutant Sti1 dependency (Reidy et al. 2018), which was not addressed in Zuehlke et al. (2017).

The second SdC suppressor, A577V, was located in the SdC region itself. In addition to its ability to suppress SdC phenotypes in cells that have Sti1, A577V also suppressed Sti1 dependency of SdC mutants (Reidy et al. 2018). Remarkably, residue A577 had also been independently studied previously. The Buchner, Virkhiver and Colombo labs reported that residue A577 was part of a long-range signaling network within Hsp90 that relayed information about the client bound state to the N-terminal domain (Morra et al. 2009; Retzlaff et al. 2009). Both A107N or A577V alone increased intrinsic ATPase rate in vitro and bound more Sba1 in vivo, demonstrating that both of these SdC suppressor mutations promoted N-terminal dimer closure (Reidy et al. 2018). Thus, SdC mutants were defective in proper dimer closure and required Sti1 to fulfill this crucial step in Hsp90 regulation.

In direct agreement with the idea that the two Sti1-dependent regions of Hsp90 relied on Sti1 for two different functions, the SdC suppressors (A107N or A577V) were unable to relieve the Sti1-dependence of the SdN mutations or rescue the growth defects of the severe SdN mutation expressed in *STI1* cells. Finally, combining the SdC suppressor A107N with the SdN suppressor E402R rescued the temperature growth defect of *sti1Δ* cells, while either alone only partially rescued. In other words, the two forward mutations imparted into Hsp90 itself the two functions of Sti1. Taken together, these findings are very strong evidence that Sti1 performs two roles in regulating Hsp90: mediating Hsp90 interaction with Hsp70 and assisting in proper N-terminal dimerization, presumably after client transfer. Figure 17.4 is a proposed model showing the effects of loss of Sti1/Hop functions on Hsp90 client capture. These two functions probably arise from the same interaction between Sti1/Hop and Hsp90. Furthermore, this study provides important in vivo validation for a large number of biochemical, biophysical and computational studies, and supports the idea first advanced by the Agard group that Sti1/Hop must be taking a more active role in Hsp90 regulation than merely serving as a bridge (Southworth and Agard 2011).

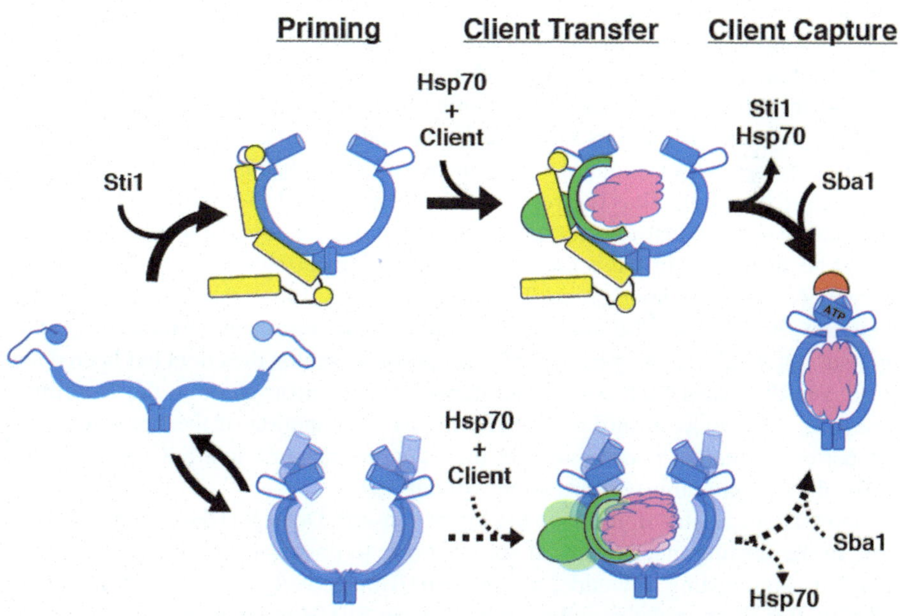

Fig. 17.4 A model for the dual roles of Hop/Sti1 in Hsp90 cycle regulation. Binding of Hop/Sti1 forces Hsp90 into the client-loading conformation, keeping Hsp90 open to accept the client from Hsp70 and then assisting Hsp90 N-terminal domain closure once the client has been successfully transferred

17.2 Conclusions

If, as described above, Sti1/Hop's role goes beyond bridging Hsp90 and Hsp70, several questions arise. First, if Sti1/Hop is so important for Hsp90 regulation, then why is it not essential? The answer to this partly lies in the fact that Hsp90 and Hsp70 can interact directly to provide enough function for cells to thrive under optimal conditions. However, *STI1* does become essential when cells are stressed. It is possible that the detrimental effects of loss of Sti1/Hop function are buffered under optimal conditions by the sheer abundance of Hsp90 in the cell. It has been shown that some point mutations in Hsp90 only give rise to a phenotype when their expression level is lowered (Jiang et al. 2013). It is reasonable to speculate that a similar phenomenon hides the full effect of loss of Sti1/Hop function. Evidence in support of this idea is the fact that *sti1Δ* cells are unable to grow when Hsp90 activity is reduced by normally sub-lethal concentrations of radicicol. Anecdotal evidence that *sti1Δ* yeast cells express even more Hsp90 than wild type cells (in which Hsp90 already comprises 1–2% of all cytosolic proteins) when grown under optimal conditions (our unpublished observations) perhaps supports the notion that loss of Sti1/Hop function is buffered by Hsp90 abundance. The fact that Sti1 functions could be bypassed by forward mutations in Hsp90 itself may also be interpreted in support of this idea (Reidy et al. 2018).

As mentioned above, residues in the SdC region were previously implicated in an interdomain signaling network (Morra et al. 2009; Retzlaff et al. 2009). Presumably this network sent information about the client state to the N-terminal ATPase domains. Mutations in this area that caused cells to depend on Sti1 were rescued by secondary mutations that increased the propensity of the N-terminal domains to dimerize. Thus this region of Hsp90 needs Sti1 to help it close. However, it remains unclear just exactly why SdC mutations need assistance in closing. It could be structural, in the sense that SdC mutations confer a mechanical impediment to N-terminal dimerization. The fact that the SdC region is located in the hinge between the M- and C-domains may support this argument. Alternatively, SdC mutant Hsp90 dimers may be unable to properly sense that the client has been delivered, and they need Sti1 to facilitate transfer of the information to the N-terminal domains. Of course, a third explanation is the combination of the first two: that a proper conformation is necessary for the signal to transmit, and SdC mutants have difficulty attaining this conformation without Sti1.

How is successful transfer of the client sensed? Does Sti1/Hop play an active role in the sensing mechanism? If so, what role to post-translational modifications play? It is interesting to wonder whether Sti1/Hop actively dissociates Hsp70 from Hsp90 after client transfer, and what happens to Sti1/Hop after the transfer is complete? The questions go on. The finding that Sti1/Hop plays a role in Hsp90 regulation beyond the bridge is an exciting development that hopefully will increase our understanding of this important cellular chaperone machine.

Acknowledgements We thank our National Institutes of Health colleagues for insightful discussions and help with the manuscript. This work was supported by the Intramural Program of the National Institutes of Health, National Institute of Diabetes and Digestive and Kidney diseases.

References

Ali MM, Roe SM, Vaughan CK et al (2006) Crystal structure of an Hsp90-nucleotide-p23/Sba1 closed chaperone complex. Nature 440:1013–1017

Beraldo FH, Soares IN, Goncalves DF et al (2013) Stress-inducible phosphoprotein 1 has unique cochaperone activity during development and regulates cellular response to ischemia via the prion protein. FASEB J 27:3594–3607

Carrigan PE, Nelson GM, Roberts PJ, Stoffer J, Riggs DL, Smith DF (2004) Multiple domains of the co-chaperone Hop are important for Hsp70 binding. J Biol Chem 279:16185–16193

Chang HC, Nathan DF, Lindquist S (1997) In vivo analysis of the Hsp90 cochaperone Sti1 (p60). Mol Cell Biol 17:318–325

Chen S, Prapapanich V, Rimerman RA, Honore B, Smith DF (1996) Interactions of p60, a mediator of progesterone receptor assembly, with heat shock proteins hsp90 and hsp70. Mol Endocrinol 10:682–693

Flom G, Weekes J, Williams JJ, Johnson JL (2006) Effect of mutation of the tetratricopeptide repeat and asparatate-proline 2 domains of Sti1 on Hsp90 signaling and interaction in *Saccharomyces cerevisiae*. Genetics 172:41–51

Flom G, Behal RH, Rosen L, Cole DG, Johnson JL (2007) Definition of the minimal fragments of Sti1 required for dimerization, interaction with Hsp70 and Hsp90 and in vivo functions. Biochem J 404:159–167

Gaiser AM, Brandt F, Richter K (2009) The non-canonical Hop protein from *Caenorhabditis elegans* exerts essential functions and forms binary complexes with either Hsc70 or Hsp90. J Mol Biol 391:621–634

Genest O, Reidy M, Street TO et al (2013) Uncovering a region of heat shock protein 90 important for client binding in E. coli and chaperone function in yeast. Mol Cell 49:464–473

Genest O, Hoskins JR, Kravats AN, Doyle SM, Wickner S (2015) Hsp70 and Hsp90 of *E. coli* directly interact for collaboration in protein remodeling. J Mol Biol 427:3877–3889

Jiang L, Mishra P, Hietpas RT, Zeldovich KB, Bolon DN (2013) Latent effects of Hsp90 mutants revealed at reduced expression levels. PLoS Genet 9:e1003600

Johnson J, Corbisier R, Stensgard B, Toft D (1996) The involvement of p23, hsp90, and immunophilins in the assembly of progesterone receptor complexes. J Steroid Biochem Mol Biol 56:31–37

Jones G, Song Y, Chung S, Masison DC (2004) Propagation of *Saccharomyces cerevisiae* [PSI+] prion is impaired by factors that regulate Hsp70 substrate binding. Mol Cell Biol 24:3928–3937

Karagoz GE, Rudiger SG (2015) Hsp90 interaction with clients. Trends Biochem Sci 40:117–125

Kravats AN, Hoskins JR, Reidy M et al (2018) Functional and physical interaction between yeast Hsp90 and Hsp70. Proc Natl Acad Sci U S A 115:E2210–E2E19

Krukenberg KA, Bottcher UM, Southworth DR, Agard DA (2009) Grp94, the endoplasmic reticulum Hsp90, has a similar solution conformation to cytosolic Hsp90 in the absence of nucleotide. Protein Sci 18:1815–1827

Lee CT, Graf C, Mayer FJ, Richter SM, Mayer MP (2012) Dynamics of the regulation of Hsp90 by the co-chaperone Sti1. EMBO J 31:1518–1528

Millson SH, Truman AW, King V, Prodromou C, Pearl LH, Piper PW (2005) A two-hybrid screen of the yeast proteome for Hsp90 interactors uncovers a novel Hsp90 chaperone requirement in the activity of a stress-activated mitogen-activated protein kinase, Slt2p (Mpk1p). Eukaryot Cell 4:849–860

Millson SH, Prodromou C, Piper PW (2010) A simple yeast-based system for analyzing inhibitor resistance in the human cancer drug targets Hsp90alpha/beta. Biochem Pharmacol 79:1581–1588

Morra G, Verkhivker G, Colombo G (2009) Modeling signal propagation mechanisms and ligand-based conformational dynamics of the Hsp90 molecular chaperone full-length dimer. PLoS Comput Biol 5:e1000323

Piper PW, Truman AW, Millson SH, Nuttall J (2006) Hsp90 chaperone control over transcriptional regulation by the yeast Slt2(Mpk1)p and human ERK5 mitogen-activated protein kinases (MAPKs). Biochem Soc Trans 34:783–785

Pratt WB, Morishima Y, Murphy M, Harrell M (2006) Chaperoning of glucocorticoid receptors. Handb Exp Pharmacol 172:111–138

Prodromou C (2016) Mechanisms of Hsp90 regulation. Biochem J 473:2439–2452

Prodromou C, Panaretou B, Chohan S et al (2000) The ATPase cycle of Hsp90 drives a molecular 'clamp' via transient dimerization of the N-terminal domains. EMBO J 19:4383–4392

Reidy M, Kumar S, Anderson DE, Masison DC (2018) Dual roles for yeast Sti1/Hop in regulating the Hsp90 chaperone cycle. Genetics 209:1139–1154

Retzlaff M, Stahl M, Eberl HC et al (2009) Hsp90 is regulated by a switch point in the C-terminal domain. EMBO Rep 10:1147–1153

Retzlaff M, Hagn F, Mitschke L et al (2010) Asymmetric activation of the hsp90 dimer by its cochaperone aha1. Mol Cell 37:344–354

Rohl A, Wengler D, Madl T et al (2015) Hsp90 regulates the dynamics of its cochaperone Sti1 and the transfer of Hsp70 between modules. Nat Commun 6:6655

Scheufler C, Brinker A, Bourenkov G et al (2000) Structure of TPR domain-peptide complexes: critical elements in the assembly of the Hsp70-Hsp90 multichaperone machine. Cell 101:199–210

Schmid AB, Lagleder S, Grawert MA et al (2012) The architecture of functional modules in the Hsp90 co-chaperone Sti1/Hop. EMBO J 31:1506–1517

Schopf FH, Biebl MM, Buchner J (2017) The HSP90 chaperone machinery. Nat Rev Mol Cell Biol 18:345–360

Song Y, Masison DC (2005) Independent regulation of Hsp70 and Hsp90 chaperones by Hsp70/Hsp90-organizing protein Sti1 (Hop1). J Biol Chem 280:34178–34185

Song HO, Lee W, An K et al (2009) *C. elegans* STI-1, the homolog of Sti1/Hop, is involved in aging and stress response. J Mol Biol 390:604–617

Southworth DR, Agard DA (2011) Client-loading conformation of the Hsp90 molecular chaperone revealed in the cryo-EM structure of the human Hsp90:Hop complex. Mol Cell 42:771–781

Vaughan CK, Piper PW, Pearl LH, Prodromou C (2009) A common conformationally coupled ATPase mechanism for yeast and human cytoplasmic HSP90s. FEBS J 276:199–209

Verba KA, Wang RY, Arakawa A et al (2016) Atomic structure of Hsp90-Cdc37-Cdk4 reveals that Hsp90 traps and stabilizes an unfolded kinase. Science 352:1542–1547

Zuehlke AD, Johnson JL (2012) Chaperoning the chaperone: a role for the co-chaperone Cpr7 in modulating Hsp90 function in *Saccharomyces cerevisiae*. Genetics 191:805–814

Zuehlke AD, Reidy M, Lin C et al (2017) An Hsp90 co-chaperone protein in yeast is functionally replaced by site-specific posttranslational modification in humans. Nat Commun 8:15328

Part III
HSP90 in Natural Products Development

Chapter 18
Hsp90: A Target for Susceptibilities and Substitutions in Biotechnological and Medicinal Application

Athanasia Warnecke, Andreas Kirschning, Daniel Landsberg, and Carsten Zeilinger

Abstract A main influencer of the chaperome is the environmental stress eliciting continuously degraded proteins. To avoid proteotoxic stress, which hinders the protein homeostasis and cell survival, most proteins are accompanied from the early existence on by heat shock proteins (HSP) and sequestrated into different routes of renaturation, de novo folding or denaturation. Therefore, the regulation of Hsp90 presence and activity is relevant for most cells and their function. In this position, Hsp90 can decide the fate between health and disease by selective refolding of denatured proteins and is a player in the evolution. From the last century on, Hsp90 was validated as a target due to its susceptibility for natural products and to make them perfect with the aim to hinder refolding and enhance the proteotoxic stress. In this review, the links between natural producer, chemical synthesis with Hsp90 as a target as well as alternative chaperoning routes by chemical compounds are illuminated.

Keywords Cancer · Chaperone · Chemical chaperones · Heat shock protein · Target

---

A. Warnecke
Clinic for Otorhinolaryngology – Head and Neck Surgery, Hannover Medical School (MHH), Hannover, Germany

A. Kirschning
Leibniz University Hannover, Institute of Organic Chemistry and Biomolecular Drug Research Center (BMWZ), Hannover, Germany

D. Landsberg · C. Zeilinger (✉)
Leibniz University Hannover, Center of Biomolecular Drug Research Center (BMWZ), Hannover, Germany
e-mail: zeilinger@cell.uni-hannover.de

A. A. A. Asea, P. Kaur (eds.), *Heat Shock Protein 90 in Human Diseases and Disorders*, Heat Shock Proteins 19,
https://doi.org/10.1007/978-3-030-23158-3_18

Abbreviations

CFTR	Cystic fibrosis transmembrane conductance regulator
HSP	Heat shock proteins
NBD	Nucleotide binding domain

18.1 Introduction

The original folding of newly synthesized proteins is so important for cell-important functions that a whole proteome section is involved, the chaperome (Wang et al. 2006; Kim et al. 2013; Joshi et al. 2018). Cellular functions such as growth, development and differentiation and organ and tissue functionality are significantly influenced by folding of proteins in several respects. On the one hand chaperones provide functional proteins at the desired time and thus are subjected to be a control station for the release of folded proteins so that various logistic stages in the provision are possible. On the other hand, chaperons re-fold worn proteins, which have lost their function, (Verba et al. 2016). All these steps make sense for economic reasons, since the costs for a new synthesis of the proteins would be much higher. The final recycling step serves to break down protein components. This concept was developed early in the evolution of cells that exist in all cells from archea to eukaryotes and has proven to be advantageous over a wide variety of selection scenarios (Geller et al. 2018; Chen et al. 2018; Starr et al. 2018). In addition, even defense strategies of some organisms are based on eliminating these central sites of a cell via natural products. HSP are catalytic mediators and produce balances of various client states. The chaperome is extremely complex in that there are many different HSP and all are highly regulated in their activity.

As a highly conserved ubiquitous protein, Hsp90 can be identified in different tissue types, body fluids and is even one of the top 100 proteins that are present in extracellular vesicles (Yáñez-Mó et al. 2015; Calderwood 2018), the main communication component between neighbouring and distant cells. It represents 1–2% of the total mammalian protein content under physiological conditions (Hoter et al. 2018). Two forms, Hsp90 alpha and beta are located in the cytoplasm, whereas the mitochondrial form of Hsp90 and a form that is present in the endoplasmatic reticulum also exist (i.e., GRP94 = 94-kDa glucose-regulated protein). The stress inducible form Hsp90 alpha shares a 85% homology with the constitutively expressed Hsp90 beta (Li and Buchner 2012; Johnson 2012). Both proteins are thought to interact with more than 10% of the human proteome. A new classification system for the human Hsp90 members has been introduced categorizing all five members under the superfamily name HspC (Kampinga et al. 2009).

In the 60s of the last century, when the function of genes on model organism begun to unravel the molecular background, a heat inducible gene activity was identified and correlated later with Hsp90 alpha as the inducible form of the gene (Ritossa 1962a, b). Later, it was shown that heat and other physiological influences as stress factors elicit the Hsp90 answer mostly in eukaryotic cell systems, but also for adaptation to changed environment required in prokaryotes (Jarosz and Lindquist 2010; Honoré et al. 2017). The activity of inducible Hsp90 is not only highly regulated on a genomic, but also on a posttranslational level and is embedded into sequences of activating/inactivating cochaperons and other regulators of posttranslational modifications (PTM). A long list of client proteins is known, but there is not an explicit specificity of client proteins known. If these candidates are accompanied by Hsp90, a higher concentration level of Hp90 should persist with the aim to hinder foldless client proteins (Echeverría et al. 2011). The step from the unfolded to the folded client contains a complex toothed sequence of single elements of the whole chaperome.

The regulation by co-chaperones and posttranslational modifications can also be an access to manipulate the function of Hsp90 by inhibitors (Franke et al. 2013; Shrestha et al. 2016). Cochaperones can exert inhibitory, activating or catalytic effects on the Hsp90 network and greatly expand its range of tasks, Fig. 18.1. The binding of co-chaperones to specific binding motifs in the Hsp90 domains causes a shift in the equilibrium between the conformations of the protein during the course of the cycle. In general, the Hsp90 is kept in an inactive state and is only activated for specific tasks controlled by ATP accessibility and lid regulation as well dimer formation (Blacklock and Verkhivker 2014), Fig. 18.2. The most important co-chaperones include Hop, p23, Cdc37 and Aha1 for allosteric modulation of the function of Hsp90 and Hsp70 (Verkhivker 2018; Ferraro et al. 2019). Hop, also referred to as STIP1 (stress-induced phosphoprotein 1), has nine TPR motifs in a total of three domains and thus binds specifically to the EEVD motif at the

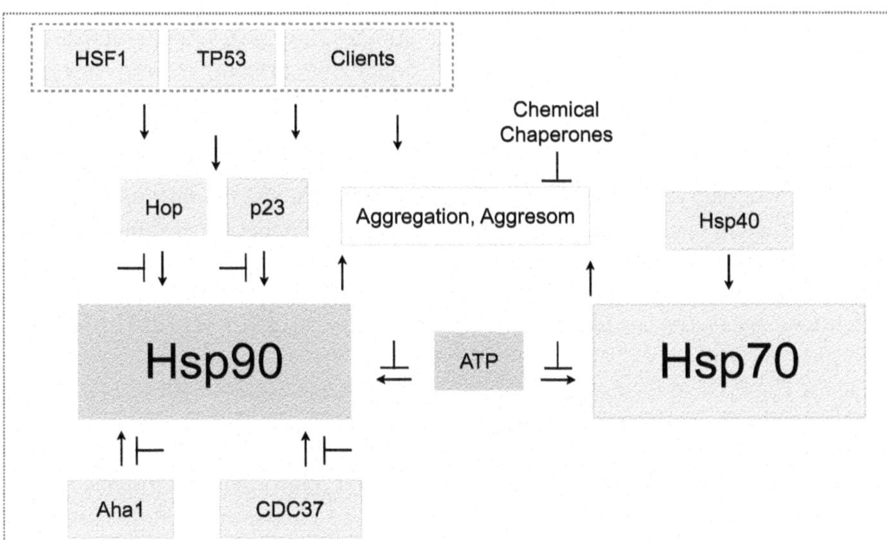

Fig. 18.1 Fate of a protein in the chaperon network with different instances. Hsp90 and Hsp70 the main driver of the protein folding machinery under control and several influences (⊥)

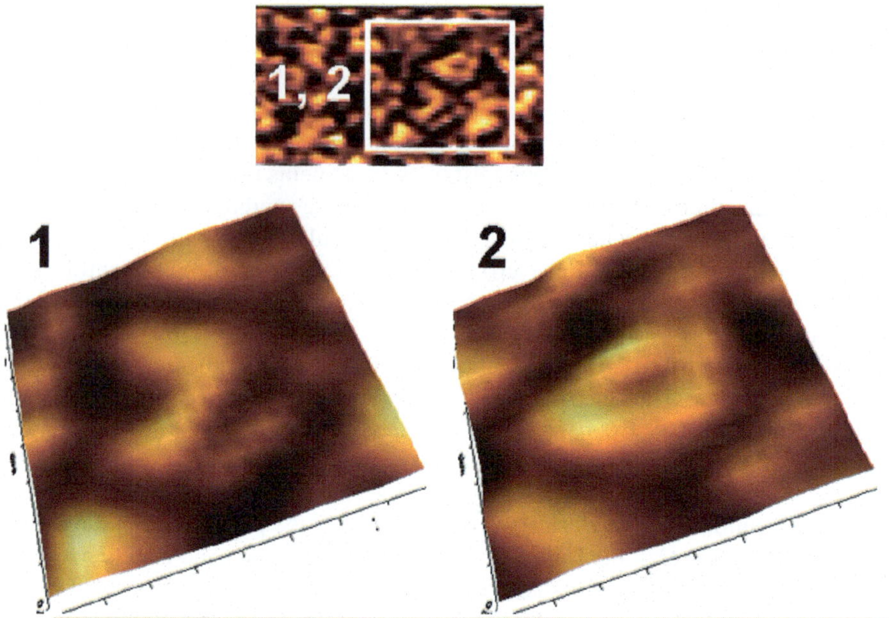

Fig. 18.2 3D high resolution AFM images of Hsp90 monolayer: A scan field of 190 × 250 nm^2 was cropped from the original scan to illustrate different Hsp90 morphologies (1, 2,) Hsp90 in the open (1) and closed (2) state. Hsp90 with a length of ~12 nm, a width of 5–7 nm and a height of ca. 2 nm picked from the overview scan

C-terminus of Hsp90. It contributes to the transfer of substrates from Hsp70 to Hsp90. In eukaryotic systems as well as in humans, it requires the participation of Sti1/Hop (stress inducible protein/Hsp90 organizing protein) thereby integrated into the multi-chaperone network and thus participates in the folding of proteins to ensure protein homeostasis (Verkhivker 2018; Ferraro et al. 2019). Hop has one binding site for Hsp90 (TPR2A) and two for Hsp70 (TPR1 and TPR2B) (Röhl et al. 2015). Newer data revealed that Hsp90 regulates tightly Sti1 binding and modulates thereby efficient client handover. Recently, it was shown that purified human Hsp90 binds directly on Hsp70 spotted on microarray devices at μM affinities (EC_{50} ~40 μM) (Mohammadi-Ostad-Kalayeh et al. 2017), Fig. 18.3. Other experimental conditions measured affinities in in similar range on yeast Hsp90/Hsp70 (EC_{50} ~13 μM) (Kravats et al. 2018; Bhattacharya et al. 2018).

Once Hsp90 is in its closed and dimerized form, it interacts with p23 (PTGES3, prostaglandin E synthase 3 protein). For this purpose, the C-terminus of p23 connects to the Hsp90 N-domain. The stabilization of Hsp90 in this conformation by p23 reduces the rate of the Hsp90 mechanism and allows ATP hydrolysis (Li and Buchner 2012; Forsythe et al. 2001). The Cdc37 co-chaperone (cell division cycle 37) has control properties, for example, in the cell division, and is essential for cell signal transduction (Vaughan et al. 2006). It works much like a tumor suppressor and inhibits specific gene expression. Furthermore, Cdc37 is able to activate DNA

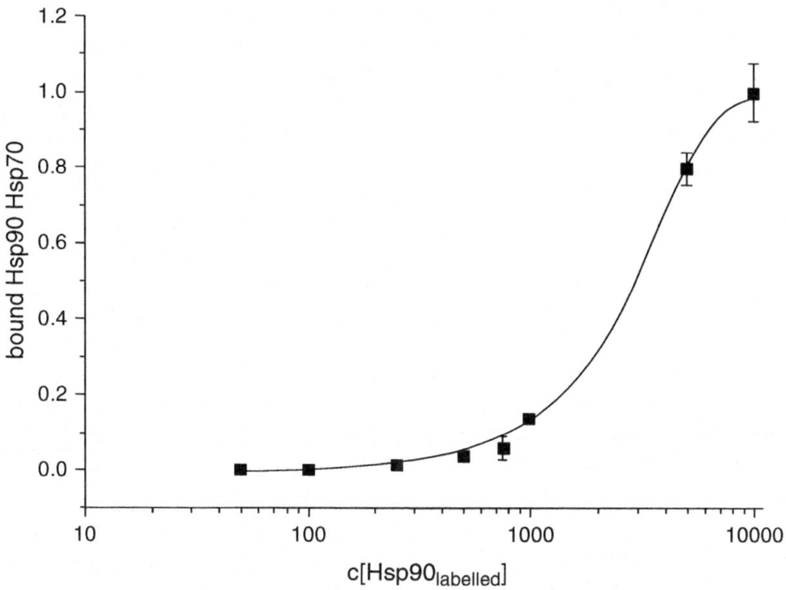

Fig. 18.3 Typical dose response curve obtained for direct Hsp90-Hsp70 interaction

repair mechanisms and participate in apoptosis induction. It binds with its C-domain to the Hsp90-N domain and is also located in tumor cells (Taipale et al. 2010; Zhao et al. 2005). Finally, mention should be made of Aha1 (activator of Hsp90 ATPase homologue 1), which promotes dimerization of the N domain by accelerating the ATPase cycle. It binds with its N domain to the central Hsp90 domain, thereby acting as an allosteric catalyst (Wandinger et al. 2008).

Posttranslational modifications include phosphorylation, acetylation, S-nitrosylation, and methylation, which affect the ATPase activity of Hsp90 (Mayer and Le Breton 2015; Sager et al. 2018). They occur after translation on the ribosome and alter the protein via chemical modification of a particular amino acid in the binding domain. Phosphorylations of tyrosine, threonine and serine are the most common. In addition, various kinases regulate the phosphorylation state of Hsp90 and its client binding. Finally, acetylation affects protein maturation and its binding with co-chaperones and reduces the ATP binding rate. The homo-dimerized Hsp90 has highly conserved binding domains, which can also be divided into three functional regions.

The N-terminal domain is about 25 kDa in size and contains an adenine-binding site (NBD) to which not only ATP but also inhibitors can bind. The 10 kDa central domain provides binding sites for co-chaperones and client proteins and stabilizes the protein (Saibil 2013). The last dimerization domain located at the C-terminus has binding sites not only for co-chaperones and inhibitors, but also for ATP. The binding motif EEVD is located in this binding domain, to which co-chaperones, like Hop, can bind to the TRP domain. Hsp90 takes on different conformations during the course of the chaperone cycle, which are caused by different factors. First, Hsp90 binds to ATP in an open, non-dimerized and nucleotide-free conformation, repositioning the "lid" segment and thus short-lived dimerization of the N domain. Hsp90 performs ATP hydrolysis by structural rearrangements from a twisted to a closed state. The binding to Aha1 is important to accelerate the conformational rearrangement and thereby increase the speed of the process. Without the co-chaperone, it is difficult for Hsp90 to achieve the proper conformation needed for ATP hydrolysis. In addition, interactions with the co-chaperones STIP1 and Cdc37, as well as with inhibitors that bind to the NBD, have the opposite effect of Aha1 and prevent the structural changes required for N-dimerization. Once successful, p23 can stabilize the closed and ATP-bound conformation and slow down the ATPase cycle to allow ATP hydrolysis. After the reaction, the N domains dissociate and lie again separated. The ATP lid opens and ADP and P_i can be released. Finally, Hsp90 is back in the initial state and the cycle is started anew (Sager et al. 2018; Saibil 2013).

18.2 HSP90 in Health and Disease

The major role of Hsp90 is to maintain and regulate proteins that are involved in apoptosis, cell cycle control, cell viability, protein folding and degradation, as well as signalling events (Saibil 2013; Dezwaan and Freeman 2008). A plethora of co-chaperones aids Hsp90 to perform their folding function in a complicated process named Hsp90 chaperone cycle (Saibil 2013; Dezwaan and Freeman 2008). Hsp90s are subject to posttranslational modifications thereby modulating their diverse functions. The role of Hsp90 in viral infection and cancer is well investigated (Trepel et al. 2010; Schuster et al. 2017). However, increasing evidence accumulates that Hsp90 may be involved in health and diseases other than infection and cancer (Taipale et al. 2010; Woodford et al. 2016).

Hsp90 interacts with a multitude of signalling pathways. Among these, protein kinases and steroid hormone receptors are most prominent. The glucocorticoid receptor is sequestered in the cytoplasm with an Hsp90-chaperone complex (Trune and Canlon 2012). Upon binding of the steroid to the glucocorticoid receptor, the Hsp90 complex is released by a conformational change of the receptor (Trune and Canlon 2012). This release unmasks the nuclear localization sequence and the receptor translocates into the nucleus (Patel et al. 2014). Among the chaperones and co-chaperones that regulate the glucocorticoid receptor, FK506 binding proteins (FKBP) are the most prominent (Fries et al. 2017). For example, intratympanic administration of dexamethasone leads to a significantly upregulated expression of FKBP in the ear (Maeda et al. 2012). Since FKBP binds FK506 (i.e., tacrolimus), an immunosuppressant that causes sensorineural hearing loss, upregulation of FKBP might be the mechanism by which dexamethasone is able to protect against FK506-mediated ototoxicity (Maeda et al. 2012). Indeed, heat shock is able to inhibit to a certain degree neomycin- and cisplatin-induced hair cell death, thus preventing ototoxic hearing loss (Cunningham and Brandon 2006). On the other hand, the HSP-chaperone complex might play a role in the mediation of glucocorticoid-induced side effects such as osteoporosis. Consequently, inhibition of Hsp90 restored glucocorticoid-induced osteoporosis by enhancing bone formation (Chen et al. 2017). Similar has been observed in the case of Cushing disease, a disease characterized by chronic elevation of glucocorticoids accompanied by overexpression of heat shock proteins (Riebold et al. 2015). Thus, misregulated glucocortoicoid receptor sensitivity was addressed with an Hsp90 inhibitor (Riebold et al. 2015). Translational modifications of Hsp90 may also result in an impaired chaperone-dependent activation of glucocorticoid receptors. An understanding of the interactions between Hsp90 and glucocorticoid receptors or other cofactors after binding of the steroid to the receptors should be the base for the use of steroids to treat different conditions.

In neurodegenerative diseases, Hsp90 may also present a target molecule for treatment. The two most prominent neurodegenerative diseases are Alzheimer and Parkinson. Both are characterized by accumulation of aggregated proteins and peptides in the extracellular and intracellular compartments (Lackie et al. 2017). Hsp90 and its co-chaperones stabilize client proteins (Sõti et al. 2005) thereby facilitating Alzheimer's and Parkinson's disease. Accumulation of the intracellular microtubule-associated protein tau into fibrillar aggregates is the hallmark of Alzheimer's disease (Morris et al. 2011). By preventing protein misfolding and aggregation, Hsp90 may be involved in Alzheimer's disease (Koren et al. 2009; Woo et al. 2017) and accumulation of tau can be diminished by the application of Hsp90 inhibitors (Roe et al. 2018). For example, dihydropyridine derivatives can compromise the ability of Hsp90 to chaperone by modulating its ATPase activity (Roe et al. 2018). Consequently, heat shock response is induced in diseased cells leading to normalization of neurofibrillary tangles, preservation of neurons, reduced tau pathology, reduced amyloid plaque, and increased dendritic spine density in a model of Alzheimer's disease (Roe et al. 2018). In Parkinson's disease, the synaptic protein synuclein accumulates and forms complex oligomers and protein aggregates, i.e., Lewy bodies (Aridon et al. 2011).

In addition to tau protein and synuclein, other proteins involved in neurodegenerative disease such as Leucine-rich repeat kinase 2, huntingtin, ataxins, protein kinase Cγ or transactive response DNA binding protein with M_r 43 kDa interact with Hsp90 (Nakazono et al. 2018; Alam et al. 2017; Inda et al. 2016; Reis et al. 2017). Thus, Hsp90 can present a therapeutic target for many neurodegenerative diseases (Nakazono et al. 2018; Inda et al. 2016; Reis et al. 2017). Identifying Hsp90 inhibitors that can be applied chronically, have a good bioavailability in order to achieve therapeutic concentrations in the brain and that are not toxic seems a promising approach. Unfortunately, any promising agent identified thus far has unacceptable side effects limiting their clinical translation (Jhaveri et al. 2014; Shi et al. 2012; Gerecitano et al. 2015; McFarland et al. 2014). On a genetic level, heat shock factor 1 is the transcription factor that regulates the expression of heat shock genes (Hooper et al. 2016). By inhibiting Hsp90, Hsf1 bound to Hsp90 is released and able to activate gene expression resulting in the upregulation of important pre- and postsynaptic proteins (Chen et al. 2014). Indeed, synaptic impairment can be reversed by the application of Hsp90 inhibitors (Thirstrup et al. 2016). Thus, Hsp90 inhibitors prevent synapse desintegration and memory loss in an Alzheimer's mouse model (Wang et al. 2017). Support of essential synaptic elements such as neuroligands, vesicle transport, synaptic scaffolding proteins, lipid rafts, synaptic spines, and axodendritic synapses is one major hallmark that can be achieved by Hsf1 due to inhibition of Hsp90 (Hooper et al. 2016). In addition to Hsp90, co-chaperones may also be involved in neurodegenerative diseases. For example, p23 binds to γ-secretases and promotes production of the beta-amyloid protein in the brain (Vetrivel et al. 2008). Silencing of the p23 gene results, similar to Hsp90 inhibitors, subverts attempts to refold proteins, thereby favouring degradation (Edkins 2015; Dickey et al. 2007). The actions of Hsp90 in neurodegenerative diseases are manifold and unravelling their mode of action may enable a precise targeting of Hsp90 for the treatment of different neurodegenerative disorders.

Another interesting aspect of Hsp90 is the canalization of disease phenotypes. For example, a high degree of phenotypic heterogeneity is observed in patients with a mutation of the transcription factor GATA3 that causes hypoparathyroidism, sensorineural deafness and renal dysplasia (HDR) syndrome. There is a growing body of evidence that Hsp90 accounts for this variability (Sheehan-Rooney et al. 2013). In cystic fibrosis, changes in the Hsp90 co-chaperone folding environment markedly altered the stability and export of cystic fibrosis transmembrane conductance regulator from the endoplasmatic reticulum (Wang et al. 2006). Maturation of some proteins that are prone to misfolding can be enhanced by manipulation of the Hsp90 chaperone system, thus, presenting in at least some mutations a promising therapeutic target.

Inherited progressive hearing loss is among others the result of alterations in autosomal dominant non-syndromic deafness genes. DFNA2 is the result of mutations in the KCNQ4 channel. Both subtypes of Hsp90 (alpha and beta) play key roles in controlling KCNQ4 homeostasis via the Hsp40-Hsp70-Hop-Hsp90 chaperone pathway and the ubiquitin- proteasome pathway (Gao et al. 2013). Even though rescue of function was not possible yet, the cell surface expression of the KCNQ4 channel using Hsp90 as a molecular target for therapeutic intervention was restored to the normal level in an in vitro model mimicking heterozygous conditions of DFNA2 patients (Gao et al. 2013). Although heat shock can also inhibit ototoxic drug- and noise-induced hair cell death (Cunningham and Brandon 2006; Gong et al. 2012), effective treatment strategies based on the modification of HSP in the inner ear have not been developed yet. A plethora of other organ systems and diseases such as diabetes or nephropathies are affected by Hsp90 and may also be subject to future molecular therapies addressing the modulation of the Hsp90 system.

18.3 HSP90 Target Validation in Pathogenic Cells

In 1970, it was reported that the soil dwelling actinomycetes *Streptomycetes hygroscopicus* produces the natural product geldanamycin (Gda), a benzoquinone ansamycin antibiotic and also herbimycin, which was first described to have a high potency against mouse leukemia (L1210) and oral carcinoma (KB) cells (DeBoer et al. 1970; Omura et al. 1979; Tanida et al. 1980). Later it was shown that Hsp90, an intracellular protein with the molecular weight of 90 kDa, is the target of Gda, which competes with ATP for the N-terminal pocket of Hsp90 and thereby inhibiting the ATPase activity of Hsp90 at very low concentrations (Sasaki et al. 1979). As a result, the recycling of unfolded proteins, which occur due to elevated stress in cancer cells, is suppressed (Srivastava et al. 1978; Moulin et al. 2005).

In the past, a "natural weapon" library has been assembled with Gda and related benzoquinone and ansamycin derivatives of fungal-derived Hsp90 inhibitors (e.g., hypothemycin, radicicol, zearalenone, and structurally related monocillins and pochonins) being key members (Winssinger and Barluenga 2007). They address Hsp90 as a target and use Hsp90 as an 'Achilles' heel (Khandelwal et al. 2016). The toxicity of Gda elicited in target plant, bacteria or human cells does not exist in the Gda producer cells of *S. hygroscopicus* (Millson et al. 2011; Piper and Millson 2012). This absence of susceptibility is due to certain amino acid exchanges in the N-terminal ATP binding domain of the bacterial Hsp90 analogue HtpG. This resistivity of HtpG demonstrates that permutations of a few amino acids enable adaptation of heat shock proteins to Hsp90 inhibitors and indicate that some amino acid exchange can induce resistance effects. An interesting effect for plant metabolites on humans was reported for green tea, which produces the catechin (−)-epigallocatechin-3-gallate (EGCG). It targets Hsp90 and shows cancer protective properties (Shirakami et al. 2012; Yin et al. 2009; Khandelwal et al. 2013). Traditionally, green tea was used a long time before this molecular knowledge was at hand. Primarily, this Hsp90 inhibitor may be a result from the rhizosphere battlefield of the plant with organisms like insects, bacteria, fungi and others, but due to cultivation by men and consequently reduced toxicity this knowledge gains new significance for human health (Raaijmakers et al. 2009). Gambogic acid (Gba), a compound used in the traditional Chinese medicine, is a candidate with anti Hsp90 activity that binds not like EGCG on the ATP-binding site (Davenport et al. 2011; Yim et al. 2016; Yue et al. 2018). It could be that a high and low affinity binding site for Gba could exist on the linker domain between N-terminal and middle domain. Novobiocin, a coumarin based antibiotic, was already in the 50s identified with antibiotic activities. The novobiocin biosynthesis clusters were identified in *Streptomyces strains* in the early 70s has also anti Hsp90 activity with binding affinity on the C-terminus to hinder dimerization of Hsp90 (Cook et al. 1956; Dlugosz and Janecka 2017; Byrd et al. 2016). The evidence that human Hsp90 is a target, is documented for several pathogenic traits and the aim is to cure the diseases by compounds with anti Hsp90 activity have been transmitted into clinical phases (Fig. 18.4). Meanwhile specific modifications gave derivatives selective for Hsp90

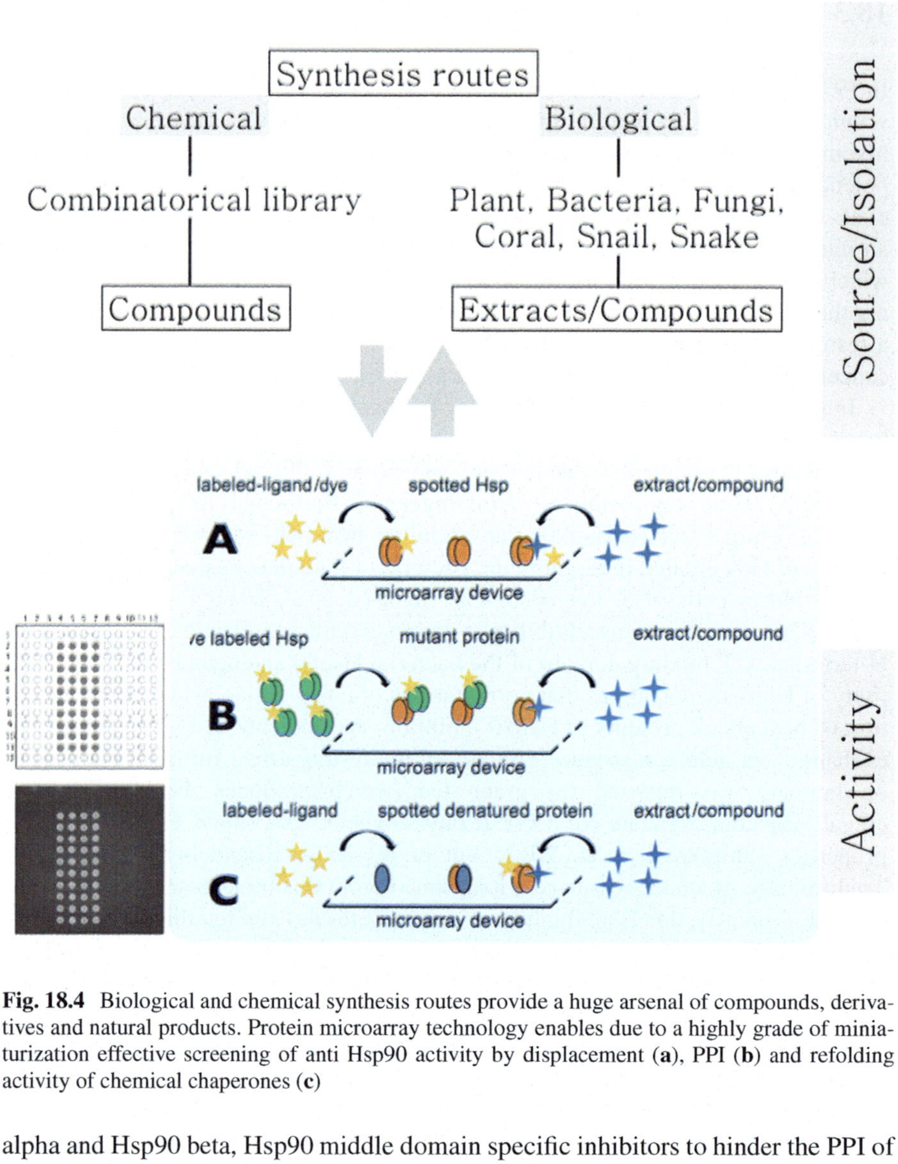

Fig. 18.4 Biological and chemical synthesis routes provide a huge arsenal of compounds, derivatives and natural products. Protein microarray technology enables due to a highly grade of miniaturization effective screening of anti Hsp90 activity by displacement (**a**), PPI (**b**) and refolding activity of chemical chaperones (**c**)

alpha and Hsp90 beta, Hsp90 middle domain specific inhibitors to hinder the PPI of Hsp90 and CDC37, or novobiocin binding site selective as C-terminal inhibitors (Li et al. 2018; Gestwicki and Shao 2019; Zhang et al. 2018; Khandelwal et al. 2018).

Although manifold different types of cancer with different genetic background exist, Hsp90 attends by mutated forms of TP53 and is accompanied with proteotoxic cellular stress. It is suggested that around 50% of all cancer types have a mutated TP53 background and increased Hsp90 levels. TP53 triggers indirectly the presence of Hsp90 in these disease pathways, therefore compounds developed to inhibit selectively increased Hsp90 levels and induce cells apoptosis. Several other players including Akt, ALK, BRAF, c-RAF, EGFR or HER2 are involved in tumorigenesis and are targets of Hsp90.

Beside in cancer, Hsp90 is a target in pathogenic cells like malaria, leishmania, trypanosoma, worms, fungi and bacteria, while no pathogen specific anti Hsp90 specific compound is known (Hombach and Clos 2014; Kanwar et al. 2017; Mohammadi-Ostad-Kalayeh et al. 2018; Posfai et al. 2018; Devaney and Gillan 2016; Schax et al. 2014). More complicated is the access of compounds against parasitic cells since the compound had to enter several membrane barriers, e.g., host membrane and parasite membranes and the target/compounds can be lost by generation cycles and can generate adaptations by multidrug resistance mechanisms, membrane surface changes or target expression and modifications. On the other hand, chaperon mediated autophagy by Hsc70 (heat shock cognate) is a housekeeping cellular strategy normally used for lysosomal degradation of protein complexes or organelles and against virus, bacteria or intracellular parasites like malaria or leishmania. However, to understand how cellular response is omitted or to control this critical point is a novel field in drug research. The arsenal from natural products and their derivatives increased in the recent years drastically, but a main hurdle, the entry into the clinical phase, remains. From all promising compounds, only a few survive due to medicinal or resource reasons, Figs. 18.4 and 18.5.

18.4 HSP90, the Cancer Target

Cancer is elicited by mutations in the genome leading to damage and unfolding of client proteins. Often folding helper proteins (chaperones) regenerate these clients. In some cases, the cellular level of incorrectly folded proteins rises together with some chaperones at the same time (Hsp90, Hsp70, small Hsp). Due to their natural rescue mechanism, these folding helpers enable the constant circuit of defective proteins and allow the survival of cancer cells thereby promoting a fatal development. On the other hand, these changes in the proteome may emerge as a pivotal point for the diagnosis and the treatment of malignancies: They are detectable as tumour markers and present targets as starting points for the development of compounds to suppress their function. Novel chemical design and synthesis strategies were used to obtain high affinity inhibitors based on natural compounds with the aim to reduce cytotoxic side effects, Fig. 18.5. Different test systems exist to investigate the potential of new chemically randomized inhibitors. In addition to cell-based tests, there are numerous in vitro test methods for Hsp90 available, which either use the ATPase, the folding activities or demonstrate the binding of inhibitors or ATP analogues via an optical marker or other biophysical techniques (e.g. retention time or use predictive structural modelling, Fig. 18.4 (Mohammadi-Ostad-Kalayeh et al. 2017). Protein microarrays can help to investigate novel compounds active on Hsp90 target-oriented screenings or to identify compounds with chaperone activity itself Fig. 18.4, (Mohammadi-Ostad-Kalayeh et al. 2017, 2018; Yue et al. 2018; Schax et al. 2014). Besides novel developments in surgical intervention

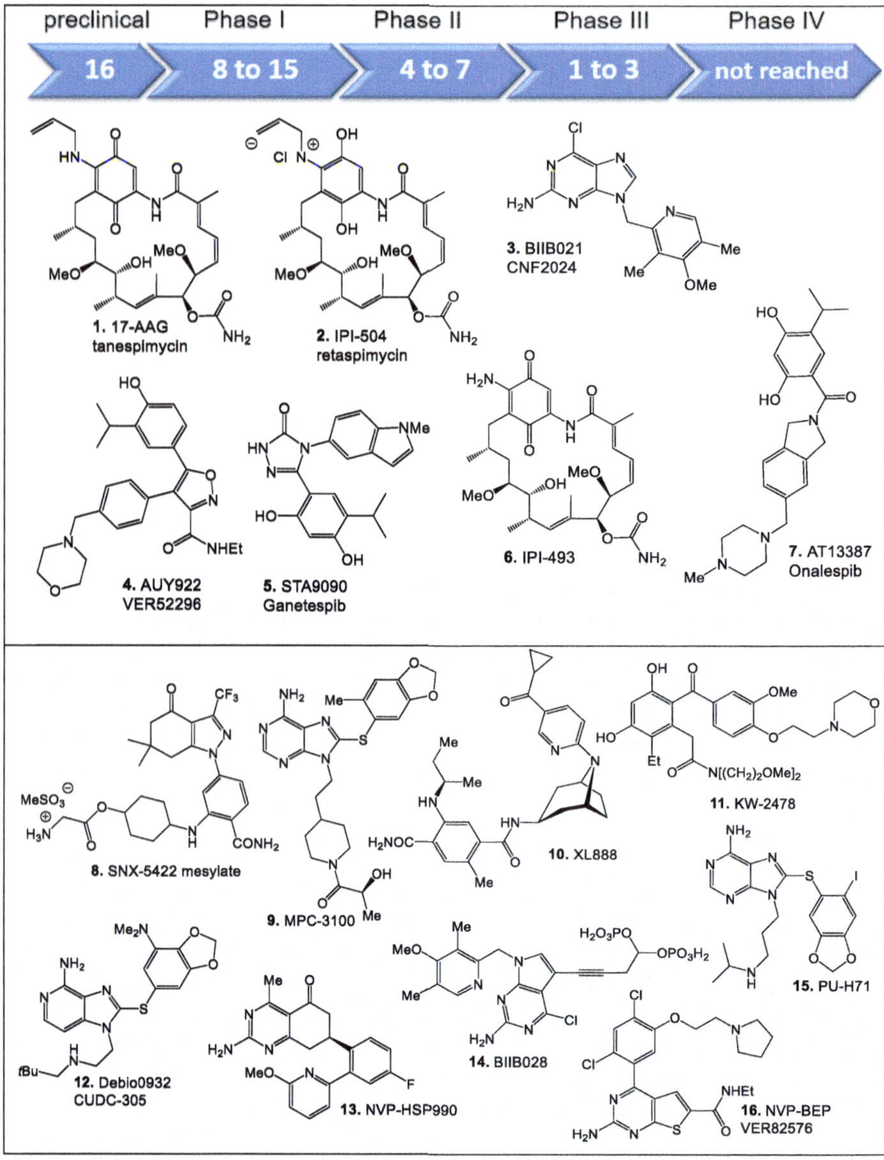

Fig. 18.5 Chemical structures of anti Hsp90 inhibitors in different clinical phases (preclinical, phase I-IV according to Jhaveri et al. (2012) and Hong et al. (2013)

accompanied by radiant exposure and several different immuno- or chemotherapies, highly selective personalized medicinal diagnostic could be a future strategy to reduce the stress for tumour patients. Inhibitors labelled with dyes or magnetic particles allow selective surgical interventions as a stain and kill application. In addition, selective and patient-tailored therapies could be developed exemplified on squamous cell carcinoma (SCC) contributing to over 90% of the malignancies in the head and neck area.

Despite good diagnostic procedures and therapy possibilities, one of every four deaths in Europe is due to cancer (https://ec.europa.eu/eurostat/web/products-eurostat-news/-/EDN-20170203-1?inheritRedirect=true). Head and neck cancer include various types of cancer: These occur in the oral cavity (tumours of the lips, tongue, floor of the mouth, palate and salivary glands), throat (pharyngeal and laryngeal), in the nose and paranasal sinuses, the lateral skull base (pinna, auditory canal, middle ear, temporal bone), the skin and the thyroid gland. Over 90% of the malignancies of the head and neck region are squamous cell carcinomas (Gregoire et al. 2010). Other tumour types are adenocarcinomas, tumours of glands, sarcomas, and other less common types of tumours.

Cancers of the head and neck area range worldwide as the sixth most common malignancy disease (Forster and Devlin 2018). Annually, about 500.000 new cases (14 million new cases of all cancer world wide and 8.2 million mortalities annually) are diagnosed and half of these patients will eventually die of this disease (Spiegelberg et al. 2014). The number of new cases per year in Germany is about 50 per 100,000 residents. The most common single type of cancer of the head and neck region is laryngeal cancer (new cases per year in Germany: 3600 men and 500 women). Malignant tumours of the oral cavity and pharynx cancer in Germany per year are about 9500 men and 3500 women. On average, men are at diagnosis 66 years old, women 70 years.

Biopsies of the suspected malignoma region are investigated in order to histologically confirm the diagnosis (Borderud et al. 2014). This is done in an intraoperative setting performing panendoscopy for thorough examination for biopsy sampling. Careful investigation of the oral cavity, nasopharynx, oropharynx, hypopharynx and larynx as well as of the trachea and esophagus is undertaken via rigid endoscopes. Biopsies are taken from suspicious mucosa and the anatomical area of the sampling is documented. Later on, histology can be assigned to each biopsy and the anatomical site of the malignant tumour is adequately documented for planning of the interdisciplinary therapy approach (Lee et al. 2018). The therapy of malignant diseases is depending on the tumour site, tumour type and tumour stage and may vary among individuals (Shah et al. 1997; Marioni et al. 2006). The surgical complete resection of most tumour entities in the head and neck region is the primary therapy option. Depending on individual tumour stage, the primary surgical resection will be supplemented by irradiation and / or chemotherapy (Zhang et al. 2013; Tamaki et al. 2018; Adelstein et al. 2003). Platinum-based drugs are used for chemotherapy and lead to the crosslinking of the DNA with subsequent apoptosis and autophagy. In order to increase the efficacy of platinum-based drugs, other chemotherapeutic agents such as taxans (Rose and Wild 2004) and 5-fluoruracil have been introduced for the treatment of HNSCC (Enokida et al. 2018). In the recent years, molecular targets have been identified for the combat of malignancies that can be used as monotherapy or in conjunction with chemotherapy or radiation (Baselga et al. 2005; Marur and Forastiere 2008).

The head and neck region is characterized by the highly complex anatomy. Here, all six sense organs are closely located to each other. In addition, many nerves (cranial nerves), blood and lymph vessels, and muscles are present (Tamaki et al. 2018). Thus, the main problems that arise in the surgical extirpation of malignant tumours in the head and neck region is detecting tumour border zones and maintaining safety margins during tumour resection. The safety margins during tumour resection, as it is observed during tumour excision in other parts of the body, would in head and neck cases harm many functionally important structures and cause a significant reduction in the quality of life for the patients. For local cancer control, the accurate detection of tumour border areas has a special role in the surgical extirpation (Mitchell et al. 2018; Abdalla et al. 2017). However, this is in infiltrative growing tumours with no sharp demarcation to the healthy tissue the main problem (Thomas Robbins et al. 2019). There is also the difficulty of the heterogeneity of the healthy tissue, such as the glandular tissue, which in contrast to the muscles cannot be seen as a homogenous structure during the procedure (Hinni et al. 2013). Therefore, the unique tumour detection, tumour marking and visualization for the surgeon would play an important role in the radical, *in sano* tumour resection and would therefore enhance (Hendriksma et al. 2018) the survival prognosis of the patients (Halicek et al. 2018).

An important cause of cancer is the instability of the genome (Dylawerska et al. 2017). A tumour arises not suddenly, but relies mostly on gene mutations, in which cells become cancerous cells and also manage to multiply and spread. The cell-physiological conditions in the tumour tissue lead sometimes to more mutations and continuous changes of the mutagenic gene pool (Loeb et al. 2003; Supek and Lehner 2017) and the corresponding proteome. This resulting change may adversely affect a therapy. Many therapies are available with strong side effects and limitations in diagnosis and localization of the tumour. However, if tumours are targeted, they produce toxins, which can lead to organ damage. With the diagnosis it is therefore unclear what molecular cause is located behind cancer, the direction of propagation is unknown and it is believed that this is also individually different. Therefore, it is important to understand what the potential targets in the range of the cancer are and which individual patterns are available. Conditionally by the special metabolic physiology, cancer cells have a proteome, but some proteins preferentially exist and can be identified as a marker (Shin et al. 2016).

18.5 Novel Strategies for Developing HSP90 as a Stress Marker

The detection of Hsp90 as a biomarker in tissue and body fluids like blood, urine or inner ear liquid could be a powerful medical tool for early diagnosis and treatment of diseases as well testing compound libraries. Thus, the early detection of biomarkers is very important in the case of cancer, cardiovascular disorders, and other pathological conditions (Polanska et al. 2014; Powers and Palecek 2012; Holton et al. 2017; Agrawal et al. 2018). Main problems are the detection limit, sensitivity and artificial quench by foreign compounds (Agrawal et al. 2018). To circumvent those limitation and to provide highest reliable sensitivities biomarkers identified by different techniques like 2D-gelelectrophoresis, ELISA, mass spectrometry (Holton et al. 2017), mass-sensing BioCD, protein array (Schax et al. 2014; Mohammadi-Ostad-Kalayeh et al. 2017), electrochemical immunoassay, SERS, flow cytometry, Europium/gold nanoparticle based detection, SPR, protein microarray, quantum dot technology, 9D DNA technology (Powers and Palecek 2012; Wu 2013). These technologies are under continuous optimization and in some cases detect picograms of biomarkers (Schmitt et al. 2018). Protein microarray technologies cover a broad application field and due to highly grade of miniaturization it is possible to testing different combinations on targets against compound libraries, Fig. 18.4 (Mohammadi-Ostad-Kalayeh et al. 2017, 2018).

18.6 Healing the Stress by Chemical Chaperones

At the beginning of this century, the decoding of the human genome promoted the development medications against point mutations (Mechler et al. 2015). Orphan drugs or chemical chaperones are pharmacological agents that target where point mutations lead to structural changes in function and are directed against rare diseases (Nourbakhsh et al. 2018; Rana and Chawla 2018). Compounds were found which selectively bind to the target proteins and stabilize them. The bound chemical compounds enable the folding of the protein into the correct three-dimensional shape. The protein can thus reach the right place and go along with the original cellular function. A treatment option for patients with rare hereditary diseases is a chance to survive or at least to improve the quality of life. Developing medicines is a lengthy and costly process that involves many risks. Research in this field has increased significantly in industrialized countries in recent years, e.g. Germany has become the world's number two after the USA (Gammie et al. 2015; Luzzatto et al. 2018). In orphan research, an average of about 5000–10,000 substances are researched and only about one drug is approved after many years. The study of rare diseases is particularly important because they allow individualized therapies and resulting drugs for patients can be of outmost importance. In each country, such research is more or less supported by the laws, for example, in Germany by the orphan drug status. There are certain advantages, such as shortening of approval times, simplification of authorization or facilitating market access with a 10-year market exclusivity (O'Sullivan et al. 2013; Tillet and Maillols-Perroy 2015). Such motivation creates new drugs and reduces costs. At present, e.g. the annual treatment cost for Ivacaftor in the year of treatment was € 268,803.98. The number of approvals showed how many projects ultimately lead to a new orphan drug being available on the market. The graph makes it clear that research in this area is particularly important. The treatment options for many people with different point mutations have improved significantly since 2000 with new drugs.

Using the example of the CFTR protein, it is briefly illustrated what significance in a therapy can achieve chemical compounds. The ABC transporter or CFTR consists of five domains, two membrane-spanning domains (MSD 1 and MSD 2), two cytoplasmic nucleotide-binding domains (NBD 1 and NBD 2) and R domain with numerous positions for phosphorylation by protein kinases (Smith et al. 2002). The two MSDs consist of six transmembrane helices (TM), which form the actual pore of the channel (Lukacs and Verkman 2012). To open the CFTR channel, ATP must bind to both NBD domains (Lukacs and Verkman 2012). Approx. 70% of all patients have ΔF508-CFTR mutation, which is caused by the deletion of a phenylalanine at position 508 (Riordan et al. 1989). Crystal structure analyzes of murine CFTR-NBD have shown that the NBD 1 mutation is localized to the surface, but results in little change in the structure (Lewis et al. 2004). Much more, Du et al. showed in 2005 that the target for degradation by proteases is due to the mutation and disrupted interaction with the NBD 2 (Du et al. 2005). The ΔF508-CFTR protein is passed through quality control in endoplasmic reticulum recognized because of his misfolding. The protein remains in the endoplasmic reticulum and is further degraded by the proteasome (Cheng et al. 1990). There are currently two approved drugs to treat cystic fibrosis. Lumacaftor increa ses the density of CFTR channel proteins on the cell surface. It acts on the protein that is expressed in too low an amount by the F508del mutation and is purged by the proteasome (Kim Chiaw et al. 2011). However, existing channel proteins are of limited functionality. Lumacaftor is one of the CFTR correctors. It has a stabilizing effect on the MSD 1 domain in the region of the N-terminus of the CFTR protein (Van Goor, et al. 2011; Gentzsch and Mall 2018). Ivacaftor improves the functionality of cell surface CFTR protein channels. The active ingredient prolongs the duration of the opening of the phosphorylation-activated channel. This influence on the defective protein increases the opening probability of the (defective) ion channel and the chloride secretion from the cells. Ivacaftor, Fig. 18.6, acts as a potentiator. It affects the ATP-dependent NBD domain for the folding function to take place (Van Goor et al. 2009). The exact mechanism of Ivacaftor and Lumacaftor has not yet been fully elucidated. Both drugs have been screened for their positive effect on CFTR protein function and have been used in the treatment of patients.

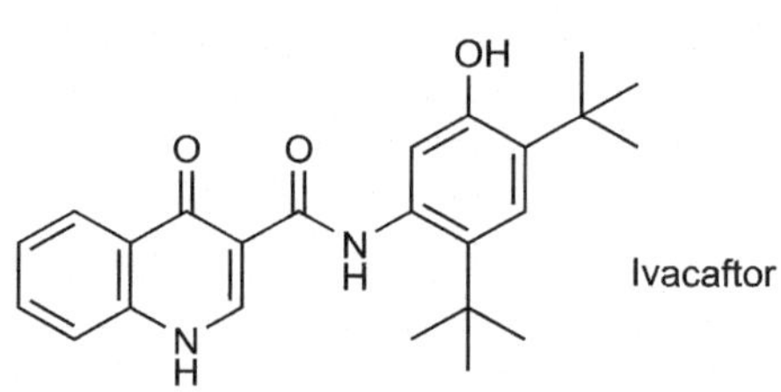

Fig. 18.6 Chemical structure of Ivacaftor

18.7 Outlook

The chaperone Hsp90 is more than a part of the folding machinery, it is a driver of the evolution, because it folds selective certain client proteins. Due to this sensitive position, Hsp90 is highly regulated and influenced by surrounding regulators mostly held in an inactive state, but the enzyme is waiting for precise activation pulses to release folded proteins. Pathogens or cancer cells use a similar survival strategy through molecular adaptation and are accompanied with different susceptibilities for drugs or natural products. The aim is to develop drugs against diseased cells by selective targeting of heat shock proteins. In several diseases like cancer, which is a folding disease due to mutations, the cellular concept runs out of frame and the Hsp90 levels become up-regulated, therefore Hsp90 can be used in some cases for the early diagnosis of liver cancer (Fu et al. 2017). In future, modern sensor technologies like microfluidics or microelectrical sensors will monitor personalized applications as well for point of care (POC) diagnostic with Hsp90 biomarker levels which could provide reliable and low-cost sensors for monitoring physiological stress status (Krüger et al. 2019). While compounds targeting Hsp90 in cancer reached different clinical phases, the antibiotic application route seems unclear and requires intense efforts. In contrast several successful therapies like antibody based therapies do not use Hsp90 as a target. Natural products, developed by a long lasting evolutionary process provided basic structures. By resynthesis and semi biosynthesis steps for several "lead"compounds, the Hsp90 function could be inhibited or at least disturbed with the aim to cure cancer and other diseases. With the increase of HSP levels in unhealthy cells, it is possible to use Hsp90 as a stress marker. Especially for head and neck cancer as a very aggressive form, Hsp90 inhibitors can be relevant to win time and to combine more personalized therapies. On the other hand it is unknown whether chemical chaperones can assume the role of client specific chaperones to compensate incorrect folded protein structures occurred by mutational effects. This could be in future a novel concept of a stop and re-functionalize therapy.

18.8 Conclusions

The chaperome is involved in the folding of newly synthesized proteins. Hsp90 is a conserved folding helper and is involved in the maintenance and regulation of proteins that are involved in apoptosis, cell cycle control, cell viability, protein folding and degradation, as well in signaling events. In cancer and inflammation, Hsp90 enables the constant circuit of defective proteins and allow the survival of cancer cells thereby promoting a fatal development. The up-regulation of Hsp90 in diseased states such as inflammation and cancer can be rapidly detected (i.e., Hsp90 may have biomarker function in several diseases) and aids in the diagnosis. In addition, it presents a suitable target for molecular therapies to cure or alleviate such diseases. In neurodegenerative diseases, Hsp90 may also present a target molecule for treatment. Another interesting aspect of Hsp90 is the canalization of disease phenotypes. This may be related to different diseases, such as cystic fibrosis, hearing loss or renal disease. Targeting Hsp90 as a therapeutic strategy in many diseases seems promising, but is currently limited due to the toxicity of the known compounds.

References

Abdalla Z, Walsh T, Thakker N, Ward CM (2017) Loss of epithelial markers is an early event in oral dysplasia and is observed within the safety margin of dysplastic and T1 OSCC biopsies. PLoS One 12:e0187449

Adelstein DJ, Li Y, Adams GL et al (2003) An intergroup phase III comparison of standard radiation therapy and two schedules of concurrent chemoradiotherapy in patients with unresectable squamous cell head and neck cancer. J Clin Oncol 21:92–98

Agrawal L, Engel KB, Greytak SR, Moore HM (2018) Understanding preanalytical variables and their effects on clinical biomarkers of oncology and immunotherapy. Semin Cancer Biol 52:26–38

Alam Q, Alam MZ, Sait KHW et al (2017) Translational shift of HSP90 as a novel therapeutic target from cancer to neurodegenerative disorders: an emerging trend in the cure of Alzheimer's and Parkinson's diseases. Curr Drug Metab 18:868–876

Aridon P, Geraci F, Turturici G, D'Amelio M, Savettieri G, Sconzo G (2011) Protective role of heat shock proteins in Parkinson's disease. Neurodegener Dis 8:155–168

Baselga J, Trigo JM, Bourhis J et al (2005) Phase II multicenter study of the antiepidermal growth factor receptor monoclonal antibody cetuximab in combination with platinum-based chemotherapy in patients with platinum-refractory metastatic and/or recurrent squamous cell carcinoma of the head and neck. J Clin Oncol 23:5568–5577

Bhattacharya K, Bernasconi L, Picard D (2018) Luminescence resonance energy transfer between genetically encoded donor and acceptor for protein-protein interaction studies in the molecular chaperone HSP70/HSP90 complexes. Sci Rep 8:2801

Blacklock K, Verkhivker GM (2014) Allosteric regulation of the Hsp90 dynamics and stability by client recruiter cochaperones: protein structure network modeling. PLoS One 9:e86547

Borderud SP, Li Y, Burkhalter JE, Sheffer CE, Ostroff JS (2014) Electronic cigarette use among patients with cancer: characteristics of electronic cigarette users and their smoking cessation outcomes. Cancer 120:3527–3535

Byrd KM, Subramanian C, Sanchez J et al (2016) Synthesis and biological evaluation of novobiocin core analogues as Hsp90 inhibitors. Chemistry 22:6921–6931

Calderwood SK (2018) Heat shock proteins and cancer: intracellular chaperones or extracellular signalling ligands? Philos Trans R Soc Lond Ser B Biol Sci 373:20160524

Chen Y, Wang B, Liu D et al (2014) Hsp90 chaperone inhibitor 17-AAG attenuates Aβ-induced synaptic toxicity and memory impairment. J Neurosci 34:2464–2470

Chen H, Xing J, Hu X et al (2017) Inhibition of heat shock protein 90 rescues glucocorticoid-induced bone loss through enhancing bone formation. J Steroid Biochem Mol Biol 171:236–246

Chen B, Feder ME, Kang L (2018) Evolution of heat-shock protein expression underlying adaptive responses to environmental stress. Mol Ecol 27:3040–3054

Cheng SH, Gregory RJ, Marshall J et al (1990) Defective intracellular transport and processing of CFTR is the molecular basis of most cystic fibrosis. Cell 63:827–834

Cook E, Eastman G, Bunn P (1956) The use of novobiocin in the management of acute pneumococcal and staphylococcal infections. Antibiot Annu 1:396–401

Cunningham LL, Brandon CS (2006) Heat shock inhibits both aminoglycoside- and cisplatin-induced sensory hair cell death. JARO – J Assoc Res Otolaryngol 7:299–307
Davenport J, Manjarrez JR, Peterson L, Krumm B, Blagg BS, Matts RL (2011) Gambogic acid, a natural product inhibitor of Hsp90. J Nat Prod 74:1085–1092
DeBoer C, Meulman PA, Wnuk RJ, Peterson DH (1970) Geldanamycin, a new antibiotic. J Antibiot (Tokyo) 23:442–447
Devaney E, Gillan V (2016) Hsp90 inhibitors in parasitic nematodes: prospects and challenges. Curr Top Med Chem 16:2805–2811
Dezwaan DC, Freeman BC (2008) HSP90. The Rosetta stone for cellular protein dynamics? Cell cycle (Georgetown, Tex) 7:1006–1012
Dickey CA, Kamal A, Lundgren K et al (2007) The high-affinity HSP90-CHIP complex recognizes and selectively degrades phosphorylated tau client proteins. J Clin Invest 117:648–658
Dlugosz A, Janecka A (2017) Novobiocin analogs as potential anticancer agents. Mini Rev Med Chem 17:728–733
Du K, Sharma M, Lukacs GL (2005) The Delta F 508 cystic fibrosis mutation impairs domain-domain interactions and arrests post- translational folding of CFTR. Nat Struct Mol Biol 12:17–25
Dylawerska A, Barczak W, Wegner A, Golusinski W, Suchorska WM (2017) Association of DNA repair genes polymorphisms and mutations with increased risk of head and neck cancer: a review. Med Oncol 34:197
Echeverría PC, Bernthaler A, Dupuis P, Mayer B, Picard D (2011) An interaction network predicted from public data as a discovery tool: application to the Hsp90 molecular chaperone machine. PLoS One 6:e26044
Edkins AL (2015) CHIP: a co-chaperone for degradation by the proteasome. Subcell Biochem 78:219–242
Enokida T, Okano S, Fujisawa T, Ueda Y, Uozumi S, Tahara M (2018) Paclitaxel plus cetuximab as 1st line chemotherapy in platinum-based chemoradiotherapy-refractory patients with squamous cell carcinoma of the head and neck. Front Oncol 8:339
Eurostat (2017) Causes of death – deaths by country of residence and occurrence. https://ec.europa.eu/eurostat/web/products-eurostat-news/-/EDN-20170203-1?inheritRedirect=true
Ferraro M, D'Annessa I, Moroni E et al (2019) Allosteric modulators of HSP90 and HSP70: dynamics meets function through structure-based drug design. J Med Chem 62:60–87
Forster MD, Devlin MJ (2018) Immune checkpoint inhibition in head and neck cancer. Front Oncol 8:310
Forsythe HL, Jarvis JL, Turner JW, Elmore LW, Holt SE (2001) Stable association of hsp90 and p23, but not hsp70, with active human telomerase. J Biol Chem 276:15571–15574
Franke J, Eichner S, Zeilinger C, Kirschning A (2013) Targeting heat-shock-protein 90 (Hsp90) by natural products. Geldanamycin, a show case in cancer therapy. Nat Prod Rep 30:1299–1323
Fries GR, Gassen NC, Rein T (2017) The FKBP51 glucocorticoid receptor co-chaperone: regulation, function, and implications in health and disease. Int J Mol Sci 18:1–31
Fu Y, Xu X, Huang D et al (2017) Plasma heat shock protein 90alpha as a biomarker for the diagnosis of liver cancer: an official, large-scale, and multicenter clinical trial. EBioMedicine 24:56–63
Gammie T, Lu CY, Babar ZU (2015) Access to orphan drugs: a comprehensive review of legislations, regulations and policies in 35 countries. PLoS One 10:e0140002
Gao Y, Yechikov S, Vazquez AE, Chen D, Nie L (2013) Distinct roles of molecular chaperones HSP90α and HSP90β in the biogenesis of KCNQ4 channels. PLoS One 8:1–10
Geller R, Pechmann S, Acevedo A, Andino R, Frydman J (2018) Hsp90 shapes protein and RNA evolution to balance trade-offs between protein stability and aggregation. Nat Commun 9:1781
Gentzsch M, Mall MA (2018) Ion channel modulators in cystic fibrosis. Chest 154:383–393
Gerecitano JF, Modi S, Rampal R et al (2015) Phase I trial of the HSP-90 inhibitor PUH71. J Clin Oncol (Meeting Abstracts) 33:253

Gestwicki JE, Shao H (2019) Inhibitors and chemical probes for molecular chaperone networks. J Biol Chem 294:2151–2161

Gong T-W, Fairfield DA, Fullarton L et al (2012) Induction of heat shock proteins by hyperthermia and noise overstimulation in hsf1 −/− mice. J Assoc Res Otolaryngol 13:29–37

Gregoire V, Lefebvre JL, Licitra L, Felip E, Group E.E. E. G. W et al (2010) Squamous cell carcinoma of the head and neck: EHNS-ESMO-ESTRO clinical practice guidelines for diagnosis, treatment and follow-up. Ann Oncologia 21:184–186

Halicek M, Little JV, Wang X, et al (2018) Optical biopsy of head and neck cancer using hyperspectral imaging and convolutional neural networks. Proc SPIE Int Soc Opt Eng Jan–Feb. 10469

Hendriksma M, Montagne MW, Langeveld TPM, Veselic M, van Benthem PPG, Sjögren EV (2018) Evaluation of surgical margin status in patients with early glottic cancer (Tis-T2) treated with transoral CO2 laser microsurgery, on local control. Eur Arch Otorhinolaryngol 275:2333–2340

Hinni ML, Ferlito A, Brandwein-Gensler MS et al (2013) Surgical margins in head and neck cancer: a contemporary review. Head Neck 35:1362–1370

Holton AB, Sinatra FL, Kreahling J, Conway AJ, Landis DA, Altiok S (2017) Microfluidic biopsy trapping device for the real-time monitoring of tumor microenvironment. PLoS One 12:e0169797

Hombach A, Clos J (2014) No stress – Hsp90 and signal transduction in Leishmania. Parasitology 141:1156–1166

Hong D, Said R, Falchook G et al (2013) Phase I study of BIIB028, a selective heat shock protein 90 inhibitor, in patients with refractory metastatic or locally advanced solid tumors. Clin Cancer Res 19:4824–4831

Honoré FA, Méjean V, Genest O (2017) Hsp90 is essential under heat stress in the bacterium *Shewanella oneidensis*. Cell Rep 19:680–687

Hooper PL, Durham HD, Török Z, Hooper PL, Crul T, Vígh L (2016) The central role of heat shock factor 1 in synaptic fidelity and memory consolidation. Cell Stress Chaperones 21:745–753

Hoter A, El-Sabban M, Naim H (2018) The HSP90 family: structure, regulation, function, and implications in health and disease. Int J Mol Sci 19:2560

Inda C, Bolaender A, Wang T, Gandu SR, Koren J (2016) Stressing out Hsp90 in neurotoxic proteinopathies. Curr Top Med Chem 16:2829–2838

Jarosz DF, Lindquist S (2010) Hsp90 and environmental stress transform the adaptive value of natural genetic variation. Science 330:1820–1824

Jhaveri K, Taldone T, Modi S, Chiosis G (2012) Advances in the clinical development of heat shock protein 90 (Hsp90) inhibitors in cancers. Biochim Biophys Acta 1823:742–755

Jhaveri K, Ochiana SO, Dunphy MP et al (2014) Heat shock protein 90 inhibitors in the treatment of cancer: current status and future directions. Expert Opin Investig Drugs 23:611–628

Johnson JL (2012) Evolution and function of diverse Hsp90 homologs and cochaperone proteins. Biochim Biophys Acta, Mol Cell Res 1823:607–613

Joshi S, Wang T, Araujo TLS, Sharma S, Brodsky JL, Chiosis G (2018) Adapting to stress – chaperome networks in cancer. Nat Rev Cancer 18:562–575

Kampinga HH, Hageman J, Vos MJ et al (2009) Guidelines for the nomenclature of the human heat shock proteins. Cell Stress Chaperones 14:105–111

Kanwar A, Eduful BJ, Barbeto L et al (2017) Synthesis and activity of a new series of antileishmanial agents. ACS Med Chem Lett 8:797–801

Khandelwal A, Hall JA, Blagg BS (2013) Synthesis and structure-activity relationships of EGCG analogues, a recently identified Hsp90 inhibitor. J Org Chem 78:7859–7884

Khandelwal A, Crowley VM, Blagg BSJ (2016) Natural product inspired N-terminal Hsp90 inhibitors: from bench to bedside? Med Res Rev 36:92–118

Khandelwal A, Kent CN, Balch M et al (2018) Structure-guided design of an Hsp90β N-terminal isoform-selective inhibitor. Nat Commun 9:425

Kim Chiaw P, Eckford PD, Bear CE (2011) Insights into the mechanisms underlying CFTR channel activity, the molecular basis for cystic fibrosis and strategies for therapy. Essays Biochem 50:233–248
Kim YE, Hipp MS, Bracher A, Hayer-Hartl M, Hartl FU (2013) Molecular chaperone functions in protein folding and proteostasis. Annu Rev Biochem 82:323–355
Koren J 3rd, Jinwal UK, Lee DC et al (2009) Chaperone signalling complexes in Alzheimer's disease. J Cell Mol Med 13:619–630
Kravats AN, Hoskins JR, Reidy M et al (2018) Functional and physical interaction between yeast Hsp90 and Hsp70. Proc Natl Acad Sci U S A 115:E2210–E2219
Krüger K, Reichel T, Zeilinger C (2019) Role of heat shock proteins 70/90 in exercise physiology, exercise immunology and their diagnostic potential in sports. J. Appl. Physiol. (1985) 126:916–927. in press
Lackie RE, Maciejewski A, Ostapchenko VG et al (2017) The Hsp70/Hsp90 chaperone machinery in neurodegenerative diseases. Front Neurosci 11:1–23
Lee DJ, Eun YG, Rho YS et al (2018) Three distinct genomic subtypes of head and neck squamous cell carcinoma associated with clinical outcomes. Oral Oncol 85:44–51
Lewis HA, Buchanan SG, Burley SK et al (2004) Structure of nucleotide- binding domain 1 of the cystic fibrosis transmembrane conductance regulator. EMBO J 23:282–293
Li J, Buchner J (2012) Structure, function and regulation of the Hsp90 machinery. Biom J 36:106–117
Li T, Jiang HL, Tong YG, Lu JJ (2018) Targeting the Hsp90-Cdc37-client protein interaction to disrupt Hsp90 chaperone machinery. J Hematol Oncol 11:59
Loeb LA, Loeb KR, Anderson JP (2003) Multiple mutations and cancer. Proc Natl Acad Sci U S A 100:776–781
Lukacs GL, Verkman AS (2012) CFTR: folding, misfolding and correcting the ΔF508 conformational defect. Trends Mol Med 18:81–91
Luzzatto L, Hyry HI, Schieppati A et al (2018) Second workshop on orphan drugs participants. Outrageous prices of orphan drugs: a call for collaboration. Lancet 392:791–794
Maeda Y, Fukushima K, Kariya S, Orita Y, Nishizaki K (2012) Intratympanic dexamethasone up-regulates Fkbp5 in the cochleae of mice in vivo. Acta Otolaryngol 132:4–9
Marioni G, Marchese-Ragona R, Cartei G, Marchese F, Staffieri A (2006) Current opinion in diagnosis and treatment of laryngeal carcinoma. Cancer Treat Rev 32:504–515
Marur S, Forastiere AA (2008) Head and neck cancer: changing epidemiology, diagnosis, and treatment. Mayo Clin Proc 83:489–501
Mayer MP, Le Breton L (2015) Hsp90: breaking the symmetry. Mol Cell 58:8–20
McFarland NR, Dimant H, Kibuuka L et al (2014) Chronic treatment with novel small molecule Hsp90 inhibitors rescues striatal dopa-mine levels but not alpha-synuclein-induced neuronal cell loss. PLoS One 9:e86048
Mechler K, Mountford WK, Hoffmann GF, Ries M (2015) Pressure for drug development in lysosomal storage disorders – a quantitative analysis thirty years beyond the US orphan drug act. Orphanet J Rare Dis 10:46
Millson SH, Chun C-S, Roe SM et al (2011) Features of the Streptomyces hygroscopicus HtpG reveal how partial geldanamycin resistance can arise by mutation to the ATP binding pocket of a eukaryotic Hsp90. FASEB J 25:3828–3837
Mitchell DA, Kanatas A, Murphy C, Chengot P, Smith AB, Ong TK (2018) Margins and survival in oral cancer. Br J Oral Maxillofac Surg 56:820–829
Mohammadi-Ostad-Kalayeh S, Hrupins V, Helmsen S et al (2017) Development of a microarray-based assay for efficient testing of new HSP70/DnaK inhibitors. Bioorg Med Chem 25:6345–6352
Mohammadi-Ostad-Kalayeh S, Stahl F, Scheper T et al (2018) Heat shock proteins revisited: using a mutasynthetically generated Reblastatin library to compare the inhibition of human and Leishmania Hsp90s. Chembiochem 19:562–574
Morris M, Maeda S, Vossel K, Mucke L (2011) The many faces of tau. Neuron 70:410–426

Moulin E, Zoete V, Barluenga S, Karplus M, Winssinger N (2005) Design, synthesis, and biological evaluation of HSP90 inhibitors based on conformational analysis of radicicol and its analogues. J Am Chem Soc 127:6999–7004
Nakazono A, Adachi N, Takahashi H et al (2018) Pharmacological induction of heat shock proteins ameliorates toxicity of mutant PKCγ in spinocerebellar ataxia type 14. J Biol Chem 293:14758–14774
Nourbakhsh F, Atabaki R, Roohbakhsh A (2018) The role of orphan G protein-coupled receptors in the modulation of pain: a review. Life Sci 212:59–69
Omura S, Iwai Y, Takahashi Y et al (1979) Herbimycin, a new antibiotic produced by a strain of Streptomyces. J Antibiot (Tokyo) 32:255–261
O'Sullivan BP, Orenstein DM, Milla CE (2013) Pricing for orphan drugs: will the market bear what society cannot? JAMA 310:1343–1344
Patel R, Williams-Dautovich J, Cummins CL (2014) Minireview: new molecular mediators of glucocorticoid receptor activity in metabolic tissues. Mol Endocrinol 28:999–1011
Piper PW, Millson SH (2012) Spotlight on the microbes that produce heat shock protein 90-targeting antibiotics. Open Biol 2:120138
Polanska H, Raudenska M, Gumulec J et al (2014) Clinical significance of head and neck squamous cell cancer biomarkers. Oral Oncol 50:168–177
Posfai D, Eubanks AL, Keim AI et al (2018) Identification of Hsp90 inhibitors with anti-plasmodium activity. Antimicrob Agents Chemother 62:e01799–e01717
Powers AD, Palecek SP (2012) Protein analytical assays for diagnosing, monitoring, and choosing treatment for cancer patients. J Healthc Eng 3:503–534
Raaijmakers JM, Paulitz TC, Steinberg C, Alabouvette C, Moënne-Loccoz Y (2009) The rhizosphere: a playground and battlefield for soilborne pathogens and beneficial microorganisms. Plant Soil 321:341–361
Rana P, Chawla S (2018) Orphan drugs: trends and issues in drug development. J Basic Clin Physiol Pharmacol 29:437–446
Reis SD, Pinho BR, Oliveira JMA (2017) Modulation of molecular chaperones in Huntington's disease and other polyglutamine disorders. Mol Neurobiol 54:5829–5854
Riebold M, Kozany C, Freiburger L et al (2015) A C-terminal HSP90 inhibitor restores glucocorticoid sensitivity and relieves a mouse allograft model of Cushing disease. Nat Med 21:276–280
Riordan JR, Rommens JM, Kerem B et al (1989) Identification of the cystic fibrosis gene: cloning and characterization of complementary DNA. Science 245:1066–1073
Ritossa F (1962a) A new puffing pattern induced by temperature shock and DNP in drosophila. Experientia 18:571–573
Ritossa F (1962b) A new puffing pattern induced by temperature shock and DNP in drosophila. Cell Mol Life Sci 18:571–573
Roe MS, Wahab B, Török Z, Horváth I, Vigh L, Prodromou C (2018) Dihydropyridines allosterically modulate Hsp90 providing a novel mechanism for heat shock protein co-induction and neuroprotection. Front Mol Biosci 5:1–14
Röhl A, Wengler D, Madl T et al (2015) Hsp90 regulates the dynamics of its cochaperone Sti1 and the transfer of Hsp70 between modules. Nat Commun 6:6655
Rose WC, Wild R (2004) Therapeutic synergy of oral taxane BMS-275183 and cetuximab versus human tumor xenografts. Clin Cancer Res 10:7413–7417
Sager RA, Woodford MR, Neckers L, Mollapour M (2018) Detecting posttranslational modifications of Hsp90. Methods Mol Biol 1709:209–219
Saibil H (2013) Chaperone machines for protein folding, unfolding and disaggregation. Nat Rev Mol Cell Biol 14:630–642
Sasaki K, Yasuda H, Onodera K (1979) Growth inhibition of virus transformed cells in vitro and antitumor activity in vivo of geldanamycin and its derivatives. J Antibiot (Tokyo) 32:849–851
Schax E, Walter JG, Märzhäuser H et al (2014) Microarray-based screening of heat shock protein inhibitors. J Biotechnol 180:1–9

Schmitt H, Roemer A, Zeilinger C et al (2018) Heat shock proteins in human perilymph: implications for cochlear implantation. Otol Neurotol 39:37–44

Schuster M, Schnell L, Feigl P et al (2017) The Hsp90 machinery facilitates the transport of diphtheria toxin into human cells. Sci Rep 7:613

Shah JP, Karnell LH, Hoffman HT et al (1997) Patterns of care for cancer of the larynx in the United States. Arch Otolaryngol Head Neck Surg 123:475–483

Sheehan-Rooney K, Swartz ME, Zhao F, Liu D, Eberhart JK (2013) Ahsa1 and Hsp90 activity confers more severe craniofacial phenotypes in a zebrafish model of hypoparathyroidism, sensorineural deafness and renal dysplasia (HDR). Dis Model Mech 6:1285–1291

Shi J, Van de Water R, Hong et al (2012) EC144 is a potent inhibitor of the heat shock protein 90. J Med Chem 55:7786–7795

Shin JM, Kamarajan P, Fenno JC, Rickard AH, Kapila YL (2016) Metabolomics of head and neck cancer: a mini-review. Front Physiol 7:526

Shirakami Y, Shimizu M, Moriwaki H (2012) Cancer chemoprevention with green tea catechins: from bench to bed. Curr Drug Targets 13:1842–1857

Shrestha L, Patel HJ, Chiosis G (2016) Chemical tools to investigate mechanisms associated with HSP90 and HSP70 in disease. Cell chemical biology 23:158–172

Smith PC, Karpowich N, Millen L et al (2002) ATP binding to the motor domain from an ABC transporter drives formation of a nucleotide sandwich dimer. Mol Cell 10:139–149

Sõti C, Nagy E, Giricz Z, Vígh L, Csermely P, Ferdinandy P (2005) Heat shock proteins as emerging therapeutic targets. Br J Pharmacol 146:769–780

Spiegelberg D, Kuku G, Selvaraju R, Nestor M (2014) Characterization of CD44 variant expression in head and neck squamous cell carcinomas. Tumour Biol 35:2053–2062

Srivastava BI, DiCioccio RA, Rinehart KL Jr, Li LH (1978) Preferential inhibition of terminal deoxynucleotidyltransferase activity among deoxyribonucleic acid polymerase activities of leukemic and normal cells by geldanamycin, streptoval C, streptovarone, and dapmavarone. Mol Pharmacol 14:442–447

Starr TN, Flynn JM, Mishra P, Bolon DNA, Thornton JW (2018) Pervasive contingency and entrenchment in a billion years of Hsp90 evolution. Proc Natl Acad Sci U S A 115:4453–4458

Supek F, Lehner B (2017) Clustered mutation signatures reveal that error-prone DNA repair targets mutations to active genes. Cell 170:534–547.e23

Taipale M, Jarosz DF, Lindquist S (2010) HSP90 at the hub of protein homeostasis. Emerging mechanistic insights. Nature reviews Mol Cell Biol 11:515–528

Tamaki A, Miles BA, Lango M, Kowalski L, Zender CA (2018) AHNS series: do you know your guidelines? Review of current knowledge on laryngeal cancer. Head Neck 40:170–181

Tanida S, Hasegawa T, Higashide E (1980) Macbecins I and II, new antitumor antibiotics. I. Producing organism, fermentation and antimicrobial activities. J Antibiot (Tokyo) 33:199–204

Thirstrup K, Sotty F, Montezinho LC et al (2016) Linking HSP90 target occupancy to HSP70 induction and efficacy in mouse brain. Pharmacol Res 104:197–205

Thomas Robbins K, Triantafyllou A, Suárez C et al (2019) Surgical margins in head and neck cancer: intra- and postoperative considerations. Auris Nasus Larynx 46:10–17

Tillet Y, Maillols-Perroy AC (2015) Orphan drugs: underrated opportunities for the developers in Europe. Therapie 70:351–357

Trepel J, Mollapour M, Giaccone G, Neckers L (2010) Targeting the dynamic HSP90 complex in cancer. Nat Rev Cancer 10:537–549

Trune DR, Canlon B (2012) Corticosteroid therapy for hearing and balance disorders. Anat Rec Adv Integr Anat Evol Biol 295:1928–1943

Van Goor F, Hadida S, Grootenhuis PD et al (2009) Rescue of CF airway epithelial cell function in vitro by a CFTR potentiator, VX-770. Proc Natl Acad Sci U S A 106:18825–18830

Van Goor F, Hadida S, Grootenhuis PD et al (2011) Correction of the F508del-CFTR protein processing defect in vitro by the investigational drug VX-809. Proc Natl Acad Sci U S A 108:18843–18848

Vaughan CK, Gohlke U, Sobott F et al (2006) Structure of an Hsp90-Cdc37-Cdk4 complex. Mol Cell 23:697–707
Verba KA, Wang RY, Arakawa A et al (2016) Atomic structure of Hsp90-Cdc37-Cdk4 reveals that Hsp90 traps and stabilizes an unfolded kinase. Science 352:1542–1547
Verkhivker GM (2018) Dynamics-based community analysis and perturbation response scanning of allosteric interaction networks in the TRAP1 chaperone structures dissect molecular linkage between conformational asymmetry and sequential ATP hydrolysis. Biochim Biophys Acta Proteins Proteom 1866:899–912
Vetrivel KS, Kodam A, Gong P et al (2008) Localization and regional distribution of p23/TMP21 in the brain. Neurobiol Dis 32:37–49
Wandinger SK, Richter K, Buchner J (2008) The Hsp90 chaperone machinery. J Biol Chem 283:18473–18477
Wang X, Venable J, LaPointe P et al (2006) Hsp90 cochaperone Aha1 downregulation rescues misfolding of CFTR in cystic fibrosis. Cell 127:803–815
Wang B, Liu Y, Huang L et al (2017) A CNS-permeable Hsp90 inhibitor rescues synaptic dysfunction and memory loss in APP-overexpressing Alzheimer's mouse model via an HSF1-mediated mechanism. Mol Psychiatry 22:990–1001
Winssinger N, Barluenga S (2007) Chemistry and biology of resorcylic acid lactones. Chem Commun 7:22–36
Woo JA, Liu T, Zhao X et al (2017) Enhanced tau pathology via RanBP9 and Hsp90/Hsc70 chaperone complexes. Hum Mol Genet 26:3973–3988
Woodford MR, Dunn DM, Ciciarelli JG, Beebe K, Neckers L, Mollapour M (2016) Targeting Hsp90 in non-cancerous maladies. Curr Top Med Chem 16:2792–2804
Wu AH (2013) Biological and analytical variation of clinical biomarker testing: implications for biomarker-guided therapy. Curr Heart Fail Rep 10:434–440
Yáñez-Mó M, Siljander PR-M, Andreu Z et al (2015) Biological properties of extracellular vesicles and their physiological functions. J Extracell Vesicles 4:27066
Yim KH, Prince TL, Qu S et al (2016) Gambogic acid identifies an isoform-specific druggable pocket in the middle domain of Hsp90β. Proc Natl Acad Sci U S A 113:E4801–E4809
Yin Z, Henry EC, Gasiewicz TA (2009) (−)-Epigallocatechin-3-gallate is a novel Hsp90 inhibitor. Biochemistry 48:336–345
Yue Q, Stahl F, Plettenburg O, Kirschning A, Warnecke A, Zeilinger C (2018) The noncompetitive effect of Gambogic acid displaces fluorescence-labeled ATP but requires ATP for binding to Hsp90/HtpG. Biochemistry 57:2601–2605
Zhang SY, Lu ZM, Luo XN et al (2013) Retrospective analysis of prognostic factors in 205 patients with laryngeal squamous cell carcinoma who underwent surgical treatment. PLoS One 8:e60157
Zhang FZ, Ho DH, Wong RH (2018) Triptolide, a HSP90 middle domain inhibitor, induces apoptosis in triple manner. Oncotarget 9:22301–22315
Zhao R, Davey M, Hsu Y-C et al (2005) Navigating the chaperone network. An integrative map of physical and genetic interactions mediated by the hsp90 chaperone. Cell 120:715–727

Chapter 19
Screening Technique for Heat Shock Protein 90 Inhibitors from Natural Products

Yue Hu, Xiao J. Zhang, Xiao T. Yang, Ying Y. Tang, Lin Y. Hu, and Dong Zhu

Abstract Heat shock protein 90 (Hsp90) has emerged as an important target in cancer therapy, in which case screening inhibitors targeted Hsp90 attracted special attention on anti-cancer research. Recent progress in the development of screening technique for Hsp90 inhibitors has taken place. Particular emphasis is focused on ligand fishing screening assay based on protein affinity such as fluorescent ligand fishing that serve as a screening platform show case. This context will focus on strategies to screen and validate potential Hsp90 inhibitors from natural products and the development of screening techniques that are currently applied for the discovery of bioactive compounds from complex mixtures.

Keywords High-throughput screening · Hsp90 · Hsp90 inhibitors · Ligand fishing · Virtual screening

Y. Hu · X. J. Zhang · X. T. Yang · Y. Y. Tang · L. Y. Hu
School of Pharmacy, Nanjing University of Chinese Medicine, Nanjing, Jiangsu, People's Republic of China

D. Zhu (✉)
School of Pharmacy, Nanjing University of Chinese Medicine, Nanjing, Jiangsu, People's Republic of China

Jiangsu Key Laboratory for Functional Substance of Chinese Medicine, Nanjing, Jiangsu, People's Republic of China

State Key Laboratory Cultivation Base for TCM Quality and Efficacy, Nanjing University of Chinese Medicine, Nanjing, Jiangsu, People's Republic of China
e-mail: dongzhu@njucm.edu.cn

A. A. A. Asea, P. Kaur (eds.), *Heat Shock Protein 90 in Human Diseases and Disorders*, Heat Shock Proteins 19,
https://doi.org/10.1007/978-3-030-23158-3_19

Abbreviations

AFI-MS	Automated flow injection ESI-MS
Aha1	Activator of Hsp90 ATPase homologue 1
Cdc37	Cell division cycle 37 homologue
CDD	C-terminal dimerisation domain
CE	Capillary electrophoresis
CEfrag	Fragment screening technology based on capillary electrophoresis
CTD	Carboxy-terminal domain
Cy3B-GM	Cy3B-labeled geldanamycin
EGCG	Epigallocatechin gallate
EI-MS	Electrospray ionization mass spectrometry
FBDD	Fragment-based drug discovery
FP	Fluorescence polarization
GA	Geldanamycin
Grp94, HSPC4	glucose-regulated protein 94
Hop	Hsc70/Hsp90-organizing protein
Hsf1	Heat shock factor 1
Hsp90	Heat shock protein 90
HSPC1	Hsp90AA1
HSPC2	Hsp90α
HSPC3	Hsp90β
HTS	High-throughput screening
MD	Middle domain
MEEVD	Met-Glu-Glu-Val-Asp
mP	Millipolarization
NMR	Nuclear magnetic resonance
NTD	Amino-terminal domain
p23	Prostaglandin E synthase 3
PTMs	Post-translational modifications
QDs	Quantum dots
QSAR	Quantitative structure-activity relationship
ROCS	Rapid overlay of chemical structures;
S. elongates	Synechococcus elongates
SCLC	Small-cell lung carcinoma
SPR	Surface plasmon resonance
TPR	Tetratricopeptide repeat
Trap-1, HSPC5	Tumor necrosis factor receptor-associated protein 1

19.1 Introduction

Cancer has been the leading cause of death with increasing incidence and mortality which is also a major public health problem worldwide, and general therapy is far from a reality yet (Chen et al. 2016). Tumors can originate from overexpressed proteins or misfolding of chaperone associated protein, which may trigger to deregulate normal physiological functions in cells (Ferrarini et al. 1992). Molecular chaperones play a crucial role in maintaining the proper folding of proteins, which is important to keep the balance between protein degradation and synthesis in cellular homeostasis. Heat shock protein 90 (Hsp90), known as a highly conserved molecular chaperone, facilitates the maturation of substrates (or clients) involved in various cellular pathways critical for cell growth, invasiveness and survival beyond protein folding (Schopf et al. 2017; Pearl et al. 2008). Hsp90 inhibition leads to the proteasome-mediated degradation of oncogenic client proteins, thus making Hsp90 a remarkable therapeutic target for cancer therapy (Hong et al. 2013). Therefore, Hsp90 inhibitors may exhibit a broad spectrum of anticancer activity.

17-AAG (tanespimycin), the first inhibitor, entered clinical trials in 1999 (Banerji et al. 2005). In 2004, a second Hsp90 inhibitor, 17-DMAG (alvespimycin), was introduced into a first-in-human study (Kaur et al. 2004). As extensive efforts in rational drug design and discovery of Hsp90 inhibitors from different origins are continually made, currently more than 17 Hsp90 inhibitors are undergoing clinical trials (Kim et al. 2009; Jhaveri et al. 2012; Neckers and Workman 2012), although there are no approved Hsp90-targeted drugs (Soga et al. 2013). Because the first and latest Hsp90 inhibitors were originated from natural products, such as geldanamycin (GA) (Prodromou et al. 1997), novobiocin (McConkey and Zhu 2008), gedunin (Brandt et al. 2008), celastrol (Hieronymus et al. 2006), epigallocatechin gallate (EGCG) (Li et al. 2009), dermbone (Hadden et al. 2007) and so on, herbal plants continued to be interested for the exploitation of Hsp90 inhibitors by virtue of their extensive research and development space.

Natural products usually contain hundreds or even thousands of components with various concentrations and properties, and only a few of them are potential inhibitors with the pharmacological activity of interest. The key is to develop high-throughput and high-selectivity methods for discovery of bioactive components in natural products. Conventional screening techniques, based on the use of high-throughput screening for the molecular structure designed on the assumption of libraries of plant-derived compounds, are obviously labor-immense and time-consuming, which are not applicable for complicated herbal extracts. Protein affinity-based ligand fishing screening assay, ascribing to direct interaction with targets (i.e. protein, enzyme and receptor), has emerged as the most efficient and convenient technology to fish out specific ligands from complex matrix (Qing et al. 2012; Lourenço et al. 2013; Singh et al. 2014). In these strategies, target proteins are usually immobilized on solid surfaces which enable to resist the environmental changes compared with solution phase-based technique, thus various types of solid supporters such as silica nanospheres (Hu et al. 2015), magnetic particles (Lourenço et al. 2013) and gold nanoparticles (New et al. 2014) were developed for bioactive molecular discovery.

Considerable progress has been made in the development of screening technique for Hsp90 inhibitors from natural products (Chini et al. 2016; Huang et al. 2016). Very recently, Zhu and coworkers established a fluorescent ligand fishing approach for the screening and identification of plant-derived Hsp90 inhibitors (Hu et al. 2018a, b; Miao et al. 2018), which not only concurrently carried out biomedical imaging and further indicated in situ screening results, but also highly improved the efficiency of specific components by largely removing the interferences and non-specific components. This chapter will focus on the strategies to screen and validate potential Hsp90 inhibitors as well as the applications for discovery of bioactive compounds from complex mixtures.

19.2 General and Structural Aspects of HSP90

Hsp90 is an evolutionarily conserved molecular chaperone, which is present in prokaryotes and eukaryotes with the exception of archaea where only small heat shock proteins exist (Taipale et al. 2010). A newer nomenclature annotates this Hsp90 family as HSPC members with five known genes in the human genome as following (Kampinga et al. 2009; Website: HUGO gene nomenclature committee): Hsp90AA1 (HSPC1), Hsp90α (HSPC2), Hsp90β (HSPC3), the glucose-regulated protein 94 (Grp94) (HSPC4), and the tumor necrosis factor receptor-associated protein 1 (Trap-1) (HSPC5). Compared with bacteria, which typically embody only a single Hsp90 gene, *Homo sapiens* contain two Hsp90 genes that encode closely related cytoplasmic isoforms: Hsp90α and Hsp90β (Schopf et al. 2017). The expression of the former is related to proliferation of cancer cells and tumor progression, whereas the latter is associated with drug resistance (McCollum et al. 2008; Whitesell and Lin 2012). Hsp90 as a homodimer and dimerization is essential for its cellular functions (Mayer and Le Breton 2015; Krukenberg et al. 2011). An Hsp90 monomer consists of four highly functional domains in Fig. 19.1a: the amino-terminal domain (NTD), which mediates binding to ATPase active domain; the carboxy-terminal domain (CTD), which is significant for Hsp90 dimerization; the middle domain (MD), which is responsible for ATP hydrolysis and the binding of Hsp90 to clients; and an unstructured, flexible, charged linker region connecting the NTD with the MD, which modulates NTD-MD contacts and impacts Hsp90 function (Schopf et al. 2017). It also contains a C-terminal Met-Glu-Glu-Val-Asp (MEEVD) motif (Buchner 1999), which plays an important role in the interaction with co-chaperones that embody tetratricopeptide repeat (TPR) domains.

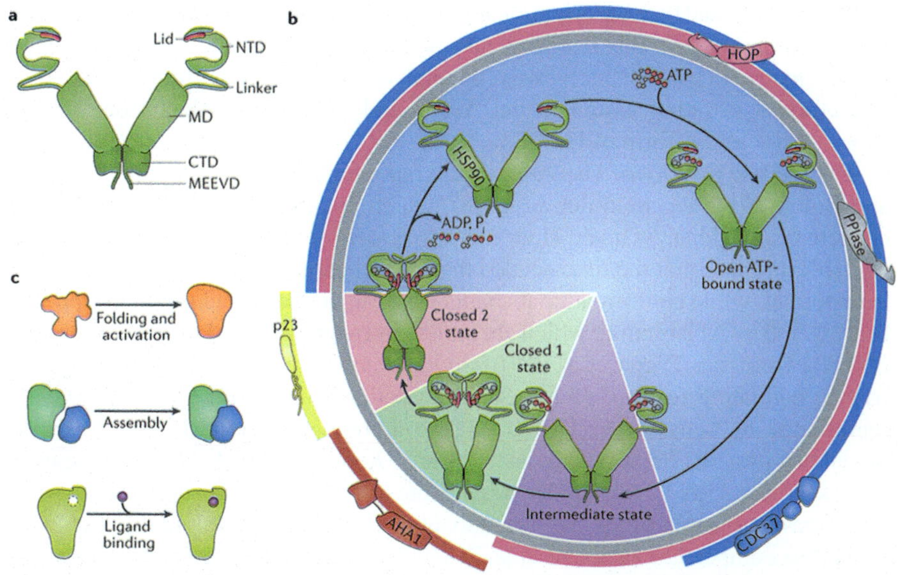

Fig. 19.1 The structure, function and conformational cycle of Hsp90. (**a**) Graphical presentation of Hsp90 domains and dimeric architecture. (**b**) The Hsp90 conformational cycle and the action of co-chaperones in different parts of the cycle. (**c**) Schematic representation of the ways in which Hsp90 can affect clients. (Reproduced from Schopf et al. 2017)

19.3 Functional Aspects of HSP90 Machinery

19.3.1 The Hsp90 Chaperone Cycle

Hsp90 chaperone adopts mainly a V-shaped open conformation ('clip') in the absence of ATP or other factors (Shiau et al. 2006), whereas it suffers from conformational change by an ATP-dependent folding cycle as shown in Fig. 19.1b (Pearl and Prodromou 2006). Current models reveal that ATP binding and hydrolysis, as well as a precisely sequenced interaction with an array of co-chaperones, subtly shift the conformational equilibrium by depressing the energy barrier between certain conformations, thus providing orientation to the Hsp90 chaperone cycle (Mickler et al. 2009; Hessling et al. 2009; Southworth and Agard 2008). During the ATP and conformational cycle, ATP binding to the undimerized (open) NTDs of Hsp90 promotes repositioning of a 'lid' segment which leads to transient dimerization of the NDs. Subsequent structural rearrangements result in the 'closed and twisted' conformation of Hsp90 (closed state 1 and 2) that is committed to ATP hydrolysis (Trepel et al. 2010). After hydrolysis of ATP, Hsp90 folds back into its original shape and the open molecular 'clip' is regenerated. Although the rate of

ATP hydrolysis and the conformational change varies with different Hsp-conformers, for the human Hsp90, this is a slow turnover rate of 3 ATP h^{-1}, and the conformational change does not necessarily depend on ATP (McLaughlin et al. 2004). Therefore, it is rather assumed that ATP merely shifts the balance between the 'open' and 'closed' form of Hsp90.

The ATPase activity of Hsp90 has been reported to be essential in vivo (Mishra and Bolon 2014; Panaretou et al. 1998), which is significantly influenced by co-chaperones (Makhnevych and Houry 2012). Recent studies based on point mutants of Hsp90 propose that the time needed in different parts of the conformational cycle is important, while conformational changes leading to the closed two state appear to be triggered by ATP binding which are sufficient for cell survival even in the absence of ATP hydrolysis (Zierer et al. 2016). In fact, more than 20 co-chaperones participate in regulating forward and backward reaction rates during the ATPase cycle, represented by activator of Hsp90 ATPase homologue 1 (Aha1), Hsc70/Hsp90-organizing protein (Hop), cell division cycle 37 homologue (Cdc37) and prostaglandin E synthase 3 (p23). Aha1 enhances the rate of ATP hydrolysis-dependent cycling by increasing the rate of the conformational alterations that result in the acquisition of ATPase competence (Retzlaff et al. 2010). Hop binds to Hsp90 and stabilizes the open conformation, thus inhibiting the ATPase activity of Hsp90 (Southworth and Agard 2011). Cdc37 binds to Hsp90 early in the cycle and is responsible for the recruitment of client kinases (Taipale et al. 2012; Aloy and Russell 2002).

19.3.2 Regulation of Hsp90 Function

As seen in Fig. 19.1c, the Hsp90 machinery evolved to facilitate protein folding, assembly of protein complexes or the binding of a ligand to its target, and the primary function of Hsp90 is regulated in many ways, including expression, post-translational modifications (PTMs), co-chaperones and client proteins. The expression of Hsp90 is induced by the stress-related transcription factor, heat shock factor 1 (Hsf1), which is also treated as an Hsp90 client (Prodromou 2016). The current model suggests that the bond of Hsp90 with Hsf1 keeps it in an inactive state, and the transcription genes will increase when chaperone proteins are no longer available for Hsf1 inhibition owing to the requirements for other functions (Voellmy and Boellmann 2007). Thus, Hsp90 directly links the stress status of the cell to the expression of HSP.

PTMs including phosphorylation, nitrosylation, acetylation and methylation, are supposed to enable fine cellular tuning for modulating Hsp90 function (Mayer and Le Breton 2015; Rohl et al. 2013). In general, phosphorylation was reported to slow down the Hsp90 conformational cycle and to hinder client maturation and the interaction with co-chaperones (Mollapour et al. 2011; Soroka et al. 2012; Mollapour and Neckers 2012). Many kinases are Hsp90 clients themselves that are reported to regulate the phosphorylation status of Hsp90 (Miyata 2009). *S*-nitrosylation at the

C-terminal domain of Hsp90 not only results in the reduction of Hsp90 ATPase activity but also influences its chaperone activity (Retzlaff et al. 2009). Acetylation can affect client maturation and co-chaperone binding, and it is reported that acylation may reduce ATP binding of Hsp90 (Scroggins et al. 2007). In addition, methylation which is regulated in several cancer cells and its catalytic activity was found to be enhanced by the interaction with Hsp90 (Hamamoto et al. 2004, 2006). The conclusion from the analysis of PTMs at different sites in Hsp90 is that, except for local effects on regulating the accessibility of binding sites, several modified sites function as allosteric switch points which regulate interdomain connection across the Hsp90 dimer (Soroka et al. 2012; Retzlaff et al. 2009).

Co-chaperones are the most important regulators that assist Hsp90 in its chaperone activity and modulate ATPase activity of Hsp90 to determine the rate of chaperone cycling. Some co-chaperones compete with each other for binding to Hsp90, whereas others can bind simultaneously or even synergistically (Schopf et al. 2017). Consistent with artificial separation, some co-chaperones function as modulators of the Hsp90 chaperone cycle and others are adaptor proteins for client recruitment. Thus, it appears that client processing integrates the functions of different co-chaperones present in the same complex.

19.4 Role in Cancer

Tumour cells are usually in a 'stressed' state due to the presence of mutain and rapid proliferation, which puts additional pressure on controlling protein equilibrium (Chen et al. 2012). Heat shock proteins are induced not only by heat but also by cellular stress situations and such as oxygen deficiency, exposure to UV radiation, or nutrient deprivation (Crevel et al. 2001; Neckers 2002; Whitesell and Lindquist 2005). These proteins, especially Hsp90, can strongly interfere with diverse cellular processes including heat shock or other stress responses, signal transduction and malignancy maintenance (Pearl and Prodromou 2000). Besides, destabilizing oncogenic mutations could enhance the dependence of tumour cells on Hsp90. Indeed, Hsp90 levels are markedly increased in tumours and high Hsp90 expression is related to a negative prognosis in breast cancer (Pick et al. 2007). Thus, Hsp90 is essential for the survival of cancer cells due to their extensive dependence on Hsp90-assisted signal pathways.

19.5 Strategies for Inhibiting HSP90 Function

Inhibiting the Hsp90 chaperone to interfere cancer cell cycles becomes possible. Thus, some compounds that can bind to different sites in Hsp90 are discovered to be potential anticancer drugs (Whitesell et al. 1994). At present, three major strategies for inhibiting Hsp90 function are being pursued. These are as follows: (a) inhibiting

ATPase activity by binding at the NTD nucleotide-binding pocket including radicicol (Soga et al. 2003), geldanamycin (Roe et al. 1999) and PU-H71 (Moulick et al. 2011); (b) altering the conformation of Hsp90 chaperone cycle with small molecules that bind to the C-terminal dimerisation domain (CDD) such as novobiocin (Marcu et al. 2000); and (c) blocking co-chaperones binding to Hsp90 such as Cdc37 which is overexpressed in cancer cells, as caused by gedunin and celastrol (Garg et al. 2016). Moreover, as mentioned above, Hsp90 activity is also regulated by relevant targeting sites for the Hsp90 with client proteins as well as sites for the inhibition of post-translational modifications. Finally, the simultaneous inhibition of Hsp90 and protein kinases was found to enable the reduction of oncogene switching in cancer models (Pashtan et al. 2008), which could indicate a promising avenue to reduce drug resistance in cancer.

19.6 Drug Discovery to Development of HSP90 Inhibitors

Extensive efforts have been undertaken for the discovery and development of new Hsp90 inhibitors suitable for clinical use in cancer patients. Currently 17 Hsp90 inhibitors have been performed clinical trials with considerable progress including improved formulations, optimum dosage and schedule, although no Hsp90-targeted agents are approved for human use (Soga et al. 2013). GA was the first identified Hsp90 inhibitor from natural product, after that, a variety of natural product inhibitors of Hsp90 have emerged, among which are herbimycin, radicicol, novobiocin, coumermycin A1, clorobiocin, EGCG, taxol, pochonin, derrubone, gedunin, and celastrol. Thus, natural products continue to play a prominent role in the discovery of Hsp90 inhibitors (Driggers et al. 2008). For the development of potential Hsp90 inhibitors from natural products, various screening techniques have been performed including library-based high-throughput screening methods, fragment-based biophysical techniques, nanocarrier-based ligand fishing and computer-aided virtual screening as summarized below for details.

19.7 Library-Based High-Throughput Screening Technique for HSP90 Inhibitors

Natural products contain hundreds or even thousands of different components, and only a few of them own pharmacological activity. The key is to develop high-throughput and high-selectivity methods for discovery of bioactive components from natural products. With improved pharmacological profiles and the rising demand for compound libraries, considerable efforts have gone into the applications of fully automated high-throughput screening devices (HTS). HTS is especially effective for finding novel inhibitors with different scaffold structures from that of a known inhibitor and with different mechanisms of action based on the assumption of chemical libraries of plant-derived compounds. The typical HTS approach utilizes a microplate for assaying and tracking target-inhibitor complexes depending on a biochemical reaction in which a substrate is functionalized with a fluorescent or radioactive label. In this section, four types of HTS techniques are introduced for Hsp90 inhibitors: enzyme-based screening, fluorescence polarization assay, chemical array-based screening, and AlphaScreen®-based screening. The HTS results are analyzed by a statistical manner and a positive assay response is called a 'hit'. In fact, the validity of HTS-derived results is associated with too many false positive responses, and the few true 'hit' was still found by laborious deliberation. Besides, pharmacokinetics of the 'hit', such as toxicity or solubility of the identified substance has not yet been stated at this stage.

19.7.1 Enzyme-Based Screening

Enzyme-based screening is based on the ATPase activity of Hsp90 (Rowlands et al. 2004). The conventional assay for Hsp90 ATPase activity is a pyruvate kinase/lactate dehydrogenase coupled-enzyme assay (Panaretou et al. 1998) which is not suitable for HTS due to low efficiency. To maintain the requirements for HTS such as sensitivity, miniaturization, and suitability for automation, various methods had been developed to measure inorganic phosphate or ATP contents, including the luciferin/luciferase assay (Lundin 2000), the *Escherichia coli* phosphate-binding protein assay (Brune et al. 1994, 1998), and the colorimetric assay (Baykov et al. 1998; Harder et al. 1994; Maehama et al. 2000). Rowlands et al. employed the colorimetric assay of inorganic phosphate for HTS (Rowlands et al. 2004), which was achieved by the formation of a phosphomolybdate complex and subsequent reaction with malachite green (Baykov et al. 1998; Harder et al. 1994; Maehama et al. 2000). In this assay, inorganic phosphate released by the formation of a phosphomolybdate complex between Hsp90 ATPase with molybdate, subsequently reacting with the dye malachite green. A distinct linear relationship exists between the absorbance at 620 nm of the phosphomolybdate-malachite green complex and the concentration of inorganic phosphate (from 0.5 to 2.8 nmol of inorganic phosphate in assay volume of 25 μL) (Rowlands et al. 2004). On this condition, Rowlands et al. screened 53,440 compounds in their collection using yeast Hsp90, and only 2 chemically related compounds (CCT018159 and CCT018158) at a hit rate of 0.3% (150 hits/53,440 screened) were confirmed to have an inhibitory activity at 40 μM (Rowlands et al. 2004).

19.7.2 Fluorescence Polarization Assay

HTS using the fluorescence polarization (FP) assay is based upon the displacement of a fluorescently labeled molecule which binds specifically to the ATP-binding site of Hsp90. The displacement is monitored by a decrease in fluorescence polarization of the fluorescent probe-Hsp90 complex when a small molecule binds competitively to the ATP-binding site. In FP assay, the magnitude of polarization lies on the Brownian rotation of fluorescently labeled molecule between absorption and emission (Lakowitz and Joseph 1999). In detail, fluorescently labeled small molecules rotate quickly during the excited state due to the shorter rotational relaxation time than its fluorescence lifetime and have a lower millipolarization (mP) value upon emission. When fluorescently labeled small molecules bind to proteins and form large complexes, the large molecules rotate little for a long rotational relaxation time, thus having a higher mP value (Fig. 19.2a). Thus, the change in mP reflects the interaction between the fluorescently labeled Hsp90 inhibitor and the Hsp90 protein (Du et al. 2007; Howes et al. 2006). In chemical screening, if a screened compounds bind to proteins competitively against fluorescently labeled small molecule, dissociated fluorescently labeled small molecules will rotate quickly, resulting in a lower mP value (Fig. 19.2b).

Du et al. developed a high-throughput fluorescence polarization screening assay by using human cancer cell lysates containing Hsp90 (Du et al. 2007). They optimized a 384-well format of the Hsp90 fluorescence polarization assay by utilizing the small-cell lung carcinoma (SCLC) cell line NCI-N417 as a source of Hsp90 and the Cy3B-labeled geldanamycin (Cy3B-GM) as a fluorescence polarization probe. In previous reports, fluorescein or BODIPY were used as the fluorophores (Howes et al. 2006; Kim et al. 2004) for screening. Auto-fluorescence compounds within the libraries of pure compounds or natural product extracts usually emit light emission at the wavelengths of these fluorophores, leading to false positives. Hence, these fluorophores limited the applicability of these assays for HTS. Consequently, Du et al. used red-shifted dyes to label fluorescence polarization probe for the HTS, thus minimizing the auto-fluorescence and reducing false-positive hits related to light scattering on account of insoluble compounds (Du et al. 2007). Then this group screened 15,157 compounds from the NIH library in a high-throughput manner. The assay resulted in 44 possible hits at a hit rate of 0.29%, which decreased the assay window by >50%, and two compounds (PubChem Substance ID 855681 and 857,148) as representative hits from this screen were identified as nucleotide-competitive inhibitors of Hsp90 (Chiosis et al. 2006), which are derivatives of the radicicol and pyrazole scaffolds, respectively. This method has the potential to identify nucleotide-competitive inhibitors of Hsp90 with a different scaffold structure from that of known inhibitors, like geldanamycin, but it is not able to identify inhibitors that carry out a different mechanism of action other than inhibition of the ATPase activity of Hsp90.

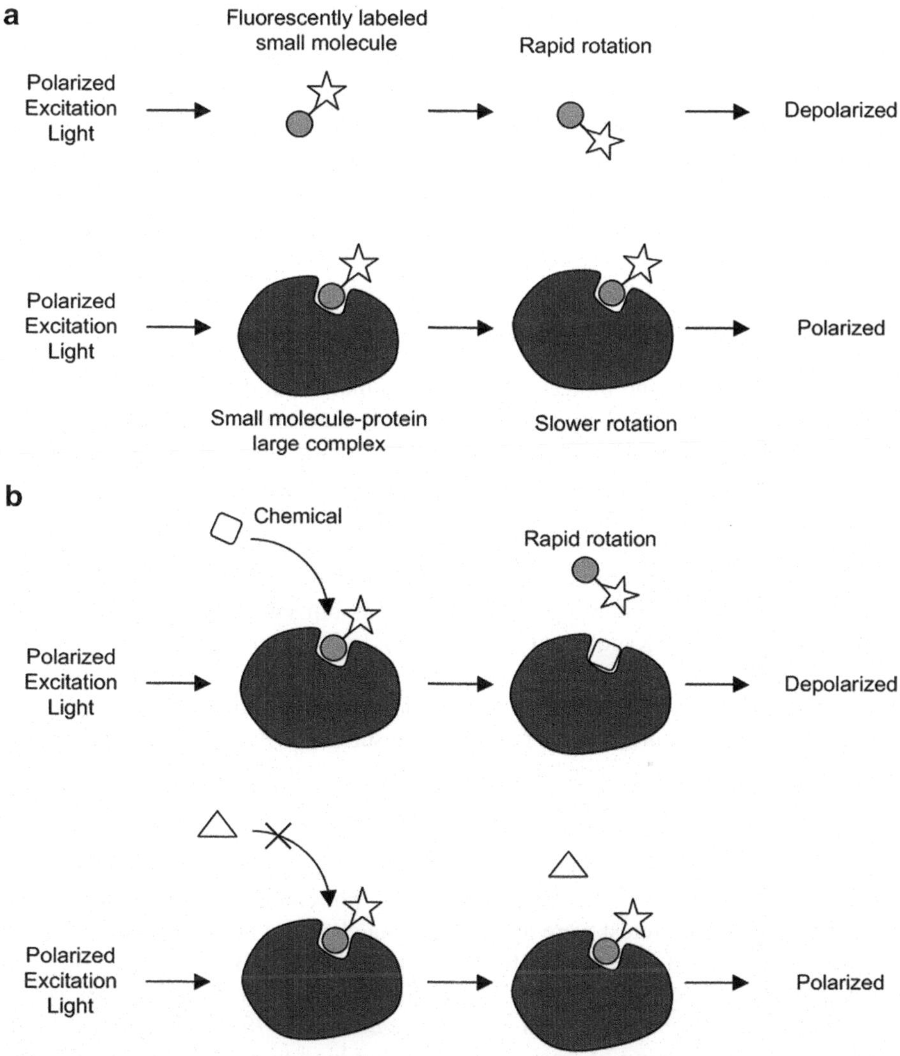

Fig. 19.2 Schematic illustration of fluorescence polarization. (**a**) Fluorescently labeled small molecules rotate quickly during the excited state and have low polarization values. When they bind to proteins and form large complexes, the large molecules rotate little during the excited state, showing high polarization values. (**b**) In chemical screening, when screened compounds bind to proteins competitively against fluorescently labeled small molecule, dissociated fluorescently labeled small molecules rotate quickly and have low polarization values. In contrast, if the screened compounds do not bind, fluorescently labeled small molecules binding to proteins have high polarization value. (Reproduced from Kondoh and Osada 2013)

19.7.3 Chemical Array Screening Based on Interaction Between Ligand and Protein

The chemical array screening is based on the binding affinity between Hsp90 and the screened compounds, which has been reported as one of the most promising approaches to find inhibitors for a protein of interest (Vegas et al. 2008; Uttamchandani et al. 2006; Ma and Horiuchi 2006; Kuruvilla et al. 2002; MacBeath et al. 1999). In general, solid surfaces are modified by certain functional groups, and compounds with a reactive functional group readily attach to the surface through printing by a selective coupling reaction (MacBeath et al. 1999; Vegas et al. 2008; Park and Shin 2002; Hergenrother et al. 2000). The resulting small molecule arrays are used to screen ligand molecules targeted to individual proteins. Several groups reported successful results in identifying ligands that regulate the biological function of proteins (Kuruvilla et al. 2002; Lee and Park 2011; Stanton et al. 2009; Koehler et al. 2003; Barnes-Seeman et al. 2003).

This selective coupling approach is obviously suitable for screening synthetic compounds, but natural products may be still difficult to be uniformly introduced onto a solid surface due to complex structure. Kanoh et al. developed a chemical array on which the natural products are immobilized in a functional group-independent manner (Kanoh et al. 2003, 2006). In this chemical array screening as shown in Fig. 19.3, small molecules are immobilized onto a solid surface by using photoreactive aryltrifluoro-methyldiazirine modification. Small molecule solutions are spotted onto the surface and coupled to the surface by UV irradiation at 365 nm after drying solvent. Then, the binding compounds are identified directly using proteins labeled with fluorochrome or indirectly via immune-staining. This photo affinity-crosslinking method would not necessitate incorporating specific functional groups among library members and is thought to be suitable for complex natural products. Several successful studies have been reported that small molecule inhibitors for proteins of interest were identified using chemical arrays fabricated by this method (Miyazaki et al. 2008, 2010; Hagiwara et al. 2010; Burger et al. 2012).

For the interest of a prokaryotic Hsp90, i.e., HtpG, from the freshwater cyanobacterium *Synechococcus elongates (S. elongates)* PCC 7942, Minagawa's group carried out the screening of small molecules that bind to HtpG by using chemical arrays of 6788 compounds immobilized in duplicate on glass slides by photo affinity-crosslinking to explore novel small molecules that regulate the HtpG function (Minagawa et al., 2011). Colistin A and polymyxin B were identified as binding strongly to HtpG and biochemical experiments to validate these compounds revealed that colistin A inhibits the chaperone activity of HtpG, but has no effect on its ATPase activity (Minagawa et al. 2011). Therefore, this chemical array method enables to discover inhibitors with different mechanisms of affinity action.

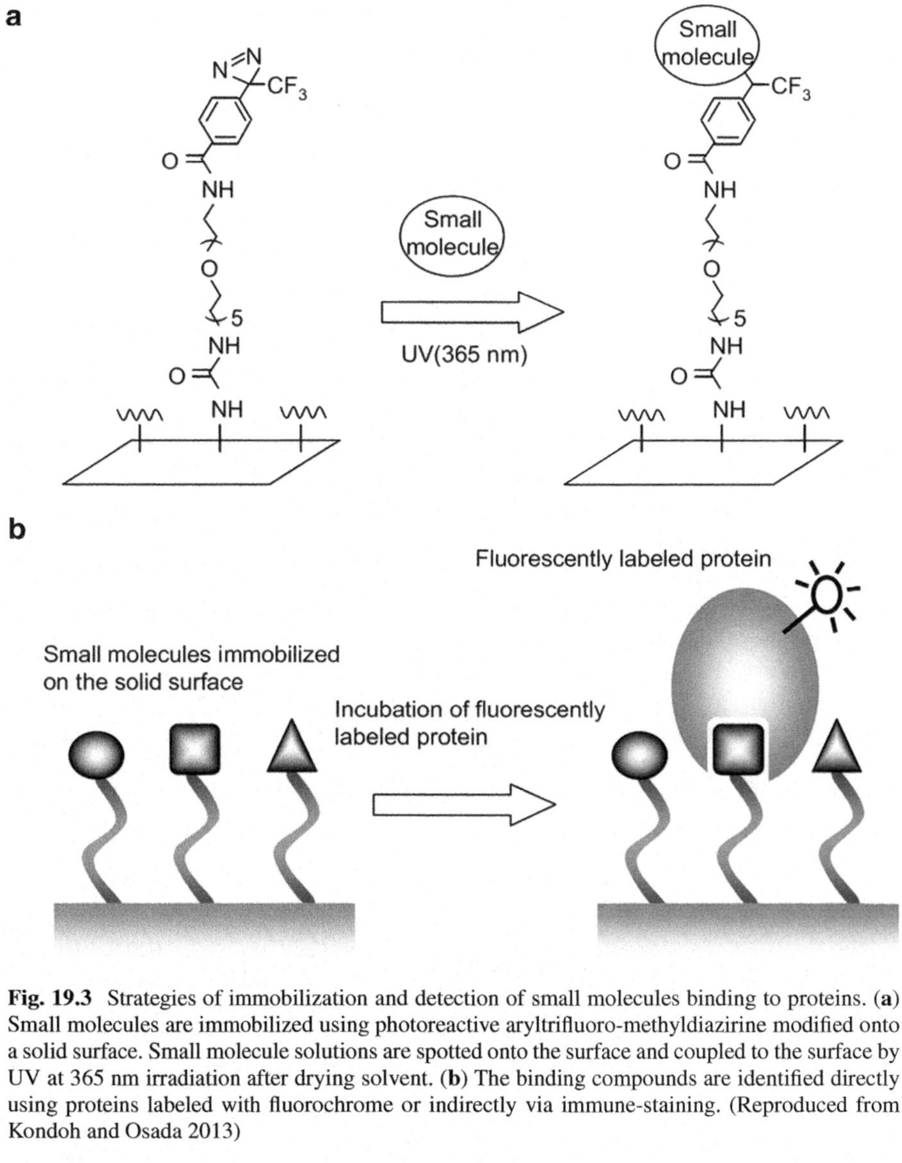

Fig. 19.3 Strategies of immobilization and detection of small molecules binding to proteins. (**a**) Small molecules are immobilized using photoreactive aryltrifluoro-methyldiazirine modified onto a solid surface. Small molecule solutions are spotted onto the surface and coupled to the surface by UV at 365 nm irradiation after drying solvent. (**b**) The binding compounds are identified directly using proteins labeled with fluorochrome or indirectly via immune-staining. (Reproduced from Kondoh and Osada 2013)

19.7.4 AlphaScreen®-Based Screening Based on Hsp90-Cochaperone Interaction

AlphaScreen® is a bead-based assay technology for identifying biomolecular interactions in a microplate format (Beaudet et al. 2001; Warner et al. 2004). This assay requires 2 types of beads: 'donor' and 'acceptor' beads. The donor beads contain a photosensitizer, phthalocyanine, which converts ambient oxygen to an excited and reactive form of O_2, singlet oxygen, upon illumination at 680 nm. Yi et al. developed a high-throughput screen based on AlphaScreen® technology to identify a novel type of Hsp90 inhibitor that interrupts its interaction with the co-chaperone HOP (Yi et al. 2009). In this AlphaScreen®-based competition screening as shown in Fig. 19.4, the donor bead is coated with streptavidin that binds biotin-labeled Hsp90 peptide, while the acceptor bead is coated with Ni-NTA that binds His6-tagged TPR2A of HOP. In the absence of inhibitors, the Hsp90 peptide-TPR2A interaction brings the donor and the acceptor beads together. Upon illumination at 680 nm, the donor bead produces excited singlet oxygen species, which can diffuse approximately 200 nm in solution and react with thioxene derivatives within the acceptor bead, thus exerting light at 520–620 nm. In the presence of small molecule inhibitors that inhibit the Hsp90 peptide-TPR2A interaction, the donor and the acceptor beads are more than 200 nm away from each other, and no signal of the acceptor bead is observed upon excitation of the donor bead.

In addition, Yi et al. previously used the AlphaScreen® competition assay in a 1536-well format to screen the NCGC compound library (76,174 compounds) for novel Hsp90 inhibitors at a dilution series of seven concentrations (Yi and Regan 2008). Three compounds remained as true positive hits (CID 460747, 653,297, and 647,501) were identified as disrupting the Hsp90-HOP interaction (Yi and Regan 2008), all of which share a 7-azapteridine ring core structure. CID 460747 binds

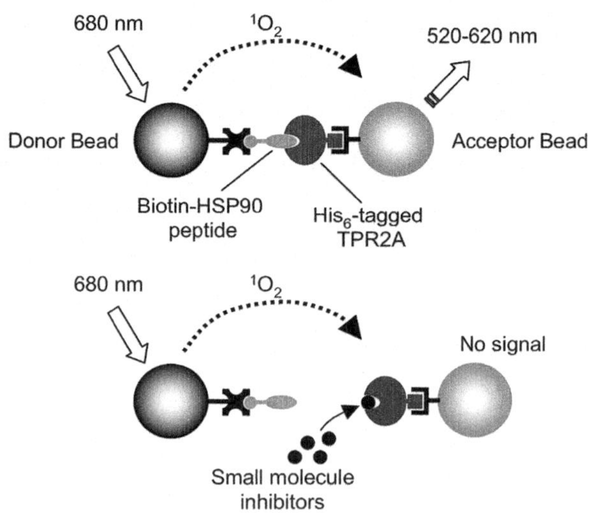

Fig. 19.4 Schematic illustration of the AlphaScreen® competition assay. (Reproduced from Kondoh and Osada 2013)

directly to TPR2A domain of HOP, inhibiting the binding of MEEVD motif of Hsp90 with TPR2A domain (Yi and Regan 2008). Interestingly, this novel type of inhibitor enables to decrease the levels of the Hsp90-dependent client proteins similarly to known Hsp90 inhibitors, such as geldanamycin, but couldn't induce the overproduction of Hsp70 as imposed by ATPase inhibitors of Hsp90 (Yi and Regan 2008, 2009; Banerji et al. 2005; Muroi et al. 2010). This result reveals that these inhibitors have a different mechanism of affinity action from that of ATPase inhibitors, which may have a potential advantage over these agents.

19.8 Fragment-Based Biophysical Techniques for Screening HSP90 Inhibitors

Fragment-based drug discovery (FBDD) is a well-established and successful drug discovery approach widely applied in the pharmaceutical community. FBDD depends on the use of sensitive biophysical techniques, such as thermal shift assays (Pantoliano et al. 2001), nuclear magnetic resonance (NMR) (Shuker et al. 1996), electrospray ionization mass spectrometry (EI-MS), capillary electrophoresis (CE) (Belenky et al. 2004), and surface plasmon resonance (SPR) (Hämäläinen et al. 2008) to detect the commonly weak affinity interactions ($K_D > 10$ μM) of the low molecular weight (<300 Da) compounds with interested targets. It plays an important role in validating fragment hits by utilizing certain techniques to help prioritize fragment follow-up.

19.8.1 Thermal Shift Assays

Thermal shift assay is a fluorescence-based method that measures the denaturation temperature (Tm) of protein samples, the basis of which relies on ligand binding for stabilizing proteins to thermal denaturation (Pantoliano et al. 2001; Lo et al. 2004; Zhang and Monsma 2010). Thermal shift assays are often applied as complementary methods to verify the hits obtained from virtual screens. Huang et al. applied this thermal shift assay to verify 20 virtual hits (Huang et al. 2016). In their research, a tenfold excess of ligand was used to ensure ligand binding to Hsp90-ND, while ADP was chosen as a positive control and gave a Tm shift of 4.5 ± 0.4 °C. Seven compounds gave positive Tm shift similar to or larger than that obtained with ADP, indicating the interaction with Hsp90-ND (Huang et al. 2016). Moreover, they also revealed that five of these compounds contained the resorcinol structure, which is a common scaffold for several Hsp90 inhibitors that are currently undergoing clinical trials (Taldone et al. 2009; Biamonte et al. 2010; Sidera and Patsavoudi 2014; Brasca et al. 2015), thus indicating the feasibility of method.

19.8.2 Protein Nuclear Magnetic Resonance Spectroscopy

Protein NMR spectroscopy is also useful in monitoring ligand binding pocket on the protein, and for the quantification of their binding affinities (K_Ds) (Fielding 2003; Williamson 2013). Huang et al. performed protein NMR spectroscopy using ^{15}N-labelled Hsp90-ND to further confirm the results obtained from previous virtual screen and thermal shift assay (Jahnke and Widmer 2004; Pellecchia et al. 2008; Stark and Powers 2011). Results revealed that four compounds bond to the ATP binding pocket of Hsp90-ND, which is in agreement with molecular modeling studies (Huang et al. 2016). Thus, in additional to screening, protein NMR experiments also allowed the mapping of the ligand binding pocket on Hsp90-ND. Moreover, they also confirmed that, whilst there is a distinct correlation between ligand binding to Hsp90 and the increase in thermal stability of the protein for some compounds, thermal shift testing is an indirect method to study protein-ligand binding that could give false positives and hence its results should always be identified by a second binding assay.

19.8.3 Electrospray Ionization Mass Spectrometry

ESI-MS is a well-established technique for studying biopolymers (nucleic acids, proteins) and their non-covalent adducts (Vivat Hannah et al. 2009; Hofstadler and Sannes-Lowery 2006), due to its capacity to directly detect ligand-target complexes in the gas phase allowing the determination of ligand-target binding in solution (Liu et al. 2009; Van den Heuvel and Heck 2004; Heck and Van Den Heuvel 2004). Different medium to high-throughput ESI-MS assays have been previously described in the literature, and some of the more recent methods make use of an automated nano-electrospray ion source for fragment-based screening (Maple et al. 2012; Vivat Hannah et al. 2009). An automated flow injection ESI-MS (AFI-MS) performs a flow injection operation mode, and samples are injected in a micro-flow by an auto-sampler which could reach the mass spectrometer ion source in few seconds. Recently, Riccardi et al. established an AFI-MS approach and employed it to screen the Nerviano Medical Sciences proprietary fragment library of approximately 2000 fragments against Hsp90 (Riccardi et al. 2015). Potential Hsp90 ligands were automatically identified by the software when a signal of sufficient intensity (300 counts in the centroid spectrum) was present at the expected molecular weight for the Hsp90–ligand complex. Five-component mixtures were tested in order to increase assay throughput, allowing the completion of the screening in 2 days and 282 compounds were found as possible hits by the primary screening carried out on these mixtures (Riccardi et al. 2015), suggesting that AFI-MS is a fast and reliable method, suitable to be used to rapidly screen fragment libraries of different sizes.

19.8.4 Capillary Electrophoresis

Capillary electrophoresis (CE) is a well-known analytical technique (Belenky et al. 2004), which has recently been extended as screening technology such as fragment screening technology based on capillary electrophoresis (CEfrag). CEfrag relies on the utilization of a known competitive ligand with high affinity to the binding site of interested target protein and detects a reduction in the interaction of the probe ligand with its target protein in the presence of test compounds (e.g., fragments) binding to the target. The main advantages that CEfrag have over other fragment screening techniques are lower protein consumption, higher throughput, and the fact that screening even can be carried out using unmodified protein in solution. CEfrag has been developed as a novel fragment screening technique enable to detect the weak affinity interactions attributing to fragments binding to their targets. Austin et al. applied this CEfrag assay for hit identification of Hsp90 ATPase inhibitors, in which case CEfrag readily detected inhibitors in the high μM to mM range (Austin et al. 2012). The hit rate obtained using the Selcia fragment library and the Hsp90 CEfrag assay (6.9%) was consistent with that obtained by others (6.1%) using NMR and the Vernalis fragment library (Brough et al. 2009), thus highlighting that CEfrag is a technique capable of detecting weak affinities.

19.8.5 Surface Plasmon Resonance

Surface plasmon resonance (SPR) has been attracted enormous attention on biosensors or chemical sensors. The fundamental principle of SPR sensor is based on constructing the resonance of a strong electromagnetic field oscillation at the interface of nanometal film and a dielectric medium with p-polarized incident light, resulting in a dark band profile in the light reflectivity at a specific wavelength and incident angle. To date, the past three decades have witnessed remarkable improvements and several implementations of the SPR sensor technology including in the area of chemical sensing, marker screening, and medical diagnostics (Šípová and Homola 2013; Piliarik et al. 2009; Homola 2008). Chini et al. reported the evaluation of the Hsp90 inhibitory activity of tetranortriterpenes by using a panel of chemical and biological approaches, including SPR measurements, biochemical and cellular assays (Chini et al. 2016). They discovered three new limonol derivatives possessed distinct anti-Hsp90 activity and also performed limited proteolysis and computational methods to identify the Hsp90 region involved in the molecular interaction with these plant small molecules (Chini et al. 2016).

19.9 Nanocarrier-Based Ligand Fishing Technique for HSP90 Inhibitors

Traditional high throughput approaches are obviously labor-immense and time-consuming, which are not applicable for complicated herbal extracts. To address these requirements, nanocarrier-based ligand fishing screening assay, has been proved to be an exciting method to screen specific ligands from complex matrix (Qing et al. 2012; Lourenço et al. 2013; Singh et al. 2014). In these strategies, the targeted protein is usually immobilized onto the surface of solid supports which enable to resist to environmental changes compared with solution phase (Liu et al. 2013; Ashtari et al. 2012), and the resulting stationary phase is applied to fish out specific compounds from different mixtures. Various types of solid supporters, such as magnetic particles (Lourenço et al. 2013), gold nanoparticles (New et al. 2014) and silica nanospheres (Hu et al. 2015), have been employed for biomolecule immobilization and drug discovery.

Marszałł et al. reported an alternative approach to the isolation of Hsp90 small ligands based upon the immobilization of Hsp90α on the surface of magnetic beads. This is the first demonstration that magnetic beads immobilized with Hsp90 can be used to fish and identify potential Hsp90 ligands in complex chemical or botanical mixtures (Marszałł et al. 2008). In their previous study, Marszałł et al. also demonstrated the effective approaches to covalently immobilizing Hsp90 on a silica-based stationary phase through either the amino- or carboxy-terminus of the protein to produce Hsp90-NT (immobilization via N-terminus) and Hsp90-(CT) (immobilization via C-terminus) columns in Fig. 19.5 and that the resulting column could be used in liquid chromatography experiments to identify small molecule Hsp90 ligands (Marszałł et al. 2008). However, on the one hand, the Hsp90-(NT) or Hsp90-(CT)

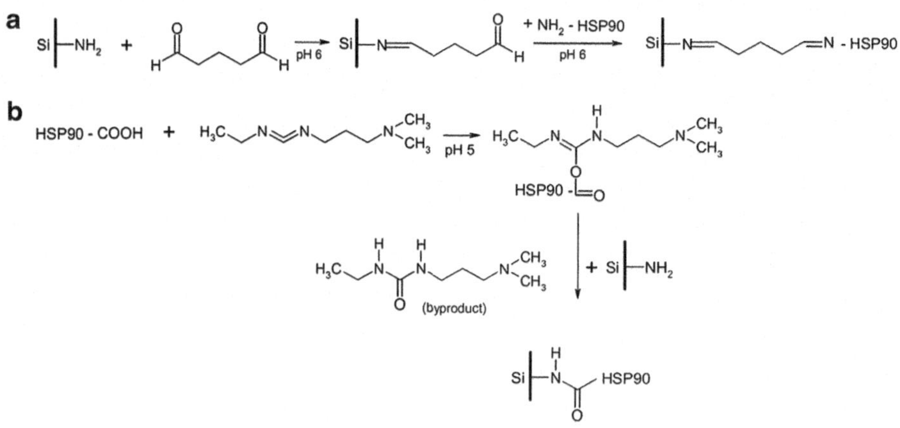

Fig. 19.5 Synthetic approaches used in the covalent immobilization of Hsp90α on an APS liquid chromatography stationary phase via the (**a**) amino terminus and (**b**) carboxyl terminus. (Reproduced from Michal et al 2008)

columns could not be readily adapted to the screening of complex matrixes such as cellular extracts or plant extracts. On the other hand, the current ligand fishing technique cannot concurrently carry out biomedical imaging, which is important to further show in situ screening results for accurate assessment bioactivity of fished ligands.

Hu et al. successfully fabricated a kind of Hsp90α functionalized fluorescent mesoporous silica-InP/ZnS quantum dots nanocomposites (i.e. Hsp90α (NT)-FQDNs), and utilized it to fish Hsp90 small ligands coupling specific enrichment with in situ biomedical imaging from *Curcuma longa* L. crude extracts as shown in Fig. 19.6 (Hu et al. 2018a). Twelve ligands were ultimately fished and identified, two of which, i.e. curcumin and curlone, reflected a strong binding effect on aliphatic hydrophobic interactions as the potential Hsp90 inhibitors associated with virtual docking (Hu et al., 2018). Moreover, Miao et al. also developed a fluorescent ligand fishing to screen and identify not only small-molecule ligands of Hsp90 from complex herbal plants but also client proteins present in a mixture of cellular extracts as seen in Fig. 19.7 (Miao et al. 2018). Twenty ligands for Hsp90 from *Alisma plantago-aquatica Linn* extracts were screened and identified by mass spectrometry, among which 14 had a certain affinity for Hsp90 on the basis of molecular docking results, and five proteins from HeLa cells were identified as potential client proteins of Hsp90 (Miao et al. 2018).

Nevertheless, these developed fluorescent ligand fishing approaches are still confronted with challenges such as the susceptible fluorescence to external environments due to the QDs only coating the surface of the mesoporous nanoparticles, thus inducing the formation of microkernel mesoporous fluorescent nanoparticles responsible for protecting QDs in further bioapplications (Zhu et al. 2012). Very recently, a multiple-layer mesoporous nanostructure composed of a protective silica

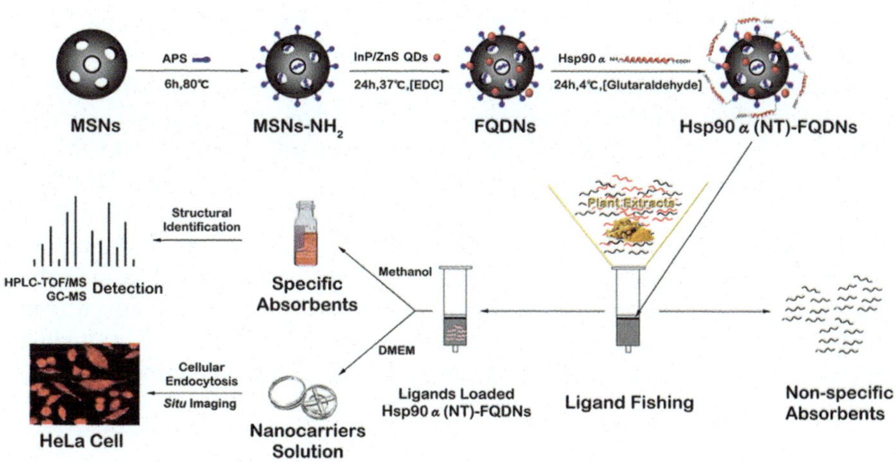

Fig. 19.6 Schematic illustration of Hsp90α functionalized nanocomposites based affinity extraction coupled with MS and in situ imaging for identification of Hsp90 inhibitors from *Curcuma longa* L. (Reproduced from Hu et al. 2018a)

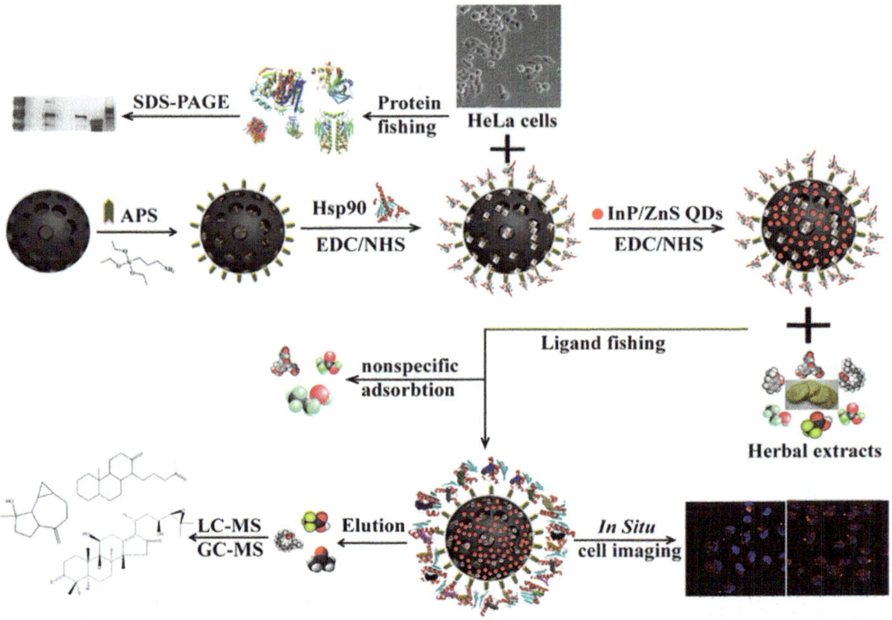

Fig. 19.7 Schematic illustration of Hsp90-coated fluorescent mesoporous silica-InP/ZnS QDs nanocomposites and their application in screening client proteins from HeLa cells and small ligands targeted Hsp90 from *Alisma plantago-aquatica Linn.* crude extract. (Reproduced from Miao et al. 2018)

coating CdTe QD core and a mesoporous silica shell, i.e., microkernel-based mesoporous (SiO_2-CdTe-SiO_2)@SiO_2 fluorescent nanoparticles (MMFNPs), was successfully fabricated as extraction sorbents and fluorescent tracers (Hu et al. 2018b). As depicted in Fig. 19.8 as-prepared MMFNPs was functionalized with Hsp90α in the presence of a cross-linking agent to obtain Hsp90α-MMFNPs for screening specific ligands from complex herbal extracts, *Tripterygium wilfordii* crude extraction. A total of 12 specific components selectively screened and identified by LC/MS and GC/MS analysis were performed to dock with the whole binding sites of Hsp90 for determining binding affinity towards Hsp90, three of which, i.e., celastrol, wilforine and demecolcine, showed a strong affinity, indicating a promising Hsp90 inhibitor (Hu et al. 2018b). These results suggest that this type of fluorescent ligand fishing approaches have great promise for the development of other plant-derived Hsp90 inhibitors and is expected to provide a facile and effective platform for in situ imaging screening and discovery of biologically active molecules in natural products.

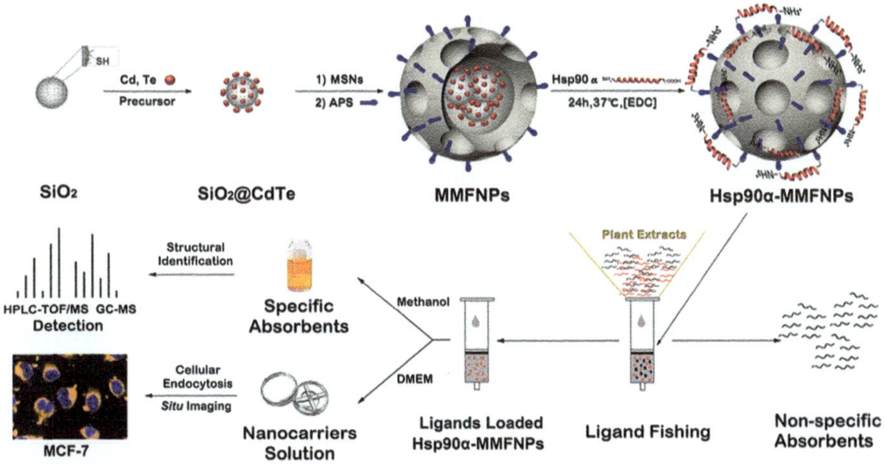

Fig. 19.8 Schematic representation of the preparation of Hsp 90MMFNPs and Its Application Based on Solid-Phase Ligand Fishing for Screening Hsp90α inhibitors from *Tripterygium wilfordii.* (Reproduced from Hu et al. 2018)

19.10 Computer-Aided Virtual Screening Technique for HSP90 Inhibitors

Virtual screening is the complementary approach of experimental high-throughput screening. It has proven to be an efficient implementation in drug design and discovery coupled with structural biology. Virtual screening refers to techniques that are utilized to search large compound databases for small molecules and further identify those structures which are most likely to bind to therapeutic receptors of interest (Willett 2006; Garcia et al. 2007; Morelli et al. 2014). In the past two decades, several molecular modelling tools such as molecular docking, pharmacophore mapping and quantitative structure-activity relationship (QSAR) have emerged as a cost effective alternative for performing virtual screening.

Barril et al. used docking-based virtual screening to discover 1-(2-phenol)-2-naphthol derivatives as a new class of Hsp90 inhibitors, whose binging mode was determined in the ATP binding site of Hsp90 by high-resolution crystal structures (Barril et al. 2005). Jia et al. explored the significant pharmacophore features required for Hsp90 inhibitors by generating a 3D-QSAR-pharmacophore model, which was further used to virtually screen out the SPECS database. Twenty-four derivatives of the most potent compounds were designed and further screened (Jia et al. 2013). Sun et al. established and validated a rapid overlay of chemical structures (ROCS) model on the basis of AT-13387 for the identification and optimization of novel Hsp90 inhibitors with tetrahydropyrido[4,3-d] pyrimidines core. Eleven compounds containing the similar scaffold from Topscience database were identified as potent Hsp90 inhibitors through shape-based screening (Sun et al. 2014). Xu et al. also reported the application of shape-based similarity screening through ROCS on the basis of CUDC-305, BIIB021, PU-H71 and PU-3 to discover Hsp90 inhibitors. A set of novel pyrazolo-pyrimidine analogues were identified and evaluated on enzyme level and cell-based level as Hsp90 inhibitors, of which compound HDI4-04 exhibited potent cytotoxicity against five human cancer cell lines (Xu et al. 2014). Caroli et al. exploited virtual screening for the discovery of novel Hsp90 inhibitors using a combination of Autodock and Surflex-Sim from the NCI Diversity Set composed of 1785 compound. The preliminary data was analyzed to screen four derivatives as hits for a subsequent medicinal chemistry optimization procedure (Caroli et al. 2014).

19.11 Conclusions

Hsp90 plays an important role in the regulation of proteins and protein complexes, and is essential for cancer cells survival. Continuous efforts have been paid for the exploitation of new Hsp90 inhibitors on anti-cancer research. Various screening techniques were developed including library-based HTS methods, fragment-based biophysical techniques, nanocarrier-based ligand fishing and computer-aided virtual screening. Recent proposed fluorescent ligand fishing approach showed not only the high efficiency and specificity for fishing ligands from herbal extraction, thus allowing easy adaptation to complex matrixes, but also characteristic optical targeting function compared with fragment-based biophysical techniques. Moreover, computer-aided virtual screening techniques are proved to be an efficient implementation towards drug design and discovery. Going forwards, a facile platform with high efficiency and selectivity is always expected to be widely applied for the discovery of biologically active molecules from complex matrix.

Acknowledgements This work was supported by the Project of National Natural Science Foundation of China (81573388). This work was also sponsored by 'Qing Lan Project of Jiangsu province' and 'Six talent peaks project of Jiangsu Province (YY-032)'. We also greatly appreciate the sponsorship of 'A Project Funded by the Priority Academic Program Development of Jiangsu Higher Education Institutions' (PAPD).

References

Aloy P, Russell RB (2002) Potential artefacts in protein-interaction networks. FEBS Lett 530:253–254

Ashtari K, Khajeh K, Fasihi J, Ashtari P, Ramazani A, Vali H (2012) Silica-encapsulated magnetic nanoparticles: enzyme immobilization and cytotoxic study. Int J Biol Macromol 50:1063–1069

Austin C, Pettit SN, Magnolo SK et al (2012) Fragment screening using capillary electrophoresis (CEfrag) for hit identification of heat shock protein 90 ATPase inhibitors. J Biomol Screen 17:868–876

Banerji U, O'Donnell A, Scurr M, Pacey S, Stapleton S, Asad Y, Simmons L, Maloney A, Raynaud F et al (2005) Phase I pharmacokinetic and pharmacodynamic study of 17-allyl-amino, 17-demethoxygeldanamycin in patients with advanced malignancies. J Clin Oncol 23:4152–4161

Barnes-Seeman D, Park SB, Koehler AN, Schreiber SL (2003) Expanding the functional group compatibility of small-molecule microarrays: discovery of novel calmodulin ligands. Angew Chem Int Ed 42:2376–2379

Barril X, Brough P, Drysdale M et al (2005) Structure-based discovery of a new class of Hsp90 inhibitors. Bioorg Med Chem Lett 15:5187–5191

Baykov AA, Evtushenko OA, Avaeva SM (1998) A malachite green procedure for orthophosphate determination and its use in alkaline phosphatase-based enzyme immunoassay. Anal Biochem 171:266–270

Beaudet L, Bédard J, Breton B, Mercuri RJ, Budarf ML (2001) Homogeneous assays for single-nucleotide polymorphism typing using AlphaScreen. Genome Res 11:600–608

Belenky A, Hughes D, Korneev A, Dunayevskiy Y (2004) Capillary electrophoretic approach to screen for enzyme inhibitors in natural extracts. J Chromatogr A 1053:247–251

Biamonte MA, Van de Water R, Arndt JW, Scannevin RH, Perret D, Lee WC (2010) Heat shock protein 90: inhibitors in clinical trials. J Med Chem 53:3–17

Brandt GE, Schmidt MD, Prisinzano TE, Blagg BS (2008) Gedunin, a novel hsp90 inhibitor: semisynthesis of derivatives and preliminary structure-activity relationships. J Med Chem 51:6495–6502

Brasca MG, Casale E, Ferguson R, Polucci P, Zuccotto F (2015) Resorcinol derivatives as Hsp90 inhibitors. U.S. patent 8993556 B2, p, 2015-03-31

Brough PA, Barril X, Borgognoni J et al (2009) Combining hit identification strategies: fragment-based and in silico approaches to orally active 2-aminothieno[2,3-d]pyrimidine inhibitors of the Hsp90 molecular chaperone. J Med Chem 52:4794–4809

Brune M, Hunter JL, Corrie JE, Webb MR (1994) Direct, real-time measurement of rapid inorganic phosphate release using a novel fluorescent probe and its application to actomyosin subfragment 1 ATPase. Biochemistry 33:8262–8271

Brune M, Hunter JL, Howell SA, Martin SR, Hazlett TL, Corrie JE, Webb MR (1998) Mechanism of inorganic phosphate interaction with phosphate binding protein from Escherichia coli. Biochemistry 37:10370–10380

Buchner J (1999) Hsp90 & co. —a holding for folding. Trends Biochem Sci 24:136–141

Burger M, Zimmermann TJ, Kondoh Y, Stege P, Watanabe N, Osada H, Waldmann H, Vetter IR (2012) Crystal structure of the predicted phospholipase LYPLAL1 reveals unexpected functional plasticity in spite of close relationship to acyl protein thioesterases. J Lipid Res 53:43–50

Caroli A, Ballante F, Wickersham RB, Corelli F, Ragno R (2014) Hsp90 inhibitors, part 2: combining ligand-based and structure-based approaches for virtual screening application. J Chem Inf Model 54:970–977

Chen G, Bradford WD, Seidel CW, Li R (2012) Hsp90 stress potentiates rapid cellular adaptation through induction of aneuploidy. Nature 482:246–250

Chen WQ, Zheng RS, Baade PD, Zhang SW, Zeng HM, Bray F, Jemal A, Yu XQ, He J (2016) Cancer statistics in China, 2015. Ca Cancer J Clin 66:115–132

Chini MG, Malafronte N, Vaccaro MC et al (2016) Identification of limonol derivatives as heat shock protein 90 (Hsp90) inhibitors through a multidisciplinary approach. Chemistry 22:13236–13250

Chiosis G, Caldas Lopes E, Solit D (2006) Hsp90 inhibitors: a chronicle from geldanamycin to today's agents. Curr Opin Investig Drugs 6:534–541

Crevel G, Bates H, Huikeshoven H, Cotterill S (2001) The Drosophila Dpit47 protein is a nuclear Hsp90 co-chaperone that interacts with DNA polymerase alpha. J Cell Sci 114:2015–2025

Driggers EM, Hale SP, Lee J, Terrett NK (2008) The exploration of macrocycles for drug discovery – an underexploited structural class. Nat Rev Drug Discov 7:608–624
Du Y, Moulick K, Rodina A, Aguirre J, Felts S, Dingledine R, Fu H, Chiosis G (2007) High-throughput screening fluorescence polarization assay for tumor-specific Hsp90. J Biomol Screen 12:915–924
Ferrarini M, Heltai S, Zocchi MR, Rugarli C (1992) Unusual expression and localization of heat-shock proteins in human tumor cells. Int J Cancer 19:613–619
Fielding L (2003) NMR methods for the determination of protein-ligand dissociation constants. Curr Top Med Chem 3:39–53
Garcia E, Hasenbank MS, Finlayson B, Yager P (2007) High-throughput screening of enzyme inhibition using an inhibitor gradient generated in a microchannel. Lab Chip 7:249–255
Garg G, Khandelwal A, Blagg BS (2016) Anticancer inhibitors of Hsp90 function: beyond the usual suspects. Adv Cancer Res 129:51–88
Hadden MK, Galam L, Gestwicki JE, Matts RL, Blagg BSJ (2007) Derrubone, an inhibitor of the Hsp90 protein folding machinery. J Nat Prod 70:2014–2018
Hagiwara K, Kondoh Y, Ueda A, amada K, Goto H, Watanabe T, Nakata T, Osada H, Aida Y (2010) Discovery of novel antiviral agents directed against the influenza a virus nucleoprotein using photo-cross-linked chemical arrays. Biochem Biophys Res Commun 394:721–727
Hämäläinen MD, Zhukov A, Ivarsson M, Fex T, Gottfries J, Karlsson R, Björsne M (2008) Label-free primary screening and affinity ranking of fragment libraries using parallel analysis of protein panels. J Biomol Screen 13:202–209
Hamamoto R, Furukawa Y, Morita M, Iimura Y, Silva FP, Li M, Yagyu R, Nakamura Y (2004) SMYD3 encodes a histone methyltransferase involved in the proliferation of cancer cells. Nat Cell Biol 6:731–740
Hamamoto R, Silva FP, Tsuge M, Nishidate T, Katagiri T, Nakamura Y, Furukawa Y (2006) Enhanced SMYD3 expression is essential for the growth of breast cancer cells. Cancer Sci 97:113–118
Harder KW, Owen P, Wong LK, Aebersold R, Clark-Lewis I, Jirik FR (1994) Characterization and kinetic analysis of the intracellular domain of human protein tyrosine phosphatase beta (HPTP beta) using synthetic phosphopeptides. Biochem J 298:395–401
Heck AJ, Van Den Heuvel RH (2004) Investigation of intact protein complexes by mass spectrometry. Mass Spectrom Rev 23:368–389
Hergenrother PJ, Depew KM, Schreiber SL (2000) Small-molecule microarrays: covalent attachment and screening of alcohol-containing small molecules on glass slides. J Am Chem Soc 122:7849–7850
Hessling M, Richter K, Buchner J (2009) Dissection of the ATP-induced conformational cycle of the molecular chaperone Hsp90. Nat Struct Mol Biol 16:287–293
Hieronymus H, Lamb J, Ross KN, Peng XP, Clement C, Rodina A, Nieto M, Du J, Stegmaier K, Raj SM, Maloney KN, Clardy J, Hahn WC, Chiosis G, Golub TR (2006) Gene expression signature-based chemical genomic prediction identifies a novel class of Hsp90 pathway modulators. Cancer Cell 10:321–330
Hofstadler SA, Sannes-Lowery KA (2006) Applications of ESI-MS in drug discovery: interrogation of noncovalent complexes. Nat Rev Drug Discov 5:585–595
Homola J (2008) Surface plasmon resonance sensors for detection of chemical and biological species. Chem Rev 108:462–493
Hong DS, Banerji U, Tavana B (2013) Targeting the molecular chaperone heat shock protein 90 (Hsp90): lessons learned and future directions. Cancer Treat Rev 39:375–387
Howes R, Barril X, Dymock BW, Grant K, Northfield CJ, Robertson AG, Surgenor A, Wayne J, Wright L, James K, Matthews T, Cheung KM, McDonald E, Workman P, Drysdale MJ (2006) A fluorescence polarization assay for inhibitors of Hsp90. Anal Biochem 350:202–213
Hu Y, Fu AC, Miao ZY, Zhang XJ, Wang TL, Kang A, Shan JJ, Zhu D, Li W (2018a) Fluorescent ligand fishing combination with in-situ imaging and characterizing to screen Hsp 90 inhibitors

from Curcuma longa L. based on InP/ZnS quantum dots embedded mesoporous nanoparticles. Talanta 178:258–267

Hu Y, Miao ZY, Zhang XJ, Yang XT, Tang YY, Yu S, Shan CX, Wen HM, Zhu D (2018b) Preparation of microkernel-based mesoporous (SiO_2−CdTe−SiO_2)@SiO_2 fluorescent nanoparticles for imaging screening and enrichment of heat shock protein 90 inhibitors from Tripterygium Wilfordii. Anal Chem 90:5678–5686

Hu Y, Shan CX, Wang J, Zhu JM, Gu CQ, Ni WT, Zhu D, Zhang AH (2015) Fabrication of functionalized SiO_2/TiO_2 nanocomposites via amidation for the fast and selective enrichment of phosphopeptides. New J Chem 39:6540–6547

Huang R, Ayine-Tora DM, Muhammad Rosdi MN, Li Y, Reynisson J, Leung IKH (2016) Virtual screening and biophysical studies lead to HSP90 inhibitors. Bioorg Med Chem Lett 27:277–281

Jahnke W, Widmer H (2004) Protein NMR in biomedical research. Cell Mol Life Sci 61:580–599

Jhaveri K, Taldone T, Modi S, Chiosis G (2012) Advances in the clinical development of heat shock protein 90 (Hsp90) inhibitors in cancers. Biochim Biophys Acta 1823:742–755

Jia J, Xu X, Liu F et al (2013) Identification, design and bio-evaluation of novel Hsp90 inhibitors by ligand-based virtual screening. PloS One 8:e59315

Kampinga HH, Hageman J, Vos MJ, Kubota H, Tanguay RM, Bruford EA, Cheetham ME, Chen B, Hightower LE (2009) Guidelines for the nomenclature of the human heat shock proteins. Cell Stress Chaperon 14:105–111

Kanoh N, Asami A, Kawatani M, Honda K, Kumashiro S, Takayama H, Simizu S, Amemiya T, Kondoh Y, Hatakeyama S, Tsuganezawa K, Utata R, Tanaka A, Yokoyama S, Tashiro H, Osada H (2006) Photo-cross-linked small molecule microarrays as chemical genomic tools for dissecting protein-ligand interactions. Chem Asian J 1:789–797

Kanoh N, Kumashiro S, Simizu S, Kondoh Y, Hatakeyama S, Tashiro H, Osada H (2003) Immobilization of natural products on glass slides by using a photoaffinity reaction and the detection of protein-small-molecule interactions. Angew Chem Int Ed 42:5584–5587

Kaur G, Belotti D, Burger AM, Fisher-Nielson K, Borsotti P, Riccardi E, Thillainathan J, Hollingshead M, Sausville EA, Giavazzi R (2004) Antiangiogenic properties of 17-(dimethylaminoethylamino)-17-demethoxygeldanamycin: an orally bioavailable heat shock protein 90 modulator. Clin Cancer Res 10:4813–4821

Kim J, Felts S, Llauger L, He H, Huezo H, Rosen N, Chiosis G (2004) Development of a fluorescence polarization assay for the molecular chaperone Hsp90. J Biomol Screen 9:375–381

Kim YS, Alarcon SV, Lee S (2009) Update on Hsp90 inhibitors in clinical trial. Curr Top Med Chem 9:1479–1492

Koehler AN, Shamji AF, Schreiber SL (2003) Discovery of an inhibitor of a transcription factor using small molecule microarrays and diversity-oriented synthesis. J Am Chem Soc 125:8420–8421

Kondoh Y, Osada H (2013) High-throughput screening identifies small molecule inhibitors of molecular chaperones. Curr Pharm Des 19:473–492

Krukenberg KA, Street TO, Lavery LA, Agard DA (2011) Conformational dynamics of the molecular chaperone Hsp90. Q Rev Biophys 44:229–255

Kuruvilla FG, Shamji AF, Sternson SM, Hergenrother PJ, Schreiber SL (2002) Dissecting glucose signalling with diversity-orienter synthesis and small-molecule microarrays. Nature 416:653–657

Lakowitz JR, Joseph R (eds) (1999) Principles of fluorescence spectroscopy, 2nd edn. Plenum, New York, pp 291–319

Lee HY, Park SB (2011) Surface modification for small-molecule microarrays and its application to the discovery of a tyrosinase inhibitor. Mol Biosyst 7:304–310

Li Y, Zhang T, Jiang Y, Lee HF, Schwartz SJ, Sun D (2009) Epigallocatechin-3-gallate inhibits Hsp90 function by impairing Hsp90 association with cochaperones in pancreatic cancer cell line Mia Paca-2. Mol Pharm 6:1152–1159

Liu LL, Shi SY, Chen XQ, Peng MJ (2013) Analysis of tyrosinase binders from Glycyrrhiza uralensis root: evaluation and comparison of tyrosinase immobilized magnetic fishing-HPLC and reverse ultrafiltration-HPLC. J Chromatogr B 932:19–25

Liu L, Bagal D, Kitova EN, Schnier PD, Klassen JS (2009) Hydrophobic proteinligand interactions preserved in the gas phase. J Am Chem Soc 131:15980–15981

Lo MC, Aulabaugh A, Jin G, Cowling R, Bard J, Malamas M, Ellestad G (2004) Evaluation of fluorescence-based thermal shift assays for hit identification in drug discovery. Anal Biochem 332:153–159

Lourenço VK, Jiang Z, Zhang X, Vieira LC, Corrêa AG, Cardoso CL, Cass QB, Moaddel R (2013) Acetylcholinesterase immobilized capillary reactors coupled to protein coated magnetic beads: a new tool for plant extract ligand screening. Talanta 116:647–652

Lundin A (2000) Use of firefly luciferace in ATP-related assays of biomass, enzymes and metabolites. Methods Enzymol 305:346–370

Ma H, Horiuchi KY (2006) Chemical microarray: a new tool for drug screening and discovery. Drug Discov Today 11:661–668

MacBeath G, Koehler AN, Schreiber SL (1999) Printing small molecules as microarrays and detecting protein-ligand interactions en masse. J Am Chem Soc 121:7967–7968

Maehama T, Taylor GS, Slama JT, Dixon JE (2000) A sensitive assay for phosphoinositide phosphatases. Anal Biochem 279:248–250

Makhnevych T, Houry WA (2012) The role of Hsp90 in protein complex assembly. Biochim Biophys Acta Mol Cell Res 1823:674–682

Maple HJ, Garlish RA, Rigau-Roca L et al (2012) Automated protein-ligand interaction screening by mass spectrometry. J Med Chem 55:837–851

Marcu MG, Schulte TW, Neckers L (2000) Novobiocin and related coumarins and depletion of heat shock protein 90-dependent signaling proteins. J Natl Cancer Inst 92:242–248

Marszałł MP, Moaddel R, Kole S, Gandhari M, Bernier M, Wainer IW (2008) Ligand and protein fishing with heat shock protein 90 coated magnetic beads. Anal Chem 80:7571–7575

Mayer MP, Le Breton L (2015) Hsp90: breaking the symmetry. Mol Cell 58:8–20

McCollum AK, TenEyck CJ, Stensgard B, Morlan BW, Ballman KV, Jenkins RB, Toft DO, Erlichman C (2008) P-glycoprotein-mediated resistance to HSP90-directed therapy is eclipsed by the heat shock response. Cancer Res 68:7419–7427

McConkey DJ, Zhu K (2008) Mechanisms of proteasome inhibitor action and resistance in cancer. Drug Resist Updat 11:164–179

McLaughlin SH, Ventouras LA, Lobbezoo B, Jackson SE (2004) Independent ATPase activity of Hsp90 subunits creates a flexible assembly platform. J Mol Biol 344:813–826

Miao ZY, Hu Y, Zhang XJ, Yang XT, Tang YY, Kang A, Zhu D (2018) Screening and identification of ligand-protein interactions using functionalized heat shock protein 90-fluorescent mesoporous silica-indium phosphide/zinc sulfide quantum dot nanocomposites. J Chromatogr A 1562:1–11

Michał P, Marszałł RM, Krzysztof J, Michel B, Irving WW (2008) Initial synthesis and characterization of an immobilized heat shock protein 90 column for online determination of binding affinities. Anal Biochem 373:313–321

Mickler M, Hessling M, Ratzke C, Buchner J, Hugel T (2009) The large conformational changes of Hsp90 are only weakly coupled to ATP hydrolysis. Nat Struct Mol Biol 16:281–286

Minagawa S, Kondoh Y, Sueoka K, Osada H, Nakamoto H (2011) Cyclic lipopeptide antibiotics bind to the N-terminal domain of the prokaryotic Hsp90 to inhibit the chaperone activity. Biochem J 435:237–246

Mishra P, Bolon DN (2014) Designed Hsp90 heterodimers reveal an asymmetric ATPase-driven mechanism in vivo. Mol Cell 53:344–350

Miyata Y (2009) Protein kinase CK2 in health and disease: CK2: the kinase controlling the Hsp90 chaperone machinery. Cell Mol Life Sci 66:1840–1849

Miyazaki I, Simizu S, Ichimiya H, Kawatani M, Osada H (2008) Robust and systematic drug screening method using chemical arrays and the protein library: identification of novel inhibitors of carbonic anhydrase II. Biosci Biotechnol Biochem 72:2739–2749
Miyazaki I, Simizu S, Okumura H, Takagi S, Osada H (2010) A small molecule inhibitor shows that pirin regulates migration of melanoma cells. Nat Chem Biol 6:667–673
Mollapour M, Neckers L (2012) Post-translational modifications of Hsp90 and their contributions to chaperone regulation. Biochim Biophys Acta 1823:648–655
Mollapour M, Tsutsumi S, Truman AW, Xu WP, Vaughan CK, Beebe K, Konstantinova A, Vourganti S, Panaretou B, Piper PW, Trepel JB, Prodromou C, Pearl LH, Neckers L (2011) Threonine 22 phosphorylation attenuates Hsp90 interaction with cochaperones and affects its chaperone activity. Mol Cell 41:672–681
Morelli L, Bernardi A, Sattin S (2014) Synthesis of potential allosteric modulators of Hsp90 by chemical glycosylation of Eupomatenoid-6. Carbohydr Res 390:33–41
Moulick K, Ahn JH, Zong H, Rodina A, Cerchietti L, Gomes DaGama EM, Caldas-Lopes E, Beebe K, Perna F, Hatzi K, Vu LP, Zhao X, Zatorska D, Taldone T, Smith-Jones P, Alpaugh M, Gross SS, Pillarsetty N, Ku T, Lewis JS, Larson SM, Levine R, Erdjument-Bromage H, Guzman ML, Nimer SD, Melnick A, Neckers L, Chiosis G (2011) Affinity-based proteomics reveal cancer-specific networks coordinated by Hsp90. Nat Chem Biol 7:818–826
Muroi M, Kazami S, Noda K, Kondo H, Takayama H, Kawatani M, Usui T, Osada H (2010) Application of proteomic profiling based on 2D-DIGE for classification of compounds according to the mechanism of action. Chem Biol 17:460–470
Neckers L (2002) Hsp90 inhibitors as novel cancer chemotherapeutic agents. Trends Mol Med 8:S55–S61
Neckers L, Workman P (2012) Hsp90 molecular chaperone inhibitors: are we there yet? Clin Cancer Res 18:64–76
New SY, Aung KM, Lim GL, Hong S, Tan SK, Lu Y, Cheung E, Su X (2014) Fast screening of ligand-protein interactions based on ligand-induced protein stabilization of gold nanoparticles. Anal Chem 86:2361–2370
Panaretou B, Prodromou C, Roe SM, O'Brien R, Ladbury JE, Piper PW, Pearl LH (1998) ATP binding and hydrolysis are essential to the function of the Hsp90 molecular chaperone in vivo. EMBO J 17:4829–4836
Pantoliano MW, Petrella EC, Kwasnoski JD, Lobanov VS, Myslik J, Graf E, Carver T, Asel E, Springer BA, Lane P, Salemme FR (2001) High-density miniaturized thermal shift assays as a general strategy for drug discovery. J Biomol Screen 6:429–440
Park S, Shin I (2002) Fabrication of carbohydrate chips for studying protein-carbohydrate interactions. Angew Chem Int Ed 41:3180–3182
Pashtan I, Tsutsumi S, Wang SQ, Xu WP, Neckers L (2008) Targeting Hsp90 prevents escape of breast cancer cells from tyrosine kinase inhibition. Cell Cycle 7:2936–2941
Pearl LH, Prodromou C (2000) Structure and in vivo function of Hsp90. Curr Opin Struct Biol 10:46–51
Pearl LH, Prodromou C (2006) Structure and mechanism of the Hsp90 molecular chaperone machinery. Ann Rev Biochem 75:271–294
Pearl LH, Prodromou C, Workman P (2008) The Hsp90 molecular chaperone: an open and shut case for treatment. Biochem J 410:439–453
Pellecchia M, Bertini I, Cowburn D et al (2008) Perspectives on NMR in drug discovery: a technique comes of age. Nat Rev Drug Discov 7:738–745
Pick E, Kluger Y, Giltnane JM, Moeder C, Camp RL, Rimm DL, Kluger HM (2007) High HSP90 expression is associated with decreased survival in breast cancer. Cancer Res 67:2932–2937
Piliarik M, Párová L, Homola J (2009) High-throughput SPR sensor for food safety. Biosens Bioelectron 24:1399–1404
Prodromou C (2016) Mechanisms of Hsp90 regulation. Biochem J 473:2439–2452

Prodromou C, Roe SM, O'Brien R, Ladbury JE, Piper PW, Pearl LH (1997) Identification and structural characterization of the ATP/ADP – binding site in the Hsp90 molecular chaperone. Cell 90:65–75

Qing LS, Tang N, Xue Y, Liang J, Liu YM, Liao X (2012) Identification of enzyme inhibitors using therapeutic target protein-magnetic nanoparticle conjugates. Anal Methods 4:1612–1615

Retzlaff M, Hagn F, Mitschke L, Hessling M, Gugel F, Kessler H, Richter K, Buchner J (2010) Asymmetric activation of the hsp90 dimer by its cochaperone aha1. Mol Cell 37:344–354

Retzlaff M, Stahl M, Eberl HC, Lagleder S, Beck J, Kessler H, Buchner J (2009) Hsp90 is regulated by a switch point in the C-terminal domain. EMBO Rep 10:1147–1153

Riccardi Sirtori F, Caronni D, Colombo M et al (2015) Establish an automated flow injection ESI-MS method for the screening of fragment based libraries: application to Hsp90. Eur J Pharm Sci 76:83–94

Roe SM, Prodromou C, O'Brien R, Ladbury JE, Piper PW, Pearl LH (1999) Structural basis for inhibition of the Hsp90 molecular chaperone by the antitumor antibiotics radicicol and geldanamycin. J Med Chem 42:260–266

Rohl A, Rohrberg J, Buchner J (2013) The chaperone Hsp90: changing partners for demanding clients. Trends Biochem Sci 38:253–262

Rowlands MG, Newbatt YM, Prodromou C, Pearl LH, Workman P, Aherne W (2004) High-throughput screening assay for inhibitors of heat-shock protein 90 ATPase activity. Anal Biochem 327:176–183

Schopf HF, Maximilian MB, Johannes B (2017) The Hsp90 chaperone machinery. Nat Rev Mol Cell Bi 18:345–360

Scroggins BT, Robzyk K, Wang D, Marcu MG, Tsutsumi S, Beebe K, Cotter RJ, Felts S, Toft D, Karnitz L, Rosen N, Neckers L (2007) An acetylation site in the middle domain of Hsp90 regulates chaperone function. Mol Cell 25:151–159

Shiau AK, Harris SF, Southworth DR, Agard DA (2006) Structural analysis of E. coli hsp90 reveals dramatic nucleotide-dependent conformational rearrangements. Cell 127:329–340

Shuker SB, Hajduk PJ, Meadows RP, Fesik SW (1996) Discovering high-affinity ligands for proteins: SAR by NMR. Science 274:1531–1534

Sidera K, Patsavoudi E (2014) HSP90 inhibitors: current development and potential in cancer therapy. Recent Pat Anticancer Drug Discov 9:1–20

Singh N, Ravichandran S, Spelman K, Fugmann SD, Moaddel R (2014) The identification of a novel SIRT6 modulator from Trigonella foenum-graecum using ligand fishing with protein coated magnetic beads. J Chromatogr B: Anal Technol Biomed Life Sci 968:105–111

Šípová H, Homola J (2013) Surface plasmon resonance sensing of nucleic acids: a review. Anal Chim Acta 773:9–23

Soga S, Akinaga S, Shiotsu Y (2013) Hsp90 inhibitors as anti-cancer agents, from basic discoveries to clinical development. Curr Pharm Design 19:366–376

Soga S, Shiotsu Y, Akinaga S, Sharma SV (2003) Development of radicicol analogues. Curr Cancer Drug Targets 3:359–369

Soroka J, Wandinger SK, Mäusbacher N, Schreiber T, Richter K, Daub H, Buchner J (2012) Conformational switching of the molecular chaperone Hsp90 via regulated phosphorylation. Mol Cell 45:517–528

Southworth DR, Agard DA (2008) Species-dependent ensembles of conserved conformational states define the Hsp90 chaperone ATPase cycle. Mol Cell 32:631–640

Southworth DR, Agard DA (2011) Client-loading conformation of the Hsp90 molecular chaperone revealed in the Cryo-EM structure of the human Hsp90: hop complex. Mol Cell 42:771–781

Stanton BZ, Peng LF, Maloof N, Nakai K, Wang X, Duffner JL, Taveras KM, Hyman JM, Lee SW, Koehler AN, Chen JK, Fox JL, Mandinova A, Schreiber SL (2009) A small molecule that binds hedgehog and blocks its signaling in human cells. Nat Chem Biol 5:154–156

Stark JL, Powers R (2011) Application of NMR and molecular docking in structure-based drug discovery. Top Curr Chem 326:1–34

Sun HP, Jia JM, Jiang F et al (2014) Identification and optimization of novel Hsp90 inhibitors with tetrahydropyrido[4,3-d]pyrimidines core through shape-based screening. Eur J Med Chem 79:399–412

Taipale M, Jarosz DF, Lindquist S (2010) HSP90 at the hub of protein homeostasis: emerging mechanistic insights. Nat Rev Mol Cell Biol 11:515–528

Taipale M, Krykbaeva I, Koeva M, Kayatekin C, Westover KD, Karras GI, Lindquist S (2012) Quantitative analysis of HSP90-client interactions reveals principles of substrate recognition. Cell 150:987–1001

Taldone T, Sun W, Chiosis G (2009) Discovery and development of heat shock protein 90 inhibitors. Bioorg Med Chem 7:2225–2235

Trepel J, Mollapour M, Giaccone G, Neckers L (2010) Targeting the dynamic HSP90 complex in cancer. Nat Rev Cancer 10:537–549

Uttamchandani M, Wang J, Yao SQ (2006) Protein and small molecule microarrays: powerful tools for high-throughput proteomics. Mol BioSyst 2:58–68

Van den Heuvel RH, Heck AJ (2004) Native protein mass spectrometry: from intact oligomers to functional machineries. Curr Opin Chem Biol 8:519–526

Vegas AJ, Fuller JH, Koehler AN (2008) Small-molecule microarrays as tools in ligand discovery. Chem Soc Rev 37:1385–1394

Vivat Hannah V, Atmanene C, Zeyer D, Van Dorsselaer A, Sanglier-Cianfèrani S (2009) Native MS: an "ESI" way to support structure- and fragment-based drug discovery. Future Med Chem 2:35–50

Voellmy R, Boellmann F (2007) Chaperone regulation of the heat shock protein response. Adv Exp Med Biol 594:89–99

Warner G, Illy C, Pedro L, Roby P, Bossé R (2004) AlphaScreen kinase HTS platforms. Curr Med Chem 11:721–730

Whitesell L, Lin NU (2012) HSP90 as a platform for the assembly of more effective cancer chemotherapy. Biochim Biophys Acta Mol Cell Res 1823:756–766

Whitesell L, Lindquist SL (2005) HSP90 and the chaperoning of cancer. Nat Rev Cancer 5:761–772

Whitesell L, Mimnaugh EG, De Costa B, Myers CE, Neckers LM (1994) Inhibition of heat shock protein HSP90-pp60v-src heteroprotein complex formation by benzoquinone ansamycins: essential role for stress proteins in oncogenic transformation. Proc Natl Acad Sci USA 91:8324–8328

Willett P (2006) Similarity-based virtual screening using 2D fingerprints. Drug Discov Today 11:1046–1053

Williamson MP (2013) Using chemical shift perturbation to characterise ligand binding. Prog Nucl Magn Reson Spectrosc 73:1–16

Xu XL, Sun HP, Liu F et al (2014) Discovery and bioevaluation of novel pyrazolopyrimidine analogs as competitive Hsp90 inhibitors through shape-based similarity screening. Mol Inform 33:293–306

Yi F, Regan L (2008) A novel class of small molecule inhibitors of Hsp90. ACS Chem Biol 3:645–654

Yi F, Zhu P, Southall N, Inglese J, Austin CP, Zheng W, Regan L (2009) An AlphaScreen-based high-throughput screen to identify inhibitors of Hsp90-cochaperone interaction. J Biomol Screen 14:273–281

Zhang R, Monsma F (2010) Fluorescence-based thermal shift assays. Curr Opin Drug Discov Devel 13:389–402

Zhu Y, Li Z, Chen M, Cooper HM, Lu GQ, Xu ZP (2012) Synthesis of robust sandwich-like SiO_2@CdTe@SiO_2 fluorescent nanoparticles for cellular imaging. Chem Mater 24:421–423

Zierer BK, Rübbelke M, Tippel F, Madl T, Schopf FH, Rutz DA, Richter K, Sattler M, Buchner J (2016) Importance of cycle timing for the function of the molecular chaperone Hsp90. Nat Struct Mol Biol 23:1020–1028

Chapter 20
Therapeutic Effects and Related Molecular Mechanisms of Celastrol, a Triterpenoid Natural Compound and Novel HSP90 Inhibitor Extracted from Plants of the Celastraceae Family

Bin Peng, Ying Wang, Yu-Ting Song, Xue Zhang, Fan-Fan Cao, Li-Min Xu, Mei Jiang, Xiao-Ling Bo, Georges Uzan, and Deng-Hai Zhang

Abstract Celastrol, a triterpenoid compound first identified in the Chinese herbal *Tripterygium Wilfordii* Hook, f. (TWHF), has attracted much attention in recent years due to its promise in multiple applications. Initially, celastrol was studied mainly as a toxic agent to treat cancer and inflammation. Celastrol's potential applications have since expanded following findings that it could protect cells due to inducing heat shock proteins via HSF-1 activation (now attributed as caused by HSP90 inhibition). In this paper, we review the therapeutic effects of celastrol towards a variety of abnormalities demonstrated in animal models, including autoimmune and inflammatory diseases, various cancers, neurodegenerative issues, and

B. Peng · Y. Wang · X. Zhang · F.-F. Cao · L.-M. Xu · D.-H. Zhang (✉)
Sino-French Cooperative Lab, Shanghai Gongli Hospital, The Second Military Medical University, Shanghai, China

Y.-T. Song
Sino-French Cooperative Lab, Shanghai Gongli Hospital, The Second Military Medical University, Shanghai, China

Postgraduate College of Ning Xiang Medical University, Yinchuan City, China

M. Jiang
Department of Neurology, Shanghai Gongli Hospital, The Second Military Medical University, Shanghai, China

X.-L. Bo
Department of Dermatology, Shanghai Gongli Hospital, The Second Military Medical University, Shanghai, China

G. Uzan
Sino-French Cooperative Lab, Shanghai Gongli Hospital, The Second Military Medical University, Shanghai, China

Inserm 1197 Research Unit, Universite Paris Sud, Paris-Saclay University, Villejuif, France

A. A. A. Asea, P. Kaur (eds.), *Heat Shock Protein 90 in Human Diseases and Disorders*, Heat Shock Proteins 19,
https://doi.org/10.1007/978-3-030-23158-3_20

chronic metabolic disorders. The cellular responses, signaling pathways and proteins reportedly affected by celastrol are also introduced. Moreover, based on celastrol's chemical structure and properties, as well as reported findings, we present a new molecular model for celastrol's direct inhibition of the HSP90 chaperone complex, which might be generalized to cover the interactions between celastrol and other targeted proteins. Finally, we suggest future work focusing on clinical translation.

Keywords Cancer · Celastrol · Heat shock protein · HSP90 inhibitor · Hydrophobicity · Inflammation · Metabolic disease · Michael addition · Molecular docking · Neurodegenerative disease · NF-κB inhibitor

Abbreviations

AKT	RAC-alpha serine/threonine-protein kinase
AML	Acute myeloid leukemia
AMPK	AMP-activated protein kinase
ATF2 or 6	Activating transcription factor 2 or 6
Cdc37	Cell division cycle 37
CDK	Cyclin-dependent kinase
c-Myc	Constitutively expressed homology with the viral gene *v-myc*
DSS	Dextran sodium sulfate
ERK	Extracellular signal-regulated kinase
HO-1	Heme oxygenase 1
HSF-1	Heat shock factor 1
HSP	Heat shock protein
HSR	Heat shock response
IKK	Inhibitor of nuclear factor kappa-B kinase
JNK	c-Jun N-terminal kinases
MAPK	Mitogen-activated protein kinase
MAX	Myc-associated factor X protein
MD2	Lymphocyte antigen 96
MMP-9	Matrix metallopeptidase 9
MnSOD	Manganese-dependent superoxide dismutase
MPO	Myeloperoxidase
mTOR	Mammalian target of rapamycin
NAD	Nicotinamide adenine dinucleotide
NF-κB	Nuclear factor kappa-light-chain-enhancer of activated B cells
NLRP3	NACHT,LRR and PYD domains-containing protein 3
NO	Nitric oxide
Nrf2	Nuclear factor (erythroid-derived 2)-like 2
PERK	Protein kinase R (PKR)-like endoplasmic reticulum kinase
PGC1α	Peroxisome proliferator-activated receptor gamma coactivator 1-alpha
PI3K	Phosphatidylinositol-4,5-bisphosphate 3-kinase

RANKL	Receptor activator of nuclear factor kappa-B ligand
ROS	Reactive oxygen species
SARM	Sterile α- and armadillo-motif-containing protein
STAT3	Signal transducer and activator of transcription 3
TGF-β1	Transforming growth factor beta 1
TLR4	Toll-like receptor 4
TNFα	Tumor necrosis factor alpha
Treg	Regulatory T cells
TWHF	*Tripterygium Wilfordii* Hook, f.
VEGF	Vascular endothelial growth factor
VEGFR	Vascular endothelial growth factor receptor

20.1 Introduction

Research into traditional medicines is a promising field of drug development. This is particularly true for plants used in Traditional Chinese Medicine, in that the effects of herbal plants in China have been tested for a long time and carefully recorded. As an example, by carefully studying the classic Chinese medical literature on the herbal plant *Artemisia apiacea*, Prof. Tu You-You identified the malaria treatment agent arteannuin, which has saved the lives of thousands of infected individuals and earned her the Nobel Prize. Celastrol, previously called tripterine, is one of these natural Chinese herbal compounds that has attracted much attention and become a leading research subject in recent years, due to its promising application in multiple clinical areas.

The story with celastrol is marvelous and inspiring, with its identification being triggered by an unexpected event and then after being neglected for about half century and with its being studied mainly as toxic agent to being used also as a protective compound against various harmful factors. In this paper, we first briefly look at the history of this compound, then summarize its therapeutic effects on a variety of disorders demonstrated in animal models, cellular responses, signaling, and the proteins reportedly affected by it. Finally, we offer some suggestions for future work. Induction of heat shock response (HSP) via activating HSF-1 (due to inhibiting HSP90 chaperone activity) is key to celastrol's effects and has dramatically affected the focus and course of research. Therefore, the molecular mechanisms for HSP90 and its co-chaperones form the core of this paper.

20.2 Brief History of Celastrol Study

Celastrol is a natural compound first extracted from the Chinese herbal plant *Triterygium wilfordii* Hook, f. (TWHF). It is now proven that celastrol exists in multiple plants, especially those of the Celastraceae family, yet TWHF applications were the major force advancing celastrol study, at least in the early stages. TWHF, called *lei gong teng*, "thunder god vine," by the Chinese, is a perennial twining vine belonging to the family Celastraceae. It is widely distributed south of the Yangtze River in China. Though TWHF has been mentioned in early Chinese herbal classic works, the effects of this plant were first clearly described by the Ming Dynasty scholar Lan Mao (1397–1479) in his medical book '*Dian Nan Ben Cao*' (Materia Medica in the South of Yun-nan Province), saying that this plant was effective for bone and tendon pain and for rheumatic diseases. Yet, TWHF was mainly used by Chinese farmers as an insecticide to protect their vegetables.

In the 1850s, TWHF was first transferred and introduced to the West by Charles Wilford, who collected this plant in Taiwan province in June 1858 and sent it to his employer, the Kew Botanical Garden, London. It was named after him and mentioned for first time in Western literature in Bentham and Hooker's "Genera plantarum" (Bentham and Hooker 1862). In the 1930s, a hillside accident, caused by harvesting TWHF roots that loosen soil and contributed to excessive erosion, brought into question whether or not TWHF cultivation should be banned. At the time, vegetable growers on the hill insisted that THWF was indispensable for killing vegetable-harming insects and Chinese entomologists and other experts begun to investigate the insecticidal effects of TWHF. The investigations produced about a dozen important papers and nearly as many brief notes on TWHF and its insecticidal use (Swingle et al. 1941). One of the key resultant achievements was identifying a red pigment, named tripterine (Chou and Mei 1936). Thereafter another red pigment, named celastrol, from the root bark of another member of the Celastraceae family, bittersweet, was identified, and these two red pigments were soon confirmed to be the same substance (Schechter and Haller 1942). Nevertheless, after then pharmacological study of celastrol was dormant for nearly a half century. Unlike celastrol study, use of TWHF in different applications continued to advance. Chinese doctors used TWHF to successfully treat leprosy in the 1960s and arthritis in the 1970s. Western researchers also became aware of the significant anti-leukemic properties of Celastraceae plants and also found TWHF extracts effective for this purpose in the 1970s. Following these advances, Prof. Qin Wan-Zhang, a famous Chinese dermatologist, and Prof. Li Ni-Shi, a famous nephrologist and member of the Chinese Academy, expanded TWHF application to SLE and nephritis. Now, TWHF and its mixed extracts have been approved for clinical applications which benefit huge numbers of patients each year. TWHF application has become so prominent in the Chinese medical field that national conferences on this plant have now been held six times. In 1980s, encouraged by the clinical efficacy of THWF application, a group of researchers (including Dr. Zhang Deng-Hai, one of this paper's authors) in China began to observe the effects of celastrol on the immune system, further awakening investigations into a molecule discovered almost half century ago. In 2003, the Morimoto group identified that celastrol is an HSP inducer (Westerheide et al. 2004), which is supported by the 2006 Golub group discovery that celastrol is a novel HSP90 inhibitor (Hieronymus et al. 2006). These outstanding works have opened up the new fields for potential celastrol application, moving from anti-inflammation and anti-cancer usage to ameliorating neurodegenerative and metabolic disorders as well.

20.3 Therapeutic Effects on Various Disorders Demonstrated In Vivo

Animal models have demonstrated that celastrol can remedy a variety of clinical disorders, including autoimmune disease, inflammations caused by endotoxin or other acute harm, various cancers, neurodegenerative disorder, and chronic metabolic diseases.

20.3.1 Autoimmune Disease and Inflammations Caused by Endotoxin or Other Acute Harm

20.3.1.1 Rheumatoid Arthritis

In the adjuvant-induced arthritis model of human RA, celastrol demonstrated potent anti-arthritic activity (Venkatesha et al. 2011). It suppressed joint inflammation, preserved joint structure, prevented bone loss and bone microarchitecture degradation, reduced bone fragility (Cascao et al. 2012, 2017; Nanjundaiah et al. 2012), and reduced biochemical hallmarks (Cascao et al. 2017). The mechanisms behind the treatment are complex, and include suppressing pro-inflammatory cytokines and chemokines, reduced STAT3 and phospho-ERK, as well as MMP-9 levels (Nanjundaiah et al. 2012; Venkatesha et al. 2011, 2012), inhibition of osteoclastic bone remodeling and reduced osteoclast numbers (Nanjundaiah et al. 2012), and reduced Th17 cells but increased Treg in the joints (Astry et al. 2015). To support this complexity, microarray analysis revealed ten differentially expressed genes and 14 pathways that constituted the "Celastrol Signature" (Yu et al. 2012). RA-related miRNAs regulation is also suggested in celastrol treatment (Dudics et al. 2018).

20.3.1.2 Systemic Lupus Erythematosus

Celastrol could reduce anti-double stranded DNA antibodies, decrease urine protein excretion, suppress the expression of collagen type IV and TGF-β1 mRNA in the murine kidney tissue (Xu et al. 2003, 2007), and protect kidneys from glomerular injury. The effects were associated with reduced interleukin-10 production and serum NO (Li et al. 2005).

20.3.1.3 Experimental Autoimmune Encephalomyelitis

Celastrol delayed disease onset, reduced clinical scores, and inhibited relapse via anti-inflammatory and anti-apoptotic effects (Yang et al. 2017). There was significant shift in the cytokines profile from Th1 towards the Th2 pattern (Abdin and Hasby 2014) and suppression of pathogenic Th17 responses (Wang et al. 2015b).

20.3.1.4 Inflammations Caused by Endotoxin

Celastrol attenuated LPS-induced splenomegaly and congestion and lung and hepatic injury (Hu et al. 2017; Wei and Wang 2017; Yu et al. 2017). Moreover, celastrol prevented circulatory failure (bradycardia and hypotension) in endotoxemic rats created by LPS challenge (Wang et al. 2015c). These therapeutic effects are based on inhibiting NF-κB activation and pro-inflammatory cytokines, inducing anti-oxidants, inhibiting NLRP3 inflammasome activation and caspase three activities (Wei and Wang 2017), and activating Nrf2/HO-1 and HSF-1/HSP70 (Wang et al. 2015c). Nur77 is reportedly needed for celastrol to act (Hu et al. 2017). However, there is one report in which Celastrol aggravates LPS-induced inflammation and injuries of liver and kidney in mice (Wu et al. 2018). The exact reasons for this contradictory result are unclear.

20.3.1.5 Inflammation Caused by Ischemia

Celastrol could protect cells from ischemic damages in the brain (Jiang et al. 2018; Li et al. 2012a), heart (Der Sarkissian et al. 2014; Li et al. 2017; Tong et al. 2018) and kidney (Chu et al. 2014; Hu et al. 2018) via its anti-oxidant and anti-inflammation effects, as well as promote an IL-33/ST2 axis-mediated microglia/macrophage M2 polarization.

20.3.1.6 Inflammation Caused by Trauma or Surgery

Celastrol or BGP-15 (a HSP70 inducer) could increase HSP70/HSP110 levels and protect cells against traumatic brain injury (Eroglu et al. 2014).Celastrol reduced surgery-induced changes in cognitive dysfunction (Wan et al. 2010) and increased retinal ganglion cell survival following crushed optic nerves via TNF-α-mediated cell death inhibition (Kyung et al. 2015). Celastrol nanoparticles inhibited corneal neovascularization induced by suturing in rats (Li et al. 2012b), partially restored the plantar incision pain-related behavior changes via reduction of proinflammatory cytokines, and markedly increased SARM (Chen et al. 2018b). Celastrol demonstrated analgesic effects in spared nerve injury animal models (Yang et al. 2014).

20.3.1.7 Inflammation Caused by Chemical Substance

Celastrol could inhibit pulmonary fibrosis in bleomycin-induced rats (Divya et al. 2016) as well as ovalbumin-induced airway inflammation, hyperresponsiveness, and tissue remodeling (Kim et al. 2009b). It attenuated DSS-induced colitis (Shaker et al. 2014; Yu et al. 2017) and improved cisplatin-induced kidney injury (Guo et al. 2017). Celastrol significantly reduced joint pain and cartilage damage induced by monoiodoacetate in osteoarthritis rats (Wang et al. 2018a). Moreover, it inhibited secretory phospholipase A2 group IIA or carrageenan-induced edema in mice by targeting multiple enzymes of the arachidonic acid pathway (Joshi et al. 2016) while it protected mice from TPA-induced ear edema, possibly by inhibiting MPO activity and production of inflammatory cytokines (Kim et al. 2009a). Celastrol demonstrated analgesic effects in carrageenan injection animal models (Yang et al. 2014). Celastrol also inhibited hearing loss in mice receiving systemic aminoglycoside treatment (Francis et al. 2011).

20.3.2 Cancers

Celastrol is a broad inhibitor towards variety of tumors as demonstrated in animal models harboring cancer cells.

20.3.2.1 Leukemia

Celastrol inhibited growth of chronic myelogenous leukemia cells in nude mouse xenografts regardless of Bcr-Abl T315I-mutant status, by downregulating the protein levels of Bcr-Abl and inducing mitochondrial-dependent apoptosis (Lu et al. 2010). Celastrol significantly impaired engraftment of acute myeloid leukemia (AML)-stem cells in sublethally irradiated mice (Hassane et al. 2008), suppressed the proliferative potential of AML cells while not affecting normal hematopoietic progenitor cells, and prolonged the survival of mice in an aggressive AML model. These effects are due to targeting Myb and C/EBPβ and thus disrupting the interaction of both these transcription factors with p300 (Coulibaly et al. 2018; Uttarkar et al. 2016). Uridine deficiency also reportedly contributes to celastrol-induced mitochondria-mediated apoptosis in human AML (Zhang et al. 2016).

20.3.2.2 Prostate Cancer

In human prostate cancer PC3 cells-loaded nude mice, celastrol strongly suppressed tumor growth (Dai et al. 2010; Pang et al. 2010; Yang et al. 2006). Tumor-inhibiting effects were also observed in mice loaded with TMPRSS2/ERG fusion gene expressing prostate cancer (Shao 2013). These effects are reportedly related to celastrol's potent proteasome inhibition (Yang et al. 2006), decreasing angiogenesis and inducing apoptosis (Dai et al. 2010; Pang et al. 2010), and suppressing the AKT/mTOR (Pang et al. 2010) and AR-ERG-NF-κB pathways (Shao 2013).

20.3.2.3 Breast Cancer

Celastrol significantly retarded the growth rate of ErbB2-overexpressed human breast cancer cells in a mouse xenograft model with only minor systemic toxicity. Mechanistically, celastrol induced the ubiquitination and degradation of ErbB2 and other HSP90 client proteins and increased the levels of ROS (Raja et al. 2011). Celastrol inhibited triple negative breast cancer cell growth in vivo by inhibiting NF-κB signaling and inducing apoptosis (Mezzanotte et al. 2014).

20.3.2.4 Lung Cancer

Celastrol binds with CIP2A (Cancerous inhibitor of PP2A) and enhances CIP2A-CHIP interaction, leading to ubiquitination/degradation of CIP2A and inhibition of lung cancer cells in vitro and in vivo (Liu et al. 2014). We showed that celastrol, combined with EGFR-TKIs, significantly suppressed in vivo growth of non-small-cell lung cancer cells with a T790 M mutation by suppressing the EGFR pathway (Wang et al. 2018b).

20.3.2.5 Hepatocellular Carcinoma

Celastrol inhibited the growth of human hepatocellular carcinoma (HCC) xenograft tumors in athymic *nu/nu* mice without any side effects (Rajendran et al. 2012), decreased growth of Hep3B cells in a xenograft tumor model (Ma et al. 2014), significantly decreased mortality, and decreased the number of tumor nodules in diethylnitrosamine-induced HCC rats (Chang et al. 2016). Celastrol's reported therapeutic effects are due to inhibiting STAT3 activation (Rajendran et al. 2012) and the expression of HIF-1α proteins (Ma et al. 2014), as well as activating intrinsic mitochondrial apoptosis (Chang et al. 2016).

20.3.2.6 Gastric Cancer

Celastrol inhibited gastric cancer growth in a xenografted mice model by inducing apoptosis and autophagy (Lee et al. 2014).

20.3.2.7 Pancreas Cancer

Celastrol significantly inhibited tumor growth in Panc-1 xenografts. Moreover, celastrol effectively suppressed tumor metastasis in a RIP1-Tag2 transgenic mouse model with pancreatic islet cell carcinogenesis. The inhibition is related to its blocking Hsp90-Cdc37 interaction (Zhang et al. 2008).

20.3.2.8 Colorectal Cancer

Celastrol ameliorated ulcerative colitis-related colorectal cancer in mice by suppressing inflammatory responses and epithelial-mesenchymal transition (Lin et al. 2015). Dietary celastrol supplementation could suppress both spontaneous and carcinogen-induced colitis-associated colorectal cancer in mice via anti-inflammation (Barker et al. 2018).

20.3.2.9 Osteosarcoma

Celastrol or parthenolide suppressed trabecular bone loss and reduced the number and size of osteolytic bone lesions following W256 injection in rats (Idris et al. 2009). Celastrol induced apoptosis and autophagy via activating the ROS/JNK signaling pathway in human osteosarcoma cells, and suppressed tumor growth in the mouse xenograft model (Li et al. 2015).

20.3.2.10 Melanoma

Celastrol and a derivative could efficiently inhibit growth of human and mouse melanoma tumors and reduced the number of lung metastases in syngeneic and xenograft mouse models (Abbas et al. 2007; Zhu et al. 2010). The underlying mechanisms were suggested as inhibiting ATF2 transcriptional activities, activating JNK and c-Jun activities (Abbas et al. 2007), and activating p38 MAPK, thus suppressing β1 integrin ligand affinity and focal adhesion formation (Zhu et al. 2010).

20.3.2.11 Glioma

Celastrol significantly reduced tumor volume in glioma xenografts models and significantly lowered the density of microvessels in tumors. This was accompanied by decreased levels of VEGFR-1 and VEGFR-2 expression (Huang et al. 2008; Zhou and Huang 2009).

20.3.2.12 Retinoblastoma

Celastrol nanoparticles inhibited the growth of retinoblastoma and induced apoptosis in retinoblastoma cells in mice (Li et al. 2012c).

20.3.3 Neurodegenerative Diseases

Celastrol can improve neurodegenerative diseases. This research has been accelerated following the discovery of celastrol's HSR induction. The therapeutic effects of celastrol on neurodegenerative diseases are due to its inflammation inhibition, anti-oxidation, and HSP induction (Allison et al. 2001; Cleren et al. 2005; Kiaei et al. 2005).

20.3.3.1 Alzheimer's Disease

Low doses of celastrol administered to Alzheimer's disease model rats significantly improved their performance in memory, learning, and psychomotor activity tests (Allison et al. 2001).

20.3.3.2 Parkinson's Disease and Huntington's Disease

Loss of dopaminergic neurons in mice induced by 1-methyl-4-phenyl-1,2,3,6- tetrahydropyridine (MPTP) in the substantia nigra pars compacta was significantly attenuated by celastrol treatment. Moreover, celastrol treatment significantly reduced depletion in dopamine concentration induced by MPTP. Similarly, celastrol significantly decreased the striatal lesion volume induced by 3-nitropropionic acid, a neurotoxin used to model Huntington's disease in rats (Cleren et al. 2005).

20.3.3.3 Amyotrophic Lateral Sclerosis (ALS)

Celastrol treatment significantly improved weight loss, motor performance, delayed disease onset, and increased neuronal number in the lumbar spinal cords in SOD1 transgenic mouse models of ALS (Kiaei et al. 2005).

20.3.4 Chronic Metabolic Diseases

Recently, there have been many reports that celastrol can effectively regulate chronic metabolic disorders, especially in obesity and diabetes.

20.3.4.1 Obesity

Celastrol has effectively reduced body weight in high fat-induced obesity animal models (Liu et al. 2015; Luo et al. 2017; Ma et al. 2015; Pfuhlmann et al. 2018; Wang et al. 2014a; Zhang et al. 2018). It reduced total cholesterol, triacylglycerol, low-density lipoprotein cholesterol and apolipoprotein B in plasma (Liu et al. 2015; Wang et al. 2014a) and increased mitochondrial function in fat and muscle. Celastrol treatment reduced hepatic steatosis (Ma et al. 2015), as well as subcutaneous and visceral fat content, liver lipid droplet formation, and hepatic intracellular triglycerides while improving glucose tolerance and insulin sensitivity (Zhang et al. 2017b). Celastrol's effectiveness as an anti-obesity treatment is suggested to be due to suppressing appetite, increasing energy expenditure via increased leptin sensitivity (Liu et al. 2015; Pfuhlmann et al. 2018), blocking binding between fatty acids and MD2 (Zhang et al. 2018), increasing antioxidant enzyme activities, inhibiting NADPH oxidase activity, and decreasing the serum levels of malondialdehyde and ROS (Wang et al. 2014a). It is also reported to suppress macrophage M1 biomarkers (Luo et al. 2017). Additionally, sirtuin 1 is reportedly needed for celastrol to ameliorate metabolic damage to the liver caused by high fat food (Zhang et al. 2017b).

20.3.4.2 Diabetes

In diabetic mice, celastrol significantly lowered fasting plasma glucose, glycated hemoglobin and homeostasis model assessment index levels, decreased body and kidney/body weight and adiposity, and reduced insulin resistance, lipid abnormalities, creatinine levels, and urinary albumin excretion. It also significantly lowered mesangial expansion and suppressed type IV collagen, PAI-1 and TGFβ1 expressions in renal tissues (Kim et al. 2013), attenuated the deterioration of skeletal muscle (Guan et al. 2016), and reduced hepatic inflammation and macrophages infiltration (Han et al. 2016). In mice with β-cell failure, celastrol reversed hyperglycemia (Weisberg et al. 2016). The anti-diabetic effects of celastrol have been attributed to anti-inflammation, anti-oxidation (Guan et al. 2016; Han et al. 2016; Kim et al. 2013), and increased AMPK phosphorylation, PGC1α, and sirtuin 3 expression levels (Guan et al. 2016).

20.3.4.3 Hypertension

Celastrol decreased both systolic and diastolic blood pressure while improving insulin sensitivity in fructose-induced hypertensive rats. Celastrol also inhibited vascular and cardiac hypertrophy, inflammatory cytokines, increased HO-1, and decreased ROS generation (Yu et al. 2010). In preeclampsia rats that induced by N^{ω}-nitro-L-arginine methyl ester hydrochloride, celastrol decreased total urine proteins, urine volume, and blood pressure, as well as urinary mRNA levels of podocin, nephrin, serum placenta growth factor, MMP-9 and soluble fms-like tyrosine kinase-1 (Xiao et al. 2018).

20.4 Heat Shock and Other Responses Affected by Celastrol

Celastrol possesses a variety of therapeutic effects based on its regulation of several important responses inside cells or bodies. These include heat shock response, inflammation, apoptosis and autophagy, endoplasmic reticulum stress, and anti-oxidation.

20.4.1 Heat Shock Response (HSR)

In 2004, celastrol was found to activate heat shock gene transcription (Westerheide et al. 2004). More recently, PRO-seq data for K562 cells revealed that HSR gene activation was the immediate and most prominent of celastrol's transcriptional effects (Dukler et al. 2017). HSR induction is important to celastrol's effects in treating neurodegenerative diseases (Brown 2007; Kalmar and Greensmith 2009; Sun et al. 2010), protecting cells and tissues from damages caused by toxic chemicals (Cleren et al. 2005), ischemia (Der Sarkissian et al. 2014), endotoxin (Wang et al. 2015c), radiation (Stankova et al. 2013), and traumatic injury (Eroglu et al. 2014; Kyung et al. 2015). However, elevated HSP, especially HSP70, might reduce celastrol's anti-cancer effects (Matokanovic et al. 2013; Peng et al. 2014).

20.4.2 Inflammation

Celastrol's inflammation inhibition was discovered in the 1980s. From then on, a variety of in vitro and in vivo models confirmed this action. Actually, celastrol can inhibit inflammation at multiple levels, such as by blocking TLR4/MD2 interaction (Han et al. 2018a; Lee et al. 2015; Li et al. 2013a), IKK/NF-κB activation (Lee et al. 2006), NLRP3 inflammasomes activation (Xin et al. 2017), vascular endothelial activation (Zhang et al. 2006), and leucocyte tissue infiltration (Xu et al. 2014). Anti-inflammation is the basis by which celastrol treats cancers and autoimmune, degenerative, and chronic metabolic disorders, as well as how it prevents damage as detailed in the previous sections on celastrol's therapeutic effects.

20.4.3 Apoptosis and Autophagy

Celastrol-induced apoptosis was first reported in a leukemia study (Nagase et al. 2003) and has been subsequently observed in celastrol treatment against many kinds of cancer cells (detailed above in the Therapeutic Effects section), while celastrol-inhibited apoptosis has been shown to protect cells from damages (Francis et al. 2011; Kalmar and Greensmith 2009; Zhang et al. 2017a). Celastrol-induced autophagy is involved in anti-tumor actions (Lee et al. 2014; Li et al. 2015; Wang et al. 2012), neuron protection (Deng et al. 2013), anti-obesity treatment (Hu and Zhang 2017) and anti-toxicity effects (Divya et al. 2017).

20.4.4 Endoplasmic Reticulum (ER) Stress

Celastrol can activate the three arms of the unfold protein response, IRE1, ATF6, and PERK (Mu et al. 2008). Inducing ER stress is involved in anti-cancer effects, as demonstrated in celastrol-treated HepG2, Bel7402 (Ren et al. 2017), human osteosarcoma cell line HOS (Chen et al. 2018c), HeLa cells (Feng et al. 2013) and oral squamous cell carcinoma (Fribley et al. 2015).

20.4.5 Anti-oxidant Response

Celastrol could inhibit NO and other ROS (Han et al. 2018b; Jin et al. 2002; Li et al. 2005). Accordingly, it could inhibit ROS-producing enzymes, including NADPH oxidase 1 (NOX1), NOX2, NOX4, and NOX5 (Jaquet et al. 2011; Whitehouse et al. 2016) and LOX-1 (Gu et al. 2013) while inducing anti-oxidant enzymes (Wang et al. 2014a), including MnSOD (Guan et al. 2016), glutathione peroxidase, glutathione reductase, and catalase (Han et al. 2018b; Jannuzzi et al. 2018), Phase II enzymes such as HO-1, glutathione-S-transferase and NADP(H): quinine oxidoreductase 1 (Divya et al. 2016; Hansen et al. 2011). Celastrol's anti-oxidant action is involved in its effects on autoimmune, disease, inflammation, and cellular protection, as detailed in the Therapeutic Effects section.

20.5 HSF-1 and Other Pathways Regulated by Celastrol

In agreement with reports that celastrol regulates multiple cellular responses, celastrol has also been reported to affect many signaling pathways, including transcription factor HSF-1, NF-κB, Nrf2 and HIF-1 pathways, as well as the upstream PI3K and MAPK.

20.5.1 HSF-1 Activation

Celastrol's inducing of heat shock response is due to HSF-1 activation (Westerheide et al. 2004). HSF-1 activation by celastrol was observed in a variety of mammary cells (Peng et al. 2014; Sharma et al. 2015; Wang et al. 2014b), as well as in yeast

(Trott et al. 2008) and Xenopus laevis (Walcott and Heikkila 2010), indicating that this is a general effect of celastrol.

20.5.2 NF-κB Activation

After the first report of celastrol inhibiting NF-κB activation and nitric oxide production (Jin et al. 2002), much work has confirmed such effects, and has been well reviewed (Chen et al. 2018a; Kashyap et al. 2018; Yadav et al. 2018).

20.5.3 Nrf2/HO-1 Pathway

Elevated HO-1 has been linked to a variety of celastrol's therapeutic effects, including in treating ischemic myocardium (Der Sarkissian et al. 2014), hypertensive (Yu et al. 2010), and bleomycin-induced pulmonary fibrosis rats (Divya et al. 2016), as well as preventing circulatory failure in endotoxemic rats (Wang et al. 2015c), inhibiting Tat-induced inflammation (Youn et al. 2014), blocking hepatitis C virus replication (Tseng et al. 2017), and inhibiting LPS-induced macrophage M1 polarization and inflammation (Luo et al. 2017). The main upstream event for inducing HO-1 was activating Nrf2 (Seo et al. 2011; Youn et al. 2014).

20.5.4 HIF-1 Pathway

Celastrol could either inhibit or induce the HIF-1 pathway. Pathway inhibition by celastrol has been reported when reducing VEGF (Huang et al. 2011), inhibiting migration and invasion of fibroblast-like synoviocytes under hypoxic conditions (Li et al. 2013b), and inhibiting human hepatoma cells Hep3B (Ma et al. 2014). However, celastrol also may improve self-renewal and differentiation of human tendon-derived stem cells, which is related to a significant upregulation of HIF-1α and downregulation of Smad7 (Wu et al. 2017).

20.5.5 PI3K/AKT

This pathway could be inhibited or activated by celastrol. Inhibition was observed in anti-tumor applications (Shrivastava et al. 2015; Wu et al. 2017) as well as in inhibiting isolated ankylosing spondylitis fibroblast proliferation and in vitro osteogenic differentiation (Zou et al. 2016). PI3K/AKT activation by celastrol was seen in endothelial barrier maintenance (Luo et al. 2016), improved glucose uptake activity (Abu Bakar and Tan 2017), myocardium protection (Tong et al. 2018), and inhibition of glucocorticoid-induced osteoporosis (Xi et al. 2018).

20.5.6 MAPK Signaling

MAPK is yet another pathway that celastrol can inhibit or activate. Its inhibition was reported when celastrol was utilized to inhibit LPS-induced inflammation (Jung et al. 2007), treat hypertensive rats (Yu et al. 2010), reduce the RANKL-induced expression of osteoclastic genes (Gan et al. 2015), inhibit poly(I:C)-induced pro-inflammatory mediators in astrocytes (An et al. 2017), and to suppress classical Hodgkin lymphoma cells proliferation (Segges et al. 2018). While MAPK pathways were activated by celastrol to cause cytoplasmic vacuolization (Wang et al. 2012).

20.6 HSP90 and Other Proteins Targeted by Celastrol

It has been discovered that celastrol can interact with several protein complexes, including the HSP90 chaperone complex, NLRP3 inflammasome, proteasome, mitochondrial respiratory chain complex, C-Myc/Max, and more than a dozen proteins. These findings support that celastrol affects multiple responses, pathways, and treat variety of disorders.

20.6.1 HSP90 and Its Co-chaperones

Celastrol can disrupt HSP90 and Cdc37 association (Zhang et al. 2009) and directly inhibit HSP90 ATPase (Peng et al. 2010). Celastrol induces the oligomerization of the chaperone via the C-terminal domain (Wyatt and Dimmer 1989) and two residuals in human HSP90 dimer are important for this protein to bind with celastrol and for binding with Cdc37 (Peng et al. 2016). HSQC NMR methods indicate that celastrol binds with co-chaperone Cdc37 and conducts a conjugate addition to cysteine residues on Cdc37 (Sreeramulu et al. 2009). Celastrol directly disrupts co-chaperone p23 function by altering its three-dimensional structure, leading to rapid formation of amyloid-like fibrils (Chadli et al. 2010).

20.6.2 NLRP3 Inflammasome

ASC (apoptosis-associated speck-like protein containing a CARD) protein oligomerizations were significantly disrupted by celastrol, and the interaction of NLRP3-ASC and ASC-caspase-1 were attenuated by celastrol (Yu et al. 2017).

20.6.3 *Proteasome*

Celastrol potently and preferentially inhibited the chymotrypsin-like activity of a purified 20S proteasome and human prostate cancer cellular 26S proteasome (Yang et al. 2006). Proteasome catalytic subunit β1 was predicted by computational analysis to be a possible direct target of celastrol and confirmed by checking the direct effects of celastrol on the activity of recombinant human proteasome subunit β1 in vitro (Feng et al. 2013).

20.6.4 *Mitochondrial Respiratory Chain Complex*

Celastrol completely inhibited H1299 cell-isolated mitochondrial respiratory chain complex I (MRC I) activity, but increased MRC complex IV activity. The activity of MRC complex II and III were decreased in H1299 cells treated with celastrol and N-acetylcysteine reversed this inhibition (Smee et al. 1990).

20.6.5 *C-Myc/Max*

Celastrol bound to and altered the quaternary structure of the pre-formed C-Myc/Max dimer and abrogated its DNA binding, thus exerting anti-leukemia effects. Interestingly, triterpenoid derivatives lacking the quinone methide showed enhanced specificity and potency against Myc, indicating that certain pharmacologically undesirable properties of celastrol such as Michael adduct formation can be eliminated while increasing selectivity and potency toward Myc and N-Myc (Wang et al. 2015a).

20.6.6 *MD2*

Celastrol suppressed LPS binding to MD2, which was reversed by thiol donors (N-acetyl-L-cysteine and dithiothreitol) (Lee et al. 2015). We found that both celastrol and palmitic acid could bind to MD2 and that celastrol could expel palmitic acid from cells, which is one of the mechanisms by which celastrol exerts anti-obesity effects (Zhang et al. 2018).

20.6.7 *ERK*

Molecular dynamics simulation showed celastrol binding to a large pocket in ERK2, which serves as the ATP-binding site, accordingly, celastrol inhibited the interaction between immunoglobulin Fc epsilon receptor I and ERK (Kim et al. 2009c).

20.6.8 IKK

Celastrol suppressed NF-κB activation by targeting cysteine 179 in IKK (Lee et al. 2006).

20.6.9 Prx I Chaperone

Celastrol could covalently bind with peroxiredoxin I through the residues Cys83 and Cys173 (Zhao et al. 2015).

20.6.10 Enoyl-Acyl Carrier Protein Reductase

Celastrol inhibited enoyl-acyl carrier protein reductase, a critical enzyme in type II fatty acid biosynthesis. Docking study suggested that celastrol may occupy both the NAD^+ and substrate-binding pocket, without requiring a bound cofactor or substrate (Tallorin et al. 2014).

20.6.11 Shp2

Celastrol inhibited Src homology region 2 domain-containing phosphatase-1 (Shp2) and a celastrol analog lacking the carboxylate group was 18-fold less potent than celastrol in Shp2 inhibition, suggesting that the carboxylate group contributes to this particular inhibition activity (Scott et al. 2011).

20.6.12 CIP2A

Celastrol bound to cancerous inhibitor of protein phosphatase 2A (CIP2A) and enhanced CIP2A-CHIP interaction, leading to ubiquitination/degradation of CIP2A and inhibition of lung cancer cells in vitro and in vivo (Liu et al. 2014).

20.6.13 Nur77

Celastrol bound to and promoted Nur77 translocation from the nucleus to mitochondria, which is suggested to be involved in celastrol's anti-inflammation effects (Hu et al. 2017).

20.6.14 Topoisomerase

Celastrol could inhibit topoisomerase II activity and induce internucleosomal DNA fragmentation and apoptosis in leukemia (Nagase et al. 2003).

20.6.15 Tubulin

Celastrol inhibited the formation of tubulin oligomers and facilitated mitotic cell death triggered by microtubule-targeting anti-cancer drugs (Jo et al. 2010).

20.6.16 Tat

Celastrol, by covalently modifying the cysteine thiols, bound transactivator of transcription (Tat) and inhibits Tat transactivation function (Narayan et al. 2011).

20.7 The Molecular Model for Celastrol's Targeting of HSP90 Complex and Other Proteins

Though celastrol can target multiple protein complexes or proteins, its inhibition of the HSP90 complex is paramount. This notion is based on three facts: HSP90 expression is high in almost all cells, HSP90 has fundamental functions in cells, and activating the heat shock response gene has been identified as the immediate and the most prominent transcriptional effect of celastrol (Dukler et al. 2017). Celastrol's inhibition of the HSP90 complex is closely related to this compound's therapeutic effects, in at least two ways. One is by reducing the levels of HSP90 clients, especially a broad swath of kinases (such as IKK, CDKs), which play an important role in anti-tumor and anti-inflammation applications. The second is by activating HSF-1 and causing heat shock response, which is how celastrol protects and/or reduces harm, especially in neurodegenerative situations. However, HSP induction might reduce celastrol's anti-cancer effects (Matokanovic et al. 2013; Peng et al. 2014).

Therefore, how celastrol inhibits the HSP90 complex has attracted the attention of several research groups including ours. The accumulated data suggests that celastrol might target HSP90 complex in multiple ways, including direct binding to the HSP90 dimer (Peng et al. 2010, 2016; Zanphorlin et al. 2014), insertion into HSP90/Cdc37 interfaces (Jiang et al. 2016; Zhang et al. 2009), direct binding to Cdc37 (Zhang et al. 2009), and direct binding to co-chaperone p23 thus blocking HSP90/p23 interaction (Chadli et al. 2010).

These findings, however, raise a further question about how celastrol could work in so many ways to target HSP90 complex, this issue becomes even more challenging when considering that celastrol has been reported to directly bind to many other proteins besides the HSP90 complex, (more than dozen of which have been mentioned above and possibly a majority of which are still awaiting identification). Here, based on the structure and properties of celastrol, and biological findings about this compound, we propose a simple working model to address this problem.

20.7.1 The Hydrophobic Pocket Binding-Based Model for Celastrol Targeting HSP90 Complex

Structurally, celastrol belongs to a subgroup of D: A-friedo-nor-oleananes characterized by a hydroxyl group in position adjacent to the carbonyl oxygen (A ring) and a chromophore that extends over the A and B ring of the triterpenoid skeleton (Fig. 20.1a). Biophysically, the C and D ring and their attached methyl groups form a hydrophobic constitute, while the two ends (rings A and E) are polar. Functionally, rings A and B could form Michael adducts with nucleophilic agents due to the quinonemethide structure on these two rings, while the C28 carboxyl could react with other molecules.

Therefore, there are three nonexclusive possible ways by which celastrol targets HSP90 complex. (1) Forming Michael adducts with nucleophilic residuals on proteins. This notion is supported in that celastrol, via its C6, forms an adduct with cysteine on Cdc37 (Sreeramulu et al. 2009), and that pretreating with a thiol-containing agent reduced celastrol's ability to induce heat shock response (Peng et al. 2010). Yet dihydrocelastrol, a derivative lacking quinonemethide, could also induce HSP, suggesting that Michael adducts are not essential for inducing HSP (Westerheide et al. 2004), albeit it might involve in celastrol binding Cdc37. (2) Interacting with proteins via C28 carboxyl groups. To support this postulation, pristimerol, a derivative with C28 carboxyl modification, lost heat shock protein-inducing effects (Westerheide et al. 2004). Moreover, we found that C28 carboxyl could form two salt bridges with two residuals on HSP90 dimers (Peng et al. 2016). Introducing polar groups on the C28 carboxyl could enhance the celastrol's disrupting influence on HSP90/Cdc37 association and increase celastrol's ability to induce HSP70 (Jiang et al. 2016). (3) Entering the hydrophobic pocket. To support this notion, celastrol could enter a hydrophobic cavity formed by the HSP90 dimer (Peng et al. 2016), and bind with co-chaperone p23 mainly in the area formed by the hydrophobic amino acids of this protein (Chadli et al. 2010). Moreover, celastrol's C ring, D ring, and part of the E ring are situated in the hydrophobic core formed on the surface of Cdc37 (Jiang et al. 2016).

In conjunction, it is clear that for celastrol to inhibit HSP90 complex in terms of inducing HSR, the quinonemethide structure-based Michael adducts are not neces-

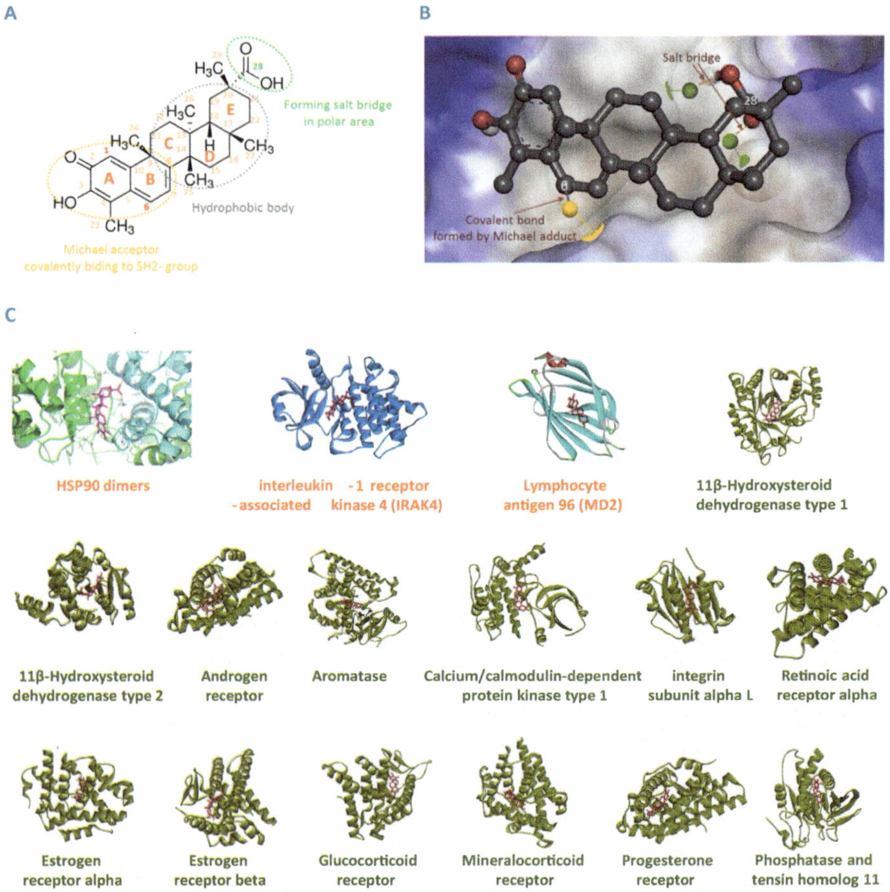

Fig. 20.1 Celastrol structure and the model for its interaction with proteins by hydrophobic pocket-based binding. (**a**) Celastrol structure. (**b**) Hydrophobic pocket-base binding model. Celastrol could enter hydrophobic pocket formed by protein or protein complex, then the C28 carboxyl could formed salt bridges with the residuals on the wall of the pocket, and the C6 could form Michele adducts with thiol groups, providing there be cysteine nearby. Nevertheless, the interaction between C6 and cysteine might be not indispensable, albeit it might enhance celastrol's binding in the pocket. (**c**) Docking prediction of entering hydrophobic pockets of some proteins by celastrol. The binding of celastrol to the first three proteins (HSP90, IRAK4 and MD2) had been biologically confirmed by us. The purple small structure is celastrol

sary, while the C28 carboxyl and the hydrophobic structure (rings C, D and part of E) are important. Based on this conclusion, we suggest a protein hydrophobic pocket-based model for celastrol's targeting of the HSP90 complex (Fig. 20.1b). Celastrol enters the hydrophobic pocket formed by the HSP90 dimer, or by HSP90 and its co-chaperones (such as Cdc37 or p23), then the C28 carboxyl group forms salt bridges with residuals in the pocket wall, enhancing the interaction between celastrol and protein. Celastrol could interact with cysteine situated in the pocket formed by Cdc37, yet this quinonemethide structure-based Michael adducts is not indispensable for disrupting HSP90/Cdc37 interaction, albeit it might enhance celastrol's binding in the pocket. Our model highlights the role of celastrol's hydrophobic sub-structure in binding, in that it is the basis by which celastrol enters the hydrophobic pocket on a protein or protein complex. To corroborate our thinking, it has been reported that hydrophobicity is important to small molecules inducing heat shock response (Ohnishi et al. 2013).

20.7.2 Generalization of the Model to Celastrol Targeting Other Proteins

Our hydrophobic pocket-based model is compatible with reports of other proteins targeted by celastrol, for example, celastrol entering the ATP-binding pocket in ERK (Kim et al. 2009c), the NAD^{+}-biding pocket in enoyl-acyl carrier protein reductase (Tallorin et al. 2014), and LPS-biding pocket in MD2 (Zhang et al. 2018). Our docking work suggests that celastrol could enter hydrophobic pockets in a variety of proteins (Fig. 20.1c). Though quinonemethide structure-based Michael adducts are not essential in celastrol inhibition of the HSP90 complex to induce HSR, our model does not preclude but rather supports the possibility that such a reaction is involved in and play important role in celastrol's targeting of other proteins or protein complexes, provided the targeted proteins have cysteine in their celastrol-binding pocket and the cysteine residual is accessible to C6 on celastrol. More specifically, there are two challenges in relating Michael adducts to celastrol's action. One is that intracellular GSH concentration (around 10 mM) (Wu et al. 1996) is higher than the usual concentration of celastrol (usually used at less than 10 μM), and could thus quickly sequester celastrol. This challenge is partially resolved by the finding that the reaction between GSH and celastrol is reversible (Zhou et al. 2016). The other challenge is to explain how celastrol limits its targets, viewing that there are a huge number of proteins with cysteine residuals in the body. To address this question, it has been suggested that nucleophiles added to the pharmacophore of celastrol in a remarkable stereospecific manner (Klaic et al. 2011). Nevertheless, our model added a novel answer to these questions: that celastrol preferentially acts on proteins' cysteine which situates in a celastrol-accessible hydrophobic pocket.

Our model, in combination with the reported structure-activity relationship, casts new light on how to translate celastrol or its derivatives into clinical application. The C28 carboxyl group is important to inducing HSP90, while the quinonemethide structure is related to cytotoxicity (Jiang et al. 2016; Westerheide et al. 2004). Therefore, if celastrol is to be used in cell protection, such as ameliorating neurodegenerative diseases, its quinonemethide structure should be abolished to reduce toxicity. On the contrary, if celastrol is to be used in anti-cancer applications, its C28 carboxyl groups should be modified to reduce its ability to induce heat shock response. For either application, the hydrophobic structure basis for celastrol-protein binding should be kept.

20.8 Conclusions

The past 30 years have witnessed exciting progress in celastrol study, making it a lead compound in treating autoimmune and various inflammatory diseases, different types of cancers, neurodegenerative, and chronic metabolic disorders. The mechanisms behind these therapeutic effects have been intensively explored, and have identified celastrol as a novel NF-κB inhibitor, HSP90 inhibitor, and proteasome inhibitor. The hydrophobic pocket-based binding, in conjunction with quinonemethide structure-based Michael adducts and C28 carboxyl-based salt bridge formation, provide the molecular basis for its direct and specific targeting of a group of proteins. It is likely that the research on this small molecule will continue to grow and accelerate with each new discovery and breakthrough. The next-step in celastrol research should be geared toward clinical translation, especially in treating neurodegenerative and chronic metabolic disorders. To do so, the structure-effect relationship should be further investigated to provide guidance for modifying this molecule to reduce its toxicity. The effects on the different organs and systems in vivo must also be well investigated before celastrol or its derivatives can be fully utilized in clinical applications.

Acknowledgements Thanks to Mr. Steven Schwab for tidying up the English expressions and grammar. This work was supported in part by the Foundation of Discipline Leader in Health Systems of the Pudong New District (No. PWRd2018-12), the National Natural Science Foundation of China (No. 81102349 and No. 81400793), Shanghai Traditional Chinese Medicine Content Construction Innovation Project (ZY3-CCCX-3-7001), Shanghai health and family planning commission General Program (No. 201540394 and No. 201540395), the Disciplines Group Construction Project of Pu Dong Health Bureau of Shanghai (No. PXZxq2017-15), the Shanghai Gongli Hospital Project (GLRb2018-01 and GLRq2017-01), and Postdoctoral Sustentation Fund of Shanghai Gongli Hospital (No. GLBH2017001).

References

Abbas S, Bhoumik A, Dahl R, Vasile S, Krajewski S, Cosford ND, Ronai ZA (2007) Preclinical studies of celastrol and acetyl isogambogic acid in melanoma. Clin Cancer Res 13:6769–6778

Abdin AA, Hasby EA (2014) Modulatory effect of celastrol on Th1/Th2 cytokines profile, TLR2 and CD3+ T-lymphocyte expression in a relapsing-remitting model of multiple sclerosis in rats. Eur J Pharmacol 742:102–112

Abu Bakar MH, Tan JS (2017) Improvement of mitochondrial function by celastrol in palmitate-treated C2C12 myotubes via activation of PI3K-Akt signaling pathway. Biomed Pharmacother = Biomed Pharmacother 93:903–912

Allison AC, Cacabelos R, Lombardi VR, Alvarez XA, Vigo C (2001) Celastrol, a potent antioxidant and anti-inflammatory drug, as a possible treatment for Alzheimer's disease. Prog Neuro-Psychopharmacol Biol Psychiatry 25:1341–1357

An SY, Youn GS, Kim H, Choi SY, Park J (2017) Celastrol suppresses expression of adhesion molecules and chemokines by inhibiting JNK-STAT1/NF-kappaB activation in poly(I:C)-stimulated astrocytes. BMB Rep 50:25–30

Astry B et al (2015) Celastrol, a Chinese herbal compound, controls autoimmune inflammation by altering the balance of pathogenic and regulatory T cells in the target organ. Clin Immunol (Orlando, Fla) 157:228–238

Barker EC, Kim BG, Yoon JH, Tochtrop GP, Letterio JJ, Choi SH (2018) Potent suppression of both spontaneous and carcinogen-induced colitis-associated colorectal cancer in mice by dietary celastrol supplementation. Carcinogenesis 39:36–46

Bentham G, Hooker JD (1862) Genera plantarum ad exempla imprimis in herbariis Kewensibus servata definita, vol I. Pars I, London, p 368

Brown IR (2007) Heat shock proteins and protection of the nervous system. Ann N Y Acad Sci 1113:147–158

Cascao R et al (2012) Effective treatment of rat adjuvant-induced arthritis by celastrol. Autoimmun Rev 11:856–862

Cascao R et al (2017) Effect of celastrol on bone structure and mechanics in arthritic rats. RMD Open 3:e000438

Chadli A et al (2010) Celastrol inhibits Hsp90 chaperoning of steroid receptors by inducing fibrillization of the co-chaperone p23. J Biol Chem 285:4224–4231

Chang W, He W, Li PP, Song SS, Yuan PF, Lu JT, Wei W (2016) Protective effects of Celastrol on diethylnitrosamine-induced hepatocellular carcinoma in rats and its mechanisms. Eur J Pharmacol 784:173–180

Chen SR, Dai Y, Zhao J, Lin L, Wang Y, Wang Y (2018a) A mechanistic overview of tiptolide and celastrol, natural products from tripterygium wilfordii Hook F. Front Pharmacol 9:104

Chen X et al (2018b) Celastrol attenuates incision-induced inflammation and pain associated with inhibition of the NF-kappaB signalling pathway via SARM. Life Sci 205:136–144

Chen Y et al (2018c) Effect and mechanisms of celastrol on the apoptosis of HOS osteosarcoma cells. Oncol Rep 40:2260–2268

Chou TQ, Mei PE (1936) Contribution a l'etude des principes de la drogue chinoise Lei-kung-teng, tripterygium wilfordii, Hook. I. Sur la matière colorante et les sucres [in French]. Chin J Phsiol 10:529–534

Chu C, He W, Kuang Y, Ren K, Gou X (2014) Celastrol protects kidney against ischemia-reperfusion-induced injury in rats. J Surg Res 186:398–407

Cleren C, Calingasan NY, Chen J, Beal MF (2005) Celastrol protects against MPTP- and 3-nitropropionic acid-induced neurotoxicity. J Neurochem 94:995–1004

Coulibaly A, Haas A, Steinmann S, Jakobs A, Schmidt TJ, Klempnauer KH (2018) The natural anti-tumor compound Celastrol targets a Myb-C/EBPbeta-p300 transcriptional module implicated in myeloid gene expression. PLoS One 13:e0190934

Dai Y et al (2010) Natural proteasome inhibitor celastrol suppresses androgen-independent prostate cancer progression by modulating apoptotic proteins and NF-kappaB. PLoS One 5:e14153

Deng YN, Shi J, Liu J, Qu QM (2013) Celastrol protects human neuroblastoma SH-SY5Y cells from rotenone-induced injury through induction of autophagy. Neurochem Int 63:1–9

Der Sarkissian S et al (2014) Celastrol protects ischaemic myocardium through a heat shock response with up-regulation of haeme oxygenase-1. Br J Pharmacol 171:5265–5279

Divya T, Dineshbabu V, Soumyakrishnan S, Sureshkumar A, Sudhandiran G (2016) Celastrol enhances Nrf2 mediated antioxidant enzymes and exhibits anti-fibrotic effect through regulation of collagen production against bleomycin-induced pulmonary fibrosis. Chem Biol Interact 246:52–62

Divya T, Sureshkumar A, Sudhandiran G (2017) Autophagy induction by celastrol augments protection against bleomycin-induced experimental pulmonary fibrosis in rats: role of adaptor protein p62/ SQSTM1. Pulm Pharmacol Ther 45:47–61

Dudics S, Venkatesha SH, Moudgil KD (2018) The micro-RNA expression profiles of autoimmune arthritis reveal novel biomarkers of the disease and therapeutic response. Int J Mol Sci 19 https://doi.org/10.3390/ijms19082293

Dukler N et al (2017) Nascent RNA sequencing reveals a dynamic global transcriptional response at genes and enhancers to the natural medicinal compound celastrol. Genome Res 27:1816–1829

Eroglu B et al (2014) Therapeutic inducers of the HSP70/HSP110 protect mice against traumatic brain injury. J Neurochem 130:626–641

Feng L et al (2013) ER stress-mediated apoptosis induced by celastrol in cancer cells and important role of glycogen synthase kinase-3beta in the signal network. Cell Death Dis 4:e715

Francis SP, Kramarenko II, Brandon CS, Lee FS, Baker TG, Cunningham LL (2011) Celastrol inhibits aminoglycoside-induced ototoxicity via heat shock protein 32. Cell Death Dis 2:e195

Fribley AM et al (2015) Celastrol induces unfolded protein response-dependent cell death in head and neck cancer. Exp Cell Res 330:412–422

Gan K, Xu L, Feng X, Zhang Q, Wang F, Zhang M, Tan W (2015) Celastrol attenuates bone erosion in collagen-induced arthritis mice and inhibits osteoclast differentiation and function in RANKL-induced RAW264.7. Int Immunopharmacol 24:239–246

Gu L et al (2013) Celastrol prevents atherosclerosis via inhibiting LOX-1 and oxidative stress. PLoS One 8:e65477

Guan Y, Cui ZJ, Sun B, Han LP, Li CJ, Chen LM (2016) Celastrol attenuates oxidative stress in the skeletal muscle of diabetic rats by regulating the AMPK-PGC1alpha-SIRT3 signaling pathway. Int J Mol Med 37:1229–1238

Guo L et al (2017) Targeted delivery of celastrol to mesangial cells is effective against mesangioproliferative glomerulonephritis. Nat Commun 8:878

Han LP, Li CJ, Sun B, Xie Y, Guan Y, Ma ZJ, Chen LM (2016) Protective effects of celastrol on diabetic liver injury via TLR4/MyD88/NF-kappaB signaling pathway in type 2 diabetic rats. J Diabetes Res 2016:2641248

Han LP, Sun B, Li CJ, Xie Y, Chen LM (2018a) Effect of celastrol on tolllike receptor 4mediated inflammatory response in free fatty acidinduced HepG2 cells. Int J Mol Med 42:2053–2061

Han XB, Tan Y, Fang YQ, Li F (2018b) Protective effects of celastrol against gamma irradiation-induced oxidative stress in human umbilical vein endothelial cells. Exp Ther Med 16:685–694

Hansen J, Palmfeldt J, Vang S, Corydon TJ, Gregersen N, Bross P (2011) Quantitative proteomics reveals cellular targets of celastrol. PLoS One 6:e26634

Hassane DC et al (2008) Discovery of agents that eradicate leukemia stem cells using an in silico screen of public gene expression data. Blood 111:5654–5662

Hieronymus H et al (2006) Gene expression signature-based chemical genomic prediction identifies a novel class of HSP90 pathway modulators. Cancer Cell 10:321–330

Hu M et al (2017) Celastrol-induced Nur77 interaction with TRAF2 alleviates inflammation by promoting mitochondrial ubiquitination and autophagy. Mol Cell 66(141–153):e146

Hu N, Zhang Y (2017) TLR4 knockout attenuated high fat diet-induced cardiac dysfunction via NF-kappaB/JNK-dependent activation of autophagy. Biochim Biophys Acta Mol basis Dis 1863:2001–2011

Hu X, Jia M, Fu Y, Zhang P, Zhang Z, Lin Q (2018) Novel low-toxic derivative of celastrol maintains protective effect against acute renal injury. ACS Omega 3:2652–2660

Huang L et al (2011) Inhibitory action of celastrol on hypoxia-mediated angiogenesis and metastasis via the HIF-1alpha pathway. Int J Mol Med 27:407–415

Huang Y, Zhou Y, Fan Y, Zhou D (2008) Celastrol inhibits the growth of human glioma xenografts in nude mice through suppressing VEGFR expression. Cancer Lett 264:101–106

Idris AI, Libouban H, Nyangoga H, Landao-Bassonga E, Chappard D, Ralston SH (2009) Pharmacologic inhibitors of IkappaB kinase suppress growth and migration of mammary car-

cinosarcoma cells in vitro and prevent osteolytic bone metastasis in vivo. Mol Cancer Ther 8:2339–2347
Jannuzzi AT, Kara M, Alpertunga B (2018) Celastrol ameliorates acetaminophen-induced oxidative stress and cytotoxicity in HepG2 cells. Hum Exp Toxicol 37:742–751
Jaquet V et al (2011) NADPH oxidase (NOX) isoforms are inhibited by celastrol with a dual mode of action. Br J Pharmacol 164:507–520
Jiang F et al (2016) Optimization and biological evaluation of celastrol derivatives as Hsp90-Cdc37 interaction disruptors with improved druglike properties. Bioorg Med Chem 24:5431–5439
Jiang M et al (2018) Celastrol treatment protects against acute ischemic stroke-induced brain injury by promoting an IL-33/ST2 axis-mediated microglia/macrophage M2 polarization. J Neuroinflammation 15:78
Jin HZ, Hwang BY, Kim HS, Lee JH, Kim YH, Lee JJ (2002) Antiinflammatory constituents of Celastrus orbiculatus inhibit the NF-kappaB activation and NO production. J Nat Prod 65:89–91
Jo H, Loison F, Hattori H, Silberstein LE, Yu H, Luo HR (2010) Natural product celastrol destabilizes tubulin heterodimer and facilitates mitotic cell death triggered by microtubule-targeting anti-cancer drugs. PLoS One 5:e10318
Joshi V et al (2016) Celastrol modulates inflammation through inhibition of the catalytic activity of mediators of arachidonic acid pathway: secretory phospholipase A2 group IIA, 5-lipoxygenase and cyclooxygenase-2. Pharmacol Res 113:265–275
Jung HW, Chung YS, Kim YS, Park YK (2007) Celastrol inhibits production of nitric oxide and proinflammatory cytokines through MAPK signal transduction and NF-kappaB in LPS-stimulated BV-2 microglial cells. Exp Mol Med 39:715–721
Kalmar B, Greensmith L (2009) Activation of the heat shock response in a primary cellular model of motoneuron neurodegeneration-evidence for neuroprotective and neurotoxic effects. Cell Mol Biol Lett 14:319–335
Kashyap D, Sharma A, Tuli HS, Sak K, Mukherjee T, Bishayee A (2018) Molecular targets of celastrol in cancer: recent trends and advancements. Crit Rev Oncol Hematol 128:70–81
Kiaei M, Kipiani K, Petri S, Chen J, Calingasan NY, Beal MF (2005) Celastrol blocks neuronal cell death and extends life in transgenic mouse model of amyotrophic lateral sclerosis. Neurodegener Dis 2:246–254
Kim JE et al (2013) Celastrol, an NF-kappaB inhibitor, improves insulin resistance and attenuates renal injury in db/db mice. PLoS One 8:e62068
Kim DH, Shin EK, Kim YH, Lee BW, Jun JG, Park JH, Kim JK (2009a) Suppression of inflammatory responses by celastrol, a quinone methide triterpenoid isolated from Celastrus regelii. Eur J Clin Investig 39:819–827
Kim DY, Park JW, Jeoung D, Ro JY (2009b) Celastrol suppresses allergen-induced airway inflammation in a mouse allergic asthma model. Eur J Pharmacol 612:98–105
Kim Y et al (2009c) Celastrol binds to ERK and inhibits FcepsilonRI signaling to exert an anti-allergic effect. Eur J Pharmacol 612:131–142
Klaic L, Trippier PC, Mishra RK, Morimoto RI, Silverman RB (2011) Remarkable stereospecific conjugate additions to the Hsp90 inhibitor celastrol. J Am Chem Soc 133:19634–19637
Kyung H, Kwong JM, Bekerman V, Gu L, Yadegari D, Caprioli J, Piri N (2015) Celastrol supports survival of retinal ganglion cells injured by optic nerve crush. Brain Res 1609:21–30
Lee JH et al (2006) Inhibition of NF-kappa B activation through targeting I kappa B kinase by celastrol, a quinone methide triterpenoid. Biochem Pharmacol 72:1311–1321
Lee HW, Jang KS, Choi HJ, Jo A, Cheong JH, Chun KH (2014) Celastrol inhibits gastric cancer growth by induction of apoptosis and autophagy. BMB Rep 47:697–702
Lee JY, Lee BH, Kim ND, Lee JY (2015) Celastrol blocks binding of lipopolysaccharides to a toll-like receptor4/myeloid differentiation factor2 complex in a thiol-dependent manner. J Ethnopharmacol 172:254–260
Li Y et al (2012a) Protective effect of celastrol in rat cerebral ischemia model: down-regulating p-JNK, p-c-Jun and NF-kappaB. Brain Res 1464:8–13

Li Z et al (2012b) Celastrol nanoparticles inhibit corneal neovascularization induced by suturing in rats. Int J Nanomedicine 7:1163–1173

Li Z et al (2012c) Antitumor activity of celastrol nanoparticles in a xenograft retinoblastoma tumor model. Int J Nanomedicine 7:2389–2398

Li G et al (2013a) Celastrol inhibits lipopolysaccharide-stimulated rheumatoid fibroblast-like synoviocyte invasion through suppression of TLR4/NF-kappaB-mediated matrix metalloproteinase-9 expression. PLoS One 8:e68905

Li GQ et al (2013b) Anti-invasive effects of celastrol in hypoxia-induced fibroblast-like synoviocyte through suppressing of HIF-1alpha/CXCR4 signaling pathway. Int Immunopharmacol 17:1028–1036

Li HY et al (2015) Celastrol induces apoptosis and autophagy via the ROS/JNK signaling pathway in human osteosarcoma cells: an in vitro and in vivo study. Cell Death Dis 6:e1604

Li H, Zhang YY, Huang XY, Sun YN, Jia YF, Li D (2005) Beneficial effect of tripterine on systemic lupus erythematosus induced by active chromatin in BALB/c mice. Eur J Pharmacol 512:231–237

Li X, Wu N, Zou L, Jia D (2017) Protective effect of celastrol on myocardial ischemia-reperfusion injury. Anatol J Cardiol 18:384–390

Lin L, Sun Y, Wang D, Zheng S, Zhang J, Zheng C (2015) Celastrol ameliorates ulcerative colitis-related colorectal cancer in mice via suppressing inflammatory responses and epithelial-mesenchymal transition. Front Pharmacol 6:320

Liu Z et al (2014) Cancerous inhibitor of PP2A is targeted by natural compound celastrol for degradation in non-small-cell lung cancer. Carcinogenesis 35:905–914

Liu J, Lee J, Salazar Hernandez MA, Mazitschek R, Ozcan U (2015) Treatment of obesity with celastrol. Cell 161:999–1011

Lu Z, Jin Y, Qiu L, Lai Y, Pan J (2010) Celastrol, a novel HSP90 inhibitor, depletes Bcr-Abl and induces apoptosis in imatinib-resistant chronic myelogenous leukemia cells harboring T315I mutation. Cancer Lett 290:182–191

Luo D, Zhao J, Rong J (2016) Plant-derived triterpene celastrol ameliorates oxygen glucose deprivation-induced disruption of endothelial barrier assembly via inducing tight junction proteins. Phytomedicine: Int J Phytother Phytopharmacol 23:1621–1628

Luo D, Guo Y, Cheng Y, Zhao J, Wang Y, Rong J (2017) Natural product celastrol suppressed macrophage M1 polarization against inflammation in diet-induced obese mice via regulating Nrf2/HO-1, MAP kinase and NF-kappaB pathways. Aging 9:2069–2082

Ma X et al (2015) Celastrol protects against obesity and metabolic dysfunction through activation of a HSF1-PGC1alpha transcriptional Axis. Cell Metab 22:695–708

Ma J, Han LZ, Liang H, Mi C, Shi H, Lee JJ, Jin X (2014) Celastrol inhibits the HIF-1alpha pathway by inhibition of mTOR/p70S6K/eIF4E and ERK1/2 phosphorylation in human hepatoma cells. Oncol Rep 32:235–242

Matokanovic M, Barisic K, Filipovic-Grcic J, Maysinger D (2013) Hsp70 silencing with siRNA in nanocarriers enhances cancer cell death induced by the inhibitor of Hsp90. Eur J Pharm Sci 50:149–158

Mezzanotte L, An N, Mol IM, Lowik CW, Kaijzel EL (2014) A new multicolor bioluminescence imaging platform to investigate NF-kappaB activity and apoptosis in human breast cancer cells. PLoS One 9:e85550

Mu TW, Ong DS, Wang YJ, Balch WE, Yates JR 3rd, Segatori L, Kelly JW (2008) Chemical and biological approaches synergize to ameliorate protein-folding diseases. Cell 134:769–781

Nagase M, Oto J, Sugiyama S, Yube K, Takaishi Y, Sakato N (2003) Apoptosis induction in HL-60 cells and inhibition of topoisomerase II by triterpene celastrol. Biosci Biotechnol Biochem 67:1883–1887

Nanjundaiah SM, Venkatesha SH, Yu H, Tong L, Stains JP, Moudgil KD (2012) Celastrus and its bioactive celastrol protect against bone damage in autoimmune arthritis by modulating osteoimmune cross-talk. J Biol Chem 287:22216–22226

Narayan V, Ravindra KC, Chiaro C, Cary D, Aggarwal BB, Henderson AJ, Prabhu KS (2011) Celastrol inhibits Tat-mediated human immunodeficiency virus (HIV) transcription and replication. J Mol Biol 410:972–983
Ohnishi K et al (2013) Non-specific protein modifications by a phytochemical induce heat shock response for self-defense. PLoS One 8:e58641
Pang X et al (2010) Celastrol suppresses angiogenesis-mediated tumor growth through inhibition of AKT/mammalian target of rapamycin pathway. Cancer Res 70:1951–1959
Peng B et al (2014) Peptide deformylase inhibitor actinonin reduces celastrol's HSP70 induction while synergizing proliferation inhibition in tumor cells. BMC Cancer 14:146
Peng B, Xu L, Cao F, Wei T, Yang C, Uzan G, Zhang D (2010) HSP90 inhibitor, celastrol, arrests human monocytic leukemia cell U937 at G0/G1 in thiol-containing agents reversible way. Mol Cancer 9:79
Peng B, Gu YJ, Wang Y, Cao FF, Zhang X, Zhang DH, Hou J (2016) Mutations Y493G and K546D in human HSP90 disrupt binding of celastrol and reduce interaction with Cdc37. FEBS Open Bio 6:729–734
Pfuhlmann K et al (2018) Celastrol induced weight loss is driven by hypophagia and independent from UCP1. Diabetes 67:2465
Raja SM et al (2011) Anticancer activity of Celastrol in combination with ErbB2-targeted therapeutics for treatment of ErbB2-overexpressing breast cancers. Cancer Biol Ther 11:263–276
Rajendran P et al (2012) Celastrol suppresses growth and induces apoptosis of human hepatocellular carcinoma through the modulation of STAT3/JAK2 signaling cascade in vitro and in vivo. Cancer Prev Res (Phila) 5:631–643
Ren B et al (2017) Celastrol induces apoptosis in hepatocellular carcinoma cells via targeting ER-stress/UPR. Oncotarget 8:93039–93050
Schechter MS, Haller HL (1942) Identify of the red pigment in the roots of tripterygium wilfordii and Celastrus scandens. J Am Chem Soc 64:182–183
Scott LM et al (2011) Shp2 protein tyrosine phosphatase inhibitor activity of estramustine phosphate and its triterpenoid analogs. Bioorg Med Chem Lett 21:730–733
Segges P et al (2018) Targeting Hodgkin and Reed-Sternberg cells with an inhibitor of heat-shock protein 90: molecular pathways of response and potential mechanisms of resistance. Int J Mol Sci 19. https://doi.org/10.3390/ijms19030836
Seo WY et al (2011) Celastrol induces expression of heme oxygenase-1 through ROS/Nrf2/ARE signaling in the HaCaT cells. Biochem Biophys Res Commun 407:535–540
Shaker ME, Ashamallah SA, Houssen ME (2014) Celastrol ameliorates murine colitis via modulating oxidative stress, inflammatory cytokines and intestinal homeostasis. Chem Biol Interact 210:26–33
Shao L et al (2013) Celastrol suppresses tumor cell growth through targeting an AR-ERG-NF-kappaB pathway in TMPRSS2/ERG fusion gene expressing prostate cancer. PLoS One 8:e58391
Sharma S et al (2015) Celastrol, an oral heat shock activator, ameliorates multiple animal disease models of cell death. Cell Stress Chaperones 20:185–201
Shrivastava S, Jeengar MK, Reddy VS, Reddy GB, Naidu VG (2015) Anticancer effect of celastrol on human triple negative breast cancer: possible involvement of oxidative stress, mitochondrial dysfunction, apoptosis and PI3K/Akt pathways. Exp Mol Pathol 98:313–327
Smee DF, Alaghamandan HA, Cottam HB, Jolley WB, Robins RK (1990) Antiviral activity of the novel immune modulator 7-thia-8-oxoguanosine. J Biol Response Mod 9:24–32
Sreeramulu S, Gande SL, Gobel M, Schwalbe H (2009) Molecular mechanism of inhibition of the human protein complex Hsp90-Cdc37, a kinome chaperone-cochaperone, by triterpene celastrol. Angew Chem Int Ed Engl 48:5853–5855
Stankova K, Ivanova K, Nikolov V, Aneva N, Georgieva R, Boteva R (2013) Proteasome inhibition protects human peripheral blood mononuclear cells from radiation-induced oxidative stress. Int J Radiat Biol 89:493–500

Sun H, Xu L, Yu P, Jiang J, Zhang G, Wang Y (2010) Synthesis and preliminary evaluation of neuroprotection of celastrol analogues in PC12 cells. Bioorg Med Chem Lett 20:3844–3847
Swingle WT, Haller EH, Swingle MC (1941) A Chinese insecticidal plant, tripterygium wilfordii, introduced into the United States. Science 93:2043–2044
Tallorin LC, Durrant JD, Nguyen QG, McCammon JA, Burkart MD (2014) Celastrol inhibits Plasmodium falciparum enoyl-acyl carrier protein reductase. Bioorg Med Chem 22:6053–6061
Tong S, Zhang L, Joseph J, Jiang X (2018) Celastrol pretreatment attenuates rat myocardial ischemia/ reperfusion injury by inhibiting high mobility group box 1 protein expression via the PI3K/Akt pathway. Biochem Biophys Res Commun 497:843–849
Trott A, West JD, Klaic L, Westerheide SD, Silverman RB, Morimoto RI, Morano KA (2008) Activation of heat shock and antioxidant responses by the natural product celastrol: transcriptional signatures of a thiol-targeted molecule. Mol Biol Cell 19:1104–1112
Tseng CK, Hsu SP, Lin CK, Wu YH, Lee JC, Young KC (2017) Celastrol inhibits hepatitis C virus replication by upregulating heme oxygenase-1 via the JNK MAPK/Nrf2 pathway in human hepatoma cells. Antivir Res 146:191–200
Uttarkar S et al (2016) Targeting acute myeloid leukemia with a small molecule inhibitor of the Myb/p300 interaction. Blood 127:1173–1182
Venkatesha SH, Yu H, Rajaiah R, Tong L, Moudgil KD (2011) Celastrus-derived celastrol suppresses autoimmune arthritis by modulating antigen-induced cellular and humoral effector responses. J Biol Chem 286:15138–15146
Venkatesha SH, Astry B, Nanjundaiah SM, Yu H, Moudgil KD (2012) Suppression of autoimmune arthritis by celastrus-derived celastrol through modulation of pro-inflammatory chemokines. Bioorg Med Chem 20:5229–5234
Walcott SE, Heikkila JJ (2010) Celastrol can inhibit proteasome activity and upregulate the expression of heat shock protein genes, hsp30 and hsp70, in Xenopus laevis A6 cells. Com Biochem Physiol A Mol Integr Physiol 156:285–293
Wan Y et al (2010) Cognitive decline following major surgery is associated with gliosis, beta-amyloid accumulation, and tau phosphorylation in old mice. Crit Care Med 38:2190–2198
Wang WB et al (2012) Paraptosis accompanied by autophagy and apoptosis was induced by celastrol, a natural compound with influence on proteasome, ER stress and Hsp90. J Cell Physiol 227:2196–2206
Wang C, Shi C, Yang X, Yang M, Sun H, Wang C (2014a) Celastrol suppresses obesity process via increasing antioxidant capacity and improving lipid metabolism. Eur J Pharmacol 744:52–58
Wang X, Cattaneo F, Ryno L, Hulleman J, Reixach N, Buxbaum JN (2014b) The systemic amyloid precursor transthyretin (TTR) behaves as a neuronal stress protein regulated by HSF1 in SH-SY5Y human neuroblastoma cells and APP23 Alzheimer's disease model mice. J Neurosci Off J Soc Neurosci 34:7253–7265
Wang H et al (2015a) Direct inhibition of c-Myc-Max heterodimers by celastrol and celastrol-inspired triterpenoids. Oncotarget 6:32380–32395
Wang Y et al (2015b) Celastrol ameliorates EAE induction by suppressing pathogenic T cell responses in the peripheral and central nervous systems. J Neuroimmune Pharmacol 10:506–516
Wang YL, Lam KK, Cheng PY, Lee YM (2015c) Celastrol prevents circulatory failure via induction of heme oxygenase-1 and heat shock protein 70 in endotoxemic rats. J Ethnopharmacol 162:168–175
Wang W, Ha C, Lin T, Wang D, Wang Y, Gong M (2018a) Celastrol attenuates pain and cartilage damage via SDF-1/CXCR4 signalling pathway in osteoarthritis rats. J Pharm Pharmacol 70:81–88
Wang Y et al (2018b) Celastrol improves the therapeutic efficacy of EGFR-TKIs for non-small-cell lung cancer by overcoming EGFR T790M drug resistance. Anti-Cancer Drugs 29:748–755
Wei Y, Wang Y (2017) Celastrol attenuates impairments associated with lipopolysaccharide-induced acute respiratory distress syndrome (ARDS) in rats. J Immunotoxicol 14:228–234
Weisberg S, Leibel R, Tortoriello DV (2016) Proteasome inhibitors, including curcumin, improve pancreatic beta-cell function and insulin sensitivity in diabetic mice. Nutr Diabetes 6:e205

Westerheide SD et al (2004) Celastrols as inducers of the heat shock response and cytoprotection. J Biol Chem 279:56053–56060

Whitehouse S, Chen PL, Greenshields AL, Nightingale M, Hoskin DW, Bedard K (2016) Resveratrol, piperine and apigenin differ in their NADPH-oxidase inhibitory and reactive oxygen species-scavenging properties. Phytomed: Int J Phytother Phytopharmacol 23:1494–1503

Wu M et al (2018) Celastrol aggravates LPS-induced inflammation and injuries of liver and kidney in mice. Am J Transl Res 10:2078–2086

Wu PH, Liu JF, Wu WL, Lanca AJ, Kalant H (1996) Development of alcohol tolerance in the rat after a single exposure to combined treatment with arginine8-vasopressin and ethanol. J Pharmacol Exp Ther 276:1283–1291

Wu T, Liu S, Wen G, Xu J, Yu Y, Chai Y (2017) Celastrol improves self-renewal and differentiation of human tendon-derived stem cells by suppressing Smad7 through hypoxia. Stem Cell Res Ther 8:274

Wyatt G, Dimmer SA (1989) A planning model for continuing education presenters. J Contin Educ Nurs 20:70–73

Xi J, Li Q, Luo X, Wang Y, Li J, Guo L, Wu G (2018) Celastrol inhibits glucocorticoidinduced osteoporosis in rat via the PI3K/AKT and Wnt signaling pathways. Mol Med Rep 18:4753

Xiao S, Zhang M, He Z, Wang D (2018) Celastrol attenuates symptoms of preeclampsia in rats by inhibiting matrix metalloproteinase-9. Eur J Pharmacol 830:33–38

Xin W, Wang Q, Zhang D, Wang C (2017) A new mechanism of inhibition of IL-1beta secretion by celastrol through the NLRP3 inflammasome pathway. Eur J Pharmacol 814:240–247

Xu LM et al (2014) Celastrol inhibits lung infiltration in differential syndrome animal models by reducing TNF-alpha and ICAM-1 levels while preserving differentiation in ATRA-induced acute promyelocytic leukemia cells. PLoS One 9:e105131

Xu X, Wu Z, Xu C, Ren Y, Ge Y (2003) Observation on serum anti-double stranded DNA antibodies of tripterine in systemic lupus erythematosus of (NZBxW)F1 mice. Ann Rheum Dis 62:377–378

Xu X, Zhong J, Wu Z, Fang Y, Xu C (2007) Effects of tripterine on mRNA expression of TGF-beta1 and collagen IV expression in BW F1 mice. Cell Biochem Funct 25:501–507

Yadav P et al (2018) Celastrol as a pentacyclic triterpenoid with chemopreventive properties. Pharm Patent Analyst 7:155–167

Yang H, Chen D, Cui QC, Yuan X, Dou QP (2006) Celastrol, a triterpene extracted from the Chinese "thunder of god vine," is a potent proteasome inhibitor and suppresses human prostate cancer growth in nude mice. Cancer Res 66:4758–4765

Yang L, Li Y, Ren J, Zhu C, Fu J, Lin D, Qiu Y (2014) Celastrol attenuates inflammatory and neuropathic pain mediated by cannabinoid receptor type 2. Int J Mol Sci 15:13637–13648

Yang H, Liu C, Jiang J, Wang Y, Zhang X (2017) Celastrol attenuates multiple sclerosis and optic neuritis in an experimental autoimmune encephalomyelitis model. Front Pharmacol 8:44

Youn GS, Kwon DJ, Ju SM, Rhim H, Bae YS, Choi SY, Park J (2014) Celastrol ameliorates HIV-1 tat-induced inflammatory responses via NF-kappaB and AP-1 inhibition and heme oxygenase-1 induction in astrocytes. Toxicol Appl Pharmacol 280:42–52

Yu X et al (2017) Celastrol ameliorates inflammation through inhibition of NLRP3 inflammasome activation. Oncotarget 8:67300–67314

Yu X, Tao W, Jiang F, Li C, Lin J, Liu C (2010) Celastrol attenuates hypertension-induced inflammation and oxidative stress in vascular smooth muscle cells via induction of heme oxygenase-1. Am J Hypertens 23:895–903

Yu H, Venkatesha SH, Moudgil KD (2012) Microarray-based gene expression profiling reveals the mediators and pathways involved in the anti-arthritic activity of Celastrus-derived Celastrol. Int Immunopharmacol 13:499–506

Zanphorlin LM, Alves FR, Ramos CH (2014) The effect of celastrol, a triterpene with antitumorigenic activity, on conformational and functional aspects of the human 90kDa heat shock protein Hsp90alpha, a chaperone implicated in the stabilization of the tumor phenotype. Biochim Biophys Acta 1840:3145–3152

Zhang DH et al (2006) Tripterine inhibits the expression of adhesion molecules in activated endothelial cells. J Leukoc Biol 80:309–319
Zhang X et al (2016) Metabolomics profiles delineate uridine deficiency contributes to mitochondria-mediated apoptosis induced by celastrol in human acute promyelocytic leukemia cells. Oncotarget 7:46557–46572
Zhang R et al (2017a) Celastrol attenuates cadmium-induced neuronal apoptosis via inhibiting ca(2+) -CaMKII-dependent Akt/mTOR pathway. J Cell Physiol 232:2145–2157
Zhang Y et al (2017b) Celastrol ameliorates liver metabolic damage caused by a high-fat diet through Sirt1. Mol Metab 6:138–147
Zhang X et al (2018) Celastrol reverses palmitic acid (PA)-caused TLR4-MD2 activation-dependent insulin resistance via disrupting MD2-related cellular binding to PA. J Cell Physiol 233:6814–6824
Zhang T, Hamza A, Cao X, Wang B, Yu S, Zhan CG, Sun D (2008) A novel Hsp90 inhibitor to disrupt Hsp90/Cdc37 complex against pancreatic cancer cells. Mol Cancer Ther 7:162–170
Zhang T, Li Y, Yu Y, Zou P, Jiang Y, Sun D (2009) Characterization of celastrol to inhibit hsp90 and cdc37 interaction. J Biol Chem 284:35381–35389
Zhao Q et al (2015) Natural products triptolide, celastrol, and withaferin A inhibit the chaperone activity of peroxiredoxin I. Chem Sci 6:4124–4130
Zhou YX, Huang YL (2009) Antiangiogenic effect of celastrol on the growth of human glioma: an in vitro and in vivo study. Chin Med J 122:1666–1673
Zhou Y, Li W, Wang M, Zhang X, Zhang H, Tong X, Xiao Y (2016) Competitive profiling of celastrol targets in human cervical cancer HeLa cells via quantitative chemical proteomics. Mol BioSyst 13:83–91
Zhu H et al (2010) Celastrol acts as a potent antimetastatic agent targeting beta1 integrin and inhibiting cell-extracellular matrix adhesion, in part via the p38 mitogen-activated protein kinase pathway. J Pharmacol Exp Ther 334:489–499
Zou YC, Yang XW, Yuan SG, Zhang P, Li YK (2016) Celastrol inhibits prostaglandin E2-induced proliferation and osteogenic differentiation of fibroblasts isolated from ankylosing spondylitis hip tissues in vitro. Drug Des Devel Ther 10:933–948

Part IV
Clinical Aspects of HSP90

Chapter 21
Hsp90 Chaperone in Disease

Luca Ferrari and Stefan G. D. Rüdiger

Abstract The molecular chaperone Hsp90 is at the heart of protein homeostasis control. A wide range of pathologies disturbs protein homeostasis, thus placing Hsp90 at the crossroads of many diseases. Here, we evaluate the impact of recent progress in understanding the molecular mechanism of Hsp90-client interactions and their role in disease. We discuss the role of Hsp90 for hormonal imbalances, cancer and neurodegenerative disorders. For each disease class we discuss implications of complexes in which Hsp90 binds to a paradigmatic client: the transcription factor Glucocorticoid Receptor, the kinase Cdk4 and the microtubule stabilizer Tau. The mechanistic insights allow us to elaborate on possible therapeutic intervention routes. Hsp90 is a druggable chaperone. Thus, understanding Hsp90 biology at molecular resolution offers an interesting approach to tackle protein-related diseases.

Keywords Kinases · Molecular chaperones · Neurodegeneration · Protein folding · Proteostasis · Steroid receptors

Abbreviations

Aha1	Activator of 90 kDa heat shock protein ATPase homolog 1
Cdk4	Cyclin-dependent kinase 4
CHIP	Carboxy terminus of Hsp70-interacting protein
FKBP51	51 kDa FK506-binding protein
GR	Glucocorticoid receptor
Grp94	94 kDa glucose-related protein

L. Ferrari · S. G. D. Rüdiger (✉)
Cellular Protein Chemistry, Bijvoet Center for Biomolecular Research, Utrecht University, Utrecht, The Netherlands

Science for Life, Utrecht University, Utrecht, The Netherlands
e-mail: s.g.d.rudiger@uu.nl

A. A. A. Asea, P. Kaur (eds.), *Heat Shock Protein 90 in Human Diseases and Disorders*, Heat Shock Proteins 19,
https://doi.org/10.1007/978-3-030-23158-3_21

Hop	Hsc70/Hsp90-organizing protein
HPA axis	Hypothalamic-pituitary-adrenal axis
Hsc70	Heat shock cognate 71 kDa protein
HSP	Heat shock protein family
Hsp40	Heat shock protein 40
Hsp70	Heat shock protein 70
Hsp90	Heat shock protein 90
Hsp90-C	Hsp90 C-terminal domain
Hsp90-M	Hsp90 middle domain
Hsp90-N	Hsp90 N-terminal domain
PPIase	Peptidyl-prolyl-cis/trans-isomerase
TPR	Tetratricopeptide repeat
Trap1	Tumor necrosis factor type 1 receptor-associated protein
XAP-2	HBV X-associated protein 2

21.1 Introduction

In the never-ending fight for adaptability, living systems and their constitutive molecules never settle, changing their internal properties to face environmental challenges (Weber 2010). A paradigmatic example of such fight is the response of a living system to temperature insults, requiring a massive, orchestrated reorganization of its internal components. This reorganization was described for the first time in fruit flies salivary glands subjected to heat shock, and aptly termed heat shock response (Ritossa 1962). Heat shock proteins orchestrate such response, navigating proteins in the turbulent waters of heat stress toward their mature folded forms, and they do so by establishing transient interactions with their assisted proteins, a mode of action that classifies them as molecular chaperones (Ellis 1987). Chaperones do not act solely in case of thermal stress. Indeed, they act in any physiological conditions or at any given environmental challenge affecting the internal protein homeostasis of living systems (Ellis 2013). At a molecular level, human diseases often create conditions challenging protein homeostasis inside cells, tissues or even entire organisms (van Oosten-Hawle and Morimoto 2014). Thus, molecular chaperones play a fundamental role in the onset and progression of any human disease (Klaips et al. 2018).

Chaperones act as a network of proteins, and a central node of this network is the molecular chaperone Hsp90 (Schopf et al. 2017; Taipale et al. 2010). This implies a major role of Hsp90 in human diseases, comprising hormone imbalances linked to metabolic stress (Patel et al. 2014), cancer (Butler et al. 2015; Jhaveri et al. 2014), neurodegenerative disorders such as Alzheimer, Huntington and Parkinson's diseases (Caron et al. 2018; Pratt et al. 2015), heart failure disorders (Ranek et al. 2018), diseases mediated by virus or parasites (Woodford et al. 2016), as in the case of Zika Virus (Pan et al. 2018) or Hepatitis B (Pei et al. 2017), and cystic fibrosis

(Ihrig and Obermann 2017). Understanding Hsp90 biology means understanding a node at crossroads of multiple diseases.

21.1.1 Hsp90: Chaperoning Disease-Relevant Proteins

Hsp90 binds to thousands of proteins, alternative named clients, with no particular common fold driving the binding (Taipale et al. 2014). Hsp90 preferentially binds to unstable proteins (Taipale et al. 2012), whose instability must be dictated by particular amino acidic patterns necessary to recognize the molecular chaperones. Such patterns emerge when we analyze structural models of Hsp90-client complexes resolved at atomic level: Hsp90 in complex with the transcription regulator glucocorticoid receptor (Kirschke et al. 2014), the kinase Cdk4 (Verba et al. 2016), the microtubule binding protein Tau (Karagöz et al. 2014) and the transport protein Transthyretin (Oroz et al. 2017). When looking at these protein complexes, common molecular recognition features arise (Karagöz and Rüdiger 2015; Radli and Rüdiger 2018). (i) Hsp90 binding surface is broad, from 50 to 170 amino acidic depending on the substrate. Hsp90 is a homodimer, each protomer is composed of three domains – the N, M and C-terminal domains – with Hsp90-N and Hsp90-M responsible for interactions with clients (Fig. 21.1). (ii) The established interactions are numerous, and the individual contributions are weak. (iii) The Hsp90 surface is heterogenous, characterized by scattered hydrophobics and charged residues. The first feature allows accommodating clients with different sizes, whereas the other two allow clients with different net charges and hydrophobic content to interact with

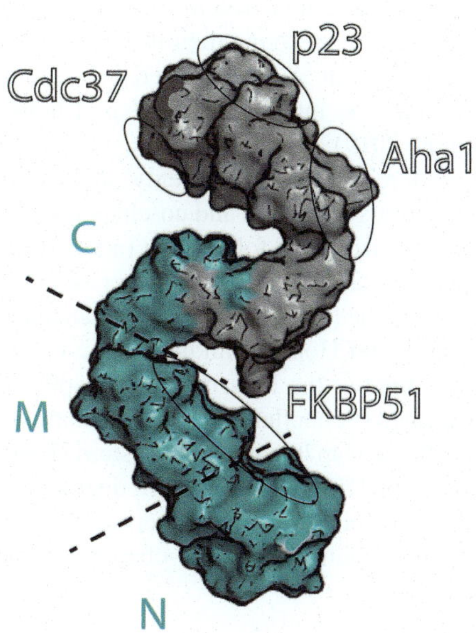

Fig. 21.1 Hsp90 structure and binding to co-chaperones. Hsp90 protomers (coloured in cyan and grey) interact via their C-terminal domains to form a homodimer. N, M and C indicate the three domains of the protomer. Circles represents docking surfaces of four representative co-chaperones, namely Cdc37, p23, Aha1 and FKBP51

Hsp90 surface. Taken together, these features explain on a structural level why Hsp90 clientele is so broad and heterogeneous.

The broad action of Hsp90 can be appreciated even further if we place this chaperon in the context of the living, dynamic environment of a cell. Hsp90 is essential in eukaryotes (Borkovich et al. 1989; Voss et al. 2000) and it is an abundant protein, accounting for ~1% of cellular protein content in non-stress conditions (Picard 2002). Hsp90 isoforms populate the major cellular compartments, with two isoforms resident in the cytoplasm (Hsp90α is the stress inducible form, while Hsp90β is the constitutive one and the main focus of this review (Schopf et al. 2017)), one in the endoplasmic reticulum (Grp94 (Melnick et al. 1992)) and one in the mitochondria (Trap1 (Felts et al. 2000)). Hsp90 centrality in the chaperones network and its broad clientele explain its involvement in otherwise unrelated diseases. Interaction of Hsp90 with disease-related clients is often regulated by co-chaperones. The co-chaperone network offers possibilities for subtle tackling of specific disease-relevant beneficiaries of Hsp90 action.

21.1.2 Hsp90 Co-chaperones: Dictating the fate of Disease-Relevant Clients

Hsp90 has a conserved role in folding of substrates in co-operation with Hsp70, a function spread throughout the evolutionary tree of life (Genest et al. 2011; Morán Luengo et al. 2018; Schumacher et al. 1996; Wegele et al. 2006). Next to this conserved role, Hsp90 acquired a regulatory role in the cytosol of eukaryotic cells, including humans (Morán Luengo et al. 2019). This means that human Hsp90 not only ensures that proteins reach their fold, but can further fine-tune their activity, for instance by delaying their interaction with a ligand (Schulke et al. 2010) or by determining their degradation (Kundrat and Regan 2010). This evolutionary acquired function is modulated by Hsp90 helpers, termed co-chaperones (Morán Luengo et al. 2019).

A recent, extensive study analyzing the folding of a family of steroid hormon receptors by Hsp90 and co-chaperones demonstrated that most of co-chaperones do not contribute to folding but rather slow down Hsp90 folding machinery (Radli and Rüdiger 2018; Sahasrabudhe et al. 2017). Co-chaperones most prominent role is to dictate the fate of clients. An elegant example is the competition of the co-chaperones CHIP and Hop for binding to the C-terminus of Hsp90. Both these co-chaperones can bind to the N-terminus of Hsp90 due to their tetratricopeptide repeat, or TPR domain (Scheufler et al. 2000; Zhang et al. 2005). Hop regulates Hsp90 interaction with the Hsp70 chaperone system (Kirschke et al. 2014; Morán Luengo et al. 2018) whilst CHIP promotes client degradation (Edkins 2015). Both co-chaperones compete for binding to the C-terminus of Hsp90, suggesting that the fate of an Hsp90 client depends on the co-chaperone winning the competition (Kundrat and Regan 2010).

More generally, there are more than 40 Hsp90 co-chaperones (an updated list can be found at https://www.picard.ch/downloads) and the fate of many clients is dictated by a subgroup of them (Cox and Johnson 2018; Picard 2002). Co-chaperones compete fiercely for binding to Hsp90 (Harst et al. 2005), and competition is divided into two leagues: from one side, co-chaperones whose docking on Hsp90 surface is dictated by the interaction between their TPR domains and the EEVD motif at the C-terminus of Hsp90; from the other, co-chaperones whose docking is TPR-independent (Schopf et al. 2017). The winners of this competition are dictated by binding affinities, steric hindrances (in the case of co-chaperones sharing the same Hsp90 docking surface, Fig. 21.1), post-translational modifications and by the levels of co-chaperones, the latter constantly changing in response to environmental cues to readily modify the fate of hsp90 clients (Assimon et al. 2015; Schopf et al. 2017). This system can be hijacked in pathology, for instance in neurodegenerative disorders. Indeed, aging changes co-chaperones and chaperones abundance in human brain, exacerbating the onset of diseases (Blair et al. 2013; Brehme et al. 2014; Labbadia and Morimoto 2015).

Co-chaperones give thus regulational power to Hsp90: based on the cellular state, the availability of ligands, co-chaperones and degradation machinery, each client may be subjected to completely opposite outcomes, from proper folding to irreversible degradation. Co-chaperones add an extra level of regulation for cytoplasmic proteins dependent on Hsp90, and this is a crucial aspect to understand Hsp90 role in disease. We discuss in detail three Hsp90 clients, to extract the mechanistic role of Hsp90 and its co-chaperones in disease. Each client will serve as paradigm for a specific set of disorders, from disorders associated to hormonal imbalances (Glucocorticoid Receptor, (GR)) to cancer (Cdk4) and neurodegeneration (Tau).

21.1.3 Hsp90 and Hormonal Imbalances

21.1.3.1 The Hsp90-Client GR Controls Hormonal Response

Hsp90 interacts with hundreds of transcription factors (Taipale et al. 2012), and among them a well characterized client is the Steroid Hormon Receptor GR. GR is the end-product of hypothalamic-pituitary-adrenal (HPA) axis, a system-wide group of molecular feedback loops present in all human tissues and governing cellular responses to steroid hormones such as cortisol (Chrousos and Kino 2009). Being so ubiquitous, imbalances in the HPA axis are associated to many disorders of different nature. HPA (and GR) imbalances can cause for instance rare diseases such as Cushing syndrome and Addison disease, two metabolic disordered with either increased or depleted levels of the GR-ligand cortisol (Chrousos 2000) but can also be linked to complex pathophysiological effects such psychosocial stress (Wust et al. 2004). HPA imbalances are also associated to neurodegeneration (Du and Pang 2015), and cancer survivors can develop depression due to HPA imbalances

(Young and Singh 2018). Thus, HPA-axis and its end-product GR are valuable targets to treat a broad group of conditions.

21.1.3.2 The Hsp90 Machinery Dictates GR Folding and Activity

Hsp90 dictates GR folding and activity, hence controlling the whole HPA-axis end signaling. GR was one of the first identified Hsp90 clients (Sanchez et al. 1987), and its association to the ligand was shown to be dependent on the presence of the molecular chaperone (Picard et al. 1990). A cohort of chaperones and co-chaperones is involved in the balance of Hsp90-mediated GR activity. Hsp90 collaborates with the Hsp70 system to allow GR to interact with glucocorticoids and implement HPA signaling (Kirschke et al. 2014). The molecular chaperone Hsp70 and its co-chaperone Hsp40 recognize GR, keeping it partially unfolded and inactive. The co-chaperone Hop links Hsp70 and Hsp90 due to its three TPR domains, allowing the downstream interaction of GR with Hsp90. While Hsp70 keeps GR unfolded, Hsp90 releases the break and dictates binding to ligand, folding and consequent downstream activity (Morán Luengo et al. 2018).

Next to the conserved folding role, Hsp90 can delay GR activation by exploiting many co-chaperones. Most of co-chaperones do not contribute to general folding of substrates, instead they slow down Hsp90 folding machinery, with different clients having their own co-chaperones signature (Radli and Rüdiger 2018; Sahasrabudhe et al. 2017). Only the co-chaperones p23 exert a general role in folding of Hsp90 clients, slowing down the Hsp90 machinery to favor client late folding events (Sahasrabudhe et al. 2017). In the case of GR, the co-chaperones FKBP51 and XAP2 delay GR translocation in the nucleus and GR signaling (Laenger et al. 2009; Schulke et al. 2010). Thus, Hsp90 controls activity of GR, with co-chaperones delaying HPA-axis signaling in a client-specific manner.

21.1.3.3 Targeting Hsp90 to Treat Hormonal Imbalances

Can we use this molecular insight to use Hsp90 as a target to treat HPA-related disorders? Hsp90 is druggable and two Hsp90 inhibitors are currently in phase II clinical trials: the inhibitor AT13387 (Onalespib) tested to treat non-small cell lung cancers and lymphomas, and the inhibitor TAS-116 tested to treat advanced gastro-intestinal stromal tumors (Miyata et al. 2013; Woodhead et al. 2010; Yuno et al. 2018). Chronical targeting of Hsp90 may be detrimental for humans, given the central role of this molecule in physiological processes too (Schopf et al. 2017; Taipale et al. 2010). To this end, more effective drugs targeting Hsp90 are being developed, for instance by targeting Hsp90 without eliciting a general stress response (Calamini et al. 2010; Neef et al. 2010). Another strategy to increase specificity and decrease off-side effects would be to target co-chaperones. An inhibitor targeting the interface of Hsp90 in complex with the co-chaperone Aha1 has been recently developed, serving as proof of principle to rationally target co-chaperones instead of Hsp90 (Stiegler et al. 2017). The inhibitor developed alters Hsp90 activity stimulated by Aha1, without altering the intrinsic ATPase activity of Hsp90. Thus HPA-mediated hormonal imbalances can be potentially tackled by targeting the Hsp90 machinery.

GR is also druggable, with the drug Mifepristone used to treat Cushing disease (Johanssen and Allolio 2007) and to alleviate stress in mice (Lesuis et al. 2018). However, there are two good reasons to prefer the targeting of The Hsp90 machinery instead of GR itself. First, promising therapies developed for GR could be expanded to other steroid hormones receptors, all relying on the Hsp90 machinery (Sahasrabudhe et al. 2017). Second, co-chaperones could be targeted to enhance specificity, as different receptors are modulated by different sets of co-chaperones (Moffatt et al. 2008; Schulke et al. 2010). As a proof of principle, small molecules targeting FKBP51:GR:Hsp90 complex have been recently developed, with the aim of treating Post Thraumatic Stress Disorders caused by HPA imbalances (Sabbagh et al. 2018). To summarize, the Hsp90 machinery dictates GR activity, making the modulation of Hsp90-dependent GR assembly a valuable strategy to fight metabolic diseases and alleviate the effect of complex pathophysiological disorders (Fig. 21.2).

21.1.4 Hsp90 and Cancer

21.1.4.1 Hsp90 Sustains Kinase-Mediated Cancer

Kinases deregulation is a common landmark of cancer, with many oncogenes coding for mutated versions of kinases (Joerger and Fersht 2007). Kinases bind to Hsp90, with the molecular chaperone preferring thermally unstable ones (Luo et al. 2017; Taipale et al. 2012). This thermodynamic preference is a crucial aspect for the development of tumors: mutated kinases are often more unstable compared to their evolutionary optimized wild-type counterparts, making them stronger Hsp90 clients. Cancerous cells become then 'addicted' to Hsp90, and indeed they are more sensitive to Hsp90 inhibition than non-cancerous cells (Isaacs et al. 2003; Whitesell et al. 1994; Whitesell and Lindquist 2005). Hsp90, together with other chaperones, sustains cancerous proliferation by reassembling into large complexes, collectively termed epichaperome (Rodina et al. 2016). Thus, Hsp90 is a key player in many forms of cancers, comprising multiple myeloma (Khong and Spencer 2011), breast cancer and lung cancer (Butler et al. 2015). In the specific case of the Hsp90-client Cdk4, deregulation of the kinase causes liposarcoma, where Cdk4 copies expand at genetic level (Barretina et al. 2010; Italiano et al. 2009), or familial melanoma, where Cdk4 is mutated to a constitutively active form (Zuo et al. 1996).

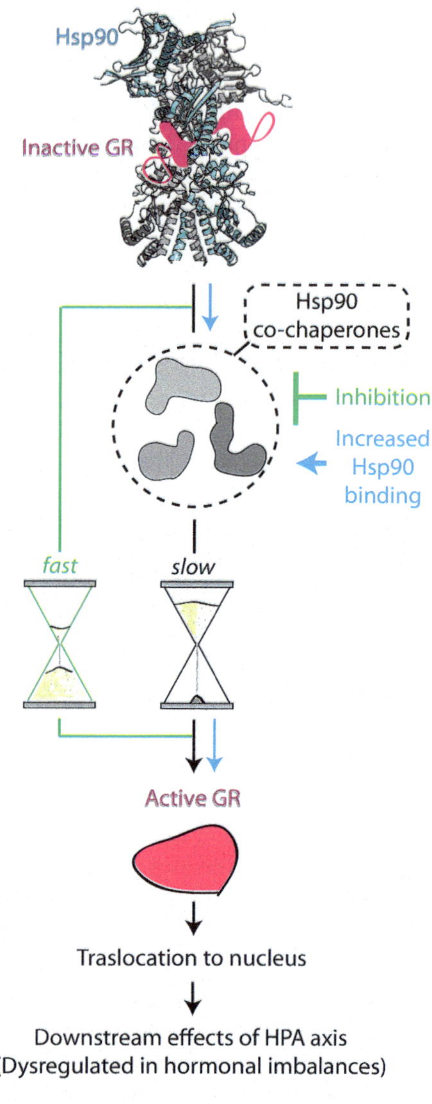

Fig. 21.2 Hsp90 controls hormone imbalances. Hsp90 homodimer is represented in blue and cyan, GR in violet. Hsp90 receives GR from the Hsp70 system (not shown) and is responsible for its downstream folding and activation. Active GR recognizes steroids and translocates to the nucleus to implement hormone-dependent signaling of the hypothalamic-pituitary-adrenal (HPA) axis. Hsp90 co-chaperones slow Hsp90 activity, delaying GR activation and signaling. Therapeutic intervention may either target co-chaperones to shortcut their inhibitory action and increase GR signaling (marked in green), or stabilize Hsp90 in complex with co-chaperones, to further delay GR signaling (marked in blue). Therapeutic intervention varies in accordance to the disease-dependent hypo- or hyper-activation of the HPA axis

21.1.4.2 Kinases Rely on the Hsp90 System for Folding and Activity

Hsp90 binding to kinases is assisted by the co-chaperones Cdc37. Together, they bind the highly conserved kinase domain, which is necessary and sufficient for a kinase to bind to the chaperon complex (Xu et al. 2001). A recent structure of the Hsp90:Cdk4:Cdc37 complex shows Hsp90 and Cdc37 stabilization of a folding client at near-atomic level (Verba et al. 2016). This structure shows how Cdc37 mimics structural features of the mature kinase, whilst Cdk4 is kept in a metastable

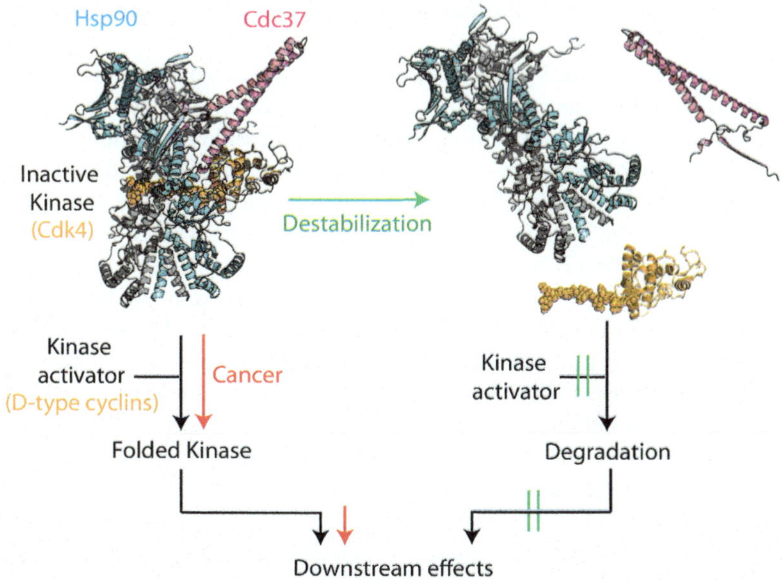

Fig. 21.3 Hsp90 stabilizes kinases promoting cancer. Hsp90 (homodimers in cyan and grey), in cooperation with Cdc37 (pink), folds a wide range of kinases, in this case Cdk4 (orange). Hsp90 and Cdc37 stabilize a partially folded kinase structure, with β-strands 4 and 5 of the kinase unravelled (atoms represented as spheres). Partially folded kinases can then react with downstream activators (e.g. enzymes modifying the kinase post translationally, such as D-type cyclins in the case of Cdk4) and in turn exert their downstream functions. Many kinases are hyper-activated in cancer and sustain growth and activity of tumor cells (extra red arrows). Therapeutic intervention (marked in green) may disrupt the Hsp90:Cdc37:kinase complex to favor kinase degradation by the proteasome system and inhibit its interaction with kinase activator and consequent downstream effects

condition. Cdc37 splits the kinases lobes at the N- and C-termini, and 'opens' the kinase by complete unfolding of the β-strands 4 and 5 (Verba et al. 2016) (Fig. 21.3). Cdc37 opening of the kinases is necessary for consequent binding of Hsp90, as shown for Fibroblast Growth Factor Receptor 3, a kinase involved in bladder cancer (Bunney et al. 2018; McConkey and Choi 2018). Stabilizing a Cdk4 folding intermediate may favor its transition between active or inactive states and accessibility to post translational modifications (Miyashita et al. 2003; Verba and Agard 2017). Indeed, Hsp90:Cdc37 binds to Cdk4, and the complex dissociates upon phosphorylation of the kinase by D-type cyclins (Stepanova et al. 1996). This highlights the evolutionary acquired function of Hsp90 as a fine-tuner of the biological activity of proteins. In the case of kinases, this function may have evolved to add an extra system to further control their activity, as kinase de-regulation is dangerously linked to cancer.

21.1.4.3 Destabilizing Hsp90:Kinase Complexes to Treat Cancer

Targeting the Hsp90:Cdc37:Cdk4 complex may prove an interesting strategy to treat Cdk4-dependent cancers. For instance, in the case of increased copies of Cdk4 causing liposarcoma, destabilization of the complex would be a valuable strategy to reduce the amount of available Cdk4, exposing Cd4 to the degradation system. As a proof of principle, targeting of the Hsp90:Cdc37:kinase complex with Whitaferin has shown promising results in the case of the kinase LRRK2, a protein linked to neurodegeneration (Narayan et al. 2015). In this case, targeting the complex reduced kinase activity, increased kinase clearance and alleviated neuronal death (Fig. 21.3). This strategy can be extended to other kinases dependent on Hsp90 and Cdc37 (Li et al. 2018; Zhang et al. 2008), potentially to the whole human kinome, as kinase dependency on Hsp90:Cdc37 is thought to be a continuum of affinities, with no clear cut between binders and non-binders (Verba and Agard 2017). Targeting the chaperone complex instead of the kinase itself may facilitate translation of therapies among different cancers caused by different Hsp90 clients.

Hsp90 targeting to treat cancer is a strategy currently under investigation: a part from the previously mentioned phase II clinical trials to treat non-small cell lung cancers lymphomas and gastrointestinal stromal tumors, there are 13 ongoing phase I clinical trials to treat chronic lymphocytic leukemia, neuroendocrine tumours, refractory hematological malignancies, solid tumors, unresectable BRAF mutated stage III/IV melanoma, advanced squamous cell carcinoma of the head and neck, advanced triple negative breast cancer, non-Hodgkin's lymphoma and myeloma (Yuno et al. 2018). Kinase deregulation is involved also in neurodegeneration, as we saw in the case of LRRK2, but plays an outstanding role in proliferation of certain viruses and parasites, capable of hijacking host Hsp90 to stabilize their own kinases (Woodford et al. 2016). This is true for instance in the case of Rous Sarcome Virus, whose transforming protein was one of the first Hsp90 clients discovered (Brugge and Darrow 1982). Thus, mastering the Hsp90:Cdc37 action on kinases may prove useful to treat a broad spectrum of diseases.

21.1.5 Hsp90 and Neurodegeneration

21.1.5.1 Protein Aggregation, Hallmark of Neurodegeneration

Hsp90 interaction with a client may also dictate its end by initiating degradation via the ubiquitin-proteasome system (Kundrat and Regan 2010). When clearance fails to be performant, proteins start to aggregate in cells. Protein aggregation is a common hallmark of many diseases, playing an outstanding role in neurodegeneration (Hartl 2017; Stoecklin and Bukau 2013). Protein aggregates may establish abnormal, toxic interactions in the cell and impair the whole protein homeostasis network (Gidalevitz et al. 2006; Hipp et al. 2014). Moreover, aggregates cannot be degraded anymore by the ubiquitin-proteasome system, further exacerbating their accumulation and their toxic effects (Bence et al. 2001). The aggregation of the Hsp90-client Tau is linked to Alzheimer's disease and other Tauopathies, comprising progressive supranuclear palsy, Pick's disease and chronic traumatic encephalopathy (Goedert et al. 2017; Lee et al. 2001; Wang and Mandelkow 2016). Hsp90 is responsible for Tau clearance, and Tau accumulation is linked to neurodegeneration, drawing a connection between the chaperone machinery and dementia.

21.1.5.2 Hsp90 Regulates Tau Turnover and Aggregation

The Hsp90:Tau complex reveals that Hsp90-N and Hsp90-M domains bind to Tau aggregation-prone regions (Karagöz et al. 2014). This interaction may inhibit Tau aggregation itself: recent data show that chaperones have different propensities to inhibit Tau aggregation, however Hsp90 activity was not tested (Mok et al. 2018). Aside from this potential buffering role, the Hsp90 machinery plays an established role in degrading Tau by bridging it to the ubiquitin-proteasome system. Many co-chaperones, with or without TPR domains, regulate Tau degradation rates (Shelton et al. 2017b). As for GR, CHIP can compete with Hop for the binding to Hsp90, dictating in this case whether Tau is degraded or linked to the Hsp70 system and kept soluble (Dickey et al. 2007; Jinwal et al. 2013; Karagöz et al. 2014; Thompson et al. 2012). FKBP51, another TPR co-chaperone, delays Tau degradation by the ubiquitin-proteasome system and favors Tau aggregation (Jinwal et al. 2010). A recent structure of Hsp90 in complex with Tau and FKBP51 shows that the co-chaperone proline isomerization domain is placed in close proximity to Tau proline-rich region, possibly favoring abnormal isomerization linked to neurodegeneration (Oroz et al. 2018) (Fig. 21.4). Aha1, a non-TPR co-chaperone, also increases Tau aggregation rates, and interestingly both Aha1 and FKBP51 levels increase in brains affected by Alzheimer's Disease (Brehme et al. 2014; Shelton et al. 2017a). As for GR and kinases, the Hsp90 machinery is a major modulator of Tau biology, and in contrast to GR and kinases Hsp90 does not influence ligand binding but rather degradation rates.

Targeting Hsp90 to Alleviate Neurodegeneration

The importance of the Hsp90 system in Tau biology and the lack of effective drugs to treat neurodegeneration make Hsp90-mediated Tau clearance an interesting approach to treat Alzheimer and other Tauopathies (Blair et al. 2014). The ratio is that reducing Tau levels may be a valuable approach to alleviate dementia, as shown in genetics studies in mice and by exploiting anti-sense nucleotides and short interference RNAs to lower Tau protein levels (DeVos and Miller 2013; Roberson et al. 2007; Wolfe 2014) (Fig. 21.4). The advantage of small chemicals over nucleotide-based therapies lies in their general ability to cross blood brain barrier more easily, a crucial factor for the designing of drugs for the central nervous system (Patel and Patel 2017). In this light, a recent study showed that chronically treating a cohort of

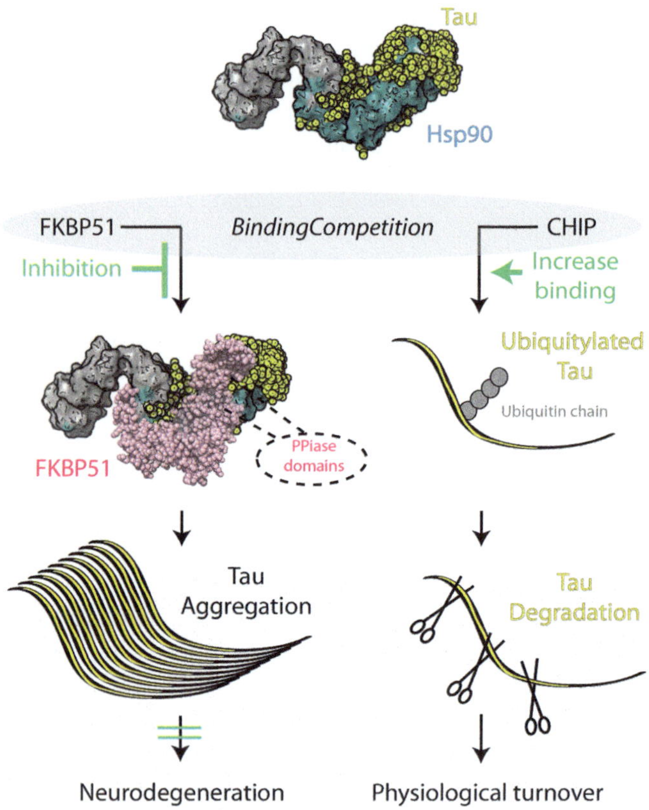

Fig. 21.4 Hsp90 control of protein aggregation in neurodegeneration. Hsp90 (homodimers in cyan and grey), in cooperation CHIP, promotes Tau physiological turnover by the proteasome (right branch of scheme). FKBP51 competes with CHIP for binding to Hsp90:Tau, reducing the amount of degraded Tau, increasing Tau soluble pool and possibly isomerizing Tau prolines via its PPIase domain (peptidyl-prolyl-cis/trans-isomerase domain), favoring overall Tau aggregation. Chaperone-based therapeutics (marked in green) shall either boost Tau turnover or disfavor interactions between Hsp90 and FKBP51 to tip the balance against protein aggregation and consequent neurodegeneration

patients with an inhibitor of FKBPs decreased the incidence of Alzheimer (Taglialatela et al. 2015), thus paving the way to possible treatments focused on the Hsp90 machinery. Other strategies to tackle neurodegeneration via the chaperon system could be also exploited. Recent studies show that the co-chaperone Cyp40 and the chaperone Hsc70 can disaggregate Tau and α-synuclein fibrils (Baker et al. 2017; Ferrari et al. 2018; Gao et al. 2015). These studies collectively imply that the chaperone machinery can be tackled from different angles to treat protein aggregation and neurodegeneration.

Hsp90 binds to other aggregation-prone proteins, such as huntingtin, α-synuclein and transthyretin, whose aggregation processes are linked respectively to Chorea Huntington, Parkinson's disease and transthyretin amyloidosis (Inda et al. 2016;

Oroz et al. 2017; Pratt et al. 2015). Thus, most of the neurodegenerative disorders rely on the Hsp90 system (Lackie et al. 2017), and this can have implications for the design of medical treatments too, as therapies targeting the Hsp90 system developed for one protein could be translated to other neurodegenerative disorders. Of note, protein aggregation can also cause certain forms of cancer, as shown for the signaling protein Axin (Anvarian et al. 2016). Thus, chaperone-based treatments against protein aggregates may cross the borders of neurodegenerative disorders and treat other types of diseases. The Hsp90 core machinery is a master regulator of protein activity and it is a versatile target to treat disease.

21.2 Conclusions

The current challenge is to bridge the molecular information obtained about Hsp90 to the pathology at the system level, with the ultimate goal of developing effective treatments. Hsp90 is a ubiquitous protein, and three crucial aspects of its machinery can allow us to treat specific sets of diseases: (i) There are more than 40 co-chaperones, allowing for targeting of unique interaction surfaces established within clients and chaperone system. (ii) As different co-chaperones add different flavors of regulational power to Hsp90, we can pinpoint one or more co-chaperones that will contribute more to a certain disease, increasing specificity. (iii) Pathological statuses are often ‘addicted’ to Hsp90 buffering role at the cellular level, thus pathological cells are often more vulnerable than physiological ones, allowing the latter to have higher fitness upon Hsp90 treatments. We can envision that a complete understanding of Hsp90 fine tuning of clients – in terms of co-chaperones, folding outcome and downstream effects – will allow us to target an even wider range of disease. Thus, the current challenge is to master combinatoriality of the Hsp90 machinery, *i.e.* to understand the output effect on a client given the thousands of possible Hsp90:co-chaperones different complexes that can assemble in a cell. The knowledge we accumulated for the main co-chaperones, such as Cdc37, p23, CHIP, Hop, must embrace also less studied co-chaperones. The combinations of effective complexes need to take into account sub-cellular compartmentalization, stress-inducible isoforms, availability of co-chaperones and post translational modifications, all influencing the assembly of the Hsp90 machinery when dealing with a specific client (Cox and Johnson 2018; Mayer and Le Breton 2015; Mollapour et al. 2014). Hsp90 buffering of protein folds is reflected not only on physiology and pathology at the system level, but on a broader scale when whole populations are considered. Indeed, Hsp90 is known to hide cryptic variations of proteins within a population, allowing these variations to reveal in time of stress, to boost adaptability (Rutherford and Lindquist 1998). The same principles applies to diseases: in the case of Fanconi anemia, Hsp90 buffers a group of protein mutants linked to mild phenotypes, which worsen upon Hsp90 inhibition (Karras et al. 2017). These findings imply that within the same disease some patients may react better to Hsp90 treatments than others. This is an important consideration when placed in the context of precision medicine, namely the designing of treatments based on the personal network of interactions and disease-variants of each patient. True mastering of Hsp90 biology will be a powerful weapon relevant for fatal diseases.

Acknowledgements We thank Magdalena Wawrzyniuk for help with PyMol. The work of the authors was supported by Marie-Curie Actions of the 7th Framework programme of the EU [Innovative Doctoral Programme "ManiFold" (No. 317371) and Initial Training Network "WntsApp" (No. 608180)], the Internationale Stichting Alzheimer Onderzoek (ISAO; project "Chaperoning Tau Aggregation"; No. 14542) and a ZonMW TOP grant ("Chaperoning Axonal Transport in neurodegenerative disease"; No. 91215084).

References

Anvarian Z, Nojima H, van Kappel EC, Madl T, Spit M, Viertler M, Jordens I, Low TY, van Scherpenzeel RC, Kuper I et al (2016) Axin cancer mutants form nanoaggregates to rewire the Wnt signaling network. Nat Struct Mol Biol 23:324–332

Assimon VA, Southworth DR, Gestwicki JE (2015) Specific binding of tetratricopeptide repeat proteins to heat shock protein 70 (Hsp70) and heat shock protein 90 (Hsp90) is regulated by affinity and phosphorylation. Biochemistry 54:7120–7131

Baker JD, Shelton LB, Zheng D, Favretto F, Nordhues BA, Darling A, Sullivan LE, Sun Z, Solanki PK, Martin MD et al (2017) Human cyclophilin 40 unravels neurotoxic amyloids. PLoS Biol 15:e2001336

Barretina J, Taylor BS, Banerji S, Ramos AH, Lagos-Quintana M, Decarolis PL, Shah K, Socci ND, Weir BA, Ho A et al (2010) Subtype-specific genomic alterations define new targets for soft-tissue sarcoma therapy. Nat Genet 42:715–721

Bence NF, Sampat RM, Kopito RR (2001) Impairment of the ubiquitin-proteasome system by protein aggregation. Science 292:1552–1555

Blair LJ, Nordhues BA, Hill SE, Scaglione KM, O'Leary JC 3rd, Fontaine SN, Breydo L, Zhang B, Li P, Wang L et al (2013) Accelerated neurodegeneration through chaperone-mediated oligomerization of tau. J Clin Invest 123:4158–4169

Blair LJ, Sabbagh JJ, Dickey CA (2014) Targeting Hsp90 and its co-chaperones to treat Alzheimer's disease. Expert Opin Ther Targets 18(10):1219–1232

Borkovich KA, Farrelly FW, Finkelstein DB, Taulien J, Lindquist S (1989) Hsp82 is an essential protein that is required in higher concentrations for growth of cells at higher temperatures. Mol Cell Biol 9:3919–3930

Brehme M, Voisine C, Rolland T, Wachi S, Soper JH, Zhu Y, Orton K, Villella A, Garza D, Vidal M, Ge H, Morimoto RI (2014) A chaperome subnetwork safeguards proteostasis in aging and neurodegenerative disease. Cell Rep 9:1135–1150

Brugge JS, Darrow D (1982) Rous sarcoma virus-induced phosphorylation of a 50,000-molecular weight cellular protein. Nature 295:250–253

Bunney TD, Inglis AJ, Sanfelice D, Farrell B, Kerr CJ, Thompson GS, Masson GR, Thiyagarajan N, Svergun DI, Williams RL, Breeze AL, Katan M (2018) Disease variants of FGFR3 reveal molecular basis for the recognition and additional roles for Cdc37 in Hsp90 chaperone system. Structure 26:446–458.e8

Butler LM, Ferraldeschi R, Armstrong HK, Centenera MM, Workman P (2015) Maximizing the therapeutic potential of Hsp90 inhibitors. Mol Cancer Res 13:1445–1451

Calamini B, Silva MC, Madoux F, Hutt DM, Khanna S, Chalfant MA, Allais C, Ouizem S, Saldanha SA, Ferguson J et al (2010) ML346: a novel modulator of proteostasis for protein conformational diseases. In: Probe reports from the NIH molecular libraries program. National Center for Biotechnology Information, Bethesda

Caron NS, Dorsey ER, Hayden MR (2018) Therapeutic approaches to Huntington disease: from the bench to the clinic. Nat Rev Drug Discov 17:729–750
Chrousos GP (2000) The role of stress and the hypothalamic-pituitary-adrenal axis in the pathogenesis of the metabolic syndrome: neuro-endocrine and target tissue-related causes. Int J Obes Relat Metab Disord 24(Suppl 2):S50–S55
Chrousos GP, Kino T (2009) Glucocorticoid signaling in the cell. Expanding clinical implications to complex human behavioral and somatic disorders. Ann N Y Acad Sci 1179:153–166
Cox MB, Johnson JL (2018) Evidence for Hsp90 co-chaperones in regulating Hsp90 function and promoting client protein folding. Methods Mol Biol 1709:397–422
DeVos SL, Miller TM (2013) Antisense oligonucleotides: treating neurodegeneration at the level of RNA. Neurotherapeutics 10:486–497
Dickey CA, Kamal A, Lundgren K, Klosak N, Bailey RM, Dunmore J, Ash P, Shoraka S, Zlatkovic J, Eckman CB et al (2007) The high-affinity HSP90-CHIP complex recognizes and selectively degrades phosphorylated tau client proteins. J Clin Invest 117:648–658
Du X, Pang TY (2015) Is dysregulation of the HPA-Axis a core pathophysiology mediating co-morbid depression in neurodegenerative diseases? Front Psych 6:32
Edkins AL (2015) CHIP: a co-chaperone for degradation by the proteasome. Subcell Biochem 78:219–242
Ellis J (1987) Proteins as molecular chaperones. Nature 328:378–379
Ellis RJ (2013) Assembly chaperones: a perspective. Philos Trans R Soc Lond Ser B Biol Sci 368:20110398
Felts SJ, Owen BA, Nguyen P, Trepel J, Donner DB, Toft DO (2000) The hsp90-related protein TRAP1 is a mitochondrial protein with distinct functional properties. J Biol Chem 275:3305–3312
Ferrari L, Geerts WJC, van Wezel M, Kos R, Konstantoulea A, van Bezouwen LS, Förster FG, Rüdiger SGD (2018) Human chaperones untangle fibrils of the Alzheimer protein Tau. Biorxiv https://www.biorxiv.org/content/10.1101/426650v1.article-info
Gao X, Carroni M, Nussbaum-Krammer C, Mogk A, Nillegoda NB, Szlachcic A, Guilbride DL, Saibil HR, Mayer MP, Bukau B (2015) Human Hsp70 disaggregase reverses Parkinson's-linked alpha-synuclein amyloid fibrils. Mol Cell 59:781–793
Genest O, Hoskins JR, Camberg JL, Doyle SM, Wickner S (2011) Heat shock protein 90 from *Escherichia coli* collaborates with the DnaK chaperone system in client protein remodeling. Proc Natl Acad Sci U S A 108:8206–8211
Gidalevitz T, Ben-Zvi A, Ho KH, Brignull HR, Morimoto RI (2006) Progressive disruption of cellular protein folding in models of polyglutamine diseases. Science 311:1471–1474
Goedert M, Eisenberg DS, Crowther RA (2017) Propagation of Tau aggregates and neurodegeneration. Annu Rev Neurosci 40:189–210
Harst A, Lin H, Obermann WM (2005) Aha1 competes with Hop, p50 and p23 for binding to the molecular chaperone Hsp90 and contributes to kinase and hormone receptor activation. Biochem J 387:789–796
Hartl FU (2017) Protein misfolding diseases. Annu Rev Biochem 86:21–26
Hipp MS, Park SH, Hartl FU (2014) Proteostasis impairment in protein-misfolding and -aggregation diseases. Trends Cell Biol 24:506–514
Ihrig V, Obermann WMJ (2017) Identifying inhibitors of the Hsp90-Aha1 protein complex, a potential target to drug cystic fibrosis, by Alpha Technology. SLAS Discov 22:923–928
Inda C, Bolaender A, Wang T, Gandu SR, Koren J 3rd (2016) Stressing out Hsp90 in neurotoxic Proteinopathies. Curr Top Med Chem 16:2829–2838
Isaacs JS, Xu W, Neckers L (2003) Heat shock protein 90 as a molecular target for cancer therapeutics. Cancer Cell 3:213–217
Italiano A, Bianchini L, Gjernes E, Keslair F, Ranchere-Vince D, Dumollard JM, Haudebourg J, Leroux A, Mainguene C, Terrier P et al (2009) Clinical and biological significance of CDK4 amplification in well-differentiated and dedifferentiated liposarcomas. Clin Cancer Res 15:5696–5703

Jhaveri K, Ochiana SO, Dunphy MP, Gerecitano JF, Corben AD, Peter RI, Janjigian YY, Gomes-DaGama EM, Koren J 3rd, Modi S, Chiosis G (2014) Heat shock protein 90 inhibitors in the treatment of cancer: current status and future directions. Expert Opin Investig Drugs 23:611–628

Jinwal UK, Koren J 3rd, Borysov SI, Schmid AB, Abisambra JF, Blair LJ, Johnson AG, Jones JR, Shults CL, O'Leary JC 3rd et al (2010) The Hsp90 cochaperone, FKBP51, increases Tau stability and polymerizes microtubules. J Neurosci 30:591–599

Jinwal UK, Koren J 3rd, Dickey CA (2013) Reconstructing the Hsp90/Tau machine. Curr Enzym Inhib 9:41–45

Joerger AC, Fersht AR (2007) Structural biology of the tumor suppressor p53 and cancer-associated mutants. Adv Cancer Res 97:1–23

Johanssen S, Allolio B (2007) Mifepristone (RU 486) in Cushing's syndrome. Eur J Endocrinol 157:561–569

Karagöz GE, Rüdiger SGD (2015) Hsp90 interaction with clients. Trends Biochem Sci 40:117–125

Karagöz GE, Duarte AMS, Akoury E, Ippel H, Biernat J, Morán Luengo T, Radli M, Didenko T, Nordhues BA, Veprintsev DB et al (2014) Hsp90-Tau complex reveals molecular basis for specificity in chaperone action. Cell 156:963–974

Karras GI, Yi S, Sahni N, Fischer M, Xie J, Vidal M, D'Andrea AD, Whitesell L, Lindquist S (2017) HSP90 shapes the consequences of human genetic variation. Cell 168:856–866.e12

Khong T, Spencer A (2011) Targeting HSP 90 induces apoptosis and inhibits critical survival and proliferation pathways in multiple myeloma. Mol Cancer Ther 10:1909–1917

Kirschke E, Goswami D, Southworth D, Griffin PR, Agard DA (2014) Glucocorticoid receptor function regulated by coordinated action of the Hsp90 and Hsp70 chaperone cycles. Cell 157:1685–1697

Klaips CL, Jayaraj GG, Hartl FU (2018) Pathways of cellular proteostasis in aging and disease. J Cell Biol 217:51–63

Kundrat L, Regan L (2010) Balance between folding and degradation for Hsp90-dependent client proteins: A key role for CHIP. Biochemistry (N. Y.) 49:7428–7438

Labbadia J, Morimoto RI (2015) The biology of proteostasis in aging and disease. Annu Rev Biochem 84:435–464

Lackie RE, Maciejewski A, Ostapchenko VG, Marques-Lopes J, Choy WY, Duennwald ML, Prado VF, Prado MAM (2017) The Hsp70/Hsp90 chaperone machinery in neurodegenerative diseases. Front Neurosci 11:254

Laenger A, Lang-Rollin I, Kozany C, Zschocke J, Zimmermann N, Ruegg J, Holsboer F, Hausch F, Rein T (2009) XAP2 inhibits glucocorticoid receptor activity in mammalian cells. FEBS Lett 583:1493–1498

Lee VM, Goedert M, Trojanowski JQ (2001) Neurodegenerative tauopathies. Annu Rev Neurosci 24:1121–1159

Lesuis SL, Weggen S, Baches S, Lucassen PJ, Krugers HJ (2018) Targeting glucocorticoid receptors prevents the effects of early life stress on amyloid pathology and cognitive performance in APP/PS1 mice. Transl Psychiatry 8:53-018–0101-2

Li T, Jiang HL, Tong YG, Lu JJ (2018) Targeting the Hsp90-Cdc37-client protein interaction to disrupt Hsp90 chaperone machinery. J Hematol Oncol 11:59-018–0602-8

Luo Q, Boczek EE, Wang Q, Buchner J, Kaila VR (2017) Hsp90 dependence of a kinase is determined by its conformational landscape. Sci Rep 7:43996

Mayer MP, Le Breton L (2015) Hsp90: breaking the symmetry. Mol Cell 58:8–20

McConkey DJ, Choi W (2018) Molecular subtypes of bladder cancer. Curr Oncol Rep 20:77-018–0727-5

Melnick J, Aviel S, Argon Y (1992) The endoplasmic reticulum stress protein GRP94, in addition to BiP, associates with unassembled immunoglobulin chains. J Biol Chem 267:21303–21306

Miyashita O, Onuchic JN, Wolynes PG (2003) Nonlinear elasticity, proteinquakes, and the energy landscapes of functional transitions in proteins. Proc Natl Acad Sci U S A 100:12570–12575

Miyata Y, Nakamoto H, Neckers L (2013) The therapeutic target Hsp90 and cancer hallmarks. Curr Pharm Des 19:347–365
Moffatt NSC, Bruinsma E, Uhl C, Obermann WMJ, Toft D (2008) Role of the cochaperone Tpr2 in Hsp90 chaperoning. Biochemistry (N. Y.) 47:8203–8213
Mok SA, Condello C, Freilich R, Gillies A, Arhar T, Oroz J, Kadavath H, Julien O, Assimon VA, Rauch JN et al (2018) Mapping interactions with the chaperone network reveals factors that protect against tau aggregation. Nat Struct Mol Biol 25:384–393
Mollapour M, Bourboulia D, Beebe K, Woodford MR, Polier S, Hoang A, Chelluri R, Li Y, Guo A, Lee MJ et al (2014) Asymmetric Hsp90 N domain SUMOylation recruits Aha1 and ATP-competitive inhibitors. Mol Cell 53:317–329
Morán Luengo T, Kityk R, Mayer MP, Rüdiger SGD (2018) Hsp90 breaks the deadlock of the Hsp70 chaperone system. Mol Cell 70:545–552
Morán Luengo T, Mayer MP, Rüdiger SGD (2019) The Hsp70-Hsp90 chaperone cascade in protein folding. Trends Cell Biol 29-2:164–177
Narayan M, Zhang J, Braswell K, Gibson C, Zitnyar A, Lee DC, Varghese-Gupta S, Jinwal UK (2015) Withaferin A regulates LRRK2 levels by interfering with the Hsp90- Cdc37 chaperone complex. Curr Aging Sci 8:259–265
Neef DW, Turski ML, Thiele DJ (2010) Modulation of heat shock transcription factor 1 as a therapeutic target for small molecule intervention in neurodegenerative disease. PLoS Biol 8:e1000291
Oroz J, Kim JH, Chang BJ, Zweckstetter M (2017) Mechanistic basis for the recognition of a misfolded protein by the molecular chaperone Hsp90. Nat Struct Mol Biol 24:407–413
Oroz J, Chang BJ, Wysoczanski P, Lee CT, Perez-Lara A, Chakraborty P, Hofele RV, Baker JD, Blair LJ, Biernat J et al (2018) Structure and pro-toxic mechanism of the human Hsp90/PPIase/Tau complex. Nat Commun 9:4532-018–406880-0
Pan T, Peng Z, Tan L, Zou F, Zhou N, Liu B, Liang L, Chen C, Liu J, Wu L et al (2018) Non-steroidal anti-inflammatory drugs (NSAIDs) potently inhibit the replication of Zika viruses by inducing the degradation of AXL. J Virol
Patel MM, Patel BM (2017) Crossing the blood-brain barrier: recent advances in drug delivery to the brain. CNS Drugs 31:109–133
Patel R, Williams-Dautovich J, Cummins CL (2014) Minireview: new molecular mediators of glucocorticoid receptor activity in metabolic tissues. Mol Endocrinol 28:999–1011
Pei Y, Wang C, Yan SF, Liu G (2017) Past, current, and future developments of therapeutic agents for treatment of chronic hepatitis B virus infection. J Med Chem 60:6461–6479
Picard D (2002) Heat-shock protein 90, a chaperone for folding and regulation. Cell Mol Life Sci 59:1640–1648
Picard D, Khursheed B, Garabedian MJ, Fortin MG, Lindquist S, Yamamoto KR (1990) Reduced levels of hsp90 compromise steroid receptor action in vivo. Nature 348:166–168
Pratt WB, Gestwicki JE, Osawa Y, Lieberman AP (2015) Targeting Hsp90/Hsp70-based protein quality control for treatment of adult onset neurodegenerative diseases. Annu Rev Pharmacol Toxicol 55:353–371
Radli M, Rüdiger SGD (2018) Dancing with the diva: Hsp90-client interactions. J Mol Biol
Ranek MJ, Stachowski MJ, Kirk JA, Willis MS (2018) The role of heat shock proteins and co-chaperones in heart failure. Philos Trans R Soc Lond Ser B Biol Sci 373 https://doi.org/10.1098/rstb.2016.0530
Ritossa F (1962) A new puffing pattern induced by temperature shock and DNP in drosophila. Experientia 18:571–573
Roberson ED, Scearce-Levie K, Palop JJ, Yan F, Cheng IH, Wu T, Gerstein H, Yu GQ, Mucke L (2007) Reducing endogenous tau ameliorates amyloid beta-induced deficits in an Alzheimer's disease mouse model. Science 316:750–754
Rodina A, Wang T, Yan P, Gomes ED, Dunphy MP, Pillarsetty N, Koren J, Gerecitano JF, Taldone T, Zong H et al (2016) The epichaperome is an integrated chaperome network that facilitates tumour survival. Nature 538:397–401

Rutherford SL, Lindquist S (1998) Hsp90 as a capacitor for morphological evolution. Nature 396:336–342
Sabbagh JJ, Cordova RA, Zheng D, Criado-Marrero M, Lemus A, Li P, Baker JD, Nordhues BA, Darling AL, Martinez-Licha C et al (2018) Targeting the FKBP51/GR/Hsp90 complex to identify functionally relevant treatments for depression and PTSD. ACS Chem Biol 13:2288–2299
Sahasrabudhe P, Rohrberg J, Biebl MM, Rutz DA, Buchner J (2017) The plasticity of the Hsp90 co-chaperone system. Mol Cell 67:947–961.e5
Sanchez ER, Meshinchi S, Tienrungroj W, Schlesinger MJ, Toft DO, Pratt WB (1987) Relationship of the 90-kDa murine heat shock protein to the untransformed and transformed states of the L cell glucocorticoid receptor. J Biol Chem 262:6986–6991
Scheufler C, Brinker A, Bourenkov G, Pegoraro S, Moroder L, Bartunik H, Hartl FU, Moarefi I (2000) Structure of TPR domain-peptide complexes: critical elements in the assembly of the Hsp70-Hsp90 multichaperone machine. Cell 101:199–210
Schopf FH, Biebl MM, Buchner J (2017) The HSP90 chaperone machinery. Nat Rev Mol Cell Biol 18:345–360
Schulke JP, Wochnik GM, Lang-Rollin I, Gassen NC, Knapp RT, Berning B, Yassouridis A, Rein T (2010) Differential impact of tetratricopeptide repeat proteins on the steroid hormone receptors. PLoS One 5:e11717
Schumacher RJ, Hansen WJ, Freeman BC, Alnemri E, Litwack G, Toft DO (1996) Cooperative action of Hsp70, Hsp90, and DnaJ proteins in protein renaturation. Biochemistry (N Y) 35:14889–14898
Shelton LB, Baker JD, Zheng D, Sullivan LE, Solanki PK, Webster JM, Sun Z, Sabbagh JJ, Nordhues BA, Koren J 3rd et al (2017a) Hsp90 activator Aha1 drives production of pathological tau aggregates. Proc Natl Acad Sci U S A 114:9707–9712
Shelton LB, Koren J 3rd, Blair LJ (2017b) Imbalances in the Hsp90 chaperone machinery: implications for Tauopathies. Front Neurosci 11:724
Stepanova L, Leng X, Parker SB, Harper JW (1996) Mammalian p50Cdc37 is a protein kinase-targeting subunit of Hsp90 that binds and stabilizes Cdk4. Genes Dev 10:1491–1502
Stiegler SC, Rubbelke M, Korotkov VS, Weiwad M, John C, Fischer G, Sieber SA, Sattler M, Buchner J (2017) A chemical compound inhibiting the Aha1-Hsp90 chaperone complex. J Biol Chem 292:17073–17083
Stoecklin G, Bukau B (2013) Telling right from wrong in life – cellular quality control. Nat Rev Mol Cell Biol 14:613–615
Taglialatela G, Rastellini C, Cicalese L (2015) Reduced incidence of dementia in solid organ transplant patients treated with calcineurin inhibitors. J Alzheimers Dis 47:329–333
Taipale M, Jarosz DF, Lindquist S (2010) HSP90 at the hub of protein homeostasis: emerging mechanistic insights. Nat Rev Mol Cell Biol 11:515–528
Taipale M, Krykbaeva I, Koeva M, Kayatekin C, Westover KD, Karras GI, Lindquist S (2012) Quantitative analysis of HSP90-client interactions reveals principles of substrate recognition. Cell 150:987–1001
Taipale M, Tucker G, Peng J, Krykbaeva I, Lin ZY, Larsen B, Choi H, Berger B, Gingras AC, Lindquist S (2014) A quantitative chaperone interaction network reveals the architecture of cellular protein homeostasis pathways. Cell 158:434–448
Thompson AD, Scaglione KM, Prensner J, Gillies AT, Chinnaiyan A, Paulson HL, Jinwal UK, Dickey CA, Gestwicki JE (2012) Analysis of the tau-associated proteome reveals that exchange of Hsp70 for Hsp90 is involved in tau degradation. ACS Chem Biol 7:1677–1686
van Oosten-Hawle P, Morimoto RI (2014) Organismal proteostasis: role of cell-nonautonomous regulation and transcellular chaperone signaling. Genes Dev 28:1533–1543
Verba KA, Agard DA (2017) How Hsp90 and Cdc37 lubricate kinase molecular switches. Trends Biochem Sci
Verba KA, Wang RY, Arakawa A, Liu Y, Shirouzu M, Yokoyama S, Agard DA (2016) Atomic structure of Hsp90-Cdc37-Cdk4 reveals that Hsp90 traps and stabilizes an unfolded kinase. Science 352:1542–1547

Voss AK, Thomas T, Gruss P (2000) Mice lacking HSP90beta fail to develop a placental labyrinth. Development 127:1–11
Wang Y, Mandelkow E (2016) Tau in physiology and pathology. Nat Rev Neurosci 17:5–21
Weber BH (2010) What is life? Defining life in the context of emergent complexity. Orig Life Evol Biosph 40:221–229
Wegele H, Wandinger SK, Schmid AB, Reinstein J, Buchner J (2006) Substrate transfer from the chaperone Hsp70 to Hsp90. J Mol Biol 356:802–811
Whitesell L, Lindquist SL (2005) HSP90 and the chaperoning of cancer. Nat Rev Cancer 5:761–772
Whitesell L, Mimnaugh EG, De Costa B, Myers CE, Neckers LM (1994) Inhibition of heat shock protein HSP90-pp60v-src heteroprotein complex formation by benzoquinone ansamycins: essential role for stress proteins in oncogenic transformation. Proc Natl Acad Sci U S A 91:8324–8328
Wolfe MS (2014) Targeting mRNA for Alzheimer's and related dementias. Scientifica (Cairo) 2014:757549
Woodford MR, Dunn DM, Ciciarelli JG, Beebe K, Neckers L, Mollapour M (2016) Targeting Hsp90 in non-cancerous maladies. Curr Top Med Chem 16:2792–2804
Woodhead AJ, Angove H, Carr MG, Chessari G, Congreve M, Coyle JE, Cosme J, Graham B, Day PJ, Downham R et al (2010) Discovery of (2,4-Dihydroxy-5-isopropylphenyl)-[5-(4-methylpiperazin-1-ylmethyl)-1,3-dihydroisoindol-2-yl]methanone (AT13387), a novel inhibitor of the molecular chaperone Hsp90 by fragment based drug design. J Med Chem 53:5956–5969
Wust S, Van Rossum EF, Federenko IS, Koper JW, Kumsta R, Hellhammer DH (2004) Common polymorphisms in the glucocorticoid receptor gene are associated with adrenocortical responses to psychosocial stress. J Clin Endocrinol Metab 89:565–573
Xu W, Mimnaugh E, Rosser MF, Nicchitta C, Marcu M, Yarden Y, Neckers L (2001) Sensitivity of mature Erbb2 to geldanamycin is conferred by its kinase domain and is mediated by the chaperone protein Hsp90. J Biol Chem 276:3702–3708
Young K, Singh G (2018) Biological mechanisms of cancer-induced depression. Front Psych 9:299
Yuno A, Lee MJ, Lee S, Tomita Y, Rekhtman D, Moore B, Trepel JB (2018) Clinical evaluation and biomarker profiling of Hsp90 inhibitors. Methods Mol Biol 1709:423–441
Zhang M, Windheim M, Roe SM, Peggie M, Cohen P, Prodromou C, Pearl LH (2005) Chaperoned ubiquitylation – crystal structures of the CHIP U box E3 ubiquitin ligase and a CHIP-Ubc13-Uev1a complex. Mol Cell 20:525–538
Zhang T, Hamza A, Cao X, Wang B, Yu S, Zhan CG, Sun D (2008) A novel Hsp90 inhibitor to disrupt Hsp90/Cdc37 complex against pancreatic cancer cells. Mol Cancer Ther 7:162–170
Zuo L, Weger J, Yang Q, Goldstein AM, Tucker MA, Walker GJ, Hayward N, Dracopoli NC (1996) Germline mutations in the p16INK4a binding domain of CDK4 in familial melanoma. Nat Genet 12:97–99

Chapter 22
Theranostic Implications of Heat Shock Proteins in Idiopathic Pulmonary Fibrosis

Ganapasam Sudhandiran, Divya Thomas, Vadivel Dineshbabu, and Soumya Krishnan

Abstract Heat shock proteins, also known as 'stress proteins' are large family of chaperones that are involved in the proper formation, maturation and maintenance of variety of proteins to promote cell survival under stress condition. Normal chaperone machinery is essential for the endogenous adaptation of several tissues in stress environment; however, altered chaperone function has been strongly associated with the molecular orchestration in the progression of several diseases. Therefore, inhibitors of molecular chaperones are emerging as attractive drug targets against various diseases. Idiopathic pulmonary fibrosis is a grave disease whose pathogenic factors remain unknown. Multiple mechanisms including TGFβ/Smad regulated epithelial mesenchymal transition and aberrant cell death pattern contribute to this disease. Recently, growing body of evidences demonstrate the involvement of large and small heat shock proteins contributing to this disease. The knowledge about the pathophysiological role of heat shock proteins has been well updated in cancers; however, its role in the initiation and progression of fibrosis is poorly addressed. Since the therapeutic options for pulmonary fibrosis are very much limited, it is noteworthy to discuss the important theranostic applications of heat shock proteins in pulmonary fibrosis which is becoming a blistering field in fibrosis research.

Keywords Apoptosis · Heat shock proteins · Idiopathic pulmonary fibrosis · Molecular chaperones · Transforming growth factor beta

---

G. Sudhandiran (✉) · D. Thomas · V. Dineshbabu · S. Krishnan
Cell Biology Laboratory, Department of Biochemistry, University of Madras, Chennai, Tamil Nadu, India
e-mail: sudhandiran@unom.ac.in

A. A. A. Asea, P. Kaur (eds.), *Heat Shock Protein 90 in Human Diseases and Disorders*, Heat Shock Proteins 19,
https://doi.org/10.1007/978-3-030-23158-3_22

Abbreviations

AEC	Alveolar epithelial cells
ECM	Extra cellular matrix
EMT	Epithelial to mesenchymal transition
Hsp	Heat shock proteins
IPF	Idiopathic pulmonary fibrosis
MAPK	Mitogen activated protein kinase
MMP	Matrix metalloproteases
ROS	Reactive oxygen species
TGFβ	Transforming growth factor beta

22.1 Introduction

Protein folding is imperative for maintenance of normal homeostasis. The scenario of protein folding in vitro under regulated conditions is favourable and rather different as compared to in vivo, where due to high protein concentration and temperature effluxes, promote extensive aggregation of proteins. The assembly of proteins, from its primary to quaternary structure is highly organized where polypeptides are arranged as functional oligomer. However, protein aggregation is too difficult to deal where unorganized, non-specific interaction of polypeptide chains are tethered contributing to the heterogenous protein clusters that leads to an unfavourable condition of proper folding (Walter and Buchner 2002). Irrespective of health and disease, integrity of proteins is essential to carry out metabolic functions. To ascertain the metabolic function, proteins need to be properly folded in their three-dimensional state, as dictated by the primary sequence of protein. Mis-folded proteins are the critical mediators of most of the disease paradigms including cancers, diabetes and neurodegenerative disorders.

22.1.1 Molecular Chaperones

Molecular chaperones span from prokaryotic to eukaryotic species are the proteins that regulate the proper folding of misfolded proteins thereby preventing its aggregation (Mayer and Bukau 1999). The folding process is spontaneous for small proteins; however certain complex proteins need an elaborate process to fold it to the native state (Mayer 2010). In this context, quality control of proteins is highly essential for maintenance of homeostasis. Molecular chaperones function in a unique manner and perform the following (a) importing proteins to the respective cellular compartment, (b) an organized mechanism of protein folding in the regions of cytosol, endoplasmic reticulum and mitochondria, (c) unstable proteins are targeted and degraded (d) deterrence of aggregation of unstable proteins (e) recognition and binding of regulatory proteins (f) repair of misfolded proteins (Feldman and Frydman 2000). In this context, heat shock proteins (Hsp) emulate appropriate folding of nascent chains, translocation of proteins preservation and reorganization of the cytoskeleton and target of malfunctioned protein for proteosomal degradation (Lindquist 1986). Hsp functions as molecular chaperons and interact with other proteins thereby reducing the possibility that these proteins interact incorrectly with each other (Richter et al. 2010). When the other proteins are in non-native state due to stress or in mis-folded architecture, Hsp recognize and bind them. How does Hsp recognize non-native protein from native one is due to the unfolded region comprising of hydrophobic amino acids, that enable chaperones in general to recognize and bind via specific peptide sequence. Hsp are oligomeric and serve as co-chaperone and as nucleotide exchange factor. They are involved in import and export process in organellar localization and works to minimize the aggregation of non-native protein and directing the aggregated proteins for prompt degradation (Feder and Hofman 1999).

In 1962, Ritossa discovered Hsp, the first time in *Drosophila malenogaster* and later was found to be expressed in all other species including prokaryotic and eukaryotic origin (Ritossa 1962). Hsp are phylogenetically highly conserved set of proteins that are induced by heat shock, though preceding studies justify that other stressors such as physiological and environmental insults could also induce these proteins. For example, a versatile set of factors such as oxidative stress insults, hypoxia, ultraviolet radiation, pathogenic infection, inflammation, malignancies and aberrant cell death program can induce Hsp. Depending on the molecular weight, Hsp are broadly classified as high molecular weight Hsp and small Hsp (Tissieres et al. 1974). The high molecular weight Hsp include Hsp110, Hsp90, Hsp70/Hsp80, Hsp60 & Hsp40. These family members have a pivotal role in regulated induction and are targeted to selective subcellular compartment (Jaattela 1999). Some major differences between high molecular weight and small Hsp (15–30 KDa) lies in their dependency to Adenosine tri phosphate (ATP). Therefore, high molecular weights Hsp are ATP dependent chaperone and need co-chaperone to orchestrate their conformation and binding to ATP, whereas small Hsp are independent of ATP. Among the high molecular weight Hsp, Hsp90 & Hsp70 are notorious and its role has been addressed in several disease settings (Mollapour and Neckers 2012). Though the role of Hsp has been well reported in ageing related diseases such as cancers and neurodegeneration, its role in fibrotic diseases is now emerging. The focus of this review is to highlight the impact of large Hsp in idiopathic pulmonary fibrosis.

22.1.2 Heat Shock Protein 90: Structure and Function

To maintain cellular proteostasis, cells encompass several mechanisms. Among the other guardians of protein quality control checks, namely ubiquitination and autophagy lysosomal degradation process that regulate protein degradation, heat shock response play an important role (Ritossa 1996). Hsp play its role till the fate of cells from folding of nascent proteins to proper degradation and modulate protein folding and repair (Kaushik and Cuervo 2015). Hsp90 is a notorious chaperone and expressed in several disease conditions (Zhao and Houry 2005). It is peculiar in several aspects. Hsp90 is evolved from prokaryotic origin to the chaperone of eukaryotes and tether with certain co-chaperones that participate in a regulated fashion (Taipale et al. 2010). Hsp90 does not bind unfolded proteins whereas it binds native like proteins. Hsp 90 is the most abundant protein present in cytosol in unstressed cells. Five types of Hsp90 in human are described (Csermely et al. 1998). Hsp90A present in cytosol, Hsp90α, Hsp90β and Hsp90B originating in the endoplasmic reticulum and Tumor Necrosis Factor Receptor-Associated Protein-1 (TRAP-1), and the mitochondrial Hsp90 (Altieri et al. 2012; Yoshida et al. 2013).

In view of the structural features of Hsp90, it possesses a 25-kDa N-terminal domain (NTD) responsible for ATP binding and hydrolysis. NTD is attached to a linker region comprising the middle domain and attribute to the co-chaperone site, ATP and substrate binding site. 50 kDa C-terminal domain with its MEEVD motif attribute to the strong dimerization interface that is indispensable for Hsp90 function. The conformational cycle of Hsp90 and its interaction with its counter substrate clients has been well reviewed (Schopf et al. 2017; Prodromou 2012). Briefly, an open conformation state was attained by Hsp90 in the absence of ATP. The activation process involves ATP which induces conformational changes, co-chaperone and through posttranslational modifications leading to a closed state (Li et al. 2012).

When cells are under stress, the expression of Hsp90 abundantly increases in cytosol, mitochondrial or endoplasmic reticulum. For instance, in malignancies of lung, breast and bladder, Hsp90 is expressed (Garrido et al. 2001; Seigneuric et al. 2011; Jego et al. 2013). Hsp90 is majorly localized in the cytoplasm (Passinen et al. 2001). Nevertheless, fragments of Hsp90 might be present in nucleus. Hsp90 modulate several signalling pathways including JAK/STAT (Hobbs et al. 2018), Mitogen activated protein kinase (MAPK) signalling (Liu et al. 2018), Transforming growth factor (TGF) signalling (Lee et al. 2016), tyrosine kinases such as Src (Haque et al. 2016; Yano et al. 2008). Several proteins are identified that depend on Hsp90 to attain their physiological state (Karagöz and Rüdiger 2015). In the case of Hsp90 and c-src, which is a cell cycle regulatory protein, the association of Hsp90 with c-src defines its functional state. Screening for natural substances that inhibit c-src has been the focus of several investigators. It has been documented that certain inhibitors that disturb src, bind through Hsp90, which are highly specific (Liu et al. 2011).

22.1.3 Heat Shock Protein 70

Hsp70 family of stress proteins are integral part of ubiquitous chaperone machinery and present in eukaryotic cells and other species. It includes inducible Hsp70, constitutively expressed Hsp70, Hsp75 from mitochondria and GRP78 originating from endoplasmic reticulum. Under unstressed condition, it functions as ATP-dependant chaperone where it interacts with nascent polypeptide, regulating their folding pattern and its transport across cell membrane (Zhang et al. 2002). However, in case of cellular stress, Hsp70 assists the stressed cells to adjust to the environment of regulating denatured proteins and its degradation (Hartl 1996). Like Hsp90, Hsp70 also possesses an N-terminal domain as ATP site, and C-terminal peptide-binding domain. Hsp70 binds hydrophobic regions in partly unfolded proteins and prevents its aggregation.

22.1.4 Idiopathic Pulmonary Fibrosis

Lung diseases are the foremost causes of death worldwide. Among various lung diseases, idiopathic pulmonary fibrosis (IPF) poses a major health burden. The incidence of IPF has increased over years and several therapeutic interventions are under investigation (Baratt et al. 2018; Raghu et al. 2006). Briefly, IPF is an aggressive disease characterized by the progressive and irreversible damage of lung tissues caused by impaired wound healing that leads to scar formation in lung that result in death of patients from respiratory failure (Wynn 2011; Ballaye and Kolb 2015). Fibrotic scar in lung is characterized by the excess accumulation of extracellular matrix (ECM) components especially interstitial collagen (Katzenstein and Myers 1998).The histological hallmarks of lung tissue of patients with IPF reveals the presence of fibroblast foci, which is formed by aggregates of activated fibroblasts that produce excessive levels of extracellular matrix proteins within the alveolar space at the site of epithelial cell loss and injury (King et al. 2011).Conventionally, IPF was believed to result from chronic inflammatory mediated response caused by the abnormal activation and accumulation of inflammatory cells. However, this view has recently been transformed and a growing body of research evidence indicates that fibrotic reaction in IPF is found to be associated with an orchestrated biological event termed as epithelial to mesenchymal transition (EMT) (Norwin et al. 2014; Sakuma 2017). Several studies elucidate that abnormally activated alveolar epithelial cells (AECs) express most of cytokines responsible for driving EMT and fibroblast activation. Among these, transforming growth factor-β (TGF-β) is a vital switch mediator that induces fibrotic response in AECs (Fernandez and Eickelberg 2012). Till now, the therapeutic opportunities to treat IPF are limited.

Pirfenidone and nintedanib were approved drugs that could retard the progression of disease. However, these drugs possesses side effects and could not completely cure the disease. Therefore, novel drug regimens are warranted to treat this disease and multiple drug targets were proposed.

22.1.4.1 Intervention of HSP90 in TGF-β/Smad Signaling Mediated Fibrosis

The TGF-β superfamily encompasses a large group of pleiotropic cytokines that regulate a wide variety of biological functions ranging from embryonic development to wound repair primarily through a canonical Smad-dependent mechanism (Massague and Chen 2000). Overproduction or potentiating its profibrotic effects leads to an aberrant wound healing response during the initiation of fibrotic processes in various organs. In mammals, three major isoforms have been identified: TGF-β1, – β2, and – β3 (Massague 2008). TGF- β1 is the isoform that is highly related with the development of IPF (Xaubet et al. 2003). Normally, it is secreted in an inactivated form through its attachment to a latency-associated peptide and to a latent TGF-β binding protein. Modulation of the TGF-β-dependent Smad pathway in animal models has provided strong evidence for a role for TGF-β in fibrotic EMT in vivo as well as in vitro (Zhang et al. 2018).Thus, one of the mechanisms of induction of fibrotic response by TGF-β in lung may be through the induction of EMT in AECs. Hsp90 have distinct role in the modulation of TGF-β1 signalling pathway. There are documented reports that inhibition of Hsp90 greatly reduces lung fibrogenesis in mice (Sibinska et al. 2017). Recent evidences suggest the direct involvement of Hsp90 and IPF (Bonniaud et al. 2018). Apart from Hsp90, the other high molecular weight Hsp70 plays a role in attenuating pulmonary fibrosis. This happens so via modulation of TGF/Smad proteins and attenuation of inflammation (Bellaye et al. 2014). The most prominent role for Hsp in fibrogenesis is their ability to modulate the TGF-β1 pathway. However, IPF subjects with antibodies specific to Hsp70 but have poor prognoses has been reported (Kahloon et al. 2013).

22.1.4.2 Modulation of Cell Death Pattern by HSP90

Over the past few decades, despite the emergence of novel cell death pathways such as autophagy and necroptosis, apoptosis remains to be the primary cell death pathway. Apoptosis, in principle is responsible for elimination of unwanted cells during development as exemplified by selective destruction of neurons, formation of interdigits, during development and play a central role in maintaining adult homeostasis (Elmore 2007). Apoptotic cell death is the predominant mode of cell death, triggered to eliminate tumour cells. Over the period, our research group as well as other investigators have investigated the pathways leading to apoptosis in several experimental models (Pandurangan et al. 2018).The death receptor or extrinsic pathway and mitochondrial or intrinsic pathway are the pathways of apoptosis. The protective role of Hsp is connected to its intrusion with apoptotic pathways. In the intrinsic pathway, mitochondria serve as an efficient organelle in subscribing pro-apoptotic molecules that involve BH3-only member of the Bcl2 family. Permeabilization of the outer mitochondrial membrane is pivotal to initiate apoptosis. To attain this, pore formation occurs through Bax. Cytochrome c resides in the inner membrane of mitochondrion and due to disruption of outer membrane, it leaks to cytosol and interacts with Apaf-1 (Apoptosis protease activated factor – 1) and trigger ATP-dependent oligomerization (Zamzami and Kroemer 2001). The oligomerized Apaf-1, then binds to the procaspase-9 and forms apoptosome which is the caspase-9 activation complex which further trigger Caspase-3 finally leading to apoptosis. In extrinsic pathway, the plasma membrane proteins spanning it known as Tumour necrosis factor receptor family such (TNFR, Fas, TRAIL etc.) are involved (Dewson and Kluck 2010; Chipuk et al. 2010). Secondly a FAS-associated death domain protein (FADD) or TNFR-associated death domainprotein (TRADD) participates followed by the activation caspases. As described in Fig. 22.1, Hsp90 positively regulates NfkB and through PI3 & Akt signaling cascades, leading to cell survival. In contrary, Hsps 90 and 70 intervene with caspase and apoptosome thereby regulating apoptosis.

The association between alveolar epithelial cell apoptosis and lung fibrosis were noted. Although the mechanisms that cause cell death in IPF are not completely understood, it is known that TGF-β plays a pivotal role as a potent inducer of apoptosis in alveolar epithelial cells (Thannickal and Horowitz 2006). In the apoptosis execution phase, TGF-β promotes the imbalance of Bcl2 family members by stimulating Bax and Bid in the murine lung. During injury, epithelial cells are exposed to a variety of stress factors that promote oxidant-antioxidant imbalances. Reactive oxygen species (ROS) promote a pro-fibrogenic microenvironment by altering the balance between MMPs and their inhibitors (tissue inhibitors of metalloproteinases [TIMPs]) and by directly or indirectly activating TGF-β. TGF-β produces ROS as part of its signalling pathway that further promote apoptosis in AEC (Liu and Desai 2015).

22.1.5 Pharmacological Inhibition of HSP90

Considering the therapeutic importance of Hsp90, it is not surprised that researchers have been attracted in identifying new scaffolds with inhibitory effect on the action of HSP90. Gefitinib, a tyrosine kinase inhibitor of the epidermal growth has been utilized for the treatment of advanced lung cancer which exhibits no response to chemotherapy (Blackledge and Averbuch 2004). Gefitinib induces IPF, through modulated expression of Hsp, whereHsp70 plays prominent role (Wang et al. 2016). Therefore, evaluation of the expression of large Hsp may be useful to evaluate drugs with potential to induce IPF. In this scenario, 17-allylamino, 17-demethoxygeldanamycin (17AAG) is reported as the first compound inhibitor of Hsp90 that entered in clinical trials (Sausville et al. 2003). However, toxicity and side effects reported in the patients enforced for the identification of novel

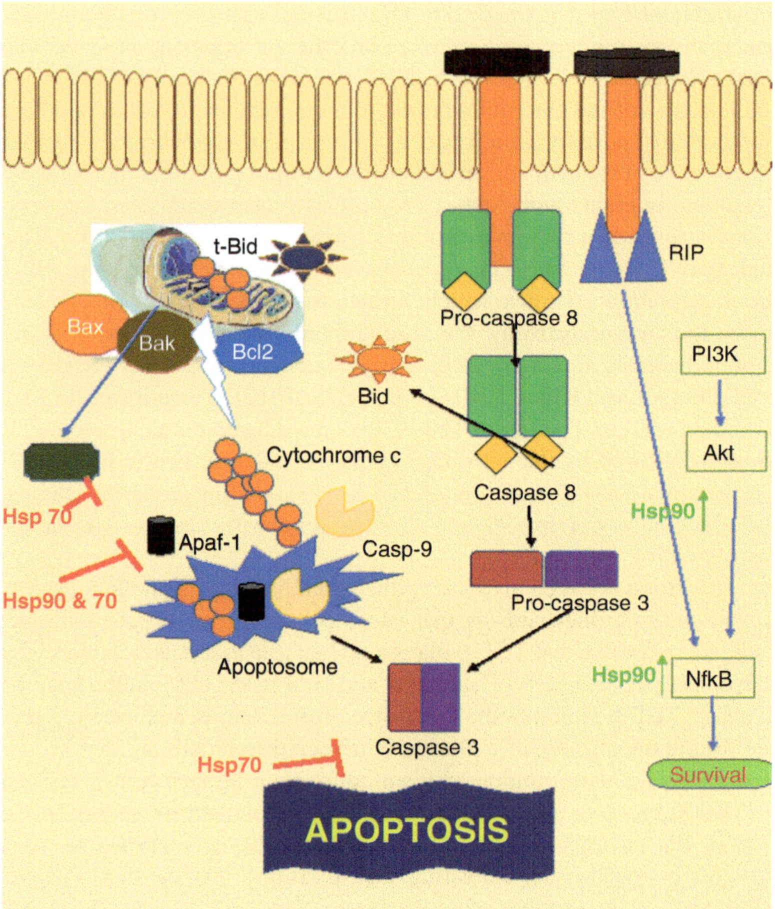

Fig. 22.1 Role of Heat shock proteins on cell death and cell survival. HSP70 and 90 prevent the oligomerization of apoptotic protease activating factor-1 (Apaf-1). HSP90 form a complex with Apaf-1 and inhibit the activation of caspase 3. Hsp70 also sequesters apoptosis-inducing factor (AIF) and thereby regulates the induction of apoptosis. On the other hand, Hsp90 promote the phosphorylation of Akt thereby promote cell survival though the positive regulation of NF-κβ

pharmacological compounds with modulatory effect on the mechanism of action of HSP90 in various diseases. Geldanamycin, a naturally occurring benzoquinone was shown to directly bind to and inhibit the function of HSP90 leading to the protein destabilization has been reported (Neckers et al. 1999). Though Geldanamycin is a polypeptide mimic that directly interact with the binding pockets of Hsp90, the poor solubility of this compound and hepatotoxicity reported in animal in the pre-clinical studies had prohibited the successful completion of clinical trial of this compound (Supko et al. 1995). Yet another potent natural product inhibitor of Hsp90 based on the structural features and binding affinity is radicicol. Its mechanism of action by binding with Hsp90 through the regulation of important oncogenic downstream signalling in 17-AAG resistant cancerous cells has been reported (Yang et al. 2004). Very recently, the effect of Hsp 90 inhibitors on the proliferation of *Leishmania donovani* have been reported (Bartsch et al. 2017). Further, Hsp90inhibitor, 17-AAG induces cytotoxicity through DNA repair mechanism and the down-regulation of Rad51 expression in lung cancer cells has been reported (Ko et al. 2012). All these studies highlight the therapeutic importance of identifying novel natural compound that target Hsp90 against various diseases. However, the use of Hsp90 inhibitors in the alterations of fibrotic condition in lung remains elusive. Very recently, Hsp90inhibitor AUY-922 was reported to block pro-fibrotic signalling, reduce fibroblast activation and improving lung function in HCl-induced pulmonary fibrosis in mice (Dimitropoulou et al. 2017). Since IPF is highly complex in nature, our research group is working on the therapeutic potential of several natural compounds in reducing the severity of the disease progression. One such compound that we found with high therapeutic potential is Celastrol, a quinine methide triterpenoid derived from Thunder God vine, a Chinese medicinal plant. For the first time, we have identified the anti-fibrotic effect of Celastrol by reducing collagen accumulation, inducing autophagy and regulating the induction of epithelial-mesenchymal transition in bleomycin-induced IPF in rats (Divya et al. 2016, 2017, 2018). In agreement with our findings, Celastrol is reported as a strong inhibitor of Hsp90 in other models. Celastrol was reported as a promising agent against gefitinib-resistant non small cell lung cancer (NSCLC) through the induction of apoptosis and Hsp90 degradation (Fan et al. 2014). The potential of Celastrol in inhibiting co-chaperone CDC37 which functions along with Hsp90to regulate protein kinases in oncogenic signalling pathways has been reported (Wei et al. 2014). Yet another study report that Celastrol synergistically act with 17-allylamino-17-demethoxy-geldanamycin (17-AAG), Hsp90 inhibitor to destroy glioblastoma cells and thereby overcoming drug resistance (Boridy et al. 2014). Growing evidences suggest the role of 17-AAG targeting cancer cells (Waza et al. 2006; Mellatyar et al. 2018). Since we observed the fascinating effects of Celastrol in regulating collagen accumulation and protects the lung from the deleterious effect of bleomycin, we sought to analyse its role in inhibiting Hsp90 during the progression of IPF. We found that Celastrol significantly reduced the expression of Hsp90 in BLM-induced rat lungs. A549 cells was transdifferentiated using TGFβ1 and the expression of Hsp90 was evaluated. The effect of celastrol was more when compared to 17-AAG in reducing the expression of HSP90 in rat lungs. Celastrol treatment caused a significant reduction in the expression of HSP90 in TGF-β trans-differentiated A549 cells, further support that Celastrol is a strong inhibitor of HSP90 (Fig. 22.2). However, more systemic evaluations are warranted in order to analyse its therapeutic index on the inhibitory effect on HSP90 against the progression of PF.

Yet another Hsp90 inhibitor is 17-DMAG (17-dimethylaminoethylamino-17-demethoxygeldanamycin). It is a geldanamycin derivative, that recognizes and ATP-binding site of Hsp90 and inhibits its action. 17-DMAG has theranostic potential to reduce fibrosis of other organs (Tomcik et al. 2014). Though the role of high molecular weight Hsp are yet to be completely elucidated in lung fibrosis, Hsp

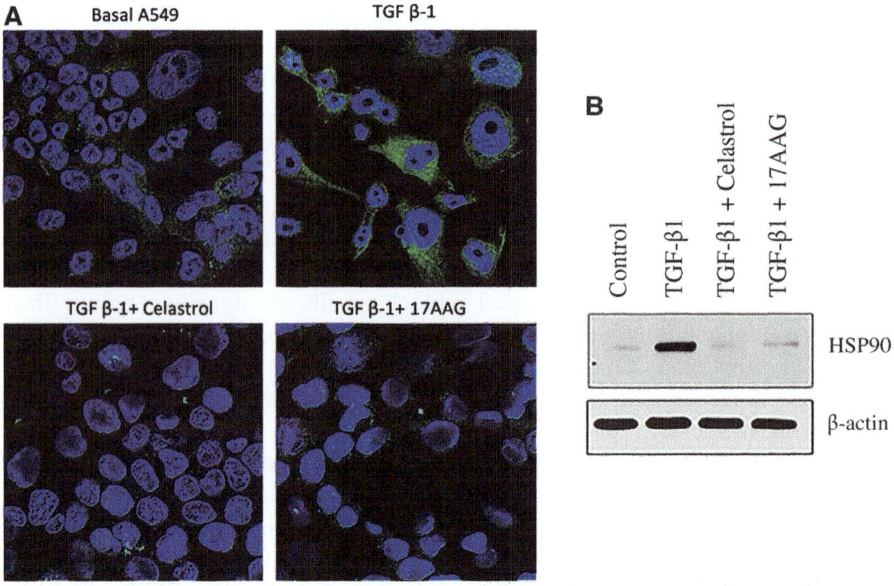

Fig. 22.2 Inhibitory effect of celastrol on the expression of Hsp90 in TGF-β1 trans-differentiated A549 cells. (**a**) Immunofluorescence analysis of Hsp90. A549 cells was administered with TGF-β1 (5 ng/ml) in serum less media. After 24 h, cells were treated with either celastrol (5 μM/ml) or 17AAG (250 nM/ml) for further 24 h. Protein expression of Hsp90 was analysed in a Confocal microscope. Hoechst 33342 was used to stain the nuclei. The fluorescence intensity was measured for Alexafluor at excitation/emission wavelength 493/520 nm and Hoechst:346/461 nm. Scale bar: 30 μm. (**b**) Western blot analysis of HSP90. After treatment of TGF-β trans-differentiated A549 cells with Celastrol or 17-AAG as mentioned above, 40 μg of proteins in cell lysate were separated on a 10% gel and were transferred to polyvinylidene difluoride (PVDF) membrane. Primary antibody specific for Hsp90 (Rabbit mAb, C45G5) and goat anti-rabbit IgG, HRP conjugated secondary antibody was used. The protein antibody complex was detected by Enhanced chemiluminesence. β-actin was used as a loading control

inhibitors are under phase I–III trials for antitumoral properties in combination with chemotherapy. The molecular events pinning the fibrosis and cancer are quite similar. It includes aberrant cell cycle proliferation, altered genetic alterations, uncontrolled cell proliferation, aberrant epithelial-mesenchymal transition etc. Interesting, in a parallel study performed in our laboratory, soy isoflavone Daidzein exerted antifibrotic effect through modulation of Hsp90 and Hsp27 through TGF/Smad regulation. The role of Small molecular weight Hsp27 in myofibroblast differentiation and instigates bleomycin induced pulmonary fibrosis has been documented (Park et al. 2016). Therefore, there is a huge possibility that such inhibitors could effectively be used as a drug target in fibrosis research.

22.2 Conclusions

Pulmonary fibrosis is a grave disease. Despite an enormous amount of research has been progressed in IPF, no definitive causative factors have been identified. In this context it would be meaningful to investigate novel antifibrotic agents. Since cancer and fibrosis are closely related, drugs to treat cancer could be expanded to IPF, as in the case of nintedanib. Though Hsp play a significant role in cancers and neurodegeneration, its role in lung fibrotic diseases is recently growing. Since Hsp90 is ubiquitous and play a major role in maintenance of homeostasis and also in would healing, its regulation is important. TGFβ signaling appear to be central in the pathogenesis of IPF. Interestingly Hsp90 targets TGF signaling and therefore modulation of TGFβ/smad signals by Hsps will provide a novel avenue in the treatment of IPF. To support this notion, the existing phase I clinical trials using inhibitors of TGFβ are promising. Hsp inhibitors might open novel avenues in discovering new drugs for the treatment of IPF.

Acknowledgements The authors thank Indian Council of Medical Research (ICMR), New Delhi, Govt. of India for a major research grant (52/21/08/BMS/) in Pulmonary fibrosis.

References

Altieri DC, Stein GS, Lian JB, Languino LR (2012) TRAP-1, the mitochondrial Hsp90. Biochim Biophys Acta (BBA)-Mol Cell Res 1823:767–773

Barratt S, Creamer A, Hayton C, Chaudhuri N (2018) Idiopathic pulmonary fibrosis (IPF): an overview. J Clin Med 7:E201

Bartsch K, Hombach-Barrigah A, Clos J (2017) Hsp90 inhibitors radicicol and geldanamycin have opposing effects on Leishmania Aha1-dependent proliferation. Cell Stress Chaperones 22:729–742

Bellaye PS, Kolb M (2015) Why do patients get idiopathic pulmonary fibrosis? Current concepts in the pathogenesis of pulmonary fibrosis. BMC Med 13:176

Bellaye PS, Burgy O, Causse S, Garrido C, Bonniaud P (2014) Heat shock proteins in fibrosis and wound healing: good or evil? Pharmacol Ther 143:119–132

Blackledge G, Averbuch S (2004) Gefitinib ('Iressa', ZD1839) and new epidermal growth factor receptor inhibitors. Br J Cancer 90:566–572

Bonniaud P, Burgy O, Garrido C (2018) Heat shock protein-90 toward theranostics: a breath of fresh air in idiopathic pulmonary fibrosis. Eur Respir J 51:pii: 1702612

Boridy S, Le PU, Petrecca K, Maysinger D (2014) Celastrol targets proteostasis and acts synergistically with a heat-shock protein 90 inhibitor to kill human glioblastoma cells. Cell Death Dis 5:e1216

Chipuk JE, Moldoveanu T, Llambi F, Parsons MJ, Green DR (2010) The BCL-2 family reunion. Mol Cell 37:299–310

Csermely P, Schnaider T, So C, Prohászka Z, Nardai G (1998) The 90-kDa molecular chaperone family: structure, function, and clinical applications. A comprehensive review. Pharmacol Ther 79:129–168

Dewson GKRM, Kluck RM (2010) Bcl-2 family-regulated apoptosis in health and disease. Cell Health Cytoskeleton 2:9–22

Dimitropoulou C, Catravas J, Barabutis N, Thangjam G (2017) Heat shock protein 90 inhibitors block pulmonary fibrosis and dysfunction in HCl-exposed mice. PA1033
Divya T, Dineshbabu V, Soumyakrishnan S, Sureshkumar A, Sudhandiran G (2016) Celastrol enhances Nrf2 mediated antioxidant enzymes and exhibits anti-fibrotic effect through regulation of collagen production against bleomycin-induced pulmonary fibrosis. Chem Biol Interact 246:52–62
Divya T, Sureshkumar A, Sudhandiran G (2017) Autophagy induction by celastrol augments protection against bleomycin-induced experimental pulmonary fibrosis in rats: role of adaptor protein p62/SQSTM1. Pulm Pharmacol Ther 45:47–61
Divya T, Velavan B, Sudhandiran G (2018) Regulation of transforming growth factor β/Smad-mediated epithelial–mesenchymal transition by celastrol provides protection against bleomycin-induced pulmonary fibrosis. Basic Clin Pharmacol Toxicol 123:122–129
Elmore S (2007) Apoptosis: a review of programmed cell death. Toxicol Pathol 35:495–516
Fan XX, Li N, Wu JL, Zhou YL, He JX, Liu L, Leung E (2014) Celastrol induces apoptosis in gefitinib-resistant non-small cell lung cancer cells via caspases-dependent pathways and Hsp90 client protein degradation. Molecules 19:3508–3522
Feder ME, Hofmann GE (1999) Heat-shock proteins, molecular chaperones, and the stress response: evolutionary and ecological physiology. Annu Rev Physiol 61:243–282
Feldman DE, Frydman J (2000) Protein folding in vivo: the importance of molecular chaperones. Curr Opin Struct Biol 10:26–33
Fernandez IE, Eickelberg O (2012) The impact of TGF-β on lung fibrosis: from targeting to biomarkers. Proc Am Thorac Soc 9:111–116
Garrido C, Gurbuxani S, Ravagnan L, Kroemer G (2001) Heat shock proteins: endogenous modulators of apoptotic cell death. Biochem Biophys Res Commun 286:433–442
Haque A, Alam Q, ZubairAlam M, Azhar EI, HussainWaliSait K, Anfinan N, Mushtaq G, Amjad Kamal M, Rasool M (2016) Current understanding of HSP90 as a novel therapeutic target: an emerging approach for the treatment of cancer. Curr Pharm Des 22:2947–2959
Hartl FU (1996) Molecular chaperones in cellular protein folding. Nature 381:571–579
Hobbs GS, Somasundara AVH, Kleppe M, Litvin R, Arcila M, Ahn J, McKenney AS, Knapp K, Ptashkin R, Weinstein H, Heinemann MH (2018) Hsp90 inhibition disrupts JAK-STAT signaling and leads to reductions in splenomegaly in patients with myeloproliferative neoplasms. Haematologica 103:e5–e9
Jäättelä M (1999) Heat shock proteins as cellular lifeguards. Ann Med 31:261–271
Jego G, Hazoumé A, Seigneuric R, Garrido C (2013) Targeting heat shock proteins in cancer. Cancer Lett 332:275–285
Kahloon RA, Xue J, Bhargava A, Csizmadia E, Otterbein L, Kass DJ, Bon J, Soejima M, Levesque MC, Lindell KO, Gibson KF (2013) Patients with idiopathic pulmonary fibrosis with antibodies to heat shock protein 70 have poor prognoses. Am J Respir Crit Care Med 187:768–775
Karagöz GE, Rüdiger SG (2015) Hsp90 interaction with clients. Trends Biochem Sci 40:117–125
Katzenstein ALA, Myers JL (1998) Idiopathic pulmonary fibrosis: clinical relevance of pathologic classification. Am J Respir Crit Care Med 157:1301–1315
Kaushik S, Cuervo AM (2015) Proteostasis and aging. Nat Med 21:1406–1415
King TE Jr, Pardo A, Selman M (2011) Idiopathic pulmonary fibrosis. Lancet 378:1949–1961
Ko JC, Chen HJ, Huang YC, Tseng SC, Weng SH, Wo TY, Huang YJ, Chiu HC, Tsai MS, Chiou RY, Lin YW (2012) HSP90 inhibition induces cytotoxicity via down-regulation of Rad51 expression and DNA repair capacity in non-small cell lung cancer cells. Regul Toxicol Pharmacol 64:415–424
Kumar Pandurangan A, Divya T, Kumar K, Dineshbabu V, Velavan B, Sudhandiran G (2018) Colorectal carcinogenesis: insights into the cell death and signal transduction pathways: a review. World J Gastrointest Oncol 10:244–259
Lee SB, Lim AR, Rah DK, Kim KS, Min HJ (2016) Modulation of heat shock protein 90 affects TGF-β-induced collagen synthesis in human dermal fibroblast cells. Tissue Cell 48:616–623

Li J, Soroka J, Buchner J (2012) The Hsp90 chaperone machinery: conformational dynamics and regulation by co-chaperones. Biochim Biophys Acta 1823:624–635
Lindquist S (1986) The heat-shock response. Annu Rev Biochem 55:1151–1191
Liu RM, Desai LP (2015) Reciprocal regulation of TGF-β and reactive oxygen species: a perverse cycle for fibrosis. Redox Biol 6:565–577
Liu X, Yan Z, Huang L, Guo M, Zhang Z, Guo C (2011) Cell surface heat shock protein 90 modulates prostate cancer cell adhesion and invasion through the integrin-β1/focal adhesion kinase/c-Srcsignaling pathway. Oncol Rep 25:1343–1351
Liu YH, Liu XM, Wang PC, Yu XX, Miao JK, Liu S, Wang YK, Du ZQ, Yang CX (2018) Heat shock protein 90α couples with the MAPK-signaling pathway to determine meiotic maturation of porcine oocytes. J Anim Sci 96:3358–3369
Massagué J (2008) TGFβ in cancer. Cell 134:215–230
Massagué J, Chen YG (2000) Controlling TGF-β signaling. Genes Dev 14:627–644
Mayer MP (2010) Gymnastics of molecular chaperones. Mol Cell 39:321–331
Mayer MP, Bukau B (1999) Molecular chaperones: the busy life of Hsp90. Curr Biol 9:R322–R325
Mellatyar H, Talaei S, Pilehvar-Soltanahmadi Y, Barzegar A, Akbarzadeh A, Shahabi A, Barekati-Mowahed M, Zarghami N (2018) Targeted cancer therapy through 17-DMAG as an Hsp90 inhibitor: overview and current state of the art. Biomed Pharmacother 102:608–617
Mollapour M, Neckers L (2012) Post-translational modifications of Hsp90 and their contributions to chaperone regulation. Biochim Biophys Acta (BBA)-Mol Cell Res 1823:648–655
Neckers L, Schulte TW, Mimnaugh E (1999) Geldanamycin as a potential anti-cancer agent: its molecular target and biochemical activity. Investig New Drugs 17:361–373
Nowrin K, Sohal SS, Peterson G, Patel R, Walters EH (2014) Epithelial-mesenchymal transition as a fundamental underlying pathogenic process in COPD airways: fibrosis, remodeling and cancer. Expert Rev Respir Med 8:547–559
Park AM, Kanai K, Itoh T, Sato T, Tsukui T, Inagaki Y, Selman M, Matsushima K, Yoshie O (2016) Heat shock protein 27 plays a pivotal role in myofibroblast differentiation and in the development of bleomycin-induced pulmonary fibrosis. PLoS One 9:e0148998
Passinen S, Valkila J, Manninen T, Syvälä H, Ylikomi T (2001) The C-terminal half of Hsp90 is responsible for its cytoplasmic localization. Eur J Biochem 268:5337–5342
Prodromou C (2012) The 'active life' of Hsp90 complexes. Biochim Biophys Acta 1823:614–623
Raghu G, Weycker D, Edelsberg J, Bradford WZ, Oster G (2006) Incidence and prevalence of idiopathic pulmonary fibrosis. Am J Respir Crit Care Med 174:810–816
Richter K, Haslbeck M, Buchner J (2010) The heat shock response: life on the verge of death. Mol Cell 40:253–266
Ritossa F (1962) A new puffing pattern induced by temperature shock and DNP in Drosophila. Experientia 18:571–573
Ritossa F (1996) Discovery of the heat shock response. Cell Stress Chaperones 1:97–98
Sakuma Y (2017) Epithelial to mesenchymal transition and its role in EGFR mutant lung adenocarcinoma and idiopathic pulmonary fibrosis. Pathol Int 67:379–388
Sausville EA, Tomaszewski JE, Ivy P (2003) Clinical development of 17-allylamino, 17-demethoxygeldanamycin. Curr Cancer Drug Targets 3:377–383
Schopf FH, Biebl MM, Buchner J (2017) The HSP90 chaperone machinery. Nat Rev Mol Cell Biol 18:345–360
Seigneuric R, Mjahed H, Gobbo J, Joly AL, Berthenet K, Shirley S, Garrido C (2011) Heat shock proteins as danger signals for cancer detection. Front Oncol 10(1):37
Sibinska Z, Tian X, Korfei M, Kojonazarov B, Kolb JS, Klepetko W, Kosanovic D, Wygrecka M, Ghofrani HA, Weissmann N, Grimminger F (2017) Amplified canonical transforming growth factor-β signalling via heat shock protein 90 in pulmonary fibrosis. Eur Respir J 49:1501941
Supko JG, Hickman RL, Grever MR, Malspeis L (1995) Preclinical pharmacologic evaluation of geldanamycin as an antitumor agent. Cancer Chemother Pharmacol 36:305–315
Taipale M, Jarosz DF, Lindquist S (2010) HSP90 at the hub of protein homeostasis: emerging mechanistic insights. Nat Rev Mol Cell Biol 11:515–528

Thannickal VJ, Horowitz JC (2006) Evolving concepts of apoptosis in idiopathic pulmonary fibrosis. Proc Am Thorac Soc 3:350–356

Tissiéres A, Mitchell HK, Tracy UM (1974) Protein synthesis in salivary glands of *Drosophila melanogaster*: relation to chromosome puffs. J Mol Biol 84:389–398

Tomcik M, Zerr P, Pitkowski J, Palumbo-Zerr K, Avouac J, Distler O, Becvar R, Senolt L, Schett G, Distler JH (2014) Heat shock protein 90 (Hsp90) inhibition targets canonical TGF-β signalling to prevent fibrosis. Ann Rheum Dis 73:1215–1222

Walter S, Buchner J (2002) Molecular chaperones—cellular machines for protein folding. Angew Chem Int Ed 41:1098–1113

Wang YQ, Shen AJ, Sun JY, Wang X, Liu HC, Zhang MM, Chen DQ, Xiong B, Shen JK, Geng MY, Zheng M (2016) Targeting Hsp90 with FS-108 circumvents gefitinib resistance in EGFR mutant non-small cell lung cancer cells. Acta Pharmacol Sin 37:1587–1596

Waza M, Adachi H, Katsuno M, Minamiyama M, Tanaka F, Sobue G (2006) Alleviating neurodegeneration by an anticancer agent: an Hsp90 inhibitor (17-AAG). Ann N Y Acad Sci 1086:21–34

Wei W, Wu S, Wang X, Sun CKW, Yang X, Yan X, Chua MS, So S (2014) Novel celastrol derivatives inhibit the growth of hepatocellular carcinoma patient-derived xenografts. Oncotarget 5:5819–5831

Wynn TA (2011) Integrating mechanisms of pulmonary fibrosis. J Exp Med 208:1339–1350

Xaubet A, Marin-Arguedas A, Lario S, Ancochea J, Morell F, Ruiz-Manzano J, Rodriguez-Becerra E, Rodriguez-Arias JM, Inigo P, Sanz S, Campistol JM (2003) Transforming growth factor-β1 gene polymorphisms are associated with disease progression in idiopathic pulmonary fibrosis. Am J Respir Crit Care Med 168:431–435

Yang ZQ, Geng X, Solit D, Pratilas CA, Rosen N, Danishefsky SJ (2004) New efficient synthesis of resorcinylic macrolides via ynolides: establishment of cycloproparadicicol as synthetically feasible preclinical anticancer agent based on Hsp90 as the target. J Am Chem Soc 126:7881–7889

Yano A, Tsutsumi S, Soga S, Lee MJ, Trepel J, Osada H, Neckers L (2008) Inhibition of Hsp90 activates osteoclast c-Srcsignaling and promotes growth of prostate carcinoma cells in bone. Proc Natl Acad Sci 105:15541–15546

Yoshida S, Tsutsumi S, Muhlebach G, Sourbier C, Lee MJ, Lee S, Vartholomaiou E, Tatokoro M, Beebe K, Miyajima N, Mohney RP (2013) Molecular chaperone TRAP1 regulates a metabolic switch between mitochondrial respiration and aerobic glycolysis. Proc Natl Acad Sci 110:E1604–E1612

Zamzami N, Kroemer G (2001) The mitochondrion in apoptosis: how Pandora's box opens. Nat Rev Mol Cell Biol 2:67–71

Zhang X, Beuron F, Freemont PS (2002) Machinery of protein folding and unfolding. Curr Opin Struct Biol 12:231–238

Zhang J, Zhang W, Zhang T, Zhou Q, Liu J, Liu Y, Kong D, Yu W, Liu R, Hai C (2018) TGF-β1 induces epithelial-to-mesenchymal transition via inhibiting mitochondrial functions in A549 cells. Free Radic Res 52:1432–1444

Zhao R, Houry WA (2005) Hsp90: a chaperone for protein folding and gene regulation. Biochem Cell Biol 83:703–710

Chapter 23
Heat Shock Protein 90 and Reproduction in Female Animals: Ovary, Oocyte and Early Embryo

Yu-Wei Yang, Lu Chen, and Cai-Xia Yang

Abstract The mammalian heat shock protein 90 (HSP90) is the most conserved member of heat shock protein family (HSP), participating in the regulation of various cellular and physiological processes, and stress adaptation and oncogenic transformation as well. In addition, in different animal reproduction system, *HSP90* plays a vital function in the folliculogenesis, oogenesis and embryo development, and also contributes to the development and progress of female reproductive diseases. This chapter focuses on the general structure, transcriptional regulation (expression and isoforms) and inhibitors of HSP90 family, and their particular roles in animal female reproductive events.

Keywords Early embryos · Female reproduction · HSP90 · Oocyte · Ovary

Abbreviations

AKT	Protein kinase B
AP1	Activator protein 1
ATP/ADP	Adenosine triphosphate/adenosine diphosphate
CDK1	Cyclin dependent kinase 1
CDK4	Cyclin dependent kinase 4
COCs	Cumulus-oocyte complexes
c-Src	Proto-oncogene tyrosine-protein kinase Src

Y.-W. Yang · C.-X. Yang (✉)
College of Animal Science, Yangtze University, Jingzhou, Hubei, China

College of Animal Science and Technology, Northeast Agricultural University, Harbin, Heilongjiang, China
e-mail: caixiayang@yangtzeu.edu.cn

L. Chen
College of Animal Science and Technology, Northeast Agricultural University, Harbin , Heilongjiang, China

A. A. A. Asea, P. Kaur (eds.), *Heat Shock Protein 90 in Human Diseases and Disorders*, Heat Shock Proteins 19,
https://doi.org/10.1007/978-3-030-23158-3_23

CTD	C-terminal domain
dpc	Day post-coital
ERBB2	Erb-B2 receptor tyrosine kinase 2
ERK	Extracellular regulated protein kinases
FLT3	FMS-Like tyrosine kinase-3
GA	Geldanamycin
GC	Glucocorticoids
GR	Glucocorticoid receptor
GRP94	94-kDa glucose-regulated protein
GV	Germinal vesicle
GVBD	Germinal vesicle breakdown
HNF4A	Hepatocyte nuclear factor 4 alpha
HOP	HSP70-HSP90 organizing protein
HSE	Heat shock element
HSF	Heat shock factor
HSP	Heat shock proteins
HSP90	Heat shock protein 90
HSP90α	Heat shock protein 90 alpha family class A member 1
HSP90β	Heat shock protein 90 alpha family class B member 1
IVM	In vitro maturation
JAK/STAT	Janus kinase-signal transducer and activator of transcription
MAPK	Mitogen-activated protein kinases
MD	Middle domain
MII	Metaphase II
Mos	MOS proto-oncogene, serine/threonine kinase
MPF	Maturation-promoting factor
NF-κB	Nuclear factor kappa-light-chain-enhancer of activated B cells
NTD	N-terminal domain
NV	Novobiocin
OGT	Beta-O-linked *N*-acetylglucosamine transferase
PB1	First polar body
PGCs	Primordial germ cells
PI3K/PTEN/AKT/mTOR	Phosphoinositide 3-kinase/phosphatase and tensin homologue/protein kinase B/mammalian target of rapamycin
PPARγ	Proliferator-activated receptor-γ
RAD	Radicicol
San A	Sansalvamide A
San A-amide	Sansalvamide A-amide
STAT5	Signal transducer and activator of transcription 5
TRAP-1	Tumor necrosis factor receptor-associated protein 1
VEGFR2	VEGF receptor 2
YTHDF2	YTH N6-methyladenosine RNA binding protein 2

23.1 Introduction

Heat shock proteins (HSP) are a family of protein chaperones, playing essential roles, especially under thermal stress, in the folding of native proteins or refolding of damaged ones, thus protecting proteins from aberrant aggregation, to maintain the normal physiological function of cells and organisms (Hoter et al. 2018). According to the molecular weights, HSP were classified into different isoforms or families, HSP100, HSP90, HSP70, HSP60 and the small molecule HSP family (Buchner 1996). As a highly conserved and abundant protein family, accounting for 1–2% of the whole cellular protein masses (Whitesell and Lindquist 2005), *HSP90* exists in all species except for Archaea (Zhang et al. 2014). Generally, in a wide variety of organisms, from bacteria to mammals, *HSP90* functions as a master regulator to regulate the maturation and activity of a variety of signaling proteins, essential not only for normal development but also for stress tolerance, and disease development as well (Pratt and Toft 2003; Taipale et al. 2010). In this chapter, we summarize the basic aspects of HSP90 biology, female reproductive processes in animals, focusing on *HSP90*'s function in female reproduction: the development of ovary, oocyte and early embryo.

23.1.1 Biology of HSP90

23.1.1.1 HSP90 Isoforms

In mammals, the HSP90 family includes four members or isoforms, HSP90α, HSP90β (in cytoplasm), 94-kDa glucose-regulated protein (GRP94, in endoplasmic reticulum, also called HSP90b1) and tumor necrosis factor receptor-associated protein 1 (TRAP-1, in mitochondria) (Hoter et al. 2018). *HSP90α* and *HSP90β* show different characteristics. Firstly, they locate on different genome regions. In the human genome, *HSP90α* locates on chromosome 14 (14q32.31), 3379 bp long and encoding 732 amino acids (84,660 Da). However, HSP90β is on the short arm of chromosome 6 (6p21.1), 2582 bp long and encoding 724 amino acids (83,264 Da). Secondly, besides the protein sequence differences, *HSP90α* and *HSP90β* have different expression patterns. The homology of amino acid consistency between HSP90α and HSP90β is about 85% (Johnson 2012). *HSP90α* expression is stress inducible, while *HSP90β* is expressed constitutively and abundantly in most tissues (Millson et al. 2007). Thirdly, though the key functions of *HSP90α* and *HSP90β* are inconsistent, functional similarities still exist to certain degree (Gao et al. 2013; Taherian et al. 2008). *HSP90β* can adjust normal and long-term cell function, while *HSP90α* might be involved in the occurrence of tumor (Li et al. 2011), the enhancement of cell cycle (Huang et al. 2014; Liu et al. 2018), and the protection of cells under stress (Islam et al. 2014).

23.1.1.2 HSP90 Expression and Regulation

Expression and Localization

Under physiological condition, HSP90 is a molecular chaperone widely and endogenously expressed at different developmental stages, showing system- and cell type-dependent patterns. In cultured rat granulosa cells, HSP90 synthesis was increased by gonadotropin treatment, suggesting the steroid hormone-dependent expression of HSP90 (Ben-Ze'ev and Amsterdam 1989). In chicken hepatoma cells, *HSP90α* transcription is up-regulated at the G1/S transition, indicating a cell cycle-dependent pattern (Jérôme et al. 1993; Szent-Gyorgyi 1995). During the differentiation of myoblasts into myotubes, HSP90α disappears and only HSP90β remains (Echeverría et al. 2016). During the hair cycle in yaks, *HSP90* mRNA and protein was expressed in a stage-dependent pattern (Yang et al. 2017). In male reproductive system, *HSP90β* mRNA and protein are expressed in mouse testis and epididymis, and its expression levels in the head, body and tail of the epididymis are higher than that in the testis, with subcellular localization of HSP90β in the cytoplasm of testicular spermatogonia, spermatocytes and round sperm (Sagare-Patil et al. 2017). *HSP90* expression in female reproductive system is summarized in section of *HSP90 And Female Reproductive Events* in this chapter.

Under stress, *HSP90* expression shows different patterns in a variety of organisms, cell types, and species. In larvae, when temperatures reach 43 °C and 45 °C, expression levels of *HSP90* in all ages are significantly higher than those under normal temperature, but not significantly different between different ages (Jiang et al. 2012). In mice, heat shock induced *HSP90* increases its expression in the lung, heart, spleen, liver and brain (Sareh et al. 2011). In primary myocardial cells of neonatal rats, *HSP90α* mRNA decreases after 360 min of heat stress (42 °C), while HSP90α protein decreases between 10 and 60 min of heat stress and increases after 360 and 480 min, with elevated signal in the cytoplasm at the end of heat stress (Islam et al. 2014). In pigs, mild heat shock (40.5 °C for 48 h) of cultured muscle satellite cells increases mRNA and protein abundance of HSP90 (Kamanga-Sollo et al. 2011). In cattle, both immediate and long term heat stress acclimation induces the elevation of *HSP90* mRNA and protein levels (Bharati et al. 2017). These findings indicate that *HSP90* expression could be induced by high temperature condition in different species, to enhance cell tolerance to thermal damages, to maintain the normal metabolism, and to improve the survival rate through adjusting to unfavorable condition.

Under pathological conditions, *HSP90* is found to be over-expressed in several types of cancers (Rappa et al. 2014). During differentiation of mouse embryonal carcinoma cells, *HSP90β* mRNA level is upregulated (Kohda et al. 1991), whereas during differentiation of rat osteoblast and HL-60 promyelocytic leukemia cells, mRNA levels of *HSP90α* and *HSP90β* are reduced (Shakoori et al. 1992). Therefore, the pattern of *HSP90* expression might be cell type-dependent in cancers.

In addition to aforementioned factors, environmental and chemical factors, and some physiological causes, could also affect *HSP90* expression (Pirkkala et al. 2001).

Regulation of HSP90 Expression

When stimulated by different stresses, the cellular expression of *HSP90* sharply increases, to enhance the cell repair function, in order to increase the tolerance and protect cells quickly and transiently (Jolly and Morimoto 2000; Morimoto 1998; Smith et al. 1998). In general, the rapid response of *HSP90* expression is mainly dependent on the heat shock element (HSE) in the upstream region of the promoter, which can be bound by the heat shock factor (HSF) (Somasundaram and Bhat 2004). Under physiological conditions, HSFs are present as monomers in the cytoplasm and nucleus (Pack and Ahn 2015). Under stressful conditions, HSFs are assembled into an active trimer structure and bind with HSE, and then initiate the transcription of *HSP90* (Metz et al. 1996). *HSP90α* is highly induced by elevated temperature as compared to the constitutively expressed *HSP90β* (Hoter et al. 2018). The foldsome containing HSP90 may serve as a repressor to bind and inactivate HSF1 (Voellmy and Boellmann 2007). In mouse oocytes, HSF1 deficiency could deplete both *HSP90α* transcript and protein levels, suggesting that HSF1 directly regulates *HSP90α* (Metchat et al. 2009). Thus, molecular factors or pathways regulating HSFs could also modulate HSP90 expression (Garbuz 2017). Recently, *HSP90* mRNA level is found to be modulated by *YTHDF2* deficiency, suggesting a new regulatory relationship between *HSP90* and *N*6-methyladenosine epigenetic modification (Yu et al. 2018).

23.1.1.3 HSP90 Inhibitors

HSP90, as a molecular chaperone, can bind to many client proteins, to maintain the normal physiological function of cells and tissues (Ferraro et al. 2018; Miao et al. 2018; Obermann 2018). The dimeric form and three functional domains (ATP-binding, protein-binding, dimerizing domain) of HSP90 protein are indispensable for exerting its function (Harris et al. 2004; Khalid and Paul 2014; Kitson et al. 2013). The molecular structure of HSP90 includes N-terminal domain (NTD), C-terminal domain (CTD) and middle domain (MD). A variable charger linker domain is existed in eukaryotes to link NTD and MD (Hoter et al. 2018). HSP90 could interact with three types of proteins (cochaperones, regulatory factors and substrate proteins) to exert its regulatory function (Ou et al. 2014). HSP90 inhibitors are mainly classified into N-terminal inhibitors, C-terminal inhibitors (Roy and Kapoor 2018, Tassone et al. 2018) and intermediate-domain inhibitors, depending on their binding to the protein binding site. Selective and specific HSP90 inhibitors are used potentially in curing cancers and neurodegenerative diseases (Soo et al. 2008; Luo et al. 2010; Chatterjee and Burns 2017).

N-Terminal Inhibitors of HSP90

ATP-binding pocket locates in the NTD of HSP90, vital for the ATPase activity, to maintain the chaperone function through binding the client protein (Hoter et al. 2018). Geldanamycin (GA), a benzoquinone ansamycin isolated from fermentation broths of streptomyces hygroscopicus, is the first HSP90 inhibitor discovered, and has been studied extensively and deeply (Whitesell et al. 1992). GA inhibits the binding of substrate protein to HSP90 by competitively binding to the ATP/ADP binding site, and leads to degradation of the substrate protein through the proteasome-ubiquitination pathway, thereby blocking the relevant signaling pathway (Soga et al. 2013). In highly metastatic human osteosarcoma 143B cells, GA treatment causes a subsequent upregulation of cytoplasmic *HSP90*, to partially account for cancer progress and drug resistance (Gorska et al. 2013). Although effective in showing preclinical anti-tumor activities (Fukuyo et al. 2010), GA has some shortcomings, such as the hepatotoxicity, poor solubility, and low in vivo stability, which limit its clinical application (Zhao et al. 2018). 17-allylamino-17-demethoxygeldanamycin (17-AAG), as the derivative of GA, is more stable and less toxic than GA (Wagatsuma et al. 2017; Whitesell and Lindquist 2005), which has been widely used in human cancer chemotherapy (Sausville et al. 2003; Singh et al. 2015; Weber et al. 2017; Xu et al. 2015) by inducing cell apoptosis and mitotic arrest (Zhao et al. 2016; Zhao et al. 2017). In canine osteosarcoma cells, 17-AAG treatment for 48 h increased *HSP90* mRNA level in one cell line, but decreased *HSP90* mRNA level in another one (Massimini et al. 2017). Our work previously showed that 17-AAG treatment of porcine cumulus-oocyte complexes (COCs) during in vitro maturation (IVM) induced the significant decrease of HSP90α protein level in oocytes collected at 24 h and 44 h of IVM, suggesting that 17-AAG could interfere with HSP90α protein expression to disturb its function (Liu et al. 2018).

Ganetespib (STA-9090), a non-geldanamycin resorcinol-containing triazolone compound, could bind also to the ATP-binding domain in the NTD of HSP90, to inhibit HSP90 activity in non-small cell lung cancer (Shimamura et al. 2012). Radicicol (RAD) is another antibiotic inhibitor produced from natural product, possibly inhibiting HSP90 activity by targeting nucleotide-binding domain (Kitson and Moody 2013). WK-88-1, a GA derivative, is a non-benzoquinone HSP90 inhibitor, obtained from genetically engineered Streptomyces hygroscopicus (Zhao et al. 2018). Reports have shown that WK-88-1 could suppress the growth of non-small cell lung cancers (Hong et al. 2014; Jang et al. 2014). As compared to GA and 17-AAG, WK-88-1 was shown to cause no detectable hepatotoxicity in mice (Zhao et al. 2018). In contrast to the action mechanism of GA and its derivatives by targeting HSP90 ATPase activity, DCZ3112, a novel derivative of triazine, could directly bind to the N-terminal region of HSP90 to interrupt the interaction of HSP90 and client protein in breast cancer, without inhibiting ATPase activity (Chen et al. 2018b). DCZ3112 can suppress the proliferation of HER2-positive breast cancer cells better than GA (Chen et al. 2018b).

C-Terminal Inhibitors of HSP90

The CTD of HSP90 contains the calmodulin binding site, the HSP90 homodimerization site and the nucleotide-binding site (Hoter et al. 2018). Novobiocin (NV), a member of the counermycin family of antibiotics, could bind to the C-terminal of HSP90 to limit its structure-activity (Audisio et al. 2014). In contrast to N-terminal inhibition of HSP90, C-terminal inhibition could avoid pro-survival signals, which provides a new and potential class of chemotherapy drug (Dlugosz and Janecka 2017). Some compounds of NV derivatives have been synthesized and their anti-proliferative activities are tested in several cancer cell lines (Kusuma et al. 2014; Wu et al. 2015).

N-Middle Inhibitors of HSP90

The main role of HSP90 MD is to combine with substrate peptide (client proteins or other synergistic molecular chaperones), and to modulate HSP90 ATPase activity (Hoter et al. 2018). Sansalvamide A (San A) is a natural product of pentadepsipeptide, produced from a marine fungus of the genus *Fusarium* (Kunicki et al. 2011). Sansalvamide A-amide (San A-amide), the peptide derivatives of San A, could bind to the HSP90 N-middle domain binding site to allosterically inhibit the binding of client protein to C-terminus of HSP90. Moreover, San A-amide does not influence the N-terminal binding of HSP90 client proteins (Vasko et al. 2010). These features of San A-amide provide the basis for the development of new anti-cancer drugs.

Other Small-Molecule HSP90 Inhibitors

In addition to the HSP90 inhibitors as described above, some other small-molecule HSP90 inhibitors have also been synthesized according to diverse scaffolds, including BIIB021 (Saif et al. 2014; Suzuki et al. 2015), VER-52296/NVP-AUY922 (Taldone et al. 2008), SNX-5422 and STA-9090 (Taldone et al. 2008). Some of them have already been used in clinic trials. In general, as compared to 17-AAG, these HSP90 inhibitors demonstrate many advantages, such as availability through synthesis, less resistance for multi-drugs, better water solubility and metabolic stability, and higher biological activity in a variety of tumors (He and Hu 2018; Taldone et al. 2008).

23.1.1.4 Biological Functions and Signaling Pathways

As a housekeeping protein, HSP90 is ubiquitously expressed and regulates the activity, turnover and trafficking of many proteins. Serving as an active and ATP-dependent molecular chaperone, HSP90 joins together with HSP70, HSP40 and other co-chaperone molecules to form the HSP90 foldsome, and to regulate the maturation process of its client proteins (Kitson et al. 2013). By chaperoning a diverse set of client proteins or interacting with different proteins, HSP90 plays a critical role in many biological processes under a wide variety of physiological, pathological and heat stress conditions (Lee et al. 2018).

Proliferation and Apoptosis

HSP90 is the ATP-consuming chaperone, and involves in post-translational protein maturation. Inhibitors targeting the ATP domain of HSP90 could interfere with HSP90 and induce the degradation of client proteins (Echeverría et al. 2016). Nuclear and cytoplasmic mRNA capping enzyme is the client of HSP90, and HSP90 inhibitor could reduce the level of capped mRNAs via declining capping enzyme (Trotman et al. 2018). Many of identified HSP90 client proteins are steroid receptors and protein kinases, involved in different cell signaling transduction pathways (Bunney et al. 2018; Eckl et al. 2015; Pratt and Toft 1997; Sima and Richter 2018). *HSP90* inhibition can disrupt many signaling cascades, thus promoting cell cycle arrest, blocking cell proliferation and inducing apoptosis (Budina-Kolomets et al. 2014).

Steroid Receptor Pathways Glucocorticoid receptor (GR) is an important signaling molecule, and can cause a wide range of physiological and pharmacological effects through binding with glucocorticoids (GC) (Hapgood et al. 2016). HSP90 functions as a key molecular chaperone for the GR signaling pathway by helping GR to form a mature conformation that could be bound by GC (Conway-Campbell et al. 2011; Shen et al. 2010). Without GC binding, the GR present in the cytosol forms a complex with the dimer of HSP90 (Kanelakis and Pratt 2003). With GC binding, HSP90 dimer binds with GC to form a receptor dimer, and to open the hydrophobic area of the GR with the collaborative assistance of chaperone HOP (HSP70-HSP90 organizing protein) (Ebong et al. 2016; Johnson et al. 1998). Then, the GC hormone-bound GR is transferred from the cytoplasm to the nucleus under the synergistic action of HSP90, immunophilin and actin. Subsequently, in the nucleus target genes are transcriptionally activated or inhibited via specific GR response elements on the chromosome bound with GC-HSP90-GR complex, or interacting with other transcription factors (AP1, NF-κB) (De et al. 2000; Tago et al. 2004). Similarly, HSP90 is also involved in the function of other receptors, such as aryl-hydrocarbon (dioxin) receptor (Kazlauskas et al. 2001), retinoid receptor (Holley and Yamamoto 1995), VEGF receptor 2 (VEGFR2) (Bruns et al. 2012), androgen receptor (Gillis et al. 2013), progesterone receptor (Patwardhan et al. 2015) and estrogen receptor (Dhamad et al. 2016; Obermann 2018).

Protein Kinase Pathways HSP90 can form protein complex with a large number of protein kinases. V-Src is the first tyrosine kinase chaperoned by HSP90 (Boczek et al. 2015; Brugge et al. 1981). More recently, *HSP90* is also found to be involved in activation of the JAK/STAT pathway (Kolosenko et al. 2014, Lin et al. 2013). In normal cells, natural conformation of HSP90 chaperone and JAK3 client protein co-exist, while in multiple myeloma, JAK3 is activated by the supermolecular chaperone structure of HSP90 (Meyer and Levine 2014). Activated JAK3 then catalyzes the phosphorylation of downstream signal transduction molecule (STAT5), subsequently regulating a series of signaling processes involved in cell proliferation and apoptosis (Li and Wang 2018; Yamada et al. 2012). In the epithelial cells of the intestinal crypt, it has been confirmed that heat shock can activate the mitogen-activated protein kinases (MAPK) pathway to mediate cell apoptosis (Chen et al. 2013). Some cell cycle-related protein kinases, cyclin-dependent kinase 4 (CDK4) and 11p110 (CDK11p110) (Mikolajczyk and Nelson 2004), greatwall kinase (Yamamoto et al. 2014) and Wee1 (Aligue et al. 1994), have also been reported to maintain their stability or activity by HSP90 and the co-chaperone CDC37/p50. In human breast cancer, inhibition of HSP90 by WK-88-1 induce G2/M cell cycle arrest via declining cell cycle factors, cyclin-dependent kinase 1 (CDK1) and cyclin B1, and other client proteins, then leading to cell apoptosis (Zhao et al. 2018). Similarly, HSP90 deficiency caused by DCZ3112 inhibitor reduces the protein kinase B (AKT) and ERK activity, inducing G1 cell cycle arrest, inhibiting cell proliferation and promoting apoptosis (Chen et al. 2018a, b). Various other kinases, such as ERBB2 (Kancha et al. 2013), c-src (Gupta et al. 2018), Phosphoinositide 3-kinase (PI3K)/AKT/mammalian target of rapamycin (mTOR) (Kim et al. 2016), NF-κB (Kim et al. 2016), and MAPK pathways (Schulte et al. 1995; Fujita et al. 2018), have been reported to be components of the HSP90 complex.

Differentiation and Development

HSP90 is reported to involve in the differentiation process of various cell types or organisms by interacting with transcription factors or co-chaperone proteins. In human induced pluripotent stem cells, HSP90β could interact directly with transcript factor HNF4A to regulate protein half-life and thereby affect hepatocyte differentiation (Jing et al. 2017). In murine 3T3-L1 preadipocytes, HSP90 could stabilize the transcription factor, the peroxisome proliferator-activated receptor-γ (PPARγ), to regulate adipose differentiation (Nguyen et al. 2013). During the differentiation of myoblasts into myotubes, HSP90β interacts with the muscle-specific HSP90 co-chaperone Aarsd1L to regulate muscle differentiation (Echeverría et al. 2016). As for the involvement of *HSP90* in oogenesis, oocyte growth and early embryo development, it will be reviewed in section of *HSP90 And Female Reproductive Events* in this chapter.

Disease Development

As a molecular chaperone and anti-apoptotic factor, previous studies have shown that HSP90 is involved in various cancers (Neckers 2007), neurodegenerative diseases (Luo et al. 2010; Ou et al. 2014), ovary autoimmunity (Pires and Khole 2009) and cystic ovarian disease (Velázquez et al. 2013). Inhibition of *HSP90* could disrupt the Janus kinase-signal transducer and activator of transcription (JAK-STAT), MAPK and PI3K/AKT signaling pathways to kill primary acute myeloid leukaemia cells expressing mutant FLT3 (FMS-Like Tyrosine kinase-3) (Al Shaer et al. 2008; Weisberg et al. 2009). In Alzheimer's disease, inhibition of *HSP90* could protect against protein toxicity via activation of HSF-1 and induction of HSP70 (Luo et al. 2010; Ou et al. 2014). Details regarding HSP90 and reproductive diseases will be summarized in section of *Under Pathological Conditions* in this chapter.

23.1.2 Female Reproductive Events in Animals

23.1.2.1 Process of Female Reproductive Events

Essential events of animal female reproduction, including ovary/follicule/oocyte development, fertilization and early embryo development, are summarized here. Main processes of ovary genesis and oogenesis in most mammals include the specification of primordial germ cells (PGCs), migration of PGCs to the genital ridges, formation of fetal ovary, folliculogenesis and oogenesis in ovary. Folliculogenesis refers to the physiological process, in which follicles develop from primordial follicles into primary follicles, secondary follicles, and tertiary follicles (Almeida et al. 2018; Zhuandi et al. 2018), accompanied by oocyte growth and maturation, granulosa cell proliferation and differentiation (King et al. 1986), transcription and deposition of maternal RNAs (De Sousa et al. 1998; Fair et al. 1997; Liu et al. 2013a), meiotic arrest and resumption (Liu et al. 2013a; Sen and Caiazza 2013). Oogenesis is a complex and highly orchestrated process in which germ cells differentiate into mature oocytes (Eppig 2001), together with the recruitment and selection of primordial follicles to further develop to antral follicles. After pellucid zone forms around oocyte during follicular development (Ernst et al. 2018), the gap junction protrudes through the pellucid zone (Allworth and Albertini 1993), to facilitate the communication of materials and signals between oocyte and cumulus cells, which plays a crucial role in oocyte growth and maturation, in order to obtain the full competence of fertilization and embryo development (de Souza et al. 2018; Elis et al. 2017). From oogenesis to puberty, nuclear of mammalian oocyte arrests at prophase of meiosis I, although oocyte increases its size with more substances deposited. Till puberty, stimuli of neuronal, hormonal and molecular signals induce mammalian fully-grown germinal vesicle (GV) oocytes to resume meiotic maturation and ovulation (Eppig 1996). Mature haploid oocytes are fertilized, and embryogenesis begins. Before implantation, mammalian fertilized oocytes could cleave and develop to blastocyst, terming as early embryo development (Godini and Fallahi 2019). These reproductive and developmental programs are orchestrated by a complex signaling network, some of which involves HSP90 function are summarized as follows.

23.1.2.2 Principle Pathways

PKA/WEE1/CDK1

The WEE1 kinase family is vital to regulate meiotic arrest in oocytes by phosphorylating Tyr 15 to inactivate CDK1 (Han et al. 2005; Nakanishi et al. 2000; Parker and Piwnica-Worms 1992). In mouse, WEE1B (WEE1 homolog 2, also known as WEE2) could be phosphorylated by protein kinase A (PKA) at Ser 15 to negatively regulate maturation promoting factor (MPF) activity and affect the first mitosis of 1-cell embryos (Liu et al. 2013b).While in pig oocytes, the phosphorylation site of WEE1B by PKA is the Ser 77, which is different from the site on Tyr 15 of mouse WEE1B (Shimaoka et al. 2011). These findings indicate that PKA/WEE1/CDK1 pathway plays roles in both oocytic meiosis and embryonic mitosis.

MPF and MAPK

The MPF complex and MAPK cascade (Mos-MEK-ERK1/2) are two key signaling pathways to regulate meiotic resumption and maturation of mammalian oocytes (Zhang et al. 2011). MPF activity is vital for oocyte meiotic resumption (Adhikari and Liu 2014) and maintenance of metaphase II (MII) arrest (Oh et al. 2013), while MAPK activity is required for germinal vesicle breakdown (GVBD) (Inoue et al. 1998; Li et al. 2008), spindle assembly (Petrunewich et al. 2009), and also metaphase II (MII) arrest (Gordo et al. 2001; Sha et al. 2017).

Cytoskeleton Organization

Cytoskeleton (actin and tubulin) assembly is important for oocyte meiotic maturation (Chen et al. 2018a; Wang et al. 2015b) and early embryo development (Howe and FitzHarris 2013; Wang et al. 2015a). HSP90 in vitro could bind actin filaments (Koyasu et al. 1986), and in vivo also interacts with cytoplasmic filaments (Csermely et al. 1998), tubulin (Weis et al. 2010), myosin (Etard et al. 2008), and other cytoplasmic skeleton components (Akner et al. 1992). Moreover, HSP90 can modulate the bundling of actin filaments (Park et al. 2007). These findings demonstrate the essential role of HSP90 in maintaining cytoskeleton architecture (Fostinis et al. 1992).

23.1.2.3 HSP90 and Female Reproductive Events

Under Physiological Conditions

Follicular Development In bovine, HSP90 protein is detected by immunostaining to be in the nucleus and cytoplasm of all types of cells in ovary, with greater amount in granulosa cells than in theca cells, showing increased expression from primary and secondary follicles to tertiary follicles. Expression of *HSP90* in atretic follicles within bovine ovary was similar to that in tertiary follicles (Velazquez et al. 2010). The increasing expression of *HSP90* during normal follicular development of bovine ovaries suggests its role in regulating follicular maturation. In pigs, HSP90 protein expresses and locates in granulosa cells, theca cells and follicles at all developmental stages within ovaries (Pennarossa et al. 2012). In rat cultured granulosa cells, *HSP90* expression increases when treating cells by gonadotropins, possibly through regulating mRNA translational efficiency, which supports *HSP90*'s regulatory role in steroid hormone action (Ben-Ze'ev and Amsterdam 1989). Hexavalent chromium treatment of rat granulosa cells downregulates HSP90 protein level and alters the translocation of HSP90 to mitochondria (Banu et al. 2011). Using atrazine to treat female rats by gavage for 30 days induces higher *HSP90* expression in follicular granulosa cells as compared to the control group, with signals detected mainly in the cytoplasm (Juliani et al. 2008). In cattle dominant follicles, HSP90 in follicular wall fragments and fluid is shown to regulate follicular maturation via inhibiting aromatase activity (Driancourt et al. 1999).

Oocyte Growth and Maturation Previous studies detected the expression and subcellular localization of *HSP90* during oogenesis, oocyte growth and maturation. *HSP90* is highly expressed in the cytoplasm of primordial germ cells in gonad of rat fetus at 13 day post-coital (dpc) (Ohsako et al. 1995). Inhibition of *HSP90* by GA during PGC migration reduces the number of PGCs in the gonads of the mouse embryo (Lejong et al. 2018). In rat ovary, *HSP90* is expressed higher in the oocyte within primordial follicle than that in the oocyte within antral follicle (Juliani et al. 2008). In mouse ovary, transcription level of *HSP90α* is relatively higher than *HSP90β* isoform and other HSP family members; HSP90α protein is concentrated in the cytoplasm of fully-grown oocytes, and HSP90β in granulosa cells (Metchat et al. 2009). Whereas, another report in mice shows that HSP90 protein is localized on plasma membrane of the major mature oocytes, and in the region opposite to the chromatin pole of some eggs (Calvert et al. 2003). In pig ovary, HSP90 protein is found in oocytes within follicles at all developmental stages (Pennarossa et al. 2012). Our work also shows that endogenous maternal *HSP90α* constantly expresses at both mRNA and protein levels in porcine maturing oocytes. HSP90α protein exists in both oocyte cytoplasm and nuclei areas, but no signal in the nucleolus region of GV oocytes, and no significant enrichment of signal in any other stained regions, either (Liu et al. 2018). The differences of *HSP90* localization could be due to species, developmental stages, and types of samples used. In different animal species, HSP90 is shown to have a diverse and vital role in the process of oogenesis, oocyte growth and maturation. During Drosophila oogenesis, HSP90 and the translational repressor Cup co-localize in the cytoplasm of the developing germ-line cells, and in vitro these two proteins interact each other, suggesting the possible role of HSP90 during female germ-line development (Pisa et al. 2009). In *Caenorhabditis elegans*, DAF-21, a homologue of HSP90, regulates the meiotic prophase/metaphase transition during oocyte development by ensuring the production of functional WEE-1.3 (Inoue et al. 2006). After translated during maturation of xenopus oocyte, Mos interacts sequentially with HSP70 and HSP90 to activate Mos via HSP90-dependent phosphorylation of Mos at Ser 3 site and subsequently activate MAPK cascade, to regulate oocyte meiotic maturation (Fisher et al. 2000). While in *Xenopus* mature oocytes, centrin (centrosomal protein) associates with the complex constituted by HSP70 and HSP90, and upon activation, the complex formed between centrin and HSP dissociates (Uzawa et al. 1995). Also, in Xenopus oocytes, both in vitro and in vivo experiments show that HSP90 inhibition reduces the nuclear export of 60S ribosomal subunits, suggesting HSP90 probably chaperone the unidirectional process of exporting mature 60S ribosome out of nuclear into cytoplasm (Schlatter et al. 2002). In mice, depleted GRP94 during oogenesis using Zp3-cre system perturbs the formation of zona pellucida, but is compatible with oocyte growth and meiotic maturation (Audouard et al. 2011). HSP90 depleted mouse oocytes induced by HSF1 knock-out delays GVBD (G2/M transition), partially blocks meiosis I and impairs asymmetrical division, showing similar developmental anomalies to 17-AAG treated oocytes, through disturbing meiotic kinases of CDK1 and MAPK (Metchat et al. 2009). Treatment of mouse COCs during IVM by hyperglycaemic mimetic glucosamine increases the level of O-GlcNAcylation catalyzed by beta-O-linked *N*-acetylglucosamine transferase (OGT). Further analysis shows the co-localization of HSP90 and OGT in both the oocytes and cumulus cells of COCs and confirms a direct interaction between these two proteins (Frank et al. 2014). Inhibition of *HSP90* during IVM of mouse COCs by 17-AAG doesn't affect the development of mature oocytes into blastocysts, while *HSP90* inhibition could restore the blastocyst development rates after hyperglycaemic mimetic glucosamine treatment during IVM, demonstrating an association between HSP90 and OGT to affect subsequent development ability of oocytes (Frank et al. 2014). In pigs, our work showed that inhibition of *HSP90* function by 17-AAG inhibitor decreases HSP90α protein level, delays GVBD, and impedes the first polar body (PB1) extrusion of oocytes, then subsequently impairs the development competence of treated mature oocytes (as indicated by the lower cleavage and higher fragmentation rates of parthenotes), showing similar developmental defects to HSP90α antibody microinjected oocytes. 17-AAG treatment of porcine maturing oocytes decreases p-ERK1/2 level, and disturbs actin/spindle assembly, indicating that *HSP90α* could couple with MAPK to regulate cytoskeleton structure and orchestrate meiotic maturation of porcine oocytes (Liu et al. 2018). Taken together, all evidences point to the important role of *HSP90* in oogenesis, oocyte growth and meiotic maturation.

Fertilization and Early Embryo Development *HSP90* is expressed constitutively throughout early embryogenesis in Xenopus (Ali et al. 1996; Taherian et al. 2010).

Depletion of GRP94 results in embryonic death on day 7 of gestation, suggesting its vital role for embryo development (Wanderling et al. 2007). To avoid embryonic lethality, using Zp3-cre system to conditional knockout oocyte-specific *GRP94* during mouse oogenesis, loss of maternal *GRP94* doesn't change follicular growth, meiotic maturation and fertilization, but blocks G2/M transition and induces abnormal mitotic spindle of zygotes during the first mitosis, leading to developmental arrest at 1-cell stage, indicating that *GPR94* is required for the first cleavage of embryonic development (Audouard et al. 2011). However, the scarce information obtained here necessitates further investigation to explore the intrinsic link between *HSP90* and embryo development.

Under Cold or Heat Shock

During mouse oogenesis, the response to heat shock is maximal in growing oocytes, decreases with oocytes growing to full size, and finally shuts off in preovulatory oocytes (Curci et al. 1991). Exposure of gilt whole ovary to 41.5 °C for 1 h boosts the level of *HSF1* and *HSF2* transcripts in oocytes, without changing *HSP90α* and *HSP90β* levels, which is consistent with the in vivo results from ovaries collected in summer and winter (Pennarossa et al. 2012). Using cryotop and solid surface vitrification methods to vitrify Yak COCs decreases the *HSP90* expression level in mature oocytes and two to eight cell embryos (Pan et al. 2015). Acute heat shock (45 °C for 2 h) to camel oocytes delays meiotic maturation, reduces ooplasmic diameter and elevates chromosomal abnormalities, together with threefold increase of *HSP90* transcripts in contrast to sevenfold increase in cumulus cells, indicating less tolerance of camel oocytes than cumulus cells to acute heat shock (Saadeldin et al. 2018). Novobiocin treatment of Xenopus oocytes followed by heat shock suggests that the initial steps of HSF1 trimerization requires ATP-binding activity of HSP90 CTD, while GA treatment followed by heat shock suggests that disassembly of activated HSF1 trimers needs the ATPase and chaperone activities of HSP90 NTD, indicating that HSF1 activity is modulated by HSP90 (Conde et al. 2009). Protein phosphatase 5 (PP5) could interact with HSF1-HSP90 complex, negatively regulating HSF1 activity after heat shock of Xenopusoocytes, but did not affect HSF1 activity in unstressed oocytes (Conde et al. 2005). Although *HSP90* transcription is active throughout Xenopus embryogenesis, it can only be induced by heat stress in late blastula, gastrula and neurula stage embryos (Ali et al. 1996). Taken together, *HSP90* is involved in regulating the response of oocytes or embryos to cold or heat stresses.

Under Pathological Conditions

In patients with infertility, HSP90 auto-antibodies have been observed (Pires 2017). Subsequently, the production of HSP90β auto-antibodies is found in human ovarian autoimmunity (Pires and Khole 2009). Furthermore, in an animal model, the deleterious effect of HSP90β auto-antibodies on ovarian function is shown, with increased number of empty and degenerated follicles (Choudhury and Khole 2015), and extensive granulosa cell death (Choudhury and Khole 2013). In bovine, HSP90 protein as detected by immunostaining shows the greatest level in cystic follicles than normal follicles at tertiary and atretic stages (Velazquez et al. 2010). The elevated *HSP90*, together with other *HSP*, might inhibit apoptosis of the follicular wall, and delay the regression of cystic follicles (Velazquez et al. 2010). *HS*P90 is also activated in stress-induced rat model of polycystic ovarian syndrome, via dysregulation of serum reproductive hormone level and steroid hormone receptor expression (Park et al. 2012, 2017).

Conclusions In conclusion, *HSP90* is closely related to homeostatic regulation of folliculogenesis, oogenesis and embryo development in different animal species. *HSP90* also contributes to the development and progress of diseases in female reproductive system. Regarding the complexity of HSP90 involved regulatory of more than 100 proteins, further investigations, possibly combined with the application of single cell biotechnology, might be performed on oocytes and early embryos for better understanding the mechanisms of *HSP90* in female reproduction, thereby providing useful strategy for improving animal reproduction and human fertility.

Acknowledgements Our original work cited herein was supported by the National Natural Science Foundation of China (grant number 31472098) and the Start-up grant from Northeast Agricultural University (C.X.Y.).

References

Adhikari D, Liu K (2014) The regulation of maturation promoting factor during prophase I arrest and meiotic entry in mammalian oocytes. Mol Cell Endocrinol 382:480–487

Akner G, Mossberg K, Sundqvist KG, Gustafsson JA, Wikström AC (1992) Evidence for reversible, non-microtubule and non-microfilament-dependent nuclear translocation of hsp90 after heat shock in human fibroblasts. Eur J Cell Biol 58:356–364

Al Shaer L, Walsby E, Gilkes A, Tonks A, Walsh V, Mills K, Burnett A, Rowntree C (2008) Heat shock protein 90 inhibition is cytotoxic to primary AML cells expressing mutant FLT3 and results in altered downstream signalling. Br J Haematol 141:483–493

Ali A, Krone PH, Pearson DS, Heikkila JJ (1996) Evaluation of stress-inducible hsp90 gene expression as a potential molecular biomarker in *Xenopus laevis*. Cell Stress Chaperones 1:62–69

Aligue R, Akhavan-Niak H, Russell P (1994) A role for Hsp90 in cell cycle control: Wee1 tyrosine kinase activity requires interaction with Hsp90. EMBO J 13:6099–6106

Allworth AE, Albertini DF (1993) Meiotic maturation in cultured bovine oocytes is accompanied by remodeling of the cumulus cell cytoskeleton. Dev Biol 158:101–112

Almeida FRCL, Costermans NGJ, Soede NM, Bunschoten A, Keijer J, Kemp B, Teerds KJ (2018) Presence of anti-müllerian hormone (AMH) during follicular development in theporcine ovary. PLoS One 13:e0197894

Audisio D, Methy-Gonnot D, Radanyi C, Renoir JM, Denis S, Sauvage F, Vergnaud-Gauduchon J, Brion JD, Messaoudi S, Alami M (2014) Synthesis and antiproliferative activity of novobiocin analogues as potential hsp90 inhibitors. Eur J Med Chem 83:498–507

Audouard C, Le Masson F, Charry C, Li Z, Christians ES (2011) Oocyte-targeted deletion reveals that hsp90b1 is needed for the completion of first mitosis in mouse zygotes. PLoS One 6:e17109
Banu SK, Stanley JA, Lee J, Stephen SD, Arosh JA, Hoyer PB, Burghardt RC (2011) Hexavalent chromium-induced apoptosis of granulosa cells involves selective sub-cellular translocation of Bcl-2 members, ERK1/2 and p53. Toxicol Appl Pharmacol 251:253–266
Ben-Ze'ev A, Amsterdam A (1989) Regulation of heat shock protein synthesis by gonadotropins in cultured granulosa cells. Endocrinology 124:2584–2594
Bharati J, Dangi SS, Bag S, Maurya VP, Singh G, Kumar P, Sarkar M (2017) Expression dynamics of HSP90 and nitric oxide synthase (NOS) isoforms during heat stress acclimation in Tharparkar cattle. Int J Biometeorol 61:1461–1469
Boczek EE, Reefschläger LG, Dehling M, Struller TJ, Häusler E, Seidl A, Kaila VR, Buchner J (2015) Conformational processing of oncogenic v-Src kinase by the molecular chaperone Hsp90. Proc Natl Acad Sci USA 112:E3189–E3198
Brugge JS, Erikson E, Erikson RL (1981) The specific interaction of the Rous sarcoma virus transforming protein, pp60src, with two cellular proteins. Cell 25:363–372
Bruns AF, Yuldasheva N, Latham AM, Bao L, Pellet-Many C, Frankel P, Stephen SL, Howell GJ, Wheatcroft SB, Kearney MT, Zachary IC, Ponnambalam S (2012) A heat-shock protein axis regulates VEGFR2 proteolysis, blood vessel development and repair. PLoS One 7:e48539
Buchner J (1996) Supervising the fold: functional principles of molecularchaperones. FASEB J 10:10–19
Budina-Kolomets A, Balaburski GM, Bondar A, Beeharry N, Yen T, Murphy ME (2014) Comparison of the activity of three different HSP70 inhibitors on apoptosis, cellcycle arrest, autophagy inhibition, and HSP90 inhibition. Cancer Biol Ther 15:194–199
Bunney TD, Inglis AJ, Sanfelice D, Farrell B, Kerr CJ, Thompson GS, Masson GR, Thiyagarajan N, Svergun DI, Williams RL, Breeze AL, Katan M (2018) Disease variants of FGFR3 reveal molecular basis for the recognition and additional roles for Cdc37 in Hsp90 chaperone system. Structure 26:446–458
Calvert ME, Digilio LC, Herr JC, Coonrod SA (2003) Oolemmal proteomics–identification of highly abundant heat shock proteins and molecular chaperones in the mature mouse egg and their localization on the plasma membrane. Reprod Biol Endocrinol 1:27
Chatterjee S, Burns TF (2017) Targeting heat shock proteins in cancer: a promising therapeutic approach. Int J Mol Sci 18:1978. https://doi.org/10.3390/ijms18091978
Chen J, Terada N, Saitoh Y, Huang Z, Ohno N, Ohno S (2013) Detection of MAPK signal transduction proteins in an ischemia/reperfusion model of mouse intestine using in vivo cryotechnique. Histochem Cell Biol 140:491–505
Chen F, Jiao XF, Zhang JY, Wu D, Ding ZM, Wang YS, Miao YL, Huo LJ (2018a) Nucleoporin35 is a novel microtubule associated protein functioning in oocyte meiotic spindle architecture. Exp Cell Res 371:435–443
Chen X, Liu P, Wang Q, Li Y, Fu L, Fu H, Zhu J, Chen Z, Zhu W, Xie C, Lou L (2018b) DCZ3112, a novel Hsp90 inhibitor, exerts potent antitumor activity against HER2-positive breast cancer through disruption of Hsp90-Cdc37 interaction. Cancer Lett 434:70–80
Choudhury A, Khole VV (2013) HSP90 antibodies: a detrimental factor responsible for ovarian dysfunction. Am J Reprod Immunol 70:372–385
Choudhury A, Khole VV (2015) Immune-mediated destruction of ovarian follicles associated with the presence of HSP90 antibodies. Mol Reprod Dev 82:81–89
Conde R, Xavier J, McLoughlin C, Chinkers M, Ovsenek N (2005) Protein phosphatase 5 is a negative modulator of heat shock factor 1. J Biol Chem 280:28989–28996
Conde R, Belak ZR, Nair M, O'Carroll RF, Ovsenek N (2009) Modulation of Hsf1 activity by novobiocin and geldanamycin. Biochem Cell Biol 87:845–851
Conway-Campbell BL, George CL, Pooley JR, Knight DM, Norman MR, Hager GL, Lightman SL (2011) The HSP90 molecular chaperone cycle regulates cyclical transcriptional dynamics of the glucocorticoid receptor and its coregulatory molecules CBP/p300 during ultradian ligand treatment. Mol Endocrinol 25:944–954

Csermely P, Schnaider T, Soti C, Prohászka Z, Nardai G (1998) The 90-kDa molecular chaperone family: structure, function, and clinical applications. A comprehensive review. Pharmacol Ther 79:129–168
Curci A, Bevilacqua A, Fiorenza MT, Mangia F (1991) Developmental regulation of heat-shock response in mouse oogenesis: identification of differentially responsive oocyte classes during Graafian follicle development. Dev Biol 144:362–368
De Bosscher K, Vanden Berghe W, Haegeman G (2000) Mechanisms of anti-inflammatoryaction and of immunosuppression by glucocorticoids: negative interference of activated glucocorticoid receptor with transcription factors. J Neuroimmunol 109:16–22
De Sousa PA, Watson AJ, Schultz GA, Bilodeau-Goeseels S (1998) Oogenetic and zygotic gene expression directing early bovine embryogenesis: a review. Mol Reprod Dev 51:112–121
de Souza DK, Salles LP, Camargo R, Gulart LVM, Costa E, Silva S, de Lima BD, Torres FAG, Rosa E, Silva AAM (2018) Effects of PI3K and FSH on steroidogenesis, viability and embryo development of the cumulus-oocyte complex after in vitroculture. Zygote 26:50–61
Dhamad AE, Zhou Z, Zhou J, Du Y (2016) Systematic proteomic identification of the heat shock proteins (Hsp) that interact with Estrogen Receptor Alpha (ERα) and biochemical characterization of the ERα-Hsp70 interaction. PLoS One 11:e0160312
Dlugosz A, Janecka A (2017) Novobiocin analogs as potential anticancer agents. Mini Rev Med Chem 17:728–733
Driancourt MA, Guet P, Reynaud K, Chadli A, Catelli MG (1999) Presence of an aromatase inhibitor, possibly heat shock protein 90 in dominant follicles of cattle. J Reprod Fertil 115:45–58
Ebong IO, Beilsten-Edmands V, Patel NA, Morgner N, Robinson CV (2016) The interchange of immunophilins leads to parallel pathways and different intermediates in the assembly of Hsp90 glucocorticoid receptor complexes. Cell Discov 2:16002
Echeverría PC, Briand PA, Picard D (2016) A remodeled Hsp90 molecular chaperone ensemble with the novel cochaperone Aarsd1 is required for muscle differentiation. Mol Cell Biol 36:1310–1321
Eckl JM, Scherr MJ, Freiburger L, Daake MA, Sattler M, Richter K (2015) Hsp90·Cdc37 complexes with protein kinases form cooperatively with multiple distinct interaction sites. J Biol Chem 290:30843–30854
Elis S, Oseikria M, Vitorino Carvalho A, Bertevello PS, Corbin E, Teixeira-Gomes AP, Lecardonnel J, Archilla C, Duranthon V, Labas V, Uzbekova S (2017) Docosahexaenoic acid mechanisms of action on the bovine oocyte-cumulus complex. J Ovarian Res 10:74
Eppig JJ (1996) Coordination of nuclear and cytoplasmic oocyte maturation in eutherian mammals. Reprod Fertil Dev 8:485–489
Eppig JJ (2001) Oocyte control of ovarian follicular development and function in mammals. Reproduction 122:829–838
Ernst EH, Franks S, Hardy K, Villesen P, Lykke-Hartmann K (2018) Granulosa cells from human primordial and primary follicles show differential global gene expression profiles. Hum Reprod 33:666–679
Etard C, Roostalu U, Strähle U (2008) Shuttling of the chaperones Unc45b and Hsp90a between the A band and the Z line of the myofibril. J Cell Biol 180:1163–1175
Fair T, Hulshof SC, Hyttel P, Greve T, Boland M (1997) Nucleus ultrastructure and transcriptional activity of bovine oocytes in preantral and early antral follicles. Mol Reprod Dev 46:208–215
Ferraro M, D'Annessa I, Moroni E, Morra G, Paladino A, Rinaldi S, Compostella F, Colombo G (2018) Allosteric modulators of HSP90 and HSP70: dynamics meets function through structure-based drug design. J Med Chem 62:60–87
Fisher DL, Mandart E, Dorée M (2000) Hsp90 is required for c-Mos activation and biphasic MAP kinase activation in Xenopus oocytes. EMBO J 19:1516–1524
Fostinis Y, Theodoropoulos PA, Gravanis A, Stournaras C (1992) Heat shock protein HSP90 and its association with the cytoskeleton: a morphological study. Biochem Cell Biol 70:779–786

Frank LA, Sutton-McDowall ML, Brown HM, Russell DL, Gilchrist RB, Thompson JG (2014) Hyperglycaemic conditions perturb mouse oocyte in vitro developmental competence via beta-O-linked glycosylation of heat shock protein 90. Hum Reprod 29:1292–1303

Fujita K, Otsuka T, Kawabata T, Kainuma S, Sakai G, Matsushima-Nishiwaki R, Kozawa O, Tokuda H (2018) HSP90 limits thrombin-stimulated IL-6 synthesis in osteoblast-like MC3T3-E1 cells: regulation of p38 MAPK. Int J Mol Med 42:2185–2192

Fukuyo Y, Hunt CR, Horikoshi N (2010) Geldanamycin and its anti-cancer activities. Cancer Lett 290:24–35

Gao Y, Yechikov S, Vazquez AE, Chen D, Nie L (2013) Distinct roles of molecular chaperones HSP90α and HSP90β in the biogenesis of KCNQ4 channels. PLoS One 8:e57282

Garbuz DG (2017) Regulation of heat shock gene expression in response to stress. Mol Biol (Mosk) 51:400–417

Gillis JL, Selth LA, Centenera MM, Townley SL, Sun S, Plymate SR, Tilley WD, Butler LM (2013) Constitutively-active androgen receptor variants function independently of the HSP90 chaperone but do not confer resistance to HSP90 inhibitors. Oncotarget 4:691–704

Godini R, Fallahi H (2019) Dynamics changes in the transcription factors during early human embryonic development. J Cell Physiol 234:6489–6502

Gordo AC, He CL, Smith S, Fissore RA (2001) Mitogen activated protein kinase plays a significant role in metaphase II arrest, spindle morphology, and maintenance of maturation promoting factor activity in bovine oocytes. Mol Reprod Dev 59:106–114

Gorska M, Marino Gammazza A, Zmijewski MA, Campanella C, Cappello F, Wasiewicz T, Kuban-Jankowska A, Daca A, Sielicka A, Popowska U, Knap N, Antoniewicz J, Wakabayashi T, Wozniak M (2013) Geldanamycin-induced osteosarcoma cell death is associated with hyperacetylation and loss of mitochondrial pool of heat shock protein 60 (hsp60). PLoS One 8:e71135

Gupta S, Lee CM, Wang JF, Parodo J, Jia SH, Hu J, Marshall JC (2018) Heat-shock protein-90 prolongs septic neutrophil survival by protecting c-Src kinase and caspase-8 from proteasomal degradation. J Leukoc Biol 103:933–944

Han SJ, Chen R, Paronetto MP, Conti M (2005) Wee1B is an oocyte-specific kinase involved in the control of meiotic arrest in the mouse. Curr Biol 15:1670–1676

Hapgood JP, Avenant C, Moliki JM (2016) Glucocorticoid-independent modulation of GR activity: implications for immunotherapy. Pharmacol Ther 165:93–113

Harris SF, Shiau AK, Agard DA (2004) The crystal structure of the carboxy-terminal dimerization domain of htpG, the Escherichia coli Hsp90, reveals a potential substrate binding site. Structure 12:1087–1097

He W, Hu H (2018) BIIB021, an Hsp90 inhibitor: a promising therapeutic strategy for blood malignancies (review). Oncol Rep 40:3–15

Holley SJ, Yamamoto KR (1995) A role for Hsp90 in retinoid receptor signal transduction. Mol Biol Cell 6:1833–1842

Hong YS, Jang WJ, Chun KS, Jeong CH (2014) Hsp90 inhibition by WK88-1 potently suppresses the growth of gefitinib-resistant H1975 cells harboring the T790M mutation in EGFR. Oncol Rep 31:2619–2624

Hoter A, El-Sabban ME, Naim HY (2018) The HSP90 family: structure, regulation, function, and implications in health and disease. Int J Mol Sci 19:2560. https://doi.org/10.3390/ijms19092560

Howe K, FitzHarris G (2013) Recent insights into spindle function in mammalian oocytes and early embryos. Biol Reprod 89:71

Huang T, Chen S, Han H, Li H, Huang Z, Zhang J, Yin Q, Wang X, Ma X, Dai P, Duan D, Zou F, Chen X (2014) Expression of Hsp90α and cyclin B1 were related to prognosis of esophageal squamous cell carcinoma and keratin pearl formation. Int J Clin Exp Pathol 7:1544–1552

Inoue M, Naito K, Nakayama T, Sato E (1998) Mitogen-activated protein kinase translocates into the germinal vesicle and induces germinal vesicle breakdown in porcine oocytes. Biol Reprod 58:130–136

Inoue T, Hirata K, Kuwana Y, Fujita M, Miwa J, Roy R, Yamaguchi Y (2006) Cell cycle control by daf-21/Hsp90 at the first meiotic prophase/metaphase boundary during oogenesis in *Caenorhabditis elegans*. Develop Growth Differ 48:25–32

Islam A, Rehana B, Zhang M, Liu ZJ, Tang S, Hartung J, Bao ED (2014) Expression of heat shock protein 90 alpha (Hsp90α) in primary neonatal rat myocardial cells exposed to various periods of heat stress in vitro. Genet Mol Res 13:2806–2816

Jang WJ, Jung SK, Kang JS, Jeong JW, Bae MK, Joo SH, Park GH, Kundu JK, Hong YS, Jeong CH (2014) Anti-tumor activity of WK88-1, a novel geldanamycin derivative, in gefitinib-resistant non-small cell lung cancers with Met amplification. Cancer Sci 105:1245–1253

Jérôme V, Vourc'h C, Baulieu EE, Catelli MG (1993) Cell cycle regulation of the chicken hsp90 alpha expression. Exp Cell Res 205:44–51

Jiang X, Zhai H, Wang L, Luo L, Sappington TW, Zhang L (2012) Cloning of the heat shock protein 90 and 70 genes from the beet armyworm, Spodoptera exigua, and expression characteristics in relation to thermal stress and development. Cell Stress Chaperones 17:67–80

Jing R, Duncan CB, Duncan SA (2017) A small-molecule screen reveals that HSP90β promotes the conversion of induced pluripotent stem cell-derived endoderm to a hepatic fate and regulates HNF4A turnover. Development 144:1764–1774

Johnson JL (2012) Evolution and function of diverse Hsp90 homologs and cochaperone proteins. Biochim Biophys Acta 1823:607–613

Johnson BD, Schumacher RJ, Ross ED, Toft DO (1998) Hop modulates Hsp70/Hsp90 interactions in protein folding. J Biol Chem 273:3679–3686

Jolly C, Morimoto RI (2000) Role of the heat shock response and molecular chaperonesin oncogenesis and cell death. J Natl Cancer Inst 92:1564–1572

Juliani CC, Silva-Zacarin EC, Santos DC, Boer PA (2008) Effects of atrazine on female Wistar rats: morphological alterations in ovarian follicles and immunocytochemical labeling of 90 kDa heat shock protein. Micron 39:607–616

Kamanga-Sollo E, Pampusch MS, White ME, Hathaway MR, Dayton WR (2011) Effects of heat stress on proliferation, protein turnover, and abundance of heat shock protein messenger ribonucleic acid in cultured porcine muscle satellite cells. J Anim Sci 89:3473–3480

Kancha RK, Bartosch N, Duyster J (2013) Analysis of conformational determinants underlying HSP90-kinase interaction. PLoS One 8:e68394

Kanelakis KC, Pratt WB (2003) Regulation of glucocorticoid receptor ligand-bindingactivity by the hsp90/hsp70-based chaperone machinery. Methods Enzymol 364:159–173

Kazlauskas A, Sundström S, Poellinger L, Pongratz I (2001) The hsp90 chaperone complex regulates intracellular localization of the dioxin receptor. Mol Cell Biol 21:2594–2607

Khalid S, Paul S (2014) Identifying a C-terminal ATP binding sites-based novel Hsp90-inhibitor in silico: a plausible therapeutic approach in Alzheimer's disease. Med Hypotheses 83:39–46

Kim SH, Kang JG, Kim CS, Ihm SH, Choi MG, Yoo HJ, Lee SJ (2016) Synergistic cytotoxicity of BIIB021 with triptolide through suppression of PI3K/Akt/mTOR and NF-κB signal pathways in thyroid carcinoma cells. Biomed Pharmacother 83:22–32

King WA, Bousquet D, Greve T, Goff AK (1986) Meiosis in bovine oocytes matured invitro and in vivo. Acta Vet Scand 27:267–279

Kitson RR, Moody CJ (2013) Learning from nature: advances in geldanamycin- and radicicol-based inhibitors of Hsp90. J Org Chem 78:5117–5141

Kitson RR, Chang CH, Xiong R, Williams HE, Davis AL, Lewis W, Dehn DL, Siegel D, Roe SM, Prodromou C, Ross D, Moody CJ (2013) Synthesis of 19-substituted geldanamycins with altered conformations and their binding to heat shock protein Hsp90. Nat Chem 5:307–314

Kohda T, Kondo K, Oishi M (1991) Cellular HSP90 (HSP86) mRNA level and in vitro differentiation of mouse embryonal carcinoma (F9) cells. FEBS Lett 290:107–110

Kolosenko I, Grander D, Tamm KP (2014) IL-6 activated JAK/STAT3 pathway and sensitivity to Hsp90 inhibitors in multiple myeloma. Curr Med Chem 21:3042–3047

Koyasu S, Nishida E, Kadowaki T, Matsuzaki F, Iida K, Harada F, Kasuga M, Sakai H, Yahara I (1986) Two mammalian heat shock proteins, HSP90 and HSP100, are actin-binding proteins. Proc Natl Acad Sci USA 83:8054–8058

Kunicki JB, Petersen MN, Alexander LD, Ardi VC, McConnell JR, McAlpine SR (2011) Synthesis and evaluation of biotinylated sansalvamide A analogs and their modulation of Hsp90. Bioorg Med Chem Lett 21:4716–4719

Kusuma BR, Khandelwal A, Gu W, Brown D, Liu W, Vielhauer G, Holzbeierlein J, Blagg BS (2014) Synthesis and biological evaluation of coumarin replacements of novobiocin as Hsp90 inhibitors. Bioorg Med Chem 22:1441–1449

Lee MG, Liu YC, Lee, YL, El-Shazly M, Lai KH, Shih SP, Ke SC, Hong MC, Du YC, Yang JC, Sung PJ, Wen ZH Lu MC (2018) Heteronemin, a marine sesterterpenoid-type metabolite, induces apoptosis in prostate LNcap cells via oxidative and ER stress combined with the inhibition of topoisomerase II and Hsp90. Mar Drugs 16, doi: https://doi.org/10.3390/md16060204

Lejong M, Choa-Duterre M, Vanmuylder N, Louryan S (2018) Geldanamycin administration reduces the amount of primordial germ cells in the mouse embryo. Morphologie 102:219–224

Li RF, Wang GF (2018) JAK/STAT5 signaling pathway inhibitor ruxolitinibreducesairway inflammation of neutrophilic asthma in mice model. Eur Rev Med Pharmacol Sci 22:835–843

Li M, Ai JS, Xu BZ, Xiong B, Yin S, Lin SL, Hou Y, Chen DY, Schatten H, Sun QY (2008) Testosterone potentially triggers meiotic resumption by activation of intra-oocyte SRC and MAPK in porcine oocytes. Biol Reprod 79:897–905

Li HL, Huang XP, Zhou XH, Ji TH, Wu ZQ, Wang ZQ, Jiang HY, Liu FR, Zhao T (2011) Correlation of seven biological factors (Hsp90a, p53, MDM2, Bcl-2, Bax, Cytochrome C, and Cleaved caspase3) with clinical outcomes of ALK+ anaplastic large-cell lymphoma. Biomed Environ Sci 24:630–641

Lin H, Kolosenko I, Björklund AC, Protsyu KD, Österborg A, Grandér D, Tamm KP (2013) An activated JAK/STAT3 pathway and CD45 expression are associated with sensitivity to Hsp90 inhibitors in multiple myeloma. Exp Cell Res 319:600–611

Liu L, Kong N, Xia G, Zhang M (2013a) Molecular control of oocyte meiotic arrest and resumption. Reprod Fertil Dev 25:463–471

Liu C, Liu Y, Liu Y, Wu D, Luan Z, Wang E, Yu B (2013b) Ser 15 of WEE1B is a potential PKA phosphorylation target in G2/M transition in one-cell stage mouse embryos. Mol Med Rep 7:1929–1937

Liu YH, Liu XM, Wang PC, Yu XX, Miao JK, Liu S, Wang YK, Du ZQ, Yang CX (2018) Heat shock protein 90α couples with the MAPK-signaling pathway to determine meiotic maturation of porcine oocytes. J Anim Sci 96:3358–3369

Luo W, Sun W, Taldone T, Rodina A, Chiosis G (2010) Heat shock protein 90 in neurodegenerative diseases. Mol Neurodegener 5:24

Massimini M, Palmieri C, De Maria R, Romanucci M, Malatesta D, De Martinis M, Maniscalco L, Ciccarelli A, Ginaldi L, Buracco P, Bongiovanni L, Della Salda L (2017) 17-AAG and apoptosis, autophagy, and mitophagy in Canine osteosarcoma cell lines. Vet Pathol 54:405–412

Metchat A, Akerfelt M, Bierkamp C, Delsinne V, Sistonen L, Alexandre H, Christians ES (2009) Mammalian heat shock factor 1 is essential for oocyte meiosis and directly regulates Hsp90 alpha expression. J Biol Chem 284:9521–9528

Metz K, Ezernieks J, Sebald W, Duschl A (1996) Interleukin-4 upregulates the heatshock protein Hsp90 alpha and enhances transcription of a reporter gene coupled toa single heat shock element. FEBS Lett 385:25–28

Meyer SC, Levine RL (2014) Molecular pathways: molecular basis for sensitivity andresistance to JAK kinase inhibitors. Clin Cancer Res 20:2051–2059

Miao W, Li L, Wang Y (2018) Identification of helicase proteins as clients for HSP90. Anal Chem 90:11751–11755

Mikolajczyk M, Nelson MA (2004) Regulation of stability of cyclin-dependent kinase CDK11p110 and a caspase-processed form, CDK11p46, by Hsp90. Biochem J 384:461–467

Millson SH, Truman AW, Rácz A, Hu B, Panaretou B, Nuttall J, Mollapour M, Söti C, Piper PW (2007) Expressed as the sole Hsp90 of yeast, the alpha and beta isoforms of human Hsp90 differ with regard to their capacities for activation of certain client proteins, whereas only Hsp90beta generates sensitivity to the Hsp90 inhibitor radicicol. FEBS J 274:4453–4463

Morimoto RI (1998) Regulation of the heat shock transcriptional response: cross talk between a family of heat shock factors, molecular chaperones, and negative regulators. Genes Dev 12:3788–3796

Nakanishi M, Ando H, Watanabe N, Kitamura K, Ito K, Okayama H, Miyamoto T, Agui T, Sasaki M (2000) Identification and characterization of human Wee1B, a new member of the Wee1 family of Cdk-inhibitory kinases. Genes Cells 5:839–847

Neckers L (2007) Heat shock protein 90: the cancer chaperone. J Biosci 32:517–530

Nguyen MT, Csermely P, Sőti C (2013) Hsp90 chaperones PPARγ and regulates differentiation and survival of 3T3-L1 adipocytes. Cell Death Differ 20:1654–1663

Obermann WMJ (2018) A motif in HSP90 and P23 that links molecular chaperones toefficient estrogen receptorα methylation by the lysine methyltransferase SMYD2. J Biol Chem 293:16479–16487

Oh JS, Susor A, Schindler K, Schultz RM, Conti M (2013) Cdc25A activity is required for the metaphase II arrest in mouse oocytes. J Cell Sci 126:1081–1085

Ohsako S, Bunick D, Hayashi Y (1995) Immunocytochemical observation of the 90 KD heat shock protein (HSP90): high expression in primordial and pre-meiotic germ cells of male and female rat gonads. J Histochem Cytochem 43:67–76

Ou JR, Tan MS, Xie AM, Yu JT, Tan L (2014) Heat shock protein 90 in Alzheimer's disease. Biomed Res Int 2014:796869. https://doi.org/10.1155/2014/796869

Pack CG, Ahn SG (2015) Heat shock-induced interactions among nuclear HSFs detected by fluorescence cross-correlation spectroscopy. Biochem Biophys Res Commun 463:303–308

Pan Y, Cui Y, Baloch AR, Fan J, He J, Zhang Y, Zheng H, Li G, Yu S (2015) Association of heat shock protein 90 with the developmental competence of immature oocytes following cryotop and solid surface vitrification in yaks (*Bos grunniens*). Cryobiology 71:33–39

Park SJ, Suetsugu S, Sagara H, Takenawa T (2007) HSP90 cross-links branched actin filaments induced by N-WASP and the Arp2/3 complex. Genes Cells 12:611–622

Park E, Cockrem JF, Han KH, Kim DH, Jung MH, Chu JP (2012) Stress-induced activation of ovarian heat shock protein 90 in a rat model of polycystic ovary syndrome. J Obstet Gynaecol Res 38:396–407

Park E, Choi CW, Kim SJ, Kim YI, Sin S, Chu JP Heo JY (2017) Hochu-ekki-to treatment improves reproductive and immune modulation in the stress-induced rat model of polycystic ovarian syndrome. Molecules 22 https://doi.org/10.3390/molecules22060978

Parker LL, Piwnica-Worms H (1992) Inactivation of the p34cdc2-cyclin B complex by the human WEE1 tyrosine kinase. Science 257:1955–1957

Patwardhan CA, Alfa E, Lu S, Chadli A (2015) Progesterone receptor chaperone complex-based high-throughput screening assay: identification of capsaicin as an inhibitor of the Hsp90 machine. J Biomol Screen 20:223–229

Pennarossa G, Maffei S, Rahman MM, Berruti G, Brevini TA, Gandolfi F (2012) Characterization of the constitutive pig ovary heat shock chaperone machinery and its response to acute thermal stress or to seasonal variations. Biol Reprod 87:119

Petrunewich MA, Trimarchi JR, Hanlan AK, Hammer MA, Baltz JM (2009) Second meiotic spindle integrity requires MEK/MAP kinase activity in mouse eggs. J Reprod Dev 55:30–38

Pires ES (2017) The unmysterious roles of HSP90: ovarian pathology and autoantibodies. Adv Anat Embryol Cell Biol 222:29–44

Pires ES, Khole VV (2009) A block in the road to fertility: autoantibodies to heat-shock protein 90-beta in human ovarian autoimmunity. FertilSteril 92:1395–1409

Pirkkala L, Nykänen P, Sistonen L (2001) Roles of the heat shock transcription factors in regulation of the heat shock response and beyond. FASEB J 15:1118–1131

Pisa V, Cozzolino M, Gargiulo S, Ottone C, Piccioni F, Monti M, Gigliotti S, Talamo F, Graziani F, Pucci P, Verrotti AC (2009) The molecular chaperone Hsp90 is a component of the cap-binding complex and interacts with the translational repressor Cup during Drosophila oogenesis. Gene 432:67–74

Pratt WB, Toft DO (1997) Steroid receptor interactions with heat shock protein and immunophilin chaperones. Endocr Rev 18:306–360

Pratt WB, Toft DO (2003) Regulation of signaling protein function and trafficking by the hsp90/hsp70-based chaperone machinery. Exp Biol Med (Maywood) 228:111–133

Rappa F, Sciume C, Lo Bello M, Bavisotto CC, Marino Gammazza A, Barone R, Campanella C, David S, Carini F, Zarcone F, Rizzuto S, Lena A, Tomasello G, Uzzo ML, Spatola GF, Bonaventura G, Leone A, Gerbino A, Cappello F, Bucchieri F, Zummo G, Farina F (2014) Comparative analysis of Hsp10 and Hsp90 expression in healthy mucosa and adenocarcinoma of the large bowel. Anticancer Res 34:4153–4159

Roy SS, Kapoor M (2018) In silico identification and analysis of the binding site for aminocmarin type inhibitors in the C-terminal domain of Hsp90. J Mol Graph Model 84:215–235

Saadeldin IM, Swelum AA, Elsafadi M, Mahmood A, Alfayez M, Alowaimer AN (2018) Differences between the tolerance of camel oocytes and cumulus cells to acute and chronic hyperthermia. J Therm Biol 74:47–54

Sagare-Patil V, Bhilawadikar R, Galvankar M, Zaveri K, Hinduja I, Modi D (2017) Progesterone requires heat shock protein 90 (HSP90) in human sperm to regulate motility and acrosome reaction. J Assist Reprod Genet 34:495–503

Saif MW, Takimoto C, Mita M, Banerji U, Lamanna N, Castro J, O'Brien S, Stogard C, Von Hoff D (2014) A phase 1, dose-escalation, pharmacokinetic and pharmacodynamic study of BIIB021 administered orally in patients with advanced solid tumors. Clin Cancer Res 20:445–455

Sareh H, Tulapurkar ME, Shah NG, Singh IS, Hasday JD (2011) Response of mice to continuous 5-day passive hyperthermia resembles human heat acclimation. Cell Stress Chaperones 16:297–307

Sausville EA, Tomaszewski JE, Ivy P (2003) Clinical development of 17-allylamino, 17-demethoxygeldanamycin. Curr Cancer Drug Targets 3:377–383

Schlatter H, Langer T, Rosmus S, Onneken ML, Fasold H (2002) A novel function for the 90 kDa heat-shock protein (Hsp90): facilitating nuclear export of 60 S ribosomal subunits. Biochem J 362:675–684

Schulte TW, Blagosklonny MV, Ingui C, Neckers L (1995) Disruption of the Raf-1-Hsp90 molecular complex results in destabilization of Raf-1 and loss of Raf-1-Ras association. J Biol Chem 270:24585–24588

Sen A, Caiazza F (2013) Oocyte maturation: a story of arrest and release. Front Biosci (Schol Ed) 5:451–477

Sha QQ, Dai XX, Dang Y, Tang F, Liu J, Zhang YL, Fan HY (2017) A MAPK cascade couples maternal mRNA translation and degradation to meiotic cell cycle progression in mouse oocytes. Development 144:452–463

Shakoori AR, Oberdorf AM, Owen TA, Weber LA, Hickey E, Stein JL, Lian JB, Stein GS (1992) Expression of heat shock genes during differentiation of mammalian osteoblasts and promyelocytic leukemia cells. J Cell Biochem 48:277–287

Shen HY, Zhao Y, Chen XY, Xiong RP, Lu JL, Chen JF, Chen LY, Zhou YG (2010) Differential alteration of heat shock protein 90 in mice modifies glucocorticoid receptor function and susceptibility to trauma. J Neurotrauma 27:373–381

Shimamura T, Perera SA, Foley KP, Sang J, Rodig SJ, Inoue T, Chen L, Li D, Carretero J, Li YC, Sinha P, Carey CD, Borgman CL, Jimenez JP, Meyerson M, Ying W, Barsoum J, Wong KK, Shapiro GI (2012) Ganetespib (STA-9090), a nongeldanamycin HSP90 inhibitor, has potent antitumor activity in in vitro and in vivo models of non-small cell lung cancer. Clin Cancer Res 18:4973–4985

Shimaoka T, Nishimura T, Kano K, Naito K (2011) Analyses of the regulatory mechanism of porcine WEE1B: the phosphorylation sites of porcine WEE1B and mouse WEE1B are different. J Reprod Dev 57:223–228
Sima S, Richter K (2018) Regulation of the Hsp90 system. Biochim Biophys Acta Mol Cell Res 1865:889–897
Singh A, Singh A, Sand JM, Bauer SJ, Hafeez BB, Meske L, Verma AK (2015) Topically applied Hsp90 inhibitor 17AAG inhibits UVR-induced cutaneous squamous cell carcinomas. J Invest J Invest Dermatol 135:1098–1107
Smith DF, Whitesell L, Katsanis E (1998) Molecular chaperones: biology and prospects for pharmacological intervention. Pharmacol Rev 50:493–514
Soga S, Akinaga S, Shiotsu Y (2013) Hsp90 inhibitors as anti-cancer agents, from basic discoveries to clinical development. Curr Pharm Des 19:366–376
Somasundaram T, Bhat SP (2004) Developmentally dictated expression of heat shock factors: exclusive expression of HSF4 in the postnatal lens and its specific interaction with alphaB-crystallin heat shock promoter. J Biol Chem 279:44497–44503
Soo ET, Yip GW, Lwin ZM, Kumar SD, Bay BH (2008) Heat shock proteins as novel therapeutic targets in cancer. In Vivo 22:311–315
Suzuki M, Takeda T, Nakagawa H, Iwata S, Watanabe T, Siddiquey MN, Goshima F, Murata T, Kawada J, Ito Y, Kojima S, Kimura H (2015) The heat shock protein 90 inhibitor BIIB021 suppresses the growth of T and natural killer cell lymphomas. Front Microbiol 6:280
Szent-Gyorgyi C (1995) A bipartite operator interacts with a heat shock element to mediate early meiotic induction of Saccharomyces cerevisiae HSP82. Mol Cell Biol 15:6754–6769
Tago K, Tsukahara F, Naruse M, Yoshioka T, Takano K (2004) Hsp90 inhibitors attenuate effect of dexamethasone on activated NF-kappaB and AP-1. Life Sci 74:1981–1992
Taherian A, Krone PH, Ovsenek N (2008) A comparison of Hsp90 alpha and Hsp90 beta interactions with cochaperones and substrates. Biochem Cell Biol 86:37–45
Taherian A, Ovsenek N, Krone PH (2010) Expression of hsp90 alpha and hsp90 beta during *Xenopus laevis* embryonic development. Iran Biomed J 14:127–135
Taipale M, Jarosz DF, Lindquist S (2010) HSP90 at the hub of protein homeostasis: emerging mechanistic insights. Nat Rev Mol Cell Biol 11:515–528
Taldone T, Gozman A, Maharaj R, Chiosis G (2008) Targeting Hsp90: small-molecule inhibitors and their clinical development. Curr Opin Pharmacol 8:370–374
Tassone G, Mangani S, Botta M, Pozzi C (2018) Probing the role of Arg97 in heat shock protein 90 N-terminal domain from the parasite Leishmania braziliensis through site-directed mutagenesis on the human counterpart. Biochim Biophys Acta Proteins Proteomics 1866:1190–1198
Trotman JB, Agana BA, Giltmier AJ, Wysocki VH, Schoenberg DR (2018) RNA-binding proteins and heat-shock protein 90 are constituents of the cytoplasmic capping enzyme interactome. J Biol Chem 293:16596–16607
Uzawa M, Grams J, Madden B, Toft D, Salisbury JL (1995) Identification of a complex between centrin and heat shock proteins in CSF-arrested Xenopus oocytes and dissociation of the complex following oocyte activation. Dev Biol 171:51–59
Vasko RC, Rodriguez RA, Cunningham CN, Ardi VC, Agard DA, McAlpine SR (2010) Mechanistic studies of Sansalvamide A-amide: an allosteric modulator of Hsp90. ACS Med Chem Lett 1:4–8
Velazquez MM, Alfaro NS, Dupuy CR, Salvetti NR, Rey F, Ortega HH (2010) Heat shock protein patterns in the bovine ovary and relation with cystic ovarian disease. Anim Reprod Sci 118:201–209
Velázquez MM, Salvetti NR, Amweg AN, Díaz PU, Matiller V, Ortega HH (2013) Changes in the expression of heat shock proteins in ovaries from bovines with cystic ovarian disease induced by ACTH. Res Vet Sci 95:1059–1067
Voellmy R, Boellmann F (2007) Chaperone regulation of the heat shock protein response. Adv Exp Med Biol 594:89–99

Wagatsuma A, Takayama Y, Hoshino T, Shiozuka M, Yamada S, Matsuda R, Mabuchi K (2017) Pharmacological targeting of HSP90 with 17-AAG induces apoptosis of myogenic cells through activation of the intrinsic pathway. Mol Cell Biochem 445:45–58

Wanderling S, Simen BB, Ostrovsky O, Ahmed NT, Vogen SM, Gidalevitz T, Argon Y (2007) GRP94 is essential for mesoderm induction and muscle development because it regulates insulin-like growth factor secretion. Mol Biol Cell 18:3764–3775

Wang QC, Liu J, Duan X, Cui XS, Kim NH, Xiong B, Sun SC (2015a) The dynamin 2 inhibitor Dynasore affects the actin filament distribution during mouse early embryo development. J Reprod Dev 61:49–53

Wang F, Zhang L, Duan X, Zhang GL, Wang ZB, Wang Q, Xiong B, Sun SC (2015b) RhoA-mediated FMNL1 regulates GM130 for actin assembly and phosphorylates MAPK for spindle formation in mouse oocyte meiosis. Cell Cycle 14:2835–2843

Weber H, Valbuena JR, Barbhuiya MA, Stein S, Kunkel H, García P, Bizama C, Riquelme I, Espinoza JA, Kurtz SE, Tyner JW, Calderon JF, Corvalán AH, Grez M, Pandey A, Leal-Rojas P, Roa JC (2017) Small molecule inhibitor screening identifified HSP90 inhibitor 17-AAG as potential therapeutic agent for gallbladder cancer. Oncotarget 8:26169–26184

Weis F, Moullintraffort L, Heichette C, Chrétien D, Garnier C (2010) The 90-kDa heat shock protein Hsp90 protects tubulin against thermal denaturation. J Biol Chem 285:9525–9534

Weisberg E, Barrett R, Liu Q, Stone R, Gray N, Griffin JD (2009) FLT3 inhibition and mechanisms of drug resistance in mutant FLT3-positive AML. Drug Resist Updat 12:81–89

Whitesell L, Lindquist SL (2005) HSP90 and the chaperoning of cancer. Nat Rev Cancer 5:761–772

Whitesell L, Shifrin SD, Schwab G, Neckers LM (1992) Benzoquinonoid ansamycins possess selective tumoricidal activity unrelated to src kinase inhibition. Cancer Res 52:1721–1728

Wu L, Chen X, Huang L, Tian J, Ke F, Xu J, Chen Y, Zheng M (2015) A Novobiocin derivative, XN4, inhibits the proliferation of chronic myeloid leukemia cells by inducing oxidative DNA damage. PLoS One 10:e0123314

Xu Y, Zhu Q, Chen D, Shen Z, Wang W, Ning G, Zhu Y (2015) The HSP90 inhibitor 17-AAG exhibits potent antitumor activity for pheochromocytoma in a xenograft model. Tumour Biol 36:5103–5108

Yamada O, Ozaki K, Akiyama M, Kawauchi K (2012) JAK-STAT and JAK-PI3K-mTORC1 pathways regulate telomerase transcriptionally and posttranslationally in ATL cells. Mol Cancer Ther 11:1112–1121

Yamamoto TM, Wang L, Fisher LA, Eckerdt FD, Peng A (2014) Regulation of greatwall kinase by protein stabilization and nuclear localization. Cell Cycle 13:3565–3575

Yang X, Cui Y, Yue J, He H, Yu C, Liu P, Liu J, Ren X, Meng Y (2017) The histological characteristics, age-related thickness change of skin, and expression of the HSPs in the skin during hair cycle in yak (*Bos grunniens*). PLoS One 12:e0176451

Yu J, Li Y, Wang T, Zhong X (2018) Modification of N6-methyladenosine RNA methylation on heat shock protein expression. PLoS One 13:e0198604

Zhang DX, Park WJ, Sun SC, Xu YN, Li YH, Cui XS, Kim NH (2011) Regulation of maternal gene expression by MEK/MAPK and MPF signaling in porcine oocytes during in vitro meiotic maturation. J Reprod Dev 57:49–56

Zhang Y, Gu S, Li C, Sang M, Wu W, Yun X, Hu X, Li B (2014) Identification and characterization of novel ER-based hsp90 gene in the red flour beetle, Tribolium castaneum. Cell Stress Chaperones 19:623–633

Zhao X, Wang J, Xiao L, Xu Q, Zhao E, Zheng X, Zheng H, Zhao S, Ding S (2016) Effects of 17-AAG on the cell cycle and apoptosis of H446 cells and the associated mechanisms. Mol Med Rep 14:1067–1074

Zhao X, Wang J, Xiao L, Xu Q, Zhao E, Zheng X, Zheng H, Zhao S, Ding S (2017) Effects of 17-allylamino-17-demethoxygeldanamycin on the induction of apoptosis and cell cycle arrest in HCT-116 cells. Oncol Lett 14:2177–2185

Zhao YR, Li HM, Zhu M, Li J, Ma T, Huo Q, Hong YS, Wu CZ (2018) Non-benzoquinone geldanamycin analog, WK-88-1, induces apoptosis in human breast cancer cell lines. J Microbiol Biotechnol 28:542–550
Zhuandi G, Tuanjie C, Luju L, Abdiryim A, Yingying D, Haoqin L, Suocheng W, Li D (2018) FSH receptor binding inhibitor restrains follicular development and possibly attenuates carcinogenesis of ovarian cancer through down-regulating expressionlevels of FSHR and ERβ in normal ovarian tissues. Gene 668:174–181

Chapter 24
Heat Shock Protein 90 in Severe Trauma

Yan Zhao and Yuan-Guo Zhou

Abstract Trauma is an important health problem in modern society. Its serious consequences, such as sepsis and multiple organ dysfunction, are the leading causes of high morbidity and mortality. The heat shock proteins (HSP) increase rapidly under a wide variety of stressful conditions to protect cell from damage. Overexpression of HSP is associated with apoptosis inhibition, inflammatory response and immunomodulation in severe trauma. Therefore, the changes in expression levels of various heat shock proteins or the differences in genetic background will affect the processes and medical outcomes of patients or animal models of diseases. Most importantly, HSP play differential and complex roles in the regulation of inflammation and immunity at different pathological stages and locations after severe trauma. In addition to the traditional therapeutic methods, strategies by the control of HSP expression and consideration of individual genetic background will pave the way for the better treatment and prediction of serious complications caused by severe trauma. In this chapter, the recent progresses of HSP in severe trauma and its complications and the possibilities of HSP as therapeutic targets will be discussed.

Keywords Gene polymorphism · Multiple organ dysfunction syndrome · Sepsis · Severe trauma · Shock

Y. Zhao · Y.-G. Zhou (✉)
Molecular Biology Center, The State Key Laboratory of Trauma, Burn and Combined Injury, Research Institute of Surgery and Daping Hospital, Third Military Medical University, Chongqing, China
e-mail: ygzhou@tmmu.edu.cn

A. A. A. Asea, P. Kaur (eds.), *Heat Shock Protein 90 in Human Diseases and Disorders*, Heat Shock Proteins 19,
https://doi.org/10.1007/978-3-030-23158-3_24

Abbreviations

DIC	Disseminated intravascular coagulation
HSP	Heat shock proteins
IL	Interleukin
MODS	Multiple organ dysfunction syndrome

24.1 Introduction

With the rapid development of modern society, severe trauma and its complications, such as multiple organ dysfunction, hemorrhagic shock, sepsis, etc., are the leading causes of death in the population under 40 years of age. It has been recognized that there are two phases following trauma, the primary injury resulted directly from an external mechanical force and the secondary injury which initiates at a delayed period. The clinical manifestations of trauma appear to be associated with both the intensity of the direct forces and the development of the secondary injuries. Appropriate inflammation after injury protects body against further damage and initiates the healing process. Prolonged inflammation, massive blood loss, tissue necrosis and wound pollution may cause serious consequences, like multiple organ dysfunction or systemic inflammatory response syndrome. In this case, damage-associated molecular patterns trigger the release of a large number of pro-inflammatory factors, which further produces waterfall release of cytokines which wound finally lead to immune dysfunction or even immune paralysis. Moreover, patients with severe trauma also present different pathological characteristics and are at different risk for serious complications due to differences in age, sex, race, genetic background, and education.

The heat Shock Proteins (HSP) are ubiquitously expressed stress molecules in all cells from bacteria to mammals. They offer the first line of defense mechanism against a variety of stressors, such as trauma, infection, fever, and intoxication. HSP are categorized according to the size of the protein, HSP have been divided into five families according on their molecular weight, including HSP110, HSP70, HSP90, HSP60, and small HSP. The synthesis of HSP can be induced not only to guard cells from stressful injuries at the early stage of trauma, but also to participate in the regulation of signaling pathways of multiple inflammatory factors and immune homeostasis at a delayed period, especially in the case of severe trauma with complications. Therefore, HSP are multifunctional molecules that connect endocrine, inflammation and immunity systems. Over the past few decades, though great advances were seen in clinical treatment in the intensive care unit, the morbidity and mortality rates among the patients with major trauma still need to be reduced. This review states the pathological changes following trauma and the roles of heat shock protein-based intervention in alleviating or preventing its complications. It is hoped that this will offer the reader a better understanding of the roles and potential applications of HSP as therapeutic targets when trauma occurs.

24.1.1 Severe Trauma and HSP

The term 'trauma' described in this chapter mainly refers to injuries induced by physical factors, such as mechanical external forces, high temperature and other factors. The severity of trauma is assessed in different criteria, and severe trauma is usually considered to be an injury with an Injury Severity Score (ISS) higher than 15. Both direct damage caused by physical stimulus and systemic stress response following trauma lead to a rapid increase in HSP expression to maintain cellular homeostasis and promote functional recovery. HSP may play various roles at different stages and locations after severe trauma (Chebotareva et al. 2017), including rescue of misfolded proteins, regulation of inflammation and inhibition of apoptosis. When cell is exposed to damaging stimuli, overactivation of apoptosis may result in the decrease even loss of normal function of various tissues, especially for the non-regenerative nerve tissue, and lead to irreversible consequences. For example, apoptosis might induce the atrophy of hippocampus and cortex and cognitive dysfunction in patients with traumatic brain injury. Thus, apoptotic injury is believed to be a significant contributor to the pathogenesis of trauma and an increasing attention was paid to explore the strategies of wound healing and functional rehabilitation by inhibition of apoptosis. Among these studies, HSP (HSP90, HSP70, and HSP27) has been suggested to be tightly related to several forms of cell death, including apoptosis in severe trauma (Arya et al. 2007; Beere 2005).

Recent studies suggest that HSP can prevent apoptosis induced by both the intrinsic and the extrinsic pathways. HSP participate in the intrinsic pathway by blocking the mitochondrial outer membrane permeabilization (MOMP), cytochrome C release, apoptosome assembly, and caspase activation, etc. It is revealed that HSP70 can interfere specifically with Apaf-1 to inhibit apoptotic complex of procaspase-9 after the release of cytochrome C from mitochondria in response to apoptotic stimuli (Saleh et al. 2000), and also affect activation of caspase-3 directly or indirectly (Li et al. 2000). Cells expressing constitutive Hsp72 were shown to have significantly reduced levels of apoptosis after heat and TNF-α stimulations, which found to be related to stress-activated protein kinase/c-Jun N-terminal kinase (SAPK/JNK) pathway (Gabai et al. 2002). These data suggest that HSP play a key role in inhibiting either proapoptotic proteins or apoptosome formation. In addition, the inflammatory response caused by trauma is closely related to the severity and the prognosis of the disease. Therefore, a large number of experiments use animal models of cecal ligation and puncture, LPS injection, traumatic brain injury, hemorrhagic shock, etc. to simulate severe infection, or organ dysfunction post trauma. HSP are increased both in injured tissue and in systemic cells, which modulates multiple inflammatory pathways following a variety of trauma. Numerous studies have shown that high levels of HSP play an active protective role in the early stages of trauma. On the one hand, HSP exerts a role in maintaining the appropriate function of protein, including stabilizing the correct structure of protein, restoring the conformation and preventing the aggregation of misfolded molecules and promoting the degradation of seriously damaged protein. On the other hand, increased HSP exert complicated roles in inflammatory responses. Thus they are tightly associated with severe complications such as sepsis and multi-organ failure.

The majority of studies used specific drugs (such as HSP90 inhibitors) or heat shock response to elevate HSP levels in advance or at a relative early stage following trauma, and attempt to uncover the underlying mechanisms in which HSP regulate inflammatory factors, apoptotic molecules, the vascular permeability, etc. The most of these studies found TLR4 and NF-κB signaling pathways are mainly responsible for their anti-inflammatory effect, whether in infectious or sterile inflammation. For instance, HSP70 is highly expressed in the area surrounding a wound after traumatic brain injury (Ondruschka et al. 2018), and the elevated HSP70 can reduce the inflammatory response after brain injury (Gu et al. 2016). HSP90 inhibitors alleviate LPS-induced inflammatory response, which is associated with decreased expressions of TLR4 (O'Neill et al. 2015) and CD14 (10) in mononuclear macrophages, and also with increased HSP70. However, the effects of HSP90 inhibitors are complex and sometimes contradictory. The inhibitors not only promote the expression of HSP70, but also inhibit some of signaling pathways of HSP90's client proteins. HSP90 is indispensable in the activation of IκB kinase, an essential activator of NF-κB and HSP90 inhibitors suppress this pathway and promote the degradation of IKK, which helps to control excessive inflammatory response (Chen et al. 2002). On the contrary, it is confirmed that heat shock response can induce the activation of NF-κB at the single-cell level (Kardynska et al. 2018). Therefore, the employing of HSP90 inhibitors should be considered entirely during experimental design and analysis. In addition to the complicated roles of HSP in cells, extracellular HSP usually presented upon severe trauma and its complications has received more attention recently (Chen et al. 2007). The prevailing opinion states that HSP in circulation indicate the serious complications or poor prognosis after severe injury. Moreover, severity of trauma is also associated with an individual's genetic background. Through the analysis of the inflammation-related genes in patients with major trauma, it was found that single nucleotide polymorphisms (SNPs) exiting in the genes of TNF-α, TNF-β, IL-1, IL-6, IL-10 as well as HSP are closely related to the occurrence and development of sepsis. In view of the multifacet actions of HSP in stressful responses, apoptosis and inflammation, the constructional and functional changes of HSP would participate in the courses of severe trauma and its serious consequences, including sepsis, MODS, shock, and disseminated intravascular coagulation (DIC), as we will discuss in the following sections.

24.1.2 Sepsis and HSP

Several factors cause the occurrence of sepsis following trauma, including the decline of the body's immune function, the mechanical damage of skin and mucosal barrier and the wound contamination. Sepsis is an excessive inflammatory response combined with infection. A large number of studies have shown that heat shock response or elevated HSP levels can reduce the severity of sepsis in animals. For example, the heat shock response and HSP90 inhibitors prolonged the survival time of sepsis mice, reduced the level of inflammatory factors, prevented the capillary leakage, and finally restored respiratory function (Antonov et al. 2008; Chatterjee et al. 2007a, 2008; Villar et al. 1994). Consistently, elevated HSP70 also alleviated the destruction of pulmonary endothelial barrier caused by pneumococcal, a Gram-positive bacteria, and alleviated pulmonary edema and prevented cells from death (Li et al. 2018). In addition, HSP also showed positive effects on the permeability of intestinal mucosal in septic mice (Wang and Hasselgren 2002). The above documents clarified that HSP play important roles in the dysfunction of pulmonary vascular barrier and intestinal mucosal barrier caused by sepsis. Therefore, investigators interfered the septic animals with direct injection of mammalian-derived HSP or gene knockout methods and found theses strategies alleviated the symptoms of sepsis in rats infected with both gram-positive and gram-negative bacteria (McConnell et al. 2011; Rozhkova et al. 2010; Vinokurov et al. 2012). Notably, the decline in HSP70 levels was found to be a risk factor for a higher incidence of sepsis in elderly patients than in adults (Heydari et al. 1994; Singh et al. 2006).

However, in the context of clinical studies, the expression of HSP in peripheral blood white cells of septic patients was often inhibited, while the level of plasmic HSP increased, which is closely related to the levels of inflammatory factors in plasma of adults and children. This is not consistent with the alteration of HSP at an early stage in trauma, which might be dependent on the severity, the stages and the injury site of the trauma, as well as the distribution of HSP. The HSP70 and HSP90 levels of monocytes and polymorphonuclear cells in patients with sepsis/septic shock fell significantly compared to those with systemic inflammatory response syndrome (SIRS) (Papadopoulos et al. 2017; Schroeder et al. 1999), but plasma Hsp90α in patients with sepsis showed an opposite change, a marked increase in its level, and it was correlated with plasma levels of Hsp72, IL-6 and IL-10 (Vardas et al. 2014). Similarly, children with sepsis had significantly higher levels of serum Hsp72 and Hsp90 than those without sepsis, but intracellular HSP in monocytes were inhibited (Spanaki et al. 2018).

24.1.3 Mods and HSP

It was found as early as 1999 that heat shock response is related to MODS (Takahashi et al. 1999). HSP were demonstrated to play an importane role in the protection of normal functions of brain, liver, kidney, heart, etc. The most widely used model is endotoxin-induced infection combined with the therapy of HSP90 inhibitors. On the one hand, HSP90 inhibitors can promote the translocation of HSF from cytoplasm into nucleus to act as a transcription factor and induce the expression of heat shock proteins, such as HSP70, HSP90, and HSP27. On the other hand, the inhibitors are involved in the modulation of proinflammatory signaling molecules by the directly interaction with IKK complex. In a rat septic model, 17-DMAG, an HSP90 Inhibitor, induced the HSF1-dependant expression of HSP70 and showed an anti-inflammatory effects and a lower mortality in MODS secondary to trauma (Wang et al. 2016).

Drugs that aim to increase the amount of HSP can reduce the levels of inflammatory factors and reduce the damage in liver and kidney. For example, 17-DMAG alleviated LPS-induced liver injury in mice (Ambade et al. 2012), and pre-elevated HSP70 can alleviate the injuries of liver and kidney in the rat model of heat stress (Chang et al. 2006). In vitro experiments also confirmed that increased HSP70 level reduced cellular damage, while RNAi against HSP70 promoted cell death (Lin et al. 2015). The beneficial effect of HSP in the liver was related to the promotion of the binding of HSF1 to the promoter of TNF-α gene which caused the drop of serum TNF-α and IL-6 and the decline of binding activity of NF-κB. The protective effect of HSP on cardiac function was also associated with HSF, as well as the increased levels of Nrf2 and glutathione in the nucleus. Several experiments revealed that increased HSP90 and HSP70 relieved endotoxin-induced heart failure and multi-organ failure (Ranek et al. 2018; Shen et al. 2017; Wang et al. 2015).

However, the high levels of plasma HSP indicate the severe injury and immune dysfunction, which often leads to bad consequences. The increase in extracellular Hsp90 in septic children predicted high risk for the occurrence of MODS (Fitrolaki et al. 2016), and the extracellular Hsp72 and Hsp90α levels was higher among non-survivors compared to survivors in ICU patients (Vardas et al. 2017). These data suggest that HSP could be associated with the occurrence and severity of MODS in trauma patients.

24.1.4 Shock and HSP

A massive hemorrhage caused by mechanical damage and secondary infection following trauma would lead to hemorrhagic and septic shock respectively and responsible for the high mortality of major trauma. It was found that there was a significant increase in HSP levels in animals with either septic or hemorrhagic shock (Su et al. 2005). It was well demonstrated that HSP could suppress the activity of induced NO synthase (iNOS), which is responsible for the synthesis of NO and closely related with the low reactivity of blood vessels, the drop of blood pressure and septic shock. It was found that HSP70 protected animals from the fatal septic shock induced by TNF-α and hypovolemic shock and correlated with the outcomes of septic shock patients (Ackermann et al. 2001; Van Molle et al. 2002). The mechanism is related to the decrease in the activity of iNOS and in NF-κB level (Kim et al. 2010). Hsp90 also has an obvious effect on the occurrence of septic shock via the formation of iNOS-Hsp90 complex (Tunctan et al. 2013). Recently it was revealed that some drugs identified effective in clinical treatment of septic shock lead to the improved expression of HSP. For example, the therapeutic effect of glutamine and ciprofloxacin on pseudomonas aeruginosa-induced septic shock was ascribed to increased levels of HSP70 and IL-10 (Mazloomi et al. 2011). Patients receiving dexamethasone showed an increased expression of HSP70 and alleviated or eliminated symptoms (Chatterjee et al. 2007b).

24.1.5 DIC and HSP

Severe trauma and infection can cause DIC, which often lead to septic shock and multiple organ dysfunction. The most common signs of DIC are prolonged prothrombin time (PT) and decreased platelets count. Recent documents showed about 55% of patients with severe sepsis had decreased platelets count, of which 80% developed DIC. By a stratification analysis, it was found that HSP70 level influenced the inflammatory response and indirectly regulated the platelets count (Bedet et al. 2018). HSP70 concentration in the blood of patients with HELLP syndrome was revealed to reflect the severity of thrombocytopenia, hemolysis, and liver injury (Madach et al. 2008). Increased HSP70 restored the prothrombin time (PT) prolongation and the decrease in platelets count and inhibited the expression of plasminogen activator inhibitor-1 (PAI-1) in severe sepsis (Wang et al. 2016). Heat shock resulted in a significant shortening of clotting time in an in vitro model induced by lipoteichoic acid (Sucker et al. 2013). Hsp70 was also identified to negatively associated with the activation of tissue factors and shortened coagulation time (Sucker et al. 2007). After heat stress, activation of tissue factors induced by c-Jun activation was inhibited and prolonged PT was recovered (Egorina et al. 2006).

24.1.6 Extracellular Heat Shock Proteins ADN Trauma

The function of extracellular HSP is a hot topic in recent years. It has been reported that extracellular HSP, including HSP60, HSP70, HSP90, and the small HSP, are released via exosomes secreted from a wide variety of cells in physiological conditions (Rashed et al. 2017), or produced by necrotic cells into interstitial fluid and bloodstream under pathological conditions such as physical trauma or stressful injury (Calderwood et al. 2007). Many of the effects of extracellular heat shock proteins and glucose regulated proteins (Grp) are mediated through TLR2, TLR4, CD40, CD91, CCR5, etc. (Calderwood et al. 2007). Thus extracellular HSP may possess the ability to act at distant sites in the body after trauma. It was found that extracellular HSP70 in the plasma of patients with multiple injuries grew tenfold after trauma and was associated with immunosuppression and poor prognosis. In another study HSP70 was regarded as a DAPM in plasma, and the increase in HSP70 level post operation suggested a poor prognosis (Leijte et al. 2018). Experiments showed that the extracellular HSP70 in wound fluid of patients with soft tissue trauma took part in immunosuppression (Flohe et al. 2007). Moreover, the mammalian HSP60 and HSP70 genes have a high homology with bacteria's, which endow them with immunoregulatory qualities (van Eden et al. 2005) and ability to promote the secretion of inflammatory factors (Radons and Multhoff 2005). In addition, gp96 promoted the secretion of IL-1 in patients with aseptic inflammation by regulating the inflammasome (Wang et al. 2018). Therefore, the plasmic HSP70 or antibody titers to HSP70 after injury or operation can be used as biomarkers for MODS and immunosuppression (Guisasola et al. 2015, 2018), as well as for myocardial ischemic injury after surgery (Santos-Junior et al. 2018).

24.1.7 Severe Trauma and Genetic Polymorphisms in HSP

Although HSP are highly conserved molecules with small differences between species, there still exit variations in the regulatory region and some coding regions of the gene. Therefore some researchers have reported that several SNPs in HSP90 and HSP70 are closely associated with severe trauma and its serious consequences. Experimental model of yeast is often employed in the research of gene variation because its genetic background is clear and easy to be observed at the phenotypic level. So far, multiple HSP90 mutants have been found to reduce the activity of GR, and these mutants showed differential effects on the function of GR (Nathan and Lindquist 1995). In a full-length scanning of HSP90 gene in caucasians, 10 and 11 mutations in Hsp90α and Hsp90β genes respectively were found (Passarino et al. 2003). After that, some researchers performed functional verification of the missense mutation found above, and indicated that some of these mutations had significant influence on GR function (MacLean et al. 2005). As a client protein of HSP90, GR is both the main effector of the hypothalamic-pituitary-adrenal axis (HPA) and a transcription factor which bind to the promoter of some pro-inflammatory factors and chemokines. Therefore, HSP90 may regulate the inflammatory effect via the signaling pathway of glucocorticoid receptor (GR) (Echeverria et al. 2009; Shen et al. 2010).

It was demonstrated that mice with different Hsp84 genotypes showed obvious differences in the severity of injury and in GR effects after acute lung injury and heat-stress injury (Zhao et al. 2010, 2018). Mutations in Hsp90β gene was found to be closely related to the patients' MODS score, but had little impact on the incidence of sepsis in patients with severe trauma (Zhao et al. 2013). In terms of HSP70, its genotypes were found to be closely correlated to MODS score and liver failure (Schroder et al. 2003), as well as the prognosis of patients with sepsis (Ramakrishna et al. 2014). It was documented that the SNPs in HSP70 were related to the decline of heat shock response in the elder (Singh et al. 2006), which may explain the high susceptibility of elderly patients to the complications, such as sepsis, after major trauma. The prognosis of pediatric patients was also related to HSP70 gene polymorphisms in a study with larger sample size (Jabandziev et al. 2014). Due to the participation of HSP in the active/inactive forms of several inflammation-related complexes simultaneously, the genetic variations in HSP are inevitably involved in the regulation of inflammation following trauma. HSP can act multiple client proteins together which making their roles in severe trauma impossible to be ignored, though trauma is regard as a complex disease caused by both pathogens and genetic factors and the SNPs in a single gene present minor effect on the process of the disease.

24.2 Conclusions

In fact, the roles of HSP in various complications after severe trauma described above are not independent and may interact with each other and aggravate the injuries. In summary, HSP exert the below three roles under severe trauma. Firstly, they function as chaperones in cytoplasm and restore misfolded molecules to the correct conformation. Secondly, HSP are key molecules in the transition of active/inactive forms of inflammatory signaling complexes. Thirdly, the immune activation of extracellular HSP affects the immunological homeostasis following trauma. All of these provide a well understanding of post-traumatic pathophysiological changes and exhibit the potential of HSP as therapeutic targets in the early intervention, development of the process, and as biomarkers of poor prognosis after severe trauma. In addition, the consideration of differences in the gene background of HSP genes would be beneficial to individualized treatment plan for trauma patients.

Acknowledgements The authors thank the State Key Laboratory of Trauma, Burn and Combined Injury. This work was supported in part by the National Natural Science Foundation of China [Grant 30470988].

References

Ackermann M, Reuter M, Flohe S, Bahrami S, Redl H, Schade FU (2001) Cytokine synthesis in the liver of endotoxin-tolerant and normal rats during hemorrhagic shock. J Endotoxin Res 7:105–112

Ambade A, Catalano D, Lim A, Mandrekar P (2012) Inhibition of heat shock protein (molecular weight 90 kDa) attenuates proinflammatory cytokines and prevents lipopolysaccharide-induced liver injury in mice. Hepatology 55:1585–1595

Antonov A, Snead C, Gorshkov B, Antonova GN, Verin AD, Catravas JD (2008) Heat shock protein 90 inhibitors protect and restore pulmonary endothelial barrier function. Am J Respir Cell Mol Biol 39:551–559

Arya R, Mallik M, Lakhotia SC (2007) Heat shock genes – integrating cell survival and death. J Biosci 32:595–610

Bedet A, Razazi K, Boissier F, Surenaud M, Hue S, Giraudier S, Brun-Buisson C, Mekontso Dessap A (2018) Mechanisms of thrombocytopenia during septic shock: a multiplex cluster analysis of endogenous sepsis mediators. Shock 49:641–648

Beere HM (2005) Death versus survival: functional interaction between the apoptotic and stress-inducible heat shock protein pathways. J Clin Invest 115:2633–2639

Calderwood SK, Mambula SS, Gray PJ Jr, Theriault JR (2007) Extracellular heat shock proteins in cell signaling. FEBS Lett 581:3689–3694

Chang CK, Chang CP, Chiu WT, Lin MT (2006) Prevention and repair of circulatory shock and cerebral ischemia/injury by various agents in experimental heatstroke. Curr Med Chem 13:3145–3154

Chatterjee A, Dimitropoulou C, Drakopanayiotakis F, Antonova G, Snead C, Cannon J, Venema RC, Catravas JD (2007a) Heat shock protein 90 inhibitors prolong survival, attenuate inflammation, and reduce lung injury in murine sepsis. Am J Respir Crit Care Med 176:667–675

Chatterjee S, Premachandran S, Shukla J, Poduval TB (2007b) Synergistic therapeutic potential of dexamethasone and L-arginine in lipopolysaccharide-induced septic shock. J Surg Res 140:99–108

Chatterjee A, Snead C, Yetik-Anacak G, Antonova G, Zeng J, Catravas JD (2008) Heat shock protein 90 inhibitors attenuate LPS-induced endothelial hyperpermeability. Am J Physiol Lung Cell Mol Physiol 294:L755–L763

Chebotareva N, Bobkova I, Shilov E (2017) Heat shock proteins and kidney disease: perspectives of HSP therapy. Cell Stress Chaperones 22:319–343

Chen G, Cao P, Goeddel DV (2002) TNF-induced recruitment and activation of the IKK complex require Cdc37 and Hsp90. Mol Cell 9:401–410

Chen Y, Voegeli TS, Liu PP, Noble EG, Currie RW (2007) Heat shock paradox and a new role of heat shock proteins and their receptors as anti-inflammation targets. Inflamm Allergy Drug Targets 6:91–100

Echeverria PC, Mazaira G, Erlejman A, Gomez-Sanchez C, Piwien Pilipuk G, Galigniana MD (2009) Nuclear import of the glucocorticoid receptor-hsp90 complex through the nuclear pore complex is mediated by its interaction with Nup62 and importin beta. Mol Cell Biol 29:4788–4797

Egorina EM, Sovershaev MA, Kondratiev TV, Olsen JO, Tveita T, Osterud B (2006) Induction of monocytic tissue factor expression after rewarming from hypothermia in vivo is counteracted by heat shock in c-Jun-dependent manner. Arterioscler Thromb Vasc Biol 26:2401–2406

Fitrolaki MD, Dimitriou H, Venihaki M, Katrinaki M, Ilia S, Briassoulis G (2016) Increased extracellular heat shock protein 90alpha in severe sepsis and SIRS associated with multiple organ failure and related to acute inflammatory-metabolic stress response in children. Medicine (Baltimore) 95:e4651

Flohe SB, Bangen JM, Flohe S, Agrawal H, Bergmann K, Schade FU (2007) Origin of immunomodulation after soft tissue trauma: potential involvement of extracellular heat-shock proteins. Shock 27:494–502

Gabai VL, Mabuchi K, Mosser DD, Sherman MY (2002) Hsp72 and stress kinase c-jun N-terminal kinase regulate the bid-dependent pathway in tumor necrosis factor-induced apoptosis. Mol Cell Biol 22:3415–3424

Gu Y, Chen J, Wang T, Zhou C, Liu Z, Ma L (2016) Hsp70 inducer, 17-allylamino-demethoxygeldanamycin, provides neuroprotection via anti-inflammatory effects in a rat model of traumatic brain injury. Exp Ther Med 12:3767–3772

Guisasola MC, Ortiz A, Chana F, Alonso B, Vaquero J (2015) Early inflammatory response in polytraumatized patients: cytokines and heat shock proteins. A pilot study. Orthop Traumatol Surg Res 101:607–611

Guisasola MC, Alonso B, Bravo B, Vaquero J, Chana F (2018) An overview of cytokines and heat shock response in polytraumatized patients. Cell Stress Chaperones 23:483–489

Heydari AR, Takahashi R, Gutsmann A, You S, Richardson A (1994) Hsp70 and aging. Experientia 50:1092–1098

Jabandziev P, Smerek M, Michalek J, Fedora M, Kosinova L, Hubacek JA, Michalek J (2014) Multiple gene-to-gene interactions in children with sepsis: a combination of five gene variants predicts outcome of life-threatening sepsis. Crit Care 18:R1

Kardynska M, Paszek A, Smieja J, Spiller D, Widlak W, White MRH, Paszek P, Kimmel M (2018) Quantitative analysis reveals crosstalk mechanisms of heat shock-induced attenuation of NF-kappaB signaling at the single cell level. PLoS Comput Biol 14:e1006130

Kim K, Kim W, Rhee JE, Jo YH, Lee JH, Kim KS, Kwon WY, Suh GJ, Lee CC, Singer AJ (2010) Induced hypothermia attenuates the acute lung injury in hemorrhagic shock. J Trauma 68:373–381

Leijte GP, Custers H, Gerretsen J, Heijne A, Roth J, Vogl T, Scheffer GJ, Pickkers P, Kox M (2018) Increased plasma levels of danger-associated molecular patterns are associated with immune suppression and postoperative infections in patients undergoing cytoreductive surgery and hyperthermic intraperitoneal chemotherapy. Front Immunol 9:663

Li CY, Lee JS, Ko YG, Kim JI, Seo JS (2000) Heat shock protein 70 inhibits apoptosis downstream of cytochrome c release and upstream of caspase-3 activation. J Biol Chem 275:25665–25671

Li X, Yu Y, Gorshkov B, Haigh S, Bordan Z, Weintraub D, Rudic RD, Chakraborty T, Barman SA, Verin AD, Su Y, Lucas R, Stepp DW, Chen F, Fulton DJR (2018) Hsp70 suppresses mitochondrial reactive oxygen species and preserves pulmonary microvascular barrier integrity following exposure to bacterial toxins. Front Immunol 9:1309

Lin KC, Lin HJ, Chang CP, Lin MT (2015) Decreasing or increasing heat shock protein 72 exacerbates or attenuates heat-induced cell death, respectively, in rat hypothalamic cells. FEBS Open Biol 5:724–730

MacLean MJ, Llordella MM, Bot N, Picard D (2005) A yeast-based assay reveals a functional defect of the Q488H polymorphism in human Hsp90alpha. Biochem Biophys Res Commun 337:133–137

Madach K, Molvarec A, Rigo J Jr, Nagy B, Penzes I, Karadi I, Prohaszka Z (2008) Elevated serum 70 kDa heat shock protein level reflects tissue damage and disease severity in the syndrome of hemolysis, elevated liver enzymes, and low platelet count. Eur J Obstet Gynecol Reprod Biol 139:133–138

Mazloomi E, Jazani NH, Sohrabpour M, Ilkhanizadeh B, Shahabi S (2011) Synergistic effects of glutamine and ciprofloxacin in reduction of Pseudomonas aeruginosa-induced septic shock severity. Int Immunopharmacol 11:2214–2219

McConnell KW, Fox AC, Clark AT, Chang NY, Dominguez JA, Farris AB, Buchman TG, Hunt CR, Coopersmith CM (2011) The role of heat shock protein 70 in mediating age-dependent mortality in sepsis. J Immunol 186:3718–3725

Nathan DF, Lindquist S (1995) Mutational analysis of Hsp90 function: interactions with a steroid receptor and a protein kinase. Mol Cell Biol 15:3917–3925

O'Neill S, Humphries D, Tse G, Marson LP, Dhaliwal K, Hughes J, Ross JA, Wigmore SJ, Harrison EM (2015) Heat shock protein 90 inhibition abrogates TLR4-mediated NF-kappaB activity and reduces renal ischemia-reperfusion injury. Sci Rep 5:12958

Ondruschka B, Rosinsky F, Trauer H, Schneider E, Dressler J, Franke H (2018) Drug- and/or trauma-induced hyperthermia? Characterization of HSP70 and myoglobin expression. PLoS One 13:e0194442

Papadopoulos P, Pistiki A, Theodorakopoulou M, Christodoulopoulou T, Damoraki G, Goukos D, Briassouli E, Dimopoulou I, Armaganidis A, Nanas S, Briassoulis G, Tsiodras S (2017) Immunoparalysis: clinical and immunological associations in SIRS and severe sepsis patients. Cytokine 92:83–92

Passarino G, Cavalleri GL, Stecconi R, Franceschi C, Altomare K, Dato S, Greco V, Luca Cavalli Sforza L, Underhill PA, de Benedictis G (2003) Molecular variation of human HSP90alpha and HSP90beta genes in Caucasians. Hum Mutat 21:554–555

Radons J, Multhoff G (2005) Immunostimulatory functions of membrane-bound and exported heat shock protein 70. Exerc Immunol Rev 11:17–33

Ramakrishna K, Pugazhendhi S, Kabeerdoss J, Peter JV (2014) Association between heat shock protein 70 gene polymorphisms and clinical outcomes in intensive care unit patients with sepsis. Indian J Crit Care Med 18:205–211

Ranek MJ, Stachowski MJ, Kirk JA, Willis MS (2018) The role of heat shock proteins and co-chaperones in heart failure. Philos Trans R Soc Lond Ser B Biol Sci 373:20160530

Rashed M, Bayraktar E, Helal G, Abd-Ellah MF, Amero P, Chavez-Reyes A, Rodriguez-Aguayo C (2017) Exosomes: from garbage bins to promising therapeutic targets. Int J Mol Sci 18:538

Rozhkova E, Yurinskaya M, Zatsepina O, Garbuz D, Karpov V, Surkov S, Murashev A, Ostrov V, Margulis B, Evgen'ev M, Vinokurov M (2010) Exogenous mammalian extracellular HSP70 reduces endotoxin manifestations at the cellular and organism levels. Ann N Y Acad Sci 1197:94–107

Saleh A, Srinivasula SM, Balkir L, Robbins PD, Alnemri ES (2000) Negative regulation of the Apaf-1 apoptosome by Hsp70. Nat Cell Biol 2:476–483

Santos-Junior VA, Lollo PCB, Cantero MA, Moura CS, Amaya-Farfan J, Morato PN (2018) Heat shock proteins: protection and potential biomarkers for ischemic injury of cardiomyocytes after surgery. Braz J Cardiovasc Surg 33:291–302

Schroder O, Schulte KM, Ostermann P, Roher HD, Ekkernkamp A, Laun RA (2003) Heat shock protein 70 genotypes HSPA1B and HSPA1L influence cytokine concentrations and interfere with outcome after major injury. Crit Care Med 31:73–79

Schroeder S, Lindemann C, Hoeft A, Putensen C, Decker D, von Ruecker AA, Stuber F (1999) Impaired inducibility of heat shock protein 70 in peripheral blood lymphocytes of patients with severe sepsis. Crit Care Med 27:1080–1084

Shen HY, Zhao Y, Chen XY, Xiong RP, Lu JL, Chen JF, Chen LY, Zhou YG (2010) Differential alteration of heat shock protein 90 in mice modifies glucocorticoid receptor function and susceptibility to trauma. J Neurotrauma 27:373–381

Shen HH, Huang SY, Cheng PY, Chu YJ, Chen SY, Lam KK, Lee YM (2017) Involvement of HSP70 and HO-1 in the protective effects of raloxifene on multiple organ dysfunction syndrome by endotoxemia in ovariectomized rats. Menopause 24:959–969

Singh R, Kolvraa S, Bross P, Jensen UB, Gregersen N, Tan Q, Knudsen C, Rattan SI (2006) Reduced heat shock response in human mononuclear cells during aging and its association with polymorphisms in HSP70 genes. Cell Stress Chaperones 11:208–215

Spanaki AM, Tavladaki T, Dimitriou H, Kozlov AV, Duvigneau JC, Meleti E, Weidinger A, Papakonstantinou E, Briassoulis G (2018) Longitudinal profiles of metabolism and bioenergetics associated with innate immune hormonal inflammatory responses and amino-acid kinetics in severe Sepsis and systemic inflammatory response syndrome in children. JPEN J Parenter Enteral Nutr 42:1061–1074

Su F, Nguyen ND, Wang Z, Cai Y, Rogiers P, Vincent JL (2005) Fever control in septic shock: beneficial or harmful? Shock 23:516–520

Sucker C, Zacharowski K, Thielmann M, Hartmann M (2007) Heat shock inhibits lipopolysaccharide-induced tissue factor activity in human whole blood. Thromb J 5:13

Sucker C, Paniczek S, Scharf RE, Litmathe J, Hartmann M (2013) Rotation thromboelastography for the detection and characterization of lipoteichoid acid-induced activation of haemostasis in an in vitro sepsis model. Perfusion 28:146–151

Takahashi T, Suzuki T, Yamasaki A, Tsukiji T, Hirakawa M, Akagi R (1999) Heat shock response in a rat model of septic multiple organ dysfunction syndrome. Nihon Yakurigaku Zasshi 114:295–302

Tunctan B, Korkmaz B, Sari AN, Kacan M, Unsal D, Serin MS, Buharalioglu CK, Sahan-Firat S, Cuez T, Schunck WH, Manthati VL, Falck JR, Malik KU (2013) Contribution of iNOS/sGC/PKG pathway, COX-2, CYP4A1, and gp91(phox) to the protective effect of 5,14-HEDGE, a 20-HETE mimetic, against vasodilation, hypotension, tachycardia, and inflammation in a rat model of septic shock. Nitric Oxide 33:18–41

van Eden W, van der Zee R, Prakken B (2005) Heat-shock proteins induce T-cell regulation of chronic inflammation. Nat Rev Immunol 5:318–330

Van Molle W, Wielockx B, Mahieu T, Takada M, Taniguchi T, Sekikawa K, Libert C (2002) HSP70 protects against TNF-induced lethal inflammatory shock. Immunity 16:685–695

Vardas K, Apostolou K, Briassouli E, Goukos D, Psarra K, Botoula E, Tsagarakis S, Magira E, Routsi C, Nanas S, Briassoulis G (2014) Early response roles for prolactin cortisol and circulating and cellular levels of heat shock proteins 72 and 90alpha in severe sepsis and SIRS. Biomed Res Int 2014:803561

Vardas K, Ilia S, Sertedaki A, Charmandari E, Briassouli E, Goukos D, Apostolou K, Psarra K, Botoula E, Tsagarakis S, Magira E, Routsi C, Stratakis CA, Nanas S, Briassoulis G (2017) Increased glucocorticoid receptor expression in sepsis is related to heat shock proteins, cytokines, and cortisol and is associated with increased mortality. Intensive Care Med Exp 5:10

Villar J, Ribeiro SP, Mullen JB, Kuliszewski M, Post M, Slutsky AS (1994) Induction of the heat shock response reduces mortality rate and organ damage in a sepsis-induced acute lung injury model. Crit Care Med 22:914–921

Vinokurov M, Ostrov V, Yurinskaya M, Garbuz D, Murashev A, Antonova O, Evgen'ev M (2012) Recombinant human Hsp70 protects against lipoteichoic acid-induced inflammation manifestations at the cellular and organismal levels. Cell Stress Chaperones 17:89–101

Wang Q, Hasselgren PO (2002) Heat shock response reduces intestinal permeability in septic mice: potential role of interleukin-10. Am J Physiol Regul Integr Comp Physiol 282:R669–R676

Wang YL, Lam KK, Cheng PY, Lee YM (2015) Celastrol prevents circulatory failure via induction of heme oxygenase-1 and heat shock protein 70 in endotoxemic rats. J Ethnopharmacol 162:168–175

Wang YL, Shen HH, Cheng PY, Chu YJ, Hwang HR, Lam KK, Lee YM (2016) 17-DMAG, an HSP90 inhibitor, ameliorates multiple organ dysfunction syndrome via induction of HSP70 in endotoxemic rats. PLoS One 11:e0155583

Wang Y, Sedlacek AL, Pawaria S, Xu H, Scott MJ, Binder RJ (2018) Cutting edge: the heat shock protein gp96 activates inflammasome-signaling platforms in APCs. J Immunol 201:2209–2214

Zhao Y, Shen HY, Chen XY, Xiong RP, Li P, Liu P, Yang N, Zhou YG (2010) Genetic variations of heat shock protein 84 in mice mediate cellular glucocorticoid response. Cell Physiol Biochem 25:359–366

Zhao Y, Tao L, Jiang D, Chen X, Li P, Ning Y, Xiong R, Liu P, Peng Y, Zhou YG (2013) The -144C/A polymorphism in the promoter of HSP90beta is associated with multiple organ dysfunction scores. PLoS One 8:e58646

Zhao Y, Xiong RP, Chen X, Li P, Ning YL, Yang N, Peng Y, Jiang YL, Zhou YG (2018) Hsp90 regulation affects the treatment of glucocorticoid for pancreatitis-induced lung injury. Mol Cell Biochem 440:189–197

Chapter 25
Hsp90: Is There an Unknown Role in Pain Neurobiology

João Dias-Ferreira and Fani L. Moreira Neto

Abstract Heat shock protein 90 (Hsp90) is a highly conserved molecular chaperone that is essential in protein homeostasis and normal cell growth and survival, assisting protein folding, protein degradation and being highly recruited in cellular stress conditions. The implication of Hsp90 in several pathological conditions is being increasingly recognized, particularly in cancer, an area of intense research on the use of different Hsp90 inhibitors as a possible therapeutic approach, as well as on the associated mechanisms of action. There is also strong indication that the Hsp90 chaperone is involved in inflammation and neuroinflammation, and neurodegeneration events. Here we summarize some scientific evidence suggesting that Hsp90, and possibly its mitochondrial-compartimentalized homologue Tumour Necrosis Factor Receptor-Associated Protein 1 (TRAP-1), participate also in pain processing mechanisms.

Keywords Hsp90 · Mitochondrial homologue · Neuroinflammation · Neuropathic pain · Oxidative stress · Pain neurobiology

Abbreviations

17-AAG	17-demethoxygeldanamycin
17-DMAG	17-Dimethylaminoethylamino-17-demethoxygeldanamycin
ATF3	Activating transcription factor 3

J. Dias-Ferreira · F. L. Moreira Neto (✉)
i3S – Instituto de Investigação e Inovação em Saúde, Universidade do Porto, Porto, Portugal

IBMC – Instituto de Biologia Molecular e Celular, Universidade do Porto, Porto, Portugal

Departamento de Biomedicina – Unidade de Biologia Experimental, Faculdade de Medicina, Universidade do Porto, Porto, Portugal
e-mail: fanineto@med.up.pt

A. A. A. Asea, P. Kaur (eds.), *Heat Shock Protein 90 in Human Diseases and Disorders*, Heat Shock Proteins 19,
https://doi.org/10.1007/978-3-030-23158-3_25

CCI	Chronic constriction injury
CGRP	Calcitonin-gene-related peptide
CHIP	Carboxyl terminus of Hsp70-interacting protein
CypD	Cyclophilin D
DAMP	Danger-associated molecular pattern
ERK	Extracellular signal-regulated kinases
GFAP	Glial fibrillary acidic protein
GHKL	**G**yrase, **H**sp90, Histidine **K**inase and Mut**L**
HSE	Heat shock element
HSF1	Heat shock factor 1
HSP	Heat shock proteins
Hsp70	Heat shock protein 70
Hsp90	Heat shock protein 90
IKK	Inhibitor of κB (IκB) Kinase complex
IL-10	Interleukin 10
IL-1β	Interleukin 1 beta
IL-6	Interleukin 6
IκB	Inhibitor of κB
JAK2	JAK/Signal Transducer and Activator of Transcription (STATs)
JNK	c-Jun N-terminal kinase
MAPKs	Mitogen-activated protein kinases
MOR	Mu-opioid receptor
mPTP	Mitochondrial permeability transition pore
NADPH	Nicotinamide adenine dinucleotide phosphate
NF-κB	Nuclear factor-kappa B
NO	Nitric oxide
NOS	NO synthases
NOX	Nicotinamide adenine dinucleotide phosphate (NADPH) oxidases
PINK-1	PTEN-induced Putative Kinase 1
PRRs	Pattern recognition receptors
RIP	Receptor interacting kinase
RNS	Reactive nitrogen species
ROS	Reactive oxygen species
siRNA	Small interference RNA
SP	Substance P
STATs	Signal transducer and activator of transcription
TAK-1/IKK	TGF-β-activated kinase-1/Inhibitor of κB (IκB) Kinase complex
TLR2	Toll-like-receptor 2
TLR4	Toll like-receptor 4
TNFα	Tumour necrosis factor alpha
TRAP-1	Tumour necrosis factor receptor-associated protein 1
β-HIVS	β-hydroxyisovalerylshikonin

25.1 Introduction

Heat shock proteins or HSP belong to a family of highly conserved proteins which were discovered accidentally in the fruit fly, following their increased gene expression in response to heat shock (Ritossa 1962; Horwich 2014). They are found in both eukaryotic and prokaryotic cells and act as molecular chaperones, therefore assisting the correct covalent folding or unfolding and the assembly or disassembly of several cellular proteins as well as the degradation of misfolded proteins (Craig 1993; Hartl 1996; Hartl and Hayer-Hartl 2002; Horwich 2014). They are also involved in intracellular trafficking, cellular signalling, and in the regulation of immune responses and of apoptotic events at the pre- and post-mitochondrial level (Didelot et al. 2006; Udono et al. 2009; Muralidharan and Mandrekar 2013; Sevin et al. 2015; Ikwegbue et al. 2017), therefore having an essential participation in protein homeostasis and normal cell growth and survival. Their expression is frequently triggered as a cytoprotective event by a number of physiological and environmental conditions that cause cellular stress and threat the cell viability, such as heat shock, hypoxia, anticancer chemotherapy, inflammation or infections, but they are also implicated in normal cellular events like development and differentiation (Neuer et al. 1998; Walsh et al. 1999; Wiesgigl and Clos 2001; Christians et al. 2003; Afzal et al. 2011; Benitez et al. 2014).

HSP in mammalian cells have been divided into two groups and have been named according to their molecular weight. The small heat shock proteins, which include Hsp27, are a group of ATP-independent chaperones, while the higher molecular weight molecular chaperones, which comprise the Hsp60, Hsp70, Hsp90, Hsp100 and Hsp110 families, have an ATP binding domain that is essential for their activity and need co-chaperones (such as Hsp40) that modulate their conformation and ATP binding (Sevin et al. 2015; Lackie et al. 2017). Each family comprises several members that may be specifically located in the nucleus, cytoplasm or defined organelles, exerting particular functions, and that can be either constitutively expressed or induced.

25.1.1 *The Hsp90 Chaperone Family*

The heat shock protein 90 (Hsp90) is a highly preserved and copious chaperone, indispensable in all the eukaryotic cells. Hsp90 is not essential in the overall process of *de novo* maturation and folding of proteins, contrarily to Hsp70 or Hsp60 (GroEL in bacteria), but simplifies the final steps in a well-known array of identified client proteins (Mayer and Bukau 1999; Young et al. 2001; Jackson 2013). Amongst these proteins are the kinases, steroid hormone receptors and transcription factors, pointing to a vast role of these chaperones in several cellular processes, as signal transduction, transportation of intracellular proteins and protein degradation (Mahalingam et al. 2009; Wayne et al. 2011). In prokaryotes like yeasts, Hsp90 is embodied by two isoforms present in the cytosol named Hsc82 and Hsp82, with the last being upregulated by 20-fold in stress heat-induced situations. In the mammalian system, there are three homologues of Hsp90 that exist compartmentalized in discrete subcellular microenvironments and organelles. The cytoplasmic Hsp90 has two isoforms, Hsp90α (Hsp90a) and Hsp90β (Hsp90b), which share around 85% of homology (Johnson 2012). Hsp90α expression is induced following various cell stressors, and Hsp90β is constitutively expressed in normal physiological conditions (Csermely et al. 1998). Inside the cell, other functional and structural homologues of Hsp90 are present, such as the Glucose-regulated protein 94 (GRP94), located in the lumen of the endoplasmic reticulum, and the mitochondrial-compartmentalized Tumour Necrosis Factor Receptor-Associated Protein 1 (TRAP-1), a protein of about ~75 kDa, initially named Hsp75. In the plant cells, the ch-Hsp90 has been described in the chloroplast. Hsp90 may also be secreted out of the cells, being known as extracellular Hsp90 (eHsp90), which seems to be related with cancer metastasis mechanisms (Subbarao Sreedhar et al. 2004; Houlihan et al. 2009; Stellas et al. 2010; Marzec et al. 2012; Sung et al. 2016). Generally, the expression of Hsp90α and Hsp90β molecular chaperones in normal physiological conditions accounts for 1–2% of all soluble cellular proteins (Csermely et al. 1998). When in the presence of various cell stressors, such as heat, heavy metals, low pH and hypoxia, the Hsp90 levels considerably rise up to tenfold (Buchner 1999; Bagatell and Whitesell 2004).

25.1.2 Ultrastructure of Hsp90, Ubiquitination and Post-translational Modifications

Hsp90 is constituted by a homodimeric functional unit with each protomer having three connected flexible portions: the N-terminal domain containing an ATP-binding site, the M-domain in the middle of the structure, and the C-domain located in the C-terminal and responsible for dimerization. The structure of Hsp90 is highly conserved between species, from bacteria to man, with a small exception in the extended portion amongst the N- and M-domains (Chen et al. 2006; Didenko et al. 2012; Jackson 2013). Hsp90 belongs to a specific category of ATPases that are associated according to their structures and the conserved motifs of their ATPase domains. This category includes the DNA **G**yrase, **H**sp90, Histidine **K**inase and Mut**L** (GHKL) protein assemblies with ATPase function conserving a ATP-binding Bergerat fold (Voisine et al. 2007; Pearl 2016). A highly preserved sequence of amino acid residues allows these regions to remain open during the ADP-bound phase and closed in the ATP-bound phase. The ATP binding domain is essential for the Hsp90 role in maintaining cell viability, since intrinsic ATPase activity triggers a conformational change in Hsp90 that is required to induce conformational changes in the client proteins (Pratt and Toft 2003). In Hsp90, the hydrolysis of the ATP molecule is attained through the M-domain catalytic residues. The C-domain portion is critical for the functional dimerization of the chaperone, while the M-domain is also responsible for helping the binding of client proteins and co-chaperones (Prodromou et al. 1999; Pullen and Bolon 2011). In fact, Hsp90 binding to many different relevant proteins, that act as its co-chaperones, yields the formation of heteromeric multichaperone complexes. These proteins participate in the inhibition or activation of Hsp90 ATPase activity and recruitment of specific client proteins, therefore regulating the chaperone function (Johnson 2012; Li et al. 2012). A great number of these co-chaperone proteins have been identified in mammals and have distinct modes of action, as previously reviewed (Li et al. 2012).

In addition to regulation by co-chaperones, Hsp90 molecules are also subject to posttranslational modifications that are of innermost importance to regulate the catalytic domains of the Hsp90 biological machinery, affecting their chaperoning abilities. These changes may occur in several amino acid residues by distinct mechanisms, such as phosphorylation, the most frequent form of posttranslational changes in Hsp90, acetylation, the second most recurrent form, as well as nitrosylation, oxidation and ubiquitination. Each one of these alterations has a subsequent outcome in the mode by which the chaperoning complexes process the maturation strategies applied to the client proteins (Mollapour and Neckers 2011; Li et al. 2012; Sager et al. 2018). Hsp90 may also be the focus of modifications by secondary products of oxidation of intracellular or extracellular molecules which yield α,β-unsaturated molecules, namely electrophile reacting molecules according to the Michael addition reaction. Exemplifying, the product of oxidation 4-hydroxynonenal undergoes chemical reactions forming adducts with Hsp90, a phenomenon connected with alterations in the expression of genes by the cell, namely those which heat shock factor 1 (HSF1) has in charge, resulting in their upregulation (Connor et al. 2011). Another chaperone, Hsp70, is apparently an intervener in this process with the gene BAG3 being induced by this complex, leading to the upgrade of the cell to a survival state (Rauch et al. 2017). Altogether, the available data indicates that Hsp90 has a preponderant role in cell protection against electrophilic reacting molecules, avoiding the usual inevitable death.

The Hsp90 molecule can also suffer an ubiquitination process that plays a role in the chaperone's regulation of in vivo functions. Ubiquitination is mediated by the co-chaperone CHIP (carboxyl terminus of Hsp70-interacting protein) in an array of residues of the amino acid lysine. These flanked regions are subjected to ubiquitin insertion in the section closest to the binding site of ATP, the protein C-terminal or the N-terminal (Kundrat and Regan 2010; Mollapour and Neckers 2012).

25.1.3 Hsp90 Conformational Cycle

The Hsp90 ATPase activity, which is required to induce large conformational rearrangements in the chaperone, and its turnover activity are extremely reduced. The values range between 1 min^{-1} in the case of yeast and 0.1 min^{-1} in the case of human beings. This is because the chaperone has an astonishing capacity to change between different conformations, adopting always a position of biochemical equilibrium. These actions are governed by the binding of the nucleotides to the active pocket, which induces a determined direction in the conformational interchanges, resulting in a nucleotide-dependent cycle of different conformations. The conformation for which the Hsp90 adopts a V shape, the apo state, is commonly known as the open configuration. By the time ATP binds to the respective pocket the chaperone initiates a sequence of conformational changes that include the relocation of the N-terminal with concomitant different rearrangement of the N- and M- domains alignment. The final step in the cycle is reached when the chaperone adopts a compact closed configuration, with the N-domains dimerized (Hessling et al. 2009; Krukenberg et al. 2011; Li et al. 2012; Vettoretti et al. 2016; Zierer et al. 2016).

The activity and normal functioning of Hsp90 may be disrupted by using small molecules known as Hsp90 inhibitors. Most of these molecules target the N-terminal domain of the chaperone, while some of them also target the C-terminal, leaving the N-terminal ATP-binding site completely intact. Although targeting each of these terminal domains appears to yield dissimilar outcomes and phenotypes, the development of such molecules is still a target of interest for in vivo research studies unravelling the role of unidentified Hsp90-related mechanisms or mediators, or even for therapeutic purposes (Georgakis and Younes 2005; Wang and McAlpine 2015a, b; Wang et al. 2017; Zuehlke et al. 2018).

25.1.4 Hsp90 Client Protein Recognition

Hsp90 chaperones have the capacity to assist the folding of numerous proteins known as clients. Until the present day, about 200 client proteins of Hsp90 were scrutinized (Wayne et al. 2011; Taipale et al. 2012). These have vast roles in the response of the organism to multiple situations as in innate immune reactions, viral threatening situations, slight alterations in the RNA molecules or the SHORT-ROOT (SHR) proteins. Further, the Hsp90 is also in charge of activation processes as well as transportation and, if necessary, degradation (Johnson 2012; Eckl and Richter 2013). Despite the tremendous efforts endured in the last years to understand Hsp90-related events there are some delicate questions that still remain, namely regarding the sites where the client proteins bind to the chaperone. Recent studies suggest these sites might be localized along the domains of the chaperone, but this raises another intricate problem related with the mechanism used by Hsp90 to recognize the client proteins. Hsp90 client proteins belong to distinct families with distinct arrangements of their amino acids' sequences. Despite some studies have pointed the αC–β4 loop in kinases as key in the process of identification of client proteins, the most important feature is related with the ability of the chaperone to recognize conformations, once there are other factors influencing further the recognition capacity, than only a sequence of polypeptides (Citri et al. 2006; García-Descalzo et al. 2011; Miao et al. 2018). This is supported by the example of Src, which has two versions of itself, vSrc and c-Src, sharing 95% of structural sequence homology but presenting alternative needs on what concerns the binding to Hsp90. The v-Src activation is only dependent on the Hsp90 chaperone while the c-Src needs to be free of Hsp90 to be activated (Duval et al. 2007; Boczek et al. 2015; Luo et al. 2017).

25.1.5 Extracellular Hsp90

A very interesting topic on Hsp90 is related with the presence of an extracellular form of the chaperone (eHsp90). Several proposals concerning the origin of this extracellular Hsp90 have been advanced but three of them have prevailed: it may result from cell secretion, cell release or it may be a cell membrane-bounded form. Indeed, it was believed for numerous years that eHsp90 would be simply the result of the release of intracellular Hsp90 by dying cells. However, the form present in the exterior milieu of the cell is Hsp90a, responsible for processes like wound regeneration and cell motility, which does not sustain that hypothesis. As a matter of fact, secretion of this eHsp90a is not accomplished by all cells, but rather by the cells subjected to stress-like events such as heat, hypoxic environment exposure, reactive oxygen species (ROS) or radiation. The mechanism by which this extracellular chaperone is secreted may be related with an innovative pathway of protein secretion by exosomes. Another interesting feature about eHsp90a is its distinct mode of action since, in contrast to the intracellular form, it does not require the activity of ATPase to proceed to its normal functioning, in what regards to its role in cell motility (Tsutsumi and Neckers 2007; Song et al. 2010; Li et al. 2013; Wong and Jay, 2016).

25.1.6 TRAP-1: The Mitochondrial Homologue of Hsp90

The Hsp90 homologue located in the mitochondria, namely in the matrix and, occasionally, in the intermembrane space, is TRAP-1, as already mentioned. Its functions range from the maintenance of the organelle to the overall integrity of the cell, ensuring that organic homeostasis is preserved by using multiple internal pathways for this purpose (Felts et al. 2000; Altieri 2013). The usual functions of TRAP-1 are uncontrolled in pathological conditions such as neurodegeneration or cancer, particularly the functions related with controlled cell death induced by ROS, maintenance of mitochondrial integrity and correct folding of proteins within a given region of a cellular organelle (Matassa et al. 2012; Agliarulo et al. 2015). The similarities between TRAP-1 and Hsp90 are of about 60%. TRAP-1 has been foreseen to have circa 6 variations in the cytosolic milieu with features matching those of Hsp90, namely the inhibition of the activity by the compounds geldanamycin and radicicol and the intrinsic ATPase activity (Hadden et al. 2006; Bartsch et al. 2017). The initiation of TRAP-1's activity is triggered by heat shock and the higher ATP binding affinity, in comparison to Hsp90, leads to a closed conformational state but without ATP hydrolysis. The measured concentrations of TRAP-1 are risen in several cancers and this complex has a relation with a component of the mitochondrial permeability transition pore (mPTP), the matrix peptidyl prolyl isomerase cyclophilin D (CypD), essential in the mitochondria-induced apoptosis process. Studies advocate the potential role of CypD as a client protein of TRAP-1 and the superior importance of this binary complex in stress-dependent cytoprotection activated by oncogene stimulation (Neckers et al. 2007; Basit et al. 2017). Recent studies demonstrated the potential of combining mitochondria-specific moieties with the HspP90 inhibitor geldanamycin to yield geldanamycin-mitochondrial matrix inhibitors. These molecules do gather inside the mitochondrial space, further encouraging the organelle to fail, and open the permeability transition pore associated with the mitochondrial milieu (Kang et al. 2010; Kang et al. 2011).

The first studies on TRAP-1 pointed to the capacity of this complex to protect the cell against mitochondrial-mediated apoptosis (Altieri 2013). With the advances of molecular biology, it was possible to discover that programmed cell death is a fate that can occur by two distinct ways: an extrinsic pathway that uses the transduction of signals received by death receptors at the surface of the cell, and an intrinsic pathway that occurs through the occasional dysfunction of a given organelle which releases death stimulating factors. This is the case of the inner mitochondrial membrane complex cytochrome c that, when in the cytosol, is a direct inducer of the mechanisms of apoptosis (Fulda and Debatin 2006; Elmore 2007). Despite of both pathways having an independent mechanism by which they act in cell homeostasis leading to the fatal course, they both do interact to regulate the fine process of programmed cell death, activating the cascade of caspases and having as consequence the disruption of the inner and fundamental structure of the cell (Lopez-Hernandez et al. 2006). In a study carried out by Masuda and collaborators, the administration of a molecule named β-hydroxyisovalerylshikonin (β-HIVS), used to treat patients with oncologic disorders (due to its mechanism of irreversible inhibition of the tyrosine kinase domain), reduced the levels of TRAP-1 in cancer cells, an event associated with the augmented apoptosis induced by mitochondria (Masuda et al. 2004; Altieri 2013). These data support a potential role for TRAP-1 in mitochondrial homeostasis, thus avoiding cell death. The TRAP-1 implication in some neurodegenerative pathologies has also been the subject of investigation. In the case of Parkinson's disease, a mechanism of preservation against cellular damages controlled by a mutated PTEN-induced Putative Kinase 1 (PINK-1) does require the mediation by TRAP-1, which is phosphorylated by PINK-1, demonstrating a similarity to Hsp90 in the sense that the activity of this complex is also managed by posttranslational alterations. These events culminate in a TRAP-1 complex that prevents mitochondrial dysfunction and ROS-induced cell apoptosis (Pridgeon et al. 2007; Zhang et al. 2013).

25.2 Role of Hsp90 in Inflammation, Neuroinflammation and Neurodegeneration

The past years in the field of Hsp90 research have been recognized by the awareness of the involvement of this chaperone in many pathological conditions. Apart from their implication in cancer-related events, also reported for other HSP (Nahleh et al. 2012; Calderwood 2018), which has led to the trial of a number of Hsp90 inhibitors as anti-tumorigenic therapeutic alternatives (Zuehlke et al. 2018), Hsp90 seems to be also implicated in some human autoimmune and inflammatory diseases, as recently reviewed (Tukaj and Wegrzyn 2016). Indeed, the list of Hsp90 protein clients includes several key mediators of the signalling pathways regulating inflammatory and immune responses, such as JAK/Signal Transducer and Activator of Transcription (STATs) (JAK2), Toll Like-Receptor 4 (TLR4) and Nuclear Factor-kappa B (NF-κB) (Sevin et al. 2015). Hsp90 also interacts and stabilizes members of the receptor interacting kinase (RIP) family, essential sensors of intracellular and extracellular stresses which are part of the signalling cascade regulating cell death and innate immunity, possibly determining if either a pathway of pro-inflammatory gene expression or apoptosis is activated in the cell (Zhang et al. 2010; Humphries et al. 2015). One of the cascades triggered by TLR4 activation requires the recruitment of RIP1, which is followed by RIP1 ubiquitination and the downstream activation of NF-κB, induced by the TGF-β-activated kinase-1/Inhibitor of κB (IκB) Kinase (TAK-1/IKK) complex. NF-κB translocation into the nucleus induces then the expression of a number of pro-inflammatory cytokines, such as interleukin 1 beta (IL-1β) and tumour necrosis factor alpha (TNFα) (Humphries et al. 2015). In addition, stimulation of RIP1 by the TNFα cytokine also causes RIP recruitment to the TNF receptor, which can also culminate in the activation of NF-κB to induce the production of pro-inflammatory factors (Zhang et al. 2010; Sevin et al. 2015). By being an Hsp90 client protein, RIP is a potential target of Hsp90 inhibitors. Indeed, many Hsp90 inhibitors have higher affinity for the ATP binding site of Hsp90 than ATP/ADP, disrupting the intrinsic ATPase activity and therefore the correct folding and stabilization of client proteins by the chaperone. The client proteins are instead directed for degradation via the ubiquitin-proteasome pathway (Zhang and Burrows 2004; Whitesell and Lindquist 2005; Sharp and Workman 2006; Tukaj and Wegrzyn 2016). In accordance, inactivation of Hsp90 leads to RIP degradation and prevents the activation of NF-κB (Zhang and Burrows 2004), thereby diminishing the production of key pro-inflammatory mediators. Additionally, Hsp90 inhibitor drugs were also shown to dissociate the IKK complex. The IKK complex has a master role in NF-κB activation, triggering its translocation into the nucleus and NF-κB-dependent gene expression of pro-inflammatory factors (Salminen et al. 2008; Israel 2010), as above mentioned. The IKK complex is composed of two protein kinases, IKKα and IKKβ, which phosphorylate the inhibitor of kB proteins (IkB) and are responsible for keeping NF-κB inactive in the cytoplasm. Once phosphorylation occurs, IkB is addressed for ubiquitination and proteasomal degradation, freeing NF-κB. A regulatory subunit devoid of catalytic activity and named IKKγ (or NEMO) is also part of the IKK complex (Salminen et al. 2008; Israel 2010). In addition to other proteins that seem to be candidate components of the IKK complex, but rather in a transitory status (Israel 2010), Hsp90 has been described as being important for stabilizing the complex together with its co-chaperone cdc37 (Salminen et al. 2008). It was verified that the dissociation of the IKK complex by Hsp90 depletion results in the inhibition of NF-κB activation and of its translocation into the nucleus (Salminen et al. 2008), therefore interfering with the expression of pro-inflammatory mediators. Another rationale for the anti-inflammatory effects of Hsp90 inhibitors concerns the activation of HSF1 (Rabindran et al. 1991), which under normal physiological conditions exists in a monomeric form. This inactive form is stabilized by the formation of a complex with cytoplasmic chaperones, including Hsp90 that is believed to work as a HSF1 major repressor, maintaining its monomeric state (Rabindran et al. 1993; Sarge et al. 1993; Westwood and Wu 1993; Zuo et al. 1995; Zou et al. 1998; Bharadwaj et al. 1999). Following cellular stress induced by heat exposure, inflammation or others, HSF1 dissociates from these molecular chaperones, migrates to the nucleus and acquires a trimeric conformation in order to promote the expression of a number of HSP genes, including Hsp70 and Hsp90, after binding to specific heat shock element (HSE) sequences in DNA (Hensold et al. 1990; Price and Calderwood 1991; Rabindran et al. 1993; Sarge et al. 1993; Westwood and Wu 1993; Bharadwaj et al. 1999; Xie et al. 2002). The role of Hsp90 as a HSF1 monomer stabilizer has been, however, recently challenged by Kijima et al. latest findings which do not support this theory and suggest instead a non-canonical role for Hsp90 in modulating HSF1 activity by removing HSF1 trimers from the heat shock elements in DNA, thus putting an end in the heat shock response (Kijima et al. 2018). Nonetheless, whatever the mechanism associating Hsp90 with HSF1 is, the use of Hsp90 inhibitors should always lead to activation of HSF1 and an increase in its transcriptional activity, causing an increased expression of factors such as Hsp70 and IL-10 that are responsible for the suppression of pro-inflammatory and the activation of anti-inflammatory genes, as already proposed (Tukaj and Wegrzyn 2016). In fact, Hsp90 inhibitors targeting the N-terminal domain were recently shown to disrupt the HSF1-Hsp90 interaction, prolonging the time of HSF1 binding to the *hsp70* gene promoter and of constitutive and inducible HSF1-mediated gene transcription (Kijima et al. 2018). Additionally, the inhibition of Hsp90 has also been associated with the selective blockade of innate and adaptive cells of the immune system in several autoimmune/inflammatory diseases (Tukaj and Wegrzyn 2016). Taking all this information together, it seems quite logical to associate Hsp90 with inflammation events and to expect an anti-inflammatory effect out of Hsp90 inhibitors. Accordingly, some Hsp90 inhibitory drugs reduced the levels of pro-inflammatory cytokines and nitric oxide (NO) in severe sepsis (Chatterjee et al. 2007), autoimmune encephalomyelitis (Dello Russo et al. 2006) and experimental stroke (Qi et al. 2014). In human vascular smooth muscle cells, the 17-demethoxygeldanamycin (17-AAG) and the 17-Dimethylaminoethylamino-17-demethoxygeldanamycin (17-DMAG) Hsp90 inhibitors were able to reduce cytokine-induced NF-κB activation, as well as the expression of chemokines induced by incubation with pro-inflammatory cytokines (Madrigal-Matute et al. 2010). In the same study, the authors showed that 17-DMAG reduced the inflammatory response and the chemokine levels in a rodent experimental model of atherosclerosis (Madrigal-Matute et al. 2010). In rheumatoid arthritis related experiments, in vitro and in vivo administration of a small molecule Hsp90 inhibitor blocked the NF-κB translocation to the nucleus, and impaired cytokine and NO production in cell cultures, while in animal models it caused an inhibition of paw swelling along with an overall improvement in the joint structures (Rice et al. 2008). Moreover, a different synthetic inhibitor was able to block the activation of antigen-specific $CD4^+$ T cells as well as attenuate the disease progression in a rodent model of collagen-induced arthritis (Yun et al. 2011).

As already mentioned and reviewed by other authors (Sevin et al. 2015; Tukaj and Wegrzyn 2016; Ikwegbue et al. 2017), there is already some evidence of the potential of Hsp90 inhibitors in the therapy of human inflammatory/autoimmune diseases. In what regards neuroinflammation in particular, data is less abundant, but the existing studies suggest that Hsp90 is also implicated in the inflammatory response associated with several pathologies of the nervous system, possibly through regulating the levels of Hsp70 (Kim et al. 2012; Li et al. 2015). In fact, in an experimental rat model of global cerebral ischemia, the administration of the 17-AAG Hsp90 inhibitor significantly reduced the cerebral infarct volume, stimulated neuroprotection against neuronal death and autophagy and reduced the expression of the pro-inflammatory cytokines IL-1β, interferon gamma (INF-γ), and TNF-α in the rat CA1 hippocampal region (Li et al. 2015). This had implications in the functional outcome, as the treated animals exhibited also improved learning and memory abilities (Li et al. 2015). The mechanisms underlying these beneficial effects seem to be related with an induced upregulation of Hsp70, which was observed upon Hsp90 inhibition by 17-AAG in Li et al. experiments (Li et al. 2015), and has been reported as an outcome of Hsp90 inhibition (Kim et al. 2012). In fact, 17-AAG administration was also able to trigger Hsp70 expression in microglia and neurons in the injured brain tissue, and promoted neuroprotection accompanied by a reduced haemorrhage volume and positive functional neurobehavioural outcomes in a mice model of traumatic brain injury (Kim et al. 2012; Kim et al. 2015). Additionally, in vitro pharmacological Hsp70 induction by a variety of Hsp90 inhibitors, including 17-AAG, resulted in cytoprotection of the cerebral vasculature against the cell death caused by pro-inflammatory insult (Kacimi and Yenari 2015).

As easily inferred from the above cited studies, the role of Hsp90 in neuroinflammation and neurological pathology has been mainly linked to the effect of its inhibition on Hsp70 increased induction and the consequent neuroprotective impact. For some time, this has been also the reasoning linking Hsp90 with neurodegenerative conditions as Huntington's, Parkinson's or Alzheimer's diseases, as well as other nervous tissue-related debilitating pathologies on which the neuronal units accumulate toxic protein aggregates or are victims of increased oxidative stress, being incapable of dealing with these states. Thus, increased levels of Hsp70 and of other molecular chaperones following Hsp90 inhibition appear to be responsible for the enhanced disaggregation and protein degradation of misfolded and accumulated protein aggregates, leading to an improvement of the general disease hallmarks in experimental models (Fujikake et al. 2008; Luo et al. 2010; Lindberg et al. 2015; Lackie et al. 2017). However, the contribution of Hsp90 to neurodegeneration might also occur through the ability of the chaperone to preserve the integrity and functional stability of abnormal neuronal proteins, leading to their accumulation (Luo et al. 2010; Lindberg et al. 2015), so that the use of specific inhibitors would derange this task, therefore having an ameliorating effect on the pathological condition (Luo et al. 2010). In fact, in the cases of Alzheimer's and Parkinson's diseases, the Hsp90 chaperone has also been found, respectively, at the sites where the Aβ-amyloid plaque resulting from the hyperphosphorylated tau protein is being generated, or in the Lewy bodies generated by α-synuclein. Using the most recent molecular biology techniques it has been possible to identify modifications in the normal patterns of protein expression as well as the post-translational modifications occurring in these proteins (Gallo 2006; Luo et al. 2010; Ou et al. 2014; Lindberg et al. 2015; Lackie et al. 2017). Some studies have even shown that the beneficial effects of Hsp90 inhibitors included directing the aberrant proteins and aggregates into the proteasome for degradation (for revision (Luo et al. 2010)) or blocking the formation of toxic aggregates (Thomas et al. 2006), and in some cases these were uncoupled from the finding of Hsp70 (or Hsp40) overt induction (Thomas et al. 2006). Additionally, when produced in higher levels within the context of neurodegeneration, Hsp90 might act as a danger-associated molecular pattern (DAMP), which is recognized by specific pattern recognition receptors (PRRs) of the innate immune system like Toll-like-receptors 2 (TLR2) and 4 (TLR4), triggering thereafter the generation of strong immune responses through the activation of inflammatory signalling cascades (Ardura-Fabregat et al. 2017). The resultant neuroinflammatory environment is now recognized as an important player in the evolution of neurodegeneration.

25.3 Neuroinflammation as a Key Event in Pain Neurobiology

Neuroinflammation is also a key event in the generation of hypersensitization responses responsible for the pain sensation. Indeed, many of the molecular players found in the inflammatory response are also found altered in the nervous tissue during the progression of chronic pain conditions (Milligan and Watkins 2009;

Clark et al. 2013; Ellis and Bennett 2013; Old et al. 2015; Ji et al. 2016; Martini and Willison 2016; Tsuda 2016; Pinho-Ribeiro et al. 2017; Lacagnina et al. 2018). As a result of a harmful event induced by mechanical, thermal or chemical stimuli to non-neural or neural tissue, the action potentials generated in primary afferent neurons will propagate not only to their central axon branches located at the dorsal horn of the spinal cord, but also towards the peripheral terminals, causing the release of neuropeptides, such as substance P (SP) and calcitonin-gene-related peptide (CGRP), and glutamate at these sites. This triggers a rapid response of the immune system that includes the recruitment of immune cells to the site of aggression and activation of local immune cells. Mast cells, macrophages, neutrophils and T-lymphocytes participate in this process, as well as Schwann cells in the event of peripheral nerve injury, releasing a vast number of pro-inflammatory molecules, such as histamine, prostaglandins, thromboxanes, leukotrienes and the cytokines IL-1β, interleukin 6 (IL-6) and TNFα, ATP from necrotic cells, neurotrophins and extracellular proteases and protons, at the lesion site, an event named neurogenic inflammation. Amid the paraphernalia of released molecules, the ones that prompt the most potent outcome are those arising from the injured tissue cells, namely ATP, bradykinin and protons (Huang et al. 2006). The molecules of this inflammatory soup are responsible for the activation of their specific receptors at the nociceptor's peripheral terminal, which culminates in the activation of different intracellular signalling cascades, increased intracellular calcium, and plastic changes in the expression of transcription factors, neurotransmitters and their receptors. As a result, the sensory neurons responsible for detecting noxious input, namely the Aδ and C nociceptors, become more responsive to lower threshold stimuli, ending up in enhanced excitability and, sometimes, spontaneous firing. This characterizes the mechanisms of peripheral sensitization to subsequent stimuli that give rise to the symptoms of hyperalgesia (increased pain from a stimulus that normally provokes pain) and allodynia (pain due to a stimulus that normally does not provoke pain) recognized in painful conditions (Gaudet et al. 2011; Gangadharan and Kuner 2013; Castro-Lopes and Neto 2014; Das 2015; Pinho-Ribeiro et al. 2017). At the spinal cord and supraspinal regions, ongoing painful stimulation also leads to central sensitization, to which contribute several mechanisms (Latremoliere and Woolf 2009; Woolf 2011), including the activation of glial cells, such as microgliocytes and astrocytes. Among other events, activated glial cells release several inflammatory mediators, contributing to the generation of a neuroinflammatory environment at central levels of the nervous system that might be neurotoxic, impacting therefore on the development and maintenance of persistent painful states (Milligan and Watkins 2009; Hansen and Malcangio 2013; Ji et al. 2013, 2016; Sofroniew 2014; Alfonso Romero-Sandoval and Sweitzer 2015; Old et al. 2015; Ren and Dubner 2016; Tsuda 2016; Zhao et al. 2017).

25.3.1 *Hsp90 Role in Pain Neurobiology: Experimental Evidence*

The activation of specific cellular signalling pathways, like those including mitogen-activated protein kinases (MAPKs), such as p38, c-Jun N-terminal kinase (JNK) and extracellular signal-regulated kinases (ERK), is a process shared by the inflammatory reaction (Sevin et al. 2015) and the plastic response to noxious input (Ji 2004; Salter 2005; Brown et al. 2008; Ji et al. 2009; Berta et al. 2016; Taves et al. 2016; Wu et al. 2016). As Hsp90 is implicated in neuroinflammation events, by having as target client proteins molecules involved in the regulation of the immune response, as previously discussed, it seems logical to hypothesize this molecular chaperone might also have some role in pain-related mechanisms. In fact, the expression of HSP is known to be enhanced after peripheral nerve injury (Costigan et al. 1998) and there is some scientific evidence suggesting Hsp90 is a molecule playing a part in pain neurobiology mechanisms, although the mechanisms involved are still elusive. One of such studies clearly showed that allodynia or hyperalgesia were attenuated in the Chronic Constriction Injury (CCI) rat model of neuropathic pain following systemic (subcutaneous) or intrathecal administration of Hsp90 inhibitors, respectively (Hutchinson et al. 2009). In the same study, Hutchinson et al. demonstrated that intrathecal morphine-induced analgesia in CCI animals, which is prevented by TLR4 activation, was also improved by intrathecal Hsp90 inhibition (Hutchinson et al. 2009). These and further studies suggested that the effects of Hsp90 inhibition on nociception might be related with a TLR4-mediated signalling, which is probably hindered by Hsp90 inhibition (Hutchinson et al. 2009; Lewis et al. 2010). Several exogenous ligands, including Hsp90, activate TLR4, therefore recruiting typical cascades of inflammatory signalling, and so this receptor is clearly associated with both inflammatory and neuropathic pain (Ji et al. 2016; Lacagnina et al. 2018). Additionally, it has already been shown that in response to lipopolysaccharide (LPS), the TLR4 receptor complex associates with both Hsp90 and Hsp70 chaperones in the cell surface, and these are required for the adequate function of the complex (Triantafilou and Triantafilou 2004; Triantafilou et al. 2008). The Hsp90 role in the opioid reaction has also been investigated by Koshimizu et al. in cell culture studies. The authors found that the adenylate cyclase overactivation that occurs as a cellular adaptation/sensitization mechanism to chronic morphine treatment was diminished following Hsp90 inhibition with Geldanamycin (Koshimizu et al. 2010), suggesting that Hsp90 might be implicated in opiate tolerance and dependence physiology. This was further supported by a series of experiments combining proteomics, network analysis and animal behaviour approaches, which showed that the Hsp90 protein levels were significantly increased in the presynaptic fractions of the striata of rats subject to escalating doses of chronic morphine administration, used as a model of opiate-induced dependence (Abul-Husn et al. 2011). Treatment with Geldanamycin immediately prior to opioid antagonist administration (naloxone precipitation) significantly attenuated the somatic behavioural signs of morphine withdrawal (Abul-Husn et al.

2011), pointing to Hsp90 as a promising target for the prevention of morphine dependence. Interestingly, most of the opiate-induced adverse effects, such as the development of morphine tolerance and/or hyperalgesia, were shown to be rescued by microglia inhibition, which also has a positive outcome on reducing nocipetive behaviours in animals models of inflammatory and neuropathic pain (Song and Zhao 2001; Raghavendra et al. 2002; Raghavendra et al. 2004; Clark et al. 2007; Cui et al. 2008; Milligan and Watkins 2009; Ji et al. 2013; Old et al. 2015; Ren and Dubner 2016; Taves et al. 2016; Tsuda 2016; Adaes et al. 2017). In accordance, Tsai and collaborators found that morphine-evoked microglia activation and migration in cell culture were prevented by administration of (+)-naloxone (Tsai et al. 2015), a compound that does not activate the traditional miu, kappa or delta opioid receptors but rather non-classical opiate receptors (Hutchinson et al. 2012; Lewis et al. 2012). Curiously, this was accompanied by an inhibition of the enhanced Hsp90 cleavage and of the decrease of intact Hsp90 expression levels induced by morphine administration (Tsai et al. 2015), thus suggesting that Hsp90 is somehow implicated in this microglial activation-mediated side effects of opiates. At this point, it is interesting to note that the events linking microglia activation with opiate side effects are likely to the mediated by activation of microglial cell surface TLR4-mediated signalling pathways (Watkins et al. 2009; Lewis et al. 2012). These data altogether point to a role of Hsp90 in pain pathophysiology, and particularly on opiate effects, that seem to be in some way related with the TLR4 signalling cascades. Curiously, a recent study using the sciatic nerve CCI model of neuropathic pain in TLR4 knockout mice has shown that microglial TLR4 is mediating autophagic impairment, therefore contributing to the development of neuropathic pain symptomatology (Piao et al. 2018). Autophagy is an intracellular degradation system that is important for the elimination of materials and for the cellular renovation and homeostasis, which, among other mechanisms, might be mediated by molecular chaperones and co-chaperones (Mizushima and Komatsu 2011). Hsp90 appears to be implicated in the regulation of autophagy, as it regulates the stability and functionality of several signalling proteins through several mechanisms that include TLR-mediated autophagy (Wang et al. 2016), and it was recently observed that Hsp90 inhibition hampers the ability of tumour cells for providing an effective autophagic response (Han et al. 2018). Thus, Hsp90 involvement in TLR4-mediated signalling in painful conditions might also occur through regulation of autophagy in those conditions of cellular stress. Additionally, other mechanisms underlying Hsp90 role in pain pathophysiology might be involved, as suggested by Lei and collaborators recent studies (Lei et al. 2017). Accordingly, while studying the mu-opioid receptor (MOR)-mediated signalling and its effects on pain, the authors have found that the consequences of Hsp90 inhibition on morphine-induced analgesia are mediated by ERK MAPK activation in the brain (Lei et al. 2017), which apparently represents an alternative signalling pathway to TLR4-mediated facilitation of Hsp90 action on opioid-induced antinociception. The authors reported that opioid-induced analgesia was effectively prevented in mice models of post-surgical pain and HIV neuropathic pain following 17-AAG intracerebroventricular administration, while in naïve animals, opioid-induced analgesia to thermal tail-flick stimuli was only slightly attenuated. In addition, Hsp90 inhibition boosted morphine-induced precipitated withdrawal, therefore increasing opioids' side effects, and all these Hsp90 blockage-induced effects were recapitulated by ERK inhibition but not by Hsp70 blockade (Lei et al. 2017).

It is currently accepted that the neuronal injury or neuroinflammation associated with pathological pain conditions are not limited to harmful events but are accompanied by induction of survival mechanisms. In a rat model of monoarthritis, a joint inflammatory painful condition, the intrathecal administration of the 17-DMAG Hsp90 inhibitor diminished the mRNA expression levels of the P_2X_3 purinergic receptor and of glial fibrillary acidic protein (GFAP, satellite glial cell activation marker), which had been both found increased in sensory ganglia of inflamed animals. These data suggest that Hsp90 might be involved in neuronal sensitization and satellite glial cell activation in sensory ganglia during painful conditions (Nascimento et al. 2017; Neto 2018). This was accompanied by the attenuation of movement-induced mechanical allodynia in the affected joint and the reversal of the monoarthritis-induced increased levels of Hsp90 mRNA and respective cleavage in dorsal root ganglia (Nascimento et al. 2017; Neto 2018). The mechanisms underlying those effects are, however, still unclear but is interesting to note that oxidative stress, which is increased in inflammatory conditions (see below), apparently leads to cleavage and deactivation of Hsp90 (Beck et al. 2009). Moreover, the activation of satellite glial cells in sensory ganglia has been proposed to occur as a rescue event in response to neuronal injury and/or stress (Ohara et al. 2009; Gu et al. 2010), and the expression of neuronal injury markers such as activating transcription factor 3 (ATF3) is associated also with survival and neuronal regeneration events (Seijffers et al. 2007; Hunt et al. 2012; Gey et al. 2016). The overexpression of ATF3 in sensory neurons is consistently found in models of neuropathic pain due to peripheral nerve injury (Tsujino et al. 2000; Hunt et al. 2012), and in osteoarthritis (Ferreira-Gomes et al. 2012; Thakur et al. 2012; Adaes et al. 2015) and monoarthritis models of joint pain this was also reported (Nascimento et al. 2011; Nascimento et al. 2014). Interestingly, partial ATF3 knockdown using small interference RNA (siRNA) in dorsal root ganglia primary cell cultures induced significant decreases in Hsp90 expression (Nascimento et al. 2017; Neto 2018), suggesting that ATF3 positively regulates Hsp90 expression in sensory ganglia. It is likely that this is related with survival mechanisms acting because of inflicted neuronal stress. In accordance, expression of ATF3 in a transgenic model of amyotrophic lateral sclerosis, a very aggressive neurodegenerative disease accompanied by pain (Chio et al. 2017), was accompanied by co-expression of Hsp70 (Vlug et al. 2005). As already mentioned, it is known that Hsp90 inhibitors exhibit neuroprotective effects against toxicity in neurodegenerative conditions probably through the activation and translocation to the nucleus of HSF1, which ultimately results in increased expression of Hsp70 and of the co-chaperone Hsp40. These promote cell proliferation and survival both by acting as co-chaperones of Hsp90 and by neutralizing the effects of a number of pro-apoptotic proteins (Triantafilou and Triantafilou 2004; Luo et al. 2010; Lackie et al. 2017). As pain is a common feature of various neurodegenerative diseases, it is conceivable that Hsp90 action during painful conditions and the effects of Hsp90 inhibition on the attenuation of nociception and neuronal sensitization are equally related with the triggering of survival pathways, although studies showing this are still missing.

25.3.2 *Oxidative Stress, Hsp90 Inhibition and TRAP-1 Mitochondrial Hsp90 Homologue in Pain Neurobiology*

There is increasing scientific evidence that molecules resulting from the oxidation of oxygen, as natural byproducts of its normal metabolism, and called reactive oxygen species (ROS), have their levels dramatically increased upon considerable cell damage and stress, an event known as oxidative stress. The same happens with molecules derived from NO and superoxide, a ROS, which are known as reactive nitrogen species (RNS) (Grace et al. 2016). Altogether, these nitroxidative species, which include superoxide, peroxynitrite and hydrogen peroxide, are produced by neurons, glia and immune cells and are implicated in several mechanisms that enhance neuroexcitability in pain pathways, by promoting cytotoxicity and neuroinflammation and thus peripheral and central sensitization (Grace et al. 2016). The most known endogenous nitroxidative species involved in pain processing are generated by nicotinamide adenine dinucleotide phosphate (NADPH) oxidases (NOX), NO synthases (NOS) and mitochondrial respiration, as already extensively reviewed (Grace et al. 2016). Nitroxidative species can lead to peripheral and central sensitization by a number of different mechanisms. Thus, they can either directly increase the excitability of spinal nociceptive neurons by a variety of pathways, activate transient receptor potential (TRP) channels located at the peripheral and central terminals and cell bodies of primary sensory neurons, induce mitochondrial dysfunction and/or regulate the production of proinflammatory mediators during pathological pain (Grace et al. 2016; Carrasco et al. 2018). Elevated levels of mitochondrial-generated ROS are frequently observed in rodent models of inflammatory and neuropathic pain and appear to be responsible for the enhanced hypersensitivity to sensory stimuli, as ROS scavengers were shown to yield significant antinociceptive effects (Kim et al. 2008, 2010; Schwartz et al. 2008, 2009). It is described that mitochondria engage into a selective mechanism of autophagy, known as mitophagy, as a result of the cell response to damage and/or stress. Mitophagy acts as defence against oxidative stress, mitochondrial dysfunction and aging (Lemasters 2005). Recent studies suggest that mitochondrial dysfunction is likely to directly affect the functioning of small-diameter axons, usually responsible for the processing of high threshold nociceptive stimuli in conditions of painful conditions. In fact, in experimental autoimmune neuritis, a model of Guillain-Barre syndrome, in which mitochondria were specifically damaged, the researchers found that in large-diameter axons healthy mitochondria moved to the damaged area, repopulating the injured axon segment where the dysfunctional mitochondria were located. In contrast, in small-diameter axons the dysfunctional mitochondria accumulated, sealing the distal part of the axon and therefore hindering the migration of functional mitochondria into the damaged distal part of the axon. These data suggested small-diameter axons are more susceptible to dysfunction and degeneration following severe mitochondrial damage in conditions of nerve injury/neurodegeneration and neuroinflammation (Sajic et al. 2018). In accordance, there are a number of studies showing that mitochondrial dysfunction is associated with the development of enhanced hypersensitivity, as reviewed (Carrasco et al. 2018). A recent paper from Toyama and collaborators reports that the continuous administration of a mitochondria–targeted peptide, known to stimulate mitochondrial respiration and restore its bioenergetics, was able to prevent the development of mechanical and cold hypersensitivity in a model of chemotherapy-induced peripheral neuropathy (Toyama et al. 2018). Additionally, in a model of diabetic peripheral neuropathy there is evidence that the neuronal mitochondrial function is decreased and that the administration of KU-32, a C-terminal Hsp90 inhibitor, to diabetic animals not only reversed diabetes-induced thermal hypoalgesia, but also enhanced mitochondrial bioenergetics in sensory neurons (Urban et al. 2010, 2012; Zhang et al. 2012). Altogether these data indicate that Hsp90 inhibition might have a beneficial effect on nociception by decreasing oxidative stress and preventing mitochondrial dysfunction, thereby improving neuronal mitochondrial bioenergetics in sensory neurons. These and other studies investigating Hsp90 inhibitors in peripheral diabetic neuropathy led to the hypothesis that selectively targeting Hsp90 chaperones to control proteostasis might be a novel and effective approach to treat diabetic complications (Dobrowsky 2016). Additionally, the supplementation with antioxidants is also being under consideration as a putative therapeutic approach for several neuropathic pain conditions (Carrasco et al. 2018).

As earlier mentioned, the mitochondrial-compartmentalized Hsp90-like chaperone is a protein of about ~75 kD, known as TRAP-1 or Hsp75 (Felts et al. 2000; Altieri 2013). Some studies in tumour cells suggest that TRAP-1 is involved in mitochondrial oxidative stress and integrity, protecting against mitochondrial apoptosis, thereby facilitating cell survival (Masgras et al. 2017). Additionally, some studies indicate that the TRAP-1 pathway of cytoprotection is deregulated in neurodegenerative conditions (Montesano Gesualdi et al. 2007; Altieri 2013). Interestingly, a retrospective study investigating bidirectional associations between mitochondrial dysfunction and different functional disorders in patients, including chronic pain, found a robust correlation between functional symptomatology and three specific very highly conserved variants in the ATPase binding domain of the TRAP-1 exon sequence (Boles et al. 2015). Thus, it is possible that TRAP-1 dysfunctional activity is also implicated in the process of hypersensitization to sensory stimuli in some types of persistent pain conditions where neuronal damage is occurring, and that the antinociceptive effects of Hsp90 inhibitors observed in those rodent pain models are associated with neuroprotective events and controlled mitochondrial oxidative stress. This hypothesis might be confirmed by the use of newly developed specific

TRAP-1 inhibitors (Seo 2015; Rondanin et al. 2018) that will allow to distinguish from Hsp90-mediated effects.

25.4 Conclusions

Hsp90 inhibitors might be a promising therapeutic tool in chronic pain, as an antinociceptive effect following their administration has been reported in some experimental models of either inflammation or nerve injury that lead to the development of persistent painful conditions. In particular, Hsp90 appears to be a potential regulator of opioid-mediated signalling, therefore having an impact on nociceptive behavioural responses and on the detrimental side effects of morphine administration, namely on dependence and tolerance. The underlying mechanisms are still unclear but might include the activation of TLR4 and/or ERK MAPK signalling cascades. Additionally, there is also indication that nitroxidative stress-mediated mechanisms might also be involved. This is supported by studies showing that oxidative stress apparently leads to increased Hsp90 cleavage and dysfunction, and that Hsp90 inhibition decreased oxidative stress and promoted mitochondrial viability. TRAP-1 (Hsp75), the mitochondrial Hsp90 homolog, might be implicated in those events. More studies using different pain models and time points of diseases duration are still needed to ascertain the real efficacy of Hsp90 inhibition as a pharmacological target to achieve antinociception and to understand the underlying mechanisms. Additionally, the putative involvement of TRAP-1 should be investigated by using a new generation of specifically targeted TRAP-1 inhibitors. This could be promising for the treatment of some specific chronic painful conditions, as for example peripheral diabetic neuropathy, where mitochondrial dysfunction has been described.

References

Abul-Husn NS, Annangudi SP, Ma'ayan A et al (2011) Chronic morphine alters the presynaptic protein profile: identification of novel molecular targets using proteomics and network analysis. PLoS One 6:e25535

Adaes S, Ferreira-Gomes J, Mendonca M et al (2015) Injury of primary afferent neurons may contribute to osteoarthritis induced pain: an experimental study using the collagenase model in rats. Osteoarthr Cartil 23:914–924

Adaes S, Almeida L, Potes CS et al (2017) Glial activation in the collagenase model of nociception associated with osteoarthritis. Mol Pain 13:1744806916688219

Afzal E, Ebrahimi M, Najafi SM et al (2011) Potential role of heat shock proteins in neural differentiation of murine embryonal carcinoma stem cells (P19). Cell Biol Int 35:713–720

Agliarulo I, Matassa DS, Amoroso MR et al (2015) TRAP1 controls cell migration of cancer cells in metabolic stress conditions: correlations with AKT/p70S6K pathways. Biochim Biophys Acta (BBA) Mol Cell Res 1853:2570–2579

Alfonso Romero-Sandoval E, Sweitzer S (2015) Nonneuronal central mechanisms of pain: glia and immune response. Prog Mol Biol Transl Sci 131:325–358
Altieri DC (2013) Hsp90 regulation of mitochondrial protein folding from organelle integrity to cellular homeostasis. Cell Mol Life Sci 70:2463–2472
Ardura-Fabregat A, Boddeke E, Boza-Serrano A et al (2017) Targeting neuroinflammation to treat Alzheimer's disease. CNS Drugs 31:1057–1082
Bagatell R, Whitesell L (2004) Altered Hsp90 function in cancer: a unique therapeutic opportunity. Mol Cancer Ther 3:1021–1030
Bartsch K, Hombach-Barrigah A, Clos J (2017) Hsp90 inhibitors radicicol and geldanamycin have opposing effects on Leishmania Aha1-dependent proliferation. Cell Stress Chaperones 22:729–742
Basit F, Van Oppen LM, Schockel L et al (2017) Mitochondrial complex I inhibition triggers a mitophagy-dependent ROS increase leading to necroptosis and ferroptosis in melanoma cells. Cell Death Dis 8:e2716
Beck R, Verrax J, Gonze T et al (2009) Hsp90 cleavage by an oxidative stress leads to its client proteins degradation and cancer cell death. Biochem Pharmacol 77:375–383
Benitez MJ, Sanchez-Ponce D, Garrido JJ et al (2014) Hsp90 activity is necessary to acquire a proper neuronal polarization. Biochim Biophys Acta 1843:245–252
Berta T, Qadri YJ, Chen G et al (2016) Microglial signaling in chronic pain with a special focus on caspase 6, p38 MAP kinase, and sex dependence. J Dent Res 95:1124–1131
Bharadwaj S, Ali A, Ovsenek N (1999) Multiple components of the HSP90 chaperone complex function in regulation of heat shock factor 1 in vivo. Mol Cell Biol 19:8033–8041
Boczek EE, Reefschläger LG, Dehling M et al (2015) Conformational processing of oncogenic v-Src kinase by the molecular chaperone Hsp90. Proc Nat Acad Sci 112:E3189–E3E98
Boles RG, Hornung HA, Moody AE et al (2015) Hurt, tired and queasy: specific variants in the ATPase domain of the TRAP1 mitochondrial chaperone are associated with common, chronic "functional" symptomatology including pain, fatigue and gastrointestinal dysmotility. Mitochondrion 23:64–70
Brown KK, Heitmeyer SA, Hookfin EB et al (2008) P38 MAP kinase inhibitors as potential therapeutics for the treatment of joint degeneration and pain associated with osteoarthritis. J Inflamm (Lond) 5:22
Buchner J (1999) Hsp90 & Co. – a holding for folding. Trends Biochem Sci 24:136–141
Calderwood SK (2018) Heat shock proteins and cancer: intracellular chaperones or extracellular signalling ligands? Philos Trans R Soc Lond Ser B Biol Sci 373
Carrasco C, Naziroglu M, Rodriguez AB et al (2018) Neuropathic pain: delving into the oxidative origin and the possible implication of transient receptor potential channels. Front Physiol 9:95
Castro-Lopes JM, Neto F (2014) Neurobiology of nociceptors. In: Raja SN, Sommer CL (eds) Pain 2014 refresher courses. 15th World Congress on Pain. IASP Press, Washington, DC, pp 407–418
Chatterjee A, Dimitropoulou C, Drakopanayiotakis F et al (2007) Heat shock protein 90 inhibitors prolong survival, attenuate inflammation, and reduce lung injury in murine sepsis. Am J Respir Crit Care Med 176:667–675
Chen B, Zhong D, Monteiro A (2006) Comparative genomics and evolution of the HSP90 family of genes across all kingdoms of organisms. BMC Genomics 7:156–156
Chio A, Mora G, Lauria G (2017) Pain in amyotrophic lateral sclerosis. Lancet Neurol 16:144–157
Christians ES, Zhou Q, Renard J et al (2003) Heat shock proteins in mammalian development. Semin Cell Dev Biol 14:283–290
Citri A, Harari D, Shohat G et al (2006) Hsp90 recognizes a common surface on client kinases. J Biol Chem 281:14361–14369
Clark AK, Yip PK, Grist J et al (2007) Inhibition of spinal microglial cathepsin S for the reversal of neuropathic pain. Proc Natl Acad Sci U S A 104:10655–10660
Clark AK, Old EA, Malcangio M (2013) Neuropathic pain and cytokines: current perspectives. J Pain Res 6:803–814

Connor RE, Marnett LJ, Liebler DC (2011) Protein-selective capture to analyze electrophile adduction of hsp90 by 4-hydroxynonenal. Chem Res Toxicol 24:1275–1282
Costigan M, Mannion RJ, Kendall G et al (1998) Heat shock protein 27: developmental regulation and expression after peripheral nerve injury. J Neurosci 18:5891–5900
Craig EA (1993) Chaperones: helpers along the pathways to protein folding. Science 260:1902–1903
Csermely P, Schnaider T, Soti C et al (1998) The 90-kDa molecular chaperone family: structure, function, and clinical applications. A comprehensive review. Pharmacol Ther 79:129–168
Cui Y, Liao XX, Liu W et al (2008) A novel role of minocycline: attenuating morphine antinociceptive tolerance by inhibition of p38 MAPK in the activated spinal microglia. Brain Behav Immun 22:114–123
Das V (2015) Chapter One – An introduction to pain pathways and pain "targets". In: Price TJ, Dussor G (eds) Prog Mol Biol Transl Sci 131:1–30. Elsevier
Dello Russo C, Polak PE, Mercado PR et al (2006) The heat-shock protein 90 inhibitor 17-allyl amino-17-demethoxygeldanamycin suppresses glial inflammatory responses and ameliorates experimental autoimmune encephalomyelitis. J Neurochem 99:1351–1362
Didelot C, Schmitt E, Brunet M et al (2006) Heat shock proteins: endogenous modulators of apoptotic cell death. Handb Exp Pharmacol 172:171–198
Didenko T, Duarte AMS, Karagöz GE et al (2012) Hsp90 structure and function studied by NMR spectroscopy. Biochim Biophys Acta 1823:636–647
Dobrowsky RT (2016) Targeting the diabetic chaperome to improve peripheral neuropathy. Curr Diab Rep 16:71
Duval M, Le Bœuf F, Huot J et al (2007) Src-mediated phosphorylation of Hsp90 in response to Vascular Endothelial Growth Factor (VEGF) is required for VEGF Receptor-2 signaling to endothelial NO synthase. Mol Biol Cell 18:4659–4668
Eckl JM, Richter K (2013) Functions of the Hsp90 chaperone system: lifting client proteins to new heights. Int J Biochem Mol Biol 4:157–165
Ellis A, Bennett DLH (2013) Neuroinflammation and the generation of neuropathic pain. Br J Anaesth 111:26–37
Elmore S (2007) Apoptosis: a review of programmed cell death. Toxicol Pathol 35:495–516
Felts SJ, Owen BL, Nguyen P et al (2000) The hsp90-related protein TRAP1 is a mitochondrial protein with distinct functional properties. J Biol Chem 275:3305–3312
Ferreira-Gomes J, Adaes S, Sousa RM et al (2012) Dose-dependent expression of neuronal injury markers during experimental osteoarthritis induced by monoiodoacetate in the rat. Mol Pain 8:50
Fujikake N, Nagai Y, Popiel HA et al (2008) Heat shock transcription factor 1-activating compounds suppress polyglutamine-induced neurodegeneration through induction of multiple molecular chaperones. J Biol Chem 283:26188–26197
Fulda S, Debatin KM (2006) Extrinsic versus intrinsic apoptosis pathways in anticancer chemotherapy. Oncogene 25:4798
Gallo KA (2006) Targeting HSP90 to halt neurodegeneration. Chem Biol 13:115–116
Gangadharan V, Kuner R (2013) Pain hypersensitivity mechanisms at a glance. Dis Model Mech 6:889–895
García-Descalzo L, Alcazar A, Baquero F et al (2011) Identification of in vivo HSP90-interacting proteins reveals modularity of HSP90 complexes is dependent on the environment in psychrophilic bacteria. Cell Stress Chaperones 16:203–218
Gaudet AD, Popovich PG, Ramer MS (2011) Wallerian degeneration: gaining perspective on inflammatory events after peripheral nerve injury. J Neuroinflammation 8:110
Georgakis GV, Younes A (2005) Heat-shock protein 90 inhibitors in cancer therapy: 17AAG and beyond. Future Oncol 1:273–281
Gey M, Wanner R, Schilling C et al (2016) Atf3 mutant mice show reduced axon regeneration and impaired regeneration-associated gene induction after peripheral nerve injury. Open Biol 6
Grace PM, Gaudet AD, Staikopoulos V et al (2016) Nitroxidative signaling mechanisms in pathological pain. Trends Neurosci 39:862–879

Gu Y, Chen Y, Zhang X et al (2010) Neuronal soma-satellite glial cell interactions in sensory ganglia and the participation of purinergic receptors. Neuron Glia Biol 6:53–62
Hadden MK, Lubbers DJ, Blagg BS (2006) Geldanamycin, radicicol, and chimeric inhibitors of the Hsp90 N-terminal ATP binding site. Curr Top Med Chem 6:1173–1182
Han J, Goldstein LA, Hou W et al (2018) HSP90 inhibition targets autophagy and induces a CASP9-dependent resistance mechanism in NSCLC. Autophagy 14:958–971
Hansen RR, Malcangio M (2013) Astrocytes—multitaskers in chronic pain. Eur J Pharmacol 716:120–128
Hartl FU (1996) Molecular chaperones in cellular protein folding. Nature 381:571–579
Hartl FU, Hayer-Hartl M (2002) Molecular chaperones in the cytosol: from nascent chain to folded protein. Science 295:1852–1858
Hensold JO, Hunt CR, Calderwood SK et al (1990) DNA binding of heat shock factor to the heat shock element is insufficient for transcriptional activation in murine erythroleukemia cells. Mol Cell Biol 10:1600–1608
Hessling M, Richter K, Buchner J (2009) Dissection of the ATP-induced conformational cycle of the molecular chaperone Hsp90. Nat Struct Mol Biol 16:287–293
Horwich AL (2014) Molecular chaperones in cellular protein folding: the birth of a field. Cell 157:285–288
Houlihan JL, Metzler JJ, Blum JS (2009) HSP90α and HSP90β isoforms selectively modulate MHC class II antigen presentation in B cells. J Immunol 182:7451–7458
Huang J, Zhang X, Mcnaughton PA (2006) Inflammatory pain: the cellular basis of heat hyperalgesia. Curr Neuropharmacol 4:197–206
Humphries F, Yang S, Wang B et al (2015) RIP kinases: key decision makers in cell death and innate immunity. Cell Death Differ 22:225–236
Hunt D, Raivich G, Anderson PN (2012) Activating transcription factor 3 and the nervous system. Front Mol Neurosci 5:7
Hutchinson MR, Ramos KM, Loram LC et al (2009) Evidence for a role of heat shock protein-90 in toll like receptor 4 mediated pain enhancement in rats. Neuroscience 164:1821–1832
Hutchinson MR, Northcutt AL, Hiranita T et al (2012) Opioid activation of toll-like receptor 4 contributes to drug reinforcement. J Neurosci 32:11187–11200
Ikwegbue PC, Masamba P, Oyinloye BE et al (2017) Roles of heat shock proteins in apoptosis, oxidative stress, human inflammatory diseases, and cancer. Pharmaceuticals (Basel) 11
Israel A (2010) The IKK complex, a central regulator of NF-kappaB activation. Cold Spring Harb Perspect Biol 2:a000158
Jackson SE (2013) Hsp90: structure and function. Top Curr Chem 328:155–240
Ji RR (2004) Peripheral and central mechanisms of inflammatory pain, with emphasis on MAP kinases. Curr Drug Targets Inflamm Allergy 3:299–303
Ji RR, Gereau RWT, Malcangio M et al (2009) MAP kinase and pain. Brain Res Rev 60:135–148
Ji RR, Berta T, Nedergaard M (2013) Glia and pain: is chronic pain a gliopathy? Pain 154(Suppl 1):S10–S28
Ji RR, Chamessian A, Zhang YQ (2016) Pain regulation by non-neuronal cells and inflammation. Science 354:572–577
Johnson JL (2012) Evolution and function of diverse Hsp90 homologs and cochaperone proteins. Biochim Biophys Acta 1823:607–613
Kacimi R, Yenari MA (2015) Pharmacologic heat shock protein 70 induction confers cytoprotection against inflammation in gliovascular cells. Glia 63:1200–1212
Kang BH, Siegelin MD, Plescia J et al (2010) Preclinical characterization of mitochondria-targeted small molecule Hsp90 inhibitors, Gamitrinibs, in advanced prostate cancer. Clin Cancer Res 16:4779–4788
Kang BH, Tavecchio M, Goel HL et al (2011) Targeted inhibition of mitochondrial Hsp90 suppresses localised and metastatic prostate cancer growth in a genetic mouse model of disease. Br J Cancer 104:629

Kijima T, Prince TL, Tigue ML et al (2018) HSP90 inhibitors disrupt a transient HSP90-HSF1 interaction and identify a noncanonical model of HSP90-mediated HSF1 regulation. Sci Rep 8:6976
Kim HY, Chung JM, Chung K (2008) Increased production of mitochondrial superoxide in the spinal cord induces pain behaviors in mice: the effect of mitochondrial electron transport complex inhibitors. Neurosci Lett 447:87–91
Kim D, You B, Jo EK et al (2010) NADPH oxidase 2-derived reactive oxygen species in spinal cord microglia contribute to peripheral nerve injury-induced neuropathic pain. Proc Natl Acad Sci U S A 107:14851–14856
Kim N, Kim JY, Yenari MA (2012) Anti-inflammatory properties and pharmacological induction of Hsp70 after brain injury. Inflammopharmacology 20:177–185
Kim N, Kim JY, Yenari MA (2015) Pharmacological induction of the 70-kDa heat shock protein protects against brain injury. Neuroscience 284:912–919
Koshimizu TA, Tsuchiya H, Tsuda H et al (2010) Inhibition of heat shock protein 90 attenuates adenylate cyclase sensitization after chronic morphine treatment. Biochem Biophys Res Commun 392:603–607
Krukenberg KA, Street TO, Lavery LA et al (2011) Conformational dynamics of the molecular chaperone Hsp90. Q Rev Biophys 44:229–255
Kundrat L, Regan L (2010) Identification of residues on Hsp70 and Hsp90 ubiquitinated by the cochaperone CHIP. J Mol Biol 395:587–594
Lacagnina MJ, Watkins LR, Grace PM (2018) Toll-like receptors and their role in persistent pain. Pharmacol Ther 184:145–158
Lackie RE, Maciejewski A, Ostapchenko VG et al (2017) The Hsp70/Hsp90 chaperone machinery in neurodegenerative diseases. Front Neurosci 11:254
Latremoliere A, Woolf CJ (2009) Central sensitization: a generator of pain hypersensitivity by central neural plasticity. J Pain 10:895–926
Lei W, Mullen N, Mccarthy S et al (2017) Heat-shock protein 90 (Hsp90) promotes opioid-induced anti-nociception by an ERK mitogen-activated protein kinase (MAPK) mechanism in mouse brain. J Biol Chem 292:10414–10428
Lemasters JJ (2005) Selective mitochondrial autophagy, or mitophagy, as a targeted defense against oxidative stress, mitochondrial dysfunction, and aging. Rejuvenation Res 8:3–5
Lewis SS, Hutchinson MR, Rezvani N et al (2010) Evidence that intrathecal morphine-3-glucuronide may cause pain enhancement via toll-like receptor 4/MD-2 and interleukin-1beta. Neuroscience 165:569–583
Lewis SS, Loram LC, Hutchinson MR et al (2012) (+)-naloxone, an opioid-inactive toll-like receptor 4 signaling inhibitor, reverses multiple models of chronic neuropathic pain in rats. J Pain 13:498–506
Li J, Soroka J, Buchner J (2012) The Hsp90 chaperone machinery: conformational dynamics and regulation by co-chaperones. Biochim Biophys Acta 1823:624–635
Li W, Tsen F, Sahu D et al (2013) Chapter Five – Extracellular Hsp90 (eHsp90) as the actual target in clinical trials: Intentionally or unintentionally. In: Jeon KW (ed) Int Rev Cell Mol Biol 303:203–35. Academic
Li J, Yang F, Guo J et al (2015) 17-AAG post-treatment ameliorates memory impairment and hippocampal CA1 neuronal autophagic death induced by transient global cerebral ischemia. Brain Res 1610:80–88
Lindberg I, Shorter J, Wiseman RL et al (2015) Chaperones in neurodegeneration. J Neurosci 35:13853–13859
Lopez-Hernandez FJ, Ortiz MA, Piedrafita FJ (2006) The extrinsic and intrinsic apoptotic pathways are differentially affected by temperature upstream of mitochondrial damage. Apoptosis 11:1339–1347
Luo W, Sun W, Taldone T et al (2010) Heat shock protein 90 in neurodegenerative diseases. Mol Neurodegener 5:24

Luo Q, Boczek EE, Wang Q et al (2017) Hsp90 dependence of a kinase is determined by its conformational landscape. Sci Rep 7:43996
Madrigal-Matute J, Lopez-Franco O, Blanco-Colio LM et al (2010) Heat shock protein 90 inhibitors attenuate inflammatory responses in atherosclerosis. Cardiovasc Res 86:330–337
Mahalingam D, Swords R, Carew JS et al (2009) Targeting HSP90 for cancer therapy. Br J Cancer 100:1523
Martini R, Willison H (2016) Neuroinflammation in the peripheral nerve: cause, modulator, or bystander in peripheral neuropathies? Glia 64:475–486
Marzec M, Eletto D, Argon Y (2012) GRP94: an HSP90-like protein specialized for protein folding and quality control in the endoplasmic reticulum. Biochim Biophys Acta 1823:774–787
Masgras I, Sanchez-Martin C, Colombo G et al (2017) The chaperone TRAP1 as a modulator of the mitochondrial adaptations in cancer cells. Front Oncol 7:58
Masuda Y, Shima G, Aiuchi T et al (2004) Involvement of tumor necrosis factor receptor-associated protein 1 (TRAP1) in apoptosis induced by beta-hydroxyisovalerylshikonin. J Biol Chem 279:42503–42515
Matassa DS, Amoroso MR, Maddalena F et al (2012) New insights into TRAP1 pathway. Am J Cancer Res 2:235–248
Mayer MP, Bukau B (1999) Molecular chaperones: the busy life of Hsp90. Curr Biol 9:R322–RR25
Miao W, Li L, Wang Y (2018) Identification of helicase proteins as clients for HSP90. Anal Chem 90:11751–11755
Milligan ED, Watkins LR (2009) Pathological and protective roles of glia in chronic pain. Nat Rev Neurosci 10:23–36
Mizushima N, Komatsu M (2011) Autophagy: renovation of cells and tissues. Cell 147:728–741
Mollapour M, Neckers L (2011) Detecting HSP90 phosphorylation. Methods Mol Biol 787:67–74
Mollapour M, Neckers L (2012) Post-translational modifications of Hsp90 and their contributions to chaperone regulation. Biochim Biophys Acta 1823:648–655
Montesano Gesualdi N, Chirico G, Pirozzi G et al (2007) Tumor necrosis factor-associated protein 1 (TRAP-1) protects cells from oxidative stress and apoptosis. Stress 10:342–350
Muralidharan S, Mandrekar P (2013) Cellular stress response and innate immune signaling: integrating pathways in host defense and inflammation. J Leukoc Biol 94:1167–1184
Nahleh Z, Tfayli A, Najm A et al (2012) Heat shock proteins in cancer: targeting the ‘chaperones’. Future Med Chem 4:927–935
Nascimento D, Pozza DH, Castro-Lopes JM et al (2011) Neuronal injury marker ATF-3 is induced in primary afferent neurons of monoarthritic rats. Neurosignals 19:210–221
Nascimento DS, Castro-Lopes JM, Moreira Neto FL (2014) Satellite glial cells surrounding primary afferent neurons are activated and proliferate during monoarthritis in rats: is there a role for ATF3? PLoS One 9:e108152
Nascimento DSM, Potes CS, Soares ML et al (2017) Drug-induced HSP90 inhibition alleviates pain in monoarthritic rats and alters the expression of new putative pain players at the DRG. Mol Neurobiol. https://doi.org/10.1007/s12035-017-0628-x
Neckers L, Kern A, Tsutsumi S (2007) Hsp90 inhibitors disrupt mitochondrial homeostasis in cancer cells. Chem Biol 14:1204–1206
Neto F (2018) Commentary on paper “drug-induced HSP90 inhibition alleviates pain in monoarthritic rats and alters the expression of new putative pain players at the DRG”. J Immunol Sci 2:17–21
Neuer A, Mele C, Liu HC et al (1998) Monoclonal antibodies to mammalian heat shock proteins impair mouse embryo development in vitro. Hum Reprod 13:987–990
Ohara PT, Vit JP, Bhargava A et al (2009) Gliopathic pain: when satellite glial cells go bad. Neuroscientist 15:450–463
Old EA, Clark AK, Malcangio M (2015) The role of glia in the spinal cord in neuropathic and inflammatory pain. Handb Exp Pharmacol 227:145–170
Ou J-R, Tan M-S, Xie A-M et al (2014) Heat shock protein 90 in Alzheimer’s disease. BioMed Res Intern 2014:7

Pearl LH (2016) Review: the HSP90 molecular chaperone—an enigmatic ATPase. Biopolymers 105:594–607

Piao Y, Gwon DH, Kang DW et al (2018) TLR4-mediated autophagic impairment contributes to neuropathic pain in chronic constriction injury mice. Mol Brain 11:11

Pinho-Ribeiro FA, Verri WA, Chiu IM (2017) Nociceptor sensory neuron-immune interactions in pain and inflammation. Trends Immunol 38:5–19

Pratt WB, Toft DO (2003) Regulation of signaling protein function and trafficking by the hsp90/hsp70-based chaperone machinery. Exp Biol Med (Maywood) 228:111–133

Price BD, Calderwood SK (1991) Ca2+ is essential for multistep activation of the heat shock factor in permeabilized cells. Mol Cell Biol 11:3365–3368

Pridgeon JW, Olzmann JA, Chin L-S et al (2007) PINK1 protects against oxidative stress by phosphorylating mitochondrial chaperone TRAP1. PLoS Biol 5:e172

Prodromou C, Siligardi G, O'brien R et al (1999) Regulation of Hsp90 ATPase activity by tetratricopeptide repeat (TPR)-domain co-chaperones. EMBO J 18:754–762

Pullen L, Bolon DN (2011) Enforced N-domain proximity stimulates Hsp90 ATPase activity and is compatible with function in vivo. J Biol Chem 286:11091–11098

Qi J, Han X, Liu HT et al (2014) 17-Dimethylaminoethylamino-17-demethoxygeldanamycin attenuates inflammatory responses in experimental stroke. Biol Pharm Bull 37:1713–1718

Rabindran SK, Giorgi G, Clos J et al (1991) Molecular cloning and expression of a human heat shock factor, HSF1. Proc Natl Acad Sci U S A 88:6906–6910

Rabindran SK, Haroun RI, Clos J et al (1993) Regulation of heat shock factor trimer formation: role of a conserved leucine zipper. Science 259:230–234

Raghavendra V, Rutkowski MD, Deleo JA (2002) The role of spinal neuroimmune activation in morphine tolerance/hyperalgesia in neuropathic and sham-operated rats. J Neurosci 22:9980–9989

Raghavendra V, Tanga FY, Deleo JA (2004) Attenuation of morphine tolerance, withdrawal-induced hyperalgesia, and associated spinal inflammatory immune responses by propentofylline in rats. Neuropsychopharmacology 29:327–334

Rauch JN, Tse E, Freilich R et al (2017) BAG3 is a modular, scaffolding protein that physically links heat shock protein 70 (Hsp70) to the small heat shock proteins. J Mol Biol 429:128–141

Ren K, Dubner R (2016) Activity-triggered tetrapartite neuron-glial interactions following peripheral injury. Curr Opin Pharmacol 26:16–25

Rice JW, Veal JM, Fadden RP et al (2008) Small molecule inhibitors of Hsp90 potently affect inflammatory disease pathways and exhibit activity in models of rheumatoid arthritis. Arthritis Rheum 58:3765–3775

Ritossa F (1962) A new puffing pattern induced by temperature shock and DNP in drosophila. Experientia 18:571–573

Rondanin R, Lettini G, Oliva P et al (2018) New TRAP1 and Hsp90 chaperone inhibitors with cationic components: preliminary studies on mitochondrial targeting. Bioorg Med Chem Lett 28:2289–2293

Sager RA, Woodford MR, Neckers L et al (2018) Detecting posttranslational modifications of Hsp90. Methods Mol Biol 1709:209–219

Sajic M, Ida KK, Canning R et al (2018) Mitochondrial damage and "plugging" of transport selectively in myelinated, small-diameter axons are major early events in peripheral neuroinflammation. J Neuroinflammation 15:61

Salminen A, Paimela T, Suuronen T et al (2008) Innate immunity meets with cellular stress at the IKK complex: regulation of the IKK complex by HSP70 and HSP90. Immunol Lett 117:9–15

Salter MW (2005) Cellular signalling pathways of spinal pain neuroplasticity as targets for analgesic development. Curr Top Med Chem 5:557–567

Sarge KD, Murphy SP, Morimoto RI (1993) Activation of heat shock gene transcription by heat shock factor 1 involves oligomerization, acquisition of DNA-binding activity, and nuclear localization and can occur in the absence of stress. Mol Cell Biol 13:1392–1407

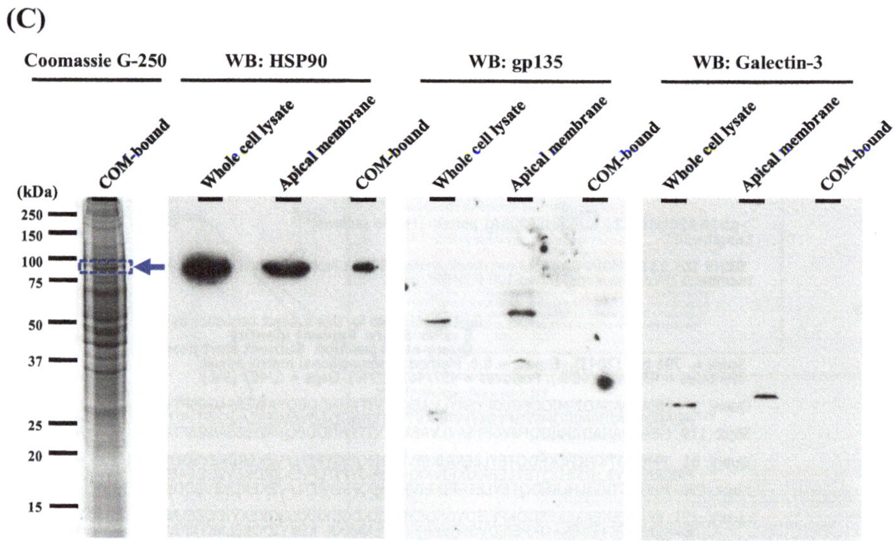

Fig. 26.1 (continued)

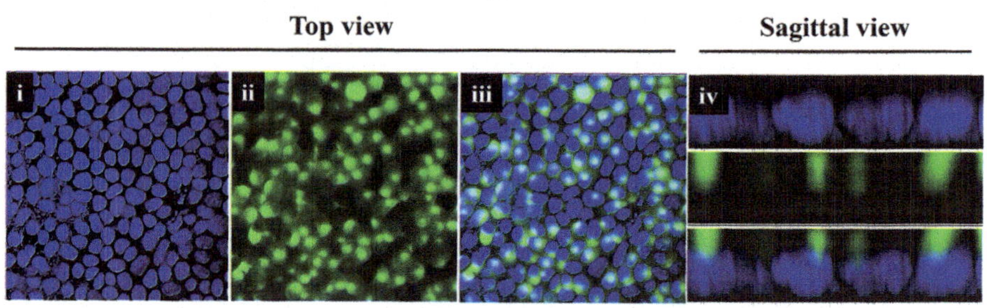

Fig. 26.2 HSP90 is also localized on apical surface of renal tubular epithelial cells. Polarized renal tubular epithelial cells were incubated with mouse monoclonal anti-HSP90 antibody without permeabilization. Subsequently, the cells were incubated with Alexa-488-conjugated goat anti-mouse antibody containing 0.1 μg/ml Hoechst dye (DNA staining for nuclear localization) and imaged by a laser-scanning confocal microscope to obtain both top view or horizontal (X-Y) section of the polarized cell monolayer (panels i–iii) and sagittal (X-Z) view from apical to basal sections of the polarized MDCK cells (panel iv). Original magnification power was 630X. The green signal represents HSP90, whereas nuclei are illustrated in blue. (Reproduced with permission from Fong-ngern et al. 2016)

26.3 HSP90-Interacting Protein Complex Is Involved in Kidney Stone Formation

To better understand the role for HSP90 in kidney stone pathogenesis, a most recent proteomics study has extensively characterized HSP90-interacting protein complex using an affinity purification followed by mass spectrometry (AP-MS) technique

(Manissorn et al. 2018). HSP90-interacting protein complex was purified from renal cells by tandem affinity purification (TAP) of the tagged HSP90. Such protein complex was then characterized by ultrahigh-resolution quadrupole time-of-flight tandem MS (Q-TOF MS/MS) (Manissorn et al. 2018). The TAP control was used to subtract non-specific binding proteins that could be co-purified by this system. TAP-tag is a highly efficient purification method that can be applied to isolate/purify protein complexes in their native forms (Puig et al. 2001). The key concept of this method is to construct the dual affinity tags to fuse with gene of the protein of interest. Thereafter, the two consecutive purification steps are performed to isolate the target protein along with its interacting partners (Li 2010). Comparing with any single-step purification, e.g. co-immunoprecipitation, the TAP-tag method has a higher specificity and higher reproducibility with minimal contaminations (Bauer and Kuster 2003; Xu et al. 2010).

Using this AP-MS technique, a total of 32 HSP90-interacting proteins could be identified from this complex, of which a half (16 proteins) are novel HSP90 interactors that had not been reported previously (Manissorn et al. 2018). Interactions of HSP90 with some of these known (i.e., HSP70, α-tubulin and β-actin) and novel (i.e., vimentin, calpain-1, and importin-β1) interactors were confirmed by co-immunoprecipitation and reciprocal co-immunoprecipitation followed by Western blotting as well as immunofluorescence study. Interestingly, α-tubulin, β-actin and vimentin have been documented as potential COM crystal receptors on apical membranes of renal tubular epithelial cells (Fong-ngern et al. 2011). Knockdown of HSP90 using siHSP90 showed significant reduction in levels of these potential COM crystal receptors together with the marked decrease of HSP90 level (Manissorn et al. 2018). Moreover, the siHSP90-transfected cells showed marked decrease in crystal-cell adhesion as compared to the non-transfected and siControl-transfected cells (Manissorn et al. 2018). Taken together, these data indicate that HSP90-interacting protein complex is involved, at least in part, in kidney stone formation (in particular, during the crystal-cell adhesion step).

26.4 COM Crystals Induce Secretion of HSP90 from Human Monocytes

Interstitial inflammation inside kidney tissue (renal parenchyma) is one of the pathological findings associated with kidney stone disease. It is also known that such interstitial inflammation can lead to fibrogenesis or fibrous change of the kidney (Holdsworth and Tipping 2007). One of the important cells that are recruited into the inflammatory site is monocyte that can then activate the inflammatory cascade by releasing several various inflammatory cytokines/chemokines (Akcay et al. 2009; Sean and Cockwell 2005; Segerer 2003; Swaminathan and Griffin 2008; Wilson et al. 2004). Interestingly, degree of monocytic cell infiltration at the inflammatory site has been shown to be correlated with the number of COM crystals

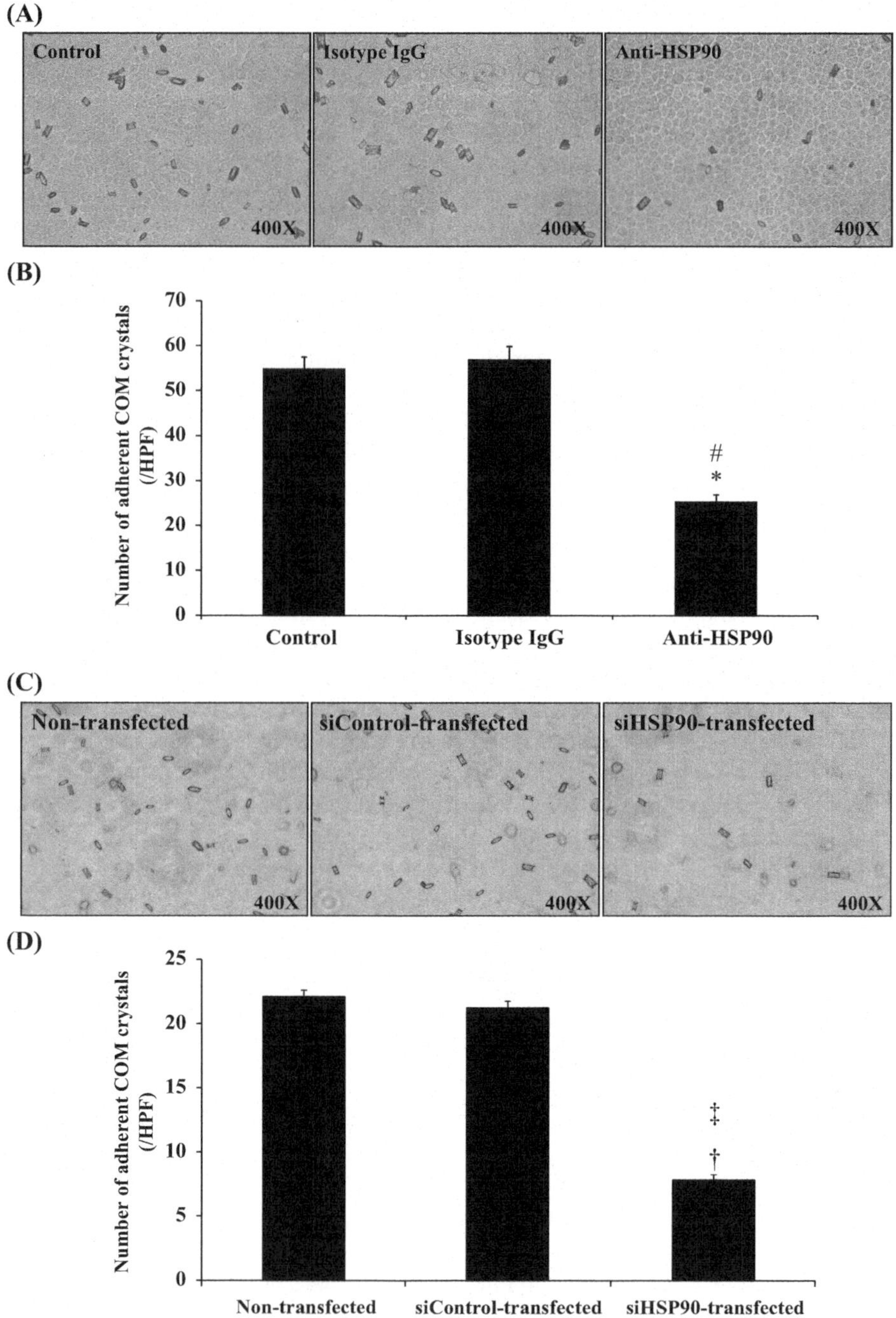

Fig. 26.3 Reduced COM crystal adhesion on renal tubular epithelial cells after neutralization and knockdown of surface HSP90. (**a**): Crystal-cell adhesion assay was performed without or with

deposited in the same locale (Okada et al. 2009, 2010). Moreover, monocytes (subsequently derived to macrophages) are frequently found surrounding the deposited COM crystals within the renal interstitium (de Water et al. 1999, 2000). It has been documented that renal tubular epithelial cells can excrete monocyte chemoattractant protein-1 (MCP-1) after exposure to COM crystals, resulting to oxidative stress and tissue inflammation (Habibzadegah-Tari et al. 2006; Umekawa et al. 2009).

In a recent proteomics study using two-dimensional gel electrophoresis followed by Q-TOF MS/MS, alterations in proteins secreted from monocytes after exposure to COM crystals have been characterized aiming for defining potential involvement of the secreted products (secretome) of human monocytes in kidney stone pathogenesis (Sintiprungrat et al. 2016). A total of 18 secreted proteins had significantly altered levels after monocytes were exposed to COM crystals as compared to the control cells. Most of them were involved in immune response and cell survival. Among the 18 significantly altered proteins secreted from monocytes, HSP90 secretion was increased after COM crystal exposure. Its increased level was validated by Western blotting. In addition, immunofluorescence study has also shown that surface HSP90 on human monocytes was significantly increased after the cells were exposed to COM crystals (Fig. 26.4). Moreover, ELISA of culture supernatants demonstrated the significantly increased level of interferon-α (IFN-α) secreted from the monocytes exposed with COM crystals (Sintiprungrat et al. 2016). These data implicate that HSP90 on the surface of human monocytes may be involved, at least in part, in the inflammatory cascade of the renal parenchyma frequently observed in kidney stone disease.

26.5 HSP90 Interacts with Filamentous Form of Actin to Form Phagosome in Human Macrophages to Eliminate COM Crystals

After migration into the crystal deposition site, monocytes can be derived to tissue macrophages that are responsible for elimination of the deposited crystals (de Water et al. 2000; Dominguez-Gutierrez et al. 2018; Kusmartsev et al. 2016; Taguchi et al.

Fig. 26.3 (continued) neutralization by mouse monoclonal anti-HSP90 antibody (specific antibody) or isotype IgG (non-specific antibody) prior to crystal-cell adhesion assay. (**b**): After rigorous wash with PBS five times to eliminate the unbound COM crystals, those remained on apical surface of the cells were counted from at least 15 random high-power fields (HPFs) under a phase contrast microscope. (**c**): Crystal-cell adhesion assay was performed in the non-transfected, siControl-transfected, and siHSP90-transfected cells. (**d**): After rigorous wash with PBS five times to eliminate the unbound COM crystals, those remained on apical surface of the cells were counted from at least 15 random HPFs under a phase contrast microscope. The quantitative data are reported as mean ± SEM. $* = p < 0.001$ *vs.* control cells; $\# = p < 0.001$ *vs.* isotype (non-specific) IgG treated cells; $\dagger = p < 0.001$ *vs.* non-transfected cells; $\ddagger = p < 0.001$ *vs.* siControl-transfected cells. (Reproduced with permission from Fong-ngern et al. 2016)

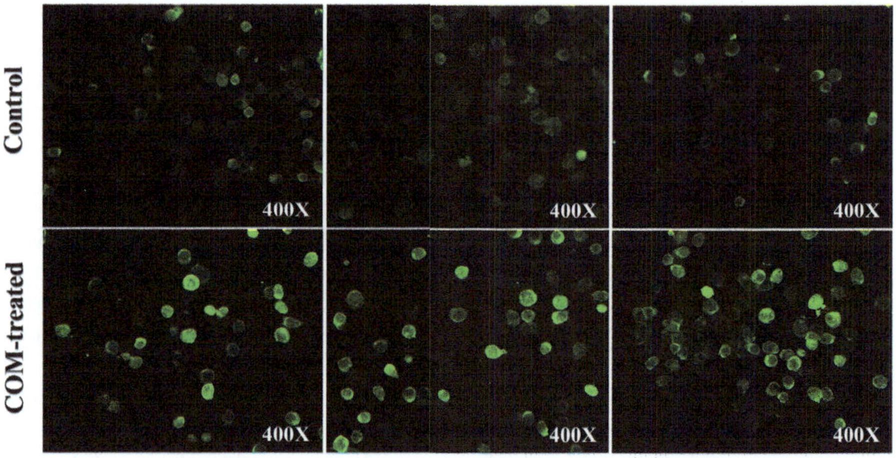

Fig. 26.4 Increased HSP90 expression on surface of human monocytes after exposure to COM crystals. U937 human monocytes were cultivated without or with COM crystals (100 μg crystals/ml medium) in serum/protein-free condition for 16 h. After washing, the cells were fixed with 1% (w/v) formaldehyde in PBS but were not permeabilized. Surface HSP90 was stained with mouse monoclonal anti-HSP90 antibody followed by goat anti-mouse IgG conjugated with Alexa Flour 488. Fluorescence images were captured from three independent experiments using a fluorescence microscope (original magnification = 400×). (Reproduced with permission from Sintiprungrat et al. 2016)

2016). After phagocytosis of COM crystals into phogosome, this vesicular microstructure fuses with lysosome (to form phagolysosome) or combines with other secretory vesicles to eliminate the degraded crystals (Chaiyarit et al. 2016; Flannagan et al. 2009; Gordon and Taylor 2005; Kinchen and Ravichandran 2008). On the other hand, the recruited tissue macrophages can worsen the inflammation cascade by overproduction of proinflammatory cytokines/chemokines, e.g., tumor necrosis factor-α (TNF-α) and interleukin-6 (IL-6), which further aggravate inflammation, oxidative damage and tissue injury (de Water et al. 2001; Khan 2004).

A role for HSP90 in phagosome formation has been evidenced in a recent proteomics study on macrophages derived from U937 human monocytes upon exposure to COM crystals (Singhto et al. 2013). Changes in cellular proteome of COM crystal-treated macrophages were characterized by two-dimensional gel electrophoresis followed by Q-TOF MS and MS/MS analyses comparing to the control cells. Using this approach, a total of 18 cellular proteins, whose levels were significantly altered, were identified from the COM crystal-treated macrophages, including those involved in cellular structure, carbohydrate metabolism, DNA/RNA processing, protein metabolism, molecular trafficking, and stress response (Singhto et al. 2013). Among these significantly altered proteins, the altered levels of ezrin, α-tubulin and β-actin in the COM crystal-treated macrophages were confirmed by Western blotting. In addition, protein-protein interactions network analysis revealed three main interaction networks of cytoskeletal proteins, chaperones/stress response proteins, and metabolic proteins. Interestingly, HSP90, one of the significantly increased proteins,

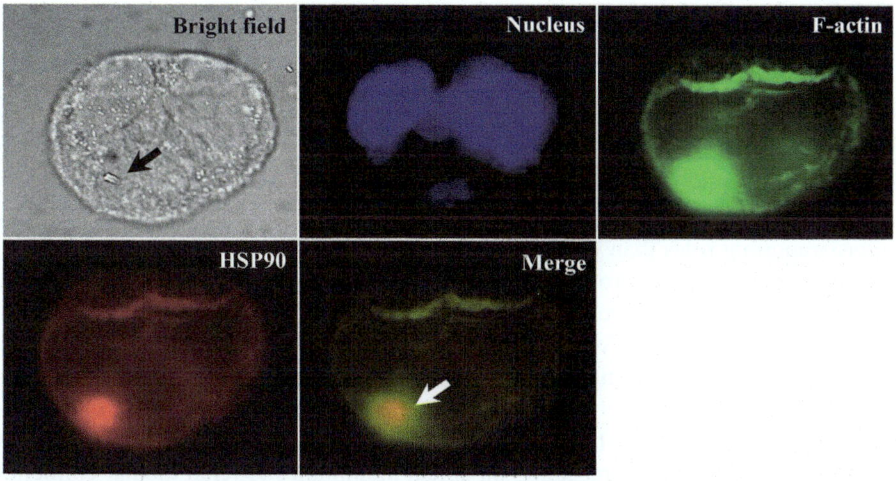

Fig. 26.5 Co-localization of HSP90 and F-actin is important for phagosome formation. After an incubation of macrophages with COM crystals, the cells were subjected to immunofluorescence co-stainings to visualize HSP90 (shown in red) together with F-actin (shown in green). Nuclei were counter-stained with Hoechst dye (shown in blue). A merged image (last panel) shows co-localization (as demonstrated by the overlapping of red and green signals that appeared as yellow) of HSP90 and F-actin on the phagosome membrane surrounding COM crystal. The arrow head indicates the locale of phagosome. Images were taken using Nikon ECLIPSE 80i fluorescence microscope (Nikon Corp.; Tokyo, Japan) with original magnification of 1000×. (Reproduced with permission from Singhto et al. 2013)

was predicted to be associated with β-actin and α-tubulin. Such prediction was then confirmed by immunofluorescence study, which showed co-localization of HSP90 and filamentous form of actin (F-actin) on the phagosome vesicular membrane in the COM crystal-treated cells (Fig. 26.5) (Singhto et al. 2013).

Functional validation was also performed using siRNA technique. The siHSP90-transfected macrophages had markedly reduced levels of HSP90, β-actin, α-tubulin and Rab5 (phagosome marker) comparing to the siControl-transfected cells (Singhto et al. 2013). These findings suggested the role for HSP90 in regulation of phagocytosis and migration of macrophages mediated by cytoskeletal proteins. Such hypothesis was then addressed by phagocytosis and migration assays. Phagocytosis assay demonstrated that the phagocytic activity of the siHSP90-transfected, COM crystal-treated macrophages was almost completely abolished comparing to the siControl-transfected, COM crystal-treated macrophages and non-transfected, COM crystal-treated macrophages (Singhto et al. 2013). Similarly, the siHSP90-transfected, COM crystal-treated macrophages had marked reduction in migration capability as compared to both siControl-transfected and non-transfected cells treated with COM crystals (Singhto et al. 2013). These findings strengthen the role for HSP in phagosome formation, phagocytic activity and migration capability of tissue macrophages.

26.6 Conclusions

Proteomics has enabled the exploratory study of the previously unknown/unexplored proteins in disease pathogenesis more feasible. In this chapter, a number of references have shown that proteomics has led to identification of HSP90 as one among several proteins that are involved in pathogenic mechanisms of kidney stone disease. Surprisingly, expression and functional data have shown that HSP90 play more important roles than we had initially anticipated. In addition to serving as the general stress response protein to regulate folding and conformation of many other proteins, HSP90 also serves as a receptor on apical surface of renal tubular epithelial cells to bind with COM crystals, leading to crystal retention inside intratubular lumen that is an important initial mechanism for kidney stone formation (Fong-ngern et al. 2016). In addition, HSP90 also forms a complex with many other proteins in renal cells (e.g., α-tubulin, β-actin and vimentin, which have been identified as potential COM crystal receptors on apical membranes of renal tubular epithelial cells (Fong-ngern et al. 2011)). Knockdown of HSP90 using siHSP90 showed significant reduction in levels of these COM crystal receptors together with the marked decrease in HSP90 level (Manissorn et al. 2018). Moreover, the siHSP90-transfected cells showed marked decrease in crystal-cell adhesion as compared to the non-transfected and siControl-transfected cells (Manissorn et al. 2018). Taken together, the data obtained from these two studies (Fong-ngern et al. 2016; Manissorn et al. 2018) indicate that HSP90 and its interacting protein complex are involved, at least in part, in kidney stone formation at the crystal-cell adhesion step.

Proteomics of human monocytes and macrophages have revealed additional significant role of HSP90 in inflammatory cascade of renal interstitium deposited with COM crystals. Monocytes treated with COM crystals had significantly increased secretion of HSP90 and IFN-α (Sintiprungrat et al. 2016), suggesting that HSP90 on the surface of human monocytes may be involved, at least in part, in inflammatory cascade of the renal parenchyma frequently observed in kidney stone disease. Moreover, macrophages derived from U937 human monocytes had increased level of cellular HSP90 upon exposure to COM crystals (Singhto et al. 2013). Interestingly, HSP90 was co-localized with F-actin on the phagosome vesicular membrane in the COM crystal-treated macrophages. siHSP90 markedly reduced levels of HSP90, β-actin, α-tubulin and Rab5 (phagosome marker) and almost completely abolished the phagocytic activity and migration capability of the COM crystal-treated macrophages (Singhto et al. 2013). These findings strengthen the role for HSP90 in phagosome formation and regulation of phagocytosis and migration of macrophages mediated by cytoskeletal proteins.

Acknowledgements This work was supported by Mahidol University research grant and the Thailand Research Fund (IRN60W0004 and IRG5980006).

References

Akcay A, Nguyen Q, Edelstein CL (2009) Mediators of inflammation in acute kidney injury. Mediat Inflamm 2009:137072

Bauer A, Kuster B (2003) Affinity purification-mass spectrometry. Powerful tools for the characterization of protein complexes. Eur J Biochem 270:570–578

Chaiyarit S, Singhto N, Thongboonkerd V (2016) Calcium oxalate monohydrate crystals internalized into renal tubular cells are degraded and dissolved by endolysosomes. Chem Biol Interact 246:30–35

de Water R, Noordermeer C, van der Kwast TH, Nizze H, Boeve ER, Kok DJ, Schroder FH (1999) Calcium oxalate nephrolithiasis: effect of renal crystal deposition on the cellular composition of the renal interstitium. Am J Kidney Dis 33:761–771

de Water R, Noordermeer C, Houtsmuller AB, Nigg AL, Stijnen T, Schroder FH, Kok DJ (2000) Role of macrophages in nephrolithiasis in rats: an analysis of the renal interstitium. Am J Kidney Dis 36:615–625

de Water R, Leenen PJ, Noordermeer C, Nigg AL, Houtsmuller AB, Kok DJ, Schroder FH (2001) Cytokine production induced by binding and processing of calcium oxalate crystals in cultured macrophages. Am J Kidney Dis 38:331–338

Dominguez-Gutierrez PR, Kusmartsev S, Canales BK, Khan SR (2018) Calcium oxalate differentiates human monocytes into inflammatory M1 macrophages. Front Immunol 9:1863

Flannagan RS, Cosio G, Grinstein S (2009) Antimicrobial mechanisms of phagocytes and bacterial evasion strategies. Nat Rev Microbiol 7:355–366

Fong-ngern K, Peerapen P, Sinchaikul S, Chen ST, Thongboonkerd V (2011) Large-scale identification of calcium oxalate monohydrate crystal-binding proteins on apical membrane of distal renal tubular epithelial cells. J Proteome Res 10:4463–4477

Fong-ngern K, Sueksakit K, Thongboonkerd V (2016) Surface heat shock protein 90 serves as a potential receptor for calcium oxalate crystal on apical membrane of renal tubular epithelial cells. J Biol Inorg Chem 21:463–474

Gordon S, Taylor PR (2005) Monocyte and macrophage heterogeneity. Nat Rev Immunol 5:953–964

Habibzadegah-Tari P, Byer KG, Khan SR (2006) Reactive oxygen species mediated calcium oxalate crystal-induced expression of MCP-1 in HK-2 cells. Urol Res 34:26–36

Holdsworth SR, Tipping PG (2007) Leukocytes in glomerular injury. Semin Immunopathol 29:355–374

Jin S, Song YC, Emili A, Sherman PM, Chan VL (2003) JlpA of Campylobacter jejuni interacts with surface-exposed heat shock protein 90alpha and triggers signalling pathways leading to the activation of NF-kappaB and p38 MAP kinase in epithelial cells. Cell Microbiol 5:165–174

Khan SR (2004) Crystal-induced inflammation of the kidneys: results from human studies, animal models, and tissue-culture studies. Clin Exp Nephrol 8:75–88

Kinchen JM, Ravichandran KS (2008) Phagosome maturation: going through the acid test. Nat Rev Mol Cell Biol 9:781–795

Kumar V, Farell G, Deganello S, Lieske JC (2003) Annexin II is present on renal epithelial cells and binds calcium oxalate monohydrate crystals. J Am Soc Nephrol 14:289–297

Kusmartsev S, Dominguez-Gutierrez PR, Canales BK, Bird VG, Vieweg J, Khan SR (2016) Calcium oxalate stone fragment and crystal phagocytosis by human macrophages. J Urol 195:1143–1151

Li Y (2010) Commonly used tag combinations for tandem affinity purification. Biotechnol Appl Biochem 55:73–83

Manissorn J, Singhto N, Thongboonkerd V (2018) Characterizations of HSP90-interacting complex in renal cells using tandem affinity purification and its potential role in kidney stone formation. Proteomics 18:XXX

Okada A, Yasui T, Hamamoto S, Hirose M, Kubota Y, Itoh Y, Tozawa K, Hayashi Y, Kohri K (2009) Genome-wide analysis of genes related to kidney stone formation and elimination in the calcium oxalate nephrolithiasis model mouse: detection of stone-preventive factors and involvement of macrophage activity. J Bone Miner Res 24:908–924

Okada A, Yasui T, Fujii Y, Niimi K, Hamamoto S, Hirose M, Kojima Y, Itoh Y, Tozawa K, Hayashi Y et al (2010) Renal macrophage migration and crystal phagocytosis via inflammatory-related gene expression during kidney stone formation and elimination in mice: detection by association analysis of stone-related gene expression and microstructural observation. J Bone Miner Res 25:2701–2711

Puig O, Caspary F, Rigaut G, Rutz B, Bouveret E, Bragado-Nilsson E, Wilm M, Seraphin B (2001) The tandem affinity purification (TAP) method: a general procedure of protein complex purification. Methods 24:218–229

Reyes-Del Valle J, Chavez-Salinas S, Medina F, Del Angel RM (2005) Heat shock protein 90 and heat shock protein 70 are components of dengue virus receptor complex in human cells. J Virol 79:4557–4567

Schubert G (2006) Stone analysis. Urol Res 34:146–150

Sean EK, Cockwell P (2005) Macrophages and progressive tubulointerstitial disease. Kidney Int 68:437–455

Segerer S (2003) The role of chemokines and chemokine receptors in progressive renal diseases. Am J Kidney Dis 41:S15–S18

Singhto N, Sintiprungrat K, Thongboonkerd V (2013) Alterations in macrophage cellular proteome induced by calcium oxalate crystals: the association of HSP90 and F-actin is important for phagosome formation. J Proteome Res 12:3561–3572

Sintiprungrat K, Singhto N, Thongboonkerd V (2016) Characterization of calcium oxalate crystal-induced changes in the secretome of U937 human monocytes. Mol BioSyst 12:879–889

Sorokina EA, Wesson JA, Kleinman JG (2004) An acidic peptide sequence of nucleolin-related protein can mediate the attachment of calcium oxalate to renal tubule cells. J Am Soc Nephrol 15:2057–2065

Swaminathan S, Griffin MD (2008) First responders: understanding monocyte-lineage traffic in the acutely injured kidney. Kidney Int 74:1509–1511

Taguchi K, Okada A, Hamamoto S, Unno R, Moritoki Y, Ando R, Mizuno K, Tozawa K, Kohri K, Yasui T (2016) M1/M2-macrophage phenotypes regulate renal calcium oxalate crystal development. Sci Rep 6:35167

Thongboonkerd V (2004) Proteomics in nephrology: current status and future directions. Am J Nephrol 24:360–378

Thongboonkerd V (2005) Proteomic analysis of renal diseases: unraveling the pathophysiology and biomarker discovery. Expert Rev Proteomics 2:349–366

Thongboonkerd V (2007a) Practical points in urinary proteomics. J Proteome Res 6:3881–3890

Thongboonkerd V (2007b) Recent progress in urinary proteomics. Proteomics Clin Appl 1:780–791

Thongboonkerd V (2008) Proteomics and kidney stone disease. Contrib Nephrol 160:142–158

Thongboonkerd V, Malasit P (2005) Renal and urinary proteomics: current applications and challenges. Proteomics 5:1033–1042

Umekawa T, Tsuji H, Uemura H, Khan SR (2009) Superoxide from NADPH oxidase as second messenger for the expression of osteopontin and monocyte chemoattractant protein-1 in renal epithelial cells exposed to calcium oxalate crystals. BJU Int 104:115–120

Verkoelen CF, Verhulst A (2007) Proposed mechanisms in renal tubular crystal retention. Kidney Int 72:13–18

Verkoelen CF, van der Boom BG, Kok DJ, Romijn JC (2000a) Sialic acid and crystal binding. Kidney Int 57:1072–1082

Verkoelen CF, van der Boom BG, Romijn JC (2000b) Identification of hyaluronan as a crystal-binding molecule at the surface of migrating and proliferating MDCK cells. Kidney Int 58:1045–1054

Vinaiphat A, Thongboonkerd V (2017) Prospects for proteomics in kidney stone disease. Expert Rev Proteomics 14:185–187

Vinaiphat A, Aluksanasuwan S, Manissorn J, Sutthimethakorn S, Thongboonkerd V (2017) Response of renal tubular cells to differential types and doses of calcium oxalate crystals: integrative proteome network analysis and functional investigations. Proteomics 17:1700192

Wilson HM, Walbaum D, Rees AJ (2004) Macrophages and the kidney. Curr Opin Nephrol Hypertens 13:285–290

Xu X, Song Y, Li Y, Chang J, Zhang H, An L (2010) The tandem affinity purification method: an efficient system for protein complex purification and protein interaction identification. Protein Expr Purif 72:149–156

Chapter 27
HSP90 et al.: Chaperome and Proteostasis Deregulation in Human Disease

Cindy Voisine and Marc Brehme

Abstract HSP90 action as a conserved chaperone is fundamental to the folding and maturation of a large fraction of structurally and functionally diverse proteins. Through its wide spectrum of clients and their connectivity within the proteome, HSP90 is a key factor in the processes that maintain cellular proteostasis, referred to as the proteostasis network (PN). Besides HSP90, hundreds of additional molecular chaperones and co-chaperones act together in complexes and intricate networks to maintain the cellular native folding environment, avoid misfolding, and redirect non-native folding intermediates back to their native state or towards clearance. The ensemble of all cellular chaperones and co-chaperones constitute the chaperome, which acts to regulate and maintain proteome functionality. Recent insights derived from systematic studies have revealed vast imbalances to cellular proteostasis reflected by chaperome deregulation from aging, to neurodegenerative diseases, and cancers. Deregulation states along with specific chaperome connectivity can represent biomarkers for certain diseases. HSP90 is not acting in isolation but as a component of a comprehensive and dynamic chaperome interactome network. Therefore, a detailed understanding of systems-level alterations of chaperome biology is fundamental towards a better understanding of HSP90 action and its therapeutic targeting in health and disease.

Keywords Chaperome · Chaperone · Co-chaperone · Heat shock protein · HSP90 · Proteostasis · Proteostasis network (PN)

C. Voisine
Department of Biology, Northeastern Illinois University, Chicago, IL, USA

M. Brehme (✉)
CB - med Center for Biomarker Research in Medicine GmbH, Graz, Austria
e-mail: marc.brehme@gmail.com

A. A. A. Asea, P. Kaur (eds.), *Heat Shock Protein 90 in Human Diseases and Disorders*, Heat Shock Proteins 19,
https://doi.org/10.1007/978-3-030-23158-3_27

Abbreviations

HSP	Heat shock protein
HSP90	Heat shock protein 90
PN	Proteostasis network

27.1 Introduction

Heat shock protein 90 (HSP90) is a highly conserved molecular chaperone that facilitates the maturation of a wide range of proteins, referred to as clients. HSP90, via its clientele, participates in a diverse range of fundamental cellular processes from transcriptional regulation, chromatin remodeling, stress response signaling, kinase activation and acts as a capacitor for evolutionary change (Jarosz and Lindquist 2010; Karras et al. 2017; Queitsch et al. 2002; Rutherford and Lindquist 1998; Sangster et al. 2008). To this end, HSP90 regulates a variety of cellular pathways impacting normal biology and disease (Taipale et al. 2010). Protein folding and maturation is a highly conserved process that enlists a number of additional molecular chaperones and co-chaperones along with HSP90.

Protein quality control mechanisms including molecular chaperones along with protein clearance mechanisms have a central role in the proteostasis network (PN) to safeguard the proteome from misfolding, aggregation and the consequences of proteotoxic stress (Balchin et al. 2016; Schmidt and Finley 2014). Molecular chaperones associate with a large and diverse set of cofactors (co-chaperones) that regulate their client specificity and functionality. With these cooperative efforts, it is apparent that individual chaperones do not function in isolation but as part of a network of cellular machines responsible for maintaining proteostasis (Brehme et al. 2014; Taipale et al. 2014).

Decades of research on the biochemical characteristics and mode of operation of molecular chaperones and their co-chaperones has highlighted the fundamental importance of HSP90 in health and disease. HSP90 along with members of the chaperome safeguard the proteome from misfolding, aggregation and proteotoxicity, maintaining proteostasis (Pearl and Prodromou 2006; Schopf et al. 2017; Taipale et al. 2010; Young et al. 2001; Zhao et al. 2005). Consequently, numerous examples exist that implicate chaperone and co-chaperone alterations as biomarkers of proteostasis collapse in aging and a series of human diseases, from neurodegenerative diseases to diverse cancers (Powers et al. 2009; Walter et al. 2011; Whitesell and Lindquist 2005). The fundamental conserved role of the human chaperome as a central component to the cellular proteostasis network (PN) and the plethora of studies implicating chaperones and co-chaperones in human diseases calls for systems-level studies of chaperome action and its deregulation in aging and disease, beyond just HSP90 (Balch et al. 2008; Brehme et al. 2014; Taipale et al. 2010).

27.2 Insights from Model Systems

Protein folding pathways are highly conserved from bacteria to humans. Molecular chaperones function by transiently binding newly synthesized or misfolded polypeptides, preventing nonproductive interactions with cellular proteins (Hartl et al. 2011; Hartl and Hayer-Hartl 2002). Through ATP dependent cycles of binding and release, these substrates fold into their functional states (Horwich et al. 2007; Liberek et al. 2008; Mayer and Bukau 2005; Saibil 2008). Molecular chaperones not only facilitate native folding but also function to protect cells from proteotoxic stressors. As a highly conserved quality control system and central arm within the proteostasis network (PN), changes in chaperome functionality affect proteome maintenance and repair capacity (Balch et al. 2008).

The chaperome consists of several families with multiple members, including the heat shock protein (HSP) families HSP100, HSP90, HSP70, HSP60; the chaperonins; HSP40 or DNAJ proteins; and the small HSP (sHSP) (Hartl and Hayer-Hartl 2002). These family names reflect the molecular weight of the original member. Prefoldins and Tetratricopeptide repeat (TPR)-domain-containing co-chaperone family members along with organellar-specific chaperones of the endoplasmic reticulum (ER) and mitochondria complete the make up of the chaperome (Kleizen and Braakman 2004; Tatsuta et al. 2005). Many studies in model systems such as yeast, flies, and worms support a role for candidates from each of the major families in proteostasis (Brehme and Voisine 2016).

Based on the high level of conservation of function, model systems have been instrumental in understanding the role of key chaperones and co-chaperones in human health and disease. Many neurodegenerative diseases such as Huntington's disease (HD), Alzheimer's disease (AD) and Parkinson's disease (PD) accumulate misfolded proteins, reflecting an imbalance in proteostatsis (Klaips et al. 2018). Model organisms expressing these disease associated proteins provide a genetic tool for identifying modulators of proteotoxicity (Brehme and Voisine 2016). Utilizing genome-wide screens in diverse model systems of protein misfolding, an unbiased sampling of genes identified key chaperones and co-chaperones critical for proteostasis maintenance (Fig. 27.1).

The majority of studies found that multiple HSP70 family members ameliorated proteotoxicity (Auluck et al. 2002; Bilen and Bonini 2007; Brehme et al. 2014; Calamini et al. 2011; Chan et al. 2002; Duennwald et al. 2012; Fernandez-Funez et al. 2000; Fonte et al. 2002; Khabirova et al. 2014; Kraemer et al. 2006; Krobitsch and Lindquist 2000; Nollen et al. 2004; Sadlish et al. 2008; Silva et al. 2011; Sun et al. 2011; Walter et al. 2011; Wang et al. 2007, 2009; Warrick et al. 1999; Zhang et al. 2010). HSP70 assists in the folding of newly synthesized polypeptides and prevents protein misfolding and aggregation. The cycle of binding and release of client proteins is coupled to ATP binding and hydrolysis. In this capacity, HSP70 maintains cellular proteostasis and is a key member of the PN. Importantly, HSP70 cooperates with additional chaperones, such as HSP40/DNAJ, to accomplish its cellular folding tasks. HSP40 family members not only bind misfolded proteins but

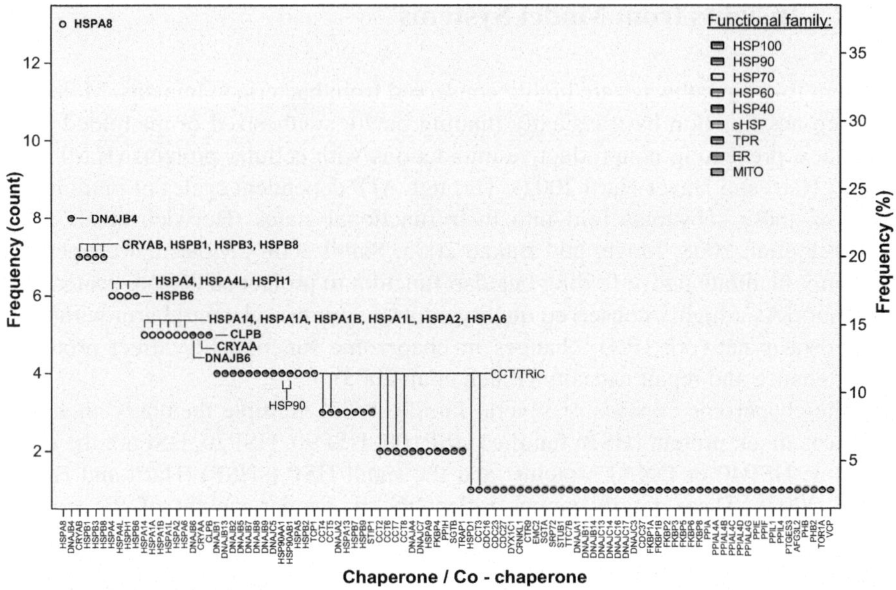

Frequency = unique observations in 35 independent studies since the release of the human genome sequence
Chaperones / co-chaperones identified ≥ 5x are labeled in bold

Fig. 27.1 Chaperone modifiers of proteotoxicity observed in model systems. Literature survey covering 35 independent studies since the release of the human genome sequence in the year 2000. (Figure adapted from Brehme and Voisine 2016)

also stimulate HSP70 ATP hydrolysis. Studies in model organisms revealed specific DNAJ proteins, DNAJB4 and DNAJB6, that modulate proteotoxicity likely by reducing HSP70 folding efficiency and HSP70 client specificity (Bilen and Bonini 2007; Duennwald et al. 2012; Fernandez-Funez et al. 2000; Kazemi-Esfarjani and Benzer 2000; Krobitsch and Lindquist 2000; Walter et al. 2011; Wang et al. 2007; Zhang et al. 2010) (Fig. 27.1). Furthermore, data from these diverse genetic studies demonstrate that small heat shock proteins (sHSP) are critical for proteostasis balance (Bilen and Bonini 2007; Calamini et al. 2011; Duennwald et al. 2012; Fonte et al. 2002; Jimenez-Sanchez et al. 2015; Silva et al. 2011; Vos et al. 2010; Walter et al. 2011; Wang et al. 2007, 2009). Small heat shock proteins are ATP independent chaperones that bind misfolded proteins, shielding them from inappropriate protein-protein interactions (Fig. 27.1). Although HSP90 was identified in a limited number of screens (Brehme et al. 2014; Butler et al. 2012; Calamini et al. 2011; Kraemer et al. 2006; Liang et al. 2008; Silva et al. 2011; van Ham et al. 2008; Walter et al. 2011), studies have demonstrated the importance of HSP90 as a centerpiece to proteostasis (Taipale et al. 2014). The large diverse clientele of HSP90 and the client connectivity within the proteome point towards HSP90 as a key factor in the PN. Genetic screens in model systems are an effective approach to identifiying key modulators of proteotoxicity but may be limited in scope. Clearly, systematic studies that uniformly assess chaperome connectivity and its dynamic alterations in physiology and disease are needed (Brehme and Voisine 2016).

27.3 The Human Chaperome Interactome

The human interactome, the integrated network of all physical interactions within the cell (Luck et al. 2017; Rolland et al. 2014) provides the foundation necessary for understanding human disease. Cataloging data from high throughput experiments on protein-protein interactions along with gene expression and co-expression data provides connectivity amongst proteins. Mapping networks formed by interactions of cellular proteins can provide insight as to how alterations in networks trigger disease. Within the human PN, a central component is represented by the chaperome (Labbadia and Morimoto 2015; Voisine et al. 2010). The chaperome corresponds to the complete ensemble of chaperones and co-chaperones that interact in a complex network of molecular folding machines to regulate the folded functional proteome through targeted client interactions (Kim et al. 2013). Several studies have harnessed the advantages of the model system yeast for the mapping of chaperone network connectivity, many of which have focused on known and central hubs such as HSP90 as entry points for the identification of physical and genetic interactions using proteomic and genomic methods. An early study using genome-wide yeast two-hybrid (Y2H) screens and affinity purification assays of HSP90 in complex with its interaction partners, coupled with synthetic genetic array technology and microarray-based chemical-genetic screens, identified HSP90's extended network. 198 putative physical interactions, including newly identified HSP90 co-chaperones (Tah1 and Pih1), and 451 genetic and chemical-genetic interactions centered around HSP90 alone, point towards the complexity of the human chaperome in its entirety (Zhao et al. 2005). A second study, using affinity purification coupled with mass-spectrometry (AP-MS), has charted a chaperone-chaperone and chaperone-client interaction network for a set of 63 yeast chaperones revealing that any given protein can interact with up to 25 different chaperones throughout its lifetime (Gong et al. 2009). These mapping approaches, however, are static representations and do not cover the anticipated dynamic changes adopted by the chaperome within the PN during physiological changes and exposure to stress. Importantly, quantitative approaches have started to assess the role of chaperone abundance (copies per cell) in their network context, investigating relationships between individual chaperones and their targets in terms of chaperone 'client throughput' (Brownridge et al. 2013). One such study followed a two-pronged approach for unbiased discovery of chaperone-cochaperone interactions by AP-MS in combination with quantitative LUMIER assays to characterize co-chaperone client specificity of HSP90 and its known co-chaperones (35 proteins) as well as HSP70 and its cofactors (13 proteins) and additional factors (68 proteins in total). Fifty four proteins were tagged (3xFLAG-V5) for AP-MS analyses and 60 proteins were fused to *Renilla* luciferase for quantitative LUMIER assays. This study yielded 468 high-confidence interactions for the 54 tagged AP-MS baits, providing a quantitative landscape of chaperone-client interaction profiles (Taipale et al. 2014). The major limitation to these approaches was the focus on a set of known chaperones and co-chaperones as entry points for interactome mapping and quantifications, disregarding the complexity of the remainder of the human chaperome and limiting the search space for interactions to the candidate baits.

As a step forward towards a more holistic understanding of the chaperome in systems-level studies, the human chaperome was assembled using structural genomics databases and interactomics databases of protein-protein interactions (PPIs). The construction of this comprehensive, annotated human chaperome compendium allows systems-level studies of chaperome biology towards an understanding of the role and implications of the chaperome in cellular proteostasis (Brehme et al. 2014). The human chaperome consists of 332 genes that partition into nine chaperone gene families, corresponding to HSP90, HSP70, HSP60, HSP40, Prefoldin, small HSP (sHSP), TPR-domain containing, and organellar-specific chaperones of the ER and mitochondria. The 332 human chaperone and co-chaperone members are attributed to highly networked communities based on functional chaperone activities, subcellular localization, and protein interactome and gene expression datasets. This compendium provided the basis for the systematic characterization of chaperome interactome alterations in human aging brains, age-onset neurodegenerative diseases (Brehme et al. 2014), and numerous cancers (Hadizadeh Esfahani et al. 2018). In a recent study that compared chaperome deregulation in cancers covered by The Cancer Genome Atlas (TCGA) consortium versus neurodegenerative diseases, 220 of the 332 human chaperome genes (nodes) were connected by 666 unique edges (PPIs). PPI data from the BioGRID, IntAct, DIP, and MINT databases with the PSI-MI annotation 'physical association' supported 222 unique interactions between 128 chaperome nodes. This knowledge of chaperome interconnectivity in the chaperone – chaperone and chaperone – co-chaperone interaction network along with integration of differential gene expression information enabled the generation of quantitative landscapes of chaperome deregulation. Future studies should aim at integrating additional functional and quantitative data towards a more complete understanding of proteostasis in disease.

27.4 Chaperome Deregulation in Aging and Neurodegenerative Diseases

Broad coverage of the chaperome and multi-level annotation with functional ontologies has enabled systems-level assessment of chaperome alterations in a broad range of complex diseases, such as major neurodegenerative diseases as well as human brain aging, which is recognized as the single most consistent and significant correlate of risk for neurodegenerative disease (Niccoli and Partridge

2012; Niccoli et al. 2017; Phillipson 2014). In the original study using the annotated human chaperome, the gene expression dynamics of 318 of its 332 members expressed in human aging brains and brains from Alzheimer's disease patients uncovered dramatically perturbed chaperome expression (Brehme et al. 2014). Overall, 32% of the chaperome, corresponding to ATP-dependent chaperone machines, were repressed, whereas only 19.5%, corresponding to ATP-independent chaperones and co-chaperones, were induced. Furthermore, chaperome dynamics correlated between aging and neurodegenerative diseases, such as Alzheimer's disease (Fig. 27.2). Considering physical interactions between brain-expressed chaperones and co-chaperones in combination with significant co-expression pairs (Pearson co-expression correlation >0.8), the study identified 1384 unique edges (interactions) between pairs of chaperones or co-chaperones, involving 192 nodes (chaperome proteins), of which 191 edges were protein-protein interactions (PPI), 1178 edges were significant co-expression interactions (COX), and 15 were both PPI and COX (PPI-COX) edges. Network link community clustering (Ahn et al. 2010) identified 40 communities comprised of only 6 induction communities with 30 interactions, as opposed to 34 repression communities with 1144 interactions, highlighting the overall robust, uniform chaperome interactome network repression in aging. These repression and induction clusters were aggravated in the brains of those with Alzheimer's, Huntington's or Parkinson's disease (Fig. 27.2). Given the strong correlation of chaperome gene expression changes between aging and neurodegenerative diseases, functional perturbation analyses were carried out in *C. elegans* and human cellular models expressing Aβ, polyglutamine, and huntingtin. Out of 219 *C. elegans* orthologs, a conserved chaperome sub-network of 16 chaperone genes, corresponding to 28 human orthologs, modulated proteotoxicity. This sub-network, which is affected in brain aging and diseases associated with protein aggregation such as Alzheimer's disease, acts as a critical proteostasis safeguard.

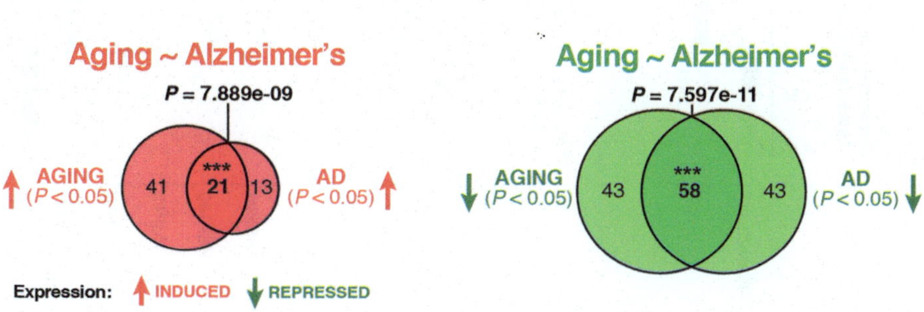

Fig. 27.2 Differential chaperome expression in aging human brains and Alzheimer's disease. Human chaperome differential induction and repression of gene expression in aging human brains and in brains from Alzheimer's disease patients significantly overlaps. (Figure adapted from Brehme et al. 2014)

27.5 Chaperome Deregulation in Cancers

Findings on systems-level proteostasis alterations by chaperome deregulation in age-onset neurodegenerative diseases has motivated further studies on its deregulation in cancers. Decades of evidence have revealed the differential expression of numerous individual chaperones in several cancers using a variety of model systems (Oromendia and Amon 2014; Valastyan and Lindquist 2014). However, a systematic and comprehensive assessment of the annotated human chaperome across a representative reference dataset of human cancers has been lacking. A recent study utilized data from the broad representation of cancer families as classified by The Cancer Genome Atlas (TCGA) compendium (Weinstein et al. 2013). Gene expression data derived from more than 10,000 patient biopsies covering 22 solid cancers provided detailed maps on the relative shifts of chaperome gene expression across the heterogeneous human cancer landscape (Hadizadeh Esfahani et al. 2018). The development of a novel Meta-PCA dimension reduction approach enabled the generation of M-scores serving as quantitative indicators of cancer-specific gene expression changes, conveniently condensing the complexity of highly multiparametric cancer transcriptomics datasets into a finite set of quantitative functional network topographies, defined by the chaperome functional families. The study confirmed HSP90 family genes as broadly upregulated across cancers and also uncovered HSP60s, Prefoldins, HSP100s, ER- and mitochondria-specific chaperones as pan-cancer induced. Interestingly, the study highlighted small heat shock proteins (sHSP) as downregulated. The chaperome deregulation in cancers shares

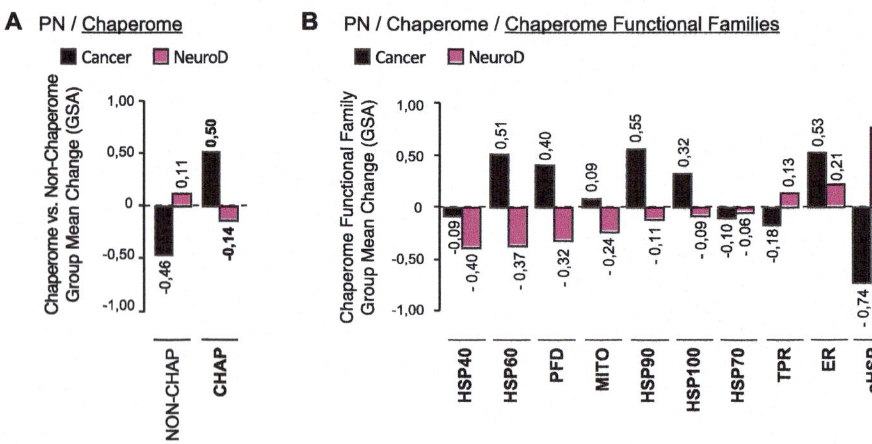

Fig. 27.3 Chaperome deregulation in cancers versus neurodegeneration. Chaperome deregulation patterns overall and especially in regards to chaperome functional sub-families is markedly opposed in cancers as compared to chaperome deregulation in neurodegenerative diseases. The side-by-side comparison shows gene expression changes of chaperome vs. non-chaperome (**a**) and chaperome functional families (**b**) in cancers (black) versus neurodegenerative diseases (NeuroD, grey). (Figure adapted from Hadizadeh Esfahani et al. 2018. Gene expression changes indicated as ΔGSA group mean changes (Hadizadeh Esfahani et al. 2018))

similarities with proteostasis alterations in stem cells, while exposing a vastly diametrically opposed chaperome deregulation in neurodegenerative diseases such as Alzheimer's, Huntington's and Parkinson's disease (Fig. 27.3).

Proteomics-based approaches have broadly assessed the role of chaperome interconnectivity in cancer. A recent study focusing on HSP90 as a target of cancer drug therapy concluded that the chaperome interactome and specifically the inherent protein complexes around HSP90 and HSP70 act as "nucleating sites" (Rodina et al. 2016). The authors refer to a highly interconnected and stable chaperome network of multimeric chaperone – co-chaperone complexes as "epi-chaperome", which they find present in over half of 'Type 1' cancers tested versus 'Type 2' cancers that have a less tightly interconnected set of chaperones. The 'epi-chaperome' can function as an enhancer of cellular survival and is a vulnerable target for therapeutic agents. The authors postulate that targeted inhibition of the 'epi-chaperome', including HSP90 specific compounds, lead to cell death in Type 1 cancers, but only partially compromises the chaperome in Type 2 cancers, resulting in cell survival. Collectively, these studies highlight the importance and benefit of systems-level, integrative and quantitative studies of chaperome network biology towards an understanding of proteostasis deregulation in cancers.

27.6 Towards a Systems-Level Understanding of Proteostasis Deregulation in Human Diseases

Based on the evidence outlined above, the entire chaperome interactome network, rather than individual chaperones, and its dynamic variations represent an important integrated correlate of the PN in human diseases. Recent evidence has shown that the effect of specific therapeutic targeting of HSP90, a central and conserved individual component, depends on its connectivity within the wider chaperome interactome. Concordantly, all proteostasis regulators must be evaluated in the context of chaperome and PN connectivity and the overall dynamics of proteostasis. Based on the importance of proteostasis in health and disease, a chaperome-targeted therapeutic intervention that adjusts proteostasis may be beneficial for a large number of diseases (Balch et al. 2008). To this end, disease-specific characteristic signatures of chaperome deregulation can serve as biomarkers of proteostasis alterations. Targeted readjustments of proteostasis through chaperome modulation represent a crucial strategy for therapeutic intervention. For example, testing HSP90 as a specific therapeutic gene target within the context of chaperome connectivity, including the conserved chaperome subnetwork that safeguards proteostasis in neurodegeneration (Brehme et al. 2014) or the epi-chaperome in cancers (Rodina et al. 2016), will lend itself towards a PN-guided therapeutic adjustment of proteostasis deficiencies. Towards this goal, further systems-level analyses and a more complete understanding of proteostasis disease biology are needed, beyond the examples discussed in this chapter. Mechanisms of PN regulation, such as HSF1-driven expression of heat shock proteins, and the interplay during adaption to external and intrinsic stressors, are incompletely understood and call for further studies across the human diseasome. For instance, the evident contrast of chaperome deregulation in neurodegenerative diseases versus cancers points towards characteristic and fundamental differences in PN deregulation between diseases (Hadizadeh Esfahani et al. 2018). Detailed maps of chaperome and PN deregulation are needed throughout the human diseasome to support the contributions of proteostasis alterations to complex disease processes. A precise knowledge of system-wide proteostasis deregulation provides potential new proteostasis biomarkers of disease and insights into global targets for therapeutic proteostasis readjustment.

27.7 Conclusions

Cellular proteostasis is fundamental to healthy aging. Alterations to the cellular proteostasis network (PN) have been observed in aging and a diverse range of human diseases. The understanding of PN deregulation lends itself for the targeted readjustment of proteostasis in these diseases and PN deregulation signatures represent useful biomarkers towards this goal. A key functional arm of the PN is represented by molecular chaperones and their respective co-chaperone factors that collectively comprise the cellular chaperome, which is in charge of regulating and maintaining the natively folded and functional proteome. Conserved key chaperone hubs such as HSP90 or HSP70 have been extensively studied and described throughout the past decades. In recent years, however, genome-scale systematic studies have highlighted the need for consideration of the entire chaperome ensemble, including less studied chaperones and co-chaperones. The quest for new pharmacologic chaperones as well as proteostasis regulators (PRs) that modulate and readjust functional PN arms in specific diseases calls for an understanding of interactome network connectivity context for both chaperome and PN wiring and the associated proteostasis dynamic changes. The systems-level understanding of proteostasis biology is inevitable towards a rational approach to therapeutic intervention in diseases of proteostasis deficiency or deregulation. The human chaperome and the PN as therapeutic target spaces in human disease thereby represent paradigmatic examples for interactome network informed poly-pharmacology through the knowledge of disease-specific network biomarker signatures. Further systems-level studies of chaperome and PN alterations in human diseases are required in order to support pharmacological intervention to proteostasis deregulation as a promising therapeutic strategy in human aging and disease.

Acknowledgements The authors would like to acknowledge Richard Morimoto, Marc Vidal, Andreas Schuppert, Julio Saez-Rodriguez, Angelina Sverchkova, and Ali Hadizadeh-Esfahani for their support and contributions to the authors' research referenced in this chapter. Figure 27.1 reproduced from Brehme and Voisine, Disease Models & Mechanisms (2016) 9, 823–838, https://doi.org/10.1242/dmm.024703 with permission of Disease Models & Mechanisms.

References

Ahn YY, Bagrow JP, Lehmann S (2010) Link communities reveal multiscale complexity in networks. Nature 466:761–764

Auluck PK, Chan HY, Trojanowski JQ, Lee VM, Bonini NM (2002) Chaperone suppression of alpha-synuclein toxicity in a *Drosophila* model for Parkinson's disease. Science (New York, NY) 295:865–868

Balch WE, Morimoto RI, Dillin A, Kelly JW (2008) Adapting proteostasis for disease intervention. Science (New York, NY) 319:916–919

Balchin D, Hayer-Hartl M, Hartl FU (2016) *In vivo* aspects of protein folding and quality control. Science (New York, NY) 353:aac4354

Bilen J, Bonini NM (2007) Genome-wide screen for modifiers of ataxin-3 neurodegeneration in *Drosophila*. PLoS Genet 3:1950–1964

Brehme M, Voisine C (2016) Model systems of protein-misfolding diseases reveal chaperone modifiers of proteotoxicity. Dis Model Mech 9:823–838

Brehme M, Voisine C, Rolland T et al (2014) A chaperome subnetwork safeguards proteostasis in aging and neurodegenerative disease. Cell Rep 9:1135–1150

Brownridge P, Lawless C, Payapilly AB et al (2013) Quantitative analysis of chaperone network throughput in budding yeast. Proteomics 13:1276–1291

Butler EK, Voigt A, Lutz AK et al (2012) The mitochondrial chaperone protein TRAP1 mitigates alpha-Synuclein toxicity. PLoS Genet 8:e1002488

Calamini B, Silva MC, Madoux F et al (2011) Small-molecule proteostasis regulators for protein conformational diseases. Nat Chem Biol 8:185–196

Chan HY, Warrick JM, Andriola I, Merry D, Bonini NM (2002) Genetic modulation of polyglutamine toxicity by protein conjugation pathways in Drosophila. Hum Mol Genet 11:2895–2904

Duennwald ML, Echeverria A, Shorter J (2012) Small heat shock proteins potentiate amyloid dissolution by protein disaggregases from yeast and humans. PLoS Biol 10:e1001346

Fernandez-Funez P, Nino-Rosales ML, de Gouyon B et al (2000) Identification of genes that modify ataxin-1-induced neurodegeneration. Nature 408:101–106

Fonte V, Kapulkin WJ, Taft A, Fluet A, Friedman D, Link CD (2002) Interaction of intracellular beta amyloid peptide with chaperone proteins. Proc Natl Acad Sci U S A 99:9439–9444

Gong Y, Kakihara Y, Krogan N et al (2009) An atlas of chaperone-protein interactions in *Saccharomyces cerevisiae*: implications to protein folding pathways in the cell. Mol Syst Biol 5:275

Hadizadeh Esfahani A, Sverchkova A, Saez-Rodriguez J, Schuppert AA, Brehme M (2018) A systematic atlas of chaperome deregulation topologies across the human cancer landscape. PLoS Comput Biol 14:e1005890

van Ham TJ, Thijssen KL, Breitling R, Hofstra RM, Plasterk RH, Nollen EA (2008) *C. elegans* model identifies genetic modifiers of alpha-synuclein inclusion formation during aging. PLoS Genet 4:e1000027

Hartl FU, Hayer-Hartl M (2002) Molecular chaperones in the cytosol: from nascent chain to folded protein. Science (New York, NY) 295:1852–1858

Hartl FU, Bracher A, Hayer-Hartl M (2011) Molecular chaperones in protein folding and proteostasis. Nature 475:324–332

Horwich AL, Fenton WA, Chapman E, Farr GW (2007) Two families of chaperonin: physiology and mechanism. Annu Rev Cell Dev Biol 23:115–145

Jarosz DF, Lindquist S (2010) Hsp90 and environmental stress transform the adaptive value of natural genetic variation. Science (New York, NY) 330:1820–1824

Jimenez-Sanchez M, Lam W, Hannus M et al (2015) siRNA screen identifies QPCT as a druggable target for Huntington's disease. Nat Chem Biol 11:347–354

Karras GI, Yi S, Sahni N et al (2017) HSP90 shapes the consequences of human genetic variation. Cell 168:856–866.e12

Kazemi-Esfarjani P, Benzer S (2000) Genetic suppression of polyglutamine toxicity in *Drosophila*. Science (New York, NY) 287:1837–1840

Khabirova E, Moloney A, Marciniak SJ et al (2014) The TRiC/CCT chaperone is implicated in Alzheimer's disease based on patient GWAS and an RNAi screen in Abeta-expressing *Caenorhabditis elegans*. PLoS One 9:e102985

Kim YE, Hipp MS, Bracher A, Hayer-Hartl M, Hartl FU (2013) Molecular chaperone functions in protein folding and proteostasis. Annu Rev Biochem 82:323–355

Klaips CL, Jayaraj GG, Hartl FU (2018) Pathways of cellular proteostasis in aging and disease. J Cell Biol 217:51–63

Kleizen B, Braakman I (2004) Protein folding and quality control in the endoplasmic reticulum. Curr Opin Cell Biol 16:343–349

Kraemer BC, Burgess JK, Chen JH, Thomas JH, Schellenberg GD (2006) Molecular pathways that influence human tau-induced pathology in *Caenorhabditis elegans*. Hum Mol Genet 15:1483–1496

Krobitsch S, Lindquist S (2000) Aggregation of huntingtin in yeast varies with the length of the polyglutamine expansion and the expression of chaperone proteins. Proc Natl Acad Sci U S A 97:1589–1594

Labbadia J, Morimoto RI (2015) The biology of proteostasis in aging and disease. Annu Rev Biochem 84:435–464

Liang J, Clark-Dixon C, Wang S et al (2008) Novel suppressors of alpha-synuclein toxicity identified using yeast. Hum Mol Genet 17:3784–3795

Liberek K, Lewandowska A, Zietkiewicz S (2008) Chaperones in control of protein disaggregation. EMBO J 27:328–335

Luck K, Sheynkman GM, Zhang I, Vidal M (2017) Proteome-scale human interactomics. Trends Biochem Sci 42:342–354

Mayer MP, Bukau B (2005) Hsp70 chaperones: cellular functions and molecular mechanism. Cell Mol Life Sci 62:670–684

Niccoli T, Partridge L (2012) Ageing as a risk factor for disease. Curr Biol 22:R741–R752

Niccoli T, Partridge L, Isaacs AM (2017) Ageing as a risk factor for ALS/FTD. Hum Mol Genet 26:R105–RR13

Nollen EA, Garcia SM, van Haaften G et al (2004) Genome-wide RNA interference screen identifies previously undescribed regulators of polyglutamine aggregation. Proc Natl Acad Sci U S A 101:6403–6408

Oromendia AB, Amon A (2014) Aneuploidy: implications for protein homeostasis and disease. Dis Model Mech 7:15–20

Pearl LH, Prodromou C (2006) Structure and mechanism of the Hsp90 molecular chaperone machinery. Annu Rev Biochem 75:271–294

Phillipson OT (2014) Management of the aging risk factor for Parkinson's disease. Neurobiol Aging 35:847–857

Powers ET, Morimoto RI, Dillin A, Kelly JW, Balch WE (2009) Biological and chemical approaches to diseases of proteostasis deficiency. Annu Rev Biochem 78:959–991

Queitsch C, Sangster TA, Lindquist S (2002) Hsp90 as a capacitor of phenotypic variation. Nature 417:618–624

Rodina A, Wang T, Yan P et al (2016) The epichaperome is an integrated chaperome network that facilitates tumour survival. Nature 538:397–401

Rolland T, Tasan M, Charloteaux B et al (2014) A proteome-scale map of the human interactome network. Cell 159:1212–1226

Rutherford SL, Lindquist S (1998) Hsp90 as a capacitor for morphological evolution. Nature 396:336–342

Sadlish H, Rampelt H, Shorter J et al (2008) Hsp110 chaperones regulate prion formation and propagation in *S. cerevisiae* by two discrete activities. PLoS One 3:e1763

Saibil HR (2008) Chaperone machines in action. Curr Opin Struct Biol 18:35–42

Sangster TA, Salathia N, Lee HN et al (2008) HSP90-buffered genetic variation is common in *Arabidopsis thaliana*. Proc Natl Acad Sci U S A 105:2969–2974
Schmidt M, Finley D (2014) Regulation of proteasome activity in health and disease. Biochim Biophys Acta 1843:13–25
Schopf FH, Biebl MM, Buchner J (2017) The HSP90 chaperone machinery. Nat Rev Mol Cell Biol 18:345–360
Silva MC, Fox S, Beam M, Thakkar H, Amaral MD, Morimoto RI (2011) A genetic screening strategy identifies novel regulators of the proteostasis network. PLoS Genet 7:e1002438
Sun Z, Diaz Z, Fang X et al (2011) Molecular determinants and genetic modifiers of aggregation and toxicity for the ALS disease protein FUS/TLS. PLoS Biol 9:e1000614
Taipale M, Jarosz DF, Lindquist S (2010) HSP90 at the hub of protein homeostasis: emerging mechanistic insights. Nat Rev Mol Cell Biol 11:515–528
Taipale M, Tucker G, Peng J et al (2014) A quantitative chaperone interaction network reveals the architecture of cellular protein homeostasis pathways. Cell 158:434–448
Tatsuta T, Model K, Langer T (2005) Formation of membrane-bound ring complexes by prohibitins in mitochondria. Mol Biol Cell 16:248–259
Valastyan JS, Lindquist S (2014) Mechanisms of protein-folding diseases at a glance. Dis Model Mech 7:9–14
Voisine C, Pedersen JS, Morimoto RI (2010) Chaperone networks: tipping the balance in protein folding diseases. Neurobiol Dis 40:12–20
Vos MJ, Zijlstra MP, Kanon B et al (2010) HSPB7 is the most potent polyQ aggregation suppressor within the HSPB family of molecular chaperones. Hum Mol Genet 19:4677–4693
Walter GM, Smith MC, Wisen S et al (2011) Ordered assembly of heat shock proteins, Hsp26, Hsp70, Hsp90, and Hsp104, on expanded polyglutamine fragments revealed by chemical probes. J Biol Chem 286:40486–40493
Wang Y, Meriin AB, Costello CE, Sherman MY (2007) Characterization of proteins associated with polyglutamine aggregates: a novel approach towards isolation of aggregates from protein conformation disorders. Prion 1:128–135
Wang J, Farr GW, Hall DH et al (2009) An ALS-linked mutant SOD1 produces a locomotor defect associated with aggregation and synaptic dysfunction when expressed in neurons of *Caenorhabditis elegans*. PLoS Genet 5:e1000350
Warrick JM, Chan HY, Gray-Board GL, Chai Y, Paulson HL, Bonini NM (1999) Suppression of polyglutamine-mediated neurodegeneration in *Drosophila* by the molecular chaperone HSP70. Nat Genet 23:425–428
Weinstein JN, Collisson EA, Mills GB et al (2013) The cancer genome atlas pan-cancer analysis project. Nat Genet 45:1113–1120
Whitesell L, Lindquist SL (2005) HSP90 and the chaperoning of cancer. Nat Rev Cancer 5:761–772
Young JC, Moarefi I, Hartl FU (2001) Hsp90: a specialized but essential protein-folding tool. J Cell Biol 154:267–273
Zhang S, Binari R, Zhou R, Perrimon N (2010) A genomewide RNA interference screen for modifiers of aggregates formation by mutant huntingtin in *Drosophila*. Genetics 184:1165–1179
Zhao R, Davey M, Hsu YC et al (2005) Navigating the chaperone network: an integrative map of physical and genetic interactions mediated by the hsp90 chaperone. Cell 120:715–727

Index

A. A. A. Asea, P. Kaur (eds.), *Heat Shock Protein 90 in Human Diseases and Disorders*, Heat Shock Proteins 19,
https://doi.org/10.1007/978-3-030-23158-3